WITHDRAWN
FROM
STOCK

HANDBOOK OF GENETICS

Volume 5

Molecular Genetics

HANDBOOK OF GENETICS

Volume 1: Bacteria, Bacteriophages, and Fungi

Volume 2: Plants, Plant Viruses, and Protists

Volume 3: Invertebrates of Genetic Interest

Volume 4: Vertebrates of Genetic Interest

Volume 5: Molecular Genetics

HANDBOOK OF GENETICS

ROBERT C. KING, EDITOR
Professor of Genetics, Department of Biological Sciences
Northwestern University, Evanston, Illinois

Volume 5
Molecular Genetics

PLENUM PRESS · NEW YORK AND LONDON

Library of Congress Cataloging in Publication Data

King, Robert C
Molecular genetics.

(His Handbook of genetics; v. 5)
Includes bibliographies and index.
1. Molecular genetics I. Title. [DNLM: 1. Genetics. QH430 H236]
QH430.K53 574.8'732 76-2444
ISBN 0-306-37615-6

A Division of Plenum Publishing Corporation
227 West 17th Street, New York, N.Y. 10011

Printed in the United States of America

Preface

Many modern geneticists attempt to elucidate the molecular basis of phenotype by utilizing a battery of techniques derived from physical chemistry on subcellular components isolated from various species of organisms. Volume 5 of the *Handbook of Genetics* provides explanations of the advantages and shortcomings of some of these revolutionary techniques, and the nonspecialist is alerted to key research papers, reviews, and reference works. Much of the text deals with the structure and functioning of the molecules bearing genetic information which reside in the nucleus and with the processing of this information by the ribosomes residing in the cytoplasm of eukaryotic cells. The mitochondria, which also live in the cytoplasm of the cells of all eukaryotes, now appear to be separate little creatures. These, as Lynn Margulis pointed out in Volume 1, are the colonial posterity of migrant prokaryotes, probably primitive bacteria that swam into the ancestral precursors of all eukaryotic cells and remained as symbionts. They have maintained themselves and their ways ever since, replicating their own DNA and transcribing an RNA quite different from that of their hosts. In a similar manner, the chloroplasts in all plants are self-replicating organelles presumably derived from the blue-green algae, with their own nucleic acids and ribosomes. Four chapters are devoted to the nucleic acids and the ribosomal components of both classes of these semi-independent lodgers. Finally, data from various sources on genetic variants of enzymes are tabulated for ready reference, and an evaluation of this information is attempted.

I am particularly appreciative of the assistance provided by Mary Bahns, Bennett Buckles, Joseph Cassidy, Patricia DeOca, Lisa Gross, and Pamela Khipple in seeing this volume to completion. My long-term collaboration with Ralph Cutler, production editor at Plenum Press, has been one of the most pleasant aspects of the project.

The 3500 pages in the five volumes of the *Handbook of Genetics* present the combined labors of its 150 dedicated contributors. These authors have digested and critically summarized a prodigious amount of information. Some measure of this effort is gained from consulting the bibliographies that close each of the 133 chapters. These contain citations totaling 14,000 publications authored by about 12,000 different scientists. Readers interested in correcting printing errors in earlier volumes should consult p. 617 in this volume.

I am grateful that, as one of the last scientific efforts during his illustrious career, Theodosius Dobzhansky was able to contribute to this series. He died near the end of his 76th year on the 18th of December, 1975, just a few weeks after the publication of Volume 3, which contained two chapters that he coauthored with Jeffrey Powell.

Robert C. King

Contributors

Anton J. M. Berns, Department of Biochemistry, University of Nijmegen, Nijmegen, The Netherlands
David P. Bloch, Department of Botany and Cell Research Institute, University of Texas, Austin, Texas
Hans Bloemendal, Department of Biochemistry, University of Nijmegen, Nijmegen, The Netherlands
Arminio Boschetti, Institute for Biochemistry, University of Berne, Berne, Switzerland
W. Yean Chooi, Whitman Laboratory, University of Chicago, Chicago, Illinois
Bertil Daneholt, Department of Histology, Karolinska Institutet, Stockholm, Sweden
Ronald A. Eckhardt, Department of Biology, Brooklyn College of The City University of New York, Brooklyn, New York
Mary G. Hamilton, Sloan-Kettering Institute for Cancer Research, New York City, New York
Rex P. Hjelm, Jr., Biophysics Laboratories, Department of Physics, Portsmouth Polytechnic Institute, Portsmouth, England
Ru Chih C. Huang, Department of Biology, The Johns Hopkins University, Baltimore Maryland
James T. Madison, U.S. Plant, Soil and Nutrition Laboratory, U.S. Department of Agriculture, Ithaca, New York
David E. Matthews, Department of Biochemistry, University of Florida, Gainesville, Florida
Kuruganti G. Murti, Laboratories of Virology, St. Jude Children's Research Hospital, Memphis, Tennessee

Margit M. K. Nass, Department of Therapeutic Research, University of Pennsylvania, School of Medicine, Philadelphia, Pennsylvania

Thomas W. O'Brien, Department of Biochemistry University of Florida, Gainesville, Florida

Robert B. Painter, Laboratory of Radiobiology, University of California, San Francisco, California

Rupi Prasad, The Universtiy of Texas, and M.D. Anderson Hospital and Tumor Institute, Texas Medical Center, Houston, Texas

Ruth Sager, Sidney Farber Cancer Center and Department of Microbiology, Harvard Medical School, Boston, Massachusetts

Gladys Schlanger, Division of Human Genetics, Cornell University Medical College, New York City, New York

Hans-Georg Schweiger, Max-Planck-Institut für Zellbiologie, Wilhelmshaven, Federal Republic of Germany

Charles R. Shaw, The University of Texas and M.D. Anderson Hospital and Tumor Institute, Texas Medical Center, Houston, Texas

Neil A. Straus, Department of Botany, University of Toronto, Toronto, Ontario, Canada

Erhard Stutz, Laboratory of Plant Biochemistry, University of Neuchâtel, Neuchâtel, Switzerland

Contents

Q. The Molecular Organization Of Chromosomes

R. Gene Transcripts

S. Chloroplasts and Mitochondria

T. Mutant Enzymes

PART Q
THE MOLECULAR ORGANIZATION OF CHROMOSOMES

1

Repeated DNA in Eukaryotes

NEIL A. STRAUS

Introduction

Based on DNA:DNA kinetics of reassociation, large fractions of eukaryotic genomes are believed to be repeated (Britten and Kohne, 1968). There have been estimates that as much as 90% of the genome of some eukaryotes is composed of repeated DNA (Straus, 1971) and that the repetition frequency of some repeated sequences may be as large as 1 million (Britten and Kohne, 1968). The filter hybridization method of Gillespie and Spiegelman (1965) has been used to show that most organisms contain multiple copies of ribosomal cistrons. Although the repetition frequency of the ribosomal cistron ranges from five copies in bacteria (Attardi and Amaldi, 1970) to thousands of copies in eukaryotes (Brown and Dawid, 1968), the sequences of DNA complementary to ribosomal RNA generally account for only a fraction of a percent of the total nuclear DNA.

This work is not intended to be a complete review of the present knowledge of the details of DNA:DNA reassociation. For those interested in more details about the reassociation reaction, there are the papers of Britten and Kohne (1968) and Wetmur and Davidson (1968) and a

NEIL A. STRAUS—Department of Botany, University of Toronto, Toronto, Ontario, Canada.

number of review articles (Marmur *et al.,* 1963; Walker, 1969; Kohne, 1970; McCarthy and Church, 1970; Kennel, 1971; Britten *et al.,* 1974).

The first section of this chapter will give the reader sufficient information to understand the meaning of repetitive DNA in the kinetic sense and what cautions should be exercised in interpretations of the data on repeated DNA. The second section focuses on the reassociation kinetics data of a number of plants and animals. By applying the information of Section I to these data, the reader will be able to obtain estimates of the amounts and repetition frequencies of repetitive DNA in these organisms.

Section I

DNA Reassociation Kinetic Parameters

DNA, as carefully isolated from organisms, consists of long fragments of double-stranded molecules. The two polynucleotide strands of this double helical structure are held together by the specific complementary pairing of bases on opposing strands. The two strands can be separated by simply heating the DNA in a solution of low or moderate salt concentration. Under the proper incubation conditions, the separated strands will realign themselves with complementary or nearly complementary sequences, forming stable paired structures (Marmur *et al.,* 1963).

Reassociation of DNA follows second-order kinetics; hence the rate of reassociation is dependent on the concentration of complementary strands in solution (Marmur *et al.,* 1963; Wetmur and Davidson, 1968). The rate of reassociation also depends on the incubation conditions. Briefly, reassociation is sensitive to temperature, salt concentration, viscosity, and DNA fragment size. A graph of the rate of reassociation vs. temperature is bell shaped, with a maximum approximately 25°C below the T_m (the temperature of half-dissociation of the native DNA, defined spectrophotometrically). Increasing salt concentration increases the rate of reassociation (Wetmur and Davidson, 1968; Britten, 1969). The rate of reassociation increases with increasing fragment size (Wetmur and Davidson, 1968). Fragment size is important when repeated DNA is being studied, and this factor will be dealt with more fully later. The rate of reassociation is decreased by increasing the viscosity of the solution of incubation (Wetmur and Davidson, 1968). Caution must be exercised since some of the above conditions are interdependent. For example, the T_m of DNA itself depends on salt concentration (Marmur and Doty, 1962), and altering the salt concentration changes the rate–temperature curve.

The rate of reassociation also depends on the DNA isolated. The rate is directly affected by the complexity of the genome studied (Britten and Kohne, 1968; Wetmur and Davidson, 1968) and to a much smaller extent by the GC content of the DNA under study (Wetmur and Davidson, 1968; Laird, 1971). For organisms containing no repeated sequences, the rate constant of reassociation is inversely proportional to the genome size (Britten and Kohne, 1968).

Methods of Monitoring Reassociation

Basically, two methods have been employed to monitor DNA:DNA reassociation kinetics. One method is to observe the time-dependent decrease in optical density at 260 nm of the DNA. Here the hypochromicity is assumed to vary linearly with the restored base pairing (Doty *et al.*, 1959). This method measures the actual amount of DNA helix that has formed. The other method is to observe the amount of DNA which binds to hydroxylapatite following the incubation period. Under certain conditions, hydroxylapatite selectively binds duplex DNA and allows single-stranded DNA to pass through (Britten and Kohne, 1968; Miyazawa and Thomas, 1965; Bernardi, 1965). The latter technique then enables one to measure the fraction of the DNA fragments which are double-stranded along some part of their length.

C_0t Plot and Repetition Frequency

DNA reassociation follows second-order kinetics. The ideal second-order rate equation is

$$dC/dt = -KC^2$$

where C is the concentration of single-stranded sequences and t is the time of incubation. This yields

$$C/C_0 = 1/(1 + KC_0t)$$

where C_0 is the initial concentration of single-stranded DNA in moles of nucleotides/liter; that is, $C = C_0$ when $t = 0$.

The progress of the reaction can be conveniently represented by plotting the fraction that remains unreacted C/C_0 against the logarithm of the product of the initial concentration and the time of incubation (C_0t). This is known as the "C_0t" plot." When the reaction is half completed, then $KC_0t = 1$ and this value of C_0t is designated the $C_0t_{1/2}$ (Britten and Kohne, 1968).

A theoretical curve (C_0t plot) of a sequence of DNA with no repetitions of any one of its parts relative to the other part is shown in Figure 1. The symmetrical S-shaped curve has an inflection point at its position of half-reaction—its $C_0t_{1/2}$. A straight line drawn to approximate the central 70% of the theoretical curve gives a ratio of 100 between the C_0t values at which the line passes through zero reaction and the C_0t value on the line for complete reaction (Figure 1). If this ratio is larger than 100, then the DNA contains some sequences that are reacting faster than other sequences (Britten and Kohne, 1968).

DNA isolated from prokaryotes reassociates with ideal second-order kinetics. The rate constant of reassociation is inversely proportional to genome size, and hence the $C_0t_{1/2}$ is proportional to genome size, as shown in Figure 2 (Britten and Kohne, 1968; Wetmur and Davidson, 1968; Laird, 1971). C_0t plots of DNA from eukaryotes, on the other hand, are unlike the ideal second-order C_0t plot. These plots indicate that some sequences are reassociating faster than others. Kinetically distinct components, when isolated from the DNA of eukaryotes, exhibit ideal second-order kinetics (Figure 2). Indeed, the last component to react (in all cases examined) reacts with ideal kinetics, and the $C_0t_{1/2}$ is that expected for the genome size of the organism in question (Britten and Kohne, 1968; Davidson and Hough, 1969; Laird, 1971; Straus, 1971). The C_0t plot of eukaryotic DNA is assumed to be a composite of a number of components, each reassociating with second-order kinetics but with different rates of reaction.

Since the DNA reassociation reaction is concentration dependent, the eukaryotic DNA sequences that reassociate rapidly must be more concentrated than those which react more slowly. Thus we assume that the

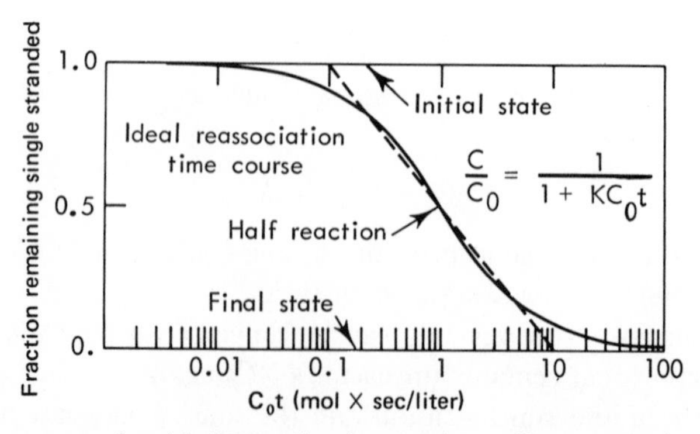

Figure 1. Time course of an ideal, second-order reaction to illustrate the features of the log C_0t plot. The equation represents the fraction of DNA which remains single-stranded after the initiation of the reaction. For this example, K is taken to be 1.0, and the fraction remaining single-stranded is plotted against the product of total concentration and time on a logarithmic scale. From Britten and Kohne (1968).

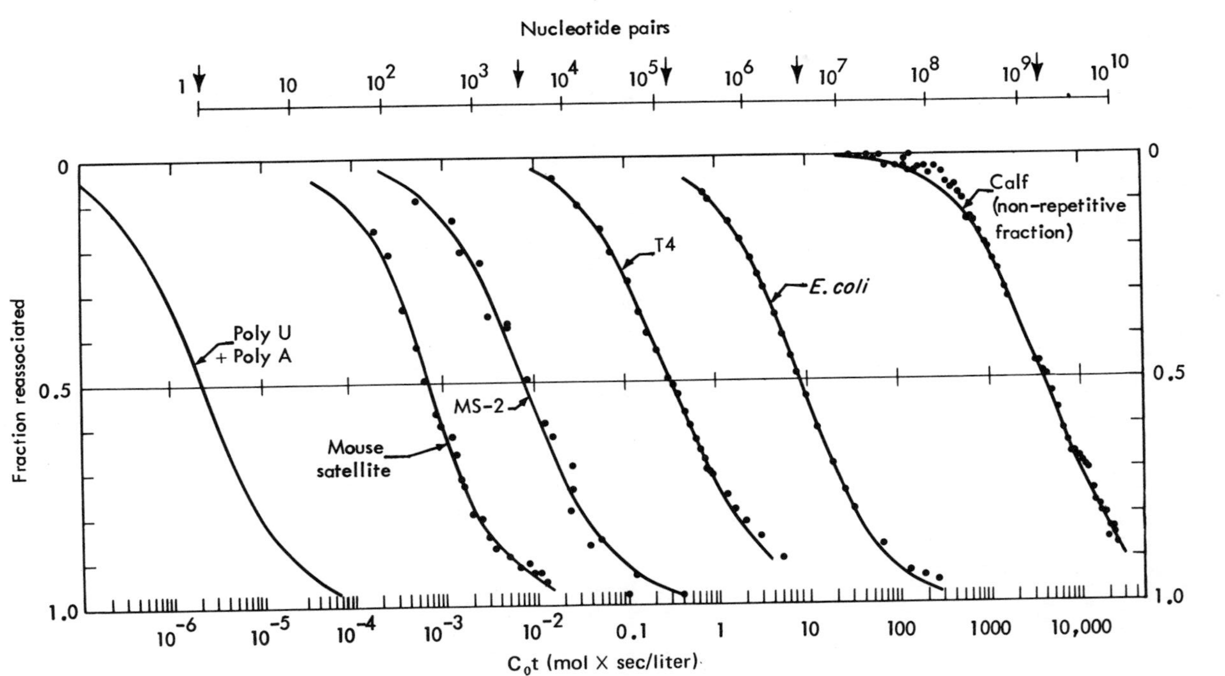

Figure 2. Reassociation of double-stranded nucleic acids from various sources. The genome size is indicated by the arrows near the upper nomographic scale. Over a factor of 10^9, this value is proportional to the C_0t required for half-reaction. The DNA was sheared and the other nucleic acids are reported to have approximately the same fragment size (about 400 nucleotides, single-stranded). Correction has been made to give the rate that would be observed at 0.18 M sodium ion concentration. No correction for temperature has been applied, as it was approximately optimum in all cases. Optical rotation was the measure of the reassociation of the calf thymus nonrepeated fraction (far right). The MS-2 RNA points were calculated from a series of measurements of the increase in ribonuclease resistance (Billeter et al., 1966). The curve (far left) for polyuridylic acid plus polyadenylic acid was estimated from the data of Ross and Sturtevant (1962). The remainder of the curves were measured by hypochromicity at 260 nm; a Zeiss spectrophotometer with a continuous recording attachment was used. From Britten and Kohne (1968).

rapidly reassociating sequences have a higher repetition frequency than the sequences reassociating with the slowest rate (Britten and Kohne, 1968; Wetmur and Davidson, 1968).

Theoretically, the $C_0t_{1/2}$ of a repeated sequence is inversely proportional to its frequency (Britten and Kohne, 1968). Thus the repetition frequency of any component is obtained by dividing the expected $C_0t_{1/2}$ of sequences present only once per haploid genome by the $C_0t_{1/2}$ of the component under investigation.

Repetitive DNA

The factors affecting the rate of renaturation of repeated DNA are assumed to be the same as those discussed earlier. However, additional complications can arise. For example, because of the interspersion of repeated sequences, unsheared DNA from eukaryotes can form large networks during renaturation. A sequence of repeated DNA can react with a related complementary sequence that is situated in a different position of the genome. Although these repeated sequences form double strands, the surrounding nucleotide sequences are unrelated and cannot react. In this way long strands are held together by relatively short double-stranded regions. Other parts of these and other strands can interact, and the result is a large network of DNA that is mostly single-stranded. Steric hindrance strongly restricts further complementary base pairing. The problem can be avoided by using DNA of smaller piece size; sheared DNA reacts more completely (Britten and Waring, 1965). A second, but related, complication is that the hydroxylapatite-monitored reassociation kinetics of repeated DNA is sensitive to fragment size. The longer the DNA fragment, the higher the chance of its containing a member of some family of repeated sequences (Britten and Davidson, 1971). Since the fraction of the genome that is repeated (based on hydroxylapatite kinetics) actually represents the fraction of fragments of DNA that contains repeated sequences, reassociation of fragment sizes that are longer than the length of the repeating units can result in a gross overestimate of the fraction of the genome that is actually repeated. Use of very small fragment size can minimize this problem, and studies of the change in rate of reassociation with different fragment sizes can lead to a better understanding of the way these repeated sequences are distributed throughout the genome (Britten and Smith, 1970; Davidson *et al.*, 1973; Graham *et al.*, 1974).

The members of most families of repeated sequences are not identical, based on thermal stability studies of the products of renaturation (Martin and Hoyer, 1966; Britten and Kohne, 1968; McCarthy and Church,

1970). Duplex structures that are paired precisely throughout their lengths have a high thermal stability; structures with various degrees of mismatching have lower thermal stabilities. Studies in a number of laboratories have shown that a depression of 1°C in T_m (ΔT_m = 1°C) corresponds to about 1–1.5% mismatching in the base pairing of the duplex structure (Bautz and Bautz, 1964; Kotuka and Baldwin, 1964; Uhlenbeck *et al.*, 1968; Laird *et al.*, 1969). Sequences of repeated DNA from some families are identical or nearly so, and the T_m of their renatured product is high. The members of other families have diverged from each other, and their renatured duplexes have lower thermal stabilities.

Cross-reaction between the distantly related members of these families will obviously be retarded if not eliminated by high incubation temperatures of reaction. Since the T_m is dependent on salt concentration and the presence of denaturing agents (McCarthy and Church, 1970; McConaughy *et al.*, 1969), the thermal stability of the reassociated product is then determined by the conditions of incubation (Church and McCarthy, 1968; McCarthy and McConaughy, 1968). Incubation conditions of lower and lower temperature and/or higher and higher salt concentration permit the reaction of more and more distantly related sequences (Martin and Hoyer, 1966). Distantly related sequences have a low melting temperature and hence a different temperature-rate dependence from precisely paired sequences of the same GC content. Correction of the estimate of repetition frequency to allow for the suppression in reassociation rate caused by mismatching is a function of the amount of mismatching and of the conditions of incubation. It has been suggested that this correction is a large one. From their studies of the reassociation of thermal fractions of renatured mouse satellite DNA, Sutton and McCallum (1971) proposed that the rate suppression was a factor of 20–50 for mismatching that results in a ΔT_m of 20°C. However, renaturation studies with hybrid sequences from different bacterial species and with deaminated bacteriophage DNA indicate that for each 10°C depression in the melting temperature due to sequence divergence, the rate of DNA reassociation is depressed by a factor of 2 after correction to the optimum temperature of incubation (Bonner and Britten, 1971; Bonner *et al.*, 1973). Similar results have been reported by Lee and Wetmur (1973) for DNA modified by reaction with chloroacetaldehyde. The same laboratory reported affects approximately half as large for DNA modifed by deamination and glyoxalation (Hutton and Wetmur, 1973). This effect of base mismatching is small when compared to the reported large differences in reassociation rate between different fractions of one eukaryotic DNA. However, these small differences may be important for hybridization measurements between DNAs of different species (Bonner *et al.*, 1973).

Section II

This section surveys DNA reassociation kinetics studies on nucleic acids of various eukaryotes. Often, for technical reasons, the DNA for reassociation studies is isolated from whole tissues on the assumption that the nucleic acid so isolated is representative of nuclear DNA sequences. The validity of such work depends on the two further assumptions that cytoplasmic DNA represents a negligible amount of the total cellular DNA and that there is no difference in the nuclear DNA of different cell types. Although such assumptions appear generally safe, there are a number of notable exceptions to this rule, examples of which follow.

Cytoplasmic DNA

Mitochondria and chloroplasts have their own specific DNA which generally represents only a few percent of the total cellular DNA (for review, see Granick and Gibor, 1967). Since the genome size of these two organelles is quite small (0.01–0.00002 pg per organelle, as compared to eukaryotic haploid genome sizes of 0.1–50 pg), DNA from these organelles would contribute to the repeated fraction of DNA if it were present as a large proportion of the total cellular DNA.

Extreme examples of excessively large amounts of mitochondrial DNA are found in the oocytes of *Rana pipiens* and *Xenopus laevis,* which contain 300–500 times as much DNA as somatic cells. This extra DNA is mitochondrial in origin (Dawid, 1965, 1966). Since the eggs of many different animals contain larger amounts of DNA than somatic cells, a relatively high abundance of mitochondrial DNA could be a general phenomenon of eggs (Dawid, 1965).

There are other examples where mitochondrial DNA represents a significant proportion of the total cellular DNA. Twenty-eight percent of the DNA isolated from whole cells of the cellular slime mold *Dictyostelium discoideum* is mitochondrial in origin and increases the estimate of repeated DNA sequences in this fungus (Firtel and Bonner, 1972). Another example of this is found in the nematode *Ascaris lumbricoides.* In the four-cell stage of this organism, 26% of the total cellular DNA is mitochondrial and is present as a light satellite on a CsCl density gradient. This cytoplasmic DNA increases the estimate of repetitive DNA sequences (Figure 3) since the DNAs of other stages of *Ascaris* development do not contain an obvious satellite and have much fewer repetitive (rapidly reassociating) sequences (Tobler *et al.,* 1972).

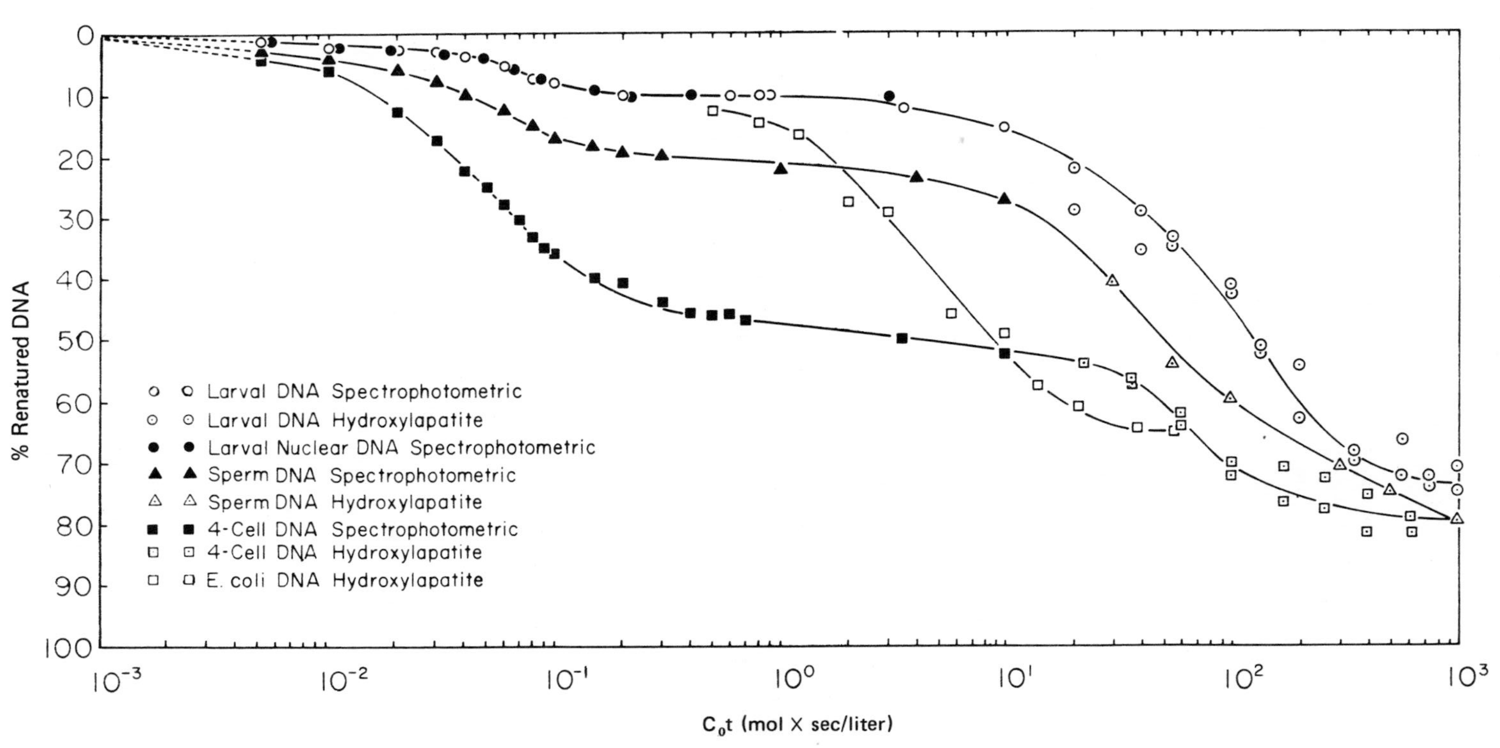

Figure 3. Kinetics of reassociation of Ascaris lumbricoides DNA. All DNAs were sheared, heat denatured, and incubated at 60° C in 0.12 M phosphate buffer, 0.001 M EDTA. C_0t points up to a C_0t of 10 were determined spectrophotometically. C_0t points above 10 were determined by hydroxylapatite binding. The data have not been normalized and have a maximum reaction of 80%. Typically, DNA reassociation should continue beyond 95%. From Tobler et al. (1972).

Alterations in Nuclear DNA

An elementary rule of cytology and genetics states that the chromosomal contents of germ-line cells and somatic cells are equivalent. However, the cytological phenomena of chromatin elimination, differential replication, and gene amplification can cause drastic differences in the DNA of some cells in specific animals.

Chromatin elimination and selective chromosome loss have been reported in a number of genera. For example, in the *Sciara,* whole chromosomes are eliminated from somatic tissues; in the genus *Ascaris,* the elimination of chromatin in somatic tissues is restricted to the terminal heterochromatin (Swanson, 1957). DNA reassociation studies show that the 27% of the DNA lost in somatic cells of *Ascaris lumbricoides* is comprised of equal amounts of repeated and unique DNA sequences (Figure 3). Since the repeated sequences initially represented 23% of the total genome, the process of chromatin elimination results in an increase in the relative amount of unique sequences over repeated sequences (Tobler *et al.,* 1972).

Differential replication has been observed during the formation of polytene chromosomes in dipterans. Presumably some regions (especially the heterochromatic centric regions) suffer from underreplication (Rudkin, 1969). Dickson *et al.* (1971) showed that DNA isolated from the salivary glands of *Drosophila hydei* contained many fewer repeated sequences than pupal or embryo DNA (Figure 4). Since a number of studies have shown that much of the repetitive DNA of *Drosophila* is restricted to centric heterochromatin (Gall *et al.,* 1971; Botchan *et al.,* 1971; Kram *et al.,* 1972), the underreplication of DNA in the centric heterochromatin is consistent with the parallel loss of repeated DNA sequences.

Amplification is a process whereby a very small part of the genome is multiplied many times while the rest of the DNA does not replicate. An extreme example of amplification is found in animal oocytes. Here the ribosomal cistrons overreplicate and become the extrachromosomal DNA reported in a large number of animal oocytes (for review, see Gall, 1969). In *Xenopus laevis,* for example, the ribosomal cistrons usually represent only 0.114% of the genome (Brown and Dawid, 1968); however, in the oocytes of *Xenopus laevis,* ribosomal cistrons may account for 7.5% of the nuclear DNA (Gall, 1969).

In spite of these selected examples, it is still probably safe to assume that the DNAs isolated from different tissues of one organism are equivalent. In plants this may be particularly true since vegetative propagation can be used to produce whole organisms complete with sex organs

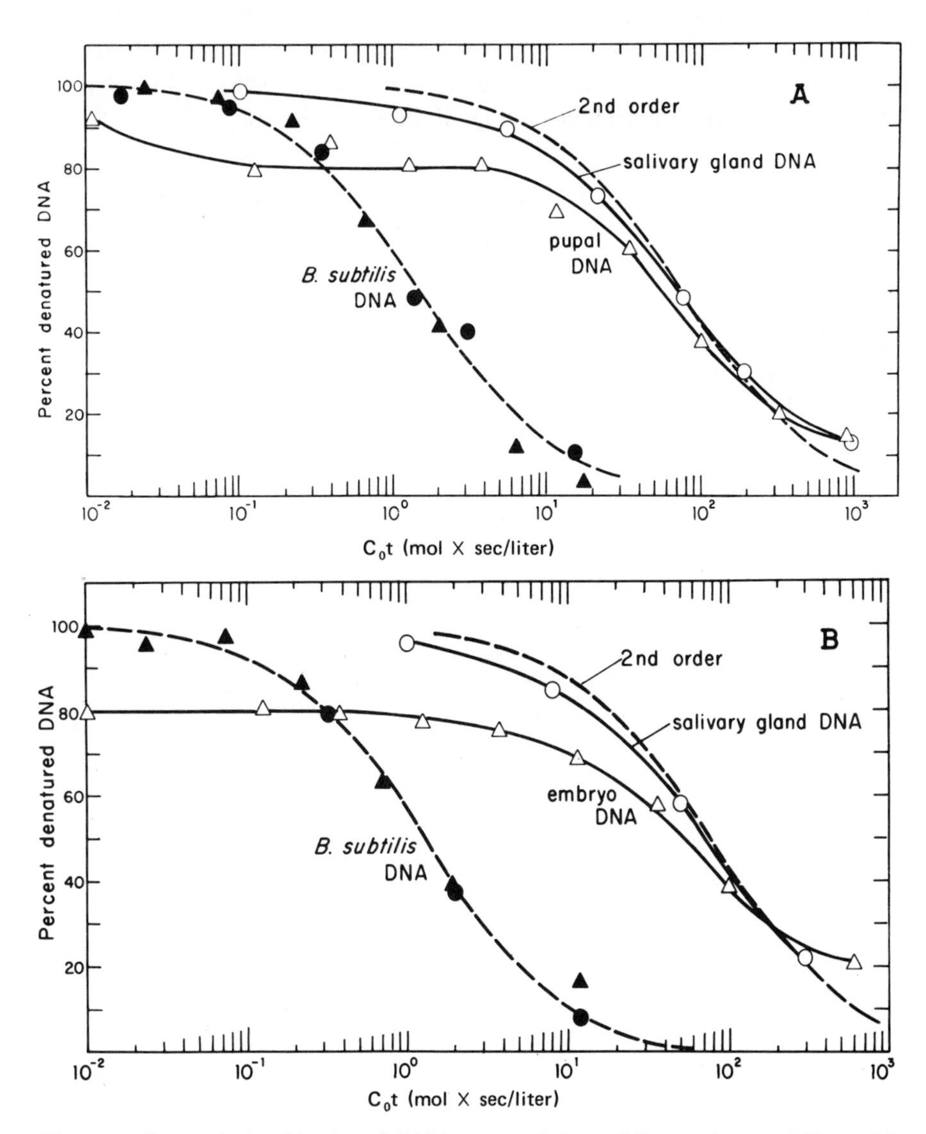

Figure 4. Reassociation kinetics of DNA extracted from different tissues of Drosophila hydei. Drosophila DNA and ^{14}C-labeled Bacillus subtilis DNA (internal standard) were sheared to a single-stranded molecular weight of about 16,500 daltons, denatured, and allowed to reassociate at 60°C in 0.12 M phosphate buffer. The extent of reassociation was measured by hydroxylapatite chromatography. From Dickson et al. (1971).

and since the same meristematic cells give rise to vegetative organs and sex organs at different stages of the life cycle. Studies involving vertebrate DNAs were unable to detect differences in the reassociation properties of the DNA of various tissues including liver, brain, spleen, kidney, and muscles (McCarthy and Hoyer, 1964; Britten and Kohne, 1968).

DNA Reassociation Kinetics Studies of Eukaryotes

If the genome size of the organism under study is known, then one can predict the rate of reassociation of the DNA sequences present only once per haploid genome. If this so-called unique component is subtracted from the C_0t plot, then the remaining faster components are assumed to be repeated DNA and the $C_0t_{1/2}$ values of reaction are assumed to be inversely proportional to their repetition frequency. One can use a computer program to fit hypothetical components to the C_0t plots (see Figure 6) or make cruder visual estimates applying the information and cautions in Sec-

TABLE 1. Estimated Amounts of Repeated DNA in the Protista and Fungi

Organism	Percent unique DNA	Repeated DNA		Reference
		Percent	Frequency	
Protozoa				
Paramecium aurelia strains 1/540, 1/90, 2/71, 8/138, 8/299	85	15	53–76	Figure 5
Tetrahymena pyriformis strains				
7/NC-651	70	30	25	Figure 5
1/D	75	25	33	
1/8, 1/D/1, 1/7	80	20	30	
1/C	85	15	28	
1/A, T[e]	95	5	—	
Acanthamoeba castellanii	70	30	130	Jantzen, 1973
Algae				
Chlamydomonas reinhardi	70	30	>1000	Wells and Sager, 1971
Fungi				
Dictyostelium discoideum (cellular slime mould)	60	28	113	Figure 6
Physarum polycephalum	58	42	15,000	Fouquet *et al.*, 1974
Neurospora crassa	80	20	60	Brooks and Huang, 1972

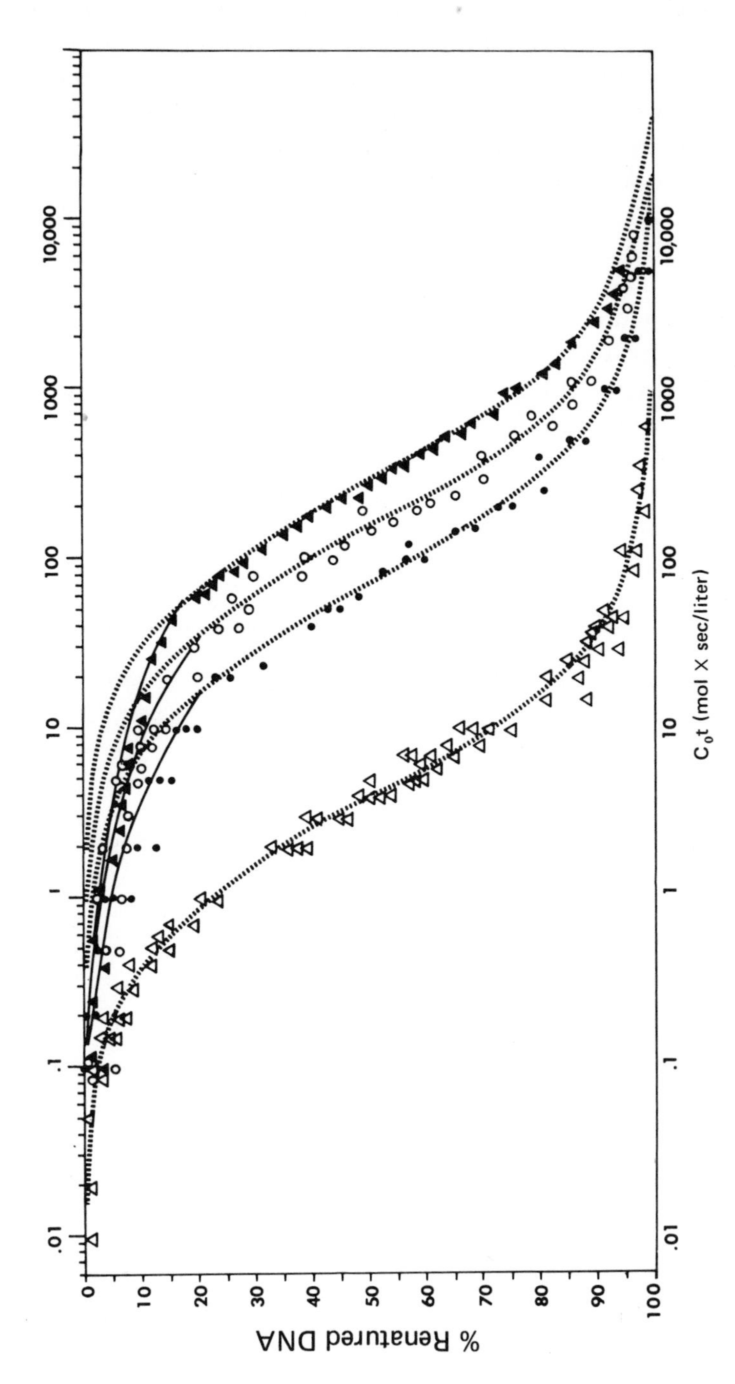

Figure 5. *Reassociation kinetics of Escherichia coli and ciliate DNA. The DNA sheared to fragments about 500 nucleotides long was reassociated in 0.12 M in phosphate buffer at 50°C and the degree of reaction was monitored by hydroxylapatite chromatography. Tetrahymena pyriformis strain 1/7 (●) has a smaller micronucleus than strain 1/8 (○); Paramecium aurelia has a larger genome still (▲). Radioactive E. coli (Δ) DNA was included as an internal standard. From Allen and Gibson (1972).*

tion I. An example of the latter form will be applied to the C_0t plot of *Triticum aestivum* in the section of the Metaphyta reassociation studies.

For convenience, the reassociation data are divided into three groups: data for the Protista (the more primitive eukaryotes), those for the Metaphyta (the more highly evolved plants), and those for the Metazoa (the more highly evolved animals).

Protista. Table 1 contains the estimates of repeated and unique fractions of various members of the Protista.

Metaphyta. The plants of Metaphyta include a large number of different organisms with a wide range of genome sizes (Rees and Jones, 1972; Bennett, 1972). Unfortunately, the reassociation kinetics of only a few plants have been studied so far. Those studied have been moderately large in genome size and all contain large amounts of repeated DNA. The reassociation kinetics of seven conifers indicate that 70–80% of the DNAs of these organisms reassociate as though they are composed of repeated sequences (Miksche and Hotta, 1973). The monocots studied are *Triticum*

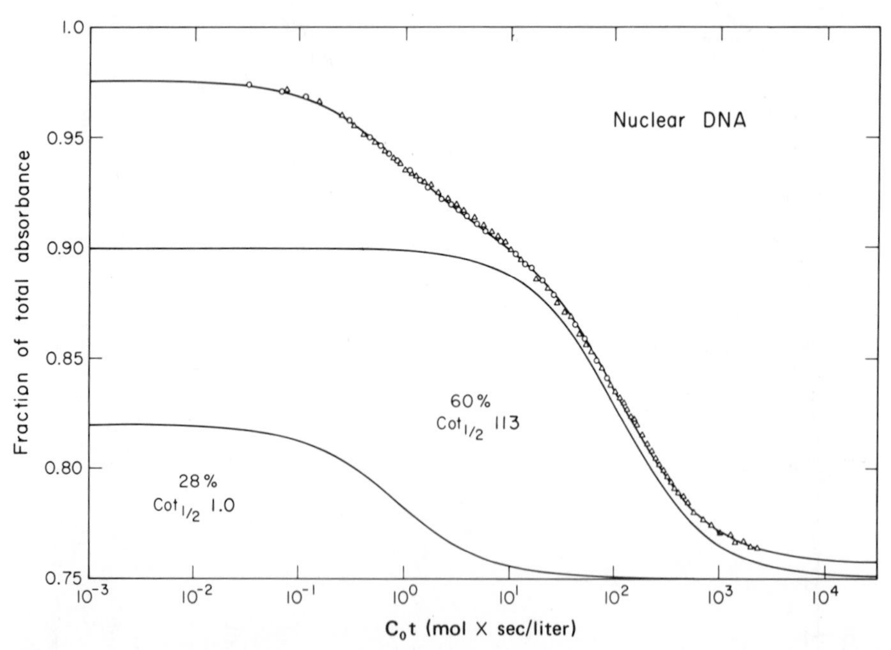

Figure 6. C_0t plot for sheared DNA of Dictyostelium discoideum (cellular slime mold). Reassociation was performed in 0.12 M phosphate buffer at 56° C (○) and in 0.24 M phosphate buffer at 60° C (Δ). The data collected from the higher salt incubation were multiplied by a correction factor to make them comparable with those of the other conditions of incubation. The data were analyzed by a computer and the three curves represent a least-squares fit of components to the data. From Firtel and Bonner (1972).

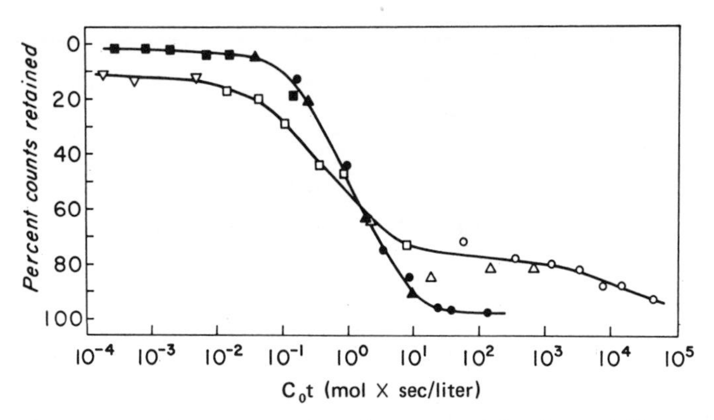

Figure 7. Kinetics of DNA reassociation of wheat and Bacillus subtilis measured with hydroxylapatite. ^{3}*H-labeled hexaploid wheat and* ^{14}*C-labeled B. subtilis (6700 cpm/μg) sheared DNAs were mixed and reassociated at 40°C in 62% formamide, 0.78 M NaCl, and* 5×10^{-3} *M Na-phosphate buffer, pH 6.8.* ●, ▲, ■, *B. subtilis DNA at 16, 4, and 0.3 μg/ml, respectively;* ○, △, □, *wheat DNA at 6000, 300, and 16 μg/ml, respectively;* ▽, *wheat DNA at 0.5 μg/ml reassociated in the absence of B. subtilis DNA. From Bendich and McCarthy (1970).*

aestivum (Figure 7) and *Secale cereale* (Bendich and McCarthy, 1970); the dicots are represented by *Vicia faba* and *Vicia sativa.*

These examples show that the plants under study have a high proportion of highly reiterated DNA sequences. For example, *Triticum aestivum* (wheat) has a genome size of 18 pg of DNA per haploid nucleus (Bennett, 1972). This is 5300 times larger than the genome size of *Bacillus subtilis* (3.4×10^{-3} pg Laird, 1971). Virtually all of the DNA of *B. subtilis* is unique. The reassociation of this DNA should approximate second-order kinetics, and this seems to be the case (Figure 7). The unique DNA component of *Triticum aestivum* is expected to react 5300 times slower than the DNA of *B. subtilis*; therefore, the $C_0t_{1/2}$ of *T. aestivum* unique DNA will be 5300 times larger than the $C_0t_{1/2}$ of *B. subtilis* DNA which is reassociating under the same conditions. Since the $C_0t_{1/2}$ of *B. subtilis* is about 1, the $C_0t_{1/2}$ of reassociation of unique DNA sequences in *T. aestivum* should be 5300 mole sec/liter. Figure 7 shows that 10% of the genome of wheat reassociates extremely rapidly. This component could have a $C_0t_{1/2}$ value of 10^{-5} or less. This means that this component either represents sequences that have intrastrand homologies and merely fold back upon reassociation, or a repeated sequence with a repetition frequency greater than $5300/10^{-5}$ or 5.3×10^8. This would represent very short sequences since 10% of the wheat genome or 1.7×10^9 nucleotide pairs (1.8 pg) must contain more than 5.3×10^8 copies of short sequence. Thus this could be shorter than three nucleotides. On the other hand, 70%

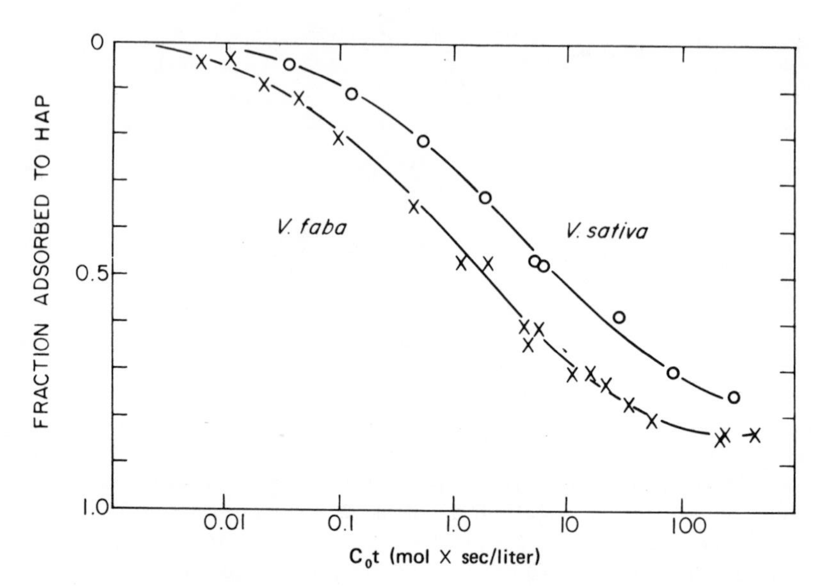

Figure 8. Reassociation kinetics of Vicia faba and Vicia sativa DNA. The DNA was sheared to a single-stranded fragment length of 400–500 nucleotides and reassociated in 0.14 M phosphate buffer at 60°C. The extent of reassociation was measured by hydroxylapatite chromatography. Under identical conditions, E. coli DNA reassociates with $C_0t_{1/2}$ value of 4.0 mole sec/liter. Based on the haploid genome size measurements for V. faba being 13–14 pg and for S. sativa being 2.7 pg (Rees and Jones, 1972; Bennett, 1972), the $C_0t_{1/2}$ of unique DNA of these two species should be 13,000 and 2400 mole sec/liter, respectively.

of the genome reassociates as though it were one component with a $C_0t_{1/2}$ value of 0.5 mole sec/liter. This would represent a component with a repetition frequency of 5300/0.5 or 10,600. The remainder of the wheat genome is probably unique DNA. Like wheat, *Secale cereale* (rye) contains a great deal of repeated DNA (about 80%). However, rye DNA is composed of a larger spectrum of components whose repetition frequencies vary from many to few (visual estimations from data of Bendich and McCarthy, 1970).

The dicots *Vicia faba* and *Vicia sativa* similarly have large proportions of their genomes present as repeated DNA (Figure 8). About 80–85% of the genome of these two plants is composed of repeated sequences. The range of repetition frequencies in *V. faba* could be 10^3–10^5 and in *V. sativa* 10^2–10^4 (Figure 8).

Metazoa. Invertebrates also contain many divergent groups with larger differences in genome size: 0.1×10^{-12} g to 6×10^{-12} g per haploid genome size (Rees and Jones, 1972). Table 2 contains estimates of

TABLE 2. Repeated DNA in Invertebrates

Animal	*Percent unique DNA*	*Repeated DNA*		*Reference*
		Percent	*Frequency*	
Ascaris lumbricoides germ-line DNA	77	23	7000–10,000	Figure 3
Ophiotrix quinquemaculata (serpent star)	45	30	30	Weinblum *et al.*, 1973
		25	600	
Psammechinus milaris (sea urchin)	60	40	100–1000	Kedes and Birnsteil, 1971
Strongylocentrotus purpuratus (sea urchin)	38	25	20–50	Britten *et al.*, 1972
		27	250	
		7	6000	
		3	Very rapid reassociation	
	50	27	10	Graham *et al.*, 1974
		19	164	
Sphaerechinus granularis (sea urchin)	30	20	3	Weinblum *et al.*, 1973
		31	100	
		19	800	
Holothuria forskali (sea cucumber)	45	25	100	Weinblum *et al.*, 1973
		30	450	
H. poli	48	24	30	Weinblum *et al.*, 1973
		28	200	
H. tubulosa	55	22	30	Weinblum *et al.*, 1973
		23	1000	
Lima sp. (marine clam)	48	28	30	Weinblum *et al.*, 1973
		24	700	
Nassaria obsoleta (marine gastropod)	38	12	20	Davidson *et al.*, 1971
		15	1000	
			Very rapid reassociation	
Phallusia mammillata (tunicate)	50	30	9	Weinblum *et al.*, 1973
		20	400	
Ciona intestinalis (tunicate)	70	30	—	Lambert and Laird, 1971;
Bombyx mori (silk worm)	55	21	500	Gage, 1974
		24	50,000	
Prosimulium multidentatum (blackfly)	56	24	1090	Sohn *et al.*, 1975
		7–10	Very rapid reassociation	
Drosophila melanogaster (fruit fly)	78	15	35	Wu *et al.*, 1972
		7	2600	
Chironomus tentans	95.5	4.5	120	Sachs and Clever, 1972

repeated DNA based on the reassociation kinetics of a number of invertebrates.

The vertebrates include fish, amphibians, reptiles, birds, and mammals. As yet no published work on reassociation kinetics exists for the Agnatha and Chondrichthyes, and only two species of Osteichthyes have been studied. The reassociation kinetics of salmon DNA indicates that about 80% of the genome of salmon reassociates as though it is composed of many components of repeated frequencies with a wide spectrum of repetition frequencies (Figure 9). In the case of *Musgurnus fossilis*, a European freshwater loach, DNA when reassociated at 66°C in 0.14 M phosphate buffer reassociates as though 50% of the DNA sequences are uniquely represented in the haploid genome, while 10% are repeated 300 times, 15% are repeated 1500 times, 15% are repeated 50,000 times, and

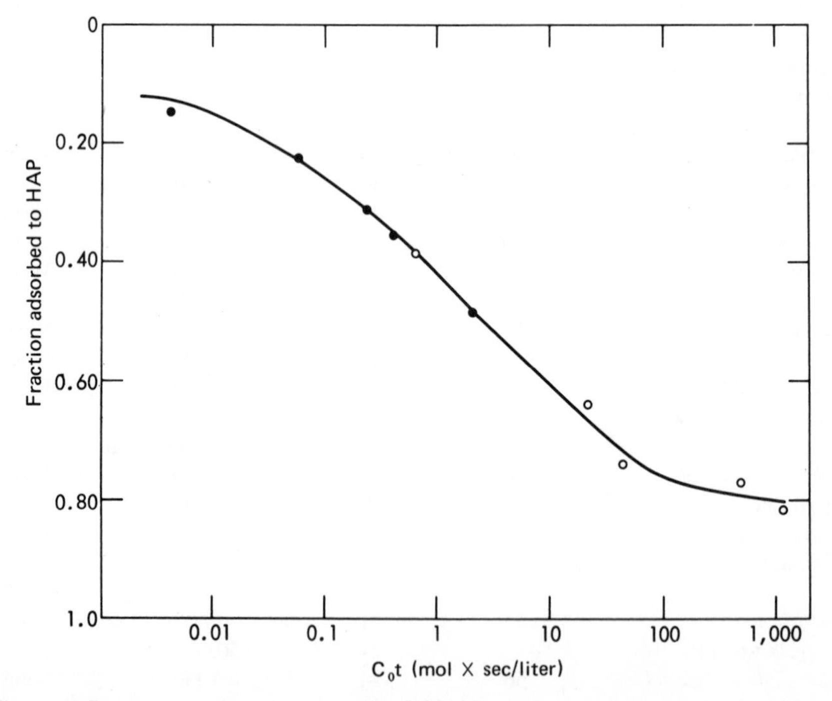

Figure 9. Reassociation kinetics of salmon DNA. Sheared denatured DNA was incubated at 50°C in 0.14 M phosphate buffer at 50°C. The extent of reassociation was monitored by hydroxylapatite chromatography. Under these conditions of reassociation, the $C_0t_{1/2}$ of E. coli is about 6.7 mole sec/liter (Straus and Bonner, 1972). The genome size of Salmo irideus is 2.5×10^{-12} g of DNA haploid genome (Rees and Jones, 1972) or 580 times larger than that of E. coli. If the salmon referred to in this diagram has a similar genome size, its $C_0t_{1/2}$ of reaction for unique DNA should be about 4000 mole sec/liter. From Britten and Kohne (1968).

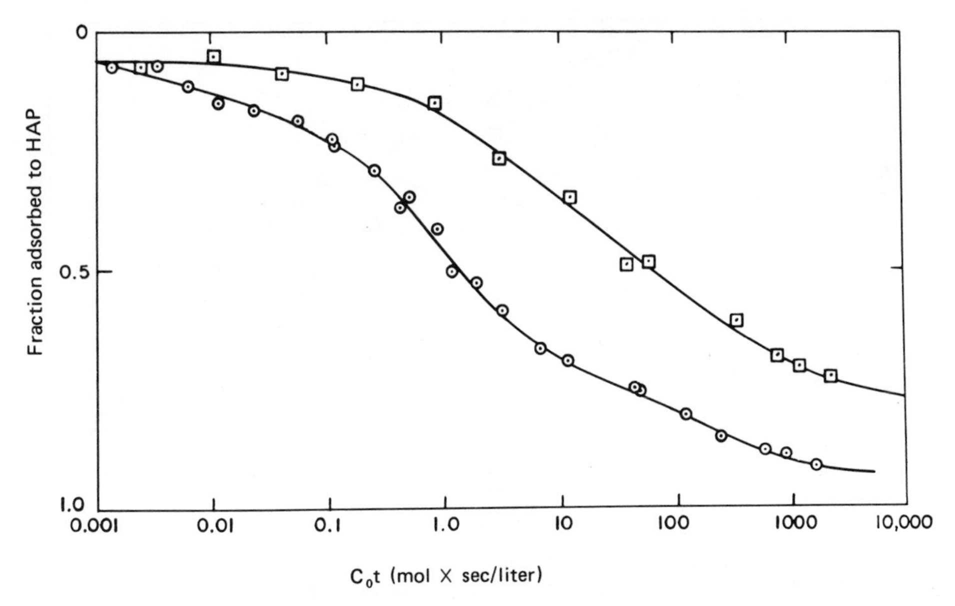

Figure 10. Renaturation of sheared Ambystoma tigrinum DNA (□, $^{A}260$ = 0.074–30.0) and Necturus maculosus DNA (○, $^{A}260$ = 0.116–23.2). "Fraction adsorbed" refers to the fraction of DNA that has renatured (double-stranded DNA adsorbs to hydroxylapatite). The solid lines are computer-plotted least-square fits of three components to the data. The incubation conditions were 0.12 M phosphate buffer at 60° C. From Straus (1971).

4% reassociate more rapidly than measured by the conditions of these experiments (Kuprijanova and Timofeeva, 1974).

Amphibians have been studied in more detail. Amphibians, for a single taxonomic class, have an impressively wide variation in genome sizes. The more primitive urodeles generally have larger genomes than the anurans. Within this group alone, there is a hundredfold spread in haploid genome size, 0.8×10^{-12} g to 80×10^{-12} g per haploid nucleus (Rees and Jones, 1972; Sexsmith, 1968).

The reassociation kinetics of a selection of DNAs from amphibians covering the complete spectrum of genome sizes is shown in Figures 10 and 11. Table 3 contains the results of computer analysis of these curves. Paralleling the wide variation in genome size, different genera of amphibians have widely differing reassociation kinetics. However, it appears that the larger genomes have greater amounts of repeated DNA and have families of repeated sequences whose repetition frequency is higher than the smaller genome amphibians.

Davidson *et al.* (1973) (Table 3) have studied the genome of *Xenopus laevis* (African clawed toad) in more detail, including an investigation into the interspersion of repetitive sequences with nonrepetitive sequences. By

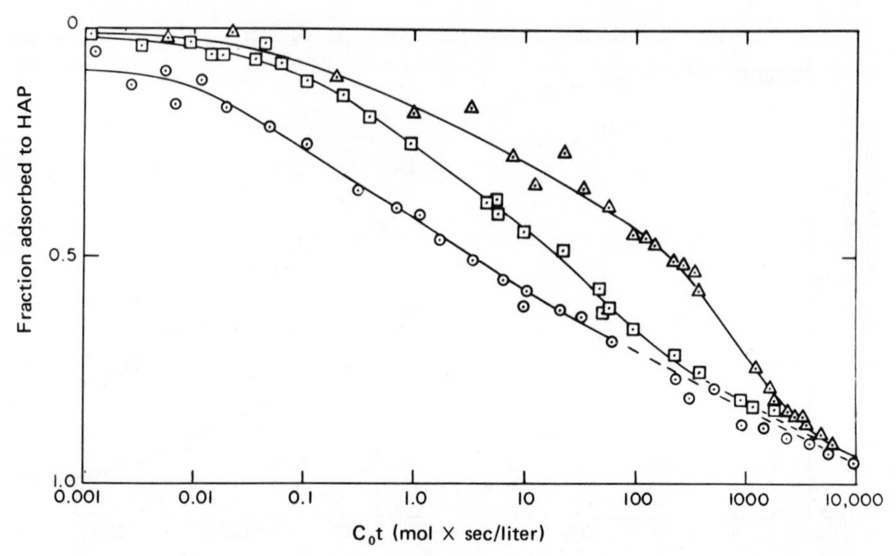

Figure 11. Renaturation of sheared Scaphiopus couchi DNA (Δ, $^{A}260 = 0.417–124$), Rana clamitans DNA (□, $^{A}260 = 0.085–116$), and Bufo marinus DNA (O, $^{A}260 = 0.103–157$). The solid lines are computer-plotted least-square fits of three components to the data. The broken lines add the theoretical unique components. The incubation conditions were 0.12 M phosphate buffer at 60° C. From Straus (1971).

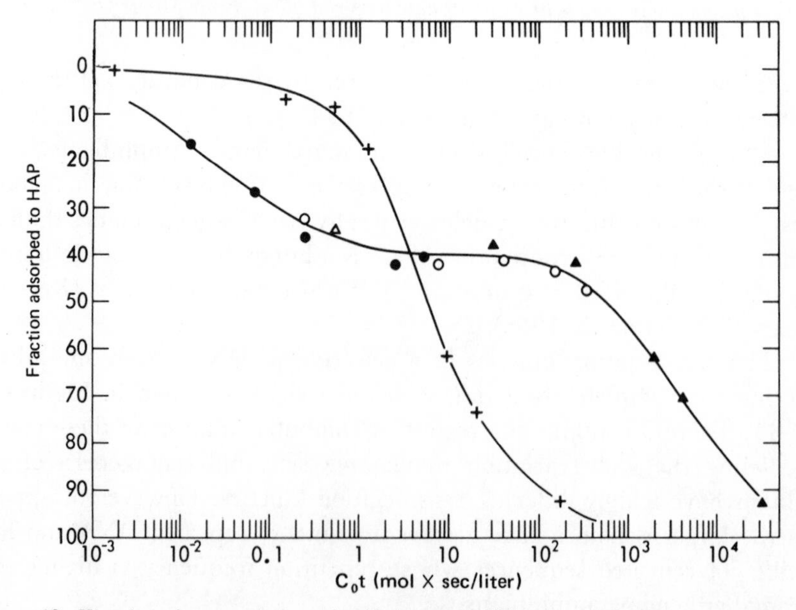

Figure 12. Kinetics of reassociation of calf thymus DNA measured with hydroxylapatite. The DNA was incubated at 60°C in 0.12 M phosphate buffer. The DNA concentrations during the reaction were (μg/ml) (Δ) 2, (●) 10, (O) 600, and (▲) 8600. Radioactively labeled E. coli DNA (X) at 43 μg/ml was present in the reaction containing calf thymus DNA at 8600 μg/ml. From Britten and Kohne (1968).

TABLE 3. *Repeated DNA in Amphibians*

Species	*Percent unique DNA*	*Repeated DNA*		*Reference*
		Percent	*Frequency*	
Scaphiopus couchi (spadefoot toad)	60	24	100	Figure 11
		4	5,400	
Bufo marinus (marine toad)	26	14	35	Figure 11
		27	1000	
		26	48,000	
Xenopus laevis (African clawed toad)	54	6	20	Davidson and Britten, 1973
		31	1,600	
		6	32,000	
		3	*f*	Davidson *et al.*, 1973
Rana clamitans (green frog)	22	40	93	Figure 11
		27	4,100	
		9	72,000	
Ambystoma tigrinum (tiger salamander)	≤24	27	54	Figure 10
		37	3,500	
		6	2,500,000	
Necturus umaculosus (mudpuppy)	≤12	20	330	Figure 10
		54	50,000	
		14	6,700,000	

reassociating tracer quantities of radioactively labeled DNA of different fragment sizes with a large excess of unlabeled DNA of fixed size, they were able to show that 50% of the *Xenopus* genome consists of repetitive sequences about 300 nucleotides long interspersed with nonrepetitive sequences 800 nucleotides long (Davidson *et al.*, 1973).

The taxonomic groups representing reptiles, birds, and mammals, unlike the amphibian group, have relatively narrow ranges in genome size. The haploid genome size of mammals is slightly less than twice that of birds (Rees and Jones, 1972; Bachmann, 1972). Table 4 contains the available reassociation data for a limited number of birds and mammals. As yet no DNA reassociation kinetics studies have been published for reptiles. Most mammalian DNAs appear to contain a fraction of very highly repeated sequences that represent 2–10% of the genome (Britten and Davidson, 1971) and a somewhat larger fraction of more slowly reassociating DNA. According to Britten and Kohne (1968), DNA from cattle, unlike that from birds and other mammals, appears to have little DNA that reassociates as though it were composed of sequences with lower repetition frequency (2–6000) (Figure 12). Again one must keep in mind that the reassociation profile depends on the experimental conditions.

TABLE 4. Repeated DNA in Birds and Mammals

Species	*Percent unique DNA*	*Repeated DNA*		*Reference*
		Percent	*Frequency*	
Gallus domesticus (domestic chicken)	70	24	120	Sanchez de Jiminez *et al.*, 1974
		3	330,000	
		3	1,100,000	
Junco hyemalis (darkeyed junco)	63	16	200	Shields and Straus, 1974
		21	370,000	
Rattus (rat)	65	19	1,500	Holmes and Bonner, 1974
		9	Very rapid reassociation	
Mus musculus (mouse)				
Incubated at 50 °C	49	13	127	Straus, unpublished
		18	7,500	
		13	Very rapid reassociation	
Incubated at 60 °C	60	14	600	Straus and Birnboim, 1974
		15	47,000	
		8	Very rapid reassociation	
Bos taurus (cattle)	55	38	60,000	Davidson and Britten, 1973
		2	1,000,000	
		3	Very rapid reassociation	
Homo sapiens	64	13	Low	Saunders *et al.*, 1972
		12	Intermediate	
		10	High	

For example, less stringent conditions of reassociation of mouse DNA allow one to observe the formation of duplex structure with less base pairing of presumably more distantly related sequences (Table 4).

Origin of Repeated Sequences

Families of repeated sequences may vary widely in their repetition frequency and in the degree to which their members are related (Section I). Britten and Kohne (1968) postulated that repeated DNA is formed by a "saltatory event"; that is, within a relatively short period of time a sequence is multiplied many, many times in one genome in the form of identical tandem repeats. With time and species divergence, mutations accumulate within the members of each of these families so that the members gradually diverge from each other. In addition, during this period of divergence, the clustering of these repeated sequences is gradually eroded as the sequences are dispersed throughout the genome.

Much of the evidence accumulated since the elaboration of the "saltation theory" has added support to Britten and Kohne's theory. Tandem multiplication of a short sequence whose base composition differs from that of the rest of the genome could result in the formation of species-specific satellite DNA. Isolated DNA consists of a population of large fragments of DNA. Generally most fragments have some average base composition which is close to that of the average base composition of total DNA. However, when a short sequence of different base composition is arranged in tandem repeats, the result is a set of fragments whose overall base composition differs from that of the remaining fragments. Such DNA will form a satellite on an isopycnic equilibrium gradient formed by centrifuging solutions of CsCl or Cs_2SO_4. The relationship between repeated DNA and satellite DNA has been extensively reviewed by Flamm (1972). Cytologically, these species-specific satellites are also clustered. *In situ* hybridization studies have located *Drosophila* satellite DNA in the centromeric heterochromatin (Gall *et al.,* 1971). Similarly, when radioactive mouse satellite DNA sequences are hybridized to chromosome preparations, autoradiography shows that these satellite sequences are located in the centromeric heterochromatin of mouse chromosomes (Pardue and Gall, 1970).

Highly repeated sequences that are clustered but have the same base composition as total DNA can be isolated on CsCl equilibrium density gradients, also. Unsheared DNA is denatured and renaturated for a short time and then centrifuged on an equilibrium density gradient. Highly repeated tandemly arranged sequences will reform double helices; however, the remainder of the DNA will either form huge networks that are mostly single stranded because of interspersion of repeated sequences (Section I) or not reassociate because of the necessity of a longer period of incubation for reassociation. Since denatured DNA has a heavier density than double-stranded DNA on a CsCl density gradient, clustered repeated sequences can be isolated as a satellite whose density is close to that of native DNA. Indeed, this method provides an alternative method of isolating sequences that normally form a satellite upon centrifugation of undenatured DNA (Britten and Waring, 1965).

The strongest evidence supporting the "saltation theory" comes from DNA:DNA hybridization studies in rodents (Rice, 1972). These studies indicate that families of repeated DNA which are restricted to one species form reassociated products of high thermal stability. In other words, the members of recently formed families are virtually identical. Families of repeated DNA whose product of reassociation had low thermal stability were found in a number of related species. These families formed hybrids of similar thermal stability when radioactive DNA from one species was hybridized to a vast excess of DNA from another species. Members of these older families of repeated DNA appear to be diverging from each other both within one species and between species, as one would expect if the saltation theory of evolution represents the real situation.

Acknowledgments

I acknowledge with gratitude the helpful suggestions of Drs. T. I. Bonner, I. R. Brown, and N. R. Rice.

Literature Cited

Allen, S. and I. Gibson, 1972 Genome amplifications and gene expression in the ciliate macronucleus. *Biochem. Genet.* **6:**293–313.

Attardi, G. and F. Amaldi, 1970 Structure and synthesis of ribosomal RNA. *Annu. Rev. Biochem.* **39:**183–226.

Bachmann, K., 1972 Genome size in mammals. *Chromosoma* **37:**85–93.

Bautz, E. K. F. and F. A. Bautz, 1964 The influence of noncomplementary bases on the stability of ordered polynucleotides. *Proc. Natl. Acad. Sci. USA* **52:**1476–1481.

Bendich, A. J. and B. J. McCarthy, 1970 DNA comparisons among barley, oats, rye and wheat. *Genetics* **65:**545–565.

Bennett, M. D., 1972 Nuclear DNA content and minimum generation time in herbaceous plants. *Proc. R. Soc. London* **181:**109–135.

Bernardi, G., 1965 Chromatography of nucleic acids on hydroxyapatite. *Nature (London)* **206:**779–783.

Billeter, M. A., C. Weissman, and R. C. Warner, 1966 Replication of viral ribonucleic acid. IX. Properties of double-stranded RNA from *Escherichia coli* infected with bacteriophage MS2. *J. Mol. Biol.* **17:**145–173.

Bonner, T. I. and R. J. Britten, 1971 The effect of sequence divergence on the rate of reassociation. *Carnegie Inst. Washington Yearb.* **70:**373–374.

Bonner, T. I., D. J. Brenner, B. R. Neufeld, and R. J. Britten, 1973 Reduction in the rate of DNA reassociation by sequence divergence. *J. Mol. Biol.* **81:**123–135.

Botchan, M., R. Kram, C. W. Schmid, and J. E. Hearst, 1971 Isolation and chromosomal localization of highly repeated DNA sequences in *Drosophila melanogaster. Proc. Natl. Acad. Sci. USA.* **68:**1125–1129.

Britten, R. J., 1969 The arithmetic of nucleic acid reassociation. *Carnegie Inst. Washington Yearb.* **67:**332–335.

Britten, R. J. and E. H. Davidson, 1971 Repetitive and non-repetitive DNA sequences and a speculation on the origins of evolutionary novelty. *Q. Rev. Biol.* **46:**111–138.

Britten, R. J. and D. E. Kohne, 1968 Repeated sequences in DNA. *Science* **161:**529–540.

Britten, R. J. and J. Smith, 1970 A bovine genome. *Carnegie Inst. Washington Yearb.* **68:**378–386.

Britten, R. J. and M. Waring, 1965 "Renaturation" of the DNA of higher organisms. *Carnegie Inst. Washington Yearb.* **64:**316–333.

Britten, R. J., D. E. Graham, and M. Henrey, 1972 Sea urchin repeated and single-copy DNA. *Carnegie Inst. Washington Yearb.* **71:**270–273.

Britten, R. J., D. E. Graham, and B. R. Neufeld, 1974 Analysis of repeating DNA by reassociation. *Methods Enzymol.* **29:**363–418.

Brooks, R. R. and P. C. Huang, 1972 Redundant DNA of *Neurospora crassa*. *Biochem. Genet.* **6**:41–49 (1972).

Brown, D. D. and I. B. Dawid, 1968 Specific gene amplification in oocytes. *Science* **160**:272–280.

Church, R. B. and B. J. McCarthy, 1968 Related base sequences in the DNA of simple and complex organisms. II. The interpretation of DNA/RNA hybridization studies with mammalian nucleic acids. *Biochem. Genet.* **2**:55–73.

Davidson, E. H. and R. J. Britten, 1973 Organization, transcription and regulation in the animal genome. *Q. Rev. Biol.* **48**:565–579.

Davidson, E. H. and B. R. Hough, 1969 High sequence diversity in the RNA synthesized at the lampbrush stage of oogenesis. *Proc. Natl. Acad. Sci. USA* **63**:342–349.

Davidson, E. H., B. R. Hough, M. E. Chamberlin, and R. J. Britten, 1971 Sequence repetition in the DNA of *Nassaria* (*Ilyanassa*) *obsoleta*. *Dev. Biol.* **25**:342–463.

Davidson, E. H., B. R. Hough, C. S. Amenson, and R. J. Britten, 1973 General interspersion of repetitive with non-repetitive sequence elements in DNA of *Xenopus*. *J. Mol. Biol.* **77**:1–23.

Dawid, I. B., 1965 Deoxyribonucleic acid in amphibian eggs. *J. Mol. Biol.* **12**:581–599.

Dawid, I. B., 1966 Evidence for the mitochondrial origin of frog egg cytoplasmic DNA. *Proc. Natl. Acad. Sci. USA* **56**:269–276.

Dickson, E., J. B. Boyd, and C. D. Laird, 1971 Sequence diversity of polytene chromosome DNA from *Drosophila hydei*. *J. Mol. Biol.* **61**:615–627.

Doty, P., H. Boedtker, J. R. Fresco, R. Haselkorn, and M. Litt, 1959 Secondary structure in ribonucleic acids. *Proc. Natl. Acad. Sci. USA* **45**:482–499.

Firtel, R. A. and J. Bonner, 1972 Characterization of the genome of the cellular slime mold *Dictyostelium discoideum*. *J. Mol. Biol.* **66**:339–361.

Flamm, W. G., 1972 Highly repetitive sequences of DNA in chromosomes. *Int. Rev. Cytol.* **32**:1–51.

Fouquet, H., B. Bierweiler, and H. W. Sauer, 1974 Reassociation kinetics of nuclear DNA from *Physarum polycephalum*. *Eur. J. Biochem.* **44**:407–410.

Gage, L. P., 1974 The *Bombyx mori* genome: Analysis by DNA reassociation kinetics. *Chromosoma* **45**:27–42.

Gall, J. G. 1969 Genes for ribosomal RNA during oogenesis, *Genetics Suppl. No. 1, Part 2* **61**:121–132.

Gall, J. G., E. H. Cohen, and M. L. Polan, 1971 Repetitive DNA sequences in *Drosophila*. *Chromosoma* **33**:319–344.

Gillespie, D. and S. Spiegelman, 1965 A quantitative assay for DNA-RNA hybrids with DNA immobilized on a membrane. *J. Mol. Biol.* **12**:829–842.

Graham, D. E., B. R. Neufeld, E. H. Davidson, and R. J. Britten, 1974 Interspersion of repetitive and non-repetitive DNA sequences in the sea urchin genome. *Cell* **1**:127–137.

Granick, S. and A. Gibor, 1967 The DNA of chloroplasts, mitochondria, and centrioles. *Prog. Nucleic Acid Res. Mol. Biol.* **6**:143–186.

Holmes, D. S. and J. Bonner, 1974 Sequence composition of rat deoxyribonucleic acid and high molecular weight nuclear ribonucleaic acid. *Biochemistry* **13**:841–848.

Hutton, J. R. and J. G. Wetmur, 1973 Effect of chemical modification on the rate of renaturation of deoxyribonucleic acid, deaminated and gluoxalated deoxyribonucleic acid. *Biochemistry* **12**:558–563.

Jantzen, H., 1973 Change of genome expression during development of *Acanthamoeba castellanii*. *Arch. Mikrobiol.* **91**:163–178.

Kedes, L. H. and M. L. Birnstiel, 1971 Reiteration and clustering of DNA sequences complementary to histone messenger RNA. *Nature (London) New Biol.* **230:**165–169.

Kennel, D. E., 1971 Principles and practices of nucleic acid hybridization. *Prog. Nucleic Acid Res. Mol. Biol.* **11:**259–301.

Kohne, D. E., 1970 Evolution of higher-organism DNA. *Q. Rev. Biophys.* **3:**327–375.

Kotuka, T. and R. L. Baldwin, 1964 Effects of nitrous acid on the dAT copolymer as a template for DNA polymerase. *J. Mol. Biol.* **9:**323–339.

Kram, R., M. Botchan, and J. E. Hearst, 1972 Arrangements of highly reiterated DNA sequences in centric heterochromatin of *Drosophila melanogaster*: Evidence of interspersed spacer DNA. *J. Mol. Biol.* **64:**103–117.

Kuprijanova, N. S. and M. J. Timofeeva, 1974 Repeated nucleotide sequences in the loach genome. *Eur. J. Biochem.* **44:**59–65.

Laird, C. D., 1971 Chromatid structure: Relationship between DNA content and nucleotide sequence diversity. *Chromosoma* **32:**378–406.

Laird, C. D., B. L. McConaughy, and B. J. McCarthy, 1969 Rate of fixation of nucleotide substitutions in evolution. *Nature* (London) **224:**149–154.

Lambert, C. C. and C. D. Laird, 1971 Molecular properties of tunicate DNA. *Biochim. Biophys. Acta* **240:**39–45.

Lee, C. H. and J. G. Wetmur, 1973 Physical studies of chloroacetaldehyde labelled fluorescent DNA. *Biochim. Biophys. Acta* **50:**879–885.

Marmur, J. and P. Doty, 1962 Determination of the base composition of deoxyribonucleic acid from its thermal denaturation temperature. *J. Mol. Biol.* **5:**109–118.

Marmur, J., R. Rownd, and C. L. Schildkraut, 1963 Denaturation and renaturation of deoxyribonucleic acid. *Prog. Nucleic Acid Res. Mol. Biol.* **1:**231–300.

Martin, M. A. and B. H. Hoyer, 1966 Thermal stabilities and species specificities of reannealed animal deoxyribonucleic acids. *Biochemistry* **5:**2706–2713.

McCarthy, B. J. and R. B. Church, 1970 The specificity of molecular hybridization reactions. *Annu. Rev. Biochem.* **39:**131–150.

McCarthy, B. J. and B. H. Hoyer, 1964 Identity of DNA and diversity of messenger RNA molecules in normal mouse tissues *Proc. Natl. Acad. Sci. USA* **52:**915–922.

McCarthy, B. J. and B. L. McConaughy, 1968 Related base sequences in the DNA of simple and complex organisms. I. DNA/DNA duplex formation and the incidence of partially related base sequences in DNA. *Biochem. Genet.* **2:**37–53.

McConaughy, B. L., C. D. Laird, and B. J. McCarthy, 1969 Nucleic acid reassociation in formamide. *Biochemistry* **8:**3289–3294.

Miksche, J. P. and Y. Hotta, 1973 DNA base composition and repetitious DNA in several conifers. *Chromosoma* **41:**29–36.

Miyazawa, Y. and C. A. Thomas, Jr., 1965 Nucleotide composition of short segments of DNA molecules. *J. Mol. Biol.* **11:**223–237.

Pardue, M. L. and J. G. Gall, 1970 Chromosomal localization of mouse satellite DNA. *Science* **168:**1356–1358.

Rees, H. and R. N. Jones, 1972 The origin of the wide species variation in nuclear DNA content. *Int. Rev. Cytol.* **32:**53–92.

Rice, N., 1972 Change in repeated DNA in evolution. In Evolution of Genetic Systems, edited by H. H. Smith. *Brookhaven Symp. Biol.* **23:**44–79.

Ross, P. D. and T. M. Sturtevant, 1962 On the kinetics and mechanism of helix formation: The two stranded poly (A + U) complex from polyriboadenylic acid and polyuridylic acid. *J. Am. Chem. Soc.* **84:**4503–4507.

Rudkin, G. T., 1969 Non-replicating DNA in *Drosophila. Genetics Suppl. No. 1, Part 2* **61:**227–238.

Sachs, R. I. and U. Clever, 1972 Unique and repetitive DNA sequences in the genome of *Chironomus tentans*. *Exp. Cell Res.* **74:**587–591.

Sanchez de Jimenez, E. F., J. L. Gomzalez, J. L. Dominguez, and E. F. Saloma, 1974 Characterization of DNA from differentiated cells: Analysis of the chicken genomic complexity. *Eur. J. Biochem.* **45:**25–29.

Saunders, G. F., S. Shirakawa, P. P. Saunders, F. E. Arrighi, and T. C. Hsu, 1972 Populations of repeated DNA sequences in the human genome. *J. Mol. Biol.* **63:**323–334.

Sexsmith, E., 1968 DNA values and karyotypes of amphibia. Ph.D. thesis, University of Toronto, Ontario.

Shields, G. F. and N. A. Straus, 1975 DNA-DNA hybridization studies of birds. *Evolution* **29:**159–166.

Sohn, U., K. H. Rothfels, and N. A. Straus, 1975 DNA-DNA hybridization studies in blackflies. *J. Mol. Evol.* (in press).

Straus, N. A., 1971 Comparative DNA renaturation kinetics in amphibians. *Proc. Natl. Acad. Sci. USA* **68:**799–802.

Straus, N. A. and H. C. Birnboim, 1974 Long pyrimidine tracts of L-cell DNA: Localization to repeated DNA. *Proc. Natl. Acad. Sci. USA* **71:**2992–2995.

Straus, N. A. and T. I. Bonner, 1972 Temperature dependence of RNA-DNA hybridization kinetics. *Biochim. Biophys. Acta* **227:**87–95.

Sutton, W. D. and M. McCallum, 1971 Mismatching and the reassociation rate of mouse satellite DNA. *Nature (London) New Biol. 232:*83–85.

Swanson, C. P., 1957 *Cytology and Cytogenetics,* 596 pp., Prentice-Hall, Englewood Cliffs, N.J.

Tobler, H., K. D. Smith, and H. Ursprung, 1972 Molecular aspects of chromatin elimination in *Ascaris lumbricoides*. *Dev. Biol.* **27:**190–203.

Uhlenbeck, O., R. Harrison, and P. Doty, 1968 Some effects of non-complementary bases on the stability of helical complexes of polyribonucleotides. In *Molecular Associations in Biology,* edited by B. Bullman, pp. 107–114, Academic Press, New York.

Walker, P. M. B., 1969 The specificity of molecular hybridization in relation to studies on higher organisms. *Prog. Nucleic Acid Res. Mol. Biol.* **9:**301:326.

Weinblum, D., U. Gungerich, M. Geisert, and R. K. Zahn, 1973 Occurrence of repetitive sequences in the DNA of some marine invertebrates. *Biochim. Biophys. Acta* **299:**231–240.

Wells, R. and R. Sager, 1971 Denaturation and the renaturation kinetics of chloroplast DNA from *Chlamydomonas reinhardi*. *J. Mol. Biol.* **58:**611–622.

Wetmur, J. G. and N. Davidson, 1968 Kinetics of renaturation of DNA. *J. Mol. Biol.* **31:**349–370.

Wu, J.-R., J. Hurn, and J. Bonner, 1972 Size and distribution of the repetitive segments of the *Drosophila* genome. *J. Mol. Biol.* **64:**211–219.

2

Cytological Localization of Repeated DNAs

RONALD A. ECKHARDT

Introduction

The discovery that the eukaryotic genome contains families of nucleotide sequences repeated to varying degrees (see Chapter 1) raises many questions concerning the distribution of these sequences within the eukaryotic cell. Based on analyses of DNA : DNA reassociation kinetics, nuclear DNA of all eukaryotes studied can be divided into three arbitrary classes according to the extent of repetition within each class. These are (1) a few sequences with a high degree of reiteration, reannealing very fast and many times appearing as satellites after isopycnic centrifugation in cesium salts, (2) moderately repeated sequences which renature with intermediate rates including the DNAs coding for ribosomal RNAs, 5 S RNAs, and transfer RNAs, and (3) sequences which reassociate at rates compatible with the genomic complexity of the organism in question. This last class contains "single-copy" DNAs presumably coding for the bulk of the messenger RNAs.

Early attempts to study the intracellular distribution of repeated DNAs primarily involved fractionating either nuclear components, size classes of chromosomes, or different types of chromatin and then examin-

---

RONALD A. ECKHARDT—Department of Biology, Brooklyn College of The City University of New York, Brooklyn, New York.

ing the DNA isolated from each fraction (Maio and Schildkraut, 1967, 1969; Schildkraut and Maio, 1968; Yasmineh and Yunis, 1970). This experimental approach was limited for the most part in that it was not possible to study nucleotide sequence distribution in single cells. The problem of cell selection, however, was resolved in 1969 with the development of the technique of "*in situ* molecular hybridization" (Gall, 1969; Gall and Pardue, 1969; John *et al.*, 1969; Buongiorno-Nardelli and Amaldi, 1970).

In situ (or cytological) hybridization involves the molecular hybridization of radioactively labeled nucleic acids to the DNA resident within cytological structures and the subsequent visualization of this hybridization by means of autoradiography. By use of this technique, it is now possible to determine the localization of many specific DNAs even within single chromosomes or certain other cytological entities. Many comprehensive review articles are available describing procedural details as well as summarizing the results of various applications of *in situ* hybridization in diverse biological situations (Gall and Pardue, 1971; Eckhardt, 1972; Pardue and Gall, 1972*a,b*; Rae, 1972; Steffensen and Wimber, 1972; Hennig, 1973; Hsu and Arrighi, 1973; Jones, 1973; Wimber and Steffensen, 1973). Here, only the general experimental design of the technique will be considered, pointing out important operational principles behind each step. This is intended to help the reader without prior knowledge of *in situ* hybridization to achieve a basic understanding of what is involved. In addition, Table 1 contains references to a variety of investigations to which the reader may wish to refer for more detailed information.

While being an extremely powerful molecular/cytogenetic investigative tool, *in situ* hybridization is relatively simple in terms of the experimental manipulations involved. The entire procedure may be divided into five basic steps.

Preparation of Slides

Obviously the quality of the finished hybrids can be no better in terms of details to be seen than the quality of the initial cytological preparations used in the procedure. A standard squash, smear, or tissue section slide is suitable, with the guiding principle being to choose the technique by which a given tissue is most properly handled. One important caution in this step is that the slides be "subbed" or coated with albumin to increase the adherence of the tissues to the glass. Moreover, the preparations should be made extremely flat. This also helps to avoid loss of tissues in subsequent treatments. A fixative consisting of 3 parts absolute ethanol to 1 part glacial acetic acid (v/v) has been most frequently used in previous investigations, but the use of other fixatives is sometimes possible. Beyond making slides for studies at the level of the light microscope, some success has been reported in preparing specimens for examination with the electron microscope. However, electron microscopic techniques have not been widely used as yet in conjunction with *in situ* hybridization and will not be discussed here (for information, see Jacob *et al.*, 1971, 1974; Croissant *et al.*, 1972; Geuskens and May, 1974).

Prehybridization Treatment and Denaturation of DNA

Various prehybridization procedures have been employed by different investigators with the overall goal of removing substances from slides which might interfere with the subsequent hybridization. However, it is usually best to limit the number of prehybridization treatments to only those necessary since many times they can distort the morphology or molecular integrity of the cytological preparations.

The use of RNAse is perhaps the most common prehybridization procedure. It is extremely valuable in investigations of the distribution of ribosomal cistrons where there can be substantial amounts of residual ribosomal RNA (rRNA) present on a slide. In this case, the residual rRNA can presumably act as a competitor for binding sites in the DNA. Hence it is beneficial to remove this RNA prior to hybridization.

To remove basic proteins from slides, many investigators have used dilute HCl rinses as a prehybridization treatment. Their concern has been that basic proteins might nonspecifically bind the challenge nucleic acid, causing incorrect results. In many instances, however, HCl rinses have proven to be unnecessary. Moreover, they can be very disruptive, causing depurination or even loss of DNA. This is especially true when NaOH is subsequently used to denature the DNA. Therefore, it is advisable to omit HCl rinses whenever possible.

Before molecular hybridization can occur, the DNA of the cytological structures must be denatured. Denaturation can be achieved by any one of a number of methods. Chemical treatments with dilute solutions of NaOH or HCl or heat treatments in low salt solutions or formamide have proven successful. The greatest problem here is to achieve DNA denaturation conditions while keeping the treatments mild enough so as not to destroy the integrity of the cytological preparations. Different tissues respond differently to each of the above treatments and the best conditions of denaturation are worked out by trial and error within the guidelines of previous work.

Molecular Hybridization

In the molecular hybridization step, a radioactively labeled nucleic acid (DNA or RNA) is placed over the cytological preparation. The slide is held under appropriately controlled conditions so that annealing of the added nucleic acid can occur with complementary regions of the denatured DNA still present in the cellular components. Conceptually, the annealing reaction is assumed to follow the same principles as for DNA: DNA or DNA: RNA hybridization on filters or in solution (see Chapter 1). Salt concentration, time and temperature of incubation, and concentration of the challenge nucleic acid are all important factors which must be controlled to form a stable hybrid. In most previous investigations, the challenge nucleic acid has been tritium-labeled to give good autoradiographic resolution, but other isotopes can also be utilized. The nucleic acid may be labeled *in vivo,* labeled *in vitro* using isolated polymerase enzymes, or chemically labeled *in vitro.* It must, however, be of high specific activity, with a practical lower limit being about 10^5 cpm/μg for most studies.

Posthybridization Treatment

After hybridization has taken place, it is necessary to remove nonspecifically bound nucleic acids yet keep the true hybrids intact. With RNA: DNA hybrids, this is conveniently done by posttreating the slides with dilute concentrations of RNAse. The conditions of digestion are similar to those used for RNA: DNA filter hybridizations (Gillespie and Spiegelman, 1965). With DNA: DNA hybrids, the nonspecifically bound DNA is most often removed by a series of hot or cold high salt washes.

Autoradiography

Standard techniques of autoradiography with stripping film or liquid emulsion are used subsequent to the posthybridization treatments to determine the binding sites of the radioactively labeled challenge nucleic acids. Exposure times are variable, of course, depending on the experiment. But the best exposure may be determined by developing test slides after various intervals of time. After photographic development of the emulsion, the slides are stained (usually with Giemsa stain), and the results are examined.

There are several important additional considerations that influence

(Text continued on page 47.)

TABLE 1. Studies on the Localization of Repeated DNAs

Organism	*Sequence studied*	*Localization*	*Reference*
Protozoa			
Stylonychia mytilus	Macronuclear DNA	Specific sites in polytene chromosomes; only slight hybridization with micronuclei	Ammermann *et al.*, 1974
Plants			
Phaseolus coccineus	Ribosomal DNA	Nucleolar organizer regions and certain other regions of the chromosomes	Avanzi *et al.*, 1971
Phaseolus vulgaris (kidney bean)		Chromosome pairs I, II, V; intranucleolar chromatin and micronucleoli	Avanzi *et al.*, 1972
		Four nucleolus-associated chromosomes and intranucleolar chromatin	Brady and Clutter, 1972
Zea mays (corn)	5 S DNA	Located near end of the long arm of chromosome 2	Wimber *et al.*, 1974
Animals: Insects			
Acheta domesticus (house cricket)	Ribosomal DNA	Amplified DNA body during pachynema; in DNA body and multiple nucleoli in diplotene-stage oocytes	Cave, 1972
	Satellite DNA	Major chromomeres of chromosomes 6 and 11	Ullman *et al.*, 1973; Lima-de-Faria, 1974
	Nonribosomal, satellite DNA	Amplified DNA body and in limited regions of the chromosomes	Cave, 1973
	Other repetitive DNAs	Along the chromosomes in addition to the DNA body	
Schistocerca gregaria (desert locust)	Highly repetitive DNA	Centromeric regions of all chromosomes; also to telomeric regions of three pairs of short chromosomes	Brown and Wilmore, 1974

TABLE 1. Continued

Organism	*Sequence studied*	*Localization*	*Reference*
Heteropeza pygmaea (gall midge)	Somatic satellite DNA	Found in the centromeric regions of somatic (S) chromosomes and germ-line limited E chromosomes	Kunz and Eckhardt, 1974
Rhynchosciara hollaenderi (fungus fly)	Ribosomal DNA	X chromosome and C chromosome; some micronucleoli	Pardue *et al.*, 1970
	Satellite DNA	Centromeric heterochromatin of all four chromosomes and telomeric regions of the A chromosome and the C chromosome	Eckhardt and Gall, 1971
	Other repetitive DNAs	Along chromosomes including areas containing satellite DNA	
Rhynchosciara angelae	DNA complementary to poly r(U)	Bound to discrete loci in the four salivary gland chromosomes	Jones *et al.*, 1973*b*
Sciara coprophila (fungus gnat)	Ribosomal DNA	Nucleolus organizer of X chromosome and some micronucleoli	Pardue *et al.*, 1970; Gerbi, 1971
Chironomus tentans	DNA complementary to RNA from Balbiani rings 1, 2, and 3	BR 2 RNA hybridizes specifically with BR 2 DNA of chromosome IV	Lambert, 1972, 1974
		BR 1 RNA hybridizes with BR 1 DNA and with BR 2 DNA	
		BR 3 RNA hybridizes with BR 2 DNA but not to BR 3 DNA under the conditions used	
	DNA complementary to various nuclear RNA fractions	RNA from BR 2 hybridizes specifically with the BR 2 region of chromosome IV	Lambert *et al.*, 1972; Lambert, 1974
		High molecular weight nuclear sap RNA hybridizes with the BR 2 region of chromosome IV	
		Low molecular weight nuclear sap RNA hybridizes diffusely with all chromosomes	

		RNA from chromosome I hybridizes diffusely with all chromosomes	
		Nucleolar RNA hybridizes specifically with nucleolus organizers found on chromosomes II and III	
	DNA complementary to cytoplasmic RNA	Found BR 2 complementary sequences in peripheral cytoplasmic RNA	Lambert, 1973, 1974
	DNA complementary to heterodisperse high molecular weight RNA	Balbiani rings 1 and 2 of chromosome IV	Sachs and Clever, 1972
Glyptotendipes barbipes	Ribosomal DNA	Three puffs on chromosomes IR, III, and IIIL	Wen *et al.*, 1974
	5 S DNA	Three sites on IIIR, two sites on IIR, and one site (Balbiani ring) on IV	
	4 S DNA	No defined bands under the conditions used	
Drosophila hydei	Ribosomal DNA	Intranucleolar chromatin	Pardue *et al.*, 1970; Alonso, 1973
	Nuclear DNA	Widespread distribution of repetitive sequences of differing degrees of repetition	Alonso *et al.*, 1974; Berendes *et al.*, 1974
	DNA complementary to total RNA	Many chromosomal sites and the nucleolus	Berendes *et al.*, 1974
Drosophila hydei, *Drosophila neohydei*, and *Drosophila pseudoneohydei*	Satellite DNA	Cross-species hybrids showed that various satellites occupy different positions within the chromosomes; generally found in α-heterochromatin of chromocenters; in some instances, widely distributed	Hennig *et al.*, 1970
Drosophila melanogaster	DNA complementary to 5 S RNA	Region 56E–F of the right arm of chromosome 2	Wimber and Steffensen, 1970; Prensky *et al.*, 1973

TABLE 1. Continued

Organism	*Sequence studied*	*Localization*	*Reference*
Drosophila melanogaster (continued)	DNA complementary to transfer RNAs	X chromosome, 25 sites; right arm of chromosome 2, 22 sites; distal half of the left arm of chromosome 2, 21 sites; some of these sites form RNA puffs during larvlal development	Steffensen and Wimber, 1971
		Lys-5 tRNA located on right arm of chromosome 2 in region 48F–49A	Grigliatti *et al.*, 1974*b*
		Lys-5 tRNA, 48F–49A; Phe-2 tRNA, 48F–49A (?); Ser-7 tRNA, no detectable bands under conditions used	Grigliatti *et al.*, 1974*a*
	Total DNA	Along chromosomes, most concentrated in chromocenter	Jones and Robertson, 1970; Rudkin and Tartof, 1974
	Highly repetitive DNA fraction	Many chromosomal sites but most concentrated in the chromocentric heterochromatin	Rae, 1970
	Low G + C fraction of DNA	Highly concentrated in a discrete region within the chromocenter; slight labeling over chromosome arms	
	G + C rich fraction of DNA	Chromosome arms with heavy labeling over chromocenter and in a region near the tip of the left arm of one of the chromosomes (later identified as region 21C–D on 2L)	
	Highly repetitive DNA fraction	Within chromocenter, particularly within a small region of the chromocenter	Botchan *et al.*, 1971
	Satellite DNA	Primarily in the chromocenter; shown to be under-replicated in polytene cells	Gall *et al.*, 1971

	Total DNA	Chromosome arms; most concentrated in chromocenter and in a small band near the tip of an autosome; also present in nucleolus	
	DNA of five isolated satellites	All satellites present in the chromocenter; three satellites located in region 21C–D on the left arm of chromosome 2	Peacock *et al.*, 1974
	Main band rapidly renaturing DNA	Heterochromatic regions of the chromosomes	Hearst *et al.*, 1974
	DNA from isolated Thomas circles	Label found over chromocenters	Schachat and Hogness, 1974
	Histone messenger DNA	Found in two adjacent bands in region 39E–40A near the centromere on the left arm of chromosome 2	Pardue *et al.*, 1972
Drosophila virilis	Satellite DNA	Most concentrated in α-heterochromatin of chromocenter but some labeling of the β-heterochromatin, particularly around the base of the X chromosome; in mitotic chromosomes, found in the heterochromatic areas of the X chromosome and autosomes, but less so in the Y chromosome; shown to be underreplicated in polytene cells	Gall *et al.*, 1971
	Other repetitive DNAs	Many chromosomal sites with heavy labeling along the X-chromosome, but only light labeling over the α- and β-heterochromatin of the chromocenter	
Animals: Crustaceans			
Cancer pagurus	A + T-rich satellite DNA	Highly dispersed within interphase nuclei from diverse sources	Chevaillier *et al.*, 1974
Gecarcinus lateralis (Bermuda land crab)	G + C-rich satellite DNA	Clustered in certain chromocenters and chromosome groups, but not in others	Musich and Skinner, 1972
	Main band DNA	General distribution in nuclei	

TABLE 1. *Continued*

Organism	*Sequence studied*	*Localization*	*Reference*
Animals: Amphibia			
Plethodon cinereus cinereus (Eastern red-backed salamander)	Satellite DNA	Centromeric regions of the chromosomes	Macgregor and Kezer, 1971
	Other repetitive DNAs	Along chromosomes; heaviest labeling over centromere regions	
	Ribosomal DNA	Nucleoli and associated chromatin	
		Short arm of the seventh-longest bivalent lampbrush chromosome near the centromere; label also found on the short arm of the fourteenth bivalent near the end	Macgregor and Kezer, 1973
	Satellite DNA	Confined to discrete regions in spermatid and mature sperm nuclei	Macgregor and Walker, 1973
	Ribosomal DNA	Near the centromere on the short arm of chromosome 7; confined to discrete regions in spermatid and mature sperm nuclei	
	Satellite DNA	Associated with centromeres of all 14 chromosomes; cross-hybridizes with other species of *Plethodon*	Macgregor *et al.*, 1973
Triturus viridescens (newt)	Total DNA	Along chromosomes with heaviest labeling over centromeric and telomeric regions	Barasacchi and Gall, 1972
	Ribosomal DNA	Only in nucleoli	
		Clustered in sperm nuclei (*Triturus cristatus*)	Macgregor and Walker, 1973
Xenopus laevis (African clawed toad)	Ribosomal DNA	Extrachromosomal nuclear cap DNA (amplified ribosomal cistrons)	Gall, 1969; John *et al.*, 1969

		Nuclear cap DNA in leptotene to pachytene stage oocytes; label also present over oogonial nucleoli	Gall and Pardue, 1969
		High-resolution electron microscopic study; results as above	Jacob *et al.*, 1971
		Clustered in sperm nuclei	Macgregor and Walker, 1973
		Studied the occurrence of amplified rDNA during germ cell development	Kalt and Gall, 1974
	Main band DNA lacking ribosomal DNA	Hybridizes with the chromosomes, but not with the amplified nuclear cap DNA	Pardue and Gall, 1969
	DNA complementary to 5 S RNA	Found in association with the telomeres of the long arm of most chromosomes; telomeres tightly clustered during meiosis, less so during interphase in most cells	Pardue *et al.*, 1973
	Ribosomal DNA	Found in the region of the secondary constriction of the largest subtelocentric chromosome	Pardue, 1974
Xenopus mulleri	Ribosomal DNA	End of the short arm of a large submetacentric chromosome	Pardue, 1974
	5 S DNA	Telomere regions of long arms of most chromosomes	
	A + T-rich satellite DNA (1.683 g/cm^3)	Hybridizes with discrete bands in the shorter arms of most chromosomes	
Animals: Birds			
Coturnix coturnix japonica (Japanese quail)	Satellite DNA	Found in microchromosomes	Brown and Jones, 1972
Gallus domesticus (chicken)	Various fractions of repetitive DNA	All fractions found in the centromeric regions of the microchromosomes; several fractions showed intense labeling over the W chromosome of females	Stefos and Arrighi, 1974

TABLE 1. *Continued*

Organism	*Sequence studied*	*Localization*	*Reference*
Animals: Mammals			
Dipodomys ordii (kangaroo rat)	HS-β satellite DNA	Centromeres of all but three pairs of chromosomes	Prescott *et al.*, 1973
	HD satellite DNA	Located predominantly in the short arms of chromosomes containing HS-β and in the centromeres of chromosome pairs lacking HS-β satellite DNA	
Microtus agrestis (European field vole)	Total repetitive DNA	Along chromosomes; heaviest labeling over X and Y chromosomes	Arrighi *et al.*, 1970
Mus musculus (mouse)	Satellite DNA	Centromeric heterochromatin of chromosomes with the possible exception of the Y chromosome; in interphase cells, found in association with condensed chromatin and nucleoli	Jones, 1970; Pardue and Gall, 1970; Jones and Robertson, 1970; Ahnström and Natarajan, 1974
		With semi-thin sectioning, shown to be present in a portion of the nuclear envelope heterochromatin, the nucleolar stalk, and the nucleolar heterochromatin	Rae and Franke, 1972
		Electron microscopic study showing satellite DNA to be found in chromatin patches, many times associated with the nuclear envelope and in association with nucleolar chromatin	Jacob *et al.*, 1974
	Other repetitive DNAs	Many chromosomal sites	Pardue and Gall, 1970
	Main band, rapidly renaturing DNA	Various regions of mouse chromosomes	Hearst *et al.*, 1974; Cech *et al.*, 1973

	Globin messenger RNA	Developed new procedures to permit the localization of messenger RNAs; globin messenger RNA was shown to be present in the cytoplasm of certain fetal liver cells and in Friend virus transformed cells	Harrison *et al.*, 1973; Conkie *et al.*, 1974; Harrison *et al.*, 1974
	DNA complementary to murine sarcoma–leukemia virus RNA	Chromocenters of interphase cells; the constitutive heterochromatin of some metaphase chromosomes	Loni and Green, 1974
Rattus norvegicus (rat)	Adenovirus-specific DNA	Found nuclear labeling in infected cells and adenovirus-transformed cells	Dunn *et al.*, 1973
Cricetulus griseus (Chinese hamster)	Ribosomal DNA	Nucleolar regions	Buongiorno-Nardelli and Amaldi, 1970
	DNA complementary to 5 S RNA	Nucleolar regions and perinucleolar dense chromatin	Amaldi and Buongiorno-Nardelli, 1971
	DNA complementary to tRNA	Nucleolar regions and perinucleolar dense chromatin	
	DNA complementary to "pulse-labeled" RNA	Diffuse labeling over nucleus.	
	Various fractions of repetitive DNA	Localized in some but not all regions of constitutive heterochromatin	Arrighi *et al.*, 1974
Mesocricetus auratus (golden hamster)	DNA complementary to poly(U)	Many diverse loci throughout chromosomes	Shenkin and Burdon, 1974
Cavia cobaya (guinea pig)	Highly repetitive DNA	Centromeric regions of the chromosomes	Natarajan and Raposa, 1974
Sylvilagus floridanus (cottontail rabbit)	SV40 viral DNA	Detected small number of SV40-transformed cells producing viral DNA; many times localized in clumps within the nucleus	Watkins, 1973
Oryctolagus cuniculus (domestic rabbit)	Shope papilloma virus DNA	Viral replication sites in keratinizing cells	Orth *et al.*, 1971
		Electron microscopic study of viral replication sites	Croissant *et al.*, 1972

TABLE 1. Continued

Organism	*Sequence studied*	*Localization*	*Reference*
Bos taurus (calf)	Four satellite DNAs	Satellite I: centromeres of all autosomes, X and Y chromosomes relatively unlabeled; satellite II: one-third of label over interstitial or telomeric regions of chromosomes, two-thirds over centromeres, X and Y least labeled; satellite III: centromeric region of all but four autosomes, X and Y relatively unlabeled; satellite IV: centromeres of most autosomes, two pairs unlabeled, X and Y chromosomes relatively unlabeled	Kurnit *et al.*, 1973
Cercopithecus aethiops (African green monkey)	α-Satellite DNA	Pericentromeric heterochromatin of all chromosomes; label found in clumps in growing interphase cells; label diffuse in confluent interphase cells	Kurnit and Maio, 1973
	Main band DNA	Throughout chromosomes	
	β-Satellite DNA	Nonrandomly scattered throughout chromosomes with the possible exception of the Y chromosome	Kurnit and Maio, 1974
	γ-Satellite DNA	Localized in pericentromeric heterochromatin of all chromosomes with the possible exception of the Y chromosome	
	SV40 viral DNA	Electron microscopic localization of viral DNA in the cytoplasm, nuclei, and nucleoli of tissue culture cells	Geuskens and May, 1974
Macaca mulatta (Rhesus monkey)	Ribosomal DNA	Achromatic region of chromosome 20	Henderson *et al.*, 1974*a*
Pan troglodytes (chimpanzee)	Ribosomal DNA	Satellited regions of chromosomes 14, 15, 17, 22, and 23	Henderson *et al.*, 1974*b*

Homo sapiens (humans)	DNA complementary to total HeLa cell RNA	Diffuse labeling over nucleus	John *et al.*, 1969
	Total DNA	Label throughout the chromosomal complement; study used ^{125}I-labeled complementary RNA	Altenburg *et al.*, 1973
	Various repetitive fractions of total DNA	In general, repetitive DNAs are more concentrated in the centromeric and telomeric regions of human chromosomes	Arrighi *et al.*, 1971
	Various repetitive fractions of total DNA	Highly repetitive DNA is more concentrated in centromeric and telomeric regions than in interstitial zones; different thermal fractions localize on different chromosomes	Saunders *et al.*, 1972*b*
	Various repetitive fractions of total DNA	Highly repetitive DNAs primarily located in centromeric and telomeric heterochromatin; intermediately repetitive DNAs are more widely scattered; chromosomes 1, 2, 3, and 16 labeled heavily with a 80–90°C thermal fraction of highly repetitive DNA; chromosome 2, an 85°C fraction; chromosome 9, an 80–95°C fraction	Hsu *et al.*, 1972
	Satellite DNA II (Corneo)	Close to the centromeres of three pairs of chromosomes [chromosome 1 of group A, a submedian centromeric chromosome from group C, and chromosome 16 (?)]	Jones and Corneo, 1971
	A repetitive main-band DNA fraction	Along many chromosomes, but not to areas containing satellite II	
	Satellite DNA III	Paracentromeric heterochromatin of the long arms of chromosome 9 and in minor concentrations in the centromeric heterochromatin of D and G group chromosomes	Jones *et al.*, 1973*a*
	A satellite DNA (1.703 g/cm^3)	Chromosome 9	Saunders *et al.*, 1972*a*

TABLE 1. Continued

Organism	Sequence studied	Localization	Reference
Homo sapiens (humans) (continued)	C_0t 0–5 DNA	Q-bands of chromosome arms, C-bands of pericentromeric regions, and T-bands of telomeric regions in several chromosomes	Sanchez and Yunis, 1974
	Ribosomal DNA	Cytologically satellited regions of chromosomes 13, 14, 15, 21, and 22	Henderson *et al.*, 1972
		Results as above using ^{125}I-labeled rRNA	Yu *et al.*, 1974
		Results as above using heterologous hybridization with *Xenopus* RNA; found rDNA to be localized in the nucleolar constrictions but not in the chromosome satellites; rDNA found in all three chromosomes 21st of Down's syndrome patients	Evans *et al.*, 1974
		Found rDNA to be contained in connecting chromatin strands between acrocentric chromosomes	Henderson *et al.*, 1973
	DNA complementary to 5 S RNA	Major site on chromosome 1; additional sites possibly on chromosomes 3, 9, 16, and 22; nucleolar label in interphase cells	Steffensen *et al.*, 1974
	DNA coding for hemoglobin messenger RNA	Possibly chromosome 2 and a B-group chromosome (either 4 or 5)	Price *et al.*, 1972
	Main-band, rapidly renaturing DNA	Many locations in different chromosomes	Hearst *et al.*, 1974
	Adenovirus type 12 DNA	Sites of viral replication; label consistently found over chromosome 1	McDougall *et al.*, 1972
	RD114 virus RNA complementary DNA	Label associated with one D-group chromosome	Price *et al.*, 1973
	Epstein-Barr viral DNA	Label found over nuclei of tumor cells; label found over many chromosomal sites of metaphase chromosomes	Wolf *et al.*, 1973; zur Hausen and Schulte-Holthausen, 1972

the general applicability of *in situ* hybridization to given research problems. Basically these present practical limits as to which sequences may be studied by means of this technique.

1. The first consideration is how much DNA can be detected using cytological hybridization? The problem here is whether enough challenge nucleic acid of sufficiently high specific activity can be bound to a region of a cellular component to be detected by the process of autoradiography. Under the most favorable combination of hybridization conditions now practiced, Gall and Pardue (1971) have estimated the limit of detection to be about 10^8 daltons of DNA. Assuming most messenger RNAs to be of an average size of about 10^5 to 10^6 daltons, one would not expect to be able to study their distribution under most circumstances. However, there are certain biological situations, such as the occurrence of polytene chromosomes in various Diptera, where the local concentrations of DNA in given chromosomal regions are increased. In such instances, it has been possible to determine the localization of large "single-copy" DNAs. But for the most part only the distribution of repetitive DNAs can be examined under present experimental conditions.

2. The other general consideration as to which sequences can be studied by *in situ* hybridization involves the problem of the effective rate of DNA : DNA or DNA : RNA hybridization. Most "single-copy" DNAs require long periods of time to form stable hybrids with the concentrations of challenge nucleic acid most commonly used in this technique. The high C_0t values necessary for the hybridization of unique DNAs subject the cytological preparations to conditions under which much resolution of structure is lost. All factors considered, again one should expect only to be able to examine the distribution of repeated DNAs in most situations using the current procedures.

Much progress has been made since 1969 in determining the localization of various repeated DNAs. Numerous individual studies have been published, many of which are included in Table 1. Likewise, much work still remains to be done before the significance of repeated sequences can be fully understood. The ready availability of the technique of *in situ* hybridization should help to alleviate this ignorance.

Literature Cited

Ahnström, G. and A. T. Natarajan, 1974 Localisation of repetitive DNA in mouse cells by *in situ* hybridization and dye binding techniques. *Hereditas* **76:**316–320.

Alonso, C., 1973 Improved conditions for *in situ* RNA/DNA hybridization. *FEBS Lett.* **31:**85–88.

Alonso, C., P. J. Helmsing, and H. D. Berendes, 1974 A comparative study of *in situ* hybridization procedures using cRNA applied to *Drosophila hydei* salivary gland chromosomes. *Exp. Cell Res.* **85:**383–390.

Altenburg, L. C., M. J. Getz, W. R. Crain, G. F. Saunders, and M. W. Shaw, 1973 ^{125}I-labeled DNA:RNA hybrids in cytological preparations. *Proc. Natl. Acad. Sci. USA* **70:**1536–1539.

Amaldi, F. and M. Buongiorno-Nardelli, 1971 Molecular hybridization of Chinese hamster 5 S, 4 S and "pulse-labelled" RNA in cytological preparations. *Exp. Cell Res.* **65:**329–334.

Ammermann, D., G. Steinbrück, L. von Berger, and W. Hennig, 1974 The development of the macronucleus in the ciliated protozoan *Stylonychia mytilus. Chromosoma* **45:**401–429.

Arrighi, F. E., T. C. Hsu, P. Saunders, and G. F. Saunders, 1970 Localization of repetitive DNA in the chromosomes of *Microtus agrestis* by means of *in situ* hybridization. *Chromosoma* **32:**224–236.

Arrighi, F. E., P. P. Saunders, G. F. Saunders, and T. C. Hsu, 1971 Distribution of repetitious DNA in human chromosomes. *Experientia* **27:**964–966.

Arrighi, F. E., T. C. Hsu, S. Pathak, and H. Sawada, 1974 The sex chromosomes of the Chinese hamster: Constitutive heterochromatin deficient in repetitive DNA sequences. *Cytogenet. Cell Genet.* **13:**268–274.

Avanzi, S., M. Buongiorno-Nardelli, P. G. Cionini, and F. D'Amato, 1971 Cytological localization of molecular hybrids between rRNA and DNA in the embryo suspensor cells of *Phaseolus coccineus. Accad. Naz. Lincei Rendic. C. Sci. Fis. Mat. Nat. Ser. VIII* **50:**357–361.

Avanzi, S., M. Durante, P. G. Cionini, and F. D'Amato, 1972 Cytological localization of ribosomal cistrons in polytene chromosomes of *Phaseolus coccineus. Chromosoma* **39:**191–203.

Barsacchi, G. and J. G. Gall, 1972 Chromosomal localization of repetitive DNA in the newt, *Triturus. J. Cell Biol.* **54:**580–591.

Berendes, H. D., C. Alonso, P. J. Helmsing, H. J. Leenders, and J. Derksen, 1974 Structure and function in the genome of *Drosophila hydei. Cold Spring Harbor Symp. Quant. Biol.* **38:**645–654.

Botchan, M., R. Kram, C. W. Schmid, and J. E. Hearst, 1971 Isolation and chromosomal localization of highly repeated DNA sequences in *Drosophila melanogaster. Proc. Natl. Acad. Sci. USA* **68:**1125–1129.

Brady, T. and M. E. Clutter, 1972 Cytolocalizaton of ribosomal cistrons in plant polytene chromosomes. *J. Cell Biol.* **53:**827–832.

Brown, A. K. and P. J. Wilmore, 1974 Location of repetitious DNA in the chromosomes of the desert locust (*Schistocerca gregaria*). *Chromosoma* **47:**379–383.

Brown, J. E. and K. W. Jones, 1972 Localisation of satellite DNA in the microchromosomes of the Japanese quail by *in situ* hybridization. *Chromosoma* **38:**313–318.

Buongiorno-Nardelli, M. and F. Amaldi, 1970 Autoradiographic detection of molecular hybrids between rRNA and DNA in tissue sections. *Nature* (*London*) **225:**946–948.

Cave, M. D., 1972 Localization of ribosomal DNA within oocytes of the house cricket, *Acheta domesticus* (Orthoptera: Gryllidae). *J. Cell Biol.* **55:**310–321.

Cave, M. D., 1973 Synthesis and characterization of amplified DNA in oocytes of the house cricket, *Acheta domesticus* (Orthoptera: Gryllidae). *Chromosoma* **42:**1–22.

Cech, T. R., A. Rosenfeld, and J. E. Hearst, 1973 Characterization of the most rapidly renaturing sequences in mouse main-band DNA. *J. Mol. Biol.* **81:**299–325.

Chevaillier, P., A. M. de Recondo, and M. Geuskens, 1974 Deoxyribonucleic acid of the crab *Cancer pagurus*. III. Intracellular localization of poly d(A-T) of *Cancer pagurus* by hybridization *in situ*. *Exp. Cell Res.* **86:**383–391.

Conkie, D., N. Affara, P. R. Harrison, J. Paul, and K. Jones, 1974 *In situ* localization of globin messenger RNA formation. II. After treatment of Friend virus-transformed mouse cells with dimethyl sulfoxide. *J. Cell Biol.* **63:**414–419.

Croissant, O., C. Dauguet, P. Jeanteur, and G. Orth, 1972 Application de la technique d'hybridation moléculaire *in situ* à la mise en évidence au microscope électronique, de la réplication végétative de l'ADN viral dans les papillomes provoqués par le virus de Shope chez de lapin cottontail. *C. R. Acad. Sci. Ser. D* **274:**614–617.

Dunn, A. R., P. H. Gallimore, K. W. Jones, and J. K. McDougall, 1973 *In situ* hybridization of adenovirus RNA and DNA. II. Detection of adenovirus-specific DNA in transformed and tumour cells. *Int. J. Cancer* **11:**628–636.

Eckhardt, R. A., 1972 Chromosomal localizaton of repetitive DNA. *Brookhaven Symp. Biol.* **23:**271–292.

Eckhardt, R. A. and J. G. Gall, 1971 Satellite DNA associated with heterochromatin in *Rhynchosciara*. *Chromosoma* **32:**407–427.

Evans, H. J., R. A. Buckland, and M. L. Pardue, 1974 Location of the genes coding for 18 S and 28 S ribosomal RNA in the human genome. *Chromosoma* **48:**405–426.

Gall, J. G., 1969 The genes for ribosomal RNA during oogenesis. *Genetics* (*Suppl.*) **61:**121–132.

Gall, J. G. and M. L. Pardue, 1969 Formation and detection of RNA-DNA hybrid molecules in cytological preparations. *Proc. Natl. Acad. Sci. USA* **63:**378–383.

Gall, J. G. and M. L. Pardue, 1971 Nucleic acid hybridization in cytological preparations. *Methods Enzymol.* **21:**(Part D) 470–480.

Gall, J. G., E. H. Cohen, and M. L. Polan, 1971 Repetitive DNA sequences in *Drosophila*. *Chromosoma* **33:**319–344.

Gerbi, S. A., 1971 Localization and characterization of the ribosomal RNA cistrons in *Sciara coprophila*. *J. Mol. Biol.* **58:**499–511.

Geuskens, M. and E. May, 1974 Ultrastructural localization of SV40 viral DNA in cells, during lytic infection, by *in situ* molecular hybridization. *Exp. Cell Res.* **87:**175–185.

Gillespie, D. and S. Spiegelman, 1965 A quantitative assay for DNA-RNA hybrids with DNA immobilized on a membrane. *J. Mol. Biol.* **12:**829–842.

Grigliatti, T. A., B. N. White, G. M. Tener, T. C. Kaufman, J. J. Holden, and D. T. Suzuki, 1974*a* Studies on the transfer RNA genes of *Drosophila*. *Cold Spring Harbor Symp. Quant. Biol.* **38:**461–474.

Grigliatti, T. A., B. N. White, G. M. Tener, T. C. Kaufman, and D. T. Suzuki, 1974*b* The localization of transfer RNA_5^{Lys} genes in *Drosophila melanogaster*. *Proc. Natl. Acad. Sci. USA* **71:**3527–3531.

Harrison, P. R., D. Conkie, J. Paul, and K. Jones, 1973 Localisation of cellular globin messenger RNA by *in situ* hybridisation to complementary DNA. *FEBS Lett.* **32:**109–112.

Harrison, P. R., D. Conkie, N. Affara, and J. Paul, 1974 *In situ* localization of globin messenger RNA formation. I. During mouse fetal liver development. *J. Cell Biol.* **63:**402–413.

Hearst, J. E., T. R. Cech, K. A. Marx, A. Rosenfeld, and J. R. Allen, 1974 Characterization of the rapidly renaturing sequences in the main CsCl density bands of *Drosophila*, mouse, and human DNA. *Cold Spring Harbor Symp. Quant. Biol.* **38:**329–339.

Henderson, A. S., D. Warburton, and K. C. Atwood, 1972 Location of ribosomal DNA in the human chromosome complement. *Proc. Natl. Acad. Sci. USA* **69:**3394–3398.

Henderson, A. S., D. Warburton, and K. C. Atwood, 1973 Ribosomal DNA connectives between human acrocentric chromosomes. *Nature (London)* **245:**95–97.

Henderson, A. S., D. Warburton, and K. C. Atwood, 1974*a* Localization of rDNA in the chromosome complement of the Rhesus (*Macaca mulatta*). *Chromosoma* **44:**367–370.

Henderson, A. S., D. Warburton, and K. C. Atwood, 1974*b* Localization of rDNA in the chimpanzee (*Pan troglodytes*) chromosome complement. *Chromosoma* **46:**435–441.

Hennig, W., 1973 Molecular hybridization of DNA and RNA *in situ*. *Int. Rev. Cytol.* **36:**1–44.

Hennig, W., I. Hennig, and H. Stein, 1970 Repeated sequences in the DNA of *Drosophila* and their localization in giant chromosomes. *Chromosoma* **32:**31–63.

Hsu, T. C. and F. E. Arrighi, 1973 Service of *in situ* nucleic acid hybridization to biology. *Nobel Symp.* **23:**307–314.

Hsu, T. C., F. E. Arrighi, and G. F. Saunders, 1972 Compositional heterogeneity of human heterochromatin. *Proc. Natl. Acad. Sci. USA* **69:**1464–1466.

Jacob, J., K. Todd, M. L. Birnstiel, and A. Bird, 1971 Molecular hybridization of ^{3}H-labelled ribosomal RNA with DNA in ultrathin sections prepared for electron microscopy. *Biochim. Biophys. Acta* **228:**761–766.

Jacob, J., K. Gillies, D. Macleod, and K. W. Jones, 1974 Molecular hybridization of mouse satellite DNA–complementary RNA in ultrathin sections prepared for electron microscopy. *J. Cell Sci.* **14:**253–261.

John, H. A., M. L. Birnstiel, and K. W. Jones, 1969 RNA-DNA hybrids at the cytological level. *Nature (London)* **223:**582–587.

Jones, K. W., 1970 Chromosomal and nuclear location of mouse satellite DNA in individual cells. *Nature (London)* **225:**912–915.

Jones, K. W., 1973 The method of *in situ* hybridization. In *New Techniques in Biophysics and Cell Biology,* Vol. 1, edited by R. H. Pain and B. J. Smith, pp. 29–66, Interscience, New York.

Jones, K. W. and G. Corneo, 1971 Location of satellite and homogeneous DNA sequences on human chromosomes. *Nature (London) New Biol.* **233:**268–271.

Jones, K. W. and F. W. Robertson, 1970 Localisation of reiterated nucleotide sequences in *Drosophila* and mouse by *in situ* hybridisation of complementary RNA. *Chromosoma* **31:**331–345.

Jones, K. W., J. Prosser, G. Corneo, and E. Ginelli, 1973*a* The chromosomal location of human satellite DNA III. *Chromosoma* **42:**445–451.

Jones, K. W., J. O. Bishop, and A. Brito-da-Cunha, 1973*b* Complex formation between poly-r(U) and various chromosomal loci in *Rhynchosciara*. *Chromosoma* **43:**375–390.

Kalt, M. R. and J. G. Gall, 1974 Observations on early germ cell development and premeiotic ribosomal DNA amplification in *Xenopus laevis*. *J. Cell Biol.* **62:**460–472.

Kunz, W. and R. A. Eckhardt, 1974 The chromosomal distribution of satellite DNA in the germ-line and somatic tissues of the gall midge, *Heteropeza pygmaea*. *Chromosoma* **47:**1–19.

Kurnit, D. M. and J. J. Maio, 1973 Subnuclear redistribution of DNA species in confluent and growing mammalian cells. *Chromosoma* **42:**23–36.

Kurnit, D. M. and J. J. Maio, 1974 Variable satellite DNA's in the African green monkey *Cercopithecus aethiops*. *Chromosoma* **45:**387–400.

Kurnit, D. M., B. R. Shafit, and J. J. Maio, 1973 Multiple satellite deoxyribonucleic acids in the calf and their relation to the sex chromosomes. *J. Mol. Biol.* **81:**273–284.

Lambert, B., 1972 Repeated DNA sequences in a Balbiani ring. *J. Mol. Biol.* **72:**65–75.

Lambert, B., 1973 Tracing of RNA from a puff in the polytene chromosomes to the cytoplasm in *Chironomus tentans* salivary gland cells. *Nature (London)* **242:**51–53.

Lambert, B., 1974 Repeated nucleotide sequences in a single puff of *Chironomus tentans* polytene chromosomes. *Cold Spring Harbor Symp. Quant. Biol.* **38:**637–644.

Lambert, B., L. Wieslander, B. Daneholt, E. Egyházi, and U. Ringborg, 1972 *In situ* demonstration of DNA hybridizing with chromosomal and nuclear sap RNA in *Chironomus tentans. J. Cell Biol.* **53:**407–418.

Lima-de-Faria, A., 1974 The molecular organization of the chromomeres of *Acheta* involved in ribosomal DNA amplification. *Cold Spring Harbor Symp. Quant. Biol.* **38:**559–571.

Loni, M. C. and M. Green, 1974 Detection and localization of virus-specific DNA by *in situ* hybridization of cells during infection and rapid transformation by the murine sarcoma-leukemia virus. *Proc. Natl. Acad. Sci. USA* **71:**3418–3422.

Macgregor, H. C. and J. Kezer, 1971 The chromosomal localization of a heavy satellite DNA in the testis of *Plethodon c. cinereus. Chromosoma* **33:**167–182.

Macgregor, H. C. and J. Kezer, 1973 The nucleolar organizer of *Plethodon cinereus cinereus* (Green). I. Location of the nucleolar organizer by *in situ* nucleic acid hybridization. *Chromosoma* **42:**415–426.

Macgregor, H. C. and M. H. Walker, 1973 The arrangement of chromosomes in nuclei of sperm from plethodontid salamanders. *Chromosoma* **40:**243–262.

Macgregor, H. C., H. Horner, C. A. Owen, and I. Parker, 1973 Observations on centromeric heterochromatin and satellite DNA in salamanders of the genus *Plethodon. Chromosoma* **43:**329–348.

Maio, J. J. and C. L. Schildkraut, 1967 Isolated mammalian metaphase chromosomes I. General characteristics of nucleic acids and proteins. *J. Mol. Biol.* **24:**29–39.

Maio, J. J. and C. L. Schildkraut, 1969 Isolated mammalian metaphase chromosomes. II. Fractionated chromosomes of mouse and Chinese hamster cells. *J. Mol. Biol.* **40:**203–216.

McDougall, J. K., A. R. Dunn, and K. W. Jones, 1972 *In situ* hybridization of adenovirus RNA and DNA. *Nature (London)* **236:**346–348.

Musich, P. R. and D. M. Skinner, 1972 A cytological study of the DNA of the Bermuda land crab, *Gecarcinus lateralis. J. Cell Biol.* **55:**184a.

Natarajan, A. T. and T. Raposa, 1974 Repetitive DNA and constitutive heterochromatin in the chromosomes of guinea pig. *Hereditas* **76:**145–147.

Orth, G., P. Jeanteur, and O. Croissant, 1971 Evidence for and localization of vegetative viral DNA replication by autoradiographic detection of RNA-DNA hybrids in sections of tumors induced by Shope papilloma virus. *Proc. Natl. Acad. Sci. USA* **68:**1876–1880.

Pardue, M. L., 1974 Localization of repeated DNA sequences in *Xenopus* chromosomes. *Cold Spring Harbor Symp. Quant. Biol.* **38:**475–482.

Pardue, M. L. and J. G. Gall, 1969 Molecular hybridization of radioactive DNA to the DNA of cytological preparations. *Proc. Natl. Acad. Sci. USA* **64:**600–604.

Pardue, M. L. and J. G. Gall, 1970 Chromosomal localization of mouse satellite DNA. *Science* **168:**1356–1358.

Pardue, M. L. and J. G. Gall, 1972*a* Chromosome structure studied by nucleic acid hybridisation in cytological preparations. In *Chromosomes Today,* Vol. 3, edited by C. D. Darlington and K. R. Lewis, pp. 47–52, Hafner Press, New York.

Pardue, M. L. and J. G. Gall, 1972*b* Molecular cytogenetics. In *Molecular Genetics and*

Developmental Biology, edited by M. Sussman, pp. 65–99, Prentice-Hall, Englewood Cliffs, N.J.

Pardue, M. L., S. A. Gerbi, R. A. Eckhardt, and J. G. Gall, 1970 Cytological localization of DNA complementary to ribosomal RNA in polytene chromosomes of *Diptera. Chromosoma* **29:**268–290.

Pardue, M. L., E. Weinberg, L. H. Kedes, and M. L. Birnstiel, 1972 Localization of sequences coding for histone messenger RNA in the chromosomes of *Drosophila melanogaster. J. Cell Biol.* **55:**199a.

Pardue, M. L., D. D. Brown, and M. L. Birnstiel, 1973 Location of the genes for 5 S ribosomal RNA in *Xenopus laevis. Chromosoma* **42:**191–203.

Peacock, W. J., D. Brutlag, E. Goldring, R. Appels, C. W. Hinton, and D. L. Lindsley, 1974 The organization of highly repeated DNA sequences in *Drosophila melanogaster* chromosomes. *Cold Spring Harbor Symp. Quant. Biol.* **38:**405–416.

Prensky, W., D. M. Steffensen, and W. L. Hughes, 1973 The use of iodinated RNA for gene localization. *Proc. Natl. Acad. Sci. USA* **70:**1860–1864.

Prescott, D. M., C. J. Bostock, F. T. Hatch, and J. A. Mazrimas, 1973 Location of satellite DNAs in the chromosomes of the kangaroo rat (*Dipodomys ordii*). *Chromosoma* **42:**205–213.

Price, P. M., J. H. Conover, and K. Hirschhorn, 1972 Chromosomal localization of human haemoglobin structural genes. *Nature* (*London*) **237:**340–342.

Price, P. M., K. Hirschhorn, N. Gabelman, and S. Waxman, 1973 *In situ* hybridization of RD114-virus RNA with human metaphase chromosomes. *Proc. Natl. Acad. Sci. USA* **70:**11–14.

Rae, P. M. M., 1970 Chromosomal distribution of rapidly reannealing DNA in *Drosophila melanogaster. Proc. Natl. Acad. Sci. USA* **67:**1018–1025.

Rae, P. M. M., 1972 The distribution of repetitive DNA sequences in chromosomes. In *Advances in Cell and Molecular Biology,* Vol. 2, edited by E. J. DuPraw, pp. 109–149, Academic Press, New York.

Rae, P. M. M. and W. W. Franke, 1972 The interphase distribution of satellite DNA-containing heterochromatin in mouse nuclei. *Chromosoma* **39:**443–456.

Rudkin, G. T. and K. D. Tartof, 1974 Repetitive DNA in polytene chromosomes of *Drosophila melanogaster. Cold Spring Harbor Symp. Quant. Biol.* **38:**397–403.

Sachs, R. I. and U. Clever, 1972 Unique and repetitive DNA sequences in the genome of *Chironomus tentans. Exp. Cell Res.* **74:**587–591.

Sanchez, O. and J. J. Yunis, 1974 The relationship between repetitive DNA and chromosomal bands in man, *Chromosoma* **48:**191–202.

Saunders, G. F., T. C. Hsu, M. J. Getz, E. L. Simes, and F. E. Arrighi, 1972*a* Locations of a human satellite DNA in human chromosomes. *Nature* (*London*) *New Biol.* **236:**244–246.

Saunders, G. F., S. Shirakawa, P. P. Saunders, F. E. Arrighi, and T. C. Hsu, 1972*b* Populations of repeated DNA sequences in the human genome. *J. Mol. Biol.* **63:**323–334.

Schachat, F. H. and D. S. Hogness, 1974 Repetitive sequences in isolated Thomas circles from *Drosophila melanogaster. Cold Spring Harbor Symp. Quant. Biol.* **38:**371–381.

Schildkraut, C. L. and J. J. Maio, 1968 Studies on the intranuclear distribution and properties of mouse satellite DNA. *Biochim. Biophys. Acta* **161:**76–93.

Shenkin, A. and R. H. Burdon, 1974 Deoxyadenylate-rich and deoxyguanylate-rich regions in mammalian DNA. *J. Mol. Biol.* **85:**19–39.

Steffensen, D. M. and D. E. Wimber, 1971 Localization of tRNA genes in the salivary chromosomes of *Drosophila* by RNA : DNA hybridization. *Genetics* **69:**163–178.

Steffensen, D. M. and D. E. Wimber, 1972 Hybridization of nucleic acids to chromosomes. *Results Probl. Cell Differ.* **3:**47–63.

Steffensen, D. M., P. Duffey, and W. Prensky, Localisation of 5 S ribosomal RNA genes on human chromosome 1. *Nature* (*London*) **252:**741–743.

Stefos, K. and F. E. Arrighi, 1974 Repetitive DNA of *Gallus domesticus* and its cytological locations. *Exp. Cell Res.* **83:**9–14.

Ullman, J. S., A. Lima-de-Faria, H. Jaworska, and T. Bryngelsson, 1973 Amplification of ribosomal DNA in *Acheta.* V. Hybridization of RNA complementary to ribosomal DNA with pachytene chromosomes. *Hereditas* **74:**13–24.

Watkins, J. F., 1973 Studies on a virogenic clone of SV40-transformed rabbit cells using cell fusion and *in situ* hybridization. *J. Gen. Virol.* **21:**69–81.

Wen, W., P. E. León, and D. R. Hague, 1974 Multiple gene sites for 5 S and 18 + 28 S RNA on chromosomes of *Glyptotendipes barbipes* (Staeger). *J. Cell Biol.* **62:**132–144.

Wimber, D. E. and D. M. Steffensen, 1970 Localization of 5 S RNA genes on *Drosophila* chromosomes by RNA-DNA hybridization. *Science* **170:**639–641.

Wimber, D. E. and D. M. Steffensen, 1973 Localization of gene function. *Annu. Rev. Genet.* **7:**205–223.

Wimber, D. E., P. A. Duffey, D. M. Steffensen, and W. Prensky, 1974 Localization of the 5 S RNA genes in *Zea mays* by RNA-DNA hybridization *in situ. Chromosoma* **47:**353–359.

Wolf, H., H. zur Hausen, and V. Becker, 1973 EB viral genomes in epithelial nasopharyngeal carcinoma cells. *Nature* (*London*) *New Biol.* **244:**245–247.

Yasmineh, W. G. and J. J. Yunis, 1970 Localization of mouse satellite DNA in constitutive heterochromatin. *Exp. Cell Res.* **59:**69–75.

Yu, M. T., L. D. Johnson, D. Vogelman, and A. S. Henderson, 1974 *In situ* hybridization of iodinated ribosomal RNA to human chromosomes. *Exp. Cell Res.* **86:**165–166.

zur Hausen, H. and H. Schulte-Holthausen, 1972 Detection of Epstein-Barr viral genomes in human tumour cells by nucleic acid hybridization. In *Oncogenesis* and *Herpesviruses,* edited by P. M. Biggs, G. de-Thé, and L. N. Payne, pp. 321–325, International Association for Research on Cancer, Lyon, France.

3

Chromosomal Proteins

Ru Chih C. Huang and
Rex P. Hjelm, Jr.

Introduction

Chromosomes of higher organisms, being the packaged genome, must contain within themselves the ability to perform certain critical functions in response to specific intra- and/or intercellular stimuli. The two most important functions are the control of transcription and cellular proliferation. That transcription is controlled at the chromosome level is evident from studies on stage-specific RNA and protein synthesis occurring during development (Bishop *et al.,* 1972; Suzuki *et al.,* 1972; Kafatos, 1973). These demonstrated that the synthesis of RNA is restricted to only a portion of the genome. This view is supported by many *in vitro* RNA synthesis experiments, including those that have documented the *in vitro* transcription of specific genes (Price and Penman, 1972; Reeder and Roeder, 1972; Marzluff *et al.,* 1973). The notion of the specific restriction of the genome implies that the chromosome synthesizes RNA only at certain controlled locations. This has been verified by direct visualization of RNA transcription by means of the electron microscope (Hamkalo and

Ru Chih C. Huang—Department of Biology, The Johns Hopkins University, Baltimore, Maryland. Rex P. Hjelm, Jr.—Biophysics Laboratories, Department of Physics, Portsmouth Polytechnic Institute, Portsmouth, England.

Miller, 1973). We understand little of the molecular basis of this control. Even less understood is the molecular mechanism underlying the chromosomal events of cell proliferation. Regulatory triggers of chromosome replication may exert their effects on the chromosome indirectly through cellular or nuclear membranes and/or through the synthesis of macromolecules essential for starting DNA synthesis. The outcome of this is the initiation of a sequence of events in which the chromosome is replicated, including its complex of regulatory machinery. The process culminates in the chromosomal events in mitosis.

Studies on isolated chromosomes have demonstrated that chromosomes consist of DNA, large amounts of proteins, and small quantities of RNA. The chromosomal proteins are categorized into two types: small, distinct, basic proteins called *histones,* and other proteins called *nonhistones.* In this chapter, we will discuss some recent information on the proteins of the chromosome and their relationship to chromosome structure and function. Their role in transcription and replication (including chromosome condensation and division) is of special interest.

Much of the work to be discussed has been carried out on isolated chromosomes. The form in which the chromosome is isolated is not the familiar entity apparent at mitosis, but rather the more diffuse material of the interphase nucleus. Such material is variously referred to as *chromatin, nucleohistone, nucleoprotein, deoxynucleoprotein,* etc. Studies on this material are necessary in determining the chemical nature of chromatin, and studies on the interrelationships of the chromosomal components are necessary in determining chromosome structure. There are some problems and limitations of working with such material. Chemically, chromatin can be defined precisely only for a specific isolation procedure. Functionally, no definition at all is possible. The latter is a serious limitation, because isolated chromosomes would be ideal for studies on how chromosomal components interact in the processes of transcription and replication. For example, it is difficult to assess to what extent isolated chromatin represents the functional genome of the cell, although, in several systems at least, transcription of specific genes by use of chromatin as template has been proven to be possible (Astrin, 1973; Richard *et al.,* 1973).

For a topic as large and important as chromosomal proteins, we can cover only a small part of the field. Several reviews on this topic have been published (Stellwagen and Cole, 1969*b*; Georgiev, 1969; Hearst and Botchan, 1970; DeLange and Smith, 1971; Elgin *et al.,* 1971), and we will refer to them and others on related subjects for more thorough discussions. The material in this review covers a part of the literature up to October 1973.

Histones

Introduction

Histones comprise the major protein component of the chromosomes of higher plants and animals. In virtually all higher plants and animals studied, chromosomes have been found to contain a weight ratio of histone to DNA that is close to 1 : 1. Since histones constitute such a major portion of the chromosome, they must play an important role in its structure and function.

The nuclei in any individual contain, qualitatively, the same genetic information. Thus a differentiated cell will utilize only a small part of the total information available in its DNA, necessitating the selective repression of large amounts of genetic material. That histones might be a genetic repressor was first suggested by Stedman and Stedman (1951). Huang and Bonner (1962) first demonstrated that histones could serve as genetic repressors by showing that histone is capable of inhibiting the intrinsic property of DNA to act as a template for polymerase-mediated RNA synthesis *in vitro*. This observation was confirmed by other workers (Allfrey *et al.,* 1963; Barr and Butler, 1963; Huang *et al.,* 1964; Marushige and Bonner, 1966; Georgiev *et al.,* 1966). However, there is no clear indication that histones act primarily as repressors *in vivo*. Various physical techniques such as sedimentation velocity, flow dichroism, rotatory diffusion, intrinsic viscosity (Zubay and Doty, 1959; Ohba, 1966*a*; Henson and Walker, 1970*a*; Wilhelm *et al.,* 1970), circular dichroism (Shih and Fasman, 1970), thermal denaturation (Bonner and Huang, 1963), X-ray diffraction (Pardon *et al.,* 1967), and electron microscopy (Ris and Kubai, 1970; DuPraw, 1970) clearly demonstrate that the structure of DNA in the nucleohistone complex is different from that of DNA free in solution. There is little doubt that most of this structure is imposed by the histones.

In some cases, structural differences are known to exist between regions of the genome that are active and inactive in the transcription of RNA. Active regions of some chromsomes are observed to have a more open structure. The RNA puffs of dipteran polytene chromsomes and the loops of lampbrush chromosomes of amphibian oocytes and *Drosophila* spermatocytes are salient examples of this phenomenon. Inactive genes are sometimes seen to exist in dense structures such as Lyonized X chromosomes (Barr bodies) and heterochromatic paternal chromosomes of mealy bugs (for reviews, see Brown, 1966; Lyon, 1968). Although these correlations may not be general (Ris and Kubai, 1970), there is some evidence that they can at least be extended to mammalian interphase nuclei, where electron microscopy has demonstrated the existence of dense and diffuse regions of the chromatin (Littau *et al.,* 1964). RNA synthesis appears to be limited to the diffuse material (Frenster *et al.,* 1963; Littau *et al.,* 1964; Chalkley and Jensen, 1968; Murphy *et al.,* 1973). Isolated chromatin can be fractionated into its dense and diffuse components (Frenster *et al.,* 1963; Duerksen and McCarthy, 1971; Murphy *et al.,* 1973), which have been shown to differ with respect to several physical parameters, especially sedimentation velocity and solubility (Chalkley and Jensen, 1968; Jensen and Chalkley, 1968; Duerksen and McCarthy, 1971). Thus the alternate hypotheses of histones as genetic repressors or as structural components of chromosomes may be equivalent.

Definition

Attempts to define histone proteins are frustrated by the inability to assign to them a distinct function. The alternative is to define them phenomenologically. Histones are recognized as basic nuclear proteins that are ultimately associated with DNA (Murray, 1964*b*). In practical terms, histones may be defined as those proteins that are extracted by dilute acid from isolated chromosomes (chromatin). Although this is a necessary condition of the definition, it is not sufficient to apply this criterion alone to classify a protein as a histone, since protamines (see Chapter 5) and nonhistone "contaminants" also share this characteristic. However, the histone proteins constitute the major part of the acid-extractable chromosomal proteins of higher plants and animals. The well-characterized histones isolated from these sources can be used to determine further properties that assist in delineating this group of proteins. Johns (1971) has listed some of these properties. Prominent among them is a high content of arginine and lysine. As a result, these proteins are highly basic, and are exceeded in this respect only by the protamines. The histones are best differentiated from the protamines by their larger size and higher lysine content.

Fractionation and Nomenclature

The techniques for the isolation and fractionation of histones have been outlined by Johns (1971). Two classical techniques deserve mention since they are still widely used and form the basis of the traditional histone nomenclature.

Total acid-extracted histone may be fractionated on the weak anion exchange resin Amberlite IRC 50 (Rasmussen *et al.*, 1962). The column is developed with a guanidinium chloride gradient, which elutes the histones in order of their relative arginine content. The first fraction contains a series of very similar histone polypeptides that are relatively rich in lysine. They are collectively known as *histone I*. The next fraction consists of two proteins, both of which are moderately rich in lysine. These proteins are termed *histone IIb1* and *histone IIb2*. The final fraction contains histones relatively rich in arginine. Two distinct histones are present here—*histone III* and *histone IV*.

The second important separation procedure utilizes differential chemical extraction (Johns and Butler, 1962; Phillips and Johns, 1965; Johns, 1964, 1967, 1971). The lysine-rich fraction, corresponding to histone I, is isolated as *fraction f1*. The two slightly lysine-rich histones, IIb1 and IIb2, are isolated as *fraction f2a2* and *fraction f2b*, respectively. The fractions corresponding to the two arginine-rich histones, III and IV, are *fraction f3* and *fraction f2a1*.

Other nomenclatures based on other fractionation procedures or properties of the histone fractions are presently in use. Johns (1971) has listed the correspondence between these various nomenclatures.

Characteristics of Calf Thymus Histones

No histones are as well characterized as those isolated from calf thymus. Four distinct histone polypeptides are found in calf thymus. Two, IIb1 (f2a2) and IIb2 (f2b), are slightly lysine rich. Histone IIb1 constitutes roughly 15% of the total histone (of calf thymus histone) and has a lysine-to-arginine ratio of 2:1. Histone IIb2 has a lysine-to-arginine ratio of 5:2 and represents 30% of the total histone complement. The other two histone proteins, histones III and IV, have a lysine-to-arginine ratio of less than 1. They are thus arginine rich, and respectively they constitute 20 and 10% of the calf thymus histone. Histone I (f1), the fraction of the lysine-rich histones, is a family of polypeptides having closely related sequences (Rall and Cole, 1971). Their lysine-to-arginine ratio is in excess of 10, and together they make up about 25% of calf thymus histone.

Total histones from calf thymus, as well as those from the other sources, are isolated from nuclei or chromatin (see review by Johns, 1971) by salt or acid extraction. The usual techniques may then be used to separate the histone types. The histone fractions isolated from thymus have, in most cases, characteristics very similar to those of the analogous fractions isolated from other plant and animal sources.

Histones are highly basic proteins of low molecular weight. The largest are the lysine-rich histones (I, f1). The lysine-rich histones from calf thymus have a molecular weight of 21,000 daltons (Teller *et al.,* 1965) corresponding to a length of 216 amino acids. The arginine-rich histones III (f3) and IV (f2a1) have molecular weights of 15,324 daltons (DeLange *et al.,* 1973) and 11,300 daltons (Phillips, 1971), respectively. Histone IIb2 (f2b) has a molecular weight of 13,774 daltons (Phillips, 1971). The molecular weight of the other slightly lysine-rich histone, IIb1 (f2a2), is 13,944 daltons.

Histones IIb1 (f2a2) (Yeoman *et al.,* 1972), IIb2 (f2b) (Iwai *et al.,* 1972), III (f3) (DeLange *et al.,* 1973), and IV (f2a1) (DeLange *et al.,* 1969*a*; Ogawa *et al.,* 1969) from calf thymus have been sequenced. These are shown in Figure 1, along with a partial sequence of the subfractions of the lysine-rich histone, I (f1) (Rall and Cole, 1971; Bustin, 1972). The most striking feature of these sequences is the distinct polarity in the distribution of the amino acids. The most important manifestation of this is the distribution of basic amino acids. In the slightly lysine-rich and arginine-rich histones, the basic amino acids (lysine, arginine, and histidine) are concentrated in the amino-terminal halves of these molecules to the exclusion of the acidic (glutamic and aspartic) amino acids. The net result is a striking, uneven distribution of charge in these polypeptides; consequently, the amino-terminal halves of the molecules are far more basic than the carboxyl-terminal halves.

To illustrate some of the unusual properties of the primary sequence of histones, we will consider the slightly lysine-rich histone IIb2 (f2b). It is instructive to arbitrarily delimit regions of the molecule. The amino-terminal end to residue 34 is one region, while residues 35–79 constitute a central region. This leaves the remaining 46 residues as the carboxyl-terminal region. Inspection of the sequence of histone IIb2 (f2b) in Figure 1 reveals that 15 of the 31 basic amino acids in this histone are in the amino-terminal end of the molecule, whereas only two of the molecule's ten acidic amino acid residues are in this region. This gives the *N*-terminal region of this histone a total charge of +13 as compared to +1 for the central 45 residues and +7 for the 46 residues in the carboxyl-terminal end. The basic amino acids contained within the amino-terminal end of the molecule are often clustered. One finds four clusters of two basic amino acid residues and one cluster of five basic residues. Other salient features are seen. Of six prolines found in IIb2, four are in the amino-terminal portion. Although the molecule contains 30 hydrophobic amino acids (valine, methionine, isoleucine, leucine, tyrosine, and phenylalanine), all but one are found in the sequence from residue 37 to the carboxyl-terminal end.

Similar characteristics may be noted for the other histones, although

```
                                   5              10              15   +         20 *
F2al (IV)[a,b]   Ac-Ser-Gly-Arg-Gly-Lys-Gly-Gly-Lys-Gly-Leu-Gly-Lys-Gly-Gly-Ala-Lys-Arg-His-Arg-Lys-
F2a2 (IIbl)[c]   Ac-Ser-Gly-Arg-Gly-Lys-Gln-Gly-Gly-Lys-Ala-Arg-Ala-Lys-Ala-Lys-Thr-Arg-Ser-Ser-Arg-
F2b (IIb2)[d]   H2N-Pro-Glu-Pro-Ala-Lys-Ser-Ala-Pro-Ala-Pro-Lys-Lys-Gly-Ser-Lys-Lys-Ala-Val-Thr-Lys-
                                                 +              *
F3 (III)[e]     H2N-Ala-Arg-Thr-Lys-Gln-Thr-Ala-Arg-Lys-Ser-Thr-Gly-Gly-Lys-Ala-Pro-Arg-Lys-Gln-Leu-
Fl (I)[f,g]      Ac-Ser-Glu-Ala-Pro-Ala-Glu-Thr-Ala-Ala-Pro-Ala-Pro-Ala-Pro-Lys-Ser-Pro-Ala-Lys-Thr-
                                  25              30              35              40
F2al (IV)          -Val-Leu-Arg-Asp-Asn-Ile-Gln-Gly-Ile-Thr-Lys-Pro-Ala-Ile-Arg-Arg-Leu-Ala-Arg-Arg-
F2a2 (IIbl)        -Ala-Gly-Leu-Gln-Phe-Pro-Val-Gly-Arg-Val-His-Arg-Leu-Leu-Arg-Lys-Gly-Asn-Tyr-Ala-
F2b (IIb2)         -Ala-Gln-Lys-Lys-Asp-Gly-Lys-Lys-Arg-Lys-Arg-Ser-Arg-Lys-Glu-Ser-Thr-Ser-Val-Tyr-
                              +               *
F3 (III)           -Ala-Thr-Lys-Ala-Ala-Arg-Lys-Ser-Ala-Pro-Ala-Thr-Gly-Gly-Val-Lys-Lys-Pro-His-Arg-
Fl (I)             -Pro-Val-Lys-Ala-Ala-Lys-Lys-Lys-Lys-Pro-Ala-Gly-Ala-Arg-Arg-Lys-Ala-Ser-Gly-Pro-
                                  45              50              55              60
F2al (IV)          -Gly-Gly-Val-Lys-Arg-Ile-Ser-Gly-Leu-Ile-Tyr-Glu-Glu-Thr-Arg-Gly-Val-Leu-Lys-Val-
F2a2 (IIbl)        -Glu-Arg-Val-Gly-Ala-Gly-Ala-Pro-Val-Tyr-Leu-Ala-Ala-Val-Leu-Glu-Tyr-Leu-Thr-Ala-
F2b (IIb2)         -Val-Tyr-Lys-Val-Leu-Lys-Gln-Val-His-Pro-Asp-Thr-Gly-Ile-Ser-Ser-Lys-Ala-Met-Gly-
F3 (III)           -Tyr-Arg-Pro-Gly-Thr-Val-Ala-Leu-Arg-Glu-Ile-Arg-Arg-Tyr-Gln-Lys-Ser-Thr-Glu-Leu-
Fl (I)             -Pro-Val-Ser-Glu-Leu-Ile-Thr-Lys-Ala-Val-Ala-Ala-Ser-Lys-Glu-Arg-Ser-Gly-Val-Ser-
                                  65              70              75              80
F2al (IV)          -Phe-Leu-Glu-Asn-Val-Ile-Arg-Asp-Ala-Val-Thr-Tyr-Thr-Glu-His-Ala-Lys-Arg-Lys-Thr-
F2a2 (IIbl)        -Glu-Ile-Leu-Glu-Leu-Ala-Gly-Asn-Ala-Ala-Arg-Asp-Asn-Lys-Lys-Thr-Arg-Ile-Ile-Pro-
F2b (IIb2)         -Ile-Met-Asn-Ser-Phe-Val-Asn-Asp-Ile-Phe-Glu-Arg-Ile-Ala-Gly-Glu-Ala-Ser-Arg-Leu-
F3 (III)           -Leu-Ile-Arg-Lys-Leu-Pro-Phe-Gln-Arg-Leu-Val-Arg-Glu-Ile-Ala-Gln-Asp-Phe-Lys-Thr-
Fl (I)             -Leu-Ala-Ala-Leu-Lys-Lys-Ala-Leu-Ala-Ala-Ala-Gly-Tyr-Asp-Val-Glu-Lys-           -
                                  85              90              95             100
F2al (IV)          -Val-Thr-Ala-Met-Asp-Val-Val-Tyr-Ala-Leu-Lys-Arg-Gln-Gly-Arg-Thr-Leu-Tyr-Gly-Phe-
F2a2 (IIbl)        -Arg-His-Leu-Gln-Leu-Ala-Ile-Arg-Asn-Asp-Glu-Glu-Leu-Asn-Lys-Leu-Leu-Gly-Lys-Val-
F2b (IIb2)         -Ala-His-Tyr-Asn-Lys-Arg-Ser-Thr-Ile-Thr-Ser-Arg-Glu-Ile-Gln-Thr-Ala-Val-Arg-Leu-
F3 (III)           -Asp-Leu-Arg-Phe-Gln-Ser-Ser-Ala-Val-Met-Ala-Leu-Gln-Glu-Ala-Lys-Glu-Ala-Tyr-Leu-
Fl (I)             -                                                                               -
                                  105             110             115             120
F2al (IV)          -Gly-Gly-COOH
F2a2 (IIbl)        -Thr-Ile-Ala-Gln-Gly-Gly-Val-Leu-Pro-Asn-Ile-Gln-Ala-Val-Leu-Leu-Pro-Lys-Lys-Thr-
F2b (IIb2)         -Leu-Leu-Pro-Gly-Glu-Leu-Ala-Lys-His-Ala-Val-Ser-Glu-Gly-Thr-Lys-Ala-Val-Thr-Lys-
F3 (III)           -Val-Gly-Leu-Phe-Glu-Asp-Thr-Asn-Leu-Cys-Ala-Ile-His Ala-Lys-Arg-Val-Thr-Ile-Met-
                                  125             130             135             140
F2a2 (IIbl)        -Glu-Ser-His-His-Lys-Ala-Lys-Gly-Lys-COOH
F2b (IIb2)         -Tyr-Thr-Ser-Ser-Lys-COOH
F3 (III)           -Pro-Lys-Asp-Ile-Gln-Leu-Ala-Arg-Arg-Ile-Arg-Gly-Glu-Arg-Ala-COOH
Fl                 -
```

*Figure 1. Sequences of calf thymus histones. [a]DeLange et al., 1969a, [b]Ogawa et al., 1969, [c]Yeoman et al., 1972, [d]Iwai et al., 1972, [e]DeLange et al., 1973, [f]Rall and Cole, 1971, [g]Bustin, 1972, +ε-N-acetyllysine, *ε-N-mono-, di-, or tri-methyllysine. Basic amino acids: Lys, Arg, His; acidic amino acids: Glu, Asp; hydrophobic amino acids: Phe, Tyr, Leu, Ile, Val, Met.*

each histone has unique properties. Histones IIb1 (f2a2) and III (f3) have fairly basic carboxyl-termini. They also have rather large hydrophobic areas which contain no basic residues (e.g., 43–70 and 100–117 in histone IIb1, and residues 85–113 in III). The two cysteines of histone III (the only calf thymus histone containing cysteine) are in the hydrophobic region.

The lysine-rich histone polypeptides are conspicuously different from the other histones. They are much larger than other histones, and other differences will be noted later. As can be seen from the illustrated amino-terminal sequence of one of the subfractions from calf thymus (CTL-3), this portion of the molecule has many of the characteristics of other histones (Rall and Cole, 1971). However, it is distinct from the other histones which have been sequenced, in that the carboxyl-terminal half of the molecule is its most basic part (Bustin and Cole, 1969*b*). Limited chymotrypsin digestion of rat thymus histone I (RTL-3) separates the molecule into four parts (Bustin and Cole, 1970). One peptide is the amino-terminal portion (residues 1–51), while two peptides come from the central 49 residues (52–100). The remaining peptide, 116 residues long, comes from the carboxyl-terminal region. As noted above, the first 51 residues make up a basic peptide containing nine prolines (total charge +7). The middle portion has a total charge of +3, has no prolines, and contains half of the hydrophobic amino acids of the molecule. Together, the first 100 residues are not unlike basic enzymes such as lysozyme or ribonuclease (Bustin and Cole, 1970). The carboxyl-terminal half of the molecule is quite different. Whereas the amino-terminal half of the molecule has a basic–acidic amino acid ratio of 1 : 6, the carboxyl-terminal half basic–acid ratio is 15 (Bustin and Cole, 1970). This portion of the molecule also contains 16 prolines. Overall, this histone sequence is unusual in that proline, lysine, and alanine account for 65% of its total amino acid content.

Histone Interactions

The uneven distribution of amino acids along the histone polypeptide suggests that each region of the molecule is involved in specific interactions. Because of the ionic nature of the histone binding with DNA, it has been suggested that the more basic part of the histone is involved in interactions with DNA while the other portion (with a composition more like basic enzymes) is involved in secondary structure and interactions with other proteins (Bustin *et al.*, 1969; DeLange *et al.*, 1969*a*). Histones in solution at low ionic strength demonstrate mostly random coiling (Bradbury *et al.*, 1965; Jirgensons and Hnilica, 1965), but take on some secondary structure when the ionic strength of the solution is raised with NaCl (Bradbury *et al.*, 1965, 1967; Jirgensons and Hnilica, 1965; Oh, 1970; Shih and Fasman, 1971). Aggregation may also occur (Edwards and Shooter, 1969; Johns, 1971). DNA induces similar effects in histones (Jirgensons and Hnilica, 1965), indicating that histones may take on secondary and higher-order structure when associated with DNA.

Following these observations and the suggestions of DeLange *et al.* (1969*a*) and Bustin *et al.* (1969), Bradbury and coworkers have determined which regions of the histone polypeptides of f1 (I), f2a1 (IV), and f2b (IIb2) can possess ordered secondary structure and/or interact with other histones (Boublík *et al.*, 1970*a,b*; Bradbury and Crane-Robinson, 1971; Bradbury *et al.*, 1972*a*; Bradbury and Rattle, 1973). They have also determined which parts might interact with DNA (Boublík *et al.*, 1971; Bradbury *et al.*, 1972*b*, 1973*a*; Bradbury and Rattle, 1972). These workers examined the proton magnetic resonance spectra (PMR) of histones in solutions of various ionic strengths. Changes in the histone secondary and higher-order interactions could be followed as the ionic strength of the solution was raised. The signals from groups that are relatively immobilized by interactions broaden, resulting in an apparent loss of area under the resonance peak. Since certain regions of the histone are richer in some amino acid residues than others, a correlation can be made between the peaks broadened and the region of the histone molecule involved in the interaction.

In a like manner, observation of which peaks broadened when histones interact with DNA determined which regions of histone are involved in the histone-DNA interaction. Lysine-rich histone–DNA interactions were studied in native deoxyribonucleoprotein (Bradbury *et al.*, 1973*a*). F2b–DNA interactions were studied in artificial histone–DNA complexes (Boublík *et al.*, 1971; Bradbury and Rattle, 1972). The results show that the areas of the histone with potential for each type of interaction are roughly those expected from the sequencing data. In the arginine-rich histone f2a1 (IV), the self-association and secondary structure region extends from approximately residue 33 to the carboxyl-terminal end at residue 102. Presumably the remainder interacts with DNA. The slightly lysine-rich histone f2b appears to have two DNA binding regions, one extending from the amino-terminal end to residue 32, the other in the region bounded by residues 102 and 125 (*C*-terminus). Residues 33–102 were shown to have the potential for protein–protein interactions. Residues 47–106 define the region of lysine-rich histone which can bind protein, while the amino-terminal end, 1–46, and the carboxyl-terminal half, 107–216, are "sticky" for DNA.

The above experiments suggest that some regions of histone interact with DNA while other regions interact with protein. There is some evidence that both the more and the less basic parts of histone bind to DNA. However, the character of the interaction for each portion is different. Histone f1 can be cleaved into two fragments at its single tyrosine (72) by *N*-bromosuccinamide (Bustin and Cole, 1969*b*). Whole f1, when bound to DNA in an artificial histone–DNA complex, is seen by circular dichroism (CD) to alter the structure of DNA significantly (Fasman *et al.*, 1970). When the *N*-bromosuccinamide fragments are used separately to make such complexes, both bind to the DNA. Only the highly basic *C*-terminal fragment (residues 73–216) affects the conformation of the DNA (Fasman *et al.*, 1971).

The temperature at which DNA melts is highly dependent on the local concentration of cations. This is due to cation shielding of repulsive forces between the negatively charged phosphate groups in the native structure of DNA. The greater the cationic concentration, the greater the shielding and the higher the temperature at which the DNA melts. Histone is a polycation. It thus stabilizes DNA. Under the proper conditions, the DNA in chromatin, or artificial histone–DNA complexes, does not melt at a single temperature. Rather, seveal denaturation transitions are observed. Each corresponds to a population of DNA in the sample which is stabilized to a different extent (Li and Bonner, 1971, Li *et al.*, 1973; Ansevin and Brown, 1971; Ansevin *et al.*, 1971; Van and Ansevin, 1973). The different portions of histone molecules differ in their cationic concentration. Thus two of the melting transitions seen in chromatin and DNA–histone complexes can be interpreted as representing two populations of DNA. One, melting at a lower temperature, is associated with the less basic portion of histone molecules; the other, melting at higher temperature, is bound to the more basic part (Li and Bonner, 1971). Thermal denaturation of the two histone–DNA complexes made with the different segments of cleaved slightly lysine-rich histone, IIb2, supports this view (Li and Bonner, 1971).

That the DNA in chromatin exists in two different environments is indicated by polylysine titration. Approximately 50% of the DNA phosphates are available to bind polylysine (Itzhaki, 1971; Clark and Felsenfeld, 1971; Itzhaki and Cooper, 1973). This result, although consistent with the above thesis of different binding to DNA by different parts of the histone, may be interpreted differently. DNAse digestion experiments on chromatin show that approximately 50% of the DNA is not protected from degradation by this means (Clark and Felsenfeld, 1971). This result and the polylysine titration result have been interpreted to indicate that some regions of the DNA in chromatin are relatively free of protein (Clark and Felsenfeld, 1971). However, there is controversy over this interpretation (Li *et al.*, 1973; Itzhaki and Cooper, 1973).

Species Specificities

Fractions similar to those isolated from calf thymus histones may be isolated from other sources by the procedures outlined above. In many cases, however, modifications must be employed to obtain fractions of a purity comparable to that of fractions isolated from calf thymus (Stellwagen and Cole, 1968; Buckingham and Stocken, 1970*a*; Dick and Johns, 1969; Oliver *et al.*, 1972*b*). Virtually every animal and plant source studied has been found to contain histones (Johns, 1971). A comparison of the character of these histones reveals some striking similarities.

Total histones isolated from plant and animal sources can be fractionated into the same five types as found in calf thymus. When homologous fractions are compared, the expected differences are seen in some cases. Surprising similarities exist, however. This is especially true in the arginine-rich histones. The now classic sequencing studies of DeLange *et al.* (1969*a,b*) on calf thymus and pea seedling histone IV (f2a1) determined that this histone from these sources is identical with the exception of two conservative amino acid replacements. Considering the evolutionary divergence between these two species, the conservation of amino acid sequence in this protein is remarkable. The extreme conservation of primary structure may reflect the close association of the protein with the highly conserved structure of a molecule such as DNA.

A highly conserved primary structure is also apparent for the other arginine-rich histone III (f3). Comparison of the sequence of this protein from the carp (Hooper *et al.*, 1973) and calf thymus (DeLange *et al.*, 1973) reveals only one amino acid difference. The calf thymus cysteine-97 is found as serine-97 in carp. Sequencing work on the chicken erythrocyte histone III (Brandt and Von Holt, 1972) supports the conclusion of conservation of the primary sequence. That the conservation is general and extends to plants is indicated by Fambrough and Bonner (1968). Fingerprints of the tryptic peptides of calf thymus and pea histone III show that 26 of the 29 peptides are identical. For comparison, 27 of the 32 tryptic peptides of histone IV from these two sources are identical (Fambrough and Bonner, 1968).

The presence of two cysteines, one at position 97, the other at 116, may be a characteristic unique to mammalian histone III (f3). Electrophoretic analysis of the behavior of histone III under thiol oxidation or disulfide reducing conditions indicates that the other sources studied including other vertebrates (Panyim *et al.*, 1970; Marzluff *et al.*, 1972), pea (Fambrough and Bonner, 1968; Fambrough, 1969), and *Drosophila melanogaster* (Panyim *et al.*, 1970) contain only one cysteine. Similar experiments indicate that small amounts of histone III with only one cysteine may be present in calf thymus (Marzluff *et al.*, 1972).

Sequences of arginine-rich histones from other sources are not available. However, the conservation of the primary sequences of these histones is evidenced by the results of other, albeit less definitive, techniques. The relatedness of histone IV from calf thymus and lobster hepatopancrease has been demonstrated by microcomplement fixation (Stollar and Ward, 1970). Comparisons of the electrophoretic pattens of histone from calf thymus and insects (Cohen and Gotchel, 1971; Panyim *et al.*, 1970; Oliver and Chalkley, 1972*a,b*; McMaster-Kaye and Kaye, 1973), angiosperms (Smith *et al.*, 1970; Fambrough and Bonner, 1968), vertebrates (Panyim *et al.*, 1971), and echinoderms (Wangh *et al.*, 1972) show identical electrophoretic mobilities of the arginine-rich histones. Curiously, electrophoretic studies, along with amino acid analysis, have revealed the presence of cysteine in histone IV of echinoderms (Subirana, 1971; Wangh *et al.*, 1972). This is unusual in that in other organisms studied only the arginine-rich histone III contains this amino acid.

The conservation of primary sequence present in the arginine-rich histones is not at all apparent in the slightly lysine-rich histones. For example, the constant electrophoretic mobility present in the arginine-rich histones is not found in the slightly lysine-rich histones. In mammals, f2a2 has a greater electrophoretic mobility than f2b, while in other vertebrates this may not be the case (Panyim *et al.*, 1971). Under the same electrophoretic conditions, the slightly lysine-rich histones of crickets (McMaster-Kaye and Kaye, 1973) and *Drosophila* (Cohen and Gotchel, 1971; Oliver and Chalkley, 1972*a,b*) show relatively different electrophoretic mobilities than calf f2b and f2a2. The slightly lysine-rich histones from all animal sources studied have electrophoretic mobilities intermediate between histones III (f3) and IV (f2a1). Thus, regardless of the differences observed, the total charge and size vary only slightly. However, in pea the slightly lysine-rich histones IIa and IIb electrophorese slower than histone III (Fambrough and Bonner, 1969; Fambrough, 1969), indicating larger size and/or less charge than the analogous fractions from animals. Furthermore, one of the slightly lysine-rich fractions from pea, IIa, has two components separable by electrophoresis (Fambrough and Bonner, 1968, 1969; Fambrough, 1969). Thus, whereas in animals slightly lysine-rich histones consist of two proteins, in pea this class of histone may consist of three proteins.

Sequence studies on the slightly lysine-rich histones of trout (Candido and Dixon, 1972*b*; Bailey and Dixon, 1973) verify the conclusion of less conserved sequence of these histones. A complete sequence of the IIb1 (f2a2) histone indicates several differences between calf and trout. It is significant that only one amino acid change has occurred in the basic part of the molecule in the evolutionary time between trout and calf (some 400 million years, Bailey and Dixon, 1973). Glutamine-6, present in calf, is a threonine in trout. In the least basic portion of the molecule starting at residue 57, ten differences are found. This is consistent with regions of the histone closely associated with DNA being conserved. A partial sequence of IIb2 (f2b) indicates even larger differences between calf and trout. In the first 20 amino acids of the histone basic region, four differences are found (Candido and Dixon, 1972*b*). None, however, change the total charge of the region.

Lysine-Rich Histone: Species and Tissue Specificities

The series of closely related polypeptides making up the lysine-rich histone exhibits the greatest species differences of any histone. Electrophoresis of total histones from a variety of vertebrates demonstrates that, while the electrophoretic mobilities of the other histones vary little, if at all, the mobility of the lysine-rich polypeptides vary appreciably between different representative species from this subphylum (Panyim *et al.*, 1971). The same result is obtained in comparison of calf with various insects including *Drosophila melanogaster* (Cohen and Gotchel, 1971; Panyim *et al.*, 1970; Oliver and Chalkley, 1972*a*) and cricket (McMaster-Kaye and Kaye, 1973). Comparisons of the electropherograms of total histones from calf and pea and the histones of different angiosperms (see Figures 1 and 2 of Smith *et al.*, 1970) also shows large differences. That these results reflect large differences in composition has been shown by amino acid analysis (Panyim *et al.*, 1971; Fambrough and Bonner, 1969; Fambrough, 1969). In *Drosophila* it has been estimated that the differences between electrophoretic mobility of the lysine-rich histone of this organism and that of calf are due to the larger molecular weight (approximately 500–1000 daltons) and lesser basicity of the *Drosophila* histone (Oliver and Chalkley, 1972*a*).

The lysine-rich histone is unique in that it consists of several subfractions of closely related but distinct primary structures (Kinkade and Cole, 1966*a,b*; Bustin and Cole, 1969*b*), differing in such a way that they can be considered to have variable and constant regions (Kinkade and Cole, 1966*a*; Bustin and Cole, 1970; Rall and Cole, 1971; Bustin, 1972). Electrophoretic analysis indicates that the number of components is different in various species of vertebrates (Panyim *et al.*, 1971), angiosperms (Figure 1 of Smith *et al.*, 1970), and *Drosophila* (Oliver and Chalkley, 1972*b*). That the number and character of the subfractions of the f1 histone are species specific among mammals has been demonstrated by chromatographic (Bustin and Cole, 1968; Kinkade, 1969; Sluyser and Hermes, 1973) and immunological (Bustin and Stollar, 1972, 1973; Hekman and Sluyser, 1973) techniques.

The relative proportions of the histone fractions vary from organ to organ within a single species (Panyim *et al.*, 1971; Fambrough *et al.*, 1968). The same holds true for the different fractions of the lysine-rich histone (Bustin and Cole, 1968; Kinkade, 1969; Sluyser and Hermes, 1973). There is some chromatographic and immunological evidence indicating that there are qualitative as well as quantitative differences between organs of the same species (Bustin and Cole, 1968; Bustin and Stollar,

1973). This conclusion has been contested (Kinkade, 1969; Hekman and Sluyser, 1973). It has been reported that each of the subfractions shows a different rate of synthesis during the pregnancy-induced development of the mammary gland (Stellwagen and Cole, 1969*a*).

Occurrence of Histones in Lower Eukaryotes

Chromosomes of lower eukaryotes demonstrate great diversity in form and mode of function. In this respect, an evolutionary progression is seen in chromosomal forms and functional modes from those that are bacterialike, such as found in dinoflagellates (Haapala and Soyer, 1973; DuPraw, 1970), to those characteristic of higher plants and animals. It is fairly certain that bacteria do not contain histones, at least in the form found in higher eukaryotes (Bonner *et al.*, 1968; Elgin *et al.*, 1971; Johns, 1971). Thus there is no *a priori* reason to assume that all eukaryotes contain histones. Correlation of the appearance of histones with the evolution of chromosome structure and function could be of significant value in determining the function of these proteins. Of no less importance is the fact that many lower eukaryotes are amenable to genetic analysis. A list of organisms for which the present of histones has been investigated is given by Johns (1971).

Early reports in which cytochemical techniques were utilized failed to discover basic, histonelike molecules in dinoflagellates (Ris, 1962; Dodge, 1964 quoted by DuPraw, 1970). Recently, however, an acid-soluble protein, with similar electrophoretic mobility on polyacrylamide gel as corn histone IV, has been extracted from chromatin of the dinoflagellate *Gyrodinium cohnii* (Rizzo and Nooden, 1972). The weight ratio of acid-soluble protein to DNA in *G. cohnii* chromatin is 0.09, less than one-tenth the relative amount found in higher eukaryotes. The structure of the dinoflagellate chromosome is very bacterialike (Haapala and Soyer, 1973). Thus it is not surprising that only small amounts of histonelike proteins are associated with its DNA. Yet, in the chromosomes of these organisms, an amount of DNA equivalent to that found in the smaller human chromosomes (Haapala and Soyer, 1973) is organized into an apparently well-ordered structure (Ris and Kubai, 1970; DuPraw, 1970). The chromosomes of dinoflagellates are permanently condensed, and it has been suggested by DuPraw (1970) that the evolution of histones corresponds to the appearance of the condensation cycle of the chromosome.

Early work on fungi also failed to detect the occurrence of histones (Dwivedi *et al.*, 1969; Leighton *et al.*, 1971). It is now apparent that at least some histones similar to those of higher plants and animals are found in *Neurospora crassa* (Hsiang and Cole, 1973) and the yeast *Sacharomyces cerevisiae* (Wintersberger *et al.,* 1973). From *N. crassa* two proteins have been isolated by acid extraction of a chromatin preparation (Hsiang and Cole, 1973). Like the dinoflagellate protein, these have a weight ratio with DNA that is only a fraction of that found in higher eukaryotes (histone–DNA ratio of 0.25). Hsiang and Cole have characterized these proteins and found that both are similar to the slightly lysine-rich histones of the calf. One of the proteins is very much like calf thymus fraction IIb1 (f2b) in amino acid composition, molecular weight (14,000), electrophoretic mobility, and behavior on exclusion chromatography. The other protein, although resembling the other slightly lysine-rich histone of the calf, IIb1 (f2a2), is smaller (molecular weight 8000). Chromatin preparations from isolated nuclei of *S. cervisiae* yield histonelike acid-soluble proteins in a mass ratio to DNA of 0.8–1.2 : 1. Three major bands are seen in urea–polyacrylamide gel electrophoresis. Although they electrophorese in the same area as mammalian histones, no bands correspond (Wintersberger *et al.,* 1973).

In some other unicellular organisms, histones have been found in amounts that are comparable to that found in plants and animals. The organisms studied include the green alga *Chlostridium ellipsoidia* (Iwai, 1964) and an amicronucleate strain of the ciliate protozoan *Tetrahymena pyriformus* (Iwai *et al.,* 1965; Hamana and Iwai, 1971). Histones have also been obtained in a 1 : 1 histone-to-DNA ratio from the plasmodial slime mold *Physarum polycephalum* (Mohberg and Rusch, 1969). The histone from *T. pyriformus* and *P. polycephalum* have been characterized. On the basis of amino acid analyses, each of five or six histones from *Tetrahymena* can be designated as being either lysine rich, slightly lysine rich, or arginine rich. Their composition, with the exception of the lysine rich histone, bears a close resemblance to the corresponding calf thymus histones. Like yeast, electrophoretic patterns of the histones from *Tetrahymena* and *Physarum* bear little resemblance to those of higher plants and animals (Mohberg and Rusch, 1969; Hamana and Iwai, 1971). In *P. polycephalum,* no one of the six discernible histone bands is seen to migrate with any band from higher animals. With *Tetrahymena,* the situation is the same. Furthermore, in this ciliate, neither exclusion chromatography, Amberlite chromatography, nor chemical fractionation by the method of Johns gives fractions compatible with those obtained from calf (with the exception of f1).

These data indicate that the conservation of histones found in higher plants and animals does not extend to the protists and fungi, some of which otherwise demonstrate a complete complement of histone proteins.

Modification of Histones

Chromosomes undergo obvious structural transformations during the cell cycle. More subtle changes occur in response to the requirements of gene activation and repression, and the replication of DNA and the chromosome. Yet, in only specialized cases are qualitative changes seen in the limited heterogeneity of histone. Whatever the function of histones, the interactions of these molecules with chromosomal components must be modified to allow the necessary structural alterations. In that no change in the histone complement occurs, structural transformations must be effected by changes in the histone character. Chemical modification of the amino acid side groups of the histone is one method of obtaining the necessary change. Three such modifications are known to occur. They are methylation, phosphorylation, and acetylation. A study of their occurrence, and the chromosomal events with which they are associated will be informative in determining their importance to histone function. Conversely, this information could give insight into the function of the histones.

Methylation

Much of the work on the biochemistry of methylation has been done by Paik and Kim. They have reviewed the subject (Paik and Kim, 1971).

Occurrence. The presence of methyl groups in histone in the form of ϵ-*N*-methyllysine was first demonstrated by Murray (1964*a*). Several other methyl derivatives of basic amino acids have been found in histone. These include ϵ-*N*-dimethyllysine (Paik and Kim, 1967), ϵ-*N*-trimethyllysine (Hempel *et al.*, 1968), ω-*N*-methylarginine (Paik and Kim, 1970*a*), and 3-methylhistidine (Gershey *et al.*, 1969).

As with the other forms of histone modification, the positions of methyl incorporation are highly specific. In calf thymus arginine-rich histones the methylated lysines are at position 20 in histone IV (f2a1) (DeLange *et al.*, 1969*a*), and at positions 14 and 27 in histone III (f3) (DeLange *et al.*, 1973).

It is now well established that methylation occurs after the synthesis of the histone polypeptide is complete. This was first elucidated by Allfrey *et al.* (1964) in isolated calf thymus nuclei. Puromycin at a dosage sufficient to block 90% of the nuclear protein synthesis is unable to block more than 50% of the incorporation of [^{14}C]methyl of methionine into ϵ-*N*-methyllysine of histone. The result was confirmed by Kim and Paik (1965), who also demonstrated that *S*-adenosylmethionine serves as a methyl donor.

Methylation Enzymes. An important implication of the observation of nuclear incorporation of methyl groups into ϵ-*N*-acetyllysine of histones (Allfrey *et al.,* 1964; Kim and Paik, 1965) is that the enzyme(s) catalyzing the transfer of a methyl group *S*-adenosylmethionine to lysines of histone are of nuclear origin. An enzyme capable of mediating such a transfer, methylase III (*S*-adenosylmethionine; protein-lysine methyl transferase, Paik and Kim, 1971) has been partially purified from calf thymus (Paik and Kim, 1970*b*). As expected, the enzyme is localized in the nucleus. *S*-Adenosylmethionine is the methyl donor. Lysine is the methyl acceptor, and the product is ϵ-*N*-monomethyllysine. The enzyme is specific for histones.

Similar enzymatic activity has been observed to be tightly bound to chromatin. The incorporation of labeled methyl groups from *S*-adenosylmethionine into histones on chromatin has been observed to occur in rat liver (Sekeris *et al.,* 1967), Ehrlich ascites (Comb *et al.,* 1966), and Krebs 2 ascites (Burdon, 1971). In liver and Ehrlich ascites the enzyme(s) remained bound to the chromatin after washing with 0.15 or 0.5 M NaCl, respectively. With the exception of the Krebs ascites, the detected modification is the monomethyllysine derivative. Mono-, di-, and trimethyllysine were detected in the Krebs ascites experiments. A histone-specific methylase was isolated from the Krebs 2 ascites chromatin by extraction with 1.5 M NaCl (Burdon and Garvin, 1971).

In the above experiments, using either isolated nuclei or chromatin, no incorporation of methyl groups into arginine was observed. Yet methyl derivatives of arginine are known to occur in histones (Paik and Kim, 1970*a*). Methylase I (S-adenosylmethionine: protein arginine methyltransferase, Paik and Kim, 1971), an enzyme which transfers methyl from *S*-adenosylmethionine to arginine yielding ω-*N*-methylarginine was isolated from calf thymus cytosol (Paik and Kim, 1968). The enzyme fraction includes endogenous histone which is preferentially methylated. Exogenous histone is specifically methylated by this preparation, but the specificity extends to other basic proteins (Paik and Kim, 1969).

It should be mentioned that another cytosol-localized methylase, methylase II, has been isolated (Liss *et al.,* 1969). The enzyme methylates the acidic amino acids (glutamic and aspartic). There is no evidence that such methyl derivatives exist in histones (Allfrey, 1971).

Functional Studies. It is apparent from work on synchronized cell populations that methylation of histones is a relatively late event in the cell cycle. The first experiments which indicated this were done on partially hepatectomized rats (Tidwell *et al.,* 1968). Regeneration of the liver proceeds with approximately 70% of the cells in synchrony. Histone and

DNA synthesis coincide. The rate of uptake of infected [^{14}C]methylmethionine peaks after histone and DNA synthetic rates have begun to decline. Thus methylation of histones does not correlate with any synthetic activity. It was suggested by the authors that methylation may be necessary for the changes in chromosome structure during mitosis.

Similar results have been obtained in synchronized cultures of Chinese hamster cells (Shepherd *et al.*, 1971*a*). Isoleucine-starved cells were released into complete medium containing [^{14}C]methylmethionine. The total incorporation of the label into ε-*N*-methyllysine was measured at various times in the cell cycle. The Johns arginine-rich histone fractions (f2a1 and f3) were seen to possess the greatest amount of methyllysine derivative at mitosis. No label in the lysine-rich histone fraction, f1, was detected.

The rates of methylation of histones relative to the synthesis of histone polypeptides were studied in HeLa cells (Borun *et al.*, 1972). Relative rates were determined by measuring the ratio of [^{3}H]methylmethionine to [^{14}C]methionine incorporated into the histone fractions. Histones were electrophoresed and the gels cut up and counted. Alternatively, the histones were hydrolyzed and the label present in the methyl-amino acids was determined. Cell synchrony was obtained either by mitotic selection or double thymidine block. The results show that the rate of methylation peaks after the rate of DNA and histone synthesis has reached a maximum. It was also found that the arginine-rich histones, f3 (III) and f2a1 (IV), are methylated at the highest rates followed by the slightly lysine-rich histones, f2b (IIb2) and f2a2 (IIb1). As in the cultured Chinese hamster cells, no f1 (I) is methylated. Observation of the rates of incorporation of label into ε-*N*-methyllysine, ε-*N*,*N*-dimethyllysine, methylarginine, and methylhistidine showed that lysine methylation occurs at the highest rates. The rates of both mono- and dimethyl derivatives of lysine peak at G_2. The methyl derivatives of arginine and histidine are labeled in f3 only. Their rates of methylation are low and have a broad maximum during S.

Turnover rates of the methyl groups relative to the histone backbone have been measured (Borun *et al.*, 1972; Byvoet *et al.*, 1972; Byvoet, 1972). It is found that the methyl groups turn over at very low rate, if at all. The data given by Borun *et al.* (1972) indicate a turnover of no more than 1–2% per hour. This indicates that methylation is a very stable modification of histone structure. Thus methylation may have little to do with the dynamics of chromosome structure.

There is little evidence to indicate how methylation might affect the structure of histones and their interactions with other chromosomal components. However, considerations of how methyl groups change the properties of amino acid side chains give three possible effects.

Methyl groups when attached to lysines change the p*K* of the ε-nitrogen. It has been found that the secondary amide (monomethyl) is more basic than the primary amide. The same holds true for the quaternary amide resulting from trimethylation. The tertiary amide of lysine is less basic than any of the amides of other methylated derivatives of lysine or of the unmodified lysine amide (Paik and Kim, 1971). Thus, with the exception of the dimethyl form, methylation of histone would enhance charge interactions between histones and other molecules, particularly DNA.

Another effect is from an increased hydrophobicity imparted to an amino acid residue that is methylated (Paik and Kim, 1971). By this mechanism, histone hydrophobic interactions with other proteins or DNA would be enhanced.

Steric effects also could play a role. With the bulky methyl groups to get in the way, the interactions between molecules could be impaired (Paik and Kim, 1971).

Which of these effects are the most pervasive? We have little idea. It should be mentioned that chemical methylation of 80% of the lysines of ribonuclease, although destroying its enzymatic activity, did not appreciably affect the enzyme's physical properties (Means and Feeney, 1968) or its susceptibility to digestion by proteases (Paik and Kim, 1972).

Phosphorylation

The incorporation of inorganic phosphate into histones and the subsequent identification of *O*-phosphoserine in these proteins was done by Ord and Stocken (1966) and Kleinsmith *et al.,* (1966*a*). That the incorporation of phosphate into histones is a real biological event has now been clearly demonstrated. However, as is typical of most matters concerning histones, we have only a vague idea of its importance to cellular processes.

Occurrence. Phosphorylation occurs on preformed histones (Kleinsmith *et al.,* 1966*a*; Stevely and Stocken, 1966; Marushige *et al.,* 1969), with ATP as the immediate phosphate donor (Allfrey, 1971). Depending on the system being studied, all, or as few as one, of the five histone fractions have been reported to incorporate phosphate. Of these, the lysine-rich (I, f1) histone has attracted the most attention, mostly because it is the one histone that is extensively and ubiquitously modified in this manner. The lysine-rich histones may be multiply phosphorylated. Multiplicities of one to five have been reported in dividing cells of trout testis (Louie and

Dixon, 1973). Singly and doubly phosphorylated f1 (I) has been observed in mouse hepatoma (HTC) (Balhorn *et al.,* 1972*a*). Thus the microheterogeneity of f1 (I) can be considerably enhanced by this modification. It is apparent that not all the subfractions of this histone are phosphorylated (Langan *et al.,* 1971; Buckingham and Stocken, 1970*b*). Also, of those subfractions that are modified in this manner, not all are phosphorylated to the same extent (Buckingham and Stocken, 1970*b*).

Langan (1968*a,b*) has isolated a tryptic peptide of f1 (I) from rat which contains 80% of the [^{32}P]phosphate incorporated into this histone fraction. The amino acid content of this peptide, which is identical for the *in vivo* and *in vitro* labeled f1 (I) (Langan, 1969*a,b*), is consistent with the known *N*-terminal f1 (I) sequence. We can identify the modified amino acid to be that shown in Figure 1 as serine-38 (Rall and Cole, 1971; Langan *et al.,* 1971). The trout f1, serine-157, may also be phosphorylated (Dixon *et al.,* 1973).

Louie *et al.* (1973) have established that all five histones in trout testes are phosphorylated, although histones f3 (III) and f2b (IIb2) are modified at very low levels. The only histone besides f1 (I) showing evidence of multiple phosphorylation is f2b (IIb2) (Louie *et al.,* 1973). The arginine-rich histone, f2a1 (IV), and the slightly lysine-rich histone, f2a2 (IIb1), are both significantly phosphorylated. Both of these contain phosphate at the *N*-terminal serine (Sung and Dixon, 1970).

Phosphorylation of histones other than f1 (I) has been shown for systems other than trout testes. Histone f2a2 (IIb1) has been shown to be phosphorylated in regenerating rat liver (Sung *et al.,* 1971; Balhorn *et al.,* 1972*e*). As in trout testes, this histone appears to be *N*-terminally phosphorylated (Sung *et al.,* 1971). Marzluff and McCarty (1972*a,b*) have detected the presence of phosphoserine in calf thymus arginine-rich histone, f3 (III). A comparison of their thermolysin peptides with the published sequence of f3 (III) (DeLange *et al.,* 1973) indicates that the phosphate is incorporated at serine-10 (see Figure 1). Both f2a1 and f2b (or f2a2) are phosphorylated at low levels in HeLa cells (Marks *et al.,* 1973).

Functional Studies. A positive relationship between the replication rate of tissue and the percentage of f1 (I) found to be phosphorylated in that tissue (Sherod *et al.,* 1970; Balhorn *et al.,* 1971, 1972*a,b,d*) has been demonstrated in rat and mouse. The amount of phosphorylated f1 was determined by utilizing inherent differences in electrophoretic mobility between phosphorylated histones and the unphosphorylated form. By this method all but 5% of the f1 (I) is found in normal rat liver to be present in a single, unphosphorylated band. In contrast, regenerating rat liver shows 50% of the lysine-rich histone in the bands corresponding to phosphorylated f1 (I) (Balhorn *et al.,* 1971, 1972*d*). Likewise, 85% of this histone is found to be phosphorylated in rapidly dividing fetal liver (Balhorn *et al.*, 1972*a*). A very nice correlation between phosphorylation and replication rates was made in different Morris hepatomas (Balhorn *et al.*, 1972*b*). Similar correlations were made for mouse tissues and their respective tumors (Balhorn *et al.*, 1972*b*). In all of the above studies, the altered electrophoretic patterns were shown to be due to phosphate modification by treating the f1 (I) preparations with *E. coli* alkaline phosphatase. In most cases, such treatment reduced the banding pattern to that of the nondividing tissue. In those cases where small residual differences remained after such treatment, they could be attributed to variations in the amounts of histone I (f1) subfractions (Sherod *et al.*, 1970; Balhorn *et al.*, 1971, 1972*a,b,d*).

The time course of f1 (I) phosphorylation relative to the cell cycle has been extensively investigated. First among these studies were those carried out on regenerating rat liver. Ord and Stocken (1967, 1968) demonstrated that an increase in the total phosphate per milligram of histone f1 (I) occurs during DNA synthesis at 22 hr after partial hepatectomy. Chalkley and collaborators (Balhorn *et al.*, 1971), using the electrophoretic system described above, studied the shift of the f1 population from unphosphorylated to phosphorylated forms and back again as the process of liver regeneration progressed. During the course of the observations, the total amount of lysine-rich histone per cell did not significantly change. The first significant increase in phosphorylated f1 (I) occurred about 20 hr after the operation. This corresponds to the onset of DNA synthesis. The amount of modified f1 (I) peaked at about 29 hr, concomitant with the first round of mitosis. The phosphate content of the histone then decreased. At the time of a second round of DNA synthesis (33 hr), there was a sharp increase of phosphorylated f1 (I). This increase peaked at about 38–40 hr, synchronous with another round of mitosis.

The concept of lysine-rich histone phosphorylation occurring during the DNA synthetic period of the cell cycle is consistent with studies on the rate of phosphate incorporation into histone I in synchronous cell populations (Balhorn *et al.*, 1972*c*; Cross and Ord, 1970, 1971; Gutierrez-Cernosek and Hnilica, 1971; Ord and Stocken, 1967; Marks *et al.*, 1973). These studies have demonstrated an increase in phosphate uptake into f1 (I) during S.

Measurements of the total content of phosphate in f1 (I) (reported as μmoles phosphate/mg histone f1) in regenerating rat liver (Ord and Stocken, 1968), PHA or dibutyryl cyclic AMP stimulated porcine lymphocytes (Cross and Ord, 1970, 1971), and regenerating rat pancreas (Fitzgerald *et al.*, 1970) have indicated that lysine-rich histone phosphorylation content reaches a maximum at S and then declines. This is at variance with the results on regenerating liver (Balhorn *et al.*, 1971). That the phosphate content of the f1 histone remains high until mitosis, as noted above, has been supported by the observations of Lake and Salzman (1972), Marks *et al:* (1973), and Shepherd *et al.* (1971*b*) in Chinese hamster cell cultures. According to Marks *et al.* (1973), the phosphate content of the lysine-rich histone remains high until G_1.

Kinetic studies on the phosphorylation of f1 (I) have been made by Oliver *et al.* (1972*a*) for randomly dividing cultured mouse hepatoma cells (HTC). A short pulse label of the culture with [^{14}C]lysine was employed to label a small population of f1 (I) molecules. The subsequent modification of this population of f1 could be followed by the usual electrophoresis technique. The cell cycle in the HTC cultures is about 24 hr. The results of this experiment show that there is a lag of at least 40 min between the synthesis of f1 (I) and its succeeding phosphorylation. There is then a gradual shift of the labeled f1 (I) back to the unphosphorylated form. F1 is again phosphorylated during the next S phase. These data show conclusively that phosphorylation of f1 is a cyclical event where both "new" and "old" f1 are phosphorylated during S. During the subsequent phases of the cell cycle, f1 is dephosphorylated with a turnover time constant ($t^{1}/_{2}$) of about 5 hr (Balhorn *et al.*, 1972*e*).

Similar results have been obtained by Dixon's lab in dividing stem cells and primary spermatocytes from trout testis (Louie and Dixon, 1973). Utilizing essentially the same experimental procedure outlined for the HTC experiments, cyclical labeling of "new" and "old" f1 was again demonstrated. The length of the f1 phosphorylation and dephosphorylation was consistent with the duration of the cell cycle—about 7 days. A lag of nearly 4 days was observed between synthesis and phosphorylation of f1.

The observed lag between histone I synthesis and the subsequent phosphorylation in the trout testis and HTC cultures indicates that the phosphorylation of lysine-rich histone is not associated with its synthesis or with its transport into the cell (Oliver *et al.*, 1972*a*).

Essentially all f1 is phosphorylated during S phase of each cell cycle (Marks *et al.*, 1973). However, if the cells are exposed to X-radiation at sufficient dosages to prevent DNA synthesis, f1 phosphorylation is also prevented (Ord and Stocken, 1967; Stevely and Stocken, 1966, 1968*a,b*). Likewise, treatment with an inhibitor of DNA synthesis, cytosine arabinoside, inhibits the phosphorylation of the lysine-rich histone by 50% (Marks *et al.*, 1973). Thus in some cells the phosphorylation of f1 is at least partially coupled to DNA synthesis, indicating a role for this modification in chromosome replication during the S phase of the cell cycle.

That the content of phosphate in the lysine-rich histone remains high until mitosis indicates that phosphorylation may be important at stages of the cell cycle later than S. This consideration, combined with the cyclical nature of the modification, has led Louie and Dixon (1973) to suggest that this modification may be important in the condensation cycle of the chromosomes.

Support for this role of phosphorylation comes from the pattern of phosphorylation of a presumed lysine-rich histone from the plasmodial slime mold *Physarum polycephalum* (Bradbury *et al.*, 1973*b*). As in higher eukaryotes, [^{3}H]lysine label is seen to incorporate into the histones of plasmodia during S. One histone showing the highest specific activity of [^{3}H]lysine is seen to be phosphorylated more extensively than the other histones. Phosphorylation of this protein does not occur at S. Rather [^{32}P]phosphate is taken up into the histone late in G_2, 20 min prior to mitosis at the time of chromosome condensation.

The cyclic nature of the phosphorylation of lysine-rich histone appears to be unique. In trout testis the kinetics of phosphorylation of histone IV (f2a1) (Louie and Dixon, 1972*a*) and the slightly lysine-rich histone, IIb1 (f2a2) (Louie *et al.*, 1973), have been studied. The experimental design of these experiments is the same as that used to study the kinetics of histone I (f1) phosphorylation. In contrast to the lysine-rich histone, these two histones appear to be phosphorylated a single time. In the case of the arginine-rich histone, IV (f2a1), the first phosphorylated species appear at 16 hr after synthesis. The percent of the phosphorylated form of the histone increases until the fourth day after synthesis, when a value of roughly 35% phosphorylated histone IV is observed. The amount of phosphorylated histone then declines over a 16-day period to basal levels. No phosphorylation of "old" histone IV (f2a1) is observable during the 7-day cell cycle. The sequence of events of IIb1 (f2a2) is similar except that the kinetics of phosphorylation and dephosphorylation are faster. Ten minutes after synthesis, portions of the IIb1 population are seen to band in the regions of the electrophoretogram corresponding to the phosphorylated molecule. The percent of IIb1 phosphorylated rises rapidly and attains a value of approximately 30% at 16 hr, whereupon the molecules are rapidly dephosphorylated. Treatment with alkaline phosphatase leaves only the electrophoretic band corresponding to unmodified histone, insuring that the changes observed in the electrophoretic pattern are due to phosphorylation.

The differences in the kinetics of phosphorylation of these histones indicate a different function of phosphorylation for each histone (Louie and Dixon, 1972*c*; Louie *et al.*, 1973). The rapid phosphorylation of IIb1 (f2a2) indicates that this modification may be necessary for the proper binding of the newly synthesized molecules into chromatin (Louie *et al.*,

1973). The kinetics indicate that IV·(fa21) is phosphorylated after it binds to the chromosome.

It is apparent from the work outlined above that histone phosphorylation has some role in the replication of the chromosome. However, there is strong evidence that at least in differentiated, nondividing tissue the phosphorylation of lysine-rich histone may be part of gene activation.

The evidence for this aspect of f1 (I) phosphorylation has come from the work of Langan (1968*a,b*, 1969*a,b*) on rat liver. Normal liver has very little phosphorylated f1 (Balhorn *et al.*, 1971, 1972*a,d*), and little [^{32}P]phosphate injected into the rat is found incorporated into liver lysine-rich histone (Langan, 1969*a*). However, when the animal is injected with glucagon, there is a twentyfold increase of uptake of [^{32}P]phosphate into f1 (Langan, 1969*a*). The increased uptake is into a single tryptic peptide. This peptide is identical to the one mentioned earlier, and the phosphorylated residue has been identified as the serine 38 referred to above (Rall and Cole, 1971; Langan *et al.*, 1971). The effects of glucagon are mediated by cyclic AMP, and the injection of cyclic AMP, or its lipid-soluble analogue, N^6-$O^{2'}$-dibutyrylcyclic AMP, causes a respective eight- or twentyfold increase in the uptake of lable into liver f1 (I) (Langan, 1969*b*). Again, the label is found in the same tryptic peptide.

A histone-specific kinase has been isolated from rat liver which incorporates phosphate into the same f1 tryptic peptide *in vitro* as is phosphorylated *in vivo* (Langan, 1968*a,b*). The activity of the preparation is stimulated by cyclic AMP (Langan, 1968*a,b*). These results suggest a means by which a hormone may effect a change in genetic expression of its target organ (Langan, 1969*a,b*).

Phosphorylation Enzymes. Langan and Smith (1967) first reported the isolation of a histone-specific protein kinase. Since that time, several preparations of histone kinases and a histone phosphatase (Meisler and Langan, 1969) have been reported. Most of these have been purified from a soluble fraction of total cell homogenates.

In many cases, the activities of the kinases have been found to be dependent on the presence of cyclic AMP (Langan, 1968*a,b;* Miyamoto *et al.*, 1969; Kuo and Greengard, 1969; Chae *et al.*, 1972; Takáts *et al.*, 1972; Faragó *et al.*, 1973; Chen and Walsh, 1971; Siebert *et al.*, 1971; Pawse *et al.*, 1971). The cyclic AMP kinases are apparently ubiquitous, having been isolated from every mammalian organ attempted, and from representative organisms of virtually every animal phylum (Kuo and Greengard, 1969). This wide distribution of the cyclic AMP dependent kinases suggests that the actions of cyclic AMP is generally mediated by these kinases (Kuo and Greengard, 1969). This supports the suggestion of

Langan on the relationship among hormones, cyclic AMP, and histone phosphorylation.

Some workers have found kinases associated with nuclei. The nuclear enzymes are only weakly associated with chromatin. Simple breakage of the nuclei (Faragó *et al.*, 1973) or washing of the chromatin with solutions of low ionic strength (0.12–0.35 M NaCl) (Chae *et al.*, 1972; Lake and Salzman, 1972) is sufficient to release the activity.

The nuclear kinase preparations are always found to be cyclic AMP independent (Lake and Salzman, 1972; Faragó *et al.*, 1973; Siebert *et al.*, 1971). However, Chen and Walsh (1971) have demonstrated that the liver cyclic AMP independent enzyme is convertible to a cyclic AMP dependent enzyme by the addition of a hepatic protein fraction. Thus preparations which demonstrate no cyclic AMP stimulation may be holoenzymes from which a regulatory subunit has been removed.

There is good indication that, at least in some of the cases, the enzymes isolated are those responsible for the phosphorylation of histone *in vivo*. First, Langan (1968*a,b,* 1969*a,b*) has demonstrated that the enzyme isolated from rat liver phosphorylates at the same residue in lysine-rich histone as is found to be phosphorylated *in vivo*. Kinase isolated from another source (Jergil and Dixon, 1969) shows this same specificity, although it may phosphorylate other residues besides (Jergil *et al.*, 1970). This is not inconsistent with the observed multiple phosphorylation of f1 (I) histone. Second, the temporal pattern of activity of a cyclic AMP dependent kinase of cytoplasmic origin, and of a cyclic AMP independent enzyme of nuclear origin from regenerating rat liver (Pawse *et al.*, 1971; Siebert *et al.*, 1971), matches the kinetics of f1 phosphorylation. Finally, the fact that some of the kinases when phosphorylating free histone favor some other histone than f1 as a substrate—contrary to the *in vivo* result—may be an artifact of using a free enzyme and free substrate (Takáts *et al.*, 1972; Faragó *et al.*, 1973; Chen and Walsh, 1971; Chae *et al.*, 1972). For example, the enzyme purified by Chae *et al.* (1972) phosphorylates free f2b preferentially. However, when this kinase is used to phosphorylate histone in chromatin, 81% of the incorporated label is found in f1 (I).

Acetylation

Occurrence. Acetyl groups are found incorporated into histones as two modified amino acids: *N*-acetylserine (Phillips, 1963; 1968) and ϵ-*N*-acetyllysine (DeLange *et al.*, 1969*a,b*; Gershey *et al.*, 1968; Vidali *et al.*, 1968). The two modifications differ in several important respects.

N-Acetylserine occurs only as the *N*-terminal amino acid of the lysine-rich histone, f1 (I) (Phillips, 1963; Rall and Cole, 1971), the slightly lysine-rich histone, f2a2 (IIb1) (Phillips, 1968; Yeoman *et al.*, 1972), and the arginine-rich histone, f2a1 (IV) (Phillips, 1968; DeLange *et al.*, 1969*a,b*). The occurrence of the *N*-acetyl group in histone IV (f2a1) (DeLange *et al.*, 1969*a,b*) and most likely histones IIb1 and I (Fambrough and Bonner, 1969) in such widely divergent species as calf and pea attests to the highly conserved nature of this "modification." As such, it must be an important part of the histone structure.

Unlike *N*-acetylserine, ε-*N*-acetyllysine never occurs at the end of the histone polypeptide. Also, any histone fraction may contain ε-*N*-acetyllysine. The occurrence of the acetylated lysyl residues in the various histone fractions and the extent of acetylation are variable and dependent on the physiological state of the system. Acetylation of lysines is a nuclear event (Allfrey *et al.*, 1964) and occurs after completion of synthesis of the histone polypeptide (Allfrey *et al.*, 1964; Marzluff and McCarty, 1970). The modification occurs enzymatically (Gallwitz, 1970*a*), with coenzyme A as the acetyl donor (Gallwitz and Sekeris, 1969*b*). The acetyl groups on the lysyl residues are observed to turn over rapidly with respect to the histone backbone (Allfrey *et al.*, 1964; Marzluff and McCarty, 1970). In contrast to this, the *N*-acetylserine modification is a cytoplasmic event occurring as part of the translation of the histone polypeptide (Liew *et al.*, 1970; Marzluff and McCarty, 1970). No turnover relative to the histone backbone is observed for this type of acetylation (Marzluff and McCarty, 1970). It is the metabolically unstable lysine modification which is suitably variable to have a role in the dynamic aspects of chromosome structure and function.

It is known that acetylation is not an obligatory state of a histone, and in any given tissue a histone may be singly or multiply acetylated, or not acetylated at all. Thus calf thymus histone IV species having 0, 1, 2, or 3 acetate groups are detectable by electrophoresis (Wangh *et al.*, 1972), while, in trout testes, species with up to four such modifications are apparent (Sung and Dixon, 1970). Histone III can be acetylated at two positions in calf thymus, and at none to four locations in trout testes. The positions at which the modifications occur are highly specific. The two which occur in calf thymus histone III are at lysines 9 and 23 (DeLange *et al.*, 1973), and those in trout testes are at 9, 14, 18, and 23 (Candido and Dixon, 1972*b*). Only one of the three potentially acetylated sites has been determined for the calf thymus arginine-rich histone IV, lysine 16 (DeLange *et al.*, 1969*a*). All four positions are known in trout testis IV. They are lysines 5, 8, 12, and 16 (Candido and Dixon, 1971). Unlike calf thymus histones, the two slightly lysine-rich histones of trout testis are acetylated. Histone IIb1 (Candido and Dixon, 1972*a*) can be acetylated at lysine 5 and histone IIb2 (Candido and Dixon, 1972*b*) at positions 5, 10, 15, and 18. As has been noted previously, no comparison of the sites of IIb2 acetylation can be made with the sequence of the analogous calf thymus histone.

Functional Studies. Histone acetylation has been seen to correlate with the onset of RNA synthesis in terminally differentiated cells which have been stimulated to divide, and in differentiating cells which produce large amounts of messenger RNA at a specific time in the normal course of development.

Human peripheral lymphocytes can be made to increase protein and nucleic acid synthesis and undergo mitosis *in vitro* by exposure to phytohemagglutinin (PHA). Likewise, differentiated liver cells of the rat are activated to divide, resulting in the regeneration of liver tissue removed by partial hepatectomy. It is seen in both cases that radiolabeled acetate is taken up into the histones of both the treated and untreated cells, but the uptake is greater in the activated cells (Pogo *et al.,* 1966, 1968). The increased incorporation of acetate into histones as acetyl groups is not due to a stimulation of acetylation. Rather it occurs as the result of a decrease of histone acetate turnover, causing a net accumulation of acetyl groups (Pogo *et al.,* 1968, 1969). The increase in histone acetylation immediately precedes or is coincident with an increase in RNA synthesis (Pogo *et al.,* 1966, 1968) and RNA polymerase activity (Pogo *et al.,* 1968, 1969). In neither case is the acetylation of histone associated with histone synthesis. The periods of acetate accumulation occur before any net increase of histone begins.

In developing chick muscle, a 2.5-fold increase in the uptake of acetate into ϵ-*N*-acetyllysine is observed concomitant with an increase of myosine mRNA synthesis (Boffa and Vidali, 1971). The increase in acetylation could be attributed to a decrease in acetate turnover. Deacetylase activities in the cell dropped as the turnover rates declined (Boffa *et al.,* 1971). The histones most extensively acetylated in this system and in the liver and lymphocyte systems are the arginine-rich histones (III and IV).

Unlike lymphocytes, PHA-treated equine granulocytes show an inhibition of incorporation of uridine into RNA (Pogo *et al.,* 1966, 1967). Consistent with a model of a causal relationship between RNA synthesis and histone acetylation, the rate of incorporation of acetate into histone is diminished in the PHA-treated cells.

Another case of inhibition of RNA synthesis being correlated with the inhibition of histone acetylation is the action of the drug aflatoxin B1.

Aflatoxin B1, a carcinogen, inhibits the action of RNA polymerase. The exact mechanism of this inhibition is unknown, but it is known not to occur by direct action on the polymerase (Edwards and Allfrey, 1973). Rats injected with this drug show unaltered uptake of acetate into liver histone for the first 15-min period of labeling. Thereafter, the turnover is greater in the aflatoxin-treated animals than in the controls. The arginine-rich histones are the only histones seen to take up acetate. The turnover of acetate for f2A1 and f3 is affected to the same extent by the drug. Neither actinomycin D, which is known to inhibit RNA polymerase by binding to the template DNA, nor aflatoxin B2, which has no effects on RNA transcription, has any effect on the kinetics of acetylation of the liver histones. Furthermore, aflatoxin B1 has no effect on the *in vitro* activity of liver deacetylases (Edwards and Allfrey, 1973). Thus aflatoxin B1 may affect some control element which is common to the histone acetate content and to RNA synthesis.

Several other studies correlate the acetylation of histones with increases in RNA synthesis (Wilhelm and McCarty, 1970*a,b*), including some involving stimulation of the liver by insulin (DeVilliers Graaff and Von Holt, 1973) and cortisol (Allfrey, 1966, 1968). However, some reports to the contrary have been given. Hydrocortisone-stimulated liver (DeVilliers Graaff and Von Holt, 1973) shows only a slight increase in acetylation. Isolated liver nuclei induced to transcribe RNA by cortisol show no increased uptake of acetate (Gallwitz and Sekeris, 1969*a*). This result is in contrast to the results of Allfrey (1966, 1968). The locality of the acetate uptake of proteins in dipteran polytene chromosomes does not correlate with the position of the RNA puffs (Cleaver and Elgaard, 1970). Finally, two cell types which differ greatly in RNA synthesis, avian reticulocytes and erythrocytes, the latter being completely inactive, show kinetics of acetate uptake which do not correspond to that expected (Sanders *et al.*, 1973).

Histone acetylation occurs in situations other than gene activation. Internal acetylation is found to be associated also with the synthesis of histone IV. Cultured Chinese hamster ovary cells (CHO) show acetylation on newly synthesized histone f2a1 (IV) during S (Shepherd *et al.*, 1971*c*; Shepherd, 1973). Newly synthesized histone IV (f2a1) is also acetylated in developing trout testes (Louie and Dixon, 1972*a*). Kinetic studies of the acetylation of this histone in trout testes indicate that histone IV appears in the diacetyl form immediately after synthesis. It is then slowly acetylated to the tetraacetate form, followed by deacetylation to the mono- and unacetylated forms. The peak of acetyl content of this arginine-rich histone peaks at 3 days and declines to basal levels 12 days after synthesis

(Louie and Dixon, 1972*a*). There is no evidence for the acetylation of "old" histone IV (the cell cycle in this system is about 7 days). It has been proposed that histone IV acetylation facilitates binding of this histone to DNA as it is incorporated into chromatin (Louie and Dixon, 1972*a*).

It is evident from the work of Candido and Dixon (1972*c*) that the acetylations of trout testes histones are not confined to the times of their synthesis. Late in spermiogenesis the histones are replaced by protamines in the chromatin of the spermatid (Dixon *et al.*, 1968; Louie and Dixon, 1972*b*). During this time no histone synthesis is evident (Louie and Dixon, 1972*b*); all the histone fractions are acetylated (Candido and Dixon, 1972*c*). It would appear that, in spermiogenesis, acetylation is important not only for histone binding but also for its release.

Acetylation Enzymes. Nohara *et al.* (1968) first isolated a protein fraction from total cell homogenate that is capable of acetylating histone using acetylcoenzyme A as the donor. Later an enzyme was purified from calf thymus nuclei (Gallwitz, 1970*b*). This preparation can be further fractionated into three enzymes having different molecular weights and histone specificities (Gallwitz and Sures, 1972).

Deacetylases have been isolated from whole calf thymus homogenate (Inoue and Fujimoto, 1969) and the soluble sap of nuclei isolated in organic solvents (Vidali *et al.*, 1972). The enzymes of the total homogenate demonstrate site specificity in that they are more effective in deacetylating histones that have been acetylated *in vitro* than histones that have been chemically acetylated with acetic anhydride (Inoue and Fujimoto, 1969).

Comments

The modifications of histones appear to be associated with distinct chromosomal events. With the possible exception of methylation, these changes are readily reversible. This observation is consistent with the interpretation that phosphorylation and acetylation modulate the character of histone interaction with other chromosomal components. It is evident that, in this capacity, phosphorylation and acetylation can serve more than one function. Thus in differentiated cells, phosphorylation may be one of the events surrounding gene activation, whereas in dividing tissue it may be important for chromosome replication or condensation. Similarly, depending on the physiology of the cell type, acetylation may be involved in any one of at least three functions. Differences among the histones (I, IV, and IIb2) in their kinetics of phosphorylation suggest differences in functions of this modification for different histone fractions.

How do the modifications affect the character of the histones involved? Both phosphorylation and acetylation affect the charge of the ϵ-amino group of lysine, phosphorylation by the introduction of negative charge and acetylation by removal of the amino positive charge. Both groups are bulky, and steric factors may come into play. Inspection of the sites of modification reveals that all of them occur in the highly basic parts of the histone molecules. Since these parts of the histones are probable regions of DNA binding, it appears that modifications affect the interactions of histones with DNA. Experimentally, this expectation is borne out. Histone f1 phosphorylated at serine-38 is found to be much less effective in inducing conformation changes on DNA than nonphosphorylated f1 (Adler *et al.*, 1971). This is all the more surprising in that this single modification changes the total charge of the molecule from about +51 to about +49, a change of roughly 4%. Phosphorylation of f1 has been shown to reduce the ability of lysine-rich histone to inhibit the transcription of DNA by RNA polymerase *in vitro* (Stevely and Stocken, 1966, 1968*b*).

A specific proposal has been put forward for the action of acetylation and phosphorylation of histone IV. Model building, using the known sequences of calf (Shih and Bonner, 1970) and trout (Sung and Dixon, 1970) histone IV, indicates that an α-helical conformation involving the first 20 amino acids would fit nicely into the major groove of DNA. The lysine residues at positions 5, 8, 12, and 16 would be favorably positioned to interact with four consecutive phosphates on one of the DNA backbones. The highly charged nature of this region of the histone is not compatible with the formation of α-helix, however (Boublík *et al.*, 1970*a,b*; Bradbury and Crane-Robinson, 1971). Acetylation as it has been demonstrated to occur is sufficient to neutralize the charge in this region, allowing the formation of α-helix (Sung and Dixon, 1970). Thus the proposed binding of histone IV could be achieved. A similar suggestion has been made in the case of histone IIb1 (Bailey and Dixon, 1973; Louie *et al.*, 1973).

The Serine-Rich Histone: Avian Erythropoiesis

It is apparent that changes in chromosome states involve small qualitative (modification) or quantitative changes in the five histone fractions. However, in some cases, new histones are associated with some chromosomal states. Such is the case in meiotic cells of the lily, *Lilium longiflorum,* and the tulip, *Tulipa gesneriana* (Sheridan and Stern, 1967). A special lysine-rich histone, I^0, is found associated with some nondividing mammalian tissue such as calf lung (Panyim and Chalkley, 1969*a,b*) and pig brain (Shaw and Huang, 1970). Best characterized among these special histones is the serine-rich histone found in nucleated avian erythrocytes. This histone is termed *histone V* or *fraction f2c*.

The serine-rich histone has been reported to occur in the nucleated erythrocytes of bird, amphibian, and fish. A histone V-like fraction has even been reported in sea urchin sperm (Paoletti and Huang, 1969). Panyim *et al.* (1971) dispute the claim that this histone occurs in any vertebrate but birds, arguing that the other reported occurrences are due to histone degradation.

Interest in this histone arises from its association with the completely inactive genome of the mature chicken erythrocyte. Electron micrographs show that the chromatin of these cells is completely condensed (Vidali *et al.*, 1973). It has been suggested that f2c plays a direct role in producing this state.

The serine-rich histone has many properties common to the lysine-rich histone. These include similar molecular weights, basic amino acid content (30%), and an asymmetrical amino acid distribution (Greenaway and Murray, 1971). A sequence of the first 31 amino acids of chicken histone V has been reported (Greenaway and Murray, 1971). Unusual in this sequence is an apparent polymorphism at residue 15 involving the occurrence of either glutamine or arginine.

Little information on the relationship of this histone to the condensed, inactive state of the chromatin exists. Studies on the appearance of f2c during erythropoiesis of adult and embryonic (primitive and definitive) cell types demonstrate that this histone is present in dividing (erythroblasts) as well as nondividing (polychromatic erythrocytes) precursors of the mature erythrocytes (Appels *et al.*, 1972; Moss *et al.*, 1973; Sotirov and Johns, 1972). Thus histone f2c is present in cells actively synthesizing RNA and DNA (Williams, 1972), and its presence is not necessarily associated with the inactive state.

Dividing erythroblasts and their immediate successors, the nondividing polychromatic erythrocytes, synthesize histone f2c (Appels and Wells, 1972). In the dividing cells, this histone makes up 6–17% of the total histone complement, and increases to 25–28% in the mature erythrocyte (Appels *et al.*, 1972; Sotirov and Johns, 1972). Only the serine-rich histone is synthesized in the polychromatic erythrocytes, and this is accompanied by significant turnover of this histone (Appels and Wells, 1972). During the maturation of erythrocytes in chicken, the total histone–DNA ratio does not change (Moss *et al.*, 1973; Appels *et al.*, 1972). Also, the histone is not seen to increase at the significant expense of another histone (Moss *et al.*, 1973; Appels *et al.*, 1972). Thus histone V does not appear to replace any other histone as the chromatin condenses and becomes inactive.

Histones and Chromosome Structure: Diversity of Function

The hydrodynamic properties of chromatin demonstrate that it is isolated as a particle of length several times shorter than the extended length of the DNA which it contains (Zubay and Doty, 1959). This conclusion implies that the DNA in chromatin is packed into a particle by some sort of coiling or folding. Studies of the X-ray diffraction of nucleohistone gels drawn into partially oriented fibers lead to the same conclusion. The occurrence of diffraction maxima at 110, 55, 37, 27, and 22 Å, which are not associated with DNA structure, and the lack of a clear, oriented set of DNA reflections are interpreted as indicating the presence of a regular coil of an elementary DNA–protein fibril in a larger coiled structure of pitch 120 Å and diameter 100 Å (Pardon and Wilkins, 1972). The same set of diffraction maxima are observed in nuclei (Luzzati and Nicholaïeff, 1959; Olins and Olins, 1972) and isolated metaphase chromosomes (Pardon *et al.,* 1973).

That extensive packing of DNA occurs in nucleohistone has also been demonstrated by electron microscopy of interphase nuclei and isolated chromatin. These observations reveal that chromatin consists of knobby fibrils 100–250 Å in diameter, depending on the state of the material being studied (Ris and Kubai, 1970; DuPraw, 1970. Dry mass measurements of these samples indicate a packing ratio (length of extended DNA to length of chromatin fibril) of approximately 56 : 1 (DuPraw, 1970). As would be expected of a coiled or folded structure, the 100 A fibrils can be unpacked by stretching. The nucleohistone fibers used for X-ray studies can be stretched to 9 times their original length (Bradbury *et al.,* 1972*c*). When this is done, the 110 Å, 55 Å, etc., series disappears and is replaced by a sharp pattern characteristic of oriented DNA (Pardon *et al.,* 1967; Bradbury *et al.,* 1972*c*). Consistent with these observations, the 100 Å fibrils when stretched are seen in the electron microscope as 25–30 Å fibrils (Bram and Ris, 1971). It is thought that that 30 Å fibril is the basic unit packed by coiling or folding into a 100 Å fibril. In contrast to the X-ray diffraction data on oriented fibers, X-ray scattering data of gels of the 100 Å fibrils are consistent with the interpretation that the 100 Å fibrils are made up of an irregular coil of the 30 Å fibril. The coil has an average pitch of 40 Å and radius of 30 Å (Bram and Ris, 1971).

Histones may be removed sequentially by increasing concentrations of salt (Ohlenbusch *et al.,* 1967). Thus at 0.6 M NaCl the lysine-rich histones are removed, followed by the slightly lysine-rich histones at 1.2 M NaCl. All histones are removed from the chromatin by 3.0 M NaCl. The nature of hydrodynamic, optical, and dye binding properties of chromatin is seen to change to that of DNA in solution as the histones are removed. It is observed, however, that in the case of optical activity (Henson and Walker, 1970*b*; Simpson and Sober, 1970; Wagner and Spelsberg, 1971; Hjelm and Huang, 1974), hydrodynamic properties (Ohba, 1966*a*, Henson and Walker, 1970*a*), binding of actinomycin D (Kleiman and Huang, 1971), and X-ray diffraction of oriented nucleohistone fibers (Murray *et al.*, 1970; Bradbury *et al.*, 1972*c*; Skidmore *et al.*, 1973) only small changes in the properties of the chromatin are observed with the removal of the lysine-rich histones. Thus it would appear that basic coiled or folded structure of the chromatin fibrils is not affected by the removal of f1. However, it is clear from the changes that occur with removal of histone I in thermal denaturation (Ohba, 1966*b*; Ohlenbusch *et al.*, 1967; Henson and Walker, 1970*a*; Wilhelm *et al.*, 1970), the number of ethidium bromide binding sites in chromatin, and the value of its binding constant (Angerer and Moudrianakis, 1972) that the lysine-rich histones are closely associated with DNA and that some sort of structure is affected by its removal.

We have remarked on the observation of the presence of diffuse and dense chromatin associated with, respectively, active and inactive regions of the genome in the nuclei of thymocytes (Littau *et al.*, 1964). Later work (Littau *et al.*, 1965) has demonstrated that the compact appearance of the dense chromatin is maintained only by the lysine-rich histones. Extraction of this class of histone caused complete disruption of the dense structure. Only the replacement of the lysine-rich histone fraction could reconstitute the dense material. Extraction by acid–ethanol of all other histones (80% of the total histone) did not alter the appearance of the dense material. It has also been demonstrated that removal of f1 changes the precipitation properties of chromatin (Bradbury *et al.*, 1972*c*). It would appear, thus, that the function of the lysine-rich histone is to maintain a packing of the 100 Å fibril into a larger structure. Perhaps this is accomplished by cross-linking of the chromatin 100 Å fibrils with f1 (Littau *et al.*, 1965; Bradbury *et al.*, 1973*a*).

That lysine-rich histones have a special function has been implied repeatedly in the discussions of the molecular biology of these molecules. Among the histone classes it shows the greatest species diversity. It is the only histone fraction with recognizable polypeptide heterogeneity and possible organ specificity. Also, the apparent nonoccurrence of internal acetylation and methylation is unique. Phosphorylation occurs most extensively and universally in this molecule and the cyclic kinetics of phosphorylation in this histone associated with chromosome replication is highly suggestive of a special function.

If the lysine-rich histones have no role in maintaining the packing of nucleoprotein into the 100 Å fibril, then this role must be relegated to the other two histone classes. The nature of the interaction of the slightly lysine-rich and arginine-rich histones with DNA appears to be different than the lysine-rich histones. The slightly lysine-rich and arginine-rich histones are dissociated from chromatin DNA at much higher salt concentrations than the lysine-rich histone, yet they are removed first if 5–7 M urea is present (Kleiman and Huang, 1972). This observation suggests a more nonpolar interaction of these histones with DNA than is present in the interaction of the lysine-rich histone with DNA. A detailed comparison of the relative role of slightly lysine-rich and arginine-rich histones is not available. It is apparent from a comparison of the hydrodynamic properties of (1) chromatin depleted of lysine-rich and slightly lysine-rich histones and (2) chromatin depleted of all its histones that the members of the slightly lysine-rich class of histone are responsible for different aspects of chromosome structure than are the members of the arginine-rich class (Ohba, 1966*a*; Henson and Walker, 1970*a*). Also, dissimilarities in evolutionary rates between these two classes of histones are suggestive of different associations of slightly lysine-rich and arginine-rich histones with the conservative structure of DNA. Differences in the occurrence and kinetics of modifications are also suggestive of unique functions.

Although we can say little about the roles of histone in the economy of the chromosome, the weight of the evidence suggests a diversity of function not recognized in the two simple hypotheses introduced at the beginning of this section. It is clear that histones are an important determinant of chromosome structure. What is not clear are the details of their roles.

Nonhistone Proteins

Isolation and Characteristics

Many different procedures for the isolation of nonhistone proteins have been published. Although a large portion of the nonhistone may be removed from chromatin or nuclei by dilute salt solutions, some are very strongly bound to the chromatin and are generally insoluble in anything but highly denaturing aqueous solutions. Thus most procedures utilize high concentrations of salt (Langan, 1967; Shaw and Huang, 1970;

Benjamine and Gellhorn, 1969; Paul and Gilmour, 1968), urea (Gronow and Griffiths, 1971), salt and urea (Bekhor *et al.*, 1969; Gilmour and Paul, 1969; Shaw and Huang, 1970), urea and guanidinim chloride (Levy *et al.*, 1972), detergent (Marushige *et al.*, 1968; Shirey and Huang, 1969), or formic acid, salt, and urea (Elgin and Bonner, 1972). Extraction with phenol has also been used (Teng *et al.*, 1971; Shelton and Neelin, 1971). If the histones have not been previously extracted with dilute acid, they may be separated from the nonhistone by the use of hydroxylapatite (MacGillivray *et al.*, 1972) or one of the ion exchange resins Bio Rex 70 (Langen, 1967; Levy *et al.*, 1972; van der Brock *et al.*, 1973), SP-Sephadex (Graziano and Huang, 1971), and QAE-Sephadex (Gilmour and Paul, 1970).

The nonhistone proteins demonstrate a wide range of affinities for chromatin. They also differ in their solubilities in the solutions used for dissociation. As a result, each of the above procedures isolates a different population of proteins. Variability as to what is isolated as nonhistone protein (NHP) is also enhanced by the imprecision of the definition of these proteins. Theoretically, the NHPs are proteins, other than histones, that are isolated as part of chromatin or of nuclei washed in dilute saline. Practically, the NHP isolated by any of the above procedures is highly dependent on the method used to obtain the nuclei (MacGillivray *et al.*, 1972) or chromatin (Bhorjee and Pederson, 1973). Added to this variability are possible contaminations by cytoplasmic and soluble nuclear proteins (Johns and Forrester, 1969) and artifacts produced by protein degradation. Furthermore, the definition of NHP suffers a functional defect because, as we have seen in the case of histone modification enzymes, some proteins important in the economy of the chromosomes may not be isolable as chromosomal constituents. These comments must be kept in mind in considering what is to follow.

The NHPs isolated by the above methods are heterogeneous with respect to their molecular weights and isolectric points. Although a limited number of major bands (13–33) are discernible by sodium dodecylsulfate (SDS) polyacrylamide electrophoresis, their relative migrations correspond to molecular weights ranging from 15,000 to over 100,000 daltons (Elgin and Bonner, 1970, 1972; Graziano and Huang, 1971; Benjamine and Gellhorn, 1969; Gronow and Griffiths, 1971; Levy *et al.*, 1972). Likewise, the p*I*s, as measured by isoelectric focusing, range from around 2.0 to above 9.0 (Elgin and Bonner, 1972; Gronow and Griffiths, 1971; Gronow and Thackrah, 1973). However, amino acid analysis which does not distinguish between acids and amides indicates that these proteins are largely acidic. Comparison of NHPs prepared from different sources shows quantitative and qualitative species (Elgin and Bonner, 1970) and tissue specificity by acrylamide gel electrophoresis (Platz *et al.*, 1970; Teng *et al.*, 1971; Elgin and Bonner, 1972; MacGillivray *et al.*, 1972; Wu *et al.*, 1973), and by immunological techniques (Chytil and Spelsberg, 1971).

The apparently limited heterogeneity observed in NHPs may be due to the limited resolution of one-dimensional electrophoretic or isoelectric focusing systems. Two-dimensional gels utilizing isoelectric focusing in one direction and SDS acrylamide gel electrophoresis in the other show 18 major spots (Barrett and Gould, 1973), but a large number of smaller spots are also seen. On these gels the qualitative differences in the patterns of NHPs from different tissues and species are far more striking than those seen on one-dimensional systems. A two-dimensional electrophoretic analysis (urea and SDS-acrylamide gels) of rat liver acid-soluble nucleolar proteins alone shows about 100 components (Orrich *et al.*, 1973). However, in this case the differences observed between liver nucleolar acid-soluble proteins and those from Novikoff hepatoma were minimal.

Another important characteristic of nonhistone proteins is that a sizable fraction of them are phosphoproteins (Langan, 1967). Most of the NHP bands discernible by polyacrylamide gel electrophoresis are observed to contain phosphate (Teng *et al.*, 1971; Platz *et al.*, 1970; Shelton *et al.*, 1972; Richter and Sekeris, 1972; Rickwood *et al.*, 1973). A soluble NHP fraction isolated by 2 M NaCl dissociation of rat liver nuclei is seen to contain 1.14 mole % phosphate (Langan, 1967). Ninety percent of the phosphate is isolated as phosphoserine, 10% as phosphothreonine. In all, 40% of the serine residues of this fraction are phosphorylated.

Functional Correlations

The amount and type of NHP found in chromatin preparations reflect the metabolic state of the nucleus. This is seen in development, where there is a positive correlation between the amount (Marushige and Ozaki, 1967; Marushige and Dixon, 1969; Dingman and Sporn, 1964) and type (Shelton and Neelin, 1971; Vidali *et al.*, 1973; Seale and Aronson, 1973) of NHPs and the RNA synthetic activity of the cells. The phosphoprotein content also is seen to increase with RNA synthesis in these and other systems (Gershey and Kleinsmith, 1969; Kleinsmith and Allfrey, 1969; Shelton *et al.*, 1972). Analysis of dense and diffuse chromatin fractions shows the same type of correlations. Diffuse chromatin has both higher RNA synthetic capabilities and larger amounts of nonhistones than the dense material (Frenster, 1965). Murphy *et al.* (1973) have shown, furthermore, that specific NHP is associated with each fraction.

Unlike histones, which are synthesized concomitantly with DNA (Robbins and Borun, 1967; Borun *et al.*, 1967; Gallwitz and Mueller, 1969; Butler and Mueller, 1973; reviewed by Hnilica *et al.*, 1971), nonhistone proteins are synthesized and transmitted to the chromatin at all times during the cell cycle (Stein and Borun, 1972; Borun and Stein, 1972). Thus the synthesis of NHP is seen as one of the earliest events in the activation of quiescent cells. increased NHP synthesis is seen in the uterus of ovariectomized rat immediately after injection of estrogen (Teng and Hamilton, 1969). That the increase is partially specific for certain proteins is seen in PHA-stimulated lymphocytes (Levy *et al.*, 1973) and in livers of adrenalectomized rats injected with hydrocortisone (Shelton and Allfrey, 1970). Increases in phosphoprotein are also seen (Kleinsmith *et al.*, 1966*b*; Ahmed and Ishida, 1971). These events are usually correlated with the increase of RNA synthesis that occurs during stimulation. However, in most cases these cells go on to divide, and the high rate of nonhistone synthesis may be associated in part with chromosomal replication (Rovera and Baserga, 1971; Baserga and Stein, 1971; Stein and Baserga, 1972).

The elegant studies of the phenol-soluble residual NHP associated with quiescent and active metabolic states of the plasmodial slime mold, *Physarum polycephalum,* by LeStourgeon and coworkers (LeStourgeon and Rusch, 1973; LeStourgeon *et al.,* 1973) indicate the association of specific nonhistones with different nuclear states. Electropherograms of residular NHP from the active and inactive cells differ at several bands. The bands that appear when the active cells differentiate into the inactive states can be shown by autoradiography to consist of newly synthesized protein. Conversely, if the quiescent cells are induced to develop into an active state, the transformation is accompanied by the disappearance of the "inactive state"-associated proteins and the reappearance of the bands which were lost during the inactivation process. Newly synthesized protein is found to be associated only with the reappearing electrophoretic bands.

Nonhistones and Genetic Control

A major aim of the study of chromosomal proteins is the elucidation of the interactions involved in genetic control. Ultimately, studies on the functional characteristics of chromatin must be made. If we wish to take advantage of the ways chromatin allows the study of the interactions in genetic control, we must face the issue of the suitability of this material for functional studies. The question is one of whether the isolation precedures used to obtain chromatin leave the basic genetic control mechanisms intact.

That different genes are expressed in the various tissues implies that if chromatin approximates an isolated, intact regulated genome then it should possess unique properties corresponding to the nuclear function of the tissue from which it was isolated. The observed tissue specificity of chromosomal NHP and the association of some of these with various nuclear states testify that some of the functional aspects of the nuclear material remain intact. The NHPs observed probably are enzymes and structural proteins necessary for the maintenance of the metabolic state of the nucleus. They most likely are not the primary effectors of genetic control. Control molecules, if like those in bacteria and virus, would not be found in large enough amounts to be detected by the usual methods (Elgin and Bonner, 1970, 1972).

Only a fraction of the genetic information contained in its DNA is used by any one cell of a higher plant or animal. Thus RNA transcribed *in vivo* has homologous base sequences to only a fraction of the total DNA as measured by DNA : RNA hybridization techniques. In isolated chromatin, this characteristic is preserved. The amount of DNA available for transcription from chromatin *in vitro* approximates that available for transcription *in vivo*. Consistent with this statement is the observation that isolated chromatin has only 10% of the sites found in DNA for the binding of exogenous RNA polymerase (Cedar and Felsenfeld, 1973). Thus at least some general characteristics of the original system are preserved in the isolated material.

There is an indication that some of the more specific features of the *in vivo* system are also left unaltered. For example, genetic control mechanisms in highly differentiated cells are preprogrammed to react in specific ways to different stimuli. The response of a target tissue such as estrogen-treated chick oviduct to the hormone progesterone is an example of such programming. A cytoplasmic protein, unique to the target tissue, binds progesterone (O'Malley *et al.,* 1972). The progesterone–receptor complex is bound to the nucleus, whereupon the events leading to the production of avidin proceed (O'Malley *et al.,* 1972). Chromatin isolated from oviduct retains the ability to bind the receptor–progesterone complex, and does so more effectively than chromatins isolated from other sources. The receptor–progesterone binding material is found among the nonhistone proteins of chick oviduct chromatin (Spelsberg *et al.,* 1972).

These results are encouraging. However, the basic criterion of the functional fidelity of chromatin is whether is is capable of synthesizing the same RNA molecules *in vitro* as *in vivo*. Present techniques severely limit an answer to this question. It is being researched, and at the moment the answer is a qualified yes. Insofar as DNA : RNA hydridization competition techniques are capable of distinguishing between different populations of highly repetitive RNA sequences, it is found that the RNAs transcribed from isolated chromatin are similar to those transcribed *in vivo* (Paul and Gilmour, 1968; Huang and Huang, 1969; Bekhor *et al.*, 1969; Smith *et al.*, 1969). RNA synthesized *in vivo* is observed to compete for the same DNA base sequences as RNA transcribed from isolated homologous chromatin. The RNAs appear to be specific for the tissue from which the chromatin was isolated in that competition between *in vivo* and *in vitro* RNAs isolated from different tissues do not compete as effectively as RNAs from the same tissue. However, Reeder (1973), using a different approach, has found that the fidelity of transcription in chromatin may be rather poor and that DNA : RNA hybridization may not be an adequate method of determining the relatedness of *in vivo* and *in vitro* RNA.

One advantage of using a material such as chromatin is the ability to manipulate material and its components so as to elucidate their functional interrelationships. Two types of manipulations have been especially fruitful in indicating the existence of genetic control elements among the nonhistone protein fractions. They are selective dissociation and selective reconstitution of chromatin.

The histone proteins can be dissociated from chromatin by the application of dilute acid (Paul and Gilmour, 1968; Seligy and Neelin, 1970), high salt (Spelsberg and Hnilica, 1971), or urea and salt at low pH (Seplsberg *et al.*, 1971), leaving behind the residual NHPs. Although the amount of DNA available for transcription in depleted chromatins is many times greater than that available in chromatin, the amount is only 50–67% of the total available to act as a template in pure DNA. Thus the NHPs may be involved in restricting the genome.

Nonhistone proteins may also act as genetic activators. With the idea that the principle of self-assembly of biological macromolecules might apply to genetic regulatory assemblies, attempts were made to reassociate partially and totally dissociated chromatins. The reassociated complexes displayed many characteristics of the undissociated material. By DNA : RNA hybridization techniques, the same amount and type of DNA is transcribed in both the undissociated and reconstituted complexes (Huang and Huang, 1969; Bekhor *et al.*, 1969; Paul and Gilmour, 1968; Gilmour and Paul, 1969; Spelsberg *et al.*, 1971, 1972).If some or all of the NHPs are not included in the reconstitution (Paul and Gilmour, 1968; Spelsberg *et al.*, 1972), or if certain of the chromosomal RNAs are destroyed (Huang and Huang, 1969; Bekhor *et al.*, 1969), the reconstituted complex does not transcribe RNA. Furthermore, hybrid complexes formed from histone and DNA from one tissue and NHP from another show RNA transcription characteristics of the NHP-donor tissue (Gilmour and Paul, 1970; Spelsberg *et al.*, 1971). Thus the NHP fraction, which includes RNA, may contain molecules which modulate template restriction by histones. The modulation, evidently, is tissue specific.

Further support for this conclusion comes from the observation that certain NHPs, notably the phosphoproteins, form insoluble complexes with histones (Langan, 1967; Marushige *et al.*, 1968; Wang, 1968). When added to histone–DNA templates, the phosphoproteins increase their RNA synthetic capacity (Langan, 1967; Wang, 1968; Spelsberg and Hnilica, 1969). If rat liver nucleoplasmic RNA is used to synthesize RNA from such a histone–DNA template, the addition of rat liver phosphorylated NHP to the system will stimulate the RNA synthesis only if homologous rat DNA is used in the template. The stimulation is abolished when the NHP fraction is treated with alkaline phosphatase (Shea and Kleinsmith, 1973). The nonhistone phosphoproteins when added to chromatin stimulate its ability to act as a template for bacterial RNA polymerase (Wang, 1971: Kamiyama and Wang, 1971; Kostraba and Wang, 1972). It has been suggested that the phosphoproteins are able to compete with the DNA phosphoproteins for binding sites on histone, thus greatly decreasing the affinity of the histone for DNA (Langan, 1967).

In order that NHPs (or RNA) act as specific genetic repressors or activators, these molecules must be highly efficient at recognizing and (possibly, but not necessarily) binding specific DNA sequences. That some phospho-NHPs have this characteristic is indicated by the observation that a small fraction of nonhistone binds only to homologous DNA (Teng *et al.*, 1971; Kleinsmith *et al.*, 1970; Kleinsmith, 1973; van der Brock *et al.*, 1973).

Conclusion

The picture that has evolved for the role of histones has been one of a general and static nature. Histones probably act as general repressors and at the same time determine the overall structure of the chromosome. In these roles, the histones show some diversity, but they appear to lack the capability to be involved by themselves in the dynamics of chromosome structure and function.

The nonhistone proteins, on the other hand, show all the attributes necessary to govern the dynamic aspects of chromosome structure and function. Unlike histones, we have seen them to possess specificities relating to the metabolic state of the nuclear material. They appear necessary to the maintenance of the state of the genetic material. We might regard the histone–DNA complex of higher eukaryotes as the substrate on which some of the NHPs work. Nonhistone proteins certainly affect the histone-DNA interaction indirectly *vis-à-vis* enzymatic modification, but as we have seen they may also intercede directly to mediate histone restriction of RNA synthesis. Interactions directly with DNA may also occur, and NHPs may act in this way as specific repressors and as structural proteins.

Acknowledgments

This work was supported in part by Grants R01CA13953 and 1R01HD08185 from the National Institutes of Health.

Literature Cited

Adler, A. J., B. Schaffhausen, T. A. Langan, and G. D. Fasman, 1971 Altered conformational effects of phosphorylated lysine-rich histone (f1) in f-1-deoxyribonucleic acid complexes: Circular dichroism and immunological studies. *Biochemistry* **10:**909–913.

Ahmed, K. and H. Ishida, 1971 Effect of testosterone on nuclear phosphoproteins of rat ventral prostate. *Mol. Pharmacol.* **7:**323–327.

Allfrey, V. G., 1966 Structural modifications of histones and their possible role in the regulation of ribonucleic acid sequences. *Can. Cancer Conf.* **6:**313–335.

Allfrey, V. G., 1968 Some observations on histone acetylation and its temporal relationship to gene activation. In *Regulatory Mechanisms for Protein Synthesis in Mammalian Cells,* edited by A. San Pietro, M. R. Lunborg, and F. T. Kenney, pp. 65–100, Academic Press, New York.

Allfrey, V. G., 1971 Functional and metabolic aspects of DNA associated proteins. In *Histones and Nucleohistones,* edited by D. M. P. Phillips, pp. 241–294, Plenum Press, New York.

Allfrey, V. G., V. C. Littau, and A. E. Mirsky, 1963 On the role of histones regulating ribonucleic acid synthesis in the cell nucleus. *Proc. Natl Acad. Sci. USA* **49:**414–421.

Allfrey, V. G., R. Faulkner, and A. E. Mirsky, 1964 Acetylation and methylation of histones and their possible role in the regulation of RNA synthesis. *Proc. Natl Acad. Sci. USA* **51:**786–793.

Angerer, L. M. and E. N. Moudrianakis, 1972 Interaction of ethidium bromide with whole and selectively deproteinized deoxynucleoproteins from calf thymus. *J. Mol. Biol.* **63:**505–521.

Ansevin, A. T. and B. W. Brown, 1971 Specificity in the association of histones with deoxyribonucleic acid: Evidence from derivative thermal denaturation profiles. *Biochemistry* **10:**1133–1142.

Ansevin, A. T., L. S. Hnilica, T. C. Spelsberg, and S. L. Kehm, 1971 Structure studies on chromatin and nucleohistones: Thermal denaturation profiles recorded in the presence of urea. *Biochemistry* **10:**4793–4803.

Appels, R. and J. R. E. Wells, 1972 Synthesis and turnover of DNA-bound histone during maturation of avian red blood cells. *J. Mol. Biol.* **70:**425–434.

Appels, R., J. R. E. Wells, and A. F. Williams, 1972 Characterization of DNA-bound histone in the cells of the avian erythropoietic series. *J. Cell Sci.* **10:**47–59.

Astrin, S. M., 1973 *In vitro* transcription of simian virus 40 sequences in SV3T3 chromatin. *Proc. Natl. Acad. Sci. USA* **70:**2304–2308.

Bailey, G. S. and G. H. Dixon, 1973 Histone IIb from rainbow trout: Comparison of amino acid sequence with calf thymus IIb1. *J. Biol. Chem.* **248:**5463–5472.

Balhorn, R., W. O. Rieke, and R. Chalkley, 1971 Rapid electrophoretic analysis for histone phosphorylation: A reinvestigation of phosphorylation of lysine-rich histone during rat liver regeneration. *Biochemistry* **10:**3952–3959.

Balhorn, R., M. Balhorn, and R. Chalkley, 1972*a* Lysine-rich histone phosphorylation and hyperplasia in the developing rat. *Dev. Biol.* **29:**199–203.

Balhorn, R., M. Balhorn, H. P. Morris, and R. Chalkley, 1972*b* Comparative high-resolution electrophoresis of tumor histones: Variation in phosphorylation as a function of cell replication rate. *Cancer Res.* **32:**1775–1784.

Balhorn, R., J. Bordwell. L. Sellers, D. Granner, and R. Chalkley, 1972*c* Histone phosphorylation and DNA synthesis are linked in synchronous cultures of HTC cells. *Biochem. Biophys. Res. Commun.* **46:**1326–1333.

Balhorn, R., R. Chalkley, and D. Granner, 1972*d* Lysine-rich histone phosphorylation: A positive correlation with cell replication. *Biochemistry* **11:**1094–1097.

Balhorn, R., D. Oliver, P. Hohmann, R. Chalkley, and D. Granner, 1972*e* Turnover of deoxyribonucleic acid, histones and lysine-rich histone phosphate in hepatoma tissue culture cells. *Biochemistry* **11:**3915–3921.

Barr, G. C. and J. A. V. Butler, 1963 Histones and gene function. *Nature (London)* **199:**1170–1172.

Barrett, T. and H. J. Gould, 1973 Tissue and species specificity of non-histone chromatin proteins. *Biochim. Biophys. Acta* **294:**165–170.

Baserga, R. and G. Stein, 1971 Nuclear acid proteins and cell proliferation. *Fed. Proc.* **30:**1752–1759.

Bekhor, I., G. M. Kung, and J. Bonner, 1969 Sequence specific interaction of DNA and chromosomal proteins. *J. Mol. Biol.* **39:**351–354.

Benjamine, W. and A. Gellhorn, 1969 Acidic proteins of mammalian nuclei: Isolation and characterization. *Proc. Natl. Acad. Sci. USA* **59:**262–268.

Bhorjee, J. S. and T. Pederson, 1973 Chromatin: Its isolation from cultured mammalian cells with particular reference to contamination by nuclear ribonucleoprotein particles. *Biochemistry* **12:**2766–2773.

Bishop, J. O., R. Pemberton, and C. Baglioni, 1972 Reiteration frequency of haemoglobin genes in the duck. *Nature (London) New Biol.* **235:**231–234.

Boffa, L. C. and G. Vidali, 1971 Acid extractable proteins from chick embryo muscle nuclei. *Biochim. Biophys. Acta* **236:**259–269.

Boffa, L. C., E. L. Gershey, and G. Vidali, 1971 Changes of the histone deacetylase activity during chick embryo muscle development. *Biochim. Biophys. Acta* **254:**135–143.

Bonner, J. and R. C. C. Huang, 1963 Properties of chromosomal nucleohistone. *J. Mol. Biol.* **6:**169–174.

Bonner, J., M. E. Dahmus, D. Fambrough, R. C. C. Huang, K. Marushige, and D. Y. H. Tuan, 1968 The biology of isolated chromatin. *Science* **159:**47–56.

Borun, T. W. and G. S. Stein, 1972 The synthesis of acidic chromosomal proteins during the cell cycle of HeLa S-3 cells. II. The kinetics of residual protein synthesis and transport. *J. Cell Biol.* **52:**308–315.

Borun, T. W., M. D. Scharff, and E. Robbins, 1967 Polyribosome associated RNA having the properties of histone messages. *Proc. Natl. Acad. Sci. USA* **58:**1977–1983.

Borun, T. W., D. Peason, and W. K. Paik, 1972 Studies of histone methylation during the HeLa S-3 cell cycle. *J. Biol. Chem.* **247**:4288–4298.

Boublik, M., E. M. Bradbury, and C. Crane-Robinson, 1970*a* An investigation of the conformational changes of histones F1 and F2a1 by proton magnetic resonance spectroscopy. *Eur. J. Biochem.* **14**:486–497.

Boublik, M., E. M. Bradbury, C. Crane-Robinson, and E. W. Johns, 1970*b* An investigation of the conformational changes of histone F2b by high resolution nuclear magnetic resonance. *Eur. J. Biochem.* **17**:151–159.

Boublik, M., E. M. Bradbury, C. Crane-Robinson, and H. W. S. Rattle, 1971 Proton magnetic resonance studies of the interactions of histones F1 and F2b with DNA. *Nature (London) New Biol.* **229**:149–150.

Bradbury, E. M. and C. Crane-Robinson, 1971 Physical and conformational studies of histones. In *Histone and Nucleohistones,* edited by D. M. P. Phillips, pp. 85–127, Plenum Press, New York.

Bradbury, E. M. and H. W. E. Rattle, 1972 Simple computer-aided approach for the analysis of the nuclear magnetic resonance spectra of histones fractions F1, F2a1, F2b cleaved halves of F2b and F2b-DNA. *Eur. J. Biochem.* **27**:270–281.

Bradbury, E. M., C. Crane-Robinson, D. M. P. Phillips, E. W. Johns, and K. Murray, 1965 Conformational investigation of histones. *Nature (London)* **205**:1315–1316.

Bradbury, E. M., C. Crane-Robinson, H. Goldman, H. W. E. Rattle, and R. M. Stephens, 1967 Spectroscopic studies of the conformation of histones and protamine. *J. Mol. Biol.* **29**:507–523.

Bradbury, E. M., P. D. Cary, C. Crane-Robinson, P. L. Riches, and E. W. Johns, 1972*a* Nuclear magnetic resonance and optical spectroscopic studies of conformation and interactions in the cleaved halves of histone F2b. *Eur. J. Biochem.* **26**:482–489.

Bradbury, E. M., C. Crane-Robinson, and E. W. Johns, 1972*b* Specific conformations and interactions in chicken erythrocyte histone F2c. *Nature (London) New Biol.* **238**:262–264.

Bradbury, E. M., H. V. Molgaard, R. M. Stephens, L. A. Bolund, and E. W. Johns, 1972*c* X-ray studies of nucleoproteins depleted of lysine-rich histone. *Eur. J. Biochem.* **31**:474–482.

Bradbury, E. M., R. G. Carpenter, and H. W. E. Rattle, 1973*a* Magnetic resonance studies of deoxyribonucleoprotein. *Nature (London)* **241**:123–126.

Bradbury, E. M., R. J. Inglis, H. R. Matthews, and N. Sarner, 1973*b* Phosphorylation of very lysine-rich histone in *Physarum polycephalum. Eur. J. Biochem.* **33**:131–139.

Bram, S. and H. Ris, 1971 On the structure of nucleohistone. *J. Mol. Biol.* **55**:325–336.

Brandt, W. F. and C. Von Holt, 1972 The complete amino acid sequence of histone F3 from chicken erythrocytes. *FEBS Lett.* **23**:357–360.

Brown, S. W., 1966 Heterochromatin. *Science* **151**:417–425.

Buckingham, R. H. and L. A. Stocken, 1970*a* Histone F1: Purification and phosphorus content. *Biochem. J.* **117**:157–160.

Buckingham, R. H. and L. A. Stocken, 1970*b* Subfractionation and incorporation of (^{32}P) phosphate *in vitro. Biochem. J.* **117**:509–512.

Burdon, R. H., 1971 Enzymatic modification of chromosomal macromolecules. I. DNA and protein methylation in mouse tumor cell chromatin. *Biochim. Biophys. Acta* **232**:359–370.

Burdon, R. H. and E. V. Garvin, 1971 Enzymatic modification of chromosomal macromolecules. II. The formation of ε-*N*-trimethyl-L-lysine by a soluble chromatin methylase. *Biochim. Biophys. Acta* **232**:371–378.

Bustin, M., 1972 Conservative amino acid replacement in the tyrosine region of the lysine-rich histones. *Eur. J. Biochem.* **29:**263–267.

Bustin, M. and R. D. Cole, 1968 Species and organ specificities in very lysine-rich histones. J. Biol. Chem. **243:**4500–4505.

Bustin, M. and R. D. Cole, 1969*a* A study of the multiplicity of lysine-rich histone. *J. Biol. Chem.* **244:**5286–5290.

Bustin, M. and R. D. Cole, 1969*b* Bisection of a lysine-rich histone by *N*-bromosuccinimide. *J. Biol. Chem.* **244:**5291–5294.

Bustin, M. and R. D. Cole, 1970 Region of high and low cationic charge in a lysine-rich histone. *J. Biol. Chem.* **245:**1458–1466.

Bustin, M. and B. D. Stollar, 1972 Immunological specificity in lysine-rich histone subfractions. *J. Biol. Chem.* **247:**5716–5722.

Bustin, M. and B. D. Stollar, 1973 Immunological relatedness of thymus and liver F1 histone subfractions. *J. Biol. Chem.* **248:**3506–3510.

Bustin, M., S. C. Cole, R. H. Stellwagen, and R. D. Cole, 1969 Histone structure: Asymmetric distribution of lysine residues in lysine-rich histones. *Science* **163:**391–392.

Butler, B. B. and G. C. Mueller, 1973 Control of histone synthesis in HeLa cells. *Biochim. Biophys. Acta* **294:** 481–496.

Byvoet, P., 1972 *In vivo* turnover and distribution of radio-*N*-methyl in arginine-rich histones from rat tissues. *Arch. Biochem. Biophys.* **152:**887–888.

Byvoet, P., G. R. Shepherd, J. M. Hardin, and B. J. Noland, 1972 The distribution and turnover of labeled methyl groups in histone fractions of cultured mammalian cells. *Arch. Biochem. Biophys.* **148:**558–567.

Candido, E. P. M. and G. H. Dixon, 1971 Site of *in vivo* acetylation of trout testis histone IV. *J. Biol. Chem.* **246:**3182–3188.

Candido, E. P. M. and G. H. Dixon, 1972*a* Acetylation of trout testis histones *in vivo:* Site of the modification in histone IIb1. *J. Biol. Chem.* **247:**3868–3873.

Candido, E. P. M. and G. H. Dixon, 1972*b* Amino-terminal sequences and site of *in vivo* acetylation of trout-testis histone III and IIb2. *Proc. Natl. Acad. Sci. USA* **69:**2015–2019.

Candido, E. P. M. and G. H. Dixon, 1972*c* Trout testis cells. III. Acetylation of histones in different cell types from developing trout testis. *J. Biol. Chem.* **247:**5506–5510.

Cedar, H. and G. Felsenfeld, 1973 Transcription of chromatin *in vitro*. *J. Mol. Biol.* **77:**237–254.

Chae, C. B., M. C. Smith, and J. L. Irvin, 1972 Effect of *in vitro* histone phosphorylation on template activity of rat liver chromatin. *Biochim. Biophys. Acta* **287:**134–153.

Chalkley, R. and R. H. Jensen, 1968 A study of the structure of isolated chromatin. *Biochemistry* **7:**4380–4388.

Chen, L.-J., and D. A. Walsh, 1971 Multiple forms of hepatic adenosine 3′:5′ monophosphate dependent protein. *Biochemistry* **10:**3614–3621.

Chytil, F. and T. C. Spelsberg, 1971 Tissue differences in antigenic properties of non-histone protein–DNA complexes. *Nature (London) New Biol.* **233:**215–218.

Clark, R. J. and G. Felsenfeld, 1971 Structure of chromatin. *Nature (London) New Biol.* **229:**101–105.

Cleaver, U. and E. G. Ellgaard, 1970 Puffing and histone acetylation in polytene chromosomes. *Science* **169:**373–374.

Cohen, L. H. and B. V. Gotchel, 1971 Histones of polytene and nonpolytene nuclei of *Drosophila melanogaster*. *J. Biol. Chem.* **246:**1841–1848.

Comb, D. G., N. Sarkar, and C. J. Pinzing, 1966 The methylation of lysine residues. *J. Biol. Chem.* **241:**1857–1862.

Cross, M. E. and M. G. Ord, 1970 Changes in the phosphorylation and thiol content of histones in phytohaemagglutinin-stimulated lymphocytes. *Biochem. J.* **118:**191–193.

Cross, M. E. and M. G. Ord, 1971 Changes in histone phosphorylation and associated early metabolic events in pig lymphocyte cultures transformed by phytohaemagglutinin or 6-*N*,2′-*O*-dibutyryladenosine 3′:5′-cyclic monophosphate. *Biochem. J.* **124:**241–248.

DeLange, R. J. and E. L. Smith, 1971 Histones; Structure and function. *Annu. Rev. Biochem.* **40:**279–314.

DeLange, R. J., D. M. Fambrough, E, L. Smith, and J. Bonner, 1969*a* Calf and pea histone IV. II. The complete amino acid sequence of calf thymus histone IV; presence of ε-*N*-acetyllysine. *J. Biol. Chem.* **244:**319–334.

DeLange, R. J., D. M. Fambrough, E. L. Smith, and J. Bonner, 1969*b* Calf and pea histone IV. III. Complete amino acid sequence of pea histone IV; comparison with the homologous calf thymus histone. *J. Biol. Chem.* **244:**5669–5679.

DeLange, R. J., J. A. Hooper, and E. L. Smith, 1973 Histone III. III. Sequence studies on the cyanogen bromide peptides; complete amino acid sequence of calf thymus histone III. *J. Biol. Chem.* **248:**3261–3273.

DeVilliers Graaf, G. and C. Von Holt, 1973 Enzymatic histone modification during the induction of tyrosine aminotransferase with insulin and hydrocortisone. *Biochim. Biophys. Acta* **299:**480–484.

Dick, C. and E. W. Johns, 1969 Histones from *Drosophila melanogaster*. *Comp. Biochem. Physiol.* **31:**529–533.

Dingman, C. W. and M. B. Sporn, 1964 Studies on chromatin. I. Isolation and characterization of nuclear complexes of deoxyribonucleic acid, ribonucleic acid and protein from embryonic and adult tissues of chicken. *J. Biol. Chem.* **239:**3483–3493.

Dixon, G. H., B. Ingles, B. Jergil, V. Ling, and K. Marushige, 1968 Protein transformations during differentiation of trout testis. *Can. Cancer Conf.* **8:**76–102.

Dixon, G. H., E. P. M. Candido, and A. J. Louie, 1973 The role of enzymatic modification in the control of histone and protamine binding to DNA. *Miami Winter Symp.* **5:**279–285.

Dodge, J. D., 1964 Chromosome structure in the Dinophyceae. II. Cytochemical studies. *Arch. Mikrobiol.* **48:**66–80.

Duerksen, J. D. and B. J. McCarthy, 1971 Distribution of deoxyribonucleic acid sequences in fractionated chromatin. *Biochemistry* **10:**1471–1478.

DuPraw, E. J., 1970 *DNA and Chromosomes,* Holt, Rinehart and Winston, New York.

Dwivedi, R. S., S. Dutta, and D. Block, 1969 Isolation and characterization of chromatin from *Neurospora crassa*. *J. Cell Biol.* **43:**51–58.

Edwards, G. S. and V. G. Allfrey, 1973 Aflatoxin B1 and actinomycin D effects on histone acetylation and deacetylation in the liver. *Biochim. Biophys. Acta* **299:**354–366.

Edwards, P. A. and K. V. Shooter, 1969 Ultracentrifugation studies of histone fractions from calf thymus deoxyribonucleoprotein. *Biochem. J.* **114:**227–235.

Elgin, S. C. R. and J. Bonner, 1970 Limited heterogeneity of the major nonhistone chromosomal proteins. *Biochemistry* **9:**4440–4447.

Elgin, S. C. R. and J. Bonner, 1972 Partial fractionation and characterization of the major nonhistone chromosomal proteins. *Biochemistry* **11:**772–781.

Elgin, S. C. R., S. Froehner, J. Smart, and J. Bonner, 1971 The biology and chemistry of chromosomal proteins. *Adv. Cell Mol. Biol.* **1:**1–62.

Fambrough, D. M., 1969 Nuclear protein fractions. In *Handbook of Molecular Cytology,* edited by A. Lima-de-Faria, pp. 438–471, North-Holland, Amsterdam.

Fambrough, D. M. and J. Bonner, 1968 Sequence homology and role of cysteine in plant and animal arginine-rich histones. *J. Biol. Chem.* **243:**4434–4439.

Fambrough, D. M. and J. Bonner, 1969 Limited molecular heterogeneity of plant histones. *Biochim. Biophys. Acta* **175:**113–122.

Fambrough, D. M., F. Fujimura, and J. Bonner, 1968 Quantitative distribution of histone components in the pea plant. *Biochemistry* **7:**575–585.

Faragó, A., F. Antoni, A. Takáts, and F. Fábián, 1973 Adenosine 3′:5′-monophosphate-dependent and independent histone kinases isolated from human tonsillar lymphocytes. *Biochim. Biophys. Acta* **297:**517–526.

Fasman, G. D., B. Scharffhaussen, L. Goldsmith, and A. Adler, 1970 Conformational changes associated with f-1 histone–DNA complexes: Circular dichroism studies. *Biochemistry* **9:**2814–2822.

Fasman, G. D., M. S. Valenzuela, and A. J. Adler, 1971 Complexes of deoxyribonucleic acid with fragments of lysine-rich histone (f-1): Circular dichroism studies. *Biochemistry* **10:**3795–3801.

Fitzgerald, P. J., W. H. March, M. G. Ord, and L. A. Stocken, 1970 Histone changes during regeneration of the pancreas. *Biochem. J.* **117:**711–714.

Frenster, J. H., 1965 Nuclear polycations as de-repressors of synthesis of ribonucleic acid. *Nature (London)* **206:**680–683.

Frenster, J. H., V. G. Allfrey, and A. E. Mirsky, 1963 Repression and active chromatin isolated from interphase lymphocytes. *Proc. Natl. Acad. Sci. USA* **50:**1026–1032.

Gallwitz, D., 1970*a* Enzymatic acetylation of HeLa cell histones in isolated nuclei *in vitro. Hoppe-Seyler's Z. Physiol. Chem.* **351:**1050–1053.

Gallwitz, D., 1970*b* Extraction and partial purification of two histone specific transacetylases from rat liver nuclei. *Biochem. Biophys. Res. Commun.* **40:**236–242.

Gallwitz, D. and G. C. Mueller, 1969 Histone synthesis *in vitro* on HeLa cell microsomes. *J. Biol. Chem.* **244:**5944–5952.

Gallwitz, D. and C. E. Sekeris, 1969*a* Stimulation of RNA polymerase activity of rat liver nuclei by cortisol *in vivo. FEBS Lett.* **3:**99–102.

Gallwitz, D. and C. E. Sekeris, 1969*b* The acetylation of rat liver nuclei *in vitro* by acetyl-CoA. *Hoppe-Seyler's Z. Physiol. Chem.* **350:**150–154.

Gallwitz, D. and I. Sures, 1972 Histone acetylation, purification and properties of three histone-specific acetyltransferases from rat thymus nuclei. *Biochim. Biophys. Acta* **263:**315–328.

Georgiev, G. P., 1969 Histone and control of gene action. *Annu. Rev. Genet.* **3:**155–180.

Georgiev, G. P., L. N. Ananieva, and J. V. Kozlov, 1966 Stepwise removal of protein from a deoxyribonucleoprotein complex and de-repression of the genome. *J. Mol. Biol.* **22:**365–371.

Gershey, E. L. and L. J. Kleinsmith, 1969 Phosphorylation of nuclear protein in avian erythrocytes. *Biochim. Biophys. Acta* **194:**519–525.

Gershey, E. L., G. Vidali, and V. G. Allfrey, 1968 Chemical studies of histone acetylation: The occurrence of ε-*N*-acetyllysine in the f2a1 histone. *J. Biol. Chem.* **243:**5018–5022.

Gershey, E. L., G. W. Haslett, G. Vidali, and V. G. Allfrey, 1969 Chemical studies of histone methylation. *J. Biol. Chem.* **244:**4871–4877.

Gilmour, R. S. and J. Paul, 1969 RNA transcribed from reconstituted nucleoprotein is similar to natural RNA. *J. Mol. Biol.* **40:**137–139.

Gilmour, R. S. and J. Paul, 1970 Role of non-histone components in determining organ specificity of rabbit chromatin. *FEBS Lett.* **9:**242–244.

Graziano, S. L. and R. C. C. Huang, 1971 Chromatographic separation of chick brain chromatin protein using a SP-Sephadex column. *Biochemistry* **10:**4770–4777.

Greenaway, P. J. and K. Murray, 1971 Heterogeneity and polymorphism in chicken erythrocyte histone fraction V. *Nature (London) New Biol.* **229:**233–238.

Gronow, M. and G. Griffiths, 1971 Rapid isolation and separation of the non-histone proteins of rat liver nuclei. *FEBS Lett.* **15:**340–344.

Gronow, M. and T. Thackrah, 1973 The nonhistone nuclear protein of some rat tissues. *Arch. Biochem. Biophys.* **158:**377–386.

Gutierrez-Cernosek, R. M. and L. S. Hnilica, 1971 Histone synthesis and phosphorylation in regenerating rat liver. *Biochim. Biophys. Acta* **247:**348–354.

Haapala, O. K. and M.-O. Soyer, 1973 Structure of dinoflagellate chromosomes. *Nature (London) New Biol.* **244:**195–197.

Hamana, K. and K. Iwai, 1971 Fractionation and characterization of *Tetrahymena* histone in comparison with mammalian histones. *J. Biochem.* **69:**1097–1111.

Hamkalo, B. A. and O. L. Miller, Jr., 1973 Electron microscopy of genetic activity. *Annu. Rev. Biochem.* **42:**379–396.

Hearst, J. E. and M. Botchan, 1970 The eukaryotic chromosome. *Annu. Rev. Biochem.* **39:**151–188.

Hekman, A. and M. Sluyser, 1973 Antigenic determinants on lysine-rich histone. *Biochim. Biophys. Acta* **295:**613–620.

Hempel, K., H. W. Langen, and L. Birkofer, 1968 ε-*N*-Trimethyllysine, eine neue Aminosäure in Histonen. *Naturwissenshaften* **55:**37–39.

Henson, P. and I. O. Walker, 1970*a* The partial dissociation of nucleohistone by salts: Hydrodynamic and deternaturation studies. *Eur. J. Biochem.* **14:**345–350.

Henson, P. and I. O. Walker, 1970*b* The partial dissociation of nucleohistone by salts: Circular dichroism and denaturation studies. *Eur. J. Biochem.* **16:**524–531.

Hjelm, R. P. and R. C. C. Huang, 1974 The role of histones in the conformation of DNA in chromatin as studied by circular dichroism. *Biochemistry* **13:**5275–5283.

Hnilica, L. S., M. E. McClure, and T. C. Spelsberg, 1971 Histone biosynthesis and the cell cycle. In *Histones and Nucleohistones,* edited by D. M. P. Phillips, pp. 187–240, Plenum Press, New York.

Hooper, J. A., E. L. Smith, K. R. Sommer, and R. Chalkley, 1973 Histone III. IV. Amino acid sequence of histone III of the testes of the carp, *Letiobus bubalis. J. Biol. Chem.* **248:**3275–3279.

Hsiang, M. W. and R. D. Cole, 1973 The isolation of histone from *Neurospora crassa. J. Biol. Chem.* **248:**2007–2013.

Huang, R. C. C. and J. Bonner, 1962 Histone: A suppressor of chromosomal RNA synthesis. *Proc. Natl. Acad. Sci. USA* **48:**1216–1222.

Huang, R. C. C. and P. C. Huang, 1969 Effect of protein-bound RNA associated with chicken embryo chromatin on template specificity of the chromatin. *J. Mol. Biol.* **39:**365–378.

Huang, R. C. C., J. Bonner, and K. Murray, 1964 Physical and biological properties of soluble nucleohistones. *J. Mol. Biol.* **8:**54–64.

Inoue, I. and D. Fujimoto, 1969 Enzymatic deacetylation of histone. *Biochem. Biophys. Res. Commun.* **36:**146–150.

Itzhaki, R. F., 1971 Studies on the accessibility of deoxyribonucleic acid in deoxyribonucleoprotein to cationic molecules. *Biochem. J.* **122:**583–592.

Itzhaki, R. F. and H. K. Cooper, 1973 Similarity of chromatin from different tissues. *J. Mol. Biol.* **75:**119–128.

Iwai, K., 1964 Histones of rice embryos and of *Chlorella.* In *Nucleohistones,* edited by J. Bonner and P. O. P. T.'so, pp. 59–65, Holden-Day, San Francisco.

Iwai, K., H. Shiomi, T. Ando, and T. Mita, 1965 Isolation of histone from *Tetrahymena. J. Biochem.* **58:**312–314.

Iwai, K., H. Hayashi, and K. Ishikawa, 1972 Calf thymus lysine- and serine-rich histone. *J. Biochem.* **72:**357–367.

Jensen, R. H. and R. Chalkley, 1968 The physical state of nucleohistone under physiological ionic strength: The effect of interaction with free nucleic acids. *Biochemistry* **7:** 4388–4395.

Jergil, B. and G. H. Dixon, 1969 Protamine kinase from rainbow trout testis. *J. Biol. Chem.* **245:**425–434.

Jergil, B., M. Sung, and G. H. Dixon, 1970 Species and tissue specific patterns of phosphorylation of very lysine-rich histones. *J. Biol. Chem.* **245:**5867–5870.

Jirgensons, B. and L. S. Hnilica, 1965 The conformational changes of calf thymus histone fractions as determined by the optical rotatory dispersion. *Biochim. Biophys. Acta* **109:**241–248.

Johns, E. W., 1964 Studies on histones. 7. Preparative methods for histone fractions from calf thymus. *Biochem. J.* **92:**55–59.

Johns, E. W., 1967 A method for the selective extraction of histone fractions f2(a) 1 and f2(a)2 from calf thymus deoxyribonucleoprotein at pH 7. *Biochem. J.* **105:**611–614.

Johns, E. W., 1971 The preparation and characterization of histones. In *Histones and Nucleohistones,* edited by D. M. P. Phillips, pp. 2–46, Plenum Press, New York.

Johns, E. W. and J. A. V. Butler, 1962 Further fractionations of histone from calf thymus. *Biochem. J.* **82:**15–18.

Johns, E. W. and S. Forrester, 1969 Studies on nuclear proteins: The binding of extra acidic protein to deoxyribonucleoprotein during the preparation of nuclear protein. *Eur. J. Biochem.* **8:**547–551.

Kafatos, F. C., 1972 The cocoonase zymogen cells of silk moths: A model for terminal cell differentiation for specific protein synthesis. *Curr. Top. Devl. Biol.* **7:**125–191.

Kamiyama, M. and T. Y. Wang, 1971 Activated transcription from rat liver chromatin by non-histone proteins. *Biochim. Biophys. Acta* **228:**563–576.

Kim, S. and W. K. Paik, 1965 Studies on the origin of ε-*N*-methyl-L-lysine in protein. *J. Biol. Chem.* **240:**4629–4634.

Kinkade, J. M., Jr., 1969 Quantitative species differences and quantitative tissue differences in the distribution of lysine-rich histones. *J. Biol. Chem.* **244:**3375–3386.

Kinkade, J. M. and R. D. Cole, 1966*a* The resolution of four lysine-rich histones derived from calf thymus. *J. Biol. Chem.* **241:**5790–5797.

Kinkade, J. M. and R. D. Cole, 1966*b* A structural comparison of different lysine-rich histone of calf thymus. *J. Biol. Chem.* **241:**5798–5805.

Kleiman, L. and R. C. C. Huang, 1971 Binding of actinomycin D to calf thymus chromatin. *J. Mol. Biol.* **55:**503–521.

Kleiman, L. and R. C. C. Huang, 1972 Reconstitution of chromatin: The sequential binding of histones to DNA in the presence of salt and urea. *J. Mol. Biol.* **64:**1–8.

Kleinsmith, L. J., 1973 Specific binding of phosphorylated nonhistone chromatin protein to deoxyribonucleic acid. *J. Biol. Chem.* **248:**5648–5653.

Kleinsmith, L. J. and V. G. Allfrey, 1969 Nuclear phosphoproteins. I. Isolation and characterization of a phosphoprotein fraction from calf thymus nuclei. *Biochim. Biophys. Acta* **175:**123–135.

Kleinsmith, L. J., V. G. Allfrey, and A. E. Mirsky, 1966*a* Phosphoprotein metabolism in isolated lymphocyte nuclei. *Proc. Natl. Acad. Sci. USA* **55:**1182–1189.

Kleinsmith, L. J., V. G. Allfrey, and A. E. Mirsky, 1966*b* Phosphorylation of nuclear protein early in the course of gene activation in lymphocytes. *Science* **154:**780–781.

Kleinsmith, L. J., J. Heidema, and A. Carroll, 1970 Specific binding of rat liver nuclear protein to DNA. *Nature (London)* **226:**1025–1026.

Kostraba, N. C. and T. Y. Wang, 1972 Differential activation of transcription of chromatin by non-histone fraction. *Biochim. Biophys. Acta* **262:**169–180.

Kuo, J. F. and P. Greengard, 1969 Cyclic nucleotide-dependent protein kinase. IV. Widespread occurrence of adenosine 3′, 5′-monophosphate-dependent protein kinase in various tissues and phyla of the animal kingdom. *Proc. Natl. Acad. Sci. USA* **64:**1349–1355.

Lake, R. S. and N. P. Salzman, 1972 Occurrence and properties of a chromatin associated Fl-histone phosphokinase in mitotic Chinese hamster cells. *Biochemistry* **11:**4817–4826.

Langan, T. A., 1967 A phosphoprotein preparation from liver nuclei and its effect on the inhibition of RNA synthesis by histones. In *Regulation of Nucleic Acid and Protein Biosynthesis,* B. B. A. Library, Vol. *10,* edited by V. V. Koningsberger and L. Bosch, pp. 233–242, Elsevier, Amsterdam.

Langan, T. A., 1968*a* Histone phosphorylation: Stimulation by adenosine 3′, 5′-monophosphate. *Science* **162:**579–580.

Langan, T. A., 1968*b* Phosphorylation of proteins of the cell nucleus. In *Regulatory Mechanisms for Protein Synthesis in Mammalian Cells,* edited by A. San Pietro, M. R. Lamborg, and F. T. Kenney, pp. 101–118, Academic Press, New York.

Langan, T. A., 1969*a* Phosphorylation of liver histone following the administration of glucagon and insulin. *Proc. Natl. Acad. Sci. USA* **64:**1276–1283.

Langan, T. A., 1969*b* Action of adenosine 3′, 5′-monophosphate-dependent histone kinase *in vitro. J. Biol. Chem.* **244:**5763–5765.

Langan, T. A. and L. K. Smith, 1967 Phosphorylation of histones and protamines by a specific protein kinase from liver. *Fed. Proc.* **26:**60 (abst. 1934).

Langan, T. A., S. C. Rall, and R. D. Cole, 1971 Variation in primary structure at a phosphorylation site in lysine-rich histones. *J. Biol. Chem.* **246:**1942–1945.

Leighton, T. J., B. C. Dill, J. J. Stock, and C. Phillips, 1971 Absence of histones from the chromosomal proteins of fungi. *Proc. Natl. Acad. Sci. USA* **68:**677–680.

LeStourgeon, W. M. and H. P. Rusch, 1973 Localization of nucleolar and chromatin residual acidic protein changes during differentiation in *Physarum polycephalum. Arch. Biochem. Biophys.* **155:**144–158.

LeStourgeon, W. M., C. Nations, and H. P. Rusch, 1973 Temporal synthesis and intranuclear accumulation of the nuclear acidic proteins during periods of chromatin reactivation in *Physarum polycephalum. Arch. Biochem. Biophys.* **159:**861–872.

Levy, R., S. Levy, S. A. Rosenberg, and R. T. Simpson, 1973 Selective stimulation of nonhistone chromatin protein synthesis in lymphoid cells by phytohemagglutinin. *Biochemistry* **12:**224–228.

Levy, S., R. T. Simpson, and H. A. Sober, 1972 Fractionation of chromatin components. *Biochemistry* **11:**1547–1553.

Li, H.-J. and J. Bonner, 1971 Interaction of histone half-molecules with deoxyribonucleic acid. *Biochemistry* **10:**1461–1470.

Li, H.-J., C. Chang, and M. Weiskopf, 1973 Helix-coil transitions in nucleoprotein-chromatin structure. *Biochemistry* **12:**1763–1772.

Liew, C. C., G. W. Haslett, and V. G. Allfrey, 1970 *N*-Acetylseryl-tRNA and polypeptide chain initiation during histone biosynthesis. *Nature* (*London*) **226:**414–417.

Liss, M., A. M. Maxam, and L. J. Cuprak, 1969 Methylation of protein by calf spleen methylase. *J. Biol. Chem.* **244:**1617–1622.

Littau, V. C., V. G. Allfrey, J. H. Frenster, and A. E. Mirsky, 1964 Active and inactive regions of nuclear chromatin as revealed by electron microscope autoradiography. *Proc. Natl. Acad. Sci. USA* **52:**93–100.

Littau, V. C., C. J. Burdick, V. G. Allfrey, and A. E. Mirsky, 1965 Role of histones in the maintenance of chromatin structure. *Proc. Natl. Acad. Sci. USA* **54:**1204–1212.

Louie, A. J. and G. H. Dixon, 1972*a* Synthesis, acetylation and phosphorylation of histone IV and its binding to DNA during spermatogenesis in trout. *Proc. Natl. Acad. Sci. USA* **69:**1975–1979.

Louie, A. J. and G. H. Dixon, 1972*b* Trout testis cells. I. Characterization by deoxyribonucleic acid and protein analysis of cells by velocity sedimentation. *J. Biol. Chem.* **247:**5490–5497.

Louie, A. J. and G. H. Dixon, 1972*c* Trout testis cells. II. Synthesis and phosphorylation of histones and protamines in different cell types. *J. Biol. Chem.* **247:**5498–5505.

Louie, A. J. and G. H. Dixon, 1973 Kinetics of phosphorylation of testis histones and their possible role in determining chromosomal structure. *Nature* (*London*) *New Biol.* **243:**154–168.

Louie, A. J., M. T. Sung, and G. H. Dixon, 1973 Modification of histones during spermatogenesis in trout. III. Levels of phosphohistone species and kinetics of phosphorylation of histone IIb1. *J. Biol. Chem.* **248:**3335–3340.

Luzzati, V. and A. Nicholaïeff, 1959 Étude par diffusion des rayons X aux petits angles des gels d'acid désoxyribonucléique et des nucléoprotéines. *J. Mol. Biol.* **1:**127–133.

Lyon, M. F., 1968 Chromosomal and subchromosomal inactivation. *Annu. Rev. Genet.* **2:**31–52.

MacGillivray, A. J., A. Cameron, R. J. Krauze, D. Rickwood, and J. Paul, 1972 The non-histone proteins of chromatin, their isolation and composition in a number of tissues. *Biochim. Biophys. Acta* **277:**384–402.

Marks, D. B., W. K. Paik, and T. W. Borun, 1973 The relationship of histone phosphorylation to deoxyribonucleic acid replication and mitosis during the HeLa S-3 cell cycle. *J. Biol. Chem.* **248:**5660–5667.

Marushige, K. and J. Bonner, 1966 Template properties of liver chromatin. *J. Mol. Biol.* **15:**160–174.

Marushige, K. and G. H. Dixon, 1969 Developmental changes in chromosomal composition and template activity during spermatogenesis in trout testis. *Dev. Biol.* **19:**397–414.

Marushige, K. and H. Ozaki, 1967 Properties of isolated chromatin from sea urchin embryo. *Dev. Biol.* **16:**474–488.

Marushige, K., D. Brutlag, and J. Bonner, 1968 Properties of chromosomal nonhistone proteins of rat liver. *Biochemistry* **7:**3149–3155.

Marushige, K., V. Ling, and G. H. Dixon, 1969 Phosphorylation of chromosomal basic proteins in maturing trout testis. *J. Biol. Chem.* **244:**5953–5958.

Marzluff, W. F. and K. S. McCarty, 1970 Two classes of histones acetylation in developing mouse mammary gland. *J. Biol. Chem.* **245:**5635–5642.

Marzluff, W. F. and K. S. McCarty, 1972*a* Structural studies of calf thymus F3 histone. I. Occurrence of cysteine, phosphoserine, and ε-*N*-acetyllysine in cyanogen bromide peptides. *Biochemistry* **11:**2672–2677.

Marzluff, W. F. and K. S. McCarty, 1972*b* Structural studies of calf thymus F3 histone. II. Occurrence of phosphoserine and ε-*N*-acetyllysine in thermolysin peptides. *Biochemistry* **11:**2677–2681.

Marzluff, W. F., L. A. Sanders, and K. S. McCarty, 1972 Two chemically and metabolically distinct forms of calf thymus histone F3. *J. Biol. Chem.* **247:**2026–2033.

Marzluff, W. F., E. C. Murphy, and R. C. C. Huang, 1973 Transcription of ribonucleic acid in isolated mouse myeloma nuclei. *Biochemistry* **12:**3440–3446.

McMaster-Kaye, R. and J. S. Kaye, 1973 An electrophoretic analysis of the histones of the house cricket. *Arch. Biochem. Biophys.* **156:**426–436.

Means, G. E. and R. E. Feeney, 1968 Reductive alkylation of amino groups in proteins. *Biochemistry* **7:**2192–2201.

Meisler, M. H. and T. A. Langan, 1969 Characterization of a phosphatase specific for phosphorylated histones and protamine. *J. Biol. Chem.* **244:**4961–4968.

Miyamoto, E., J. F. Kuo, and P. Greengard, 1969 Cyclic nucleotide-dependent protein kinases. *J. Biol. Chem.* **244:**6395–6402.

Mohberg, J. and H. P. Rusch, 1969 Isolation of nuclear histones from the mixomycete, *Physarum polycephalum*. *Arch. Biochem. Biophys.* **134:**577–589.

Moss, B. A., W. G. Joyce, and V. M. Ingram, 1973 Histones in chick embryonic erythropoiesis. *J. Biol. Chem.* **248:**1025–1031.

Murphy, E. C., S. H. Hall, J. H. Shepherd, and R. S. Weiser, 1973 Fractionation of mouse myeloma chromatin. *Biochemistry* **12:**3843–3852.

Murray, K., 1964*a* The occurrence of ε-*N*-methyllysine in histones. *Biochemistry* **3:**10–15.

Murray, K., 1964*b* Histone nomenclature. In *The Nucleohistones,* edited by J. Bonner and P. Ts'o, pp. 15–20, Holden-Day, San Francisco.

Murray, K., E. M. Bradbury, C. Crane-Robinson, R. M. Stephens, A. J. Haydon, and A. R. Peacocke, 1970 The dissociation of chicken erythrocyte deoxyribonucleoprotein and some properties of its partial nucleoproteins. *Biochem. J.* **120:**859–871.

Nohara, H., T. Takahashi, and K. Ogata, 1968 Enzymatic acetylation of histones and some chemical characteristics of their acetyl groups. *Biochim. Biophys. Acta* **154:**529–539.

Ogawa, Y., G. Quagliarotti, J. J. Jordon, C. W. Taylor, W. C. Starbuck, and H. Busch, 1969 Structural analysis of the glycine-rich, arginine-rich histone. *J. Biol. Chem.* **244:**4387–4392.

Oh, Y. H., 1970 Spectroscopic studies of five purified histones from calf thymus. *J. Biol. Chem.* **245:**6404–6416.

Ohba, Y., 1966*a* Structure of nucleohistone. I. Hydrodynamic behaviour. *Biochim. Biophys. Acta* **123:**76–83.

Ohba, Y., 1966*b* Structure of nucleohistone. II. Thermal denaturation. *Biochim. Biophys. Acta* **123:**84–90.

Ohlenbusch, H. H., B. O. Oliver, D. Tuan, and N. Davidson, 1967 Selective dissociation of histones from calf thymus nucleoprotein. *J. Mol. Biol.* **25:**299–315.

Olins, D. E. and A. L. Olins, 1972 Physical studies of isolated eucaryotic nuclei. *J. Cell Biol.* **53:**715–736.

Oliver, D. R. and R., Chalkley, 1972*a* An electrophoretic analysis of *Drosophila* histones. *Exp. Cell Res.* **73:**295–302.

Oliver, D. R. and R. Chalkley, 1972*b* An electrophoretic analysis of *Drosophila* histones. II. Comparison of larval and adult histone patterns in two species of *Drosophila. Exp. Cell Res.* **73:**303–310.

Oliver, D. R., R. Balhorn, and R. Chalkley, 1972*a* Molecular nature of F1 histone phosphorylation in cultured hepatoma cells. *Biochemistry* **11:**3921–3925.

Oliver, D. R., K. R. Sommer, S. Panyim, S. Spiker, and R. Chalkley, 1972*b* A modified procedure for fractionating histones. *Biochem. J.* **129:**349–353.

O'Malley, B. W., T. C. Spelsberg, W. T. Schrader, F. Chytil, and A. W. Steggles, 1972 Mechanisms of interaction of a hormone-receptor complex with the genome of a eukaryote target cell. *Nature (London)* **235:**141–144.

Ord, M. G. and L. A. Stocken, 1966 Metabolic properties of histones from rat liver and thymus gland. *Biochem. J.* **98:**888–897.

Ord, M. G. and L. A. Stocken, 1967 Changes in the phosphorylation of histones during liver regeneration. *Biochem. J.* **103:**5P.

Ord, M. G. and L. A. Stocken, 1968 Variations in the phosphate content and thiol/disulphide ratio of nucleohistones during the cell cycle. *Biochem. J.* **107:**403–410.

Orrich, L. R., M. O. J. Olson, and H. Busch, 1973 Comparison of nucleolar proteins of normal rat liver and Novikoff hepatoma ascites cells by two-dimensional polyacrylamide gel electrophoresis. *Proc. Natl. Acad. Sci. USA* **70:**1316–1320.

Paik, W. K. and S. Kim, 1967 ε-*N*-Dimethyllysine in histones. *Biochem. Biophys. Res. Commun.* **27:**479–483.

Paik, W. K. and S. Kim, 1968 Methylase. I. Purification and properties of the enzyme. *J. Biol. Chem.* **243:**2108–2114.

Paik, W. K. and S. Kim, 1969 Enzymatic methylation of histone. *Arch. Biochem. Biophys.* **134:**632–637.

Paik, W. K. and S. Kim, 1970*a* ω-*N*-Methylarginine in protein. *J. Biol. Chem.* **245:**88–92.

Paik, W. K. and S. Kim, 1970*b* Purification of protein methylase. III from calf thymus nuclei. *J. Biol. Chem.* **245:**6010–6015.

Paik, W. K. and S. Kim, 1971 Protein methylation. *Science* **174:**114–119.

Paik, W. K. and S. Kim, 1972 Effect of methylation on susceptibility of protein to proteolytic enzymes. *Biochemistry* **11:**2589–2593.

Panyim, S. and R. Chalkley, 1969*a* The heterogeneity of histones. I. A quantitative analysis of calf histones in very long polyacrylamide gels. *Biochemistry* **8:**3972–3979.

Panyim, S. and R. Chalkley, 1969*b* A new histone found only in mammalian tissues with little cell division. *Biochem. Biophys. Res. Commun.* **37:**1042–1049.

Panyim, S., R. Chalkley, S. Spiker, and D. Oliver, 1970 Constant electrophoretic mobility of the cysteine-containing histone in plants and animals. *Biochim. Biophys. Acta* **214:**216–221.

Panyim, S., D. Bilek, and R. Chalkley, 1971 An electrophoretic comparison of vertebrate histones. *J. Biol. Chem.* **246:**4206–4215.

Paoletti, R. A. and R. C. C. Huang, 1969 Characterization of sea urchin sperm chromatin and its basic proteins. *Biochemistry* **8:**1615–1625.

Pardon, J. F. and M. H. F. Wilkins, 1972 A super-coil model for nucleohistone. *J. Mol. Biol.* **68:**115–124.

Pardon, J. F., M. H. F. Wilkins, and B. M. Richards, 1967 Super-helical model for nucleohistone. *Nature (London)* **215:**508–509.

Pardon, J. F., B. M. Richards, L. G. Skinner, and C. H. Ochey, 1973 X-ray diffraction from isolated metaphase chromosomes. *J. Mol. Biol.* **76:**267–270.

Paul, J. and R. S. Gilmour, 1968 Organ-specific restriction of transcription in mammalian chromatin. *J. Mol. Biol.* **34:**305–316.

Pawse, A. R., M. G. Ord, and L. A. Stocken, 1971 Histone kinase and cell division. *Biochem. J.* **122:**713–719.

Phillips, D. M. P., 1963 Presence of acetyl groups in histones. *Biochem. J.* **87:**258–263.

Phillips, D. M. P., 1968 *N*-Terminal acetyl-peptide from two calf thymus histones. *Biochem. J.* **107:**135–138.

Phillips, D. M. P., 1971 The primary structure of histones and protamines. In *Histones and Nucleohistones,* edited by D. M. P. Phillips, pp. 47–84, Plenum Press, New York.

Phillips, D. M. P. and E. W. Johns, 1965 A fractionation of the histones of group F2a from calf thymus. *Biochem. J.* **94:**127–130.

Platz, R. D., V. M. Kish, and L. J. Kleinsmith, 1970 Tissue specificity of non-histone chromatin phosphoproteins. *FEBS Lett.* **12:**38–40.

Pogo, B. G. T., V. G. Allfrey, and A. E. Mirsky, 1966 RNA synthesis and histone acetylation during the course of gene activation in lymphocytes. *Proc. Natl. Acad. Sci. USA* **55:**805–812.

Pogo, B. G. T., V. G. Allfrey, and A. E. Mirsky, 1967 The effect of phytohemagglutinin on ribonucleic acid synthesis and histone acetylation in equine leukocytes. *J. Cell Biol.* **35:**477–482.

Pogo, B. G. T., A. O. Pogo, V. G. Allfrey, and A. E. Mirsky, 1968 Changing patterns of histone acetylation and RNA synthesis in regeneration of the liver. *Proc. Natl. Acad. Sci. USA* **59:**1337–1344.

Pogo, B. G. T., A. O. Pogo, and V. G. Allfrey, 1969 Histone acetylation and RNA synthesis in rat liver regeneration. *Genetics Suppl.* **62:**373–379.

Price, R. and S. Penman, 1972 A distinct RNA polymerase activity, synthesizing 5.5s, 5s and 4s RNA in nuclei from adenovirus 2-infected HeLa cells. *J. Mol. Biol.* **70:**435–450.

Rall, S. C. and R. D. Cole, 1971 Amino acid sequence variability of the amino terminal region of lysine-rich histones. *J. Biol. Chem.* **246:**7175–7196.

Rasmussen, R. S., K. Murray, and J. M. Luck, 1962 On the complexity of calf thymus histone. *Biochemistry* **1:**79–89.

Reeder, R. H., 1973 Transcription of chromatin by bacterial RNA polymerase. *J. Mol. Biol.* **80:**229–241.

Reeder, R. H. and R. G. Roeder, 1972 Ribosomal RNA synthesis in isolated nuclei. *J. Mol. Biol.* **67:**433–441.

Richard, A., H. Cedar, and G. Felsenfeld, 1973 Synthesis of globin ribonucleic acid from duck reticulocyte chromatin *in vitro*. *Proc. Natl. Acad. Sci. USA* **70:**2029–2032.

Richter, K. H. and C. E. Sekeris, 1972 Isolation and partial purification of non-histone chromosomal proteins from rat liver, thymus, and kidney. *Arch. Biochem. Biophys.* **148:**44–53.

Rickwood, D., P. G. Riches, and A. J. MacGillivray, 1973 Studies of the *in vitro* phosphorylation of chromatin nonhistone proteins in isolated nuclei. *Biochim. Biophys. Acta* **299:**162–171.

Ris, H., 1962 Interpretation of ultrastructure in the cell nucleus. In *Interpretation of Ultrastructure,* edited by R. J. Harris, pp. 69–88, Academic Press, New York.

Ris, H. and D. F. Kubai, 1970 Chromosome structure. *Annu. Rev. Genet.* **4:**263–294.

Rizzo, P. J. and L. D. Nooden, 1972 Chromosomal proteins in the dinoflagellate alga *Gyrodinium cohnii*. *Science* **176**:796–797.

Robins, E. and T. W. Borun, 1967 The cytoplasmic synthesis of histone in HeLa cell and its temporal relationship to DNA replication. *Proc. Natl. Acad. Sci. USA* **57**:409–416.

Rovera, G. and R. Baserga, 1971 Early changes in the synthesis of acidic nuclear proteins in human diploid fibroblasts stimulated to synthesize DNA by changing the medium. *J. Cell. Physiol.* **77**:201–212.

Sanders, L. A., N. M. Schecter, and K. S. McCarty, 1973 A comparative study of histone acetylation, histone deacetylation, and ribonucleic acid synthesis in avian reticulocytes and erythrocytes. *Biochemistry* **12**:783–791.

Seale, R. L. and A. I. Aronson, 1973 Chromatin-associated protein of the developing sea urchin embryo. I. Kinetics of synthesis and characterization of non-histone proteins. *J. Mol. Biol.* **75**:633–645.

Sekeris, C. E., K. E. Sekeri and D. Gallwitz, 1967 The methylation of the histones of rat liver nuclei *in vitro*. *Hoppe-Seyler's Z. Physiol. Chem.* **348**:1660–1666.

Seligy, V. L. and J. M. Neelin, 1970 Transcription properties of stepwise acid-extracted chicken erythrocyte chromatin. *Biochim. Biophys. Acta* **213**:380–390.

Shaw, L. M. J. and R. C. C. Huang, 1970 A description of two procedures which avoid the use of extreme pH conditions for the resolution of components isolated from chromatins prepared from pig cerebellar and pituitary nuclei. *Biochemistry* **9**:4530–4542.

Shea, M. and L. J. Kleinsmith, 1973 Template specific stimulation of RNA synthesis by phosphorylated non-histone chromatin proteins. *Biochem. Biophys. Res. Commun.* **50**:473–477.

Shelton, K. R. and V. G. Allfrey, 1970 Selective synthesis of a nuclear acidic protein in liver cells stimulated by cortisol. *Nature* (*London*) **228**:132–134.

Shelton, K. R. and J. M. Neelin, 1971 Nuclear residual proteins from goose erythroid cells and liver. *Biochemistry* **10**:2342–2348.

Shelton, K. R., V. L. Seligy, and J. M. Neelin, 1972 Phosphate incorporation into "nuclear" residual proteins of goose erythrocytes. *Arch. Biochem. Biophys.* **153**:375–383.

Shepherd, G. R., 1973 Evidence for the biological coupling of biosynthesis and internal acetylation of histone fraction f2a1 (IV) in cultured mammalian cells. *Biochim. Biophys. Acta* **299**:485–491.

Shepherd, G. R., J. M. Hardin, and B. J. Noland, 1971*a* Methylation of lysine residues of histone fractions in synchronized mammalian cells. *Arch. Biochem. Biophys.* **143**:1–5.

Shepherd, G. R., B. J. Noland, and J. M. Hardin, 1971*b* Histone phosphorylation in synchronized mammalian cell cultures. *Arch. Biochem. Biophys.* **142**:299–302.

Shepherd, G. R., J. M. Noland, and J. M. Hardin, 1971*c* Histone acetylation in synchronized mammalian cell cultures. *Biochim. Biophys. Acta.* **228**:544–549.

Sheridan, W. F. and H. Stern, 1967 Histones of meiosis. *Exp. Cell Res.* **45**:323–335.

Sherod, D., G. Johnson, and R. Chalkley, 1970 Phosphorylation of mouse ascites tumor cell lysine-rich histone. *Biochemistry* **9**:4611–4615.

Shih, T. Y. and J. Bonner, 1970 Thermal denaturation and template properties of DNA complexes with purified histone fractions. *J. Mol. Biol.* **48**:469–487.

Shih, T. Y. and G. D. Fasman, 1970 Conformation of DNA in chromatin: A circular dichroism study. *J. Mol. Biol.* **52**:125–129.

Shih, T. Y. and G. D. Fasman, 1971 Circular dichroism studies of nucleic acid complexes with arginine-rich histone IV (f2a1). *Biochemistry* **10**:1675–1683.

Shirey, T. and R. C. C. Huang, 1969 Use of sodium dodecyl sulfate, alone, to separate chromatin proteins from deoxyribonucleoprotein of *Arbacia punctulata* sperm chromatin. *Biochemistry* **8:**4138–4148.

Siebert, G., M. G. Ord, and L. A. Stocken, 1971 Histone phosphokinase activity in nuclear and cytoplasmic cell fractions from normal and regenerating rat livers. *Biochem. J.* **122:**721–725.

Simpson, R. T. and H. A. Sober, 1970 Circular dichroism of calf liver nucleohistone. *Biochemistry* **9:**3103–3110.

Skidmore, C., I. O. Walker, J. F. Pardon, and B. M. Richards, 1973 The structure of partially depleted nucleohistones. *FEBS Lett.* **32:**175–178.

Sluyser, M. and Y. Hermes, 1973 A rat-specific lysine-rich histone. *Biochem. Biophys. Acta* **295:**605–612.

Smith, E. L., R. J. DeLange, and J. Bonner, 1970 Chemistry and biology of the histones. *Physiol. Rev.* **50:**159–170.

Smith, K. D., R. B. Church, and B. J. McCarthy, 1969 Template specificity of isolated chromatin. *Biochemistry* **8:**4271–4277.

Sotirov, N. and E. W. Johns, 1972 Quantitative differences in the content of the histone f2c between chicken erythrocytes and erythroblasts. *Exp. Cell Res.* **73:**13–16.

Spelsberg, T. C. and L. Hnilica, 1969 The effects of acidic proteins and RNA on the histone inhibitions of the DNA-dependent RNA synthesis *in vitro*. *Biochim. Biophys. Acta* **195:**63–75.

Spelsberg, T. C. and L. S. Hnilica, 1971 Proteins of chromatin in template restriction. II. Specificity of RNA synthesis. *Biochim. Biophys. Acta* **228:**212–222.

Spelsberg, T. C., L. S. Hnilica, and A. T. Ansevin, 1971 Proteins of chromatin in template restriction. III. The macromolecules in specific restriction. *Biochim. Biophys. Acta* **228:**550–562.

Spelsberg, T. C., A. W. Steggles, F. Chytil, and B. W. O'Malley, 1972 Progesterone-binding components of chick oviduct. IV. Exchange of progesterone-binding capacity from target to non-target tissue chromatins. *J. Biol. Chem.* **247:**1368–1374.

Stedman, E. and E. Stedman, 1951 The basic proteins of cell nuclei. *Philos. Trans. R. Soc. Lond. Ser. B Biol. Sci.* **235:**565–595.

Stein, G. and R. Baserga, 1972 Nuclear proteins and the cell cycle. *Adv. Cancer Res.* **15:**287–330.

Stein, G. and T. W. Borun, 1972 The synthesis of acidic chromosomal proteins during the cell cycle of HeLa S-3 cells. I. The accelerated accumulation of acidic residular proteins before the initiation of DNA replication. *J. Cell Biol.* **52:**292–307.

Stellwagen, R. H. and R. D. Cole, 1968 Danger of contamination in chromatographically prepared arginine-rich histone. *J. Biol. Chem.* **243:**4452–4455.

Stellwagen, R. H. and R. D. Cole, 1969*a* Histone biosynthesis in the mammary gland during development and lactation. *J. Biol. Chem.* **244:**4878–4887.

Stellwagen, R. H. and R. D. Cole, 1969*b* Chromosomal proteins. *Annu. Rev. Biochem.* **38:**951–990.

Stevely, W. S. and L. A. Stocken, 1966 Phosphorylation of rat thymus histone. *Biochem. J.* **100:**20–21C.

Stevely, W. S. and L. A. Stocken, 1968*a* Histone phosphorylation and cell division. *Biochem. J.* **109:**24–25P.

Stevely, W. S. and L. A. Stocken, 1968*b* Variations in the phosphate content of histone F1 in normal and irradiated tissue. *Biochem. J.* **110:**187–191.

Stollar, B. D. and M. Ward, 1970 Rabbit antibodies to histone fractions as specific reagents for preparative and comparative studies. *J. Biol. Chem.* **245:**1261–1266.

Subirana, J. A., 1971 Specific aggregation of histone fractions (presence of cysteine in F2a1 from echinoderms). *FEBS Lett.* **16:**133–136.

Sung, M. T. and G. H. Dixon, 1970 Modification of histones during spermatogenesis in trout: A molecular mechanism for altering histone binding to DNA. *Proc. Natl. Acad. Sci. USA* **67:**1616–1623.

Sung, M. T., G. H. Dixon, and O. Smithies, 1971 Phosphorylation and synthesis of histone in regeneration rat liver. *J. Biol. Chem.* **246:**1358–1364.

Suzuki, Y., L. P. Gage, and D. D. Brown, 1972 The genes for silk fibroin in *Bombyx mori. J. Mol. Biol.* **70:**637–649.

Takáts, A., A. Faragó, and F. Antoni, 1972 Adenosine 3′,5′-monophosphate dependent protein kinase in the lacrymal gland. *Biochim. Biophys. Acta* **268:**77–80.

Teller, D. C., J. M. Kinkade, and R. C. Cole, 1965 Molecular weight of lysine-rich histone. *Biochem. Biophys. Res. Commun.* **20:**739–744.

Teng, C.-S. and T. H. Hamilton, 1969 Role of chromatin in estrogen action in the uterus. II. Hormone-induced synthesis of non-histone acidic proteins which restore histone-inhibited DNA-dependent RNA synthesis. *Proc. Natl. Acad. Sci. USA* **63:**465–472.

Teng, C.-S., C. T. Teng, and V. G. Allfrey, 1971 Studies of nuclear acidic proteins: Evidence for their phosphorylation, tissue specificity, selective binding to deoxyribonucleic acid, and stimulatory effects on transcription. *J. Biol. Chem.* **246:**3597–3609.

Tidwell, T., V. G. Allfrey, and A. E. Mirsky, 1968 The methylation of histones during regeneration of the liver. *J. Biol. Chem.* **243:**707–715.

Van, N. T. and A. T. Ansevin, 1973 Ion-induced splitting in thermal denaturation profiles of F1 nucleohistone. *Biochim. Biophys. Acta* **299:**367–373.

van der Brock, H. W. J., L. D. Noodén, J. S. Sevall, and J. Bonner, 1973 Isolation, purification and fractionation of nonhistone chromosomal proteins. *Biochemistry* **12:**229–236.

Vidali, G., E. L. Gershey, and V. G. Allfrey, 1968 Chemical studies on histone acetylation: The distribution of ε-*N*-acetyllysine in calf thymus histones. *J. Biol. Chem.* **243:**6361–6366.

Vidali, G., L. C. Boffa, and V. G. Allfrey, 1972 Properties of an acidic histone-binding protein fraction from cell nuclei. *J. Biol. Chem.* **247:**7365–7372.

Vidali, G., L. C. Boffa, V. C. Littau, K. M. Allfrey, and V. G. Allfrey, 1973 Changes in nuclear acidic protein complement of red blood cells during embryonic development. *J. Biol. Chem.* **248:**4065–4068.

Wagner, T. and T. C. Spelsberg, 1971 Aspects of chromosomal structure. I. Circular dichroism studies. *Biochemistry* **10:**2599–2605.

Wang, T. Y., 1968 Restoration of histone-inhibited DNA-dependent RNA synthesis by acidic chromatin proteins. *Exp. Cell Res.* **53:**288–291.

Wang, T. Y., 1971 Tissue specificity of non-histone chromosomal proteins. *Exp. Cell Res.* **69:**217–219.

Wangh, L., A. Ruiz-Carrillo, and V. G. Allfrey, 1972 Separation and analysis of histone subfractions differing in the degree of acetylation: Some correlations with genetic activity in development. *Arch. Biochem. Biophys.* **150:**44–56.

Wilhelm, F. X., M. H. Champagne, and M. P. Daune, 1970 Conformation du DNA dans la nucléoprotéine. *Eur. J. Biochem.* **15:**321–330.

Wilhelm, J. A. and K. S. McCarty, 1970*a* Partial characterization of the histones and histone acetylation in cell cultures. *Cancer Res.* **30:**409–417.

Wilhelm, J. A. and K. S. McCarty, 1970*b* The uptake and turnover of acetate in HeLa cell histone fractions. *Cancer Res.* **30:**418–425.

Williams, A. F., 1972 DNA synthesis in purified populations of avian erythroid cells. *J. Cell Sci.* **10:**27–46.

Wintersberger, U., P. Smith, and K. Letnansky, 1973 Yeast chromatin: Preparation from isolated nuclei. Histone composition and transcription capacity. *Eur. J. Biochem.* **33:**123–130.

Wu, F. C., S. C. R. Elgin, and L. E. Hood, 1973 Non-histone chromosomal proteins of rat tissue: A comparative study by gel electrophoresis. *Biochemistry* **12:**2792–2797.

Yeoman, L. C., M. O. J. Olsen, N. Sugano, J. J. Jordon, C. W. Taylor, W. C. Starbuck, and H. Busch, 1972 Amino acid sequence of the center of the arginine-lysine-rich histone from calf thymus: The total sequence. *J. Biol. Chem.* **247:**6018–6023.

Zubay, G. and P. Doty, 1959 The isolation and properties of deoxyribonucleoprotein particles containing single nucleic acid molecules. *J. Mol. Biol.* **1:**1–20.

4

Organization of Genetic Material in the Macronucleus of Hypotrichous Ciliates

KURUGANTI G. MURTI

Introduction

Hypotrichous ciliates (e.g., *Euplotes, Oxytricha, Stylonychia*), like most other ciliated protozoans, possess two morphologically and functionally distinct nuclei, macronuclei and micronuclei. The macronuclei are large, divide amitotically, and function as somatic nuclei with distinct nucleoli and active ribonucleic acid (RNA) synthesis. Micronuclei are small, divide by mitosis, and have little or no RNA synthesis. The primary function of a micronucleus is to give rise to a new macronucleus each time the cells undergo conjugation.

Conjugation in ciliated protozoans occurs under certain conditions such as decreased food supply. During conjugation, two cells unite to form conjugal pairs. Within each conjugant the macronuclei begin to degenerate

KURUGANTI G. MURTI—Laboratories of Virology, St. Jude Children's Research Hospital, Memphis, Tennessee.

and the micronuclei undergo meiotic divisions. All the haploid micronuclear division products in each cell degenerate except for two, the stationary pronucleus and the migratory pronucleus. A mutual exchange of the migratory pronucleus occurs between the two conjugants, and the exchanged pronucleus fuses with the resident stationary pronucleus. The net result of this process is a hybrid, diploid *synkaryon* in each cell. The cells now separate and the synkaryon undergoes mitotic divisions. Of the resulting division products, which are genetically identical, some remain as such to become the new micronuclei, while others (designated as the macronuclear *anlagen*) go through a maturation process to become the new macronuclei. The old macronuclei totally degenerate during this process.

The formation of a mature macronucleus from the macronuclear anlage appears to differ in different groups of ciliates. In holotrichous ciliates such as *Paramecium* and *Tetrahymena,* the macronuclear anlagen become mature macronuclei after continuous deoxyribonucleic acid (DNA) replication (Woodard *et al.,* 1966; Johansson and Zech, unpublished; Jurand *et al.,* 1964). In these ciliates, the macronuclei possibly contain the same genetic information as the micronuclei, but in many copies ("polyploid," Raikov, 1968). In *Tetrahymena,* a comparison of the renaturation kinetics, melting patterns, and buoyant densities of the DNA purified from vegetative macro- and micronuclei has revealed few qualitative differences (Yao and Gorovsky, private communication).

In hypotrichous ciliates such as *Euplotes, Oxytricha,* and *Stylonychia,* the conversion of a macronuclear anlage to the mature macronucleus is a complex process and occurs in a series of steps. These steps have been well documented in *S. mytilus* (Ammermann, 1965) and *E. aediculatus* (Ammermann, 1971), but many observations suggest that similar events occur in all other hypotrichs as well. Briefly, the events are as follows. The macronuclear anlage, which has the same genetic complement as the micronucleus, undergoes extensive DNA replication resulting in the formation of the polytene chromosomes. After the formation of the polytene chromosomes, the DNA synthesis ceases and the polytene chromosomes are destroyed. At this time, a major portion (80% in *Euplotes,* over 95% in *Stylonychia,* Ammermann, 1971) of the DNA is degraded to acid-soluble products that are excreted into the medium (Ammermann, 1969, 1971). Electron microscopy of the macronuclear anlage at the time of DNA destruction has revealed that the polytene chromosomes are cut up into short segments by "membranous" partitions that appear at the interband regions (Kloetzel, 1970; Murti, 1973). Subsequently, each band with the adjacent interband is totally enclosed by membranes to form a physically independent "vesicle" (Ammermann, 1971). The anlage now contains thousands of vesicles, each of which contains a band plus an interband of the polytene chromosome. Within each vesicle, a major portion of the DNA is degraded (Prescott and Murti, 1973) and eventually eliminated from the anlage (Ammermann, 1969). After the destruction of DNA, the vesicles disappear, and the anlage synthesizes DNA by means of replication bands (a mode of replication commonly encountered in the mature macronucleus of hypotrichous ciliates). During this second period of DNA synthesis, the DNA content of the anlage increases to the level of the mature macronucleus. The old macronuclei totally degenerate during this process.

It has been suggested (Rao and Ammermann, 1970; Kloetzel, 1970) that the above series of events produces a mature macronucleus that contains only a portion of the gene complement of the micronucleus amplified many times. Direct experimental proof for this hypothesis was provided by comparisons of the buoyant density, melting pattern, and renaturation kinetics of the DNAs obtained from the mature micro- and macronuclei of *Oxytricha** (Bostock and Prescott, 1972; Lauth *et al.*, 1976). These studies have revealed that the macronucleus contains very few of the DNA components (less than 10%) present in the micronucleus.

Electron microscopic and sedimentation analyses have revealed another major difference between the vegetative micro- and macronuclear DNAs of *Oxytricha.* Whereas the micronucleus contained large DNA molecules (over 45.0 μm), the macronuclear DNA is composed of small molecules (0.2–2.2 μm, Prescott *et al.*, 1971; Murti, 1972). The small size of the macronuclear DNA is believed to be the result of the selective genetic diminution (i.e., breakdown of the polytene chromosomes and the subsequent degradation of DNA in the vesicles) that occurs in the macronuclear development. The above studies lead to two conclusions. First, the macronucleus in hypotrichous ciliates cannot be considered as a polyploid version of the micronucleus, since it contains only a small portion of the genetic material of the micronucleus amplified many times. Second, the macronucleus contains many small pieces of DNA, an organization of genetic material apparently restricted to hypotrichous ciliates among eukaryotes.

In summary, evidence in recent years suggests that the hypotrichous ciliates have a complex pattern of macronuclear development that results in a mature macronucleus with a uniquely organized genetic material. A detailed analysis of the development and structure of the macronucleus

* Publications from Prescott's laboratory have referred to this cell as *Stylonychia mytilus.* Dr. A. C. Borror of the Department of Zoology, University of New Hampshire, Durham, has recently identified the cell as *Oxytricha fallax.*

may lead to an understanding of the peculiar mode of macronuclear replication (by means of replication bands, Turner, 1930) and macronuclear aging (death of vegetative cells after specific number of macronuclear divisions, Maupas, 1888; Preer, 1968; Ammermann, 1971) commonly seen among hypotrichs. It is the purpose of this chapter to provide a detailed analysis of the development, replication, and division of the hypotrich macronucleus.

Conjugation

Conjugation in hypotrichous ciliates follows the same basic plan as that of other ciliates. A schematic diagram of conjugation in *Stylonychia* constructed from the light microscopic studies of Ammermann (1965) is presented in Figure 1. The vegetative cells of *Stylonychia* each contain two macronuclei and four micronuclei, but in the figure only two micronuclei (Mi) and a macronucleus (MA) are represented, for the sake of simplicity. Briefly, the events of conjugation are as follows. After the cells have formed conjugal pairs, the micronuclei undergo three successive prezygotic (two meiotic followed by one mitotic) divisions in each cell. During this time, the macronucleus begins to disintegrate by breaking up into four pieces. Of the eight haploid micronuclear division products in each cell, all disintegrate except for two, and these two are designated as the stationary pronucleus and the migratory pronucleus. The migratory pronucleus of one cell migrates across the cytoplasmic bridge that links the two cells and fuses with the stationary pronucleus of the other cell to form the diploid zygote nucleus or synkaryon (Sy). After the formation of the synkaryon, the conjugants separate. The synkaryon undergoes two successive postzygotic (mitotic) divisions to produce four genetically identical nuclei. Of these, two become the new micronuclei, the third becomes the presumptive macronucleus (or macronuclear anlage), and the fourth degenerates. The macronuclear anlage (Ma) eventually becomes a macronucleus. Finally, the two micronuclei and the newly formed macronucleus divide in the absence of cytokinesis to produce the nuclear constitution of the vegetative cell.

Macronuclear Development

In *E. aediculatus* and *S. mytilus,* Ammermann (1971) has studied the formation of the mature macronucleus from the macronuclear anlage. According to him, the macronuclear development occurs in three stages:

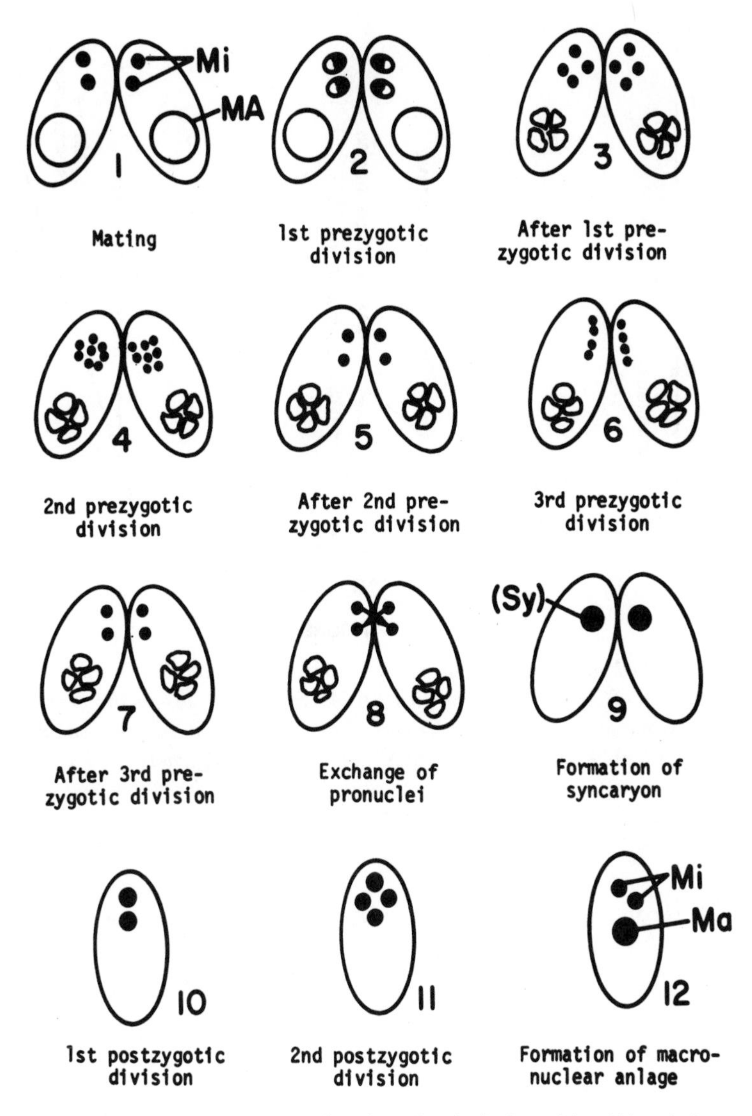

Figure 1. Schematic diagram of conjugation in Stylonychia. Abbreviations: Mi, micronucleus; MA, macronucleus; Ma, macronuclear anlage; Sy, synkaryon. See text for details.

1. A first period of DNA synthesis which results in the formation of polytene chromosomes.
2. The period in which the polytene chromosomes are degraded and a major portion of the DNA previously synthesized is lost from the anlage (so called DNA-poor stage, Ammermann, 1965, 1971).

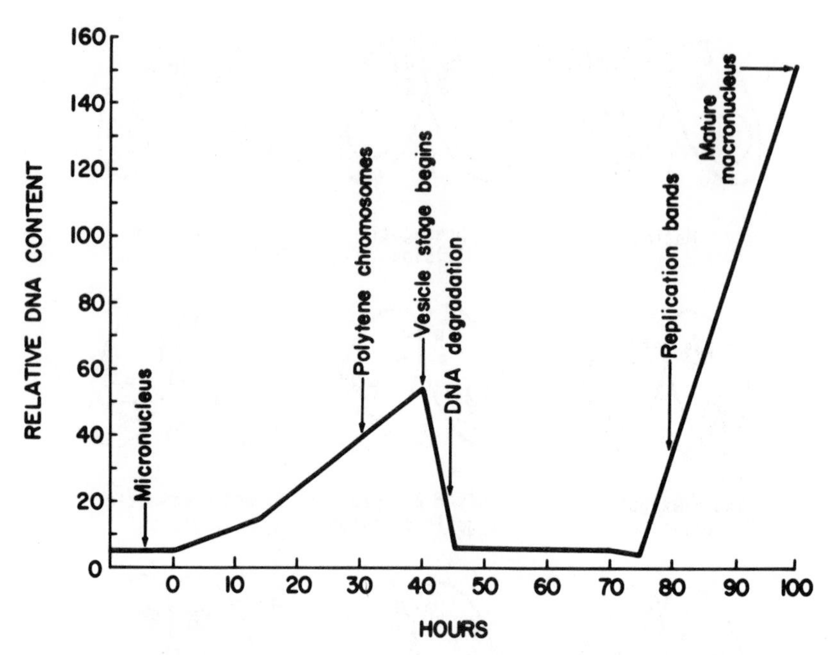

Figure 2. Schematic representation of the course of DNA changes in the developing macronucleus of Stylonychia. The first DNA synthesis results in the formation of polytene chromosomes. The polytene chromosomes are cut up into individual bands, and the bands are enclosed in the vesicles. Within the vesicles, most of the DNA is degraded. Finally, the remaining DNA replicates by replication bands and a mature macronucleus is formed. Redrawn from Ammermann (1965).

3. A second period of DNA synthesis in which the fraction of DNA that remained undegraded (in stage 2) is replicated many times to produce the mature macronucleus. The DNA synthesis in this stage occurs by means of replication bands, a process of replication that is typical of the mature macronucleus.

A schematic diagram of the course of DNA changes in the developing macronucleus of *S. mytilus* studied by microspectrometry (Ammermann, 1965) is given in Figure 2. In *O. fallax* the pattern of DNA changes in the macronuclear anlage is similar to that of *S. mytilus* (Prescott, private communication). The DNA changes in the anlage in *E. aediculatus* are also similar except for the quantitative differences in DNA replication and DNA degradation (Ammermann, 1971). It appears that the general process of macronuclear development shown in Figure 2 is true for all the hypotrichs. The salient features in macronuclear development are now described.

Polytene Chromosomes

The first period of DNA synthesis in the macronuclear anlage results in the formation of polytene chromosomes. At this time the DNA content of the anlage is increased up to 15 times (*S. mytilus,* Ammermann, 1971), or 30 times (*O. fallax,* Prescott, unpublished), or about 53 times (*E. aediculatus,* Ammermann, 1971) over the DNA content of the diploid micronucleus. With the formation of the polytene chromosomes, the DNA synthesis stops.

Polytene chromosomes have been observed in the macronuclear anlagen of hypotrichs *S. mytilus* (Ammermann, 1965; Jareño *et al.,* 1972), *S. muscorum* (Alonso and Perez-Silva, 1965), *S. notophora* (Sapra and Dass, 1970), *O. fallax* (Prescott, unpublished), *E. patella* (Turner, 1930), *E. aediculatus* (Rao and Ammermann, 1970; Ammermann, 1971), *Euplotes* sp. (Ris and Ruffolo, unpublished), *Kahlia* sp. (Rao, 1966), *Paraurostyla* sp. (Heumann, unpublished), *Urostyla* sp. (Ammermann, private communication), *Keronopsis rubra* (Ruthmann, 1972), *Holosticha* sp. (Prescott, unpublished), *Steinia candens,* and *Opisthotricha* sp. (Alonso and Perez-Silva, 1967). The hypotrich polytene chromosomes resemble those in the tissues of dipteran (insect) larvae in that they are much larger than the normal chromosomes and contain dense bands that alternate with less dense interbands (Figure 3). However, there are certain differences between the two kinds of polytene chromosomes. In dipterans, there is a haploid number of polytene chromosomes, each consisting of two homologous chromosomes in close association (Balbiani, 1881). At specific times during the growth of the insect larvae, localized swellings ("puffs") appear at the banded regions. Extensive RNA synthesis has been demonstrated in the puffed regions (*Drosophila,* Berendes, 1968).

The number of polytene chromosomes in the hypotrich macronuclear anlage has not been established. In Ammermann's (1971) observations on *S. mytilus,* the number varied from 25 to 50. This number appears to be dependent on the cytological techniques used in preparative procedure (Ammermann, 1971). When the isolated anlage is transferred to a hypotonic (swelling) medium, the nucleus bursts, releasing polytene chromosomes that stay together as a single clump. Further handling of the polytene chromosomes with glass needles or micropipettes often results in the breaking of the polytene chromosomes. Therefore, it is possible that the observed polytene chromosomes are in fact the breakdown products of much larger chromosomes. It is interesting to note that before polytenization the macronuclear anlage (which is still diploid) contains 250–300 chromosomes (Ammermann *et al.,* 1974), and after polytenization the

Figure 3. Electron micrograph of polytene chromosomes from the macronuclear anlage of Oxytricha. The preparation was obtained by using a modification of Miller's technique (Miller and Bakken, 1972). Arrows indicate a "puffed" region. ×5625.

number of polytene chromosomes ranges between 25 and 50. It appears that most of the micronuclear chromosomes are not polytenized during the macronuclear development in *S. mytilus*. The fate of the unpolytenized chromosomes is not known, but presumably they are broken down and eliminated from the macronuclear anlage during subsequent development. In *E. aediculatus,* the number of polytene chromosomes (approximately 150) is larger than the number of diploid micronuclear chromosomes (50–60). Ammermann (1971) contends that the observed number of polytene chromosomes in this cell is due to the crude cytological techniques that cause fragmentation of the polytene chromosomes.

The hypotrich polytene chromosomes show certain deviations in structure from the dipteran polytene chromosomes. First, the homologous chromosomes are not paired in the hypotrich polytene chromosomes. By careful spreading of the anlage of *S. mytilus,* Ammermann (1971) has demonstrated that the two homologous chromosomes (identifiable by their heterochromatic bands at specific locations) are distinctly separate from one another. In addition, no longitudinal splits in the polytene chromosomes (that could be interpreted as pairing gaps) have been observed. Second, the hypotrich polytene chromosomes neither contain "puffs" (Ammermann *et al.,* 1974) nor show evidence of transcriptional activity (Rao and Ammermann, 1970). In *S. mytilus,* Jareño *et al.* (1972) have observed structures resembling puffs in the polytene chromosomes. These puffs, however, seem to resemble "DNA puffs" seen in the polytene chromosomes of *Rhynchosciara* (Pavan and Breuer, 1955). In our own observation with *O. fallax,* we have seen puffs in the polytene chromosomes. These puffs appear as localized swellings at the banded regions (Figure 4a). These swellings contain diffuse material which in turn is composed of loops of DNA–protein fibers (Figure 4b). Whether or not these regions possess transcriptional activity is yet to be determined.

Degradation of Polytene Chromosomes

When their formation is complete, the polytene chromosomes become transected by "membranous" partitions that appear at the interband regions (Kloetzel, 1970; Murti, 1973). These "membranes" are composed of a single diffuse layer measuring 100 Å and do not have the trilaminar structure typical of lipoprotein membranes. It is possible that these membranes are composed of protein. The formation of these membranes is concomitant with the polytenization of the chromosomes (Murti, 1973). The membranes in the macronuclear anlage originate from a diffuse ma-

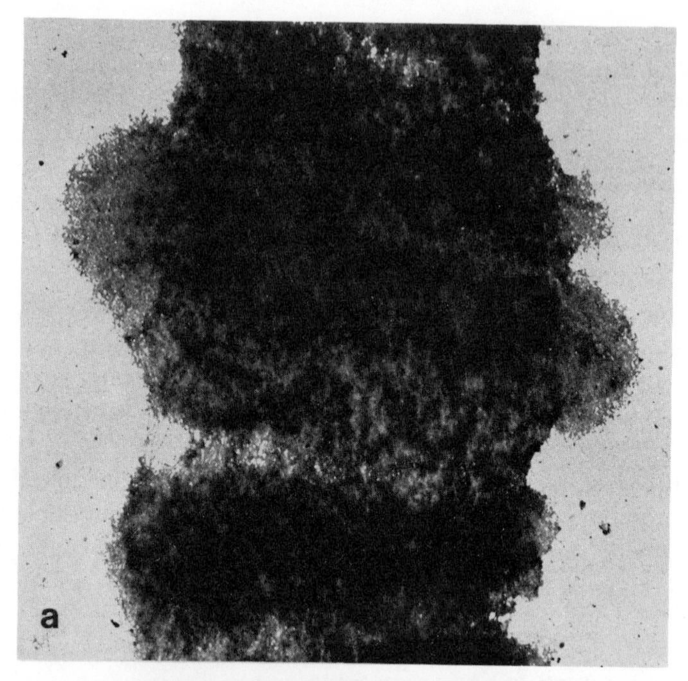

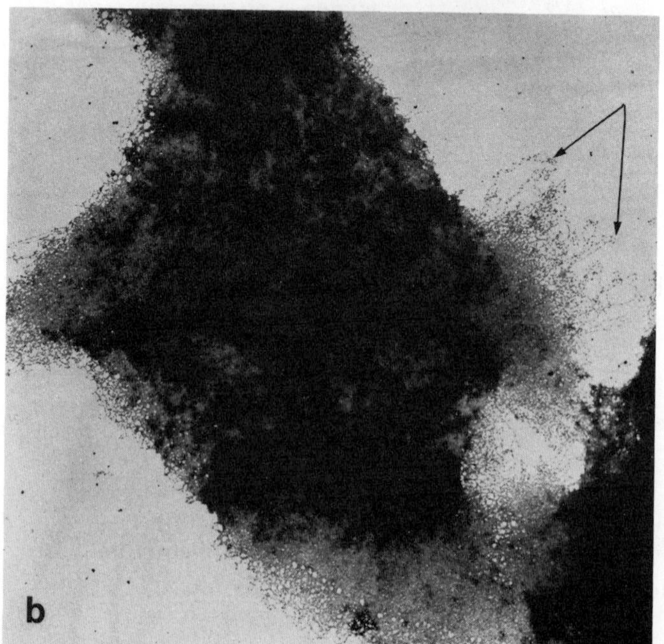

Figure 4. (a) A "puffed" region in the polytene chromosomes. ×8000. *(b) An extended puff in which loops of DNA–protein fibers are seen (arrow).* ×8000.

terial seen in the early anlage (Murti, 1973). The diffuse material in the early anlage is concentrated at several points. In later stages it is seen in the interior of the anlage, outlining the developing polytene chromosomes. As the polytene chromosomes reach the maximum degree of polyteny, the diffuse material condenses to give the membrane appearance and extends into the interband regions (Figure 5). Finally, each band plus the adjacent interband of the polytene chromosomes is completely enclosed by the membranes to form a large number of vesicles (Figure 6). The anlage now contains thousands of physically independent (Ammermann, 1971) vesicles, each of which contains a band plus the adjacent interband of the polytene chromosome.

The vesicle stage is followed by shrinkage of the anlage. The transection of the polytene chromosomes, formation of vesicles, and shrinkage are coincident with progressive reduction in the DNA content of the anlage (Ammermann, 1971; Murti, 1973). In *S. mytilus,* over 90% of the DNA of the polytene chromosomes is degraded to acid-soluble products that permanently leave the anlage (Ammermann, 1969). In *E. aediculatus,* approximately 80% of the DNA is destroyed.

The breakdown of DNA in the anlage may involve two steps. First, the continuity of the polytene chromosomes is interrupted either before or while the membranes extend into the interbands. Ammermann (1971) has provided evidence for this hypothesis by disrupting the anlage (of a stage after polytene chromosome breakdown) in distilled water. He observed the release from the anlage of a large number of physically independent granules (presumably the independent bands or vesicles seen in the electron microscope). At the molecular level, the breaks in the continuity of the polytene chromosomes may be accomplished by a deoxyribonuclease (DNase) that has its activity restricted to interbands. Such restriction endonucleases that are site-specific in their action on DNA have been found in bacterial systems (Kelly and Smith, 1970). Second, after the enclosure of the short segments of the polytene chromosomes in vesicles, a further degradation of DNA may occur in the vesicles. There are two possibilities concerning the reduction of DNA in the vesicles. Either all of the DNA in 90% of the vesicles is destroyed, or over 90% of the DNA in each vesicle is destroyed. The latter possibility is favored by electron microscope–autoradiographic studies (Prescott and Murti, 1973) in which the anlagen were labeled with [^{3}H]thymidine during the first period of DNA synthesis and the label was traced in the vesicles.

After the destruction and elimination of DNA, the vesicles disappear, and the fraction of DNA remaining is evenly distributed in the anlage (Murti, 1973). Subsequently, this remaining DNA is replicated many

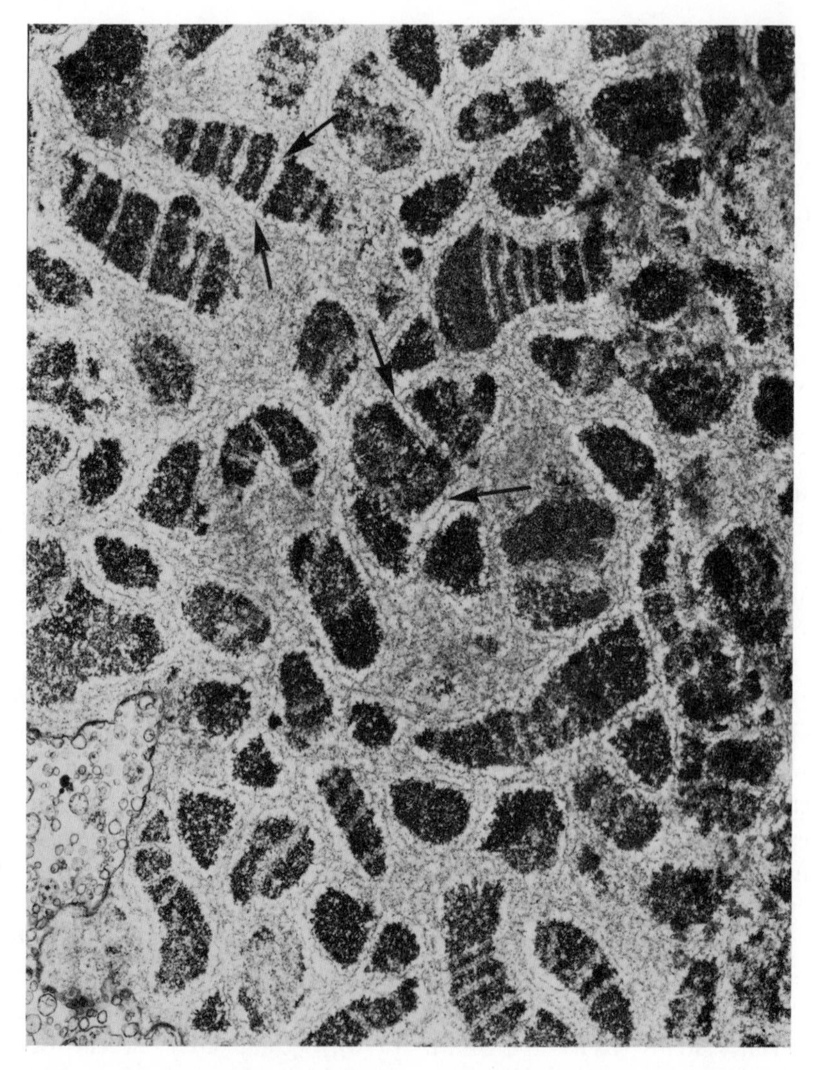

Figure 5. Electron micrograph of a thin section of the macronuclear anlage of Oxytricha at the time of polytene chromosome breakdown. Membranous material is seen around the polytene chromosomes and in the interbands (arrows). ×7900.

times, and a mature macronucleus is constituted. This completes the construction of the macronucleus, and the cell is now ready for its first postconjugational cell division.

Significance of DNA Destruction

The events in the macronuclear anlage (i.e., destruction of polytene chromosomes and elimination of most of DNA) result in an elimination of micronuclear DNA sequences from the macronucleus (Bostock and Prescott, 1972; Prescott and Murti, 1973). Both micronucleus and macronuclear anlage are derived from the synkaryon and are genetically identical. But the mature macronuclei and micronuclei show qualitative differences in the composition of their genetic material. Thus, in *O. fallax,* buoyant density studies have revealed that the micronuclear DNA contains at least four density components, whereas the macronuclear DNA consists of a single density component (Bostock and Prescott, 1972). The difference in

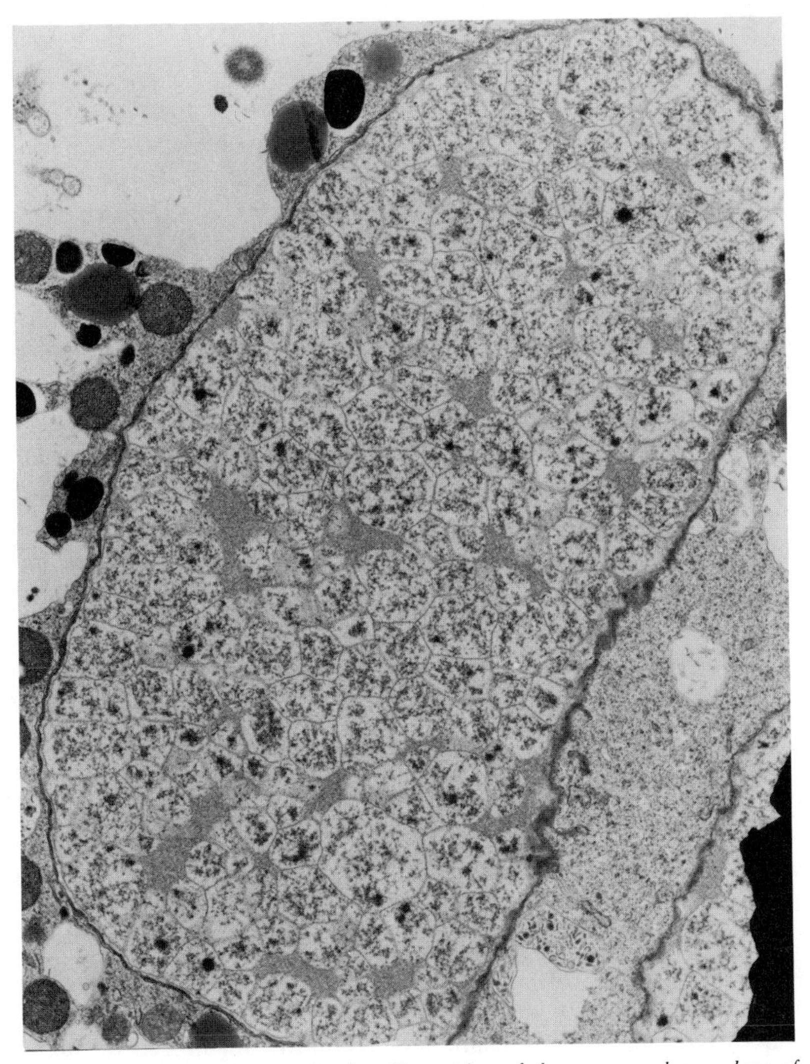

Figure 6. Electron micrograph of a thin section of the macronuclear anlage of Oxytricha during DNA destruction phase. The bands of polytene chromosomes enclosed in the vesicles are being destroyed. ×9400.

the DNA composition of the macro- and micronuclei are also reflected in the melting pattern of the DNAs obtained from these nuclei. Macronuclear DNA melts as if it were a single component, whereas micronuclear DNA melts as if it contained several components of different base composition (Bostock and Prescott, 1972). A study of the renaturation kinetics of the micro- and macronuclear DNAs showed that macronuclear DNA consists of a single DNA component with a complexity about 13 times greater than that of the DNA of *E. coli*; the micronuclear DNA contains at least two components, a rapidly renaturing (repetitious) fraction and a slowly renaturing fraction (Lauth *et al.*, 1976).

Qualitative differences begin to appear in the DNAs of the micronucleus and the anlage as the anlage enters the first period of DNA synthesis. Most of the micronuclear chromosomes are not polytenized in the anlage (Ammermann *et al.*, 1974). The DNA of the polytene chromosomes contains a single buoyant density component that has a peak density slightly less than the peak density for macronuclear DNA (Spear and Prescott, 1974). This component is believed to be composed of macronuclear DNA and a portion of micronuclear DNA sequences. So it appears that most of the micronuclear DNA sequences are "eliminated" from the anlage by nonreplication during the formation of the polytene chromosomes (Spear and Prescott, 1974; Lauth *et al.*, 1976). The situation resembles the nonreplication of satellite DNAs in the formation of polytene chromosomes in *Drosophila* (Gall *et al.*, 1971). A final elimination of the micronuclear DNA sequences from the anlage presumably occurs when segments of polytene chromosomes are destroyed in the vesicles. In conclusion, the formation and destruction of the polytene chromosomes in the anlage appears to achieve selective elimination of micronuclear DNA sequences from the macronucleus.

The function of the DNA that is eliminated from the original micronuclear genome during the formation of the macronucleus is not known. Such DNA might have a role in the various steps of conjugation (mating, meiosis, crossing over, etc.) and is eliminated from the macronucleus since the macronucleus does not participate in conjugation. In addition, features such as the formation of chromosomes and mitotic mode of division present in the micronucleus are not seen in the macronucleus. The macronucleus contains no detectable chromosomes (see next section) and divides by amitosis. It is possible that some of the DNA in the micronucleus is involved in maintaining the linear continuity of the chromosomes; such DNA is not required in the macronucleus.

There have been other examples of elimination of genetic material from the somatic nuclei. In the embryos of *Ascaris megalocephala*, the germ cells contain four large chromosomes, whereas somatic cells have about 60 much smaller chromosomes. In the course of differentiation of somatic cells from the zygote, specific portions of the chromosomes are discarded from the cell, and the rest are fragmented into smaller structures that behave as chromosomes in subsequent mitotic divisions (Lin, 1954). In certain insects (*Cecidomyidae,* gall midges), the number of chromosomes in the germ cells of females is much higher than the numer in somatic cells of the same individual. The classical example is *Miastor,* in which germ cells have 29 chromosomes and the somatic cells only six. During embryonic development of these insects, whole chromosomes are eliminated from the somatic cells (Nicklas, 1959; Painter, 1966). By cytological staining techniques, it has been shown in both *Ascaris* and *Miastor* that the eliminated portion of the genetic material is "heterochromatic." It is believed that heterochromatin is eliminated from somatic cells because of its "nonfunctional" nature, but the question of its function in germ cells is yet to be understood. Thus the selective gene diminution remains a poorly understood process in cell biology. A molecular approach to this problem is possible in simpler organisms such as hypotrichous ciliates.

Structure of Genetic Material in the Mature Macronucleus

Electron microscopic and biochemical studies of the purified macronuclear DNA have shown that the macronuclear DNA in *O. fallax* and *E. eurystomus* is composed of unusually small molecules (Prescott *et al.,* 1971). In a sucrose gradient, the macronuclear DNA consists of two components, a major component that sediments at 10 S and a minor component that sediments at 14 S. In the electron microscope, most of the molecules measure between 0.2 and 1.6 μm with an average length of 0.75 μm. The minor size class consists of DNA pieces 2.0–2.2 μm in length. These two size classes with average molecular weights of 1.6×10^6 and 4.0×10^6 daltons presumably correspond to the 10 S and 14 S fractions seen in sucrose gradient. Purifed micronuclear DNA, on the other hand, sediments with S values ranging from about 30 to greater than 100 (Prescott and Murti, 1973) and measures over 45.0 μm (Murti, 1972).

The small molecules of DNA in the macronucleus are believed to represent the state of macronuclear DNA *in vivo.* First, control experiments indicate that these molecules are not the products of mechanical shearing or enzymatic degradation (of larger DNA molecules) that may occur during purification procedures (Prescott *et al.,* 1971). Second, the low molecular weight DNA seems to be a feature of the hypotrich (*O. fallax*, *E. eurystomus*, and *Paraurostyla*, Prescott *et al.*, 1971; Prescott and Davern, unpublished) macronucleus. In the developing macronucleus of these cells, polytene chromosomes are transected in the interbands in a way that would greatly reduce the length of the original micronuclear DNA. The absence of a major portion of micronuclear DNA sequences in the macronucleus suggests that the selective DNA destruction process that occurs in the developing macronucleus might generate the low molecular weight DNA in the mature macronucleus. In holotrichs such as *Tetrahymena*, the macronucleus contains DNA of relatively high molecular weight (7×10^7 daltons, Murti, 1972; extrachromosomal ribosomal genes have a molecular weight of 12.6×10^6, Gall, 1974). In these cells, the macronuclear anlage contains neither polytene chromosomes nor a DNA destruction phase, and the mature macronucleus is an amplified version of the micronucleus (Murti, 1973; Yao and Gorovsky, private communication).

The small molecules of macronuclear DNA may represent individual genes (Prescott *et al.*, 1971; Prescott and Murti, 1973). The main fraction of DNA consists of molecules with an average length of 0.75 μm, which is sufficient to code for 750 amino acids. Such a molecule would represent an unusually large polypeptide, but it is likely that a significant portion of each piece of DNA serves for functions other than amino acid coding, e.g, control regions for transcription and replication. Some of the larger pieces of DNA may, however, code for two to several polypeptides. The minor class of DNA consists of molecules of about 2 μm (4×10^6 daltons). These may represent ribosomal genes. In *Euplotes* and *Oxytricha*, ribosomal RNA precursor is about 34 S (or 2×10^6 daltons, Prescott and Lauth, private communication). The synthesis of such a molecule would require a template of duplex DNA of about 4×10^6 daltons. Hybridization experiments (Prescott and Murti, 1973) have in fact confirmed that the minor class of DNA is enriched for ribosomal genes.

In the light of the above observations, it can be concluded that the hypotrich macronucleus is a bag that contains gene-sized DNA molecules (Prescott *et al.*, 1971). Presumably, each DNA molecule represents a single transcription and replication unit. This hypothesis is attractive since it provides some explanations for the peculiar mode of macronuclear replication (by replication bands) and macronuclear aging seen in hypotrichs.

Electron microscopic observations on the thin sections of the vegetative macronucleus of hypotrichs have shown that the chromatin is organized in the form of chromatin granules of varying dimensions (0.1–1 μm in diameter), and in these studies there is no evidence for the presence of individual free-floating DNA molecules (Kluss, 1962; Ringertz *et al.*, 1967; Kloetzel, 1970; Murti, 1973). It is possible, however, that the individual DNA molecules in the macronucleus are aggregated with protein and RNA to form the chromatin granules. Recently, we have had success in obtaining whole-mount preparations of well-spread macronuclei for electron microscopy. The procedure consists of centrifuging the contents of well-dispersed macronuclei onto electron microscope grids in the manner described by Miller (Miller and Bakken, 1972). In such preparations we found that each chromatin granule is composed of a large number of small (0.2–3.0 μm) DNA–protein fibers that are extremely active in transcription (Murti, unpublished).

A model explaining the events in macronuclear development is presented in Figure 7. The DNA of the micronucleus (the early anlage is identical to the micronucleus in its genetic constitution) is of high molecular weight (Murti, 1972; Prescott and Murti, 1973) and presumably is arranged in a series of condensed regions separated by noncondensed regions. The noncondensed regions contain individual genes separated by stretches of "spacer" DNA of the condensed regions, a situation similar to the gene–spacer arrangement described for the genes coding for ribosomal RNA (Birnstiel *et al.*, 1968; Dawid *et al.*, 1970; Miller and Beatty, 1969), 5 S RNA (Brown and Sugimoto, 1973), and tRNA (Clarkson *et al.*, 1973) in amphibians. During polytenization of the micronuclear chromosomes, the condensed regions (containing spacers) become visible as the bands of the polytene chromosomes, and the noncondensed regions (containing genes) form the interbands (see Crick, 1971). When polytenization in the developing macronucleus is complete, the DNA is cut into short pieces by endonucleases, and membranes partitions appear at the interbands. Next, each band plus an adjacent interband becomes enclosed by membrane to form a vesicle. Electron microscopic observations have shown that a complete band is enclosed in the vesicle, but it is not clear whether the vesicles also enclose a single, complete interband or a portion of the interband or portions of the two adjacent interbands. During DNA destruction phase, most of the DNA in each vesicle is destroyed (Prescott and Murti, 1973). The process may represent total destruction of the band (spacer) DNA, possibly by sequence-specific nucleases. After DNA destruction, the pieces of DNA that are left in each vesicle are believed to be gene copies. The final process of macronuclear development consists of multiple replications of the gene copies preserved in each vesicle.

The hypothesis that the macronuclear DNA sequences in the micronuclear DNA are separated by large stretches of spacer DNA is borne out by the following electron microscopic observations in *O. fallax*. Each

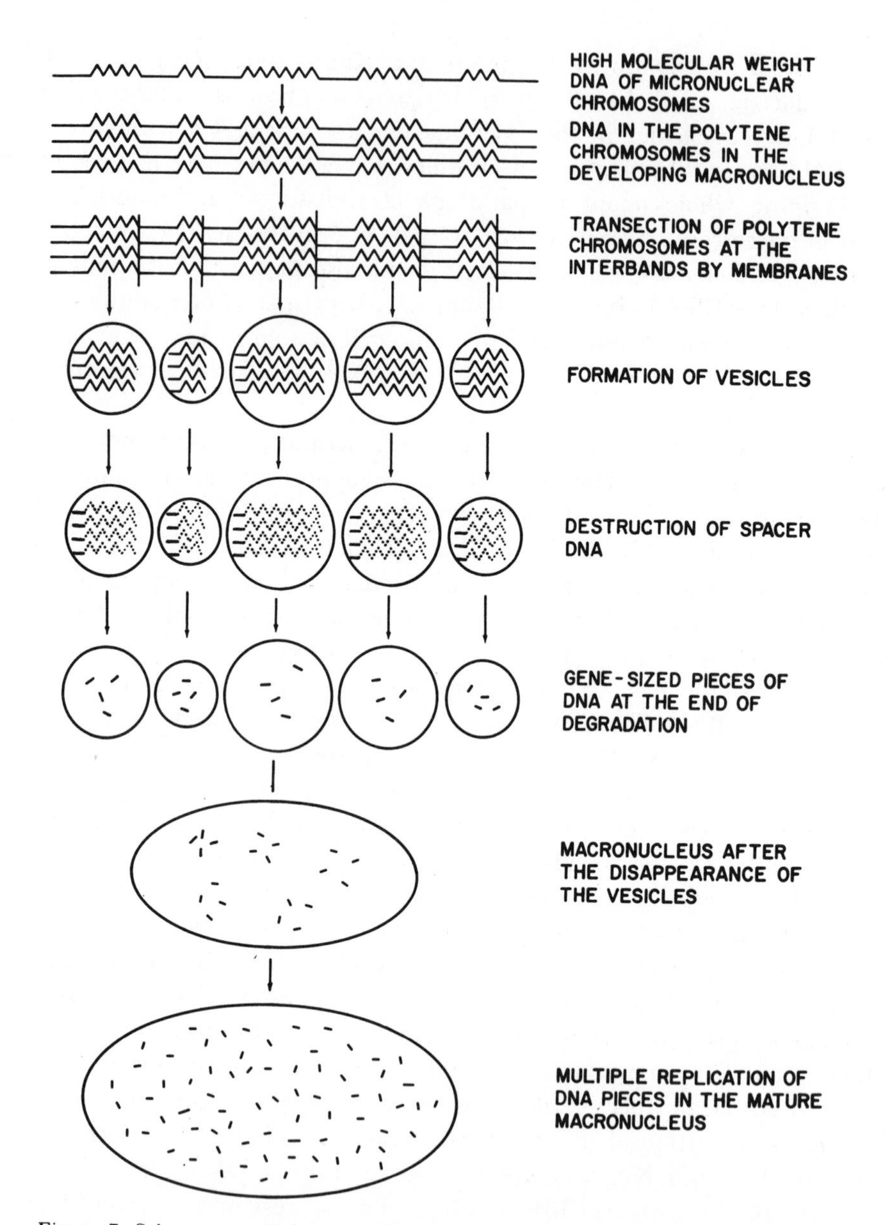

Figure 7. Scheme to explain the derivation of the gene-sized pieces of DNA in the mature macronucleus. See text for details.

piece of macronuclear DNA is known to melt at one end when partially denatured by heat or alkali (Murti *et al.*, 1972; Prescott and Murti, 1973). These unstable ends are believed to contain DNA sequences that are richer in adenine and thymidine than the rest of the molecule. When the high molecular weight micronuclear DNA is subjected to similar conditions of denaturation, each denatured site appears to be separated by large stretches of native DNA of length between 1 and 34 μm (Murti and Prescott, unpublished). The total number of denatured sites divided by the total length of DNA examined yields an average length of 4.7 μm of double-stranded DNA per denatured site in the micronucleus. The same calculation for macronuclear DNA yields an average of 0.7 μm of double-stranded DNA per denatured site (one denatured site per each piece of DNA in the macronucleus). This means that on the average about 85% (4.0 μm) of the native DNA between the denatured sites is destroyed during the vesicle stage in macronuclear development. Another observation that leads to the same conclusion is the binding pattern of RNA polymerase (from *Bacillus subtilis*) to macro- and micronuclear DNAs. Each piece of macronuclear DNA binds a single molecule of RNA polymerase at one end when the binding experiment was performed in the absence of pyrimidine deoxynucleoside triphosphates to prevent transcription (Murti *et al.*, 1972). It is not known whether the bacterial polymerase identifies the natural binding site of homologous polymerase to macronuclear DNA. When the binding experiment was performed using micronuclear DNA (Murti and Prescott, unpublished), the bound polymerase molecules were seen to be separated by large stretches of native DNA (average length of about 6 μm), enough to account for the spacer.

DNA Replication in the Macronucleus

In the hypotrich macronucleus, the DNA synthesis starts at one or two specific points and proceeds in the form of replication bands (Turner, 1930; see also Raikov, 1968, for a review). In *Euplotes* and *Paraurostyla*, two replication bands appear simultaneously at the ends of the macronucleus, move to the middle, fuse, and then disappear. In *Aspidisca*, the direction of movement of the replication band is exactly opposite (i.e., the replication bands start at the middle of the macronucleus and travel to the ends). When there are several macronuclei, replication bands appear simultaneously in all, even when their number exceeds a hundred, as in *Urostyla grandis*. Only one replication band that travels from one end of the macronucleus to the other is seen in the macronuclei of *Dysteria*, *Chlamydodon*, *Spirochona*, *Gastrostyla*, *Kahlia*, *Stylonychia*, and *Oxytricha*.

Replication bands in a hypotrich macronucleus (*Paraurostyla*) are shown in Figure 8. Structurally, the replication band consists of two distinct regions. A forward zone (FZ) composed of a large number of small chromatin particles (Cp) and a rear zone (RZ) that consists of a mass of

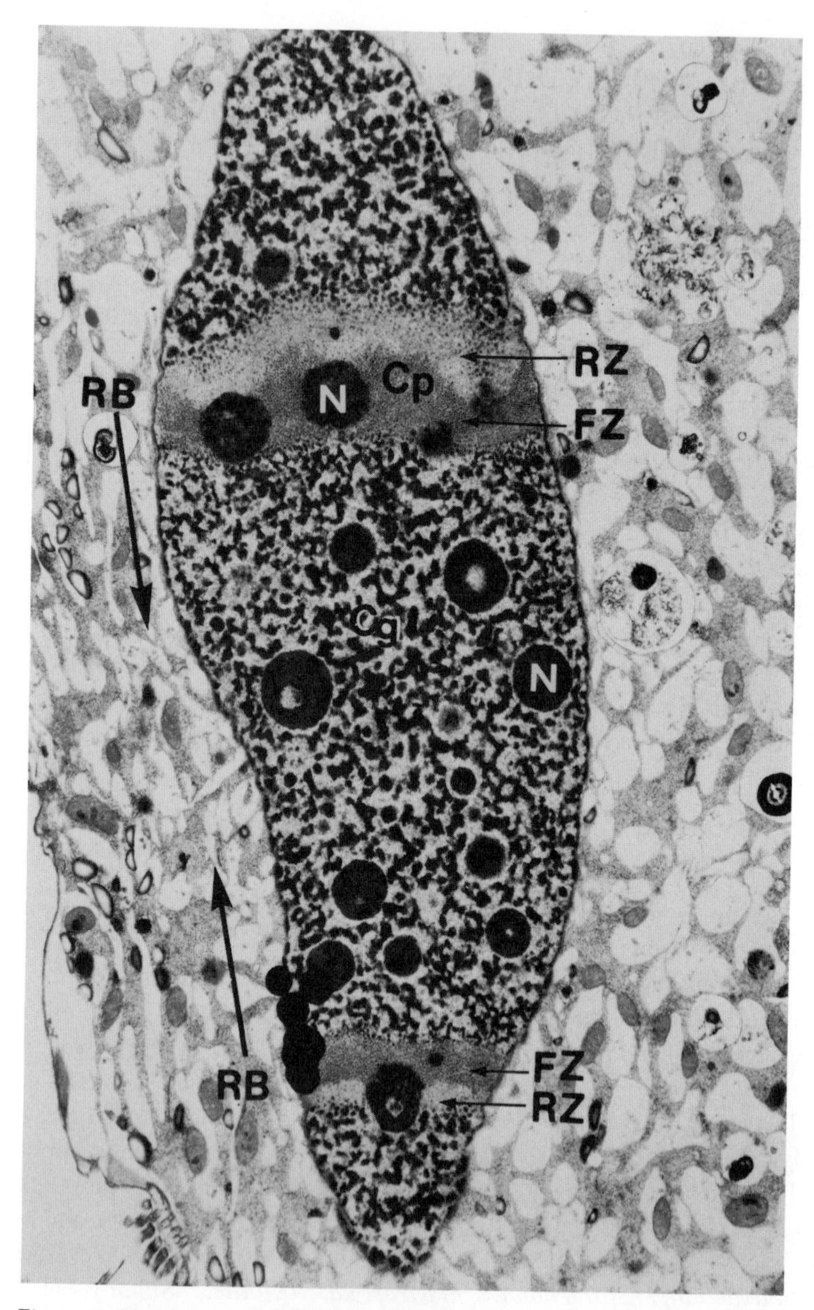

Figure 8. Electron micrograph of a thin section of the macronucleus of Paraurostyla showing replication bands (RB). The two replication bands originate at the ends of the macronucleus and proceed to the middle (the directions are indicated by arrows). Abbreviations: FZ, forward zone; RZ, rear zone; Cp, chromatin particles; Cg, chromatin granules; N, nucleolus. ×10,000.

finer (50 Å) and less clearly resolvable fibers (Kluss, 1962; Raikov, 1968). Autoradiographic studies (Gall, 1959; Kimball and Prescott, 1962; Stevens, 1963; Ammermann, 1971) have shown that DNA synthesis is restricted to the rear zone of the replication band. Behind the rear zone, the replicated fibrils reaggregate to form chromatin granules (Cg) that are usually seen in the vegetative macronucleus.

A schematic diagram of the replication band based on the available information on the structural organization of the genetic material in the macronucleus is provided in Figure 9. As already stated, the chromatin granules in the vegetative macronucleus consist of aggregates of individual particles. Each individual particle is believed to contain a single molecule of DNA with the attached RNA and protein. In the nonreplicated state, all the individual particles are held together into a chromatin granule. As the replication band approaches, the chromatin granules disaggregate into individual particles (forward zone). In the replication zone, the particles unfold into single DNA molecules (with the release of RNA and protein) that replicate either unidirectionally (by a fork) or bidirectionally (by a "bubble"). After replication, the individual DNA molecules fold back into particles, presumably because of addition of RNA and protein. The particles subsequently reaggregate to form the chromatin granules.

The model presented above may represent an oversimplification of the structure of the replication band. Many more studies are needed to understand the structure and function of the replication band. The structural features so far described in the replication band suggest that this type of DNA synthesis provides a means to replicate large numbers of individual molecules of DNA in the macronucleus.

Division of the Macronucleus

The mature macronucleus in hypotrichs, as in other ciliates, divides by amitosis. The amitotic mode of division differs from the basic plan of mitosis. In mitosis, the movement of sister chromosomes to opposite poles of the mitotic spindle is brought about by spindle fibers (Harris, 1962; Kane, 1962). In amitosis, neither a distinct organization of chromosomes nor the formation of a spindle is seen (Raikov, 1968). The macronucleus seems to divide randomly during amitosis.

In hypotrichs, the number of divisions that a macronucleus can undergo appears to be limited (Ammermann, 1971). Clones of *Stylonychia* become senile and die after 1–1½ years (400–500 divisions) of vegetative growth. The cells, however, can be revived by conjugation. Since the macronucleus takes care of somatic functions, the cell death after prolonged

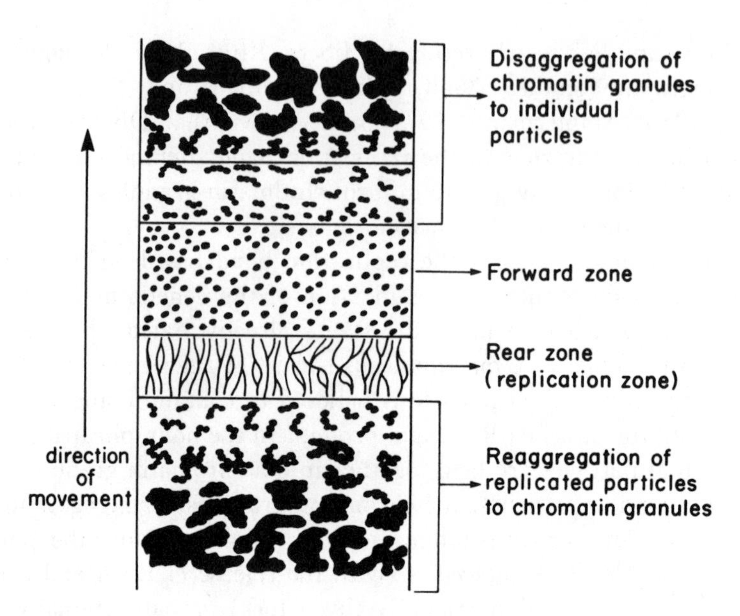

Figure 9. Model of the replication band. See text for details.

vegetative growth is possibly due to macronuclear aging. The process can be reversed by conjugation, because conjugation produces new macronuclei in the cell.

The macronuclear aging is related to the structure and organization of the genetic material in the macronucleus. If the macronucleus contains many copies of genetic information (whether chromosomes, portions of chromosomes, or individual genes) that are not distributed in an organized fashion during amitosis, there is still a good chance of the daughter macronuclei receiving complete genetic information during the earlier divisions of clonal growth. The number of divisions that a macronucleus can undergo without becoming aneuploid depends on the number of genetic elements (whether chromosomes, portions of chromosomes, or genes) required for normal growth and survival, and the extent of their multiplicity.

Using Kimura's formula for aging

$$t = \frac{4m}{n-1}(0.39n - 0.35 - 2.3 \log 10\Omega)$$

where m is the degree of multiplicity, n is the number of genetic elements in the nucleus, and Ω is the probability that after t generations the cells die through loss of an element, Ammermann (1971) has calculated the possible number of viable generations (the period in which the daughter macronuclei would receive complete genetic information) if the macronucleus contained the micronuclear chromosomes in many copies (n = 150, m =

60, Ω = 99%). He arrived at a number of 100, which is far below the known number of viable generations, i.e., 400–500. He concluded that if the macronucleus contains selected parts of micronuclear chromosomes at a higher degree of multiplicity (approximately 240), then the macronucleus can divide as many as 400 times without becoming aneuploid. Evidence summarized in this chapter strongly suggests that indeed the hypotrich macronucleus contains many copies of individual genes that represent qualitatively a small portion of the micronuclear genome.

Summary

In hypotrichous ciliates, the macronucleus contains a minor fraction of the micronuclear genome, although it develops from the micronucleus following conjugation. In the developing macronucleus, the micronuclear DNA sequences are eliminated. The result of this process is a mature macronucleus that contains many gene-sized DNA molecules. These molecules replicate and transcribe independently in the mature macronucleus. Random distribution of these molecules during amitosis leads to macronuclear aging and eventually to cell death during vegetative growth.

Acknowledgments

The author is grateful to Dr. D. M. Prescott for help, encouragement, and criticism. This work was supported by Grants GB-32232 from the National Science Foundation and RO1 GM 19199-01 CBY from the National Institutes of General Medical Science to Dr. D. M. Prescott.

Literature Cited

Alonso, P. and J. Perez-Silva, 1965 Giant chromosomes in protozoa. *Nature* (*London*) **205**:313–314.

Alonso, P. and J. Perez-Silva, 1967 Apareamiento somático y politenización en los cromasomas gigantes de ciliados hipotricos. *Bol. Real Soc. Espan. Hist. Natl. Secc. Biol.* **65**:469–475.

Ammermann, D., 1965 Cytologische und genetische Untersuchungen an dem Ciliaten *Stylonychia mytilus* Ehrenberg. *Arch. Protistenk.* **108**:109–152.

Ammermann, D., 1969 Release of DNA breakdown products into the culture medium of *Stylonychia mytilus* exconjugants (Protozoa, Ciliata) during the destruction of the polytene chromosomes. *J. Cell Biol.* **40**:576–577.

Ammermann, D., 1971 Morphology and development of the macronuclei of the ciliates *Stylonychia mytilus* and *Euplotes aediculatus*. *Chromosoma* **33**:209–238.

Ammermann, D., G. Steinbrück, L. Von Berger, and W. Hennig, 1974 The development of the macronucleus in the ciliated protozoan *Stylonychia mytilus*. *Chromosoma* **45:**401–429.

Balbiani, E. G., 1881 Sur la structure du noyau des callules salivaires chez les larves de *Chironomus*. *Zool. Anz.* **4:**637–641.

Berendes, H. D., 1968 Factors involved in the expression of gene activity in polytene chromosomes. *Chromosoma* **24:**418–437.

Birnstiel, M., J. Speirs, I. Purdom, K. Jones, and U. E. Loening, 1968 Properties and composition of the isolated ribosomal DNA satellite of *Xenopus laevis*. *Nature (London)* **219:**454–463.

Bostock, C. J. and D. M. Prescott, 1972 Evidence of gene diminution during the formation of the macronucleus in the protozoan, *Stylonychia*. *Proc. Natl. Acad. Sci. USA* **69:**139–142.

Brown, D. D., and K. Sugimoto, 1973 5 S DNAs of *Xenopus laevis* and *Xenopus mulleri:* Evolution of a gene family. *J. Mol. Biol.* **78:**397–415.

Clarkson, S. G., M. L. Birnsteil, and I. E. Purdom, 1973 Clustering of transfer RNA genes of *Xenopus laevis*. *J. Mol. Biol.* **79:**411–429.

Crick, F., 1971 General model for the chromosomes of higher organisms. *Nature (London)* **234:**25–27.

Dawid, I. B., D. D. Brown, and R. H. Reeder, 1970 Composition and structure of chromosomal and amplified ribosomal DNAs of *Xenopus laevis*. *J. Mol. Biol.* **51:**341–360.

Gall, J. G., 1959 Macronuclear duplication in the ciliated protozoan Euplotes. *J. Biophys. Biochem. Cytol.* **5:**295–308.

Gall, J. G., 1974 Free ribosomal RNA genes in the macronucleus of *Tetrahymena*. *Proc. Natl. Acad. Sci. USA* **71:**3078–3081.

Gall, J. G., E. H. Cohen, and M. L. Polan, 1971 Repetitive DNA sequences in *Drosophila Chromosoma* **33:**319–344.

Harris, P., 1962 Some structural aspects of the mitotic apparatus in sea urchin embryo. *J. Cell Biol.* **14:**475–487.

Jareño, M. A., P. Alonso, and J. Perez-Silva, 1972 Identification of some puffed regions in the polytene chromosomes of *Stylonychia mytilus*. *Protistologica* **8:**237–243.

Jurand, A., G. Beale, and M. Young, 1964 Studies on the macronucleus of *Paramecium aurelia*. II. Development of macronuclear anlagen. *J. Protozool.* **11:**491–497.

Kane, R. E., 1962 The mitotic apparatus: Fine structure of the isolated unit. *J. Cell Biol.* **15:**279–287.

Kelly, T. J. and H. O. Smith, 1970 A restriction enzyme from *Hemophilus influenzae*. II. Base sequence of the recognition site. *J. Mol. Biol.* **51:**393–409.

Kimball, R. F. and D. M. Prescott, 1962 Deoxyribonucleic acid synthesis and distribution during growth and amitosis of the macronucleus of *Euplotes*. *J. Protozool.* **9:**88–92.

Kloetzel, J. A., 1970 Compartmentalization of the developing macronucleus following conjugation in *Stylonychia* and *Euplotes*. *J. Cell Biol.* **47:**395–407.

Kluss, B. C., 1962 Electron microscopy of the macronucleus of *Euplotes eurystomus*. *J. Cell Biol.* **13:**462–465.

Lauth, M. R., B. B. Spear, J. Heumann, and D. M. Prescott, 1976 DNA of ciliated protozoa: DNA sequence diminution during macronuclear development of *Oxytricha*. *Cell* **7:**67–74.

Lin, T. P., 1954 The chromosomal cycle in *Parascaris equorum* (*Ascaris megalocephala*): Oogenesis and diminution. *Chromosoma* **6:**175–198.

Maupas, E., 1888 Recherches expérimentales sur la multiplication des infusoires ciliés. *Arch. Zool. Exp. Gen.* **6:**165–277.

Miller, O. L., Jr. and A. Bakken, 1972 Morphological studies of transcription. *Acta Endocrinol. Suppl.* **168:**155–177.

Miller, O. L., Jr. and B. R. Beatty, 1969 Visualization of nucleolar genes. *Science* **164:**955–957.

Murti, K. G., 1972 An electron microscopic study of the structure and function of the genetic material in two ciliated protozoans. Ph.D. thesis, Department of Molecular, Cellular, and Developmental Biology, University of Colorado, Boulder.

Murti, K. G., 1973 Electron microscopic observations on the macronuclear development of *Stylonychia mytilus* and *Tetrahymena pyriformis* (Ciliophora–Protozoa). *J. Cell Sci.* **13:**479–509.

Murti, K. G., D. M. Prescott, and J. J. Pène, 1972 DNA of ciliated protozoa. III. Binding of RNA polymerase and denaturation at the ends of the low molecular weight DNA in *Stylonychia. J. Mol. Biol.* **68:**413–416.

Nicklas, R. B., 1959 An experimental and descriptive study of chromosome elimination in *Miastor* sp. *Chromosoma* **10:**301–336.

Painter, T. S., 1966 The role of E chromosome in cecidomyidae. *Proc. Natl. Acad. Sci. USA* **56:**853–855.

Pavan, C. and M. Breuer, 1955 Differences in nucleic acid content of the loci in polytene chromosomes of *Rhynchosciara angelae* according to tissue and larval stages. In *Symposium on Cell Secretion,* edited by G. Schreiber, pp. 90–99, Belo Horizonte, Brazil.

Preer, J. R., 1968 Genetics of Protozoa. In *Research in Protozoology,* Vol. 3, edited by Tse-Tuan Chen, pp. 234–237, Pergamon Press, New York.

Prescott, D. M. and K. G. Murti, 1973 Chromosome structure in ciliated protozoans. *Cold Spring Harbor Symp. Quant. Biol.* **38:**609–618.

Prescott, D. M., C. J. Bostock, K. G. Murti, M. R. Lauth, and E. Gamow, 1971 DNA of ciliated protozoa. I. Electron microscopic and sedimentation analyses of macronuclear and micronuclear DNA of *Stylonychia mytilus. Chromosoma* **34:**355–366.

Raikov, I., 1968 The macronucleus of ciliates. In *Research in Protozoology,* Vol. 3, edited by T. T. Chen, pp. 1–128, Pergamon Press, New York.

Rao, M. V. N., 1966 Conjugation in *Kahlia* sp. with special reference to meiosis and endomitosis. *J. Protozool.* **13:**565–573.

Rao, M. V. N. and D. Ammermann, 1970 Polytene chromosomes and nucleic acid metabolism during macronuclear development in *Euplotes. Chromosoma* **29:**246–254.

Ringertz, N., J. Ericsson, and O. Nilsson, 1967 Macronuclear chromatin structure in *Euplotes. Exp. Cell Res.* **48:**97–117.

Ruthmann, A., 1972 Division and formation of the macronuclei of *Keronopsis rubra. J. Protozool.* **19:**661–666.

Sapra, G. R. and C. M. S. Dass, 1970 Organization and development of the macronuclear anlage in *Stylonychia notophora* Stokes. *J. Cell Sci.* **6:**351–363.

Spear, B. B. and D. M. Prescott, 1974 Under-replication of satellite DNA in polytene chromosomes of a protozoan, *Oxytricha. J. Cell Biol.* **63:**327a.

Stevens, A. R., 1963 Electron microscope radioautography of DNA and RNA synthesis in *Euplotes eurystomus. J. Cell Biol.* **19:**67A.

Turner, J. P., 1930 Division and conjugation in *Euplotes patella* Ehrbg., with special reference to the nuclear phenomena. *Univ. Calif. Publ. Zool.* **33:**193–258.

Woodard, J., M. Woodard, B. Gelber, and H. Swift, 1966 Cytochemical studies of conjugation in *Paramecium aurelia. Exp. Cell Res.* **41:**55–63.

5

Histones of Sperm

DAVID P. BLOCH

Introduction

In "A Catalog of Sperm Histones" (Bloch, 1969), organisms were grouped into five categories according to the basic proteins of their sperm: a *Salmo* type containing easily extractable "monoprotamines" (Kossel, 1928) whose sole basic amino acid is arginine, a mammalian type containing a very basic "protaminelike" protein that nevertheless binds strongly to the sperm head and can be extracted only after breaking disulfide linkages (Henricks and Mayer, 1965), a *Mytilus* type easily extractable and intermediate in composition between the monoprotamines and somatic histones (also designated di- and triprotamines by Kossel, 1928), a *Rana* type showing no discernible differences from somatic histones, and a crab type perhaps containing histones in the cytoplasm but none in the nucleus.

A search of the literature and a cursory cytochemical survey of sperm proteins of a number of species failed to reveal any relationship between protein type and sperm function. Neither nuclear activity, conditions of fertilization (external or internal, saltwater or freshwater), sperm longevity, nuclear condensation, cleavage pattern of the fertilized egg, or change in totipotency of the sperm nucleus, among other things, offered a ready explanation for the protein change during sperm development (Bloch, 1969).

DAVID P. BLOCH—Department of Botany and Cell Research Institute, University of Texas at Austin, Austin, Texas.

A hypothesis to explain the extreme nonconservatism of sperm proteins was proposed: In organisms with separate sexes, where sex determination had become chromosomally based and sperm functions localized on the heteromorphic regions of the sex chromosomes, thereby protected from crossing over, mutations might accumulate. Since the sperm nucleus is inert, the function of its histones may be rudimentary and the restraints that ordinarily cause histones to be evolutionarily conservative may not operate. If the only requirement were basicity, one might expect an increase in arginine content. Since there are more triplets that code for arginine than for lysine and histidine, the code would provide a statistical trap, and genetic drift would be the motivating force behind the accumulation of arginine.

This chapter contains an updated version of the table presented in 1969 categorizing organisms on the basis of the protein of their sperm, and tables giving the amino acid compositions of some known sperm proteins and the sequences of some protamines. Recent work on the synthesis of protamines in trout is discussed. Some of the views concerning histone composition in mammalian sperms are reexamined.

Table 1 lists a number of organisms whose sperm head proteins have been studied. The protein types are indicated in the table as follows:

1. *Salmo* type, or monoprotamine.
2. Mammalian type: arginine rich, highly basic, containing —SH groups or associated with —SH-containing proteins.
3. *Mytilus* type, intermediate between monoprotamines and somatic histones.
4. *Rana* type, no apparent difference between sperm and somatic histones.
5. Crab type, no basic protein in the nucleus.

Amino Acid Composition of Some Sperm Proteins

Table 2 gives the amino acid compositions of the basic proteins extracted from sperm. Where available, values are given in number of residues per molecule. Otherwise, values are in mole % ratios.

Amino Acid Sequence of Protamines

The protamines of a number of species consist of several components that are similar in composition. The sequence of three clupeines, Z, YI, and YII, are shown in Table 3, along with those of a salmine, three iridines, and four thynnines. The Z configuration of clupeine was depicted

TABLE 1. *Sperm Head Proteins of Some Different Species of Animals and Plants*[a]

Phylum, class, and species	*Common name*	*Protein type*	*Reference*
Nemathelminths			
Ascaris lumbricoides	—	2	Bloch, 1969
Plathyhelminths			
Planaria	—	4	Bloch, 1969
Echinodermata[b]			
Arbacia lixula	Sea urchin	3	Palau *et al.*, 1969
A. punctulata	Sea urchin	3 or 4	Hamer, 1955
Asterias forbesi	Starfish	3 or 4	Hamer, 1955
A. glacialis	Starfish	3 or 4	Vendrely and Vendrely, 1966
A. tenuispina	Starfish	3 or 4	Subirana and Palau, 1968
Astropecten aurantiacus	Starfish	3 or 4	Kossel, 1928
Brissopsis lyrifera	Sea urchin	3 or 4	Hultin and Herne, 1949
Echinarachnius parma	Sand dollar	3 or 4	Paolleti and Huang, 1969
Echinaster sepositus	Starfish	3 or 4	Subirana and Palau, 1968
Echinocardium cordatum	Sea urchin	3 or 4	Hultin and Herne, 1949
Echinocyanus pusillus	Sea urchin	3 or 4	Vendrely and Vendrely, 1966
Echinus esculentus	Sea urchin	3 or 4	Kossel, 1928
E. acutus	Sea urchin	3 or 4	Kossel and Staudt, 1926
Holothuria polii	Sea cucumber	3 or 4	Subirana and Palau, 1968
H. tubulosa	Sea cucumber	4	Subirana, 1970
Lytechinus pictus	Sea urchin	3 or 4	Bloch, 1969
Ophiothrix fragilis	Brittle star	3 or 4	Subirana and Palau, 1968
Pateria miniata	Starfish	3 or 4	Bloch, 1969
Sphaerechinus granularis	Sea urchin	3 or 4	Subirana and Palau, 1968
Strongylocentrotus drobächiensis	Sea urchin	3 or 4	Gineitis *et al.*, 1970
S. intermedius	Sea urchin	3 or 4	Gineitis *et al.*, 1970
S. lividus	Sea urchin	3 or 4	Kossel, 1928
S. nudus	Sea urchin	3 or 4	Gineitis *et al.*, 1970

TABLE 1. Continued

Phylum, class, and species	Common name	Protein type	Reference
Annelida			
Chaetopterus variopedatus	Parchment worm	4	
Eudistylia polymorpha	Feather duster	3	Bloch, 1969
Lumbricus terrestris	Earth worm	1	
Urechis caupo	Euchuroid worm	3_c	Das *et al.*, 1967
Arthropoda			
Crustaceae			
Cancer borealis	Crab	5	
C. irroratus	Crab	5	
C. magister	Crab	5	Langreth, 1969
C. productus	Crab	5	
Carcinus maenus	Crab	5_c	Chevaillier, 1967
Emerita analoga	Mole crab	5_c	Bloch, 1966; Vaughn, 1968
Eupagerus bernhardus	Hermit crab	5_c	Chevaillier, 1967
Lepas	Gooseneck barnacle	4	Bloch, 1969
Libinia emarginata	Spider crab	5_c	Vaughn and Hinsch, 1972
Nephrops norvegicus	Crayfish (saltwater)	5_c	Chevaillier, 1967
Procambus clarkii	Crayfish (freshwater)	5_c	Bloch, 1969
Myriapoda			
Armadillidium vulgare	Pillbug	$1^{an.}$	Bloch, 1969
Lithobius forficatus	Centipede	1?	Descamps, 1969
Insecta			
Acheta domestica	Cricket	$2^{an.-}$	Kaye, 1958; Kaye and McMaster-Kaye, 1966
Apis mellifera	Bee	$4^{is.}$	Bloch, 1966, 1969
Chortophaga viridifasciata	Grasshopper	$2^{an.-}$	Bloch and Brack, 1964
Drosophila melanogaster	Fruit fly	2	Das *et al.*, 1964

Pseudococcus obscura	Coccid	2	Berlowitz, 1965
Rehnia spinosus	Katydid	$2^{\text{an.}-}$	Bloch, 1969
Schistocerca gregaria	Grasshopper	2	Das *et al.*, 1965
Scuderia	Katydid	2	Bloch, 1969
Steatococcus	Coccid	$2^{\text{an.}-}$	Moses, private communication
Arachinida			
Lycosa carolinensis	Wolf spider	$1^{\text{an.}-}$	Bloch, 1969
Mollusca			
Amphineura			
Chiton olivaceus	Chiton	3	Subirana *et al.*, 1973
Cryptochiton stellerii	Chiton	3	
Gastropoda			
Crepidula fornicata	Slipper	2	Bloch, 1969
C. plana	Slipper	1	
Gibbula divaricata		3	Subirana *et al.*, 1973
Haliotus tuberculata		3	
Helix aspersa	Land snail	$1^{\text{an.}-}$	Bloch and Hew, 1960*a,b*
Nucella lapillus	Whelk	1 or 2	Walker and McGregor, 1968
Patella coerulea	Limpet	3	Hultin and Herne, 1949
P. vulgata	Limpet	3	
Pelecypoda			
Mytilus californicus	Mussel	3	Bloch, 1966
M. edulis	Mussel	3	Subirana *et al.*, 1973
Spisula solidissima	Surf clam	1?	Bloch, 1969
Cephalopoda			
Eledone cirrosa		4?	Subirana *et al.*, 1973
Loligo pealii	Squid	3	Hamer, 1955
L. opalescens	Squid	$1^{\text{an.}-}$	Bloch, 1962
Octopus vulgaris	Octopus	3	Subirana *et al.*, 1973

TABLE 1. Continued

Phylum, class, and species	*Common name*	*Protein type*	*Reference*
Chordata (Vertebrates)			
Elasmobranches			
Centrophorus granulosus	Shark	3	Kossel, 1928
Dasyatis sabina	Ray	3	Bloch, 1969
Hydrolagus colliei	Ratfish	2	
Raja rhina	Longnose skate	2	Bols and Kasinsky, 1974
Squalus acanthias	Spiny dogfish	2	
Teleosts (Acipenseriformes)			
Acipenser guldenstadii	Sturgeon	3	
A. huso	Sturgeon	3	Lisitzuin and Aleksandrovskaya, 1936
A. stellatus	Sturgeon	3	
A. sturio	Sturgeon	3	Felix, 1952
Teleosts (Batrachiformes)			
Opsanus beta	Toadfish	4	Bloch, 1969
Teleosts (Beloniformes)			
Cypselurus agoo	Flying fish	3	Yamakawa *et al.*, 1923[c]
Teleosts (Clupeiformes)			
Amblygaster immaculatus	Sardine	3	Yamakawa and Ibuka, 1926[c]
Clupeus harengus	Herring	1	Felix, 1952
C. palaseii	Herring	1	Ando *et al.*, 1953
Coregonus albus	Whitefish	1	Kossel, 1928
C. lavaretus		1	Waldschmidt-Leitz and Gutermann, 1961
C. macrophthalamus	Whitefish	1	Kossel, 1928
Gymnosarda vagaus		3	Yamakawa and Nokata, 1926[c]
Katsuwonis altivelis	Salmon	1	Inafuku, 1951

Oncorhynchus kisutch	Coho salmon	1	Dixon and Smith, 1968
O. nerka	Sockeye salmon	3	Yamakawa and Nokata, 1926[c]
O. tschawytscha	Salmon	1	Alfert, 1956; Callanan *et al.*, 1957
O. keta	Salmon	1	Ando *et al.*, 1957
Plecoglossus pelanus	Salmon	1	Inafuku, 1951
Salmo fontinalis	Trout	1	Felix, 1952
S. irrideus	Trout	1	
S. lacustris	Sea trout	3	Klezkowski, 1946[c]
S. salar	Trout	1	Felix, 1952
S. truto	Trout	1	
Salvelinus alpinus	Sable	1	Waldschmidt-Leitz and Gutermann, 1961
S. fontinalis	Canadian char	1	
S. namaycush	Trout	1	Kossel, 1928
Sardinia coerulea	Sardine	3	Dunn, 1926; Hirohata, 1929
Thynnus alalonga		1	Kossel and Staudt, 1926
T. thynnus	Tuna	1	Kossel, 1928
Truta fario	Trout	1	Kossel, 1928
Teleosts (Cypriniformes)			
Barbus fluviatilis	Barbel	3	Kossel and Schenck, 1928
Carrassius auratus	Goldfish	4	Zirkin, 1971
Cyprinus carpio	Carp	4	Kossel, 1928
Leuciscus rutilus	Barbel	3	Kossel and Staudt, 1926
Tinca tinca	Tench	4	Vendrely and Vendrely, 1966
Teleosts (Gadiformes)			
Gaddus marrhua	Cod	3	Kossel, 1928
Lota vulgaris	Burbot	3	
Teleosts (Cyprinodontiformes)			
Lebistes reticulates	Guppy	$1^{an.-}$	Bloch, 1969
Xiphophorus helleri	Swordtail	$1^{an.-}$	
Teleosts (Perciformes)			
Crenilabrus pavo	Wrasse	3	Kossel, 1928

TABLE 1. Continued

Phylum, class, and species	*Common name*	*Protein type*	*Reference*
Cyclopterus lumpus	Lumpsucker	1	Morkowin, 1899
Esox lucius	Pike	3	Vendrely and Vendrely, 1966
Lateolabrax japonicus		3	Yamakawa *et al.*, 1916[c]
Luciperca sandra	Sander	3	Lisitzuin and Aleksandrovskaya, 1933
Lutianus vitta		1?	Yamakawa *et al.*, 1923[c]
Mugil cephalus	Mullet	1	Hirohata, 1929
M. japonicus	Mullet	1	
Pelamys sarda	Spanish mackerel	1	Kossel, 1913
Perca flavescens	Perch	3	Kossel, 1928
Sagenichthys ancylodon	Croaker	1	
Sciaena schlegeli		3	Yamakawa *et al.*, 1916[c]
Scomber japonicus	Japanese mackerel	1?	Yamakawa and Yoshimoto, 1926[c]
S. scomber	Mackerel	1	Kuroda, 1951
Scombremorus niphonius	Spanish mackerel	3	Yamakawa *et al.*, 1916[c]
Scombropus boops		3	
Seriola aureovittata	Yellowtail	3	
Stenotomus chrysops	Porgie	4	Bloch, 1969
Stereolepis ishinaga	Sea bass	3	Yamakawa *et al.*, 1916[c]
Stizostedion vitreum	Perch	3	Kossel, 1913
Xiphias gladius	Swordfish	1	
Teleosts (tetraodontiformes)			
Spheroides rubripes	Puffer	1	Kuroda, 1951
Spheroides pardalis		3	Yamakawa *et al.*, 1913[c]
Amphibia			
Amphiuma means	Congo eel	1	Bloch, 1969
Bufo americanus	Toad	1 or 3	Bols and Kasinsky, 1973
B. boreus	Toad	4?	Bols and Kasinsky, 1972

B. vulgaris	Toad	3	Bloch, 1969
Hyla regilla	Frog	3	
H. versicolor	Frog	3	Bols and Kasinsky, 1972
Rana palustris	Frog	4	
R. pipiens	Frog	4	Zirkin and Wolfe, 1970; Bols and Kasinsky, 1972
R. pretiosa	Frog	4	Bols and Kasinsky, 1972
Pleurodeles waltheii	Salamander	1	Picheral, 1970
Triturus viridescens	Newt	2[is.]	Bloch, 1969
Xenopus laevis	South African clawed frog	3	Bols and Kasinsky, 1973
Reptilia			
Holbrookia texana	Lizard	1[an.–]	Bloch, 1969
Natrix natrix	Snake	1	Sud, 1961
Aves			
Gallus domesticus	Domestic fowl	1	Fischer and Kreuzer, 1953; Nakano *et al.*, 1973
Mammalia (metatheria)			
Didelphus virginiana	Opossum	1	Bloch, 1969
Smithopsis crassacaudata	Marsupial rat	1	
Mammalia (eutheria)			
Bos bovis	Bull	2	Hendricks and Mayer, 1965
Canis familiaria	Dog	2	Dallam and Thomas, 1953
Cavia esperia	Guinea pig	2	Alfert, 1958
Homo sapiens	Man	2	Dallam and Thomas, 1953
Mus musculus	Mouse	2,3	Lam and Bruce, 1971
Ovis aries	Ram	2	Dallam and Thomas, 1953
Rattus norvegicus	Rat	2	Vaughn, 1966
Sus scrofa	Boar	2	Henricks and Mayer, 1965
Plants			
Nitella	Stonewort	4	Bloch, 1969
Sphaerocarpus texana	Liverwort	1	

TABLE 1. Continued

Phylum, class, and species	*Common name*	*Protein type*	*Reference*
Lycopodium	Club moss	1 (in spores)	D'Alcontres, 1953
Allium cepa	Onion	4 (in pollen)	Rasch and Woodard, 1959
Ginkgo biloba	Ginkgo	4	Bloch, 1969
Hippiastrum belladonna	Amaryllis	4 (in pollen)	
Lilium henryi	Lily	4 (in pollen)	Rasch and Woodard, 1959
Rhoeo discolor			
Tradescantia paludosa		4 (in pollen)	
Vicia faba	Broad bean	4 (in pollen)	
Zea mays	Corn	4 (in pollen)	

[a] In the column "Protein type," 1 indicates *Salmo* type; 2, mouse or *Chortophaga* type; 3, *Mytilus*; 4, *Rana*, and 5, crab type. The superscript "an." indicates that the sperm head is known to be anisotropic or that evidence of chromatin orientation has been obtained by electron microscopy. The superscript "is." shows known isotropy, or amorphous structure as seen with the electron microscope. The subscript "c" denotes the presence of cytoplasmic basic proteins in the sperm.
[b] The echinoderm sperm histones are usually classed as intermediate. However, for most it is not known to what extent the difference in amino acid composition may be due to quantitative differences in the proportions of typical histones, rather than to qualitative differences in the histones themselves.
[c] Original reference cited in Ando *et al.* (1973). In this the authors refer to mono-, di- and triprotamines, containing arginine, arginine plus either lysine or histidine, and all three basic amino acids. The di- and triprotamines are designated type 3 in the table presented here.

by Black and Dixon (1967) as an ancestral one that gave rise to YI, then YII, through a series of mutations. Fitch (1971) pointed out that Z might be derived more simply by unequal crossover between genes YI and YII. Whether Z is ancestral, a crossover product, or neither remains a matter or speculation.

Note that the sequences shown in Table 3 permit comparison of intergeneric differences (*Oncorhyncus, Salmo, Clupeus, Thynnus*) and developmental differences (the several molecules obtained from each of the species).

Heterogeneity of Sperm Histones

The heterogeneity of the sperm basic proteins in a given species has been noted on several occasions. The staining changes undergone by differentiating spermatids indicate changes in overall chromatin composition during development (Bloch, 1969). The extraction of highly basic "intermediate"-type histones from the immature spermatids of *Loligo opalescens,* whose mature stages contain simple monoprotamines (Bloch, 1962), and the sequential incorporation of labeled amino acids into proteins of differing electrophoretic mobility from mussel testes (Bloch, 1966) both suggest that heterogeneity of the testes proteins reflects sequential replacement by a succession of histones as sperm development proceeds. It would be interesting to know whether the three clupeine molecules occur in the same sperm cells or whether each is derived from cells in different stages of development.

Variations in the patterns of changes that occur among different organisms (somatic → intermediate → protamine in squid and salmon, somatic → intermediate in mussel, no change in frog) suggest that clues to the evolutionary trend might be sought in detailed analyses of the developmental changes leading to protamine in the organisms where the change is this extreme. The similarities between the "evolutionarily intermediate" proteins of the mature sperm of mussel and the "developmentally intermediate" proteins of the spermatids of *Loligo* and salmon, both of whose sperm eventually go the whole route, suggest a relationship between ontogeny and phylogeny. Clearly more comparative studies such as those of Ando and his collaborators are needed to unravel the mysteries of the evolution of these proteins and also their developmental roles.

The Synthesis of Protamines

Dixon and his coworkers have used the rainbow trout (*Salmo gairdnerii,* or *S. irideus*) for an extensive analysis of protamine (iridine)

TABLE 2. Amino Acid Compositions

	Annelida		**Arthropoda**	**Echinodermata**				
	Urechis caupo, acrosome (*Das et al., 1967*)	*Urechis caupo sperm head (ibid)* (*Das et al., 1967*)	*Libinia emarginata* (*Vaughn and Hinsch, 1972*)	*Arbacia glacialis* (*Vendrely and Vendrely, 1966*)	*Arbacia lixula* $\phi 1$ (*Palau et al., 1969*)	*same* $\phi 2b$	*same* $\phi 2a2$	*same* $\phi 2a$
Ala	4.1	11.4	5.9	14.5	23.3	13.1	14.1	8.2
NH_3								
Arg	21.3	36.9	2.8	7.9	11.7	10.4	9.3	13.0
Asp	4.9	0.9	11.6	7.8	2.9	4.9	7.1	6.0
Cys	0.4							
Glu	4.1	0.8	15.3	12.4	2.6	6.3	8.5	7.7
Gly	15.2	2.7	6.1	6.2	4.9	7.2	11.4	14.9
His	2.1	0.1	1.4	2.0	1.2	1.8	1.9	2.4
Ile	2.1	0.3	3.4	1.4	3.4	3.3	3.9	5.4
Leu	3.1	0.4	4.7	8.1	1.9	4.6	11.2	8.8
Lys	23.2	19.2	8.1	10.6	24.6	15.5	11.3	10.3
Met			0.5	1.5				
Phe	1.4	0.2	3.5	3.1	0.6	1.4	1.9	2.3
Pro	2.2	0.9	6.1	6.0	8.0	6.0	3.9	1.6
Ser	7.4	22.4	6.6	4.4	7.4	10.9	4.2	2.8
Thr	2.7	1.6	11.6	5.7	2.9	5.8	3.1	5.7
Trypt								
Tyr	1.1	0.2	1.1	3.0	1.1	2.6	2.5	3.4
Val	4.8	2.0	6.5	5.6	3.3	5.7	5.6	7.3

[a] Protein from immature sperm cells or from elsewhere in the testes than from known mature sperm cells.

[b] Data given as number of residues per molecule, otherwise as mole %.

[c] Values in parentheses include more than one amino acid, e.g. (Ile, Leu, Val) and (Asp, Glu).

of Basic Proteins Extracted from Sperm

same φ3	*A. punctulata* α (*Paoletti and Huang, 1969*)	*same* β	*same* γ	*same* δ	*same* ε	*Asteria forbesi* (*Hamer, 1955*)	*Asteria tenuispina* (*Subirana and Palau, 1968*)	*Astrospectin aurantiacus* (*Subirana and Palau, 1968*)
13.7	13.1	12.4	19.3	11.8	13.6	15.1	14.8	14.5
						5.9		
11.5	14.4	15.3	10.3	11.6	10.7	9.1	10.0	8.0
5.1	7.6	4.6	2.8	5.7	5.3	4.4	(15.6)	(17.0)
	tr	1.5	tr	2.0	1.0			
11.8	11.8	11.1	3.1	8.4	6.4	6.1	( )	( )
7.9	8.9	12.3	7.1	10.1	11.8	6.7	8.4	8.5
1.4	tr	tr	0.3	1.2	1.7	1.5	0.6	1.5
4.4	3.4	3.8	3.0	3.6	3.0		(18.5)	(15.8)
10.0	7.2	8.4	1.9	7.5	4.7	8.8	( )	( )
9.3	10.6	6.7	27.5	11.9	14.3	14.4	13.0	14.3
	5.1	6.0	1.5	1.3	4.1	2.1		
2.5	tr	0.7	0.6	1.9	0.9	1.4	1.4	1.8
4.6	tr	2.0	8.4	4.1	4.0	6.4	5.1	5.7
4.7	7.2	6.7	5.7	6.4	7.4	5.8	5.3	5.8
5.6	4.6	5.8	2.4	4.9	4.2	4.7	5.3	5.8
2.1	tr	0.7	0.7	1.4	0.9	1.7	1.9	1.9
5.4	5.9	4.0	3.3	6.3	5.5	5.9	( )	( )

TABLE 2.

	Brissopsis lyrifera (*Hultin and Herne, 1949*)	*Echinarachnius parma* (*Hamer, 1955*)	*Echinaster sepositus* (*Subirana and Palau, 1968*)	*Echinocardium cordatum* (*Hultin and Herne, 1949*)	*Echinocyamus pusillus* (*Vendrely and Vendrely, 1966*)	*Holothuria polii* (*Subirana and Palau, 1968*)	*Holothuria tubulosa f1* (*Subirana and Palau, 1968*)	*same f2*
Ala	17.3	15.0	12.2	20.8	13.7	17.5	23.4	14.2
NH_3		5.9						
Arg	9.5	9.1	8.2	11.8	10.5	10.8	2.2	8.1
Asp		4.4	(15.3)		6.0	(13.2)	4.8	5.7
Cys							0.0	0.0
Glu		6.1	( )		8.9	( )	7.8	8.2
Gly	17.7	6.7	8.0	17.7	7.8	6.9	5.3	8.2
His	0.8	1.5	1.5	0.6	1.7	0.3	0.2	1.9
Ile	3.2		(19.1)	3.0	4.2	(17.7)	2.9	4.9
Leu	4.0	8.8	( )	2.1	7.2	( )	2.9	6.6
Lys	17.7	14.4	13.5	21.6	12.1	14.5	30.3	14.4
Met		2.1			1.3		0.2	1.6
Phe		1.4	2.5		2.4	1.4	0.7	1.9
Pro	6.7	6.5	5.0	8.5	3.2	4.8	5.2	4.2
Ser	15.6	5.8	6.1	7.1	5.6	5.6	4.0	6.1
Thr	3.7	4.7	5.9	3.5	5.1	5.6	4.6	5.8
Trypt								
Tyr		1.7	2.7		2.7	1.4	0.4	2.8
Val	3.7	5.9	( )	3.5	7.4	( )	5.2	5.3

[a] Protein from immature sperm cells or from elsewhere in the testes than from known mature sperm cells.

[b] Data given as number of residues per molecule, otherwise as mole %.

[c] Values in parentheses include more than one amino acid, e.g. (Ile, Leu, Val) and (Asp, Glu).

Continued

same f2a	*same f3*	*Ophiotrix fragilis* (Subirana and Palau, 1968)	*Paracentrotus lividus* (Vendrely and Vendrely, 1966)	*Patella coerulea* (Hultin and Herne, 1949)	*P. vulgata* (Hultin and Herne, 1949)	*Sphaerechinus granularis* (Subirana and Palau, 1968)	*Strongylocentrotus nudus* (Gineitis et al., 1970)	*Strongylocentrotus drobachiensis* (Gineitis et al., 1970)
11.4	13.3	12.8	13.9	5.6	12.4	15.0	9.16	11.63
11.5	10.3	11.6	11.1	8.6	25.0	13.5	12.95	12.83
6.0	5.7	(15.7)	7.7			(12.6)	4.8	4.96
0.3	0.1							
8.3	10.8	( )	10.2			( )	7.44	7.67
12.9	9.5	8.3	7.2	9.4	9.5	8.6	9.45	9.5
2.8	2.0	1.5	1.9	0.0	0.6	0.9	1.57	1.54
5.1	5.6	(17.0)	3.9	0.6	1.2	(16.8)	4.8	4.56
9.8	9.9	( )	7.7	3.8	3.5	( )	6.58	6.85
10.7	9.3	11.4	10.8	8.9	19.9	12.0	13.1	13.4
0.4	0.4		1.4					0.45
1.9	2.5	2.3	2.7			1.1	2.65	2.57
2.6	3.0	4.6		4.0	9.7	4.7	3.44	4.37
2.6	4.0	6.7	6.2	13.3	9.3	7.0	5.58	5.65
4.3	5.1	5.4	5.4	2.2	4.3	5.7	5.0	5.34
3.0	2.8	2.2	3.0			1.7	1.80	1.94
6.4	5.3	( )	6.6	3.8	4.5	( )	12.38	6.6

TABLE 2.

	Strongylocentrotus intermedius (*Gineitis et al.*, 1970)	**Mollusca** *Chiton olivaceus* (*Subirana et al.*, 1973)	*Cryptochiton stellerii* (*Subirana et al.*, 1973)	*Eledone cirrosa* (*Subirana et al.*, 1973)	*Gibbula divericata* (*Subirana et al.*, 1973)	*Haliotis tuberculata* (*Subirana et al.*, 1973)	*Loligo opalescens* spermatid[a] (*Bloch*, 1962)	*same* sperm 1
Ala	13.39	8.5	10.5	5.1	9.6	11.2	6.0	0.7
NH_3								
Arg	14.96	37.8	33.5	4.7	56.3	44.4	47.5	70.3
Asp	3.58	2.0	3.5	5.3	0.3	0.8	3.1	0.7
Cys		0.0	0.0	12.8	0.0	0.0		
Glu	8.3	2.0	3.3	6.9	0.2	1.0	3.0	0.9
Gly	7.63	6.9	6.4	8.6	4.9	3.4	5.3	2.2
His	1.57	0.3	0.8	8.1	tr	0.1	0.9	1.6
Ile	3.9	0.9	1.7	2.8	0.1	0.5	1.0	0.2
Leu	5.41	2.1	3.2	3.6	tr	1.0	5.0	1.2
Lys	16.66	15.4	13.5	5.4	5.8	11.8	8.5	4.3
Met		0.3	0.3	1.4	tr	0.1	1.3	0.1
Phe	2.01	0.6	0.6	1.3	tr	0.2		
Pro	5.03	4.8	4.4	20.1	0.0	1.4	2.5	1.9
Ser	5.91	10.5	6.5	4.7	17.1	16.2	6.5	8.8
Thr	4.46	3.5	4.5	4.1	2.0	2.5	3.1	1.7
Trypt								
Tyr	1.07	0.9	1.2	1.6	tr	0.4	3.4	4.8
Val	5.6	3.4	6.1	3.6	3.5	4.6	2.8	0.4

[a] Protein from immature sperm cells or from elsewhere in the testes than from known mature sperm cells.

[b] Data given as number of residues per molecule, otherwise as mole %.

[c] Values in parentheses include more than one amino acid, e.g. (Ile, Leu, Val) and (Asp, Glu).

Continued

same sperm 2	*Loligo pealeii* ("ext. 2") (Subirana *et al.*, 1973)	*Loligo pealeii* ("ext. 3") (Subirana *et al.*, 1973)	*Mytilus californicus* [b] α (Bloch 1966)	*same* γ	*same* δ	*Mytilus edulis*, whole testes histones (Subirana *et al.*, 1973)	*Mytilus edulis* fraction I	*same* fraction 3
0.3	0.0	tr	12.3	12.9	12.5	12.8	13.7	18.1
77.5	75.2	77.1	8.3	29.4	7.9	19.3	30.0	6.0
0.4	0.6	0.3	5.8	0.6	4.7	2.8	0.3	2.3
	0.0	0.0	0.5	0.4		tr,	0.0	0.0
0.5	0.7	0.1	3.5	0.6	4.3	2.8	0.2	1.6
0.4	1.6	0.5	7.2	6.6	5.9	6.2	6.3	2.1
1.5	0.1	tr	.9	0.1	0.9	0.6	tr	0.4
0.2	1.2	tr	3.5	0.3	2.5	1.5	0.1	0.6
0.2	1.7	tr	5.0	0.4	3.8	2.4	0.2	0.8
1.8	0.8	1.0	22.0	23.1	29.0	21.3	24.3	42.9
	0.0	0.0	0.8		0.7	0.5	0.0	0.2
	0.2	0.0	1.4	0.3	1.6	0.7	0.0	0.3
2.1	2.0	2.7	6.8	4.7	6.8	5.7	3.9	9.6
9.6	8.2	9.7	12.7	16.2	9.7	15.0	15.9	10.5
1.4	0.3	0.1	5.1	2.8	4.1	4.7	4.0	2.6
4.2	7.0	8.4	0.8	0.2	1.7	0.6	0.0	0.0
	0.5	tr	4.0	1.2	4.0	2.6	0.9	1.9

TABLE 2.

	Mytilus edulis fraction 2a (Subirana et al., 1973)	same fraction 2b	*Octopus vulgaris* (Subirana et al., 1973)	*Patella vulgata* (Subirana et al., 1973)	*Spisula solidissima* (Subirana et al., 1973)	**Chordata, Teleosts** *Acipenser stellatus* (Kaverzneva and Rakhmatulina, 1970)	*Acipenser sturio*[b] (Felix, 1952, 1955)	*Clupeus harengus* YI[b] (Ando and Suzuki, 1967)
Ala	10.1	10.7	1.8	6.9	12.1	2.35	5	2
NH_3								
Arg	11.6	9.2	54.2	43.0	25.5	60.8	35	20
Asp	6.6	7.2	1.8	1.7	1.6	0.3		
Cys	tr	0.0	0.0	0.0	0.1			
Glu	8.2	8.1	1.1	0.7	0.9	1.62	1	
Gly	12.5	7.0	15.7	6.9	4.1	3.21	2	1
His	1.8	2.2	3.1	0.0	0.3	9.6	7	
Ile	5.0	4.3	0.7	0.7	0.8		2	1
Leu	9.4	6.3	0.7	3.9	1.9	1.96		
Lys	11.1	12.9	11.7	9.9	28.8	13.2	9	
Met	0.8	1.5	0.0	tr	1.2			
Phe	2.1	1.7	1.0	0.6	0.4			
Pro	2.6	4.1	0.0	2.2	2.2	0.73		2
Ser	3.4	7.7	5.6	14.7	13.8	3.9	3	3
Thr	4.5	6.1	0.3	2.5	3.3	2.26	1	2
Trypt								
Tyr	3.3	3.4	1.0	0.6	0.2			
Val	6.7	6.2	1.2	5.5	2.9			

[a] Protein from immature sperm cells or from elsewhere in the testes than from known mature sperm cells.
[b] Data given as number of residues per molecule, otherwise as mole %.
[c] Values in parentheses include more than one amino acid, e.g. (Ile, Leu, Val) and (Asp, Glu).

Continued

same YII[b]	*same* Z[b]	*Clupeus palasii* (Yamashima, 1969, in Ando et al., 1973)	*Coregonus lavaretus* (Waldschmidt-Leitz and Gutermann, 1961)	*Cyprineus carpio* (Vendrely and Vendrely, 1966)	*Esox lucius* (Vendrely and Vendrely, 1966)	*Mugil japonicus* (Ota et al., 1966, cited in Ando et al., 1973)	*Oncorhyncus keta* AI[b] (Ando and Watanabe, 1969)	*Salmo fontinalis*[b] (Felix, 1952, 1955)
2	2	7.4	1.0	12.0	6.7	6.1		2
			1.8					
20	21	65.5	65.0	7.8	44.0	63.3	21	50
				5.4	1.5	0.9		
				9.1	2.2	3.2		
		1.3	6.9	7.5	7.3		2	2
				0.9				
		0.8		4.6	0.7	3.8		1
				8.6	1.7			
				13.4	6.5			2
				0.9	1.5	4.1		
				2.5	0:5			
3	2	7.3	10.4	6.7	10.5	9.4	3	5
2	3	7.9	8.4	6.0	10.5	3.1	4	3
1		3.4	1.0	6.5	1.4	3.0		
				3.0	0.5			
2	2	4.4	6.1	5.7	4.5	2.7	2	5

TABLE 2.

	Salmo irideus Ia[b] *(Ando and Watanabe, 1969)*	*same Ib*[b]	*same II*[b]	*Salmo salar*[b] *(Felix, 1952, 1955)*	*Salmo truta*[b] *(Felix, 1952, 1955)*	*Salvelinus alpinus (Waldschmidt-Leitz and Gutermann, 1961)*	*Salvelinus fontinalis (Waldschmidt-Leitz and Gutermann, 1961)*	*Scomber scomber (Kuroda, 1951)*
Ala			1	2	2	0.9	0.9	10.8
NH_3						2.7	1.8	
Arg	22	22	21	55	50	66.7	67.5	64.3
Asp								
Cys								
Glu								
Gly	2	2	2	3	2	5.9	6.3	
His								
Ile								
Leu		1						1.8
Lys								
Met								
Phe								
Pro	3	3	2	5	5	8.9	9.0	10.8
Ser	4	4	4	5	3	9.2	9.3	5.8
Thr								2.6
Trypt								
Tyr								
Val	2	1	2	4	5	5.9	6.3	4.7

[a] Protein from immature sperm cells or from elsewhere in the testes than from known mature sperm cells.
[b] Data given as number of residues per molecule, otherwise as mole %.
[c] Values in parentheses include more than one amino acid, e.g. (Ile, Leu, Val) and (Asp, Glu).

Continued

Spheroides rubripes (Kuroda, 1951)	*Thynnus thynnus* YI[b] (Bretzel, 1973)	*same* Y2	*same* ZI	*same* Z2	**Aves** *Gallus domesticus*[b] (Fischer and Kreuzer, 1953)	*Gallus domesticus* GI (Nakano *et al.*, 1973)	*same* GII	*same* GIII
14.5	3	3	1	2	5		1.9	3.1
59.4	21	21	22	22	42+1	14±1	64.3	63.8
	1	1			1		0.5	
					1	3	9.8	7.4
2.5					1			
3.1	2	2	2	2	5		2.4	4.4
14.5	2	2	3	3	5	3	11.3	14.8
2.8	1	1	1	1	2		0.8	1.1
	1	1	1	1		2	6.3	4.1
3.3	3	3	4	3	3	1	2.8	1.1

TABLE 2. *Continued*

	Gallus domesticus GIV (Nakano et al., 1973)	same GV	same GVI	same GVII	same GVIII	**Mammalia** *Bos taurus*[b] (Coelingh et al., 1972)	*Mus musculus*[b] (Lam and Bruce, 1971)	*Rattus rattus* testis histone (Kistler et al., 1973)	*Rattus rattus* sperm histone (Kistler et al., 1973)	*Sus scrofa* (Henricks and Mayer, 1965)
Ala	2.1	2	2	3.0	4.0	1	4	3.6	1.5	5.4
NH_3										7.5
Arg	65.7	25±1	29±1	54.4	59.0	22	10	20.4	65.0	22.7
Asp							2	7.4	0	5.0
Cys						5		0	9.4	8.1
Glu	0.6						2	0	0	5.8
Gly	5.8	3	3	11.1	8.3	2	5	11.1	tr	4.6
His						1	2	5.4	tr	2.1
Ile								0	0	3.4
Leu						1	1	5.5	0	4.4
Lys							10	19.5	3.4	3.3
Met						1		1.2	0	1.0
Phe						1		0	2.3	1.5
Pro	4.9	2	2	4.5	3.6			2.9	0	4.9
Ser	16.1	8	9	17.5	15.1	2	4	14.4	9.5	5.8
Thr	1.6	1	1	1.0	1.9	1	2	3.7	2.2	5.8
Trypt								0	0	
Tyr	3.1	2	2	6.0	6.6	1		2.9	5.1	3.9
Val	0.3			2.6	1.6	1		1.7	0	4.7

[a] Protein from immature sperm cells or from elsewhere in the testes than from known mature sperm cells.
[b] Data given as number of residues per molecule, otherwise as mole %.
[c] Values in parentheses include more than one amino acid, e.g. (Ile, Leu, Val) and (Asp, Glu).

synthesis during spermiogenesis. Their findings, in brief, are these. Protamines are synthesized during a late spermatid stage (Ling *et al.*, 1969) as indicated by a shift in the ratio of arginine to lysine incorporation from less than 5 to 30–150. The synthesis occurs in the cytoplasm on disomes (Ling *et al.*, 1969), a class of short polysomes that becomes prevalent at this stage (Ling and Dixon, 1970). Synthesis uses a stable message, as indicated by lack of immediate effect of actinomycin D on protamine synthesis. The newly synthesized protamine becomes phosphorylated in the cytoplasm, in contrast to the histones, which are phosphorylated while in the nucleus long after their synthesis (Marushige *et al.*, 1970). Iridine phosphorylation entails a kinase (Jergil and Dixon, 1970), involves cyclic AMP, and occurs at one, two, three, or all four of the serines (Sanders and Dixon, 1972). The phosphorylated protamines enter the nucleus, combine with chromatin as phosphorylated proteins, and later become dephosphorylated (Marushige and Dixon, 1971). Chromatin can be isolated during these stages and separated into histone chromatin and protamine chromatin, taking advantage of the insolubility of the latter. The protamine chromatin contains some histone remnants, indicating that replacement is accompanied by breakdown of the histone. In physical studies on isolated chromatin, the lysine-rich histone, which incidentally is the most loosely bound histone, is the last to be replaced by protamine (Marushige and Dixon, 1969). Also of interest is the fact that replacement of protamine occurs after the largest decline in RNA synthesis, so that this inactivity seems not to be a consequence of replacement by protamine. The protamines themselves are heterogeneous, consisting of three closely related but dissimilar proteins (Ling *et al.*, 1971). Three of these have been well characterized by Ando and Wantanabe (1969) (see table 3). The prevalence of the three changes with testes maturation, the later ones either replacing the earlier, or perhaps comprising the sperm cells that mature later in the season (Ling *et al.*, 1971). Ling and Dixon (1970) offer an interesting suggestion to explain the stability of protamine message. They propose that a high GCX RNA (GCX being a triplet coding for arginine, which makes up 70% of the protamine) could form extensive helical structures by base pairing. Such RNA might be resistant to degradation by the cellular RNAses.

Mammalian Sperm Histones

Lam and Bruce's findings (1971) that a much simplified histone devoid of sulfur-containing amino acids can be extracted from the mouse,

TABLE 3. Amino Acid Sequences of Some Fish

	Sequence
Clupeine (Ando and Suzuki, 1967)	
YI:	Ala-Arg-Arg-Arg-Arg-Ser————Ser-Ser-Arg-Pro-Ile-Arg-Arg-Arg-Arg—
Z:	Ala-Arg-Arg-Arg-Arg-Ser-Arg-Arg-Ala-Ser-Arg-Pro-Val-Arg-Arg-Arg-Arg—
YII:	Pro-Arg-Arg-Arg————Thr-Arg-Arg-Ala-Ser-Arg-Pro-Val-Arg-Arg-Arg-Arg—
Salmine (Ando and Watanabe, 1969)	
AI:	Pro-Arg-Arg-Arg-Arg————Ser-Ser-Ser-Arg-Pro-Val-Arg-Arg-Arg-Arg-A
Iridine (Ando and Watanabe, 1969)	
Ia:	Pro-Arg-Arg-Arg-Arg————Ser-Ser-Ser-Arg-Pro-Val-Arg-Arg-Arg-Arg-A
Ib:	Pro-Arg-Arg-Arg-Arg-Arg-Arg-Ser-Ser-Ser-Arg-Pro-Ile-Arg-Arg-Arg-Arg—
II:	Pro-Arg-Arg-Arg-Arg————Ser-Ser-Ser-Arg-Pro-Val-Arg-Arg-Arg-Arg—
Thinnine (Bretzel, 1973)	
Y1:	Pro-Arg-Arg-Arg-Arg————Glu-Ala-Ser-Arg-Pro-Val-Arg-Arg-Arg-Arg-A
Y2:	Pro-Arg-Arg-Arg-Arg————Gln-Ala-Ser-Arg-Pro-Val-Arg-Arg-Arg-Arg-A
Z1:	Pro-Arg-Arg-Arg-Arg————Arg-Ser-Ser-Arg-Pro-Val-Arg-Arg-Arg-Arg-A
Z2:	Pro-Arg-Arg-Arg-Arg————Arg-Ser-Ser-Arg-Pro-Val-Arg-Arg-Arg-Arg-A
Bull sperm (Coelingh *et al.*, 1972)	
	Ala-Arg-Tyr-Arg-Cys-Cys-Leu-Thr-His-Ser-Gly-Ser-Arg-Cys-Arg-Arg-Arg-Arg-A

[a] The spacing between some of the amino acids in the diagram are to facilitate comparison of the various fish proteins. The spaces span positions that may represent additions or deletions occurring

under conditions that loosen disulfide linkages, indicates that the apparent "keratin" nature of mammalian histones may be the result of physical entrapment of a more typical sperm histone in a complex held together by disulfide linkages. In the mouse the sulfur is undoubtedly contributed by another protein. Whether such sulfur-containing histones exist in addition to the histone extracted by Lam and Bruce and whether a similar situation holds for other mammals remain to be seen. The recent findings of Coelingh *et al.* (1972) show that the basic protein of bull sperm is very rich in cysteine residues. The sequence is shown in Table 3.

Cytochemical analysis of the sperm of *Smithopsis,* a ratlike marsupial, indicated the presence of a protamine (see Table 1). The opossum had previously been shown to contain a protamine. It should be interesting to learn whether the marsupials will provide a consistent exception to the mammalian situation.

Protamines and A Sperm Protein from Bull[a]

```
Pro-Arg-Arg-Arg——————Thr-Thr-Arg-Arg-Arg-Arg——————Ala-Gly-Arg-Arg-Arg-Arg
Pro-Arg-Arg——————————Val-Ser-Arg-Arg-Arg-Arg——————Ala——————Arg-Arg-Arg-Arg
Pro-Arg-Arg——————————Val-Ser-Arg-Arg-Arg-Arg——————Ala——————Arg-Arg-Arg-Arg-

Pro-Arg——————————————Val-Ser-Arg-Arg-Arg-Arg-Arg-Arg-Gly-Gly-Arg-Arg-Arg-Arg

Pro-Arg-Arg——————————Val-Ser-Arg-Arg-Arg-Arg-Arg-Arg-Gly-Gly-Arg-Arg-Arg-Arg
Pro-Arg-Arg——————————Val-Ser-Arg-Arg-Arg-Arg-Arg——————Gly-Gly-Arg-Arg-Arg-Arg
Ala-Arg-Arg——————————Val-Ser-Arg-Arg-Arg-Arg-Arg-Arg-Gly-Gly-Arg-Arg-Arg-Arg

Tyr-Arg-Arg-Ser-Thr-Ala-Ala-Arg-Arg-Arg-Arg-Arg——————Val-Val-Arg-Arg-Arg-Arg-
Tyr-Arg-Arg-Ser-Thr-Ala-Ala-Arg-Arg-Arg-Arg-Arg——————Val-Val-Arg-Arg-Arg-Arg
Tyr-Arg-Arg-Ser-Thr-Val-Ala-Arg-Arg-Arg-Arg-Arg——————Val-Val-Arg-Arg-Arg-Arg
Tyr-Arg-Arg-Ser-Thr-Ala-Ala-Arg-Arg-Arg-Arg-Arg——————Val-Val-Arg-Arg-Arg-Arg

Arg-Arg-Cys-Arg-Arg-Arg-Arg-Arg-Arg-Phe-Gly-Arg-Arg-Arg-Arg-Arg-Arg-Val-Cys
```

during evolution. The amino acids in boxes show developmental variants within a species. These might have arisen in each case by a single base substitution in one of the members. The shaded areas indicate large blocks common to all 11 of the fish protamines.

Acknowledgments

This work was supported by a grant from the United States Public Health Service, No. GM 09654, and done under tenure of a USPHS Career Development Award.

Literature Cited

Alfert, M., 1956 Chemical differentiation of nuclear proteins during spermatogenesis in the salmon. *J. Biophys. Biochem. Cytol.* **2:**109–114.

Alfert, M., 1958 Cytochemische Untersuchung an basischen Kernproteinen warend der Gametenbildung, Befruchtung, und Entwicklung. *Ges. Physiol. Chem. Colloq.* **9:**73–84.

Ando, T. and K. Suzuki, 1967 The amino acid sequence of the third component of clupeine. *Biochim. Biophys. Acta* **10:**375–377.

Ando, T. and S. Watanabe, 1969 A new method for fractionation of protamines and the amino acid sequences of salmine and three components of iridine. *Int. J. Protein Res.* **1:**221–224.

Ando, T., K. Iwai, M. Yamasaki, C. Hashimoto, M. Kimura, S. Ishii, and T. Tamura, 1953 Further notes on protamines. *Bull. Chem. Soc. Jap.* **26:**406–407.

Ando, T., S. Ishii, M. Yamasaki, K. Iwai, C. Hashimoto, and F. Sawada, 1957 Studies of protamines. I. Amino acid composition and homogeneity of clupeine, salmine, and iridine. *J. Biochem.* **44:**275–288.

Ando, T., M. Yamasaki, and K. Suzuki, 1973 *Protamines—Isolation, Characterization, Structure and Function.* Springer-Verlag, Berlin.

Berlowitz, L., 1965 Analysis of histone *in situ* in developmentally inactivated chromatin. *Proc. Natl. Acad. Sci. USA* **54:**476–480.

Black, J. A. and G. H. Dixon, 1967 Evolution of protamine: A further example of partial gene duplication. *Nature* (*London*) **216:**152–154.

Bloch, D. P., 1962 Synthetic processes in the cell nucleus. I. Histone synthesis in non-replicating chromosomes. *J. Histochem. Cytochem.* **10:**137–144.

Bloch, D. P., 1966 Cytochemistry of the histones. *Protoplasmatologia* **5:**1–56.

Bloch, D. P., 1969 A catalog of sperm histones. *Genetics Suppl.* **63:**93–111.

Bloch, D. P. and S. D. Brack, 1964 Evidence for the cytoplasmic synthesis of nuclear histone during spermiogenesis in the grasshopper *Chortophaga viridifasciata* (De Geer). *J. Cell Biol.* **22:**327–340.

Bloch, D. P. and H. Y. C. Hew, 1960*a* Schedule of spermatogenesis in the pulmonate snail *Helix aspersa,* with special reference to histone transition. *J. Biophys. Biochem. Cytol.* **7:**515–532.

Bloch, D. P. and H. Y. C. Hew, 1960*b* Changes in nuclear histones during fertilization and early embryonic development in the pulmonate snail *Helix aspera. J. Biophys. Biochem. Cytol.* **8:**69–81.

Bols, N. C. and H. E. Kasinsky, 1972 Basic protein composition of anuran sperm: A cytochemical study. *Can. J. Zool.* **50:**171–177.

Bols, N. C. and H. E. Kasinsky, 1973 An electrophoretic comparison of histones in anuran testes. *Can. J. Zool.* **51:**203–208.

Bols, N. C. and H. E. Kasinsky, 1974 Cytochemistry of sperm histones in three cartilaginous fishes. *Can J. Zool.* **52:**437–439.

Bretzel, G., 1973 Über Thynnin, das Protamin des Thunfisches. Die Aminosaüre Sequenz von Thynnin. XIV. Mitteilungen über die Struktur der Protamine im der Untersuchungsreihe von E. Waldschmidt-Leitz und Mitarbeiten. *Hoppe-Seyler's Z. Physiol. Chem.* **354**–543–549.

Callanan, M. J., W. R. Carroll, and E. R. Mitchell, 1957 Physical and chemical properties of protamine from the sperm of salmon (*Oncorhyncus tschawytscha*). *J. Biol. Chem.* **229:**279–287.

Chevaillier, P., 1967 Mise en évidence et étude cytochimique d'une protéine basique extranucléaire dans les spermatozoides de crustaces decapodes. *J. Cell Biol.* **32:**547–556.

Coelingh, J. P., C. H. Monfoort, T. H. Rozijn, J. A. Gevers Leuven, R. Schophof, F. P. Steyn-Parve, G. Barunitzer, B. Schrank, and A. Ruhfus, 1972 The complete amino acid sequence of the basic nuclear protein of bull spermatozoa. *Biochim. Biophys. Acta* **285:**1–14.

D'Alcontres, G. S., 1953 Acerca del la presencia de protaminas en al polen. *Acta Cientia Venez.* **4:**23–24.

Dallam, R. D. and L. E. Thomas, 1953 Chemical studies on mammalian sperm. *Biochem. Biophys. Acta* **11:**79–89.

Das, C. C., H. Gay, and B. P. Kaufmann, 1964 Histone–protein transition in *Drosophila melanogaster.* I. Changes during spermatogenesis. *Exp. Cell Res.* **35:**507–514.

Das, N. K., E. P. Siegel, and M. Alfert, 1965 Synthetic activities during spermatogenesis in the locust. *J. Cell Biol.* **25:**387–395.

Das, N. K., J. Micou-Eastwood, and M. Alfert, 1967 Cytochemical and biochemical properties of basic proteins of *Urechis acrosomes. J. Cell Biol.* **35:**455–458.

Descamps, M., 1969 Étude cytochimique de la spermatogenèse chez *Lithobius forficatus* L. *Histochemie* **20:**46–57.

Dixon, G. H. and M. Smith, 1968 Nucleic acid and protamines in salmon testes. *Prog. Nucleic Acid Res. Mol. Biol.* **8:**9–34.

Dunn, M. S., 1926 Basic proteins. I. The nitrogen distribution and the percentages of some amino acids in the protamines of the sardine *Sardinia caerulea. J. Biol. Chem.* **70:**697–703.

Felix, K., 1952 Zur Chemie des Zellkerns. *Experientia* **8:**312–318.

Felix, K., 1955 Protamines, nucleoprotamines, and nuclei. *Am. Sci.* **43:**431–449.

Fischer, H. and L. Kreuzer, 1953 Über Gallin. *Z. Physiol. Chem.* **293:**176–182.

Fitch, W. M., 1971 Evolution of clupeine Z, a probable crossover product. *Nature (London) New Biol.* **229:**245–247.

Gineitis, A. A., I. A. Vinogradova, I. V. Volkova, and V. I. Vorobyev, 1970 Specificity of histone removed from sperms of different species of sea urchin. *Tsitologia* **7:**1132–1136.

Hamer, D., 1955 The composition of the basic proteins of echinoderm sperm. *Biol. Bull.* **108:**35–39.

Henricks, D. M. and D. T. Mayer, 1965 Isolation and characterization of basic keratin-like protein from mammalian spermatozoa. *Exp. Cell Res.* **40:**402–412.

Hirohata, R., 1929 Studies on protamine (I). *J. Biochem.* **10:**251–258.

Hultin, T. and R. Herne, 1949 Amino acid analysis of a basic protein fraction from sperm nuclei of some different invertebrates. *Ark. Kem. Minerol. Geol.* **26A(20):**1–8.

Inafuku, Z., 1951 Protamine. *Kyushu Mem. Med. Soc. Fukuoka* **2:**33–39.

Jergil, B. and Dixon, G. H., 1970 Protamine kinase from rainbow trout testis. *J. Biol. Chem.* **245:**425–434.

Kaverzneva, F. D. and A. Z. Rakhmatulina, 1970 Isolation and characteristics of protamine from *Acipenser stellatus. Khim. Prir. Soedin.* **6:**119–123.

Kaye, J. S., 1958 Changes in the fine structure of nuclei during spermiogenesis. *J. Morphol.* **103:**311–329.

Kaye, J. S. and R. McMaster-Kaye, 1966 The fine structure and chemical composition of nuclei during spermiogenesis of the house cricket. *J. Cell Biol.* **31:**159–179.

Kistler, W. S., M. E. Geroch, and H. G. Williams-Ashman, 1973 Specific basic proteins from mammalian testes. *J. Biol. Chem.* **248:**4532–4543.

Kossel, A., 1913 Weitere Mitteilungen über die Proteine der Fischspermien. *Z. Physiol. Chem.* **88:**183–185.

Kossel, A. 1928 *The Protamines and Histones,* Longmans Green, London.

Kossel, A. and E. G. Schenck, 1928 Untersuchungen über die basischen Eiweissstoff: Eine Beitrag zu ihrer Entwicklungsgeschichte. *Z. Physiol. Chem.* **173:**278–308.

Kossel, A. and W. Staudt, 1926 Zur Kentniss der basischen Proteine. *Z. Physiol. Chem.* **159:**172–178.

Kuroda, Y., 1951 Studies on protamine. III. *J. Biochem.* **38:**115–118.

Lam, D. M. K. and W. R. Bruce, 1971 Protamine synthesis in mouse. *J. Cell. Physiol.* **78:**13–24.

Langreth, S. G., 1969 Spermiogenesis in *Cancer* crabs. *J. Cell Biol.* **43:**575–609.

Ling, V. and G. H. Dixon, 1970 The biosynthesis of protamine in trout testis. II. Polysome patterns and protein synthetic activities during testis maturation. *J. Biol. Chem.* **245:**3035–3042.

Ling, V., J. P. Trevithick, and G. H. Dixon, 1969 The biosynthesis of protamine in trout testis. I. Intracellular site of synthesis. *Can. J. Biochem.* **47:**51–60.

Ling, V., B. Jergil, and G. H. Dixon, 1971 The biosynthesis of protamine in trout testis. III. Characterization of protamine components and their synthesis during testis development. *J. Biol. Chem.* **246:**1168–1176.

Lisitzuin, M. A. and N. S. Aleksandrovskaya, 1933 Über Protamine einer Fischarten. *Z. Physiol. Chem.* **221:**156–164.

Lisitzuin, M. A. and N. S. Aleksandrovskaya, 1936 Über die chemische Zusammensetzung der Storprotamine. *Z. Physiol. Chem.* **238:**54–58.

Marushige, K. and G. H. Dixon, 1969 Developmental changes in chromosomal composition and template activity during spermatogenesis in trout testis. *Dev. Biol.* **19:**397–414.

Marushige, K. and G. H. Dixon, 1971 Transformation of trout testis chromatin. *J. Biol. Chem.* **246:**5799–5805.

Marushige, K., V. Ling, and G. H. Dixon, 1970 Phosphorylation of chromosomal basic proteins in maturing trout testis. *J. Biol. Chem.* **244:**5953–5958.

Morkowin, N., 1899 Ein Beitrag zur Kenntnis der Protamine *Z. Physiol. Chem.* **28:**313–317.

Nakano, M., T. Tobita, and T. Ando, 1973 Studies on a protamine (Galline) from fowl sperm. I. Fractionation and some characterization. *Int. J. Peptide Prot. Res.* **5:**149–159.

Palau, J., A. Ruiz Carrillo, and J. A. Subirana, 1969 Histones from the sperm of the sea urchin *Arbacia lixula*. *Eur. J. Biochem.* **7:**209–213.

Paolleti, R. A. and R. C. Huang, 1969 Characterization of sea urchin sperm chromatin and its basic proteins. *Biochemistry* **8:**1615–1624.

Picheral, B., 1970 Nature et evolution des protéines basiques au cours de la spermiogenèse chez *Pleurodeles waltii Michah.*, amphibiene Urodèle *Histochemie* **23:**189–206.

Rasch, E. and J. W. Woodard, 1959 Basic proteins of plant nuclei during normal and pathological cell growth. *J. Biophys. Biochem. Cyto.* **6:**263–276.

Sanders, M. M. and G. H. Dixon, 1972 The biosynthesis of protamine in trout testis. IV. Sites of phosphorylation. *J. Biol. Chem.* **247:**851–855.

Subirana, J., 1970 Nuclear proteins from a somatic and germinal tissue of the echinoderm *Holothuria tubulosa*. *Exp. Cell Res.* **63:**253–260.

Subirana, J. A. and J. Palau, 1968 Histone-like proteins from the sperm of echinoderms. *Exp. Cell Res.* **53:**471–477.

Subirana, J. A., C. Cozcolluela, and J. Palau, 1973 Protamines and other basic proteins from spermatozoa of molluscs. *Biochim. Biophys. Acta* **317:**364–379.

Sud, B. N., 1961 Morphological and cytochemical studies of the chromatoid body in the grass snake *Natrix natrix*. *Quart. J. Microsc. Sci.* **102:**51–58.

Vaughn, J. C., 1966 Relationship of the "*sphère chromatophile*" to the fate of displaced histones following histone transition in rat spermiogenesis. *J. Cell Biol.* **31**:257–278.

Vaughn, J. C., 1968 Changing nuclear histone patterns during development. I. Fertilization and early cheavage in the crab *Emerita analoga. J. Histochem. Cytochem.* **16**:473–479.

Vaughn, J. C. and G. W. Hinsch, 1972 Isolation and characterization of chromatin and DNA from the sperm of the spider crab, *Libinia emarginata. J. Cell Sci.* **11**:131–152.

Vendrely, R. and C. Vendrely, 1966 Biochemistry of histones and protamines. *Protoplasmatologia* **5**:1–88.

Waldschmidt-Leitz, E. and H. Gutermann, 1961 Über die Struktur der Protamine. IV. Vergleich der Protamine aus Salmonidenarten. *Z. Physiol. Chem.* **323**:98–104.

Walker, M. and H. C. MacGregor, 1968 Spermatogenesis and the structure of the mature sperm in *Nucella Lapillus* (L). *J. Cell Sci.* **3**:95–104.

Zirkin, B. R., 1971 The fine structure of nuclei in mature sperm. *J. Ultrastruct. Res.* **36**:237–248.

Zirkin, B. and S. Wolfe, 1970 The protein composition of nuclei during spermiogenesis in the leopard frog, *Rana pipiens. Chromosoma* **31**: 231–240.

6

Organization and Size of Replicons

ROBERT B. PAINTER

Introduction

The term *replicon* as applied to eukaryotes is probably used incorrectly. In Jacob and Brenner's original definition (1963), the replicon had autonomous control over its own replication, and initiation occurred at a membrane site. The evidence as it now stands strongly suggests that duplication of units of DNA replication, in mammalian cells at least, is controlled primarily by factors external to those units. The term will be used in this chapter with the following definition: a *eukaryotic replicon* is a segment of DNA containing exactly one site, called the origin, at which initiation of DNA replication begins and from which both parental strands are duplicated. This operational definition, which is an elaboration on one proposed by Blumenthal *et al.* (1973), differs from the original one for prokaryotes not only in that it lacks an indication of control mechanisms but also in that it does not implicate any cellular structures outside of DNA as being involved in initiation.

ROBERT B. PAINTER—Laboratory of Radiobiology, University of California, San Francisco, California.

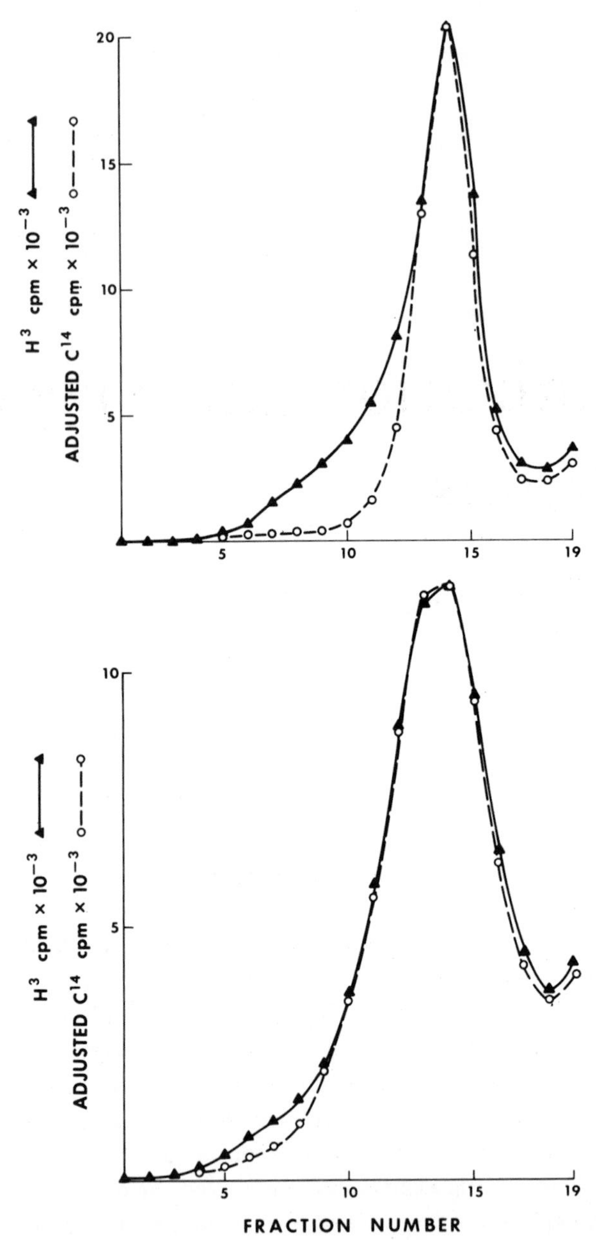

Figure 1. Isopycnic analysis of HeLa S 3 DNA labeled for 30 min with $[^3H]$dTrd and then for 2 hr with BrdUrd and sheared (upper panel) to produce B (number-average molecular weight) of 1.3×10^7 and F (fraction of tritium at densities greater than normal) of 0.225 or with ultrasound (lower panel) to produce B of 0.26×10^7 and F of 0.26. ▲, *^{3}H radioactivity;* ○, *^{14}C radioactivity (adjusted). The ultrasound F value is considered the minimum possible and is primarily due to pool mixing of $[^3H]$dTrd and BrdUrd. This is subtracted from the F value for the 12,000 rpm shear to give F_{net}. From B and F_{net}, L, the average molecular weight of DNA labeled during the pulse with $[^3H]$dTrd, can be estimated (see text). Since about one-twelfth of the total DNA on the average must be made in 30 min (S period $\simeq$360 min), the total number of sites replicating DNA $\simeq$ DNA molecular weight per cell $\div$ 12·L. Reprinted from Painter and Schaefer (1969b) with the permission of the authors and Academic Press, Inc.*

The Number and Size of Eukaryotic Replicons

Eukaryotic cells have many DNA replicating units in each chromosome, as was evident from early autoradiograms of DNA in whole chromosomes. The first indication of this derived from work by Taylor (1960), who showed that Chinese hamster cells that had been pulse-labeled with [^{3}H]thymidine ([^{3}H]dTrd) exhibited multiple clusters of grains over many of their chromosomes. This kind of analysis has since been performed on many kinds of mammalian cells with essentially the same results. The resolving power of autoradiography at the whole chromosome level, however, is much too low to permit an estimate of the total number of replicating units per chromosome.

The first indications that there is a very high number of replicons per chromosome came from two techniques. One was DNA fiber autoradiography, which Cairns (1966) used to show that the length of labeled regions of HeLa DNA molecules increased with increasing times of incubation with [^{3}H]dTrd. He estimated that "at least 100 sites of duplication" per average chromosome were in operation. The second method, an equilibrium density gradient technique, was used by Painter *et al.* (1966) to demonstrate that 10^3–10^4 HeLa cell replicons were in operation at any instant. In this method, cells are incubated with [^{3}H]dTrd for 10–30 min and then washed and shifted to a medium containing the density-labeling analogue of thymidine, bromodeoxyuridine (BrdUrd). After lysis of the cell, fragments are produced that have the radioactive label (^{3}H) at one end and the density label (BrdUrd) at the other. By determining the average molecular weight of the DNA in the gradient and measuring the amount of tritium at densities greater than that of unsubstituted DNA, a rough estimate of the average molecular weight of the segment of DNA labeled by the [^{3}H]dTrd can be made; from this it is possible to estimate the number of sites per cell with ^{3}H label (Figure 1). Somewhat later, using a method similar to that of Painter *et al.* (1966), Okada (1968) estimated the total number of replicating units in L5178Y cells to be about 10^5 per cell, and Taylor (1968), using another equilibrium density gradient method, estimated "several hundred" in the long arm of the X chromosome of Chinese hamster cells.

Actually, none of these estimates gave any solid information on the total number of replicons per cell. Both Cairns' method and that of Painter *et al.* yielded data on the number of replicons in operation at any instant but did not indicate what fraction of the total this was. Okada's estimate, which turns out to be quite close to the real number, actually derived from an improper use of sucrose gradients. The first relatively accurate estimate of the number of replicons per mammalian cell came from the work of

Huberman and Riggs (1968). They used fiber autoradiography but elaborated on the procedure in an extremely elegant way. They gave not only pulses of [^{3}H]dTrd but also pulse-chases, and they measured the distances between centers of adjacent "hot spots" (Figure 2). The histograms resulting from these measurements gave an estimate of the sizes of replicons in the Chinese hamster and HeLa cells they used; the average size was 30–40 μm. They pointed out that the method discriminates against small replicons, both because of inability to resolve short clusters of grains and because during the incubation with [^{3}H]dTrd small adjoining replicons can join to form labeled regions that would be analyzed as larger replicons. Thus the true average size of the replicon in these cells is probably smaller than 30 μm. Using 20–30 μm as the average, the number of replicons per cell can be estimated as $(1.5–2.0) \times 10^5$.

Another method that can be used to estimate replicon size and number is simply to pulse for a few minutes with [^{3}H]dTrd, lyse the cells on top of an alkaline sucrose gradient, centrifuge, and determine the distribution of sizes of DNA in the resulting velocity gradient. When this is done, the pattern of radioactivity is similar to that seen in Figure 3: the curve rises sharply at low S values, peaks at 30–40 S, and descends much less sharply through higher S values. If the cells are preincubated with [^{14}C]dTrd (before the pulse with [^{3}H]dTrd) to label parental strands, the ^{14}C-labeled DNA sediments near the bottom of the gradient where molecules of 130–140 S are found. This shows that the molecules pulse-labeled with [^{3}H]dTrd have not been subjected to mechanical shear and establishes that their distribution in the gradient truly reflects the distribution of size of nascent molecules within the cell. These results for mouse L5178Y cells have been repeated for HeLa cells (Habener *et al.*, 1969*b*; Gautschi *et al.*, 1973) and for several other cell lines, including Chinese hamster CHO (Gautschi and Kern, 1974), human WI38, and mouse P815 (Gautschi, J. M. Clarkson, and Painter, unpublished results), and indicate that the most frequent sizes for replicons are 15–30 μm. The result for L5178Y cells differs markedly from that reported by Lehmann and Ormerod (1971), who estimated the average size of the replicon to be 200 μm by analyzing their sucrose gradients after a 10-min incubation of L5178Y cells with [^{3}H]dTrd. No satisfactory explanation for this discrepancy exists, but it is possible that a large fraction of replicons joined together to form higher molecule weight molecules during the procedures that Lehmann and Ormerod used. In general, therefore, this method yields estimates of replicon size, and thus number of replicons per cell, that are similar to those derived from fiber autoradiography. A third method for estimating replicon size is an elaboration of the equilibrium density gradient method of Painter *et al.* (1966). Quantitation of this method led

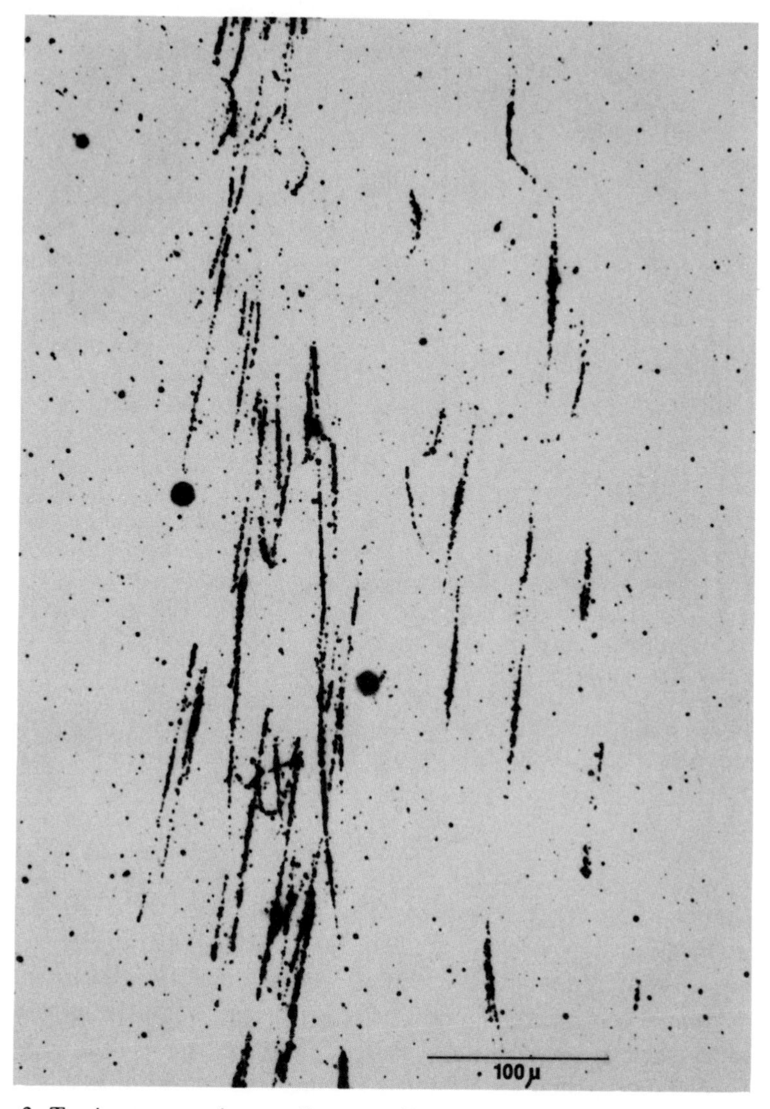

Figure 2. Tandem arrays of autoradiograms. Chinese hamster B14FAF28 fibroblast cells were grown as monolayer cultures on plastic petri dishes in Eagle's medium supplemented with 10% calf serum. After 12 hr pretreatment with fluorodeoxyuridine (FdUrd) (0.1 μg/ml), [^{3}H]dTrd (18 Ci/mmol) was added to 0.5 μg/ml. Thirty minutes later, the radioactive medium was removed and incubation was continued for 1 hr. The cells were then harvested by trypsinization and diluted to 1 × 10^4 cells/ml in isotonic saline containing 0.1 μg/ml FdUrd. Next they were diluted tenfold into 1.0 M sucrose, 0.05 M NaCl, 0.01 M EDTA, pH 8.0, lysed by dialysis against this mixture plus 1% sodium dodecylsulfate, and dialyzed further against 0.05 M NaCl, 0.005 M EDTA, pH 8.0. The released DNA was trapped on Millipore VM filters which had served as dialysis membranes and then was subjected to autoradiography. Exposure time with Kodak AR-10 Autoradiographic Stripping Film (Eastman Kodak Co.) was 4 months. Reprinted from Huberman and Riggs (1968) with the permission of the authors and Academic Press, Inc.

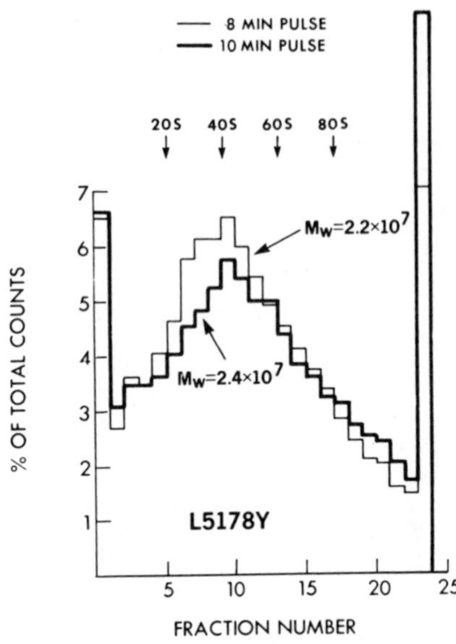

Figure 3. Sedimentation profiles for DNA from mouse L5178Y cells. An asynchronous culture was incubated for either 8 or 10 min with [³H]dTrd. Crushed ice made from saline was then added to the cell suspension. The ice-cold mixture was centrifuged and resuspended in ice-cold saline so that about 10^5 cells in 0.5 ml could be added to a lysis layer (0.5 N NaOH, 0.02 M EDTA) on top of a 5–20% sucrose gradient (35 ml). After 3 hr lysis, the tubes were centrifuged for 10 hr at 18,000 rpm in an SW27 rotor (Beckman) and fractions were collected, precipitated with cold 4% perchloric acid, and filtered through Whatman GF/C filters. After being washed and dried, the filters were added to a scintillation counting fluid and their radioactivity was determined with a Tri-Carb liquid scintillation spectrometer (Packard). The high amounts of radioactivity in the first and last fractions are artifacts due to wall effects and are not due to the sedimentation properties of the molecules containing them.

to derivation of the equation (J. L. Roti Roti, Gautschi, and Painter, unpublished)

$$L = \frac{B}{2F}\left(1 - \frac{B}{2R}\right)$$

where L is the average length of DNA labeled by [^{3}H]dTrd, B is the average length of DNA in the gradient (a measurable parameter), F is the fraction of tritium found at densities greater than unsubstituted DNA (another measurable parameter), and R is the average length of replicons. By preparing a large number of cells all labeled at the same time with [^{3}H]dTrd, mixing them well, and shearing aliquots to different extents, different values of B, and therefore of F, are generated. Then, because L is the same among all the aliquots,

$$R = \frac{B_1^2F_2 - B_2^2F_1}{2(B_1F_2 - B_2F_1)}$$

where B_1 and F_1 derive from one aliquot and B_2 and F_2 from another one, sheared to a different extent. Results from this method give average values for HeLa and WI38 replicons of 15–20 μm (unpublished results), in good agreement with the other methods.

The size of replicons in eukaryotic cells other than mammalian cells has also been estimated. Indeed, it is mainly from work on nonmammalian sources that the concept that replicon size is not necessarily a fixed parameter has come. The most complete analysis was performed by Blumenthal *et al.* (1973), who analyzed the DNA of *Drosophila melanogaster*, both in cleavage nuclei and in somatic cell cultures. For the embryo work it was possible to examine the DNA directly with the electron microscope, since the S period in *D. melanogaster* nuclei is only 3 min long, thus allowing a great number of growing points to be viewed in each nucleus at any instant. Since strand separation accompanies replication at each site of chain elongation, loops of DNA are formed (Figure 4). The average distance between origins of adjacent loops is about 2.5 μm, and so this is the average size (with little deviation) of the "replicon" in these cleavage nuclei. In the somatic cell nuclei in culture, however, where fiber autoradiography was used for analysis, the minimum average origin-to-origin distance is 9.5 μm, and there seem to be multiples of this value, with the other major class of origin-to-origin distances at about 19 μm. At least two models can be used to explain these observations. In one, origins are uniformly fixed at intervals of about 9.5 μm along the DNA fiber, but the probability is much less than 1 that each acts as an initiation site during each round of DNA replication. In the second model, favored by Blumenthal *et al.* (1973), some of the potential origins are at positions (condensed regions and/or regions already covered by chromosomal protein) which prevent their use as initiation sites.

McFarlane and Callan (1973) interpreted their fiber autoradiographic data from chick embryo cells to indicate an average initiation interval of 63 μm with a range of 25–145 μm. Their measurements (Figure 5), however, fall rather neatly into modes of about 30–35 μm, 40–45 μm, and 65–70 μm and suggest a unit of 15–17 μm as the basic replicon size in these cells. Since 30 min, a rather long time, was used for the incubation with [^{3}H]dTrd, it is very possible that the unit replicon length of 15–17 μm was not represented because of joining of adjacent units during the pulse, especially since this investigation (among others) also furnishes evidence for rather good synchronization of initiation within clusters of replicons.

Callan's earlier work (1972) with fiber autoradiography was actually the first to suggest that replicon size may vary from tissue to tissue. First he determined the origin-to-origin distances in somatic cells of *Xenopus laevis*, synchronized with fluorodeoxyuridine, to be 60 μm. With somatic cells of *Triturus cristatus carnifex*, which contain 10 times as much DNA as *Xenopus*, the distance between initiation sites was much greater, perhaps 10 times as much as in *Xenopus*. In meiotic cells (spermatocytes)

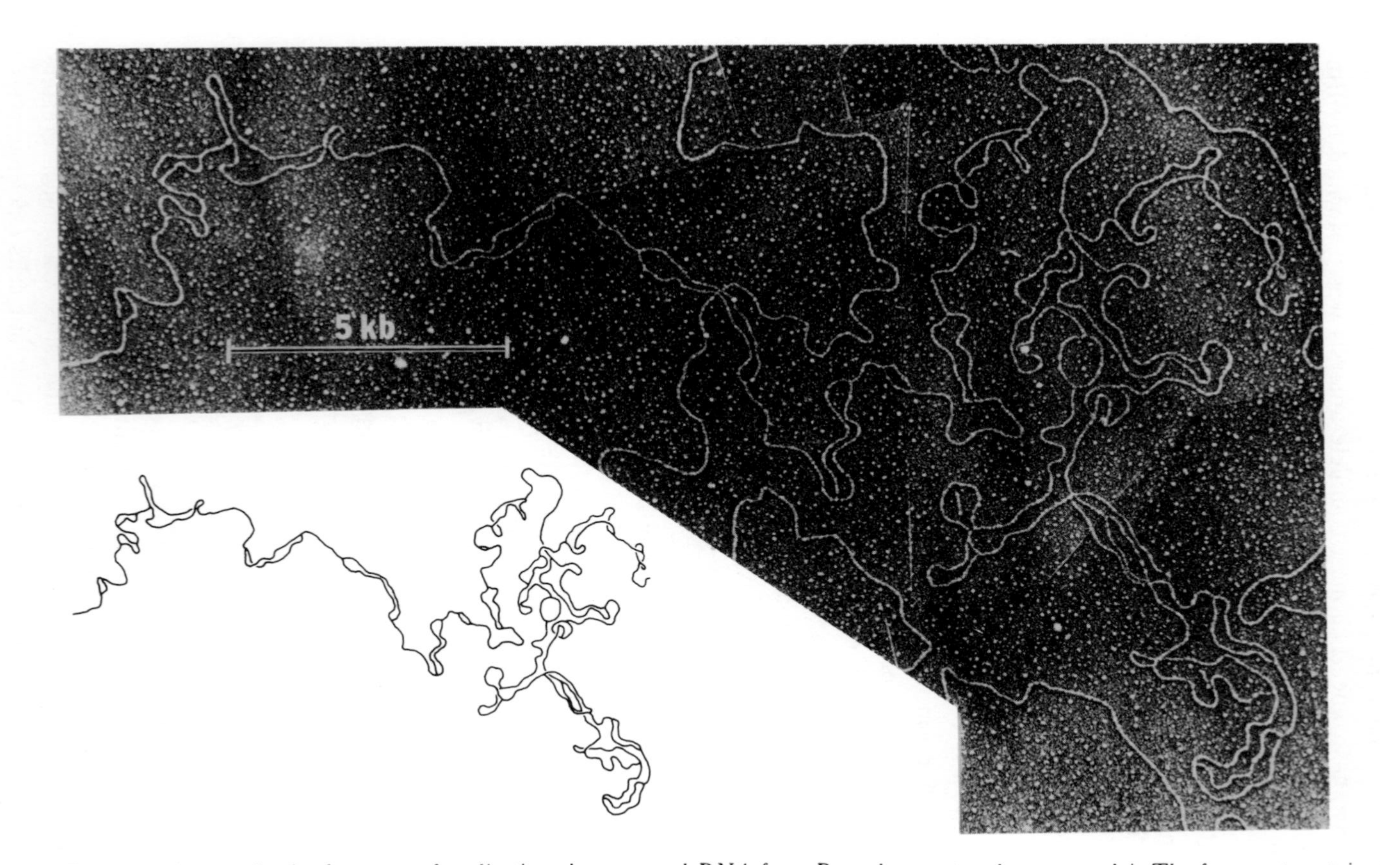

Figure 4. Electron micrograph of a fragment of replicating chromosomal DNA from *D. melanogaster* cleavage nuclei. The fragment contains 23 loops ("eye forms") in a total length of 12 μm. (5 kb = length of DNA fiber containing 5000 nucleotides ≃ 0.8 μm.) Reprinted from Kriegstein and Hogness (1974) with the permission of the authors.

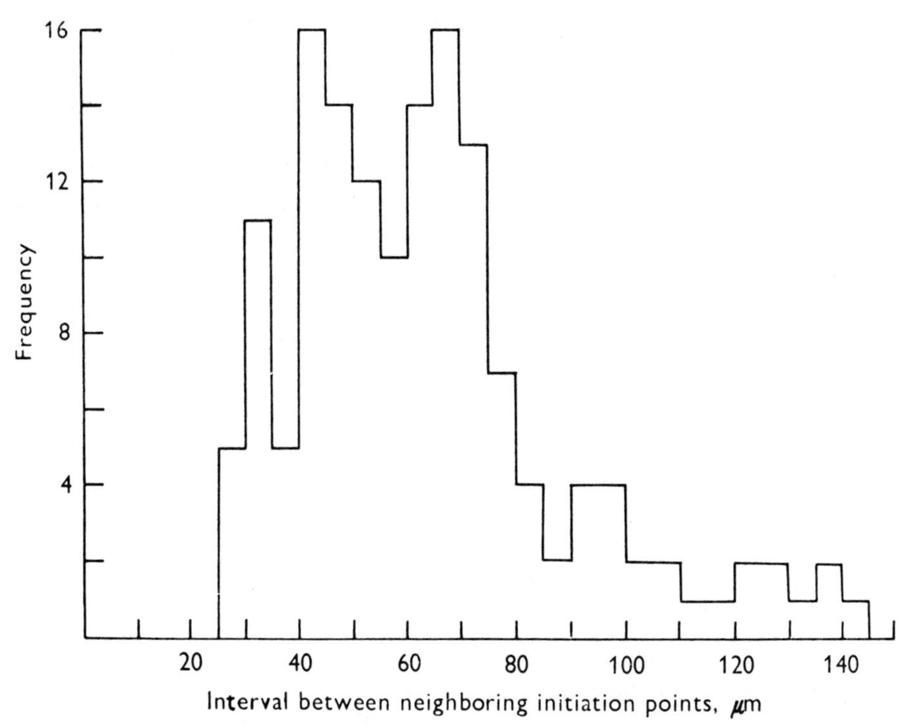

Figure 5. Frequency distribution of initiation point intervals from DNA fiber autoradiograms derived from chicken cells in culture. The cells were treated with FdUrd, labeled with [^{3}H]dTrd for 30 min, and chased for 30 min. Reprinted from McFarlane and Callan (1973) with the permission of the authors and R. J. Scaer Co.

from *Triturus,* however, even though the total lengths of some labeled segments were very long, there was no evidence for initiation or termination events in the autoradiograms observed, indicating that the average distance between initiations was very much longer than that observed for the somatic cells. Thus, although no accurate estimate for replicon size in *Triturus* was made, it is apparent that it varies by about an order of magnitude between somatic and meiotic cells. It is possible that in the latter case the replicon size is the size of the chromosomal DNA.

Initiation

The initiation of replicon operation is a variable that can be changed both by nature and by experimentation. The works of Callan (1972) in *Triturus* and of Blumenthal *et al.* (1973) in *Drosophila* show that from tissue to tissue within an organism the number of DNA initiation events can vary by 1–2 orders of magnitude. This variability, of course, is accompanied by corresponding changes in S time and cell generation time. The mechanism by which nucleotide sequences act as sites for initiation in the cells of one tissue but not in the cells of another tissue is not understood. Callan (1972) has suggested that a class or classes of proteins that recognize certain initiation sites in rapidly growing cells are not present in more slowly growing cells or that macromolecular configurations, including the association of structural proteins, differ in such a way that access is occluded in the slower-growing cells. Other explanations are possible, of course. For instance, the function of methylation of DNA in eukaryotes is unknown; this modification of DNA may act to alter sites to or from a condition that makes them accessible to initiation proteins. In this explanation the same polymerases and other proteins required for initiation are always present, but the number of modified sites would differ from tissue to tissue.

Gautschi *et al.* (1973) showed that experimental manipulation of cells in culture could change the pattern of replicon operation. They found that 2,4-dinitrophenol increased the average initiation-to-initiation distance; this observation strongly implies that, even within a single cell type, selection of initiation sites is not fixed. Although little is known about the mechanism by which this alteration in sites occurs, it is obvious that further study along these lines may lead to a better understanding of the mechanisms for initiation of DNA synthesis in eukaryotes.

Chain Elongation

Mechanism

There is no doubt that DNA chain elongation in eukaryotes involves discontinuous synthesis. This process involves the formation of a polynucleotide ahead of and separate from the nascent chain of DNA. The elongation of this polynucleotide and its joining to the nascent chain cause the further extension of the chain. Iteration of the process is the basis of DNA chain elongation. The maximum size of these polynucleotides appears to be about 4 S, or roughly 200 nucleotides long (Schandl and Taylor, 1969; Nuzzo *et al.*, 1970; Hyodo *et al.*, 1971; Fox *et al.*, 1973; Gautschi, 1974), much shorter than the "Okazaki fragments," which are about 1000 nucleotides long in bacteria (Okazaki *et al.*, 1968). There are still many doubts about several features of this discontinuous synthesis. First, there is still disagreement among workers in the field about the secondary structure of the 4 S fragments. Early evidence (Painter and Schaefer, 1969*a*; Schandl and Taylor, 1969) indicated that a large proportion of these fragments appeared as single-stranded DNA in mammalian cells under conditions where no denaturation should have occurred. Earlier results of Tsukada *et al.* (1968) suggested that newly made DNA in rat liver was only double-stranded. These results were resolved with those of Painter and Schaefer (1969*a*) and of Schandl and Taylor (1969) by Habener *et al.* (1969*a*), who showed that the configuration of newly made DNA depended on extraction conditions. Subsequent work has indicated that the single-stranded state probably occurs (Nuzzo *et al.*, 1970; Fox *et al.*, 1973; Wanka, 1973), although negative reports persist (Berger and Irvin, 1970; Gautschi, 1974).

A second point of controversy is whether all or only half of DNA is made by the discontinuous mode. Because of the technical difficulties associated with isolating newly made DNA, this question is an extremely difficult one to answer and there are data supporting both ideas (Hershey and Taylor, 1973; Gautschi, 1974).

The other point about discontinuous synthesis in eukaryotes that remains in doubt is the involvement of RNA as a primer for the formation of the small fragments. A short report by Waqar and Huberman (1973) indicated that RNA is associated with Okazaki fragments in slime molds. The best evidence that RNA is involved in eukaryotic DNA synthesis, however, is that of Fox *et al.* (1973), who found that newly made single-stranded DNA from human lymphocyte cultures assumed a position in Cs_2SO_4 equilibrium density gradients expected for a combination RNA-DNA molecule. A great deal of difficulty has been encountered by other groups in detecting such RNA-containing DNA molecules. For example, Gautschi (1974) has not been able to detect this association and favors the very transient existence of a recyclable, tRNA-like molecule as the primer.

Two problems with the experiments of Fox *et al.* (1973) should be mentioned before their results are accepted completely. One is that because transformed human lymphocytes were used, the role of a virus in the demonstration of RNA-DNA molecules must be considered. The second is that no attempt was made to disprove the possibility that the radioactivity at intermediate density was in an RNA-DNA double-stranded hybrid molecule. Thus the role and even the existence of RNA as an intermediate of eukaryotic DNA replication remain uncertain.

Direction

The work of Huberman and Riggs (1968) was the first to indicate that replication of DNA in Chinese hamster replicons occurred in opposite directions along the parental DNA from an initiation site. This interpretation was challenged by Lark *et al.* (1971), who found results similar to those of Huberman and Riggs when using the same protocol for labeling but conflicting results when using a "reverse" protocol (i.e., low specific activity [^{3}H]dTrd followed by high specific activity [^{3}H]dTrd) to label the DNA to be analyzed by fiber autoradiography. But Weintraub (1972) used the lability of BrdUrd-containing DNA to ultraviolet light to show that DNA replication was bidirectional in developing chick red blood cells. Huberman and Tsai (1973) then repeated the experiments of Lark *et al.* (1971), except that they used specific activities that were more different from one another than the ones Lark *et al.* used, and confirmed Huberman and Riggs' original conclusion. It is now generally accepted that bidirectional DNA replication is the rule because it has been found in several systems, including λ phage (Schnös and Inman, 1970), *Escherichia coli* (Prescott and Kuempel, 1972), and *Drosophila* (Kriegstein and Hogness, 1974).

Rate

Chain growth probably occurs primarily by fork displacement in opposite directions, so the former occurs twice as fast as the latter. Because most methods now in use measure the rate of movement of forks, we will refer to this parameter in this section. Among mammalian cells there seems to be a rather uniform rate of DNA fork displacement at 37°C. The first estimate by Cairns (1966) of about 0.5 μm/min for HeLa is remarkably close to estimates for other mammalian cells. Huberman and Riggs (1968) estimated 0.5–2.5 μm/min for Chinese hamster (B14FAF) cells using fiber autoradiography and Taylor (1968) estimated 1 μm/min for the same cells using equilibrium density gradients and the equations of Bonhoeffer and Gierer (1963). Using another equilibrium density gradient technique, Painter and Schaefer (1969*b*) estimated rates from 0.5 to about 1.8 μm/min for five different cell lines: HeLa S3, Chinese hamster V79, mouse L, human diploid WI38, and rabbit brain cell (CBL). It is interesting to note that the brain cell line has an S time of about 25 hr, or about 4 times as long as the others, and yet the rate of fork displacement is very similar. Lehmann and Ormerod (1970) used a sucrose gradient method to show the rate of fork displacement in mouse L5178Y cells to be 0.6–0.9 μm/min.

There is only scanty evidence for variation in the rate of fork displacement within the S period. Huberman and Riggs (1968) favored the concept that it varies by a factor of perhaps 5 (0.5–2.5 μm/min) at any instant in the S period but that the average rate does not change from one part of the S period of the Chinese hamster cell to another. Painter and Schaefer (1971), however, produced evidence that HeLa cells in early S had an average rate of fork displacement about one-third that of cells in the middle of S. Nevertheless, all evidence thus far accumulated suggests that the rather large fluctuations in overall rate of cellular DNA synthesis observed in many cells during the S period (Terasima and Tolmach, 1963; Remington and Klevecz, 1973) derive from the different numbers of replicons in operation at different times in S and that variability in rate of fork displacement influences overall synthesis rate only in a minor way.

In eukaryotes other than mammalian cells, there has been evidence of rather large variations in rate of fork displacement. Callan (1972) gives values of about 0.15 μm/min for *Xenopus* somatic cells and about 0.33 μm/min for *Triturus* somatic and meiotic cells, all at 25°C. He determined a value of about 0.5 μm/min at 37°C for chick embryo cells (McFarlane and Callan, 1973), which is very similar to the value of 0.7 μm/min that Weintraub and Holtzer (1972) found for developing chick red blood cells. Using equilibrium density gradient analysis and the equations of Bonhoeffer and Gierer (1963), Hyodo and Flickinger (1973) obtained low values for rate of fork displacement in frog embryos, varying from about 0.1 μm/min down to as low as 0.02 μm/min, depending on the tissue and the position in the S period. For both rapidly dividing cleavage nuclei and the slowly growing somatic cells of *Drosophila melangogaster,* the rate of fork displacement is about 1.0 μm/min (Blumenthal *et al.*, 1973). Cordeiro and Meneghini (1973) measured this parameter in the polytene chromosomes of salivary gland cells from another dipteran (*Rhyncosciara angelae*) and found the very low rate of 0.025 μm/min. It would be of interest to measure the rate of fork displacement in somatic cells as well as in the polytene chromosomes of *R. angelae* because the limited evidence now available indicates that there is little difference in rate of fork displacement between different kinds of cells dividing at greatly different rates within the same organism.

In summary, there seems to be a rather uniform rate of fork displacement among many different kinds of mammalian cells, but in other eukaryotic cells this does not necessarily hold. For instance, in *Drosophila* grown at 25°C the rate is almost the same as in mammalian cells, whereas in frog cells grown at 20°C the rate is as low as 0.02 μm/min. It is interesting, however, that the rate in eukaryotic cells appears never to exceed about 2 μm/min, suggesting that the complexity of eukaryotic chromosomes severely restricts this parameter, which in prokaryotic cells can be an order of magnitude higher (at least 15 μm/min, based on Cairns', 1963, results and assuming bidirectional growth).

Termination

Essentially nothing is known about termination of replicon operation. It is possible that termination occurs most of the time simply by the merging of growing points from adjacent replicons and that the site for this joining is not fixed. The data showing that initiation sites are different in different tissues (Callan, 1972; Blumenthal *et al.*, 1973) and that the distance between initiation sites can be modified experimentally (Gautschi *et al.*, 1973) imply that termination sites are also different under different conditions and thus not fixed. Blumenthal *et al.* (1973) point out that this concept makes the existence of inversions, translocations, and deletions much easier to explain, since fixed termini transferred by one of these mechanisms would result in regions that could not be replicated. Data are so sparse, however, that one cannot disregard the possibility that some or all termination sites are fixed.

Control of Replicon Operation

There is at present little insight into the processes that control the operation of eukaryotic replicons. It is well established that inhibition of protein synthesis before S phase will block or delay the onset of DNA synthesis (Mueller *et al.*, 1962; Schneiderman *et al.*, 1971). Once the S phase has started, however, the action of protein synthesis inhibitors is varied and controversial. In the slime mold *Physarum polycephalum* it seems clear that cycloheximide inhibits DNA replication in several steps, leading to the conclusion that new proteins must be formed several times during the S period in order to initiate each of the several rounds of synthesis that occur (Muldoon *et al.*, 1971). Somewhat similar results have been reported for mammalian cells (Hyodo *et al.*, 1971; Fujiwara, 1972). Other work, however, indicates that cycloheximide acts in avian and mammalian cells to inhibit the rate of chain growth almost immediately (within 30–100 sec) (Painter, 1970; Weintraub and Holtzer, 1972; Gautschi and Kern, 1973), but with little effect on initiation. Results with another protein synthesis inhibitor, puromycin, are even more controversial. Hand and Tamm (1972) and Hori and Lark (1973), using mouse L cells and Chinese hamster ovary cells, respectively, have interpreted their results with fiber autoradiography as indicating that puromycin inhibits initiation in Chinese hamster cells without affecting rate of fork displacement. Gautschi

(1974), however, used density gradient and velocity gradient analysis to show that puromycin inhibited the rate of fork displacement in HeLa and Chinese hamster ovary cells. Thus conflicting interpretations have resulted from work with the same cell line.

The basis for this difference may lie in the different techniques used. Molecular autoradiography is a powerful and valuable tool and it is esthetically satisfying because one can "see" the molecules of interest. On the other hand, it is an extremely selective procedure in which more than 99% of all the information is unusable and discarded; the usable 1% or less is assumed to be a representative sample. This technique is also especially vulnerable in the area of distinguishing between effects on rate of fork displacement and effects on delayed initiation. It is my prejudice that biochemical techniques using equilibrium density gradients and velocity gradients, which measure the average value for the total population tested, yield more representative information. Therefore, I favor the concept that protein synthesis inhibitors act as do DNA synthesis inhibitors (Painter, 1970; Amaldi *et al.,* 1972; Gautschi *et al.,* 1973) to inhibit immediately the rate of fork displacement in mammalian cells with changes in initiation and termination not important in a first approximation.

The very reproducible DNA replication patterns observable in whole chromosomes during the S period [e.g., in Chinese hamster cells (Taylor, 1960) or in human cells (German, 1964)] make it clear that control of chromosomal DNA replication is very tight at this level. These observations, combined with the evidence for clustering of replicons observed in several systems (Huberman and Riggs, 1968; Lark *et al.,* 1971; Amaldi *et al.,* 1972; Blumenthal *et al.,* 1973), suggest that the main level of control is over groups or clusters of replicons. Within clusters, however, the evidence of Gautschi *et al.* (1973) suggests that initiation and termination sites can be experimentally changed. Along with experiments by Amaldi *et al.* (1973), which showed poor reproducibility of the timing of initiation at identical sites from one S phase to another, this evidence supports the concept that, in contrast to the stringent control of initiation at the level of whole clusters, there is relatively loose control of replication within the clusters.

There is much evidence that euchromatin replicates early in the S phase of many cells with heterochromatin replication almost invariably late (Tobia *et al.,* 1970; Flamm *et al.,* 1971; Bostock *et al.,* 1972), so there must be an inherent control responsible for this sequence of replication events. How the duplication of replicons relates to the replication of clusters and to the replication of heterochromatin and euchromatin, and how any of this is involved in chromosomal duplication, is still very, very obscure.

Acknowledgment

This work was supported by the U.S. Atomic Energy Commission.

Literature Cited

Amaldi, F., F. Carnevali, L. Leoni, and D. Mariotti, 1972 Replicon origins in Chinese hamster cell DNA. I. Labeling procedure and preliminary observations. *Exp. Cell Res.* **74:**367–374.

Amaldi, F., M. Buongiorno-Nardelli, F. Carnevali, L. Leoni, D. Mariotti, and M. Pomponi, 1973 Replicon origins in Chinese hamster cell DNA. II. Reproducibility. *Exp. Cell Res.* **80:**79–87.

Berger, H., Jr. and J. L. Irvin, 1970 Changes in physical state of DNA during replication in regenerating liver of the rat. *Proc. Natl. Acad. Sci. USA* **65:**152–159.

Blumenthal, A. B., H. J. Kriegstein, and D. S. Hogness, 1973 The units of DNA replication in *Drosophila melanogaster* chromosomes. *Cold Spring Harbor Symp. Quant. Biol.* **38:**205–223.

Bonhoeffer, F. and A. Gierer, 1963 On the growth mechanism of the bacterial chromosome. *J. Mol. Biol.* **7:**534–540.

Bostock, C. J., D. M. Prescott, and F. T. Hatch, 1972 Timing of replication of the satellite and main band DNAs in cells of the kangaroo rat (*Dipodomys ordii*). *Exp. Cell Res.* **74:**487–495.

Cairns, J., 1963 The bacterial chromosome and its manner of replication as seen by autoradiography. *J. Mol. Biol.* **6:**208–213.

Cairns, J., 1966 Autoradiography of HeLa cell DNA. *J. Mol. Biol.* **15:**372–373.

Callan, H. G., 1972 Replication of DNA in the chromosomes of eukaryotes. *Proc. R. Soc. London Ser. B Biol. Sci.* **181:**19–41.

Cordeiro, M. and R. Meneghini, 1973 The rate of DNA replication in the polytene chromosomes of *Rhynchosciara angelae*. *J. Mol. Biol.* **78:**261–274.

Flamm, W. G., N. J. Bernheim, and P. E. Brubaker, 1971 Density gradient analysis of newly replicated DNA from synchronized mouse lymphoma cells. *Exp. Cell Res.* **64:**97–104.

Fox, R. M., J. Mendelsohn, E. Barbosa, and M. Goulian, 1973 RNA in nascent DNA from cultured human lymphocytes. *Nature (London) New Biol.* **245:**234–237.

Fujiwara, Y., 1972 Effect of cycloheximide on regulatory protein for initiating mammalian DNA replication at the nuclear membrane. *Cancer Res.* **32:**2089–2095.

Gautschi, J. R., 1974 Effects of puromycin on DNA chain elongation in mammalian cells. *J. Mol. Biol.* **84:**223–229.

Gautschi, J. R. and R. M. Kern, 1973 DNA replication in mammalian cells in the presence of cycloheximide. *Exp. Cell Res.* **80:**15–26.

Gautschi, J. R., R. M. Kern, and R. B. Painter, 1973 Modification of replicon operation in HeLa cells by 2,4-dinitrophenol. *J. Mol. Biol.* **80:**393–403.

German, J., 1964 The pattern of DNA synthesis in the chromosomes of human blood cells. *J. Cell Biol.* **20:**37–55.

Habener, J. F., B. S. Bynum, and J. Shack, 1969*a* Unique secondary structure of newly replicated HeLa DNA. *Biochim. Biophys. Acta* **186:**412–414.

Habener, J. F., B. S. Bynum, and J. Shack, 1969*b* Changes in the sedimentation properties of HeLa cell DNA and nucleoprotein during replication. *Biochim. Biophys. Acta* **195**:484–493.

Hand, R. and I. Tamm, 1972 Rate of DNA chain growth in mammalian cells infected with cytocidal RNA viruses. *Virology* **47**:331–337.

Hershey, H. V. and J. H. Taylor, 1973 Discontinuous DNA replication *in vitro:* Two distinct size classes of intermediates. *J. Cell Biol.* **59**:140*a*.

Hori, T. and K. G. Lark, 1973 Effect of puromycin on DNA replication in Chinese hamster cells. *J. Mol. Biol.* **77**:391–404.

Huberman, J. A. and A. D. Riggs, 1968 On the mechanism of DNA replication in mammalian chromosomes. *J. Mol. Biol.* **32**:327–341.

Huberman, J. A. and A. Tsai, 1973 Direction of DNA replication in mammalian cells. *J. Mol. Biol.* **75**:5–12.

Hyodo, M. and R. A. Flickinger, 1973 Replicon growth rates during DNA replication in developing frog embryos. *Biochim. Biophys. Acta* **299**:24–33.

Hyodo, M., H. Koyama, and T. Ono, 1971 Intermediate fragments of newly replicated DNA in mammalian cells. II. Effect of cycloheximide on DNA chain elongation. *Exp. Cell Res.* **67**:461–463.

Jacob, F. and S. Brenner, 1963 *Sur la régulation de la synthèse* du DNA *chez les bactéries: L'hypothèse du réplicon. C. R. Acad. Sci.* **256**:298–300.

Kriegstein, H. J. and D. S. Hogness, 1974 Mechanism of DNA replication in *Drosophila* chromosomes: Structure of replication forks and evidence for bidirectionality. *Proc. Natl. Acad. Sci. USA* **71**:135–139.

Lark, K. G., R. Consigli, and A. Toliver, 1971 DNA replication in Chinese hamster cells: Evidence for a single replication fork per replicon. *J. Mol. Biol.* **58**:873–875.

Lehmann, A. R. and M. G. Ormerod, 1970 The replication of DNA in murine lymphoma cells (L5178Y). I. Rate of replication. *Biochim. Biophys. Acta* **204**:128–143.

Lehmann, A. R. and M. G. Ormerod, 1971 The replication of DNA in murine lymphoma cells (L5178Y). II. Size of replicating units. *Biochim. Biophys. Acta* **272**:191–201.

McFarlane, P. W. and H. G. Callan, 1973 DNA replication in the chromosomes of the chicken, *Gallus domesticus. J. Cell Sci.* **13**:821–839.

Mueller, G. C., K. Kajiwara, E. Stubblefield, and R. R. Rueckert, 1962 Molecular events in the reproduction of animal cells. I. The effect of puromycin on the duplication of DNA. *Cancer Res.* **22**:1084–1090.

Muldoon, J. J., T. E. Evans, O. F. Nygaard, and H. H. Evans, 1971 Control of DNA replication by protein synthesis at defined times during the S period in *Physarum polycephalum. Biochim. Biophys. Acta* **247**:310–321.

Nuzzo, F., A. Brega, and A. Falaschi, 1970 DNA replication in mammalian cells. I. The size of newly synthesized helices. *Proc. Natl. Acad. Sci. USA* **65**:1017–1024.

Okada, S., 1968 Replicating units (replicons) of DNA in cultured mammalian cells. *Biophys. J.* **8**:650–664.

Okazaki, R., T. Okazaki, K. Sakabe, K. Sugimoto, R. Kainuma, A. Sugino, and N. Iwatsuki, 1968 *In vivo* mechanism of DNA chain growth. *Cold Spring Harbor Symp. Quant. Biol.* **33**:129–142.

Painter, R. B., 1970 The molecular basis of changes in rate of mammalian DNA synthesis. *J. Cell Biol.* **47**:153*a*.

Painter, R. B. and A. Schaefer, 1969*a* State of newly synthesized HeLa DNA. *Nature (London)* **221**:1215–1217.

Painter, R. B. and A. W. Schaefer, 1969*b* Rate of synthesis along replicons of different kinds of mammalian cells. *J. Mol. Biol.* **45:**467–479.

Painter, R. B. and A. W. Schaefer, 1971 Variation in the rate of DNA chain growth through the S phase in HeLa cells. *J. Mol. Biol.* **58:**289–295.

Painter, R. B., D. A. Jermany, and R. E. Rasmussen, 1966 A method to determine the number of DNA replicating units in cultured mammalian cells. *J. Mol. Biol.* **17:**47–56.

Prescott, D. M. and P. L. Kuempel, 1972 Bidirectional replication of the chromosome in *Escherichia coli. Proc. Natl. Acad. Sci. USA* **69:**2842–2845.

Remington, J. A. and R. R. Klevecz, 1973 Families of replicating units in cultured hamster fibroblasts. *Exp. Cell Res.* **76:**410–418.

Schandl, E. K. and J. H. Taylor, 1969 Early events in the replication and integration of DNA into mammalian chromosomes. *Biochem. Biophys. Res. Commun.* **34:**291–300.

Schneiderman, M. H., W. C. Dewey, and D. P. Highfield, 1971 Inhibition of DNA synthesis in synchronized Chinese hamster cells treated in G1 with cycloheximide. *Exp. Cell Res.* **67:**147–155.

Schnös, M. and R. B. Inman, 1970 Position of branch points in replicating λ DNA. *J. Mol. Biol.* **51:**61–73.

Taylor, J. H., 1960 Asynchronous duplication of chromosomes in cultured cells of Chinese hamster. *J. Biophys. Biochem. Cytol.* **7:**455–464.

Taylor, J. H., 1968 Rates of chain growth and units of replication in DNA of mammalian chromosomes. *J. Mol. Biol.* **31:**579–594.

Terasima, T. and L. J. Tolmach, 1963 Growth and nucleic acid synthesis in synchronously dividing populations of HeLa cells. *Exp. Cell Res.* **30:**344–362.

Tobia, A. M., C. L. Schildkraut, and J. J. Maio, 1970 Deoxyribonucleic acid replication in synchronized cultured mammalian cells. I. Time of synthesis of molecules of different average guanine + cytosine content. *J. Mol. Biol.* **54:**499–515.

Tsukada, K., T. Moriyama, W. E. Lynch, and I. Lieberman, 1968 Polydeoxynucleotide intermediates in DNA replication in regenerating liver. *Nature* (*London*) **220:**162–164.

Wanka, F., 1973 Separation of rapidly labeled intermediates of DNA synthesis in mammalian cells. *Biochem. Biophys. Res. Commun.* **54:**1410–1417.

Waqar, M. A. and J. A. Huberman, 1973 Evidence for the attachment of RNA to pulse-labeled DNA in the slime mold, *Physarum polycephalum. Biochem. Biophys. Res. Commun.* **51:**174–180.

Weintraub, H., 1972 Bi-directional initiation of DNA synthesis in developing chick erythroblasts. *Nature* (*London*) *New Biol.* **236:**195–197.

Weintraub, H. and H. Holtzer, 1972 Fine control of DNA synthesis in developing chick red blood cells. *J. Mol. Biol.* **66:**13–35.

PART R
GENE TRANSCRIPTS

7

Nuclear RNA

BERTIL DANEHOLT

Introduction

The cell nucleus supplies the cytoplasm with a variety of RNA species (e.g., see Weinberg, 1973). RNA molecules are not only synthesized but also undergo certain maturation steps within the nucleus before their delivery into the cytoplasm. This processing of RNA molecules was first demonstrated for ribosomal RNA. The two ribosomal RNA species originate from a common precursor molecule. This primary transcript is methylated and then cleaved during a multistep process so that the two ribosomal RNA species are finally generated. Apart from the true ribosomal RNA sequences, other sequences are also present: in HeLa cells these sequences constitute as much as half the primary transcript (Weinberg and Penman, 1970). The precursor for transfer RNA is also somewhat longer than mature transfer tRNA. This primary transcript is modified and trimmed to the size of mature transfer RNA. Earlier, this process was thought to take place in the cytoplasm. However, there is now good evidence suggesting that, at least to a large extent, it takes place within the nucleus (Egyházi *et al.*, 1969; Egyházi and Edström, 1972). Compared to that for ribosomal RNA and transfer RNA precursors, there has been considerably less solid information about the behavior of the messenger RNA precursors in the nucleus. For a long time, heterogeneous, high molecular weight RNA (HnRNA) has been regarded as a likely precursor to messenger RNA (mRNA). Again a cleavage process has been invoked, because HnRNA has a higher average molecular weight than

BERTIL DANEHOLT—Department of Histology, Karolinska Institutet, Stockholm, Sweden.

mRNA. During the most recent years, the relationship between HnRNA and mRNA has been extensively studied. Support for a precursor–product relationship has been obtained, and different kinds of processing steps, e.g., cleavage and addition of a polyadenylic acid sequence, seem to be involved. The majority of all RNA molecules exported to cytoplasm participate in protein synthesis. It has, however, also been proposed that some protein synthesis takes place within the nucleus (see, e.g., Allfrey, 1970) and consequently these various RNA species could already be functionally active inside the nucleus. Transcription–translation complexes, like those recorded in prokaryotic cells (Miller *et al.,* 1970), cannot be found in eukaryotic cells (Miller and Bakken, 1972), yet minor intranuclear protein synthesis is difficult to exclude. It can, however, be stated that the principal role of the nucleus in RNA metabolism is to generate and deliver completed RNA products to cytoplasm, where they exert their function in the protein synthesizing machinery.

When discussing nuclear RNA, it is useful to divide it into ribosomal RNA including ribosomal 5 S RNA, heterogeneous nuclear RNA (HnRNA), transfer RNA, and other low molecular weight RNAs. The synthesis of preribosomal RNA and its subsequent processing into the two major ribosomal RNA species as well as the generation and metabolism of the ribosomal 5 S RNA are considered in this volume along with the generation of ribosomes (Chapter 11). Transfer RNA and its characteristics will also be dealt with separately (Chapter 10). I will therefore mainly concentrate my presentation on HnRNA and also to some extent deal with nuclear, low molecular weight RNA. These two latter topics have previously been discussed more or less extensively in reviews on various aspects of RNA metabolism. HnRNA has been treated by Darnell *et al.* (1973), Davidson and Britten (1973), Lewin (1975*a,b*), and Weinberg (1973), while low molecular weight RNA has been discussed by Burdon (1971), Busch *et al.* (1971), and Weinberg (1973).

The Size Range of Primary Transcripts

HnRNA has been considered to cover a broad molecular size range ever since its recognition in the 1960s (Attardi *et al.,* 1966; Georgiev and Lerman, 1964; Houssais and Attardi, 1966; Perry, 1962; Scherrer *et al.,* 1963, 1966; Warner *et al.,* 1966). In most higher eukaryotes it is comprised of molecules in the range of 400–50,000 nucleotides, while in some lower eukaryotes, such as *Dictyostelium* (Lodish *et al.,* 1973), *Tetrahymena* (Prescott *et al.,* 1971*a*), and *Amoeba* (Prescott *et al.,* 1971*b*), the range is more limited, and molecules larger than 6000 nucleotides are essentially absent. Since it is known that RNA can form artifactual aggregates, particularly when phenol extraction is involved (Bramwell, 1972; Macnaughton *et al.,* 1974; McKnight and Schimke, 1974), it has been questioned whether, in fact, molecules of giant size exist at all in eukaryotes. This issue has therefore been investigated in great detail by visualization of HnRNA molecules in the electron microscope under denaturing conditions (Granboulan and Scherrer, 1969; Holmes and Bonner, 1973). These studies provide hard evidence for the existence of molecules of very high molecular weights (up to 40,000–50,000 nucleotides in length). The size range estimates for the largest HnRNA molecules given in the earlier studies of HnRNA seem therefore to be essentially correct. It is even possible that the present biochemical techniques cannot keep the highest molecular weight RNA undegraded. To take an extreme example, the electron micrographs of transcription products in amphibian oocytes (Miller and Hamkalo, 1972) indicate that the primary transcripts are probably of at least one order of magnitude larger than the largest HnRNA molecules recorded by biochemical means. It can therefore be concluded that in eukaryotes some primary transcripts have to be very long (at least on the order of 50,000 nucleotides). However, it is more difficult to establish a minimum size of the primary transcripts, as smaller molecules in the HnRNA population might also very well represent molecules under synthesis (growing RNA molecules) or molecules that have already been reduced in size (processed RNA molecules; see pp. 199–201.

Electron micrographs have nicely demonstrated the presence of growing RNA molecules in eukaryotic nuclei (for review, see Miller and Bakken, 1972). Usually this type of RNA is regarded as negligible when compared to the total amount of HnRNA in the nucleus. It has, however, been difficult to distinguish growing RNA molecules from completed molecules released from the DNA template, and to directly determine the actual proportions of each. In insect nuclei containing giant polytene chromosomes, this can be efficiently achieved, as chromosomes or chromosome segments can be separated from the surrounding nuclear sap by microdissection. The polytene chromosomes of *Chironomus* are particularly well studied, and cytological maps are presented in the chapter by K. Hägele in Volume 3 of this series. In two large, RNA-producing chromosome regions, Balbiani rings 1 and 2 (BR 1 and 2), almost all HnRNA molecules are likely to be growing ones (Daneholt, 1975; Egyházi, 1975). When completed, the BR products (75 S RNA) are immediately released from the template into the nuclear sap. It has been established that after a 20-min pulse, RNA within the Balbiani rings, i.e., nascent RNA, contains at least 10 times more radioactivity than 75 S RNA present in the nuclear sap (Egyházi, 1975). These data therefore suggest that growing RNA molecules might constitute a larger proportion of total HnRNA than generally appreciated, particularly when short labeling times are employed. Furthermore, most HnRNA sequences are only present in five to ten copies (for discussion, see Lewin, 1975*b*). At least a few copies of most of these sequences may be parts of growing molecules, as each transcription complex is likely to contain several growing molecules (Hamkalo *et al.*, 1973). Therefore, again it is indicated that contributions from growing RNA molecules should not be considered negligible when total HnRNA is discussed and that probably only a certain fraction of HnRNA (perhaps less than 50%) should be regarded as molecules actually released from the template and ready for transfer to cytoplasm.

For the two main reasons given above (the presence of growing and processed molecules), studies of total HnRNA have not been conclusive as to the minimum size of the primary transcription products. Recently, however, a new approach to the problem of the size of the primary transcripts has been successful and support for the presence of small as well as large primary transcripts has been obtained. By different technical means it has been feasible to select defined transcripts from the complex HnRNA population of molecules. A method that evidently will be frequently used in future studies has been used by Macnaughton *et al.* (1974). They isolated globin mRNA and synthesized the corresponding complementary DNA (cDNA) by reverse transcriptase and then used this cDNA as a probe to demonstrate the presence and size of HnRNA, containing globin mRNA sequences. They found that the nuclear globin sequences almost exclusively could be found in a symmetrical 14 S RNA peak (corresponding to about 2000 nucleotides). A similar technique was used by McKnight and Schimke (1974), who found that *all* ovalbumin-coding RNA was of 18 S size (2000–2500 nucleotides). Quite a different procedure was used by Stevens and Williamson (1973). They could isolate a precursor to mRNA for immunoglobin heavy chain by selective precipitation of the precursor by an antibody–antigen complex. The size of the precursor was estimated to be 2.4×10^6 daltons (about 8000 nucleotides). Finally, it has been possible in insect salivary gland cells to extract a specific giant HnRNA species (75 S RNA corresponding to about 50,000 nucleotides) from a discrete chromosome region (Balbiani ring 2), isolated by microdissection (Daneholt, 1972). This giant RNA molecule is a putative precursor for salivary polypeptide mRNA (see pp. 200–201). The ovalbumin mRNA precursor and 75 S RNA in BR 2 are likely to be the primary transcription products. In the other two experimental systems it is more difficult to rule out the possibility of intermediate products, although no evidence in favor of such an alternative was found. Somewhat contradictory experiments concerning the globin mRNA precursor have been reported by Imaizumi *et al.* (1973) and Spohr *et al.* (1974). They found that globin sequences were present in 14 S RNA but also that a minor amount of globin sequences sedimented more rapidly in a dimethylsulfoxide gradient without forming a distinct peak. From the studies by McKnight and Schimke (1974) it seems necessary to carry out repeated centrifugation and denaturation steps until all mRNA sequences artifactually cosedimenting with giant HnRNA have been removed. Since both Macnaughton *et al.* (1974) and Spohr *et al.* (1974) found that most globin sequences constituted a distinct 14 S RNA peak, at the present time it seems more appropriate to stress the significance of this particular RNA species than to regard the existence of a more giant RNA precursor to 14 S as proven. In conclusion, it can be stated from studies on total HnRNA as well as on defined primary transcripts that in higher eukaryotes there are primary transcripts of very high molecular weight (up to at least 50,000 nucleotides) but also of considerably smaller size (2000 nucleotides or even smaller).

Structure and Sequence Content

Being a putative mRNA precursor, HnRNA is likely to be synthesized from a large number of genome regions. It is therefore consistent that the base composition of HnRNA is similar to that of total DNA, in sharp contrast to ribosomal RNA, which is transcribed from a very restricted part of the genome (e.g., Darnell, 1968). Moreover, hybridization experiments have shown that HnRNA contains sequences corresponding to as much as 5–25% of the unique part of the genome (Hahn and Laird, 1971; Holmes and Bonner, 1974*a*; Turner and Laird, 1973). Complexity measurements of unique HnRNA (or total RNA) sequences also demonstrate that considerable complexity differences exist between tissues of the same organism (Brown and Church, 1972; Grouse *et al.*, 1972; Hahn and Laird, 1971; Kohne and Byers, 1973). Remarkable differences have been recorded, for example, between adult mouse brain and liver (18% and 6% of the unique DNA involved in transcription) (Hahn and Laird, 1971). Direct sequence homology comparisons have also been carried out in mouse tissues (Brown and Church, 1972; Grouse *et al.*, 1972) and again there is good evidence for tissue-specific sequences in HnRNA. It is also interesting in this context that mouse tissues (Church and Brown, 1972) as well as *Dictyostelium* (Firtel, 1972) display considerable differences in sequence content at various developmental stages. It is, however, also appropriate to stress that HnRNA molecules from different tissues (or developmental stages) are likely to have a great number of HnRNA sequences in common (e.g., Brown and Church, 1972).

Apart from unique sequences, HnRNA also contains sequences complementary to the repetitive part of the genome (for review, see Davidson and Britten, 1973). It has been estimated that the repetitive sequences comprise about 10% of the sequences in HnRNA (Church and McCarthy, 1967; Melli *et al.*, 1971; Pagoulatos and Darnell, 1970; Scherrer *et al.*, 1970; Shearer and McCarthy, 1967, 1970; Smith *et al.*, 1974; Sullivan, 1968). The average degree of reiteration is 100–1000, but there is a wide variation in these figures, as shown by Darnell *et al.* (1970) and Pagoulatos and Darnell (1970). Although there are HnRNA molecules that to a large extent consist of repetitive sequences (e.g., the BR 2 transcript, according to Lambert, 1972), it seems to be the rule that only a minor part of the HnRNA molecule consists of repetitive RNA. It has been directly demonstrated by hybridization experiments in HeLa cells (Darnell and Balint, 1970) and sea urchins (Smith *et al.*, 1974) that repetitive sequences are usually linked to unique sequences. Accordingly, HnRNA hybridized to repetitive sequences displays a high degree of RNAse sensitivity (Smith *et al.*, 1974). A similar conclusion on the arrangement of repetitive and nonrepetitive sequences in HnRNA was reached by Holmes and Bonner (1974*b*) when they analyzed the hybridization properties of unsheared and sheared HnRNA. Although comprising only 10–15% of HnRNA, the repetitive sequences are present in more than 70% of HnRNA molecules longer than 15,000 nucleotides as a consequence of the interspersed arrangement. This organization of the unique and repetitive sequences in HnRNA is in agreement with the general structural pattern of DNA obtained from DNA reassociation experiments. In experiments of various kinds, Davidson and Britten (1973) found that in *Xenopus,* sea urchin, and calf about 50% of the DNA consists of short repetitive DNA sequences (about 300 nucleotides long) separated from each other by unique sequences (700–1100 nucleotides in length). Furthermore, evidence has been presented to suggest that the repetitive sequences are preferentially located toward the 5′ ends of the HnRNA molecules (Molloy *et al.*, 1974). This was shown in experiments in which HnRNA was treated with alkali and then the 3′ ends carrying poly(A) were collected. It was found that the hybridizing ability was low when the poly(A)-terminated fragments were 3000 nucleotides or shorter but was increased 3–4 times when fragments of

8000 nucleotides (or longer) were analyzed. Some of the repetitive sequences form double-stranded regions in the HnRNA molecules (Jelinek and Darnell, 1972; Jelinek *et al.*, 1974).

Although not conclusive, competition hybridization experiments suggest that there are considerable differences in sequence homology between repetitive RNA isolated from various tissues and developmental stages (Church and McCarthy, 1967, 1970; Davidson *et al.*, 1968; Glisin *et al.*, 1966; McCarthy and Hoyer, 1964). One particularly striking example is the lack of sequence homology between oocytes and blastulas in *Xenopus*. It also seems as if different tissues have quite different repetitive sequence complexities. It can therefore be concluded, for repetitive as well as for unique sequences, that there are major qualitative differences between various tissues. It is not known whether the tissue-specific repetitive sequences are parts of HnRNA molecules containing the tissue-specific unique sequences.

During recent years a series of structural features have been revealed, the significance of which can still only be matters of speculation. In agreement with what can be expected from primary transcripts, HnRNA contains terminal nucleoside triphosphates in the 5′ end (Ryskov and Georgiev, 1970; Georgiev *et al.*, 1972). Furthermore, HnRNA is methylated, although to a much lesser extent than ribosomal or transfer RNA (Perry and Kelley, 1974). Within HnRNA molecules there are short stretches of oligouridylic acid [oligo(U)] (Burdon and Shenkin, 1972; Molloy *et al.*, 1972) and oligoadenylic acid [oligo(A)] (Nakazato *et al.*, 1974). Both the oligo(U) (Molloy *et al.*, 1974) and the oligo(A) sequence (Nakazato *et al.*, 1974) are mainly located toward the 5′ end of the HnRNA molecules. A long poly(A) sequence is frequently present in the 3′ terminus of HnRNA. The latter sequence is added posttranscriptionally and will be discussed separately (see p. 201). RNAse-resistant, double-stranded RNA was recorded in nuclei by Montagnier (1968) and somewhat later shown to be present as part of HnRNA molecules (Jelinek and Darnell, 1972; Ryskov *et al.*, 1972, 1973). These double-stranded sequences represent only a minor portion of total HnRNA 1–3%), are located mainly toward the 5′ ends of HnRNA, and are of repetitive sequence composition (Harel and Montaignier, 1971; Jelinek and Darnell, 1972; Jelinek *et al.*, 1974; Kimball and Duesberg, 1971).

To sum up, HnRNA harbors unique as well as repetitive sequences, representing a large portion of the genome. Some of these sequences have been shown to be specific to certain tissues or, alternatively, to certain developmental stages. The characterization of HnRNA in terms of sequence content and structural properties such as methylation and presence of poly(A) will provide the basis for investigations as to the significance of HnRNA.

Addition of Poly(A)

The presence of poly(A) in HnRNA sequences was established some years ago (Edmonds and Caramela, 1969; Lim and Cañellakis, 1970). These sequences are 100–200 nucleotides in length and covalently linked to HnRNA at its 3′ end (for review, see Darnell *et al.*, 1973; Weinberg, 1973). DNA templates corresponding to these sequences have, however, not been detected in the genome. This has been convincingly shown for certain viruses. Nuclear virus genomes (adenovirus, SV40) lack poly(dA:dT) segments that might code for poly(A) sequences, but still virus-specific sequences can be recorded in poly(A)-containing HnRNA (and mRNA) (Philipson *et al.*, 1971; Wall *et al.*, 1972; Weinberg *et al.*, 1972). The same situation probably prevails in mammalian cells (Birnboim *et al.*, 1973; Bishop *et al.*, 1974), where at least an insufficient number of poly(dA:dT) segments exist in DNA to account for the transcription of poly(A) with each HnRNA molecule. Posttranscriptional addition of poly(A) was therefore suggested and nicely verified in a series of experiments. It could be shown that, while the synthesis of HnRNA was strongly inhibited by actinomycin D, poly(A) synthesis occurred essentially unchanged for 1–2 min (Darnell *et al.*, 1971; Jelinek *et al.*, 1973). Furthermore, HnRNA molecules, but not poly(A), could be synthesized in the presence of cordycepin (3′-deoxyadenosine) (Darnell *et al.*, 1971; Mendecki *et al.*, 1972; Philipson *et al.*, 1971). Nakazato *et al.* (1974) were able to demonstrate that, although 3′-terminal poly(A) synthesis was largely inhibited by cordycepin, the synthesis of oligo(A) tracts located internally in HnRNA proceeded as normal, suggesting that cordycepin would interfere with neither adenosine metabolism nor transcription of HnRNA. These various observations strongly support the idea that poly(A) is added to HnRNA in a process separate from transcription of HnRNA. As no free poly(A) sequences can be recorded in the nucleus even after very short pulses (Edmonds *et al.*, 1971; Jelinek *et al.*, 1973), it has been assumed that the HnRNA molecules act as primers for the synthesis of poly(A) by the stepwise addition of single adenylate residues.

It should be noted that not all HnRNA molecules in the HnRNA population carry a poly(A) segment; in fact, only about 30% do (Greenberg and Perry, 1972; Jelinek *et al.*, 1973). This does not necessarily mean that only some transcripts are polyadenylated. First of all, as indicated above, a large portion of the HnRNA population might represent growing RNA molecules to which the addition of poly(A) to a finished 3′ end cannot take place (synthesis proceeds from the 5′ end of the RNA molecules). It might also be that some HnRNA molecules have not yet obtained their poly(A) tail, and, finally, some HnRNA molecules might represent cleavage products from the 5′ ends (an *in vivo* or preparative process) of a previously giant poly(A)-containing molecule. Therefore, the number of transcripts that become polyadenylated is evidently uncertain. On the other hand, cleavage fragments from the same molecules might all be polyadenylated, which might lead to an overestimation of the number of primary transcripts that obtain a poly(A) sequence. More work has to be carried out until it can be stated whether all primary transcripts do get a poly(A) tail and whether 3′ ends, exposed by cleavage of primary transcripts, also get polyadenylated. The poly(A) sequence has received a lot of attention during recent years because of its possible role in the transfer of mRNA sequences to cytoplasm (see p. 201).

Lifetime and Fate of HnRNA

A short lifetime for HnRNA was indicated from the observation that HnRNA is the major labeled RNA fraction in short pulse experiments, although it constitutes only 3% of total cellular RNA (Darnell, 1968; Soeiro *et al.*, 1968). When the relative synthetic rates of preribosomal and HnRNA were measured during short pulses (10 and 20 min), it was concluded that HnRNA has a mean half-life of only 3 min (Soeiro *et al.*, 1968). This very low value has, however, been questioned (e.g., Brandhorst and McConkey, 1974). The time-course of HnRNA synthesis in HeLa cells was studied by Penman *et al.* (1968). They measured the accumulation of activity in HnRNA in the presence of a low dose of actinomycin D and could show that a plateau was rapidly established with a life span of HnRNA of about 60 min. The rapid turnover of HnRNA was also shown in pulse experiments, followed by actinomycin D chases. In these studies, the half-life of HnRNA was estimated to 10–30 min (Attardi *et al.*, 1966; Houssais and Attardi, 1966; Scherrer *et al.*, 1966). Since actinomycin D causes an artificial degradation of HnRNA, in particular the nascent part (Egyházi, 1974*a*), these estimates are probably minimum values for half-lives. Cold chase experiments have also been tried, but because of the difficulty in excluding reutilization of label in these experiments they have been less efficient, at least in certain types of cells (e.g., in HeLa cells, according to Warner *et al.,* 1966). However, in other cell types, such as avian erythroblasts, there has been some success (Spohr *et al.,* 1974). It was established again that the turnover of giant HnRNA molecules is rapid (half-life of about 30 min). In experiments by Brandhorst and McConkey (1974), specific activities of the pools were also measured, and the half-life of HnRNA in the L cell nucleus was determined to be 23 min. It can therefore be stated that, in general, the half-lives of HnRNA is within the range 10–30 min. It should also be noted that half-lives of at least this order have to be expected in order to be compatible with recent determinations of the time necessary for the transcription process (for review, see Daneholt, 1975).

There are two major alternatives to account for the short lifetimes for HnRNA in the nucleus: either the molecules are delivered to cytoplasm or they are degraded within the nucleus. The rapidly labeled HnRNA was quickly regarded as a likely precursor to mRNA in cytoplasm, since both HnRNA and mRNA had a DNA-like base composition and a heterogeneous size distribution (for review, see Darnell, 1968). Moreover, label appeared in HnRNA very rapidly, in contrast to the lag period before radioactivity appeared as mRNA in cytoplasm. This delay is usually on the order of 15–20 min (Penman *et al.,* 1968; Schochetman and Perry, 1972), but the time can be considerably shorter, e.g., only some minutes for histone mRNAs (Adesnik and Darnell, 1972; Schochetman and Perry, 1972), or longer, e.g., at least 45 min for 75 S RNA in *Chironomus* salivary gland cells (Daneholt and Hosick, 1973; Edström and Tanguay, 1974). When the kinetic characteristics of HnRNA were compared in detail to those of cytoplasmic mRNA, it was, however, inferred that a simple precursor–product relationship between the two molecular species could not be prevalent (Attardi *et al.,* 1966; Brandhorst and Humphreys, 1971; Brandhorst and McConkey, 1974; Houssais and Attardi, 1966; Penman *et al.,* 1968; Roberts and Newman, 1966; Scherrer *et al.,* 1966; Soeiro *et al.,* 1966, 1968). Conclusive data were difficult to ascertain, mainly because of the extensive degradation of HnRNA (see p. 202). From the rate of accumulation of label in HnRNA and mRNA, it seems as if only a minor part of HnRNA can in fact be utilized as mRNA, the remainder then being rapidly degraded. The amount of degradation has been recently estimated to be on the order of 85% in sea urchins (Brandhorst and Humphreys, 1971, 1972) and as much as 98% in mammalian cells (Brandhorst and McConkey, 1974). In the beginning, the arguments in favor of a mRNA precursor role for HnRNA were rather weak, but more recent studies clearly suggest such a role, and I will discuss this alternative pathway prior to discussing the degradation process.

Delivery of mRNA Sequences to Cytoplasm

Investigations that show that mRNA sequences are present in HnRNA have provided strong support for the idea that HnRNA is a precursor to mRNA. The first mRNA sequences that could be directly recorded in HnRNA were virus-specific sequences in transformed cells. In studies on SV40 virus transformed cells (Lindberg and Darnell, 1970; Tonegawa *et al.*, 1970; Wall and Darnell, 1971; Wall *et al.*, 1973), it was possible to establish that the viral sequences were present in cellular HnRNA as well as in mRNA. The HnRNA molecules containing viral sequences were heterogeneous in size (covering the whole HnRNA size range) and also contained host cell sequences. The virus-specific mRNA was recorded as discrete RNA species, lacking host cell sequences, and of considerably lower molecular weights than HnRNA. These studies clearly indicated that HnRNA can act as a precursor. Whether the largest HnRNA molecules deliver sequences to cytoplasm could not be determined, as a proper chase of sequences from larger to smaller molecules could not be performed. It can, however, be argued that viral sequences are atypical sequences. It is therefore pleasing that techniques are now available to test HnRNA for its content of ordinary mRNA sequences (Lewin, 1975*b*). At the present time, only certain specific mRNA sequences have been demonstrated in HnRNA (see below). It seems reasonable to assume that HnRNA will prove to contain all the different mRNA sequences that are present in cytoplasm.

Since a precursor–product relationship between HnRNA and mRNA appears likely, it is worth comparing the two molecular populations in some detail. It has been argued that HnRNA must be considerably diminished in size (cleavage and/or degradation) prior to the delivery of the mRNA sequences (perhaps a size reduction of 10–50 times) (e.g., see Weinberg, 1973). First, it has been observed that HnRNA covers not only the mRNA size range (400–4000 nucleotides) but also molecules of much higher molecular weights (up to 50,00 nucleotides) and has a higher average molecular weight than mRNA. It is, however, possible that this difference in size range has been overemphasized to some extent, as messenger RNA molecules of very high molecular weights (corresponding in size to giant HnRNA) have been recorded in cytoplasm in many different systems (Brandhorst and Humphreys, 1972; Daneholt, *et al.*, 1969; Giudice *et al.*, 1972; Kumar and Lindberg, 1972; Ovchinnikov *et al.*, 1969; Sconzo *et al.*, 1974; Suzuki and Brown, 1972). It has been estimated for HeLa cells that as much as 25% of the mRNA molecules are at least 2800 nucleotides long (Davidson and Britten, 1973). These results indicate that very high molecular weight HnRNA might be transferred into cytoplasm without a considerable size reduction. A striking example is the transport of 75 S from BR 2 into cytoplasm (see below). Another argument for a considerable size diminution has been obtained from chase experiments (e.g., Spohr *et al.,* 1974). The activity in the highest molecular weight HnRNA decreases more rapidly than in the lower molecular weight HnRNA regions during a chase. This result is of course consistent with a cleavage process, but as a decrease in one region has not been accompanied by an absolute increase in activity in another region the evidence is not conclusive. A difference in turnover rates might provide a sufficient explanation. It is evident that kinetic information on total HnRNA and mRNA cannot provide strong support for a cleavage mechanism. In conclusion, the available information on the molecular size ranges of HnRNA and mRNA does not necessitate a major reduction in size (10 times or more) of the giant HnRNA molecules and a subsequent delivery of mRNA of relatively small size to cytoplasm.

It can be predicted that the analysis of the relation between HnRNA and mRNA will be considerably simplified when defined mRNA sequences can be studied. It is therefore important that it recently became feasible to investigate the transfer of some specific mRNA sequences from nucleus to cytoplasm in a few eukaryotic systems:

Globin mRNA. A defined globin precursor, 14 S RNA ($6–7 \times 10^5$ daltons), has been detected in the nucleus, while the finished globin mRNA (9.5 S, corresponding to 2×10^5 daltons) cannot be recorded there (Macnaughton *et al.,* 1974). It therefore seems as if 14 S RNA is cleaved just prior to or after the transfer of the mRNA sequence of cytoplasm. The fate of the remaining two-thirds of the precursor is not known.

Immunoglobin Heavy Chain mRNA. The putative nuclear precursor has an estimated molecular weight of 2.4×10^6 daltons, while the finished mRNA in cytoplasm has a molecular weight of at least 6.0×10^5 daltons (Stevens and Williamson, 1973). Again, there seems to be a processing event, but not to a particularly large extent.

Ovalbumin mRNA. It was feasible to demonstrate that all molecules containing ovalbumin mRNA sequences are confined to one particular species, an 18 S molecule (McKnight and Schimke, 1974). If there had been one RNA molecule per cell of another size containing this particular sequence, it should have been detected during the experiment. Therefore it seems likely that a processing of ovalbumin mRNA does not take place, or, alternatively, the precursor has to be very similar in size to the mRNA molecule.

Balbiani Ring 2 RNA. Balbiani ring 2 RNA, a giant RNA (75 S), is synthesized in one particular chromosome region, Balbiani ring 2, in the salivary glands of the dipteran *Chironomus tentans* (Daneholt, 1972) and has been followed from its site of synthesis via nuclear sap and into cytoplasm (Daneholt and Hosick, 1973; Lambert and Edström, 1974). Recently it has also been shown that this molecule enters polysomes and therefore probably functions as mRNA (Daneholt and Wieslander, to be published). Indirect evidence suggests that BR 2 RNA contains mRNA sequences for the major product of these cells, the salivary gland polypeptides (for discussion, see Daneholt, 1974). Also, in this particular case there does not seem to be a major size reduction of the primary transcript, although because of the problems of properly deciding the molecular weight of this giant RNA molecule, a minor size reduction cannot be excluded. The possibility that a major segment of the molecule is lost has been excluded.

In the coming years it will be possible to isolate and characterize several defined molecules which are precursors to mRNA and describe their further fate in considerable detail. From the information available just now, it seems clear that cleavage of primary transcripts can occur, but the transcripts are perhaps not in general reduced in size to such a large extent as earlier assumed. In certain cases (e.g., ovalbumin mRNA), the primary transcripts might be transferred to cytoplasm without being cleaved at all. It should also be pointed out that in primitive eukaryotic cells, e.g., in *Dictyostelium* (Lodish *et al.*, 1973), there are no giant RNA molecules and consequently little or no reduction in size.

It can very well be anticipated that noncoding sequences in HnRNA are involved in the delivery of mRNA sequences to cytoplasm (for discussion, see p. 208). The poly(A) sequence is the most intensively studied, noncoding sequence in HnRNA, and it has been proposed that the addition of this sequence to an HnRNA molecule might save an adjacent mRNA sequence from degradation and mediate the transfer of this mRNA sequence to cytoplasm (for review, see Darnell *et al.*, 1973). It is well known that mRNA, like HnRNA, contains poly(A) tails (e.g.; Darnell *et al.*, 1971). Moreover, it has been shown that poly(A) is synthesized within the nucleus and delivered from the nucleus to the cytoplasm (Jelinek *et al.*, 1973). In short pulses (45 sec) 90% of labeled poly(A) is found in the nucleus (Jelinek *et al.*, 1973), while after long periods (150 min) more than 80% resides in cytoplasm (Darnell *et al.*, 1971). When the addition of poly(A) to HnRNA is blocked by cordycepin, no poly(A)-containing mRNA appears in cytoplasm (Adesnik *et al.*, 1972; Darnell *et al.*, 1971; Jelinek *et al.*, 1973; Lee *et al.*, 1971; Mendecki *et al.*, 1972; Penman *et al.*, 1970). These early results were therefore in good agreement with the proposed role for poly(A) in the transfer process, but more recent investigations suggest a more complex situation. First, it has been discovered that not all poly(A) sequences linked to HnRNA are exported from nucleus to cytoplasm but some are degraded within the nucleus (Perry *et al.*, 1974). Furthermore, the kinetics of the labeling of poly(A) in the nucleus and cytoplasm indicated that there is probably also a concurrent cytoplasmic synthesis of poly(A) (Perry *et al.*, 1974). Such a synthesis has been convincingly demonstrated in sea urchin embryos (Slater and Slater, 1974; Slater *et al.*, 1972; Wilt, 1973). In this context, it is also interesting to note that poly(A) synthesis can occur also in mitochondria (Perlman *et al.*, 1973) as well as in cytoplasmic viruses (Eaton *et al.*, 1972; Kates, 1970; Kates and Beeson, 1970; Yogo and Wimmer, 1972). Various observations therefore suggest as concluded by Perry *et al.* (1974) that there cannot be a simple precursor–product relationship between nuclear poly(A) and cytoplasm poly(A). The cordycepin experiments suggest that some HnRNA molecules need a poly(A) tail in order to be able to deliver mRNA sequences to cytoplasm. However, poly(A) cannot be the only factor involved in the transfer of mRNA sequences from nucleus to cytoplasm, as poly(A) is evidently not needed for transport of a certain part of the mRNA population [unless a transient addition of poly(A) is assumed]. It was initially suggested that all mRNA except histone mRNA contains poly(A) (for review, see Darnell *et al.*, 1973; Weinberg, 1973), but it is now clear that there are two distinct populations of mRNA molecules, one lacking and one containing poly(A) (Milcarek *et al.*, 1974; Nemer *et al.*, 1974). The two species have about the same average size and show no or little cross-hybridization. It is obvious that more detailed information on the poly(A) metabolism and the fate of HnRNA and mRNA sequences linked to poly(A) is needed in order to establish a role for poly(A) in the delivery of mRNA sequences from nucleus to cytoplasm.

Intranuclear Degradation of HnRNA

As pointed out previously, kinetic studies suggest that most HnRNA is rapidly degraded within the nucleus. The information on this process is still very restricted, but some relevant data are now available. One important issue is, of course, whether there are specific sequences that are degraded, i.e., whether there are nuclear-specific sequences in HnRNA. It can now be stated that there are both unique sequences and short repetitive sequences of the interspersed type that never leave the nucleus. The complexity of unique HnRNA sequences is considerably greater than that of the corresponding mRNA sequences, e.g., 10 times more in sea urchin gastrulas (Galau *et al.*, 1974; Smith *et al.*, 1974), at least 5 times in Friend mouse cells (Birnie *et al.*, 1974; Getz *et al.*, 1975), and 4 times in

L cells (Hames and Perry, quoted by Lewin, 1975*b*). Also, the number of different repetitive sequences is considerably greater in HnRNA than in mRNA (Arion and Georgiev, 1967; Scherrer *et al.,* 1970; Shearer and McCarthy, 1967). In some extreme cases, mRNA lacks repetitive sequences altogether; one example is sea urchin gastrula mRNA (Galau *et al.,* 1974; Goldberg *et al.,* 1973), and another is L cells (Greenberg and Perry, 1971). Here it is quite evident that the repetitive sequences present in HnRNA are not transferred to mRNA in cytoplasm. It should also be noted that when defined mRNA species have been studied no repetitive sequences have been recorded (see Davidson and Britten, 1973). Although there seem to be exceptions, e.g., *Xenopus* neurula (Crippa *et al.,* 1973; Dina *et al.,* 1973) and *Dictyostelium* (Lodish *et al.,* 1973), it can be concluded that, in general, short repetitive sequences of the interspersed type are not delivered to the cytoplasm (as shown for HeLa cells, L cells, rat myoblasts, and sea urchin embryos). Longer repetitive sequences, probably corresponding to repeated structural genes, have, however, been recorded in mRNA (for discussion, see Lewin, 1975*b*). Finally, it should be added that the short poly(U) (Molloy *et al.,* 1972) and the short internal poly(A) sequences in HnRNA (Nakazato *et al.,* 1974) do not appear at all (or to a very low extent in mRNA) and therefore are likely to be degraded within the nucleus.

It will become of great interest to establish the intranuclear location for the cleavage and degradation processes, which modify the primary transcripts. They can, of course, take place when the RNA molecules are still growing (as frequently occurs in prokaryotes), when they reside in the nuclear sap, or possibly when they are in transit from the nucleus into the cytoplasm. At the present time, it cannot be firmly stated when HnRNA is modified, but there are certain observations that might hint at the possibility that cleavage and possibly also degradation of HnRNA are already taking place on the DNA template. When the molecules once have been released into the nuclear sap, they are possibly more stable and prone to enter the cytoplasm. It has been observed in the insect *Chironomus* that the HnRNA pattern of chromosomal RNA is quite different from the HnRNA pattern in the nuclear sap, while, on the other hand, there are striking similarities between the spectrum of nuclear sap and cytoplasmic heterogeneous RNA (Daneholt and Hosick, 1973; Egyházi, 1974*b*). It is evident that discrete RNA species that will later on appear in cytoplasm, e.g., 75 S RNA (Daneholt and Svedhem, 1971; Daneholt and Hosick, 1973) and 35 S RNA (Egyházi, 1974*b*), are enriched in nuclear sap compared to their presence on the chromosomes. A gradual accumulation within the nucleus of a particular HnRNA population, with a high metabolic stability and a size spectrum more similar to that of mRNA in cytoplasm than to that of the most rapidly labeled HnRNA, has also been reported by Georgiev (1967) and Scherrer *et al.* (1970). It is interesting that Price *et al.* (1974) have detected an HnRNA fraction with mRNA size characteristics and with a high poly(A) content. This fraction seems to be less intimately bound to the DNA template than most HnRNA, and consequently can very well correspond to the nuclear sap fraction in *Chironomus,* while the rest of HnRNA can be mainly growing molecules. These studies therefore suggest that HnRNA is processed (cleaved and/or degraded) when it is associated with the DNA template, while when released from the template it is ready for transfer to cytoplasm without any further major size changes. Addition of poly(A) might, of course, take place on the template (just after the molecules have been completed or processed), in the nuclear sap, or even to some extent in the cytoplasm.

The very nature of the degradation process is still unknown, but several suggestions have been made that relate the rapid degradation of HnRNA to the regulation of gene expression (for discussion, see p. 207).

Nuclear RNA of Low Molecular Weight

Apart from ribosomal 5 S RNA and transfer RNA (see appropriate chapters in this volume), there are a series of discrete RNA species of low molecular weight (less than 30 nucleotides) in the nucleus. These RNA species have been extensively studied (for review, see Weinberg, 1973). There are at least 11 fractions present, probably in general with a low complexity (Hellung-Larsen and Fredriksen, 1971; Larsen *et al.,* 1969; Moriyama *et al.,* 1969; Prestayko *et al.,* 1970, 1971; Weinberg and Penman, 1969). Some display characteristic fingerprints, and three of them have even been sequenced by Busch and coworkers (Reddy *et al.,* 1974; Ro-Choi *et al.,* 1972; Shibata *et al.,* 1973). Apart from one species (Ro-Choi *et al.,* 1972), they are extensively methylated (Hellung-Larsen and Fredriksen, 1972; Weinberg and Penman, 1968; Zapisek *et al.,* 1969). It is interesting that these RNA fractions are metabolically stable (half-life of at least 1 day) and are not present outside the nucleus. Two of the species are located in the nucleoli and appear to take part in the maturation of 28 S RNA. The remaining species are probably extranucleolar. Their structure and metabolic properties suggest that they cannot be nonselective degradation products of HnRNA (e.g., Reddy *et al.,* 1974), but it is not excluded that they might nevertheless have some relation to HnRNA and its transcription or processing.

An RNA species smaller than transfer RNA has been described and extensively studied (see Holmes *et al.,* 1972, for review). Most investigators now seem to agree that this RNA (designated chromosomal RNA, cRNA) appears as a result of degradation, mainly of HnRNA (for discussion, see Weinberg, 1973). If proper precautions are taken against degradation, no RNA smaller than tRNA can be recorded, e.g., in HeLa cell nuclei (Weinberg, 1973) and in total RNA from *Chironomus* polytene chromosomes (Egyházi *et al.,* 1968).

Concluding Remarks

The transcription units in eukaryotes are probably of very different sizes, as implied by the size range of the primary transcripts (see p. 190). It is, of course, a delicate matter to determine the size of a transcription unit (used here to mean the DNA segment between the sites for initiation and termination of the RNA synthesis) just from the size of the recorded RNA product. This has been amply demonstrated for prokaryotes, in which RNA molecules can be processed or degraded while still attached to their DNA template (e.g., Morse *et al.,* 1969). Such mechanisms will lead to underestimates of the sizes of the transcription units. It can, however, be safely concluded for eukaryotes that there must at least be some transcription units of very large sizes (up to 50,000 nucleotide pairs), because of the presence of giant transcripts. It has even been argued that these very large transcripts might be the rule and that most or all mRNA species are generated from such large transcripts. Such an idea (e.g., see the model proposed by Georgiev, 1969) is pleasing from the genetic point of view. Recent data suggest that in *Drosophila* a genetic unit corresponds to a whole chromomere (Judd *et al.,* 1972) (average size about 30,000 nucleotide pairs according to Rudkin, 1965). Therefore, transcription of the entire chromomere into one single giant transcript appears an attractive possibility. In one particular case, the Balbiani ring 2 system, it has been possible to make a rough comparison between the size of the RNA product (75 S RNA) and the size of the corresponding chromomere. It was inferred that at least a considerable part of the BR 2 chromomere, and maybe the whole, is transcribed into one single giant transcript (Daneholt, 1972). At the present time, the relevant question is therefore not whether transcription units of giant size exist, but rather how frequent they are. Recent studies on specific primary transcripts (e.g., the ovalbumin mRNA precursor) clearly suggest that considerably shorter transcription units do exist (2000 nucleotide pairs or even shorter). When attempts are made to estimate the most frequent transcription unit size from an HnRNA distribution (e.g., in gradients), it is, of course, important to remember that the distribution is usually expressed as a radioactivity distribution rather than as a frequency distribution of molecular lengths. If proper adjustments are made for the molecular sizes in the various parts of the HnRNA distribution, the proportion of smaller molecules becomes quite impressive. It is not excluded that the presence of very large transcription units has been overemphasized and that smaller units might be more frequent than earlier anticipated. Investigations of transcription complexes with the Miller spreading technique for electron microscopy (e.g., Miller and Bakken, 1972) should provide detailed information as to the size range and the abundance of different size classes. The electron microscope technique should give proper size determinations even if processing of growing molecules does occur before the primary transcripts are completed.

It is evident, mainly from hybridization experiments, that as a rule a large portion (5–25%) of the eukaryotic genome is transcribed into HnRNA (see p. 193). The impressive figure of 30% has been measured in *Drosophila* during pupal metamorphosis (Turner and Laird, 1973). Even if the average size of the transcription units is not known, it is obvious that hundreds or rather thousands of separate transcription units have to be active per haploid genome. For example, in the case of mouse brain there are at least as many as 6000 active units, even if it is assumed that the average primary transcript is as large as 5×10^4 nucleotides (Hahn and Laird, 1971).

The expression of at least some of the transcription units is probably regulated on the transcriptional level. This can be inferred from the observations that there are sequences in HnRNA that are only present in certain tissues or developmental stages (see p. 193). These biochemical results are in excellent agreement with cytological evidence obtained from studies of polytene chromosomes in insect cells. A great number of the chromosome regions are active in transcription. Pelling (1964) showed that in each cell as many as 10% of all chromosome bands are synthesizing RNA. Direct biochemical analyses of giant chromosomes have shown that this RNA is mainly of the HnRNA type (Daneholt, 1975; Daneholt *et al.*, 1969; Pelling, 1970). While it has been shown that certain chromosome sites active in transcription (the so-called puffs) are common to many tissues and developmental stages, other puffs are tissue or developmental stage-specific (for review, see Daneholt, 1974). The polytene chromosome studies are therefore in agreement with the view that the sequence differences in HnRNA from different tissues (or developmental stages) are mainly due to a transcriptional rather than posttranscriptional regulation of HnRNA synthesis. Of course, it is difficult to rigorously exclude the possibility that posttranscriptional regulation, closely coupled to the transcription process, does play a role.

It is established that HnRNA contains mRNA sequences (see p. 199), and consequently at least some and probably most transcription units harbor genetic information in the conventional sense. It is, however, not possible to state whether *all* HnRNA transcription units contain amino acid coding sequences, or whether, in general, the transcription units that do contain amino acid coding sequences are mono- or polycistronic. Evidently, because of their small size, some primary transcripts have to be monocistronic (e.g., the ovalbumin mRNA precursor), like the vast majority of mRNA (for discussion, see Davidson and Britten, 1973). The giant size of other primary transcripts implies that a great number (let us say 10–50) of average-sized mRNA sequences could very well be present in each of these giant transcripts. However, there is still no convincing evidence that the highest molecular weight HnRNA is cleaved into much smaller segments, one or more of which are exported to the cytoplasm as messengers. On the contrary, it can be stated that some giant HnRNA molecules are delivered to the cytoplasm and probably function as mRNA. It is worth pointing out that at least some of these giant RNA species are required to code for very large protein subunits (see compilation of subunit sizes by Klotz (1970)) and, hence, again are likely to be monocistronic. Messenger RNA for the silk fibroin is, for example, 16,000 nucleotides in length and the coding length needed for the fibroin is 14,000 nucleotides (Lizardi *et al.*, 1975). Even if all giant mRNA molecules in the cytoplasm turn out to be monocistronic, it is still, of course, conceivable that there are polycistronic transcription units containing several structural and/or regulatory genes.

There is little doubt that a regulation of gene expression takes place on the transcriptional level, but it is quite possible that posttranscriptional regulation might operate as well. Then, as pointed out by Darnell *et al.* (1973), there has to be a production of more mRNA sequences than are actually used as messengers, the surplus of sequences being either accumulated or degraded. It is therefore evident that the extensive intranuclear degradation recorded in eukaryotes can be understood on the basis of a posttranscriptional regulation process, but proof for such a mechanism is lacking. More information is needed on the RNA sequences that are degraded within the nucleus. First, it has to be established whether mRNA sequences are degraded at all in the nucleus (*cf.* the cascade hypothesis presented by Scherrer and Marcaud, 1968). If so, is one particular population of mRNA sequences degraded while another is delivered to the cytoplasm, or, alternatively, do a certain number of copies from each mRNA sequence leave the nucleus, while the remaining copies are degraded? It is true that there are nuclear-specific sequences, which would argue in favor of the first alternative, but this question has to be answered in terms of mRNA sequences and not only in terms of total RNA sequences. Furthermore, it can be expected that the elucidation of the structure and metabolism of the noncoding segments of HnRNA might contribute to a better understanding of a putative posttranscriptional regulation and perhaps also transcriptional regulation and translational modulation (see Darnell *et al.*, 1973).

It has frequently been proposed that noncoding sequences in HnRNA represent either regulatory segments in DNA involved in regulation of transcription or alternatively different kinds of recognition sites involved in posttranscriptional regulation or translational modulation. The first group of sequences can be represented by the repetitive sequences of the interspersed type. They have been proposed to reflect sequences in DNA serving as binding sites for molecules (proteins or RNA) regulating the transcription process (Britten and Davidson, 1969; Davidson and Britten, 1973). It is true that the interspersed repetitive sequences in DNA are preferentially linked to amino acid coding sequences (Davidson *et al.*, 1975). However, direct evidence for the suggested role as binding sites for regulatory molecules is still not available. It should be noted in this context that DNA segments binding repressor molecules can be transcribed in prokaryotes (e.g., in the *lac* operon in *Escherichia coli,* according to Gilbert *et al.,* 1973). Other structural properties of HnRNA have been invoked in posttranscriptional regulation. When Perry and Kelley (1974) discovered that HnRNA as well as mRNA is methylated, it was logical from the known characteristics of the ribosomal maturation pathway (e.g., Weinberg, 1973) to suggest that methylation of certain sequences might save those from being degraded and enable them to reach the cytoplasm and function as messengers. A similar role has been proposed for the poly(A) sequence (see p. 201). An addition of poly(A) seems to be necessary in order to achieve a proper maturation and transport of mRNA from nucleus to cytoplasm. This is at least true for a certain part of the mRNA population. However, there are still many perplexing observations on poly(A) metabolism that prevent firm conclusions as to its role in posttranscriptional regulation (see p. 202). Perhaps the hypotheses on a posttranscriptional regulation of gene expression can not be efficiently tested until the pathways for the mRNA sequences have been elucidated in considerable detail.

The progress in the research of HnRNA and its functional role in gene expression can be expected to be very rapid in the near future. This will mainly be a consequence of the development of hybridization techniques for unique RNA sequences, the recognition in HnRNA of several useful markers [such as methylated sequences and poly(A)], and finally the availability of specific primary transcripts. It can be anticipated that the main structural features will soon be revealed, and these will be the basis for further experimentation on the regulatory mechanisms governing gene expression in eukaryotes.

Acknowledgments

The author's work was supported by the Swedish Cancer Society, Magnus Bergvalls Stiftelse, and Karolinska Institutet (Reservations-anslaget). I am also indebted to Dr. Steven T. Case for revising the English text and to Miss Hannele Jansson for typing the manuscript.

Literature Cited

Adesnik, M. and J. E. Darnell, 1972 Biogenesis and characterization of histone messenger RNA in HeLa cells. *J. Mol. Biol.* **67**:397–406.

Adesnik, M., M. Salditt, W. Thomas, and J. E. Darnell, 1972 Evidence that all messenger RNA molecules (except histone messenger RNA) contain poly(A) sequences and that the poly(A) has a nuclear function. *J. Mol. Biol.* **71**:21–30.

Allfrey, V. G., 1970 Biosynthetic reactions in the cell nucleus. In *Aspects of Protein Biosynthesis,* edited by C. B. Anfinsen, Jr., Part A, pp. 247–365, Academic Press, New York.

Arion, V. J. and G. P. Georgiev, 1967 On the functional heterogeneity of chromosomal information. *Dokl. Akad. Nauk USSR* **172**:716–719.

Attardi, G., H. Parnas, M.-I. H. Hwang, and B. Attardi, 1966 Giant size rapidly labeled nuclear ribonucleic acid and cytoplasmic messenger ribonucleic acid in immature duck erythrocytes. *J. Mol. Biol.* **20**:145–182.

Birnboim, H. C., R. E. J. Mitchel, and N. A. Straus, 1973 Analysis of long pyrimidine polynucleotides in HeLa cell nuclear DNA: Absence of polydeoxythymidylate. *Proc. Natl. Acad. Sci. USA* **70**:2189–2192.

Birnie, G. D., E. MacPhail, B. D. Young, M. J. Getz, and J. Paul, 1974 The diversity of the messenger RNA population in growing Friend cells. *Cell Differ.* **3**:221–232.

Bishop, J. O., M. Rosbash, and D. Evans, 1974 Polynucleotide sequences in eukaryotic DNA and RNA that form ribonuclease-resistant complexes with polyuridylic acid. *J. Mol. Biol.* **85**:75–86.

Bramwell, M. E., 1972 A comparison of gel electrophoresis and density gradient centrifugation of heterogeneous nuclear RNA. *Biochim. Biophys. Acta* **281**:329–337.

Brandhorst, B. P. and T. Humphreys, 1971 Synthesis and decay rates of major classes of deoxyribonucleic acid like ribonucleic acid in sea urchin embryos. *Biochemistry* **10**:877–881.

Brandhorst, B. P. and T. Humphreys, 1972 Stabilities of nuclear and messenger RNA molecules in sea urchin embryos. *J. Cell Biol.* **53**:474–482.

Brandhorst, B. P. and E. H. McConkey, 1974 Stability of nuclear RNA in mammalian cells. *J. Mol. Biol.* **85**:451–463.

Britten, R. J. and E. H. Davidson, 1969 Gene regulation for higher cells: A theory. *Science* **165**:349–357.

Brown, I. R. and R. B. Church, 1972 Transcription of non-repeated DNA during mouse and rabbit development. *Dev. Biol.* **29:**73–85.

Burdon, R. H., 1971 Ribonucleic acid maturation in animal cells. *Prog. Nucleic Acid Res. Mol. Biol.* **11:**33–79.

Burdon, R. H. and A. Shenkin, 1972 Uridylate-rich sequences in rapidly labelled RNA of mammalian cells. *FEBS Lett.* **24:**11–15.

Busch, H., T. S. Ro-Choi, A. W. Prestayko, H. Shibata, S. T. Crooke, S. M. El-Khatib, Y. C. Choi, and C. M. Mauritzen, 1971 Low-molecular-weight nuclear RNAs. *Perspect. Biol. Med.* **15:**117–139.

Church, R. B. and I. R. Brown, 1972 Tissue specificity of genetic transcription. In *Results and Problems in Differentiation,* Vol. 3, edited by H. Ursprung, pp. 11–24, Springer-Verlag, New York.

Church, R. B. and B. J. McCarthy, 1967 Ribonucleic acid synthesis in regenerating and embryonic liver. II. The synthesis of RNA during embryonic liver development and its relationship to regenerating liver. *J. Mol. Biol.* **23:**477–486.

Church, R. B. and B. J. McCarthy, 1970 Unstable nuclear RNA synthesis following estrogen stimulation. *Biochim. Biophys. Acta* **199:**103–114.

Crippa, M., I. Meza, and D. Dina, 1973 Sequence arrangement in mRNA: Presence of poly(A) and identification of a repetitive fragment at the 5′ end. *Cold Spring Harbor Symp. Quant. Biol.* **38:**933–942.

Daneholt, B., 1972 Giant RNA transcript in a Balbiani ring. *Nature (London) New Biol.* **240:**229–232.

Daneholt, B., 1974 Transfer of genetic information in polytene cells. *Int. Rev. Cytol. Suppl.* **4:**417–462.

Daneholt, B., 1975 Transcription in polytene chromosomes. *Cell* **4:**1–9.

Daneholt, B. and H. Hosick, 1973 Evidence for transport of 75 S RNA from a discrete chromosome region via nuclear sap to cytoplasm in *Chironomus tentans. Proc. Natl. Acad. Sci. USA* **70:**442–446.

Daneholt, B. and L. Svedhem, 1971 Differential representation of chromosomal HnRNA in nuclear sap. *Exp. Cell Res.* **67:**263–272.

Daneholt, B., J.-E. Edström, E. Egyházi, B. Lambert, and U. Ringborg, 1969 Chromosomal RNA synthesis in polytene chromosomes of *Chironomus tentans. Chromosoma* **28:**399–417.

Darnell, J. E., 1968 Ribonucleic acids from animal cells. *Bacteriol. Rev.* **32:**262–290.

Darnell, J. E. and R. Balint, 1970 The distribution of rapidly hybridizing RNA from HeLa cells. *J. Cell. Physiol.* **76:**349–357.

Darnell, J. E., G. N. Pagoulatos, U. Lindberg, and R. Balint, 1970 Studies on the relationship of mRNA to heterogeneous, nuclear RNA in mammalian cells. *Cold Spring Harbor Symp. Quant. Biol.* **35:**555–560.

Darnell, J. E., L. Philipson, R. Wall, and M. Adesnik, 1971 Polyadenylic acid sequences: Role in conversion of nuclear RNA into messenger RNA. *Science* **174:**507–510.

Darnell, J. E., W. R. Jelinek, and G. R. Molloy, 1973 Biogenesis of messenger RNA: Genetic regulation in mammalian cells. *Science* **181:**1215–1221.

Davidson, E. H. and R. J. Britten, 1973 Organization, transcription, and regulation in the animal genome. *Rev. Biol.* **48:**565–613.

Davidson, E. H., M. Crippa, and A. E. Mirsky, 1968 Evidence for the appearance of novel gene products during amphibian blastulation. *Proc. Natl. Acad. Sci. USA* **60:**152–159.

Davidson, E. H., B. R. Hough, W. H. Klein, and R. J. Britten, 1975 Structural genes adjacent to interspersed repetitive DNA sequences. *Cell* **4:**217–238.

Dina, D., M. Crippa, and E. Beccari, 1973 Hybridization properties, and sequence arrangement in a population of mRNAs. *Nature (London) New Biol.* **242:**101–105.

Eaton, B. T., T. P. Donaghue, and P. Faulkner, 1972 Presence of poly(A) in the polyribosome-associated RNA of sindbis-infected BHK cells. *Nature (London) New Biol.* **238:**109–111.

Edmonds, M. and M. G. Caramela, 1969 The isolation and characterization of adenosine monophosphate-rich polynucleotides synthesized by Ehrlich ascites cells. *J. Biol. Chem.* **244:**1314–1324.

Edmonds, M., M. H. Vaughan, and H. Nakazato, 1971 Polyadenylic acid sequences in the heterogeneous nuclear RNA and rapidly-labeled polyribosomal RNA of HeLa cells: Possible evidence for a precursor relationship. *Proc. Natl. Acad. Sci. USA* **68:**1336–1340.

Edström, J.-E. and R. Tanguay, 1974 Cytoplasmic ribonucleic acids with messenger characteristics in salivary gland cells of *Chironomus tentans. J. Mol. Biol.* **84:**569–583.

Egyházi, E. 1974*a* Actinomycin D and RNA transport, *Nature (London)* **250:**221–222.

Egyházi, E. 1974*b* Processing of heterogeneous chromosomal RNA: Studies with an initiation inhibitor. *Proc. 9th FEBS Meet. (Budapest)*, pp. 57–62.

Egyházi, E., 1975 Inhibition of Balbiani ring RNA synthesis at initiation level. *Proc. Natl. Acad. Sci. USA,* **72:**947–950.

Egyházi, E. and J.-E. Edström, 1972 Evidence for transport of 4 S RNA from the nucleus to the cytoplasm in salivary gland cells of *Chironomus tentans. Biochem. Biophys. Res. Commun.* **46:**1551–1556.

Egyházi, E., U. Ringborg, B. Daneholt, and B. Lambert, 1968 Extraction and fractionation of low molecular weight RNA on the microscale. *Nature (London)* **220:**1036–1037.

Egyházi, E., B. Daneholt, J.-E. Edström, B. Lambert, and U. Ringborg, 1969 Low molecular weight RNA in cell components of *Chironomus tentans* salivary glands. *J. Mol. Biol.* **44:**517–532.

Firtel, R. A., 1972 Changes in the expression of single-copy DNA during development of the cellular slime mold *Dictyostelium discoideum. J. Mol. Biol.* **66:**363–377.

Galau, G. A., R. J. Britten, and E. H. Davidson, 1974 A measurement of the sequence complexity of polysomal messenger RNA in sea urchin embryos. *Cell* **2:**9–20.

Georgiev, G. P., 1967 The nature and biosynthesis of nuclear ribonucleic acids. *Prog. Nucleic Acid Res.* **6:**259–351.

Georgiev, G. P., 1969 On the structural organization of operon and the regulation of RNA synthesis in animal cells. *J. Theor. Biol.* **25:**473–490.

Georgiev, G. P. and M. I. Lerman, 1964 Separation and some properties of distinct classes of newly-formed ribonucleic acid from animal cells. *Biochim. Biophys. Acta* **91:**678–680.

Georgiev, G. P., A. P. Ryskov, C. Coutelle, V. L. Mantieva, and E. R. Avakyan, 1972 On the structure of transcriptional unit in mammalian cells. *Biochim. Biophys. Acta* **259:**259–283.

Getz, M. J., G. D. Birnie, B. D., Young, E. MacPhail, and J. Paul, 1975 A kinetic estimation of base sequence complexity of nuclear poly(A)-containing RNA in mouse Friend cells. *Cell* **4:**121–129.

Gilbert, W., N. Maizels, and A. Maxam, 1973 Sequences of controlling regions of the lactose operon. *Cold Spring Harbor Symp. Quant. Biol.* **38:**845–855.

Giudice, G., G. Sconzo, F. Raurirez, and I. Albanese, 1972 Giant RNA is also found in the cytoplasm in sea urchin embryos. *Biochim. Biophys. Acta* **262:**401–403.

Glisin, V. R., M. V. Glisin, and P. Doty, 1966 The nature of messenger RNA in the early stages of sea urchin development. *Proc. Natl. Acad. Sci. USA* **56:**285–289.

Goldberg, R. B., G. A. Galau, R. J. Britten, and E. H. Davidson, 1973 Nonrepetitive DNA sequence representation in sea urchin embryo messenger RNA. *Proc. Natl. Acad. Sci. USA* **70:**3516–3520.

Granboulan, N. and K. Scherrer, 1969 Visualisation in the electron microscope and size of RNA from animal cells. *Eur. J. Biochem.* **9:**1–20.

Greenberg, J. R. and R. P. Perry, 1971 Hybridization properties of DNA sequences directing the synthesis of messenger RNA and heterogeneous nuclear RNA. *J. Cell Biol.* **50:**774–786.

Greenberg, J. R. and R. P. Perry, 1972 Relative occurrence of polyadenylic acid sequences in messenger and heterogeneous nuclear RNA of L cells as determined by poly(U)-hydroxylapatite chromatography. *J. Mol. Biol.* **72:**91–98.

Grouse, L., M.-D. Chilton, and B. J. McCarthy, 1972 Hybridization of ribonucleic acid with unique sequences of mouse deoxyribonucleic acid. *Biochemistry* **11:**798–805.

Hägele, K., 1975 *Chironomus.* In *Handbook of Genetics,* Vol. 3, edited by R. C. King, pp. 269–278, Plenum Press, New York.

Hahn, W. E. and C. D. Laird, 1971 Transcription of nonrepeated DNA in mouse brain. *Science* **173:**158–160.

Hamkalo, B. A., O. L. Miller, Jr., and A. H. Bakken, 1973 Ultrastructural aspects of genetic activity. In *Molecular Cytogenetics,* edited by B. A. Hamkalo and J. Papaconstantinou, pp. 315–323, Plenum Press, New York.

Harel, L. and L. Montagnier, 1971 Homology of double stranded RNA from rat liver cells with the cellular genome. *Nature (London) New Biol.* **229:**106–108.

Hellung-Larsen, P. and S. Fredriksen, 1971 Correlation between separation of low molecular weight RNA on MAK columns and polyacrylamide gels. *Anal. Biochem.* **40:**227–232.

Hellung-Larsen, P. and S. Fredriksen, 1972 Small molecular weight RNA components in Ehrlich ascites tumor cells. *Biochim. Biophys. Acta* **262:**290–308.

Holmes, D. S. and J. Bonner, 1973 Preparation, molecular weight, base composition, and secondary structure of giant nuclear ribonucleic acid. *Biochemistry* **12:**2330–2338.

Holmes, D. S. and J. Bonner, 1974*a* Sequence composition of rat nuclear deoxyribonucleic acid and high molecular weight nuclear ribonucleic acid. *Biochemistry* **13:**841–849.

Holmes, D. S. and J. Bonner, 1974*b* Interspersion of repetitive and single-copy sequences in nuclear ribonucleic acid of high molecular weight. *Proc. Natl. Acad. Sci. USA* **71:**1108–1112.

Holmes, D. S., J. E. Mayfield, G. Sander, and J. Bonner, 1972 Chromosomal RNA: Its properties. *Science* **177:**72:74.

Houssais, J.-F. and G. Attardi, 1966 High molecular weight nonribosomal-type nuclear RNA and cytoplasmic messenger RNA in HeLa cells. *Proc. Natl. Acad. Sci. USA* **56:**616–623.

Imaizumi, T., H. Diggelmann, and K. Scherrer, 1973 Demonstration of globin messenger sequences in giant nuclear precursors of messenger RNA of avian erythroblasts. *Proc. Natl. Acad. Sci. USA* **70:**1122–1126.

Jelinek, W. and J. E. Darnell, 1972 Double-stranded regions in heterogeneous nuclear RNA from HeLa cells. *Proc. Natl. Acad. Sci. USA* **69:**2537–2541.

Jelinek, W., M. Adesnik, M. Salditt, D. Sheiness, R. Wall, G. Molloy, L. Philipson, and J. E. Darnell, 1973 Further evidence on the nuclear origin and transfer to the cytoplasm of polyadenylic acid sequences in mammalian cell RNA. *J. Mol. Biol.* **75:**515–532.

Jelinek, W., G. Molloy, R. Fernandez-Munoz, M. Salditt, and J. E. Darnell, 1974 Secondary structure in heterogeneous nuclear RNA: Involvement of regions from repeated DNA sites. *J. Mol. Biol.* **82:**361–370.

Judd, B. H., M. W. Shen, and T. C. Kaufman, 1972 The anatomy and function of a segment of the X chromosome of *Drosophila melanogaster. Genetics* **71:**139–156.

Kates, J., 1970 Transcription of the vaccinia virus genome and the occurrence of polyriboadenylic acid sequences in messenger RNA. *Cold Spring Harbor Symp. Quant. Biol.* **35:**743–752.

Kates, J. and J. Beeson, 1970 Ribonucleic acid synthesis in vaccinia virus. II. Synthesis of polyriboadenylic acid. *J. Mol. Biol.* **50:**19–33.

Kimball, P. C. and P. H. Duesberg, 1971 Virus interference by cellular double-stranded ribonucleic acid. *J. Virol.* **7:**697–706.

Klotz, I. M., 1970 Protein subunits. In *Handbook of Biochemistry. Selected Data for Molecular Biology,* edited by H. A. Sober, pp. C47–49, Chemical Rubber Co., Cleveland.

Kohne, D. E. and M. J. Byers, 1973 Amplification and evolution of deoxyribonucleic acid sequences expressed as ribonucleic acid. *Biochemistry* **12:**2373–2379.

Kumar, A. and U. Lindberg, 1972 Characterization of messenger ribonucleoprotein and messenger RNA from KB cells. *Proc. Natl. Acad. Sci. USA* **69:**681–685.

Lambert, B., 1972 Repeated DNA sequences in a Balbiani ring. *J. Mol. Biol.* **72:**65–75.

Lambert, B. and J.-E. Edström, 1974 Balbiani ring nucleotide sequences in cytoplasmic 75 S RNA of *Chironomus tentans* salivary gland cells. *Mol. Biol. Rep.* **1:**457–464.

Larsen, C. J., F. Galibert, A. Hampe, and M. Boiron, 1969 Etude des RNA nucléaires de faible poids moléculaire de la cellule KB. *Bull. Soc. Chim. Biol.* **51:**649–668.

Lee, S. Y., J. Mendecki, and G. Brawerman, 1971 A polynucleotide segment rich in adenylic acid in the rapidly-labeled polyribosomal RNA component of mouse sarcoma 180 ascites cells. *Proc. Natl. Acad. Sci. USA* **68:**1331–1335.

Lewin, B. 1975*a* Units of transcription and translation: The relationship between heterogeneous nuclear RNA and messenger RNA. *Cell* **4:**11–20.

Lewin, B., 1975*b* Units of transcription and translation: Sequence components of heterogeneous nuclear RNA and messenger RNA. *Cell* **4:**77–93.

Lim, L. and E. S. Canellakis, 1970 Adenine-rich polymer associated with rabbit reticulocyte messenger RNA. *Nature (London)* **227:**710–712.

Lindberg, U. and J. E. Darnell, 1970 SV40-specific RNA in the nucleus and polyribosomes of transformed cells. *Proc. Natl. Acad. Sci. USA* **65:**1089–1096.

Lizardi, P. M., R. Williamson, and D. D. Brown, 1975 The size of fibroin messenger RNA and its polyadenylic acid content. *Cell* **4:**199–205.

Lodish, H. F., R. A. Firtel, and A. Jacobson, 1973 Transcription and structure of the genome of the cellular slime mold *Dictyostelium discoideum. Cold Spring Harbor Symp. Quant. Biol.* **38:**899–914.

Macnaughton, M., K. B. Kreeman, and J. O. Bishop, 1974 A precursor to hemoglobin mRNA in nuclei of immature duck red blood cells. *Cell* **1:**117–125.

McCarthy, B. J. and B. H. Hoyer, 1964 . Identity of DNA and diversity of messenger RNA molecules in normal mouse tissues. *Proc. Natl. Acad. Sci. USA* **52:**915–922.

McKnight, G. S. and R. T. Schimke, 1974 Ovalbumin messenger RNA: Evidence that the initial product of transcription is the same size as polysomal ovalbumin messenger. *Proc. Natl. Acad. Sci. USA* **71:**4327–4331.

Melli, M., C. Whitfield, K. V. Rao, M. Richardson, and J. O. Bishop, 1971 DNA-RNA hybridization in vast DNA excess. *Nature (London) New Biol.* **231:**8–12.

Mendecki, J., Y. Lee, and G. Brawerman, 1972 Characteristics of the polyadenylic acid segment associated with messenger ribonucleic acid in mouse sarcoma 180 ascites cells. *Biochemistry* **11:**792–798.

Milcarek, C., R. Price, and S. Penman, 1974 The metabolism of a poly(A) minus mRNA fraction in HeLa cells. *Cell* **3:**1–10.

Miller, O. L., Jr. and A. H. Bakken, 1972 Morphological studies of transcription. *Acta Endocrinol.* **168**–155–177.

Miller, O. L., Jr. and B. A. Hamkalo, 1972 Visualization of RNA synthesis on chromosomes. *Int. Rev. Cytol.* **33:**1–25.

Miller, O. L., Jr., B. A. Hamkalo, and C. A. Thomas, Jr., 1970 Visualization of bacterial genes in action. *Science* **169:**392–395.

Molloy, G. R., W. L. Thomas, and J. E. Darnell, 1972 Occurrence of uridylate-rich oligonucleotide regions in heterogeneous nuclear RNA of HeLa cells. *Proc. Natl. Acad. Sci. USA* **69:**3684–3688.

Molloy, G. R., W. Jelinek, M. Salditt, and J. E. Darnell, 1974 Arrangement of specific oligonucleotides within poly(A) terminated HnRNA molecules. *Cell* **1:**43–53.

Montagnier, L., 1968 Présence d'un acide ribonucléique en double chaîne dans les cellules animales. *C. R. Acad. Sci. D.* **267:**1417–1420.

Moriyama, Y., J. L. Hodnett, A. W. Prestayko, and H. Busch, 1969 Studies on the nuclear 4 to 7 S RNA of the Novikoff hepatoma. *J. Mol. Biol.* **39:**335–349.

Morse, D. E., R. Mosteller, R. F. Baker, and C. Yanofsky, 1969 Direction of *in vivo* degradation of tryptophan messenger RNA-A correction. *Nature (London)* **223:**40–43.

Nakazato, H., M. Edmonds, and D. W. Kopp, 1974 Differential metabolism of large and small poly(A) sequences in the heterogeneous nuclear RNA of HeLa cells. *Proc. Natl. Acad. Sci. USA* **71:**200–204.

Nemer, M., M. Graham, and L. M. Dubroff, 1974 Co-existence of non-histone messenger RNA species lacking and containing polyadenylic acid in sea urchin embryos. *J. Mol. Biol.* **89:**435–454.

Ovchinnikov, L. P., M. A. Ajtkhozhin, T. F. Bystrova, and A. S. Spirin, 1969 Informosomes of loach embryos (*Misgurnus fossilis* L.). 1. Sedimentation and density characteristics. *Mol. Biol. SSSR* **3:**449–464.

Pagoulatos, G. N. and J. E. Darnell, 1970 Fractionation of heterogeneous nuclear RNA: Rates of hybridization and chromosomal distribution of reiterated sequences. *J. Mol. Biol.* **54:**517–535.

Pelling, C., 1964 Ribonukleinsäure-Synthese der Riesenchromosomen. *Chromosoma* **15:**71–122.

Pelling, C., 1970 Puff RNA in polytene chromosomes. *Cold Spring Harbor Symp. Quant. Biol.* **35:**521–531.

Penman, S., C. Vesco, and M. Penman, 1968 Localization and kinetics of formation of nuclear heterodisperse RNA, cytoplasmic heterodisperse RNA and polyribosome-associated messenger RNA in HeLa cells. *J. Mol. Biol.* **34:**49–69.

Penman, S., M. Rosbash, and M. Penman, 1970 Messenger and heterogeneous nuclear

RNA in HeLa cells: Differential inhibition by cordycepin. *Proc. Natl. Acad. Sci. USA* **67:**1878–1885.

Perlman, S., H. T. Abelson, and S. Penman, 1973 Mitochondrial protein synthesis: RNA with the properties of eukaryotic messenger RNA. *Proc. Natl. Acad. Sci. USA* **70:**350–353.

Perry, R. P., 1962 The cellular sites of synthesis of ribosomal and 4 S RNA. *Proc. Natl. Acad. Sci. USA* **48:**2179–2186.

Perry, R. P. and D. E. Kelley, 1974 Existence of methylated messenger RNA in mouse L cells. *Cell* **1:**37–42.

Perry, R. P., D. E. Kelley, and J. LaTorre, 1974 Synthesis and turnover of nuclear and cytoplasmic polyadenylic acid in mouse L cells. *J. Mol. Biol.* **82:**315–331.

Philipson, L., R. Wall, G. Glickman, and J. E. Darnell, 1971 Addition of polyadenylate sequences to virus-specific RNA during adenovirus replication. *Proc. Natl. Acad. Sci. USA* **68:**2806–2809.

Prescott, D. M., C. Bostock, E. Gamow, and M. R. Lauth, 1971*a* Characterization of rapidly labeled RNA in *Tetrahymena pyriformis*. *Exp. Cell Res.* **67:**124–129.

Prescott, D. M., A. R. Stevens, and M. R. Lauth, 1971*b* Characterization of nuclear RNA synthesis in *Amoeba proteus*. *Exp. Cell Res.* **64:**145–156.

Prestayko, A. W., M. Tonato, and H. Busch, 1970 Low molecular weight RNA associated with 28 S nucleolar RNA. *J. Mol. Biol.* **47:**505–515.

Prestayko, A. W., M. Tonato, B. C. Lewis, and H. Busch, 1971 Heterogeneity of nucleolar U3 ribonucleic acid of the Novikoff hepatoma. *J. Biol. Chem.* **246:**182–188.

Price, R. P., L. Ransom, and S. Penman, 1974 Identification of a small subfraction of hnRNA with the characteristics of a precursor to mRNA. *Cell* **2:**253–258.

Reddy, R., T. S. Ro-Choi, D. Hennig, and H. Busch, 1974 Primary sequence of U-1 nuclear ribonucleic acid of Novikoff hepatoma ascites cells. *J. Biol. Chem.* **249:**6486–6494.

Roberts, W. K. and J. F. E. Newman, 1966 Use of low concentrations of actinomycin D in the study of RNA synthesis in Ehrlich ascites cells. *J. Mol. Biol.* **20:**62–73.

Ro-Choi, T. S., R. Reddy, D. Hennig, T. Takano, C. W. Taylor, and H. Busch, 1972 Nucleotide sequence of 4.5 S ribonucleic acid$_I$ of Novikoff hepatoma cell nuclei. *J. Biol. Chem.* **247:**3205–3222.

Rudkin, G. T., 1965 The relative mutabilities of DNA in regions of the X chromosome of *Drosophila melanogaster*. *Genetics* **52:**665–681.

Ryskov, A. P. and G. P. Georgiev, 1970 Polyphosphate groups at the 5′-ends of nuclear d-RNA fractions. *FEBS Lett.* **8:**186–188.

Ryskov, A. P., V. R. Farashyan, and G. P. Georgiev, 1972 Ribonuclease-stable base sequences specific exclusively for giant dRNA. *Biochim. Biophys. Acta* **262:**568–572.

Ryskov, A. P., G. F. Saunders, V. R. Farashyan, and G. P. Georgiev, 1973 Double-helical regions in nuclear precursor of mRNA (pre-mRNA). *Biochim. Biophys. Acta* **312:**152–164.

Scherrer, K. and L. Marcaud, 1968 Messenger RNA in avian erythroblasts at the transcriptional and translational levels and the problem of regulation in animal cells. *J. Cell Physiol.* **72:**Suppl. 1, 181–212.

Scherrer, K., H. Latham, and J.-E. Darnell, 1963 Demonstration of an unstable RNA and of a precursor to ribosomal RNA in HeLa cells. *Proc. Natl. Acad. Sci. USA* **49:**240–248.

Scherrer, K., I. M. London, and F. Gros, 1966 Patterns of RNA metabolism in a differentiated cell: A rapidly labeled unstable 60 S RNA with messenger properties in duck erythroblasts. *Proc. Natl. Acad. Sci. USA* **56:**1571–1578.

Scherrer, K., G. Spohr, N. Granboulan, C. Morel, J. Grosclaude, and C. Chezzi, 1970 Nuclear and cytoplasmic messenger-like RNA and their relation to the active messenger RNA in polyribosomes of HeLa cells. *Cold Spring Harbor Symp. Quant. Biol.* **35:**539–554.

Schochetman, G. and R. P. Perry, 1972 Early appearance of histone messenger RNA in polyribosomes of cultured L cells. *J. Mol. Biol.* **63:**591–596.

Sconzo, G., I. Albanese, A. M. Rinaldi, G. F. Lo Presti, and G. Giudice, 1974 Cytoplasmic giant RNA in sea urchin embryos. II. Physico-chemical characterization. *Cell Diff.* **3:**297–304.

Shearer, R. W. and B. J. McCarthy, 1967 Evidence for ribonucleic acid molecules restricted to the cell nucleus. *Biochemistry* **6:**283–289.

Shearer, R. W. and B. J. McCarthy, 1970 Characterization of RNA molecules restricted to the nucleus in mouse L-cells. *J. Cell. Physiol.* **75:**97–107.

Shibata, H., R. Reddy, D. Hennig, T. S. Ro-Choi, and H. Busch, 1973 Nucleotide sequence of nuclear U2 RNA. *Fed. Proc.* **32:**664.

Slater, I. and D. W. Slater, 1974 Polyadenylation and transcription following fertilization. *Proc. Natl. Acad. Sci. USA* **71:**1103–1107.

Slater, D. W., I. Slater, and D. Gillespie, 1972 Post-fertilization synthesis of polyadenylic acid in sea urchin embryos. *Nature (London)* **240:**333–337.

Smith, M. J., B. R. Hough, M. E. Chamberlin, and E. H. Davidson, 1974 Repetitive and non-repetitive sequences in sea urchin heterogeneous nuclear RNA. J. Mol. Biol. **85:**103–126.

Soeiro, R., H. C. Birnboim, and J. E. Darnell, 1966 Rapidly labeled HeLa cell nuclear RNA. II. Base composition and cellular localization of a heterogeneous RNA fraction. *J. Mol. Biol.* **19:**362–372.

Soeiro, R., M. H. Vaughan, J. R. Warner, and J. E. Darnell, 1968 The turnover of nuclear DNA-like RNA in HeLa cells. *J. Cell Biol.* **39:**112–118.

Spohr, G., T. Imaizumi, and K Scherrer, 1974 Synthesis and processing of nuclear precursor-messenger RNA in avian erythroblasts and HeLa cells. *Proc. Natl. Acad. Sci. USA* **71:**5009–5013.

Stevens, R. H. and A. R. Williamson, 1973 Isolation of precursor coding for immunoglobulin heavy chain. *Nature (London) New Biol.* **245:**101–104.

Sullivan, D. T., 1968 Molecular hybridization used to characterize the RNA synthesized by isolated bovine thymus nuclei. *Proc. Natl. Acad. Sci. USA* **59:**846–853.

Suzuki, Y. and D. D. Brown, 1972 Isolation and identification of the messenger RNA for silk fibroin from *Bombyx mori. J. Mol. Biol.* **63:**409–429.

Tonegawa, S., G. Walter, A. Bernardini, and R. Dulbecco, 1970 Transcription of the SV 40 genome in transformed cells and during lytic infection. *Cold Spring harbor Symp. Quant. Biol.* **35:**823–831.

Turner, S. H. and C. D. Laird, 1973 Diversity of RNA sequences in *Drosophila melanogaster. Biochem. Genet.* **10:**263–274.

Wall, R. and J. E. Darnell, 1971 Presence of cell and virus specific sequences in the same molecules of nuclear RNA from virus transformed cells. *Nature (London) New Biol.* **232:**73–76.

Wall, R., L. Philipson, and J. E. Darnell, 1972 Processing of adenovirus specific nuclear RNA during virus replication. *Virology* **50:**27–34.

Wall, R., J. Weber, Z. Gage, and J. E. Darnell, 1973 Production of viral mRNA in adenovirus-transformed cells by the post-transcriptional processing of heterogeneous nuclear RNA containing viral and cell sequences. *J. Virol.* **11:**953–960.

Warner, J. R., R. Soeiro, H. C. Birnboim, M. Girard, and J. E. Darnell, 1966 Rapidly labeled HeLa cell nuclear RNA. I. Identification by zone sedimentation of a heterogeneous fraction separate from ribosomal precursor RNA. *J. Mol. Biol.* **19:**349–361.

Weinberg, R. A., 1973 Nuclear RNA metabolism. *Ann. Rev. Biochem.* **42:**329–354.

Weinberg, R. A. and S. Penman, 1968 Small molecular weight monodisperse nuclear RNA. *J. Mol. Biol.* **38:**289–304.

Weinberg, R. A. and S. Penman, 1969 Metabolism of small molecular weight monodisperse nuclear RNA. *Biochim. Biophys. Acta* **190:**10–29.

Weinberg, R. A. and S. Penman, 1970 Processing of 45 S nucleolar RNA. *J. Mol. Biol.* **47:**169–178.

Weinberg, R. A., Z. Ben-Ishai, and J. E. Newbold, 1972 Poly A associated with SV 40 messenger RNA. *Nature (London) New Biol.* **238:**111–113.

Wilt, F. H., 1973 Polyadenylation of maternal RNA of sea urchin eggs after fertilization. *Proc. Natl. Acad. Sci. USA* **70:**2345–2349.

Yogo, Y. and E. Wimmer, 1972 Polyadenylic acid at the 3′-terminus of poliovirus RNA. *Proc. Natl. Acad. Sci. USA* **69:**1877–1882.

Zapisek, W. F., A. G. Saponara, and M. D. Enger, 1969 Low molecular weight methylated ribonucleic acid species from Chinese hamster ovary cells. I. Isolation and characterization. *Biochemistry* **8:**1170–1182.

8

RNA Transcription and Ribosomal Protein Assembly in *Drosophila melanogaster*

W. Yean Chooi

Introduction

Definition of the structure of primary transcription units of RNA and their corresponding translation units is crucial to an understanding of the organization and regulation of the eukaryotic genome. During the last decade, great progress was made in determination of the structure of transcription units in prokaryotes. However, the organization of transcription in eukaryotes appears to be more complex than that in prokaryotes. The rapid progress made in the analysis of the structure of the operon in bacteria depended on combined genetic and biochemical analyses. In eukaryotes, however, the most convenient systems for genetic studies (e.g., *Drosophila*) are often not readily amenable to conventional biochemical techniques. For this reason, the majority of the scientific effort has been directed toward other eukaryotic systems. This chapter is devoted mainly

---

W. Yean Chooi—Whitman Laboratory, University of Chicago, Chicago, Illinois. After August 1976, Department of Zoology, Indiana University, Bloomington, Indiana.

to a discussion of some aspects of transcription and translation in *D. melanogaster* and their relationship to the organization of the *Drosophila* genome. Particular emphasis is placed on certain molecular and electron microscopic properties of RNA transcription and maturation. The more general aspects of these topics will only be briefly covered here, as these have been dealt with either in other chapters of this handbook series or in recent reviews.

This chapter is divided into three parts. The first deals with the nontranscribed spacers in ribosomal DNA, the second is concerned with the *in vivo* assembly of ribosomal proteins, and the last part discusses nonribosomal RNA transcription.

The Genes for Ribosomal RNA*

The genes for rRNA are unique in that they are associated with one or more distinctive organelles called nucleoli. The ability to form a nucleolus is probably present at a specific chromosomal locus called the nucleolar organizer. McClintock (1934) was the first to show that the nucleolar organizer of *Zea mays* could be split by a translocation and that each piece was capable of organizing a separate nucleolus. This work represented the first demonstration that the genes for rRNA are present in multiple copies.

Since then, several independent lines of evidence have demonstrated quite convincingly that the nucleolus is the site of rRNA synthesis and partial ribosome assembly. The anucleolate mutant of *Xenopus laevis* lacks the ability to organize a nucleolus (Elsdale *et al.,* 1958) and does not have rDNA (Wallace and Birnstiel, 1966). In the homozygous state such anucleolate mutants are unable to synthesize any rRNA and consequently die (Brown and Gurdon, 1964). Ritossa and Spiegelman (1965) were able to study *Drosophila* stocks that contained from one to four doses of the nucleolar organizer by means of nucleic acid hybridization. They found that the amount of rRNA that could hybridize specifically to the DNA

* The abbreviations used throughout the text are as follows: rRNA, ribosomal RNA; rDNA, ribosomal DNA which contains the ribosomal transcription unit and the intervening nontranscribed spacer DNA; spacer, nontranscribed spacer (spacers of mean lengths 1.3 μm and 0.8 μm are referred to as long spacers and short spacers, respectively); rRNP, ribosomal ribonucleoprotein; non-rRNP, nonribosomal ribonucleoprotein; HnRNA, heterogeneous, high molecular weight RNA; IgG, immunoglobulin G. Papain catalyzes the proteolysis of IgG into two antigen-binding fragments (Fab) and a crystallizable fragment (Fc). The Fab fragments each contain a single antibody site, a complete L chain, and a portion of an H chain.

from each genotype was related to the number of nucleolar organizers present.

When nucleoli are extracted with dithiothreitol in the presence of ethylenediamine tetraacetic acid, the main products are two species of rRNP which sediment at 80 S and 55 S. The 80 S nucleolar particles contain 45 S RNA, 5 S RNA, ribosomal proteins and nucleolar proteins (Warner and Soeiro, 1967; Kumar and Warner, 1972). The detailed composition of these proteins have not been resolved. The 55 S nucleolar particle contains 32 S RNA, 5 S RNA and two classes of proteins. One class appears to be newly formed ribosomal proteins (Shepard and Maden, 1972), while the other class appears to be different from ribosomal proteins and may be nucleolar specific (Kumar and Warner, 1972; Tsurugi *et al.*, 1973).

Recently the technique of *in situ* hybridization has demonstrated that rDNA is located within the nucleolus (Pardue and Gall, 1969; Pardue *et al.*, 1970; Gerbi, 1971; Amaldi and Buongiorno-Nardelli, 1971). The most intriguing observations are that the rDNAs have also been localized at chromosomal loci other than nucleoli (Pardue *et al.*, 1970; Avanzi *et al.*, 1972). Have these rDNAs diverged in nucleotide sequence from the nucleolar rDNAs? Their presence is reminiscent of the nucleolar dominance phenomenon seen in some interspecific crosses of *Xenopus* and the tissue specificity of the transcription of particular 5 S genes in *Xenopus*.

In hybrids between *X. laevis* and *X. mulleri*, only the *X. laevis* rDNA appears to be transcribed (Honjo and Reeder, 1973) regardless of whether *X. laevis* is the male or female parent. In hybrids that contain a complete set of chromosomes from each species (except that *X. laevis* rDNA is absent and only *X. mulleri* rDNA is present) only *X. mulleri* rDNA is expressed, but the time of turn on is subject to maternal influence. If the female parent is *X. laevis*, turn on is delayed beyond the gastrula stage. In the reverse cross there appears to be no delay in turn on at gastrulation.

The nucleotide sequences of the 5 S RNAs from two different tissues of *X. laevis* have been determined (Ford and Southern, 1973; Brownlee *et al.*, 1974). Two distinctive types of 5 S RNA species are present; one synthesized principally in the oocyte and the other in somatic tissues. The oocyte 5 S molecule differs in six bases from the somatic type. What is the nature of the phenomenon that controls the selective expression of the two 5 S RNA loci in somatic tissues and oocytes of *Xenopus*? What evolutionary mechanisms are responsible for the divergence of the two sets of the 5 S RNA genes? At this time, the answers to these questions can only be speculative. However, such phenomena are probably not unique to *Xe-*

nopus. A similar situation is seen in the control of selective expression of the rRNA genes in *D. melanogaster* oogenesis and embryogenesis and in the apparent independent evolution of the X-linked and Y-linked rRNA loci (see p. 229).

In all the eukaryotes examined, the genes for rRNA are present in multiple copies (see review by Birnstiel *et al.*, 1971; Bostock, 1971) and are arranged in tandem repeats (Birnstiel *et al.*, 1968; Dawid *et al.*, 1970). Each repeat contains a sequence that is transcribed as a single precursor molecule and a sequence that is not transcribed (nontranscribed spacer). In *Xenopus* the gene transcript is 40 S, and within each gene region is a sequence for 28 S RNA, a sequence for 18 S RNA and two smaller regions (transcribed spacer). While all the gene regions within an organism appear identical, there is some heterogeneity in the nontranscribed region (Morrow *et al.*, 1974). Considerable divergences in nucleotide sequences of nontranscribed spacers are also found between closely related organisms (Brown *et al.*, 1972). The transcribed and nontranscribed spacers in *X. laevis* and *X. mulleri* show only 10% cross homology, although the 18 S and 28 S molecules cannot be distinguished between the two species. This observation raises the question of how repeated genes can be kept homogeneneously uniform within an organism and yet allow diversity to accumulate in the nontranscribed spacers. This problem is not restricted to rRNA cistrons alone. How are repeats such as the ribosomal 5 S RNA (e.g., 24,000 copies in *X. laevis*) or simple sequence DNA (e.g., the 12 million copies of satellite I from *D. virilis*) kept identical or "rectified" and kept from accumulating a high degree of nucleotide diversity? Several hypotheses have been advanced in attempts to resolve this dilemma (see p. 235).

What is the sequence of arrangement of the 28 S, 18 S and the spacers on the rDNA molecule? Wellauer and Dawid (1974) have approached this question by making use of the characteristic regions in rDNA and rRNA from HeLa cells which reproducibly form double stranded loops under certain spreading conditions for electron microscopy. By partial digestion with a 3′ exonuclease they were able to establish the polarity of these RNA sequences. They concluded that the 28 S sequence is located on the extreme 5′ end of the 40 S precursor, followed by the transcribed spacer, the 18 S sequence and then a short transcribed spacer on the 3′ end. This assignment of the RNA sequences is in conflict with three others (Reeder and Brown, 1970; Hecht and Birnstiel, 1972; Hackett and Sauerbier, 1975). The question of the polarity of the individual rRNA sequences is still unresolved. A different approach (immune electron microscopy) has been used in the re-examination of this question in *D. melanogaster* (see p. 239).

The 18 S and 28 S rRNA of eukaryotes are synthesized initially as part of a single large precursor molecule, which is subsequently processed to yield the two mature rRNA molecules. Some ribosomal proteins are complexed with the nascent rRNA precursors while they are still localized within the nucleolus (Warner, 1974). However, it is not known which of the ribosomal proteins take part in the formation of nucleolar-specific, nascent RNP particles. Little is known about the topography of the ribosomal proteins on the rRNA precursors and the sequence of the assembly of the different proteins on the nascent rRNA precursors. Recently, the sequence of assembly of a few ribosomal proteins to nascent rRNA still attached to the DNA template has been analyzed in *D. melanogaster* (see p. 239).

Electron microscopists have long been able to distinguish between the fibrillar and granular regions of the nucleolus in sections (Busch and Smetana, 1970). From EM autoradiography, Granboulan and Granboulan (1965) suggested that the rRNA precursors were transcribed from the DNA in the fibrillar region and assembled with protein in the granular region. From his electron micrographs of active transcription units, Miller *et al.* (1970) identified denser staining regions of RNP fibrils as representing regions of RNA and protein interaction.

One of the most dramatic events in rDNA replication is the process of amplification. Amplification refers to the extrachromosomal replication of rDNA during oogenesis in animals (Brown and Dawid, 1968; see reviews by Gall, 1969; Birnstiel *et al.*, 1971; MacGregor, 1972). In general, rDNA amplification seems to occur in species that synthesize and store large amounts of ribosomes in their oocytes. The mechanism by which rDNA amplification occurs is unknown. Two theories have been proposed, the reverse transcriptase theory of Crippa and Tocchini-Valentini (1971) and the concept of a DNA replication from a DNA template. The main controversy seems to center around the mechanism whereby the first extrachromosomal copy of the rDNA is produced. Crippa and Tocchini-Valentini (1971) suggested that a DNA-dependent RNA polymerase must first transcribe at least one copy of the entire chromosomal rDNA cistron (18 S, 28 S and spacers). This RNA transcript is copied by reverse transcriptase to yield DNA copies, and these DNA copies are then joined by ligase to form the amplified rDNA. Crippa and Tocchini-Valentini reported finding evidence for a DNA–RNA hybrid molecule sedimenting at 27 S. However, Bird *et al.* (1973) have not been able to detect these structures. DNA structures resembling rolling circles have recently been found in amplified rDNA (Hourcade *et al.*, 1974), which suggests that some parts of the amplification process could involve this mechanism. Rolling circles were first detected in viral replication (Gilbert and Dressler,

1968). The data of Hourcade *et al.* suggest that the first extrachromosomal template probably arises directly by DNA synthesis from the chromosome.

There appear to be changes besides amplification that also result in a modification of the number of copies of rRNA cistrons. An example of such a phenomenon is *magnification,* a term coined by Ritossa (1968) to describe the increase in the number of ribosomal cistrons found in *D. melanogaster* males homozygous for the bobbed mutation (see p. 237). A related phenomenon is called rDNA compensation. Like rDNA magnification it involves an increase in rDNA levels in *Drosophila* which have a deficiency for rDNA. However, there are some important differences between these phenomena. While magnification is restricted to males, compensation can take place in either males or females (Tartof, 1971). In addition, changes in rDNA content due to magnification are inherited (Ritossa, 1968; Henderson and Ritossa, 1970), whereas those due to compensation are not (Tartof, 1971, 1973).

The Genes for HnRNA

It is generally recognized that many eukaryotic organisms contain far more DNA than would be required to specify the proteins required for structural and metabolic functions if each message was carried only once. The amount of DNA per nucleus is 30–20,000 times greater than that present in a bacterium (Britten and Kohne, 1968; Britten and Davidson, 1971). What is the nature of this seemingly vast excess of DNA? Investigations on the molecular arrangement of nucleotide sequences and nucleotide sequence diversity in eukaryotic chromosomes are therefore essential for a more complete understanding of the structure and function of chromosomes. The definitions of the structures of primary transcription units and translation units can in principle help in the resolution of this problem. *Drosophila* is a particularly useful organism for such studies, since the banded morphology and high DNA content of the polytene chromosomes provide an unusual opportunity to relate cytological units and functional transcription units.

When the genetic significance of the giant polytene chromosomes in Dipterans was first recognized forty years ago, most cytogeneticists were partial towards the hypothesis that each individual polytene chromosome band was associated with a single gene. Although many attempts have been made since then to either validate or discredit this one-band–one-gene hypothesis the question still remains unresolved.

An average size chromomere is about 20 times larger than an average size bacterial gene (see p. 249). Such large amounts of DNA in an average chromomere is difficult to reconcile with single gene functions, for each band would have enough DNA to code for 20 to 30 average cistrons. It is true that bands vary in size, the smallest having fewer than 5000 base pairs and the largest perhaps 100,000 base pairs. Besides, there is some dispute as to whether the number of bands can be counted accurately (see review by Lefevre, 1974). However, even if these factors are taken into consideration, it still does not solve the problem of why a gene in *D. melanogaster* requires an average of 30 times the number of nucleotides that are required in bacteria or phage. A different aspect of this problem may be responsible for the C value paradox (Callan, 1967). There are many closely related species that differ significantly in the amount of DNA per haploid cell (C value). For example, within the class Amphibia, the range in genome sizes varies over two orders of magnitude (Britten and Davidson, 1971).

Many speculations have been proposed in attempts to resolve this dilemma of the seemingly excessive amounts of DNA per band or per function. One of the most attractive theoretical hypotheses advanced was the possibility that the chromomere represented repetitive copies of structural cistrons. Callan and Loyd (1960), following their studies on lampbrush chromosomes in salamanders, advanced the *master–slave* concept of cistronic reiteration. According to the hypothesis, the chromosome consists of families of serially repeated genes. A special member of each family is the master; all the other members are slaves. The master gene specifies the sequence of every slave gene by a process called rectification. Because of rectification, only mutational alterations in the master gene are ever detected. Difficulties in visualizing an accurate and persistent mechanism for rectification and for maintaining constancy in the number of copies in the face of mutation and recombination have led other authors to propose derivative models (Georgiev, 1969; Crick, 1971; Paul, 1972; Whitehouse, 1973; Thomas, 1974; Smith, 1974).

The attempts which have been made toward the resolution of the structure of a band in *Drosophila* include a range of techniques from cytogenetics, biochemistry, and electron microscopy. Most cytogenetic evidence points to the possibility that one band could be equated with one complementation group. Judd and his co-workers (1972) did extensive genetic analyses of the zeste-white region (3A1–3C2) of the X chromosome of *D. melanogaster*. They analyzed 130 lethal mutations which were shown to fall into 16 complementation groups. Each of these complementation groups could be attributed to one of the 15–16 bands in this region. Similar conclusions were reached by Hochman (1974) from his cytogenetic analysis of lethals, semilethals, or sterile mutants from the 4th chromosome of *D. melanogaster*. His analysis of about 200 mutations revealed the presence of 37 essential loci. In addition to these there are 6 other separate loci discovered by earlier workers (Muller, 1965; see also Lindsley and Grell, 1968). Altogether there are at least 43 distinct functions on chromosome 4, which has about 50–60 bands. Lefevre (1974) has also examined X-ray induced lethals and semilethals of the X chromosome, and his results are also compatible with the one-band–one-function hypothesis.

Biochemical evidence relating to this problem is still inconclusive. Most of the efforts have been directed toward the relationship between mRNA and the genome by trying to define the structure of the primary transcription unit. In this way the unit of transcription can be compared with both the unit of translation and with the organization of the genome. This approach is based on the premise that transcriptional control depends on the organization of the nucleotide sequences in the DNA. Since an extensive review of HnRNA is presented in Chapter 7 of this volume, this chapter will be restricted to the problem in *Drosophila*.

Davidson and co-workers (Davidson *et al.,* 1973*a,* 1974; Graham *et al.,* 1974; Goldberg *et al.,* 1975) have demonstrated that there are at least two distinct forms of DNA sequence organization in eukaryotes. One is the *Xenopus* pattern in which repeated sequences each about 300 nucleotides long alternate with single copy sequences which vary in length from about 800 nucleotides to several thousand nucleotides. However, a very different form of interspersion pattern appears to exist in *D. melanogaster.* A large portion of the middle repetitive sequences are interspersed with single copy sequences. The middle repetitive sequences, which vary in length from 500 to 13,000 base pairs (the average is 5600 base pairs), alternate with single copy sequences of lengths greater than 13,000 base pairs. The significance of the difference in interspersion pattern between *D. melanogaster* and other eukaryotes cannot be assessed until the functions of these components have been resolved.

Biochemical determinations (Lengyel and Penman, 1975) have shown that the average size of pulse-labeled HnRNA from tissue culture cells of *D. melanogaster* is 1.4×10^6 daltons, a size too small to represent the transcript of all the DNA in an average band. These results are at variance with data obtained from analyses of electron micrographs of nonribosomal transcription units which show that band-sized RNA transcripts do exist (see p. 249). It may be that current biochemical techniques are incapable of preserving high molecular weight HnRNA molecules, especially in insects where a high concentration of nucleases prevail. Until further and more extensive characterizations are available, no conclusions can be drawn about the size of an average transcription unit.

Heterogeneity in the Nontranscribed Spacers of Ribosomal DNA

The unique characteristics of the ribosomal RNA cistrons in both prokaryotes and eukaryotes have captured the attention of biochemists, geneticists, and electron microscopists. A diversity of biophysical techniques ranging from nucleic acid hybridization to nucleic acid sequencing have been employed in the elucidation of their arrangement, localization, replication, transcription, structure, and regulation. Comprehensive surveys of the literature concerned with rRNA cistrons have been published recently (Birnstiel *et al.,* 1971; Bostock, 1971).

A complicating feature of the rDNA of *D. melanogaster* which will be dicussed in this section concerns the heterogeneity in the nontranscribed spacers of the ribosomal DNA observed in the different sex chromosomes and at different developmental stages. The implications of such findings are very important, especially in view of the fact that during the past few years much effort has been made to integrate the concept of gene reiteration with molecular models of eukaryotic chromosome structure, function, and evolution (Callan, 1967; Britten and Kohne, 1968; Beermann, 1972; Georgiev, 1969; Thomas, 1970, 1974; Smith, 1974).

The Ultrastructure of the Transcription Units of Ribosomal RNA

Figure 1A illustrates a typical set of tandemly arranged, active, rRNA transcription units obtained from the ovarian nurse cells of *D. melanogaster.* Similar transcription units have been visualized in amphibians (Miller and Beatty, 1969), HeLa cells (Miller and Bakken, 1972), insects (Hamkalo *et al.,* 1973; Trendelenburg, 1974), and algae (Trendelenburg *et al.,* 1974; Spring *et al.,* 1974).

When chromatin is spread on a film and viewed under the electron microscope, two main regions are observed (see Figure 1), one covered with fibrils (matrix) and one free of fibrils (spacer). In general, the fibrils in the matrix region appear to form a single gradient in terms of their length. These matrix regions were first identified by Miller and Beatty (1969) as active rRNA cistrons. This conclusion was based mainly on two criteria. The first was that the length of the matrix region approximated roughly the length of the rRNA precursor of the organism studied. The second criterion was that the alternation of matrix with spacer regions appeared to reflect the renaturation kinetics of rDNA of the organism under study. Recently (see p. 239) more direct evidence relating to the identification of these putative

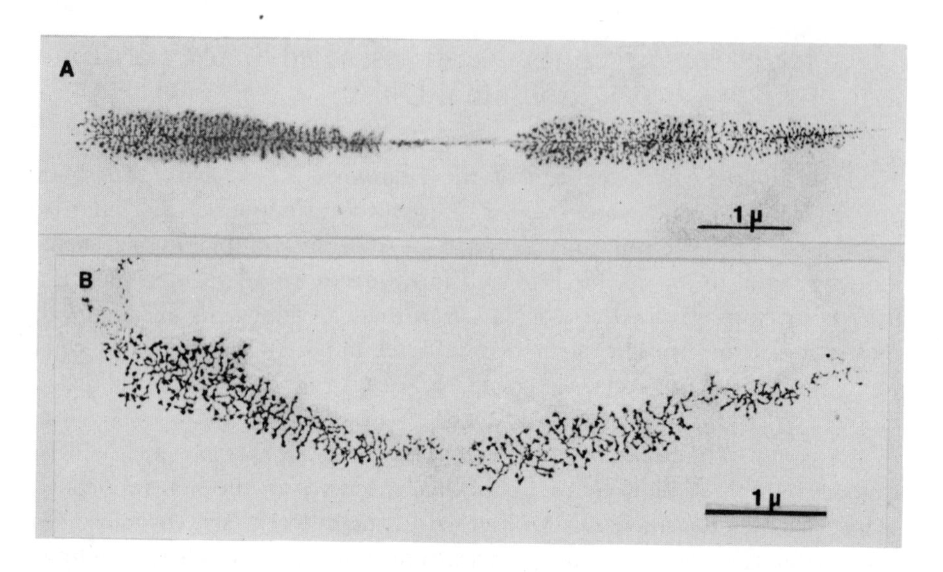

Figure 1. A: Electron micrograph of two rRNA genes from a nurse cell of a stage 10 egg chamber of D. melanogaster. B: The rRNA genes from a blastoderm embryo. The methods of preparation and staining procedures used were as follows: Embryos were collected in 35-mm petri dishes which were partially filled with semigelled agar (5%). The chorion of each embryo was removed by gently rolling it on the sticky side of a strip of Scotch tape. Staging of embryos was done while the dechorionated embryos were viewed under a light microscope. Ovaries were dissected from 3- to 4-day adult D. melanogaster females. Staging of egg chambers was done according to the conventions summarized by King (1970). From each egg chamber the 15 nurse cells were isolated with forceps in 0.1 M KCl, pH 8–9, transferred by braking pipette to a fresh 0.1 M KCl solution, and ruptured with forceps. The nurse cells (in about 5 μl) were transferred by pipette to 100 μl pH 9 water and allowed to disperse for up to 40 min. Centrifugation of cell lysates onto grids was performed for 5 min at 1020 × g in a Sorvall HB4 rotor (4°C). After centrifugation the grids were washed with 0.4% Kodak Photo Flo and air-dried. Grid chambers, sucrose–formalin, and Kodak Photo-Flo 200 solutions were prepared and used as described by Miller and Bakken (1972) and Chooi and Laird (1976). Grids were stained for 1 min with 1% phosphotungstic acid (pH unadjusted) in 70% ethanol, rinsed in 95% ethanol for 5 sec, and air-dried. Grids were then restained for 1 min with 1% uranyl acetate in 70% ethanol, rinsed in 95% ethanol for 5 sec, and air-dried.

rRNA cistrons has emerged. The fibrils seen in the matrix regions of transcription units have been referred to as *ribonucleoprotein fibrils,* and the gradient of fibrils has been interpreted as reflecting the progress of transcription. It should be noted that the sizes of the RNP fibrils do not correspond directly to their expected sizes as calculated from their positions on the gradient; they are always shorter. This observation has been interpreted by many authors as being due to a foreshortening caused by configurational changes and not the result of cleavage, although there is no conclusive evidence supporting either conclusion solely. The third section of this chapter describes some preliminary evidence that bears directly on these questions.

The transcription units in *D. melanogaster* such as those shown in Figure 1 have also been interpreted as rRNA transcription units, although they are, on the average, shorter than what one would expect from the size of the rRNA precursor. According to the biochemical estimates of Perry *et al.* (1970), the size of the rRNA precursor in *D. melanogaster* is about 38 S, which should correspond to a matrix length of about 3.2 μm. The length of the rRNA transcription units in *D. melanogaster* is 2.66 $\pm$ 0.41 μm.

Specificity of the rRNA Spacer Sequences of the X and Y Chromosomes in *D. melanogaster*

During oogenesis in *D. melanogaster,* the oocyte nucleus synthesizes little RNA, while the nurse cells are responsible for virtually all the RNA contributed to the oocyte (King, 1970). During previtellogenesis and vitellogenesis in *D. melanogaster,* the nurse cells undergo endoreplication and increase in size. At about stage 10 (King, 1970), some nurse cells reach a maximum level of about 1024C, although it is uncertain if the rRNA genes are replicated to the same degree. Massive transport of RNA, ribosomes and mitochondria takes place from the nurse cells into the oocyte (King, 1970).

Abundant syntheses of rRNA and non-rRNA have been observed to take place at about stage 9–10 during oogenesis and at the blastoderm stage of embryogenesis (Chooi and Laird, in preparation; Lamb and Laird, 1975). Egg chambers and embryos from these stages were mainly used for the studies on rRNA and non-rRNA transcription presented in this chapter.

When rRNA genes from nurse cells of Oregon R, wild-type *D. melanogaster* were studied in the electron microscope it was observed that the spacer regions had a mean length of about 1.3 μm or approximately half the length of the transcribed region (Figure 1A). Surprisingly, however, when rRNA genes from wild-type *D. melanogaster* embryos were examined using the same technique the spacers were found to be shorter (Figure 1B). The spacers had a mean length of about 0.8 μm or about a third the length of the transcribed region. This distinct difference in spacer lengths observed during two separate stages of development suggested some further experiments.

In *D. melanogaster* the ribosomal genes are found at two separate chromosomal loci. There is one nucleolar organizer on the X chromosome and one on the Y chromosome (Heitz, 1933; Kaufmann, 1933, 1934). The question arises as to whether there is selective expression or selective amplification of a particular set of ribosomal genes during oogenesis and embryogenesis. Initially, the speculation was that this difference in spacer length could be accounted for by either of two simple hypotheses: (1) that the long ribosomal spacers were specific to the rRNA genes on the X chromosome, while the short spacers were specific to the rRNA genes on the Y chromosome, or (2) that there were two types of spacers on the X-linked rRNA genes, that the genes with the long spacers were specifically expressed during oogenesis, that the genes with the short spacers were expressed during embryogenesis, and that the Y-linked rRNA genes had only short spacers.

If the first hypothesis is correct, then while long spacers are observed during oogenesis, both long and short spacers should be observed during embryogenesis, since both XX and XY embryos were examined. A bimodal distribution expected from the presence of both types of spacers from XX and XY embryos was not apparent (Figure 2), and therefore the first hypothesis is not tenable.

In order to try to resolve this difference in spacer lengths, *D. melanogaster* females with rRNA genes of only one of the sex chromosome types were generated. Spacers in flies that had only the X chromosome rDNA or only the Y chromosome rDNA were examined during oogenesis and embryogenesis. In the nurse cells of females with the genotype $sc^4sc^8/sc^4sc^8/y^+Y$ [where sc^4sc^8 = In (1) $sc^{4L}sc^{8R}$ and $y^+Y = sc^8Y$; see Lindsley and Grell, 1968], the spacer to matrix ratios of the Y-linked rRNA genes were about 0.3 (see Figures 3 and 4). Flies with the genotype $sc^4sc^8/sc^4sc^8/y^+Y$ possess two X chromosomes in which the nucleolar organizers have been eliminated and one Y chromosome which has a functional nucleolar organizer. Spacer to matrix ratios of about 0.3 were also obtained for rRNA genes from nurse cells of flies with the genotype $sc^4sc^8/sc^4sc^8/Y$, where the Y chromosome is derived from a strain different from that in $sc^4sc^8/sc^4sc^8/y^+Y$. These observations indicate that when only the rRNA genes on the Y chromosome were present only short spacers were observed in nurse cell nuclei. However, the possibility that rRNA genes with long spacers are expressed during embryogenesis in the absence of the X-linked genes is not excluded.

To investigate this problem, rRNA spacers from the F_1 from the cross $sc^4sc^8/sc^4sc^8/Y \times sc^4sc^8/Y$ were examined. In the F_1 of such a cross, zygotes of several different genotypes are produced: (1) zygotes with the genotypes sc^4sc^8/sc^4sc^8 and $sc^4sc^8/sc^4sc^8/sc^4sc^8$ (these chromosomes do not have nucleolar organizers, and therefore the embryos die before cellular

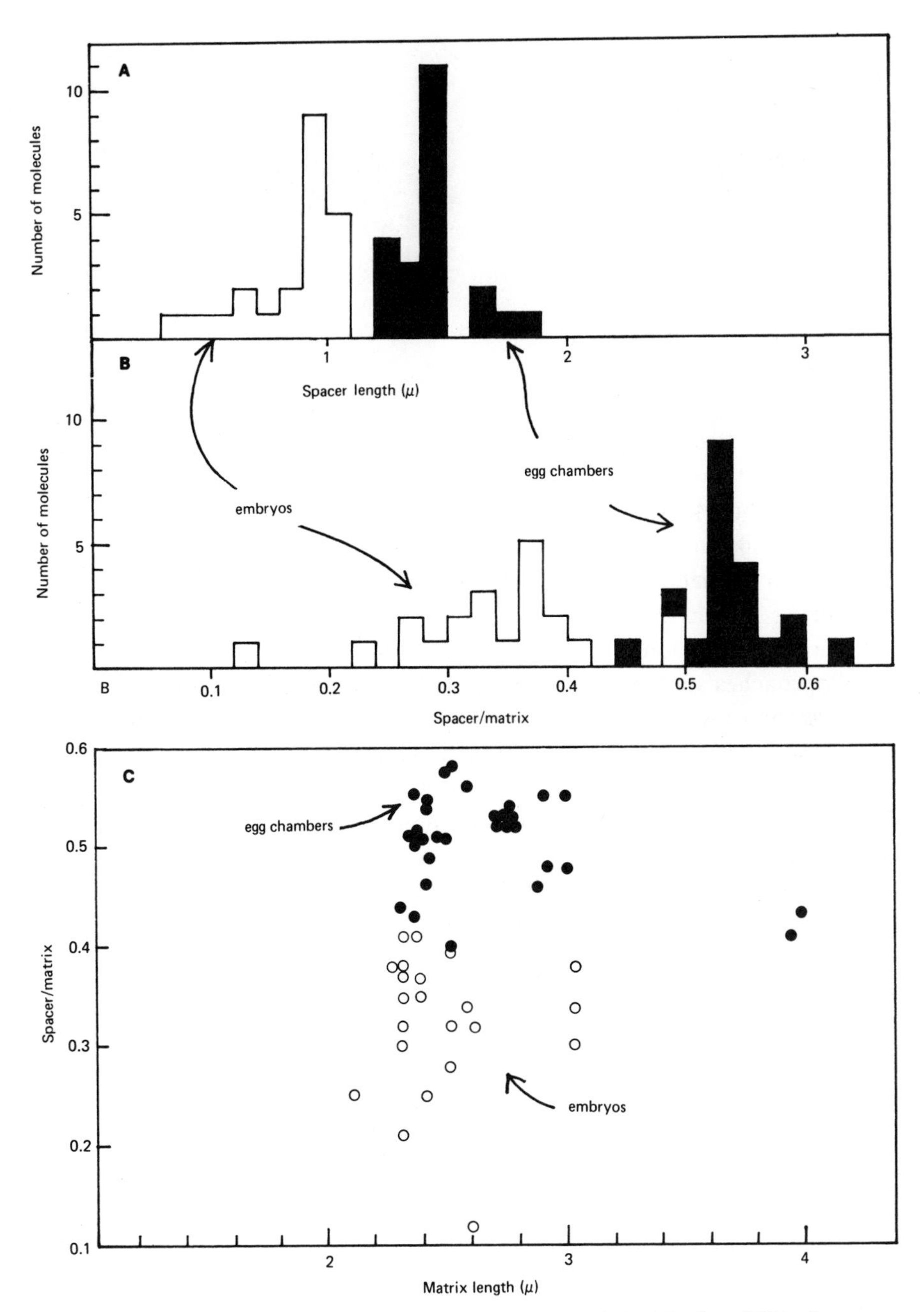

Figure 2. A: Histogram showing the distributions of spacer lengths for rDNAs from nurse cells and embryos of D. melanogaster. For all the data presented, spacer lengths were measured only from preparations where at least the two adjacent rRNA cistrons were present. B: Histogram showing the distributions of the ratios of spacer to matrix lengths from rRNA genes taken from nurse cells and embryos. C: Plot of matrix lengths against ratios of spacer to matrix lengths.

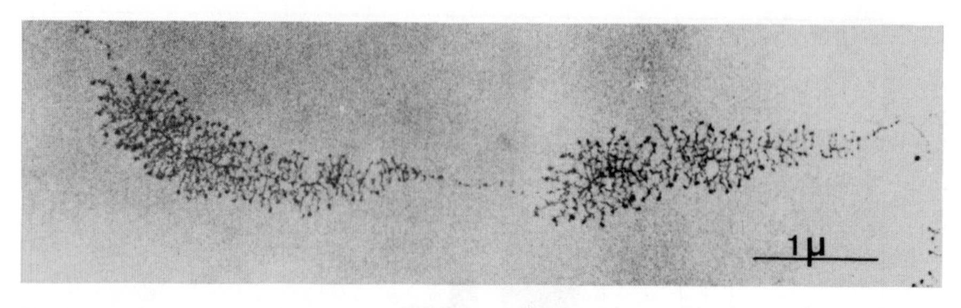

Figure 3. Electron micrograph of rRNA genes from a stage 10 nurse cell of a $sc^4sc^8/sc^4sc^8/y^+y$ female.

blastoderm), (2) females with the genotype sc^4sc^8/sc^4sc^8/Y, and (3) males with the genotypes sc^4sc^8/Y and sc^4sc^8/Y/Y. Regardless of the sex, all the viable F_1 embryos from such a cross would carry rRNA genes from only the Y chromosome. All the rRNA spacers examined from such a cross were similar in size to those seen in wild-type embryos. The data indicate that during both oogenesis and embryogenesis the Y-linked rRNA genes that were transcriptionally active had only short spacers.

Approaches such as those described for the rRNA spacers of different Y chromosomes were also made for different X chromosomes. When nurse cells of the wild-type strain of Canton X were examined, it was found that only long spacers were present. From the evidence presented, it seems likely that the rDNA on the X chromosome contains both types of spacers and that the genes with the long spacers are preferentially transcribed during oogenesis but the genes with the short spacers are preferentially transcribed during embryogenesis.

Alternatively, one could argue that the rRNA genes with long spacers on the X chromosome are also expressed during embryogenesis, but they were not detected because through sampling errors only embryos carrying XXY and/or XY chromosomes were examined and for some unknown reason only the Y chromosome was preferentially transcribed on these. This is unlikely, since the data presented in Figure 2 were obtained from at least ten randomly selected embryos, and many more embryos were examined subsequently with similar results.

In general, whenever a large number of genes are arranged in tandem on a single piece of chromatin, all of them exhibit the same direction of transcription. Occasionally, adjacent genes exhibit opposite directions of transcription (Figure 5A). In addition to the significant differences in spacer lengths observed between the two sex chromosomes (Figures 2 and 4), variations in spacer lengths between adjacent ribosomal cistrons on the same stretch of rDNA have also been observed (Figure 5B,C). It may be

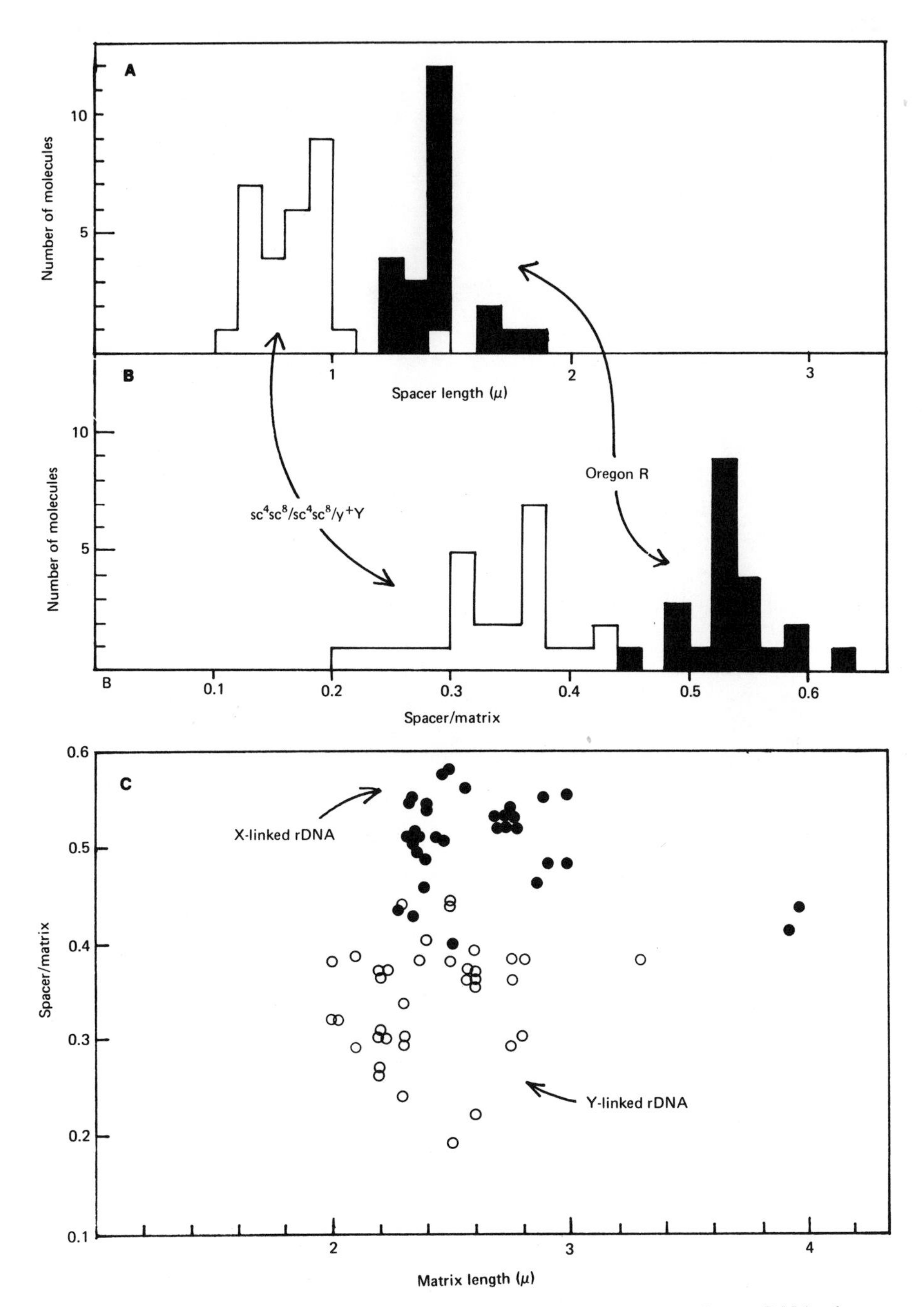

Figure 4. A: Histogram showing the distributions of spacer lengths from rDNA of nurse cell nuclei of D. melanogaster containing X-linked rDNA (Oregon R) or Y-linked rDNA ($sc^4sc^8/sc^4sc^8/Y$)*. B: Histogram showing the distributions of the ratios of spacer to matrix from rRNA genes isolated from nurse cell nuclei of Orgeon R and* $sc^4sc^8/sc^4sc^8/y^+Y$ *females. C: Plot of matrix lengths against the ratios of spacer to matrix for the two classes of rDNA.*

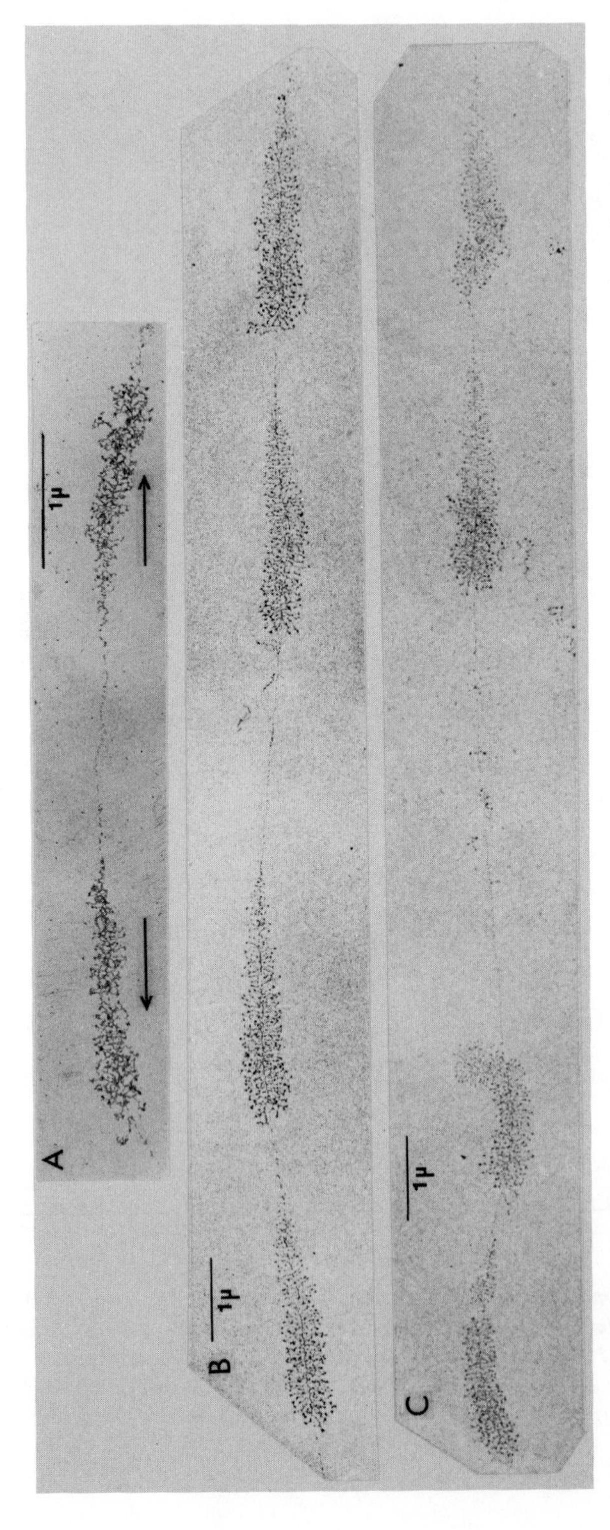

Figure 5. A: Electron micrograph of rRNA genes from a stage 10 nurse cell of a $sc^4sc^8/sc^4sc^8/Y$ female. The adjacent genes on the chromatin show opposite directions of transcription (arrows). The spacer to matrix ratio of rDNA is 0.9, which is about 3 times greater than the expected mean from the Y chromosome. B, C: rRNA genes from a stage 10 nurse cell of a $sc^4sc^8/FM7$ female. Heterogeneity in spacer lengths between adjacent genes is clearly demonstrated.

argued that such length measurements are liable to error owing to the problem of stretching. To test if this had affected my interpretations, the data were carefully quantitated. Figures 2C and 4C demonstrate that while there are variations in spacer lengths, some of which can be attributed to stretching, it may nevertheless be concluded that (1) the spacer to matrix ratio of rRNA genes in nurse cells is significantly higher than that in embryos, although similar variations in matrix lengths occur in both, and (2) that the spacer to matrix ratio of the rRNA genes in the X chromosome is significantly greater than that in the Y chromosome, although similar variations in spacer length occur in both sets of genes.

The data at hand are consistent with the second hypothesis, namely that the rRNA genes on the X chromosome have two kinds of spacers and that one set is preferentially transcribed during oogenesis while the other is preferentially transcribed during embryogenesis. The Y chromosome, on the contrary, has only short spacers. This feature of selective transcription of one gene locus over another gene locus (which produces primarily the "same" functional product) is also found in the transcription of the 5 S RNA genes in *Xenopus* (see p. 221). However, at this stage in the investigation it is not clear whether there is selective amplification or selective expression of these two classes of genes. Further characterizations—for example, with restriction enzyme EcoR1 (Hedgepeth *et al.*, 1972)—could further test the validity of this hypothesis and extend these tentative conclusions.

General Implications of Heterogeneity in the Nontranscribed Spacers of Ribosomal DNA

Early biochemical experiments using nucleic acid hybridization techniques have demonstrated that although there are large numbers of rRNA cistrons in higher eukaryotes, these cistrons show a complexity close to that of a single family of genes (Birnstiel *et al.*, 1969). For example, *Xenopus* rDNA renatures with a rate expected of a basic mass of 10^7 daltons, a value in good agreement with the proposed structural repeat length of the 28 S and 18 S tandem cistron together with the associated spacer DNA. Since the fidelity of base pairing in these nucleic acid hybridization experiments was surprisingly high, this suggested that the repetition within any one species was strikingly similar. Electron microscopic studies, including denaturation mapping (Wensinck and Brown, 1971; Brown *et al.*, 1972) and heteroduplex analysis (Forsheit *et al.*, 1974), also failed to detect size heterogeneity in rDNA. However, recent experiments with EcoR1 restriction endonucleases suggested that some heterogeneity was present in rDNA from *X. laevis* (Morrow *et al.*, 1974) (see below).

The ribosomal cistron DNA sequences are now known to be extremely conservative throughout evolution, and there is no conclusive evidence to indicate that there exists intercistronic divergence in any one organism. Maden and Tartof (1974), for example, have shown that there are no detectable differences between the rRNAs from the X and Y chromosomes of *D. melanogaster*. However, if the heterogeneity in eukaryotic rRNA is relatively small and is random, it may not be fully revealed even by nucleotide sequence analysis.

Many organisms contain hundreds or thousands of similar rRNA cistrons. While there is general agreement that the nucleotide sequences of ribosomal cistrons are rather well conserved within an organism, there is also good evidence to indicate that the spacer sequences exhibit some significant heterogeneity (Wellauer *et al.*, 1974; Morrow *et al.*, 1974; Chooi and Laird, in preparation). This raises the question of the mechanism which protects the hundreds of similar rRNA genes from rapid divergence through random mutational change and, at the same time, which permits the spacers to diverge. However, it is important not to forget that major changes in the structure and activity of the rDNA have clearly occurred during evolution, the most impressive of which is the rapid increase in the size of the ribosomal transcription unit during eukaryotic evolution. To try to account for this apparent homogeneity in the rRNA cistrons alone, it has been suggested that the cell contains special corrective or rectification mechanisms (Callan, 1967; Thomas, 1970, 1974) (see p. 225). Alternatively, random mutations may be distributed throughout the rRNA cistrons by unequal crossing over. In such a case, the spacer may play a key role as a point for genetic exchange. Clustering of the rRNA cistrons on one chromosomal locus clearly facilitates this process. Since RNA is the final product of transcription, the maintenance of base sequences in the rRNA cistron region must be more stringent than that in the spacer region. Therefore, a majority of successful mutations are those that accumulate in the spacer regions, whereas mutations in the transcribed regions are not tolerated.

Smith (1974) proposed a scheme involving differential crossing over and demonstrated on theoretical grounds that one could easily generate the accumulation of mutations in the spacer portion while maintaining intrachromosomal homogeneity in the gene portion if two assumptions were made: (1) that mutations in the ribosomal cistronic DNA would generally be deleterious and therefore that an rRNA cistron harboring such a mutation would rapidly be eliminated by natural selection, and (2) that many mutations in the spacers are generally selectively neutral and therefore that an rRNA cistron harboring a neutral mutation would not be eliminated by natural selection. These neutral mutations would therefore accumulate.

The demonstration of heterogeneity in spacer sequences between species and within species of *Xenopus* (Brown *et al.*, 1972; Wellauer *et al.*, 1974; Wellauer and Reeder, 1975) as well as the demonstration of heterogeneity in spacers between adjacent genes and between genes from different loci of the same organism (Chooi and Laird, in preparation) seems to support this hypothesis.

Male *Drosophila* with a bobbed mutation on one sex chromosome and a deficient or deleted bobbed locus on the other have been shown to pass on to the progeny a bobbed locus with an increase in rDNA. If the F_1 males have sex chromosomes identical to those of their fathers, their bobbed phenotype is less severe compared with that of their fathers. These F_1 males in turn pass on extra rDNA to their progeny. However, the males of the F_2 generation are nonbobbed, both in phenotype and in rDNA content despite carrying the same sex chromosome as their severely bobbed grandfathers. Ritossa (1968) referred to this increase in rDNA and the phenotypic reversion as *magnification*. Magnification is restricted to males, and the changes in rDNA content due to magnification are inherited (Ritossa, 1968; Henderson and Ritossa, 1970). Magnification may be interpreted as another interesting aspect of the rectification phenomenon. This concept may make the parallel evolution of rDNA located on different chromosomal loci (e.g., on different sex chromosomes as in the case of *D. melanogaster*) more likely.

Assembly of Ribosomal Proteins

The *E. coli* ribosome is composed of two subunits which were named according to their sedimentation coefficients. The smaller 30 S subunit contains one 16 S RNA molecule and 21 proteins. The larger 50 S subunit consists of one 23 S RNA molecule, one 5 S RNA molecule, and about 34 proteins. Prokaryotic ribosomes, especially *E. coli* ribosomes, have been the subject of intensive research. Characterizations of the primary structure of the ribosomal proteins, the structure of the ribosomal RNA, the interaction between proteins and RNA, and the topographical arrangement of these proteins in the ribosome have led to a somewhat more coherent assessment of their structure and function. In contrast, much less is known about the more complex eukaryotic ribosomes. Their complexity is manifested in their tendency to have higher molecular weight rRNAs, an increase in the number of proteins in the ribosome (54 in prokaryotes vs. about 70 in eukaryotes), and a considerable increase in the percentage of proteins in the ribosome (35% in bacteria vs. 50% in eukaryotes, see Chapter 11).

Protein–RNA Interactions in Prokaryotes

In *E. coli,* protein–RNA interactions have been studied by two main methods in an attempt to locate the binding sites of proteins on their respective RNA molecules. The first method was developed by Zimmermann and his coworkers (Zimmermann *et al,* 1972, 1974; Muto *et al.,* 1974), who showed that mild digestion of the 16 S RNA produced large RNA fragments to which certain proteins could be reattached. This permitted a partial localization of proteins along the 16 S RNA. In this type of approach, the proteins are not covalently attached to the RNA, and therefore special care must be taken to ensure that the observed results are genuinely specific.

The second method involves the digestion by nuclease of a complex between single proteins and the RNA with a view to isolating a region of RNA which is protected from digestion by the presence of the bound protein. This method has yielded the most detailed information on the binding sites of proteins on the 16 S RNA (Ungewickell *et al.,* 1975; Zimmermann *et al.,* 1972, 1975; Schaup *et al.,* 1971, 1973; Schaup and Kurland, 1972), the 5 S RNA (Gray *et al.,* 1972, 1973), and the 23 S RNA (Branlant *et al.,* 1973, 1975). However, this method also has the disadvantage that it is difficult to produce protected ribonucleoprotein fragments containing proteins which bind weakly.

Assembly Mapping in Prokaryotes

One of the most interesting aspects of ribosome research was the building of the assembly map by Nomura and his coworkers (Mizushima and Nomura, 1970). Using individual purified 30 S ribosomal proteins, they studied the assembly of 30 S subunits. Under the conditions of assembly, only seven of the proteins bound to the RNA. Certain other proteins became bound after some of the first seven proteins were bound (Nomura, 1972; Mizushima and Nomura, 1970). The remaining proteins required the presence of proteins in both of the above groups in order to become bound. In this manner, the sequence of addition of proteins to the 16 S RNA was worked out, and this led in turn to the construction of the assembly map.

Some of the interactions described in the assembly map have been confirmed by other workers (Green and Kurland, 1973). Although these interactions need not necessarily be a reflection of the protein neighborhoods within the 30 S particle, there has been good agreement between both those proteins which are found together in crosslinked pairs in ribonucleoprotein fragments and those proteins which are related in the assembly map. Therefore, it is reasonable to conclude that most, if not all, of the assembly interactions are indeed direct reflections of the ribosomal topography.

However, it is uncertain whether or not the assembly map corresponds to the *in vivo* sequence of assembly. There does not appear to be good agreement between the assembly map and the estimated order of addition of proteins *in vivo* (Marvaldi *et al.*, 1972). The 30 S particles can be formed by reconstitution from ribosomal proteins and precursor 16 S RNA (Wireman and Sypherd, 1974; Mangiarotti *et al.*, 1974), but the energy of activation is lowered when proteins from nascent (as opposed to completed) ribosomes are used (Mangiarotti *et al.*, 1975). This observation suggests that *in vivo* assembly may be facilitated by a recycling factor.

A successful total assembly of the 50 S particle has been described by Nierhaus and Dohme (1974). The lack of an assembly map of the 50 S particle has retarded the progress on the topography of the larger subunit.

In Vivo Assembly Mapping in *D. melanogaster*

Ribosome manufacture in eukaryotes is generally believed to occur in the nucleolus, although the entire ribosome is probably not assembled there. There is little conclusive evidence to indicate which steps of the ribosome assembly take place inside the nucleolus (see review by Warner, 1974). Nascent ribonucleoprotein particles comprising the ribosomal RNA precursors, as well as newly synthesized ribosomal proteins, and 5 S RNA have been isolated from nucleoli (Warner and Soeiro, 1967). It has been suggested that in mammals the steps involved in the processing of the 45 S complex take place within the nascent ribonucleoprotein particles. The ribosomal proteins associate with the ribosomal section of the 45 S RNA, leaving the rest accessible to nuclease degradation (Maden, 1968).

Other attempts have also been made to try to resolve the question of whether certain nucleolar-specific proteins are absent from the more mature ribonucleoprotein particles or mature ribosomes. It has been inferred from two-dimensional gel electrophoresis that there may be a number of proteins in nucleolar RNA that are not present in ribosomal proteins (Tsurugi *et al.*, 1973; Kumar and Subramaniam, 1975). However, the question of possible contamination of ribosomal proteins with absorbed extraneous soluble proteins during the preparation of ribosomal subunits has not been resolved. To date, no further conclusive evidence has emerged.

The main point of discussion in this section is concerned with the assembly of ribosomal proteins in *D. melanogaster* during the process of ribosomal RNA transcription. Are ribosomal proteins complexed with nascent rRNA precursor molecules during their process of maturation? If so, what is the sequence of assembly of the different ribosomal proteins? Are there differences in ribosomal proteins present in nucleolar nascent ribonucleoprotein particles compared with those in cytoplasmic mature ribosomes? The approach used in the study of ribosomal protein assembly in *D. melanogaster* which will be described here is that of *immune electron microscopy* (see below). Although the method has not been tested widely, it appears likely that it will be useful and novel for the study of the localization, orientation, and polarity of particular ribosomal RNA molecules and proteins. An essential prerequisite for the electron microscopic determination of assembly is the availability of pure ribosomal proteins, and these also should be detectable under the electron microscope.

The separation and purification of all the 55 ribosomal proteins from *E. coli* have been achieved (see review by Wittmann, 1974). The current techniques employed in the isolation and characterization of total ribosomal proteins from eukaryotic ribosomal subunits are sufficiently good to provide reasonably "clean" proteins. Purification of individual eukaryotic proteins has, however, proved to be considerably more difficult, and up to now few have been separated and characterized (see review by Wool and Stöffler, 1974).

In the study described here, antibodies against *E. coli* and rat liver ribosomal proteins were used. IgGs or Fabs were covalently linked to ferritin in order to facilitate a more precise localization of the antigen–antibody complex. A number of criteria were used to evaluate the ferritin–antibody conjugates used. One criterion was that the antibody in the conjugate should retain most of its immunological activity. All the IgGs and Fabs in the conjugates were still immunologically active when tested. A second criterion was that the conjugate should be relatively free of antibodies which were not linked to ferritin. Otherwise, the unconjugated antibodies would bind to antigenic sites in the specimen and competitively prevent the binding of the labeled antibody. The conjugates did not contain any free Fabs or free IgGs, because these were effectively removed by gel filtration after the coupling reaction. The gel filtration step also removed the ferritin–ferritin polymers from the final product. Their removal was important, since such polymers tended to be relatively insoluble and to adsorb nonspecifically.

The preliminary results which will be presented below represent attempts to localize the following: (1) antibodies against total 60 S subunit proteins from rat liver (anti TP 60), (2) antibodies against total 50 S subunit proteins from *E. coli* (anti TP 50), (3) antibodies against 30 S subunit proteins from *E. coli* (anti TP 30), and (4) antibodies against protein L18 from *E. coli* (anti L18).

The ease at which rRNA transcription units can be visualized in the electron microscope (see p. 226) has provided an exceptional advantage in this approach to the resolution of the *in vivo* assembly of ribosomal proteins. The electron micrographs of rRNA transcription in *D. melanogaster* (Figures 1 and 3) display several features that prove highly advantageous in such a study. Since there is a gradient of RNP fibrils along the transcription unit, the orientations of the initiation and termination regions of the transcription unit and the 5′ and 3′ termini of the growing ribonucleoprotein fibrils can easily be distinguished. Since there are about 100 ribonucleoprotein fibrils on each active transcription unit which has a mean length of about 2.6 μm, it should be theoretically possible to distinguish the sequence and pattern of assembly of proteins that are situated at least 80 nucleotides apart on the RNA.

The reactions obtained between antibodies to *E. coli* ribosomal proteins and rat liver ribosomal proteins with *Drosophila* ribosomal proteins are shown in Figures 6–9. Initially it was unclear whether the optimal conditions for electron microscopy would also be conducive to the specificity of the antibody–antigen complex. Figure 6 demonstrates very clearly that the experimental conditions allowed for both visualization of transcription units and antigen–antibody complex specificity. When antibodies to ribosomal proteins were added to cell lysates, they were found attached to ribosomes or to ribosomal transcription units, and binding to other cell components like HnRNA and chromatin was not apparent.

A common effect of IgGs on cell lysates was the production of large complexes or "lattices" of ribosomes or rRNA transcription units. One type of lattice I observed was composed of three or more ribosomes linked by IgG molecules. Each IgG has two reactive sites, and it can bind to a certain determinant on two separate ribosomes simultaneously, thereby producing a dimer. Further linkage of these dimers by more IgGs would lead to the formation of lattices. Similar lattices have also been reported for *E. coli* ribosomes complexed with specific antibodies (Wabl, 1974).

Ferritin-labeled IgGs to total proteins from the 60 S subunit of rat liver were found attached mainly to the terminal (5′) knobs of the RNP fibrils (Figure 7A). For simplicity, the length of the total rRNA cistron is referred to arbitrarily as 1 unit, with the initiation point (5′) of the transcription unit as the 0 position. The IgGs raised against TP 60 (rat liver) were located unambiguously from a mean distance of about 0.3 unit all the way to the end of the transcription unit. More than one row of ferritin-labeled IgGs was observed along the transcription unit. At least an additional row (arrow) was present toward the latter part of the transcription unit. This observation suggests that different ribosomal proteins are added at least in two and possibly more steps during RNA synthesis.

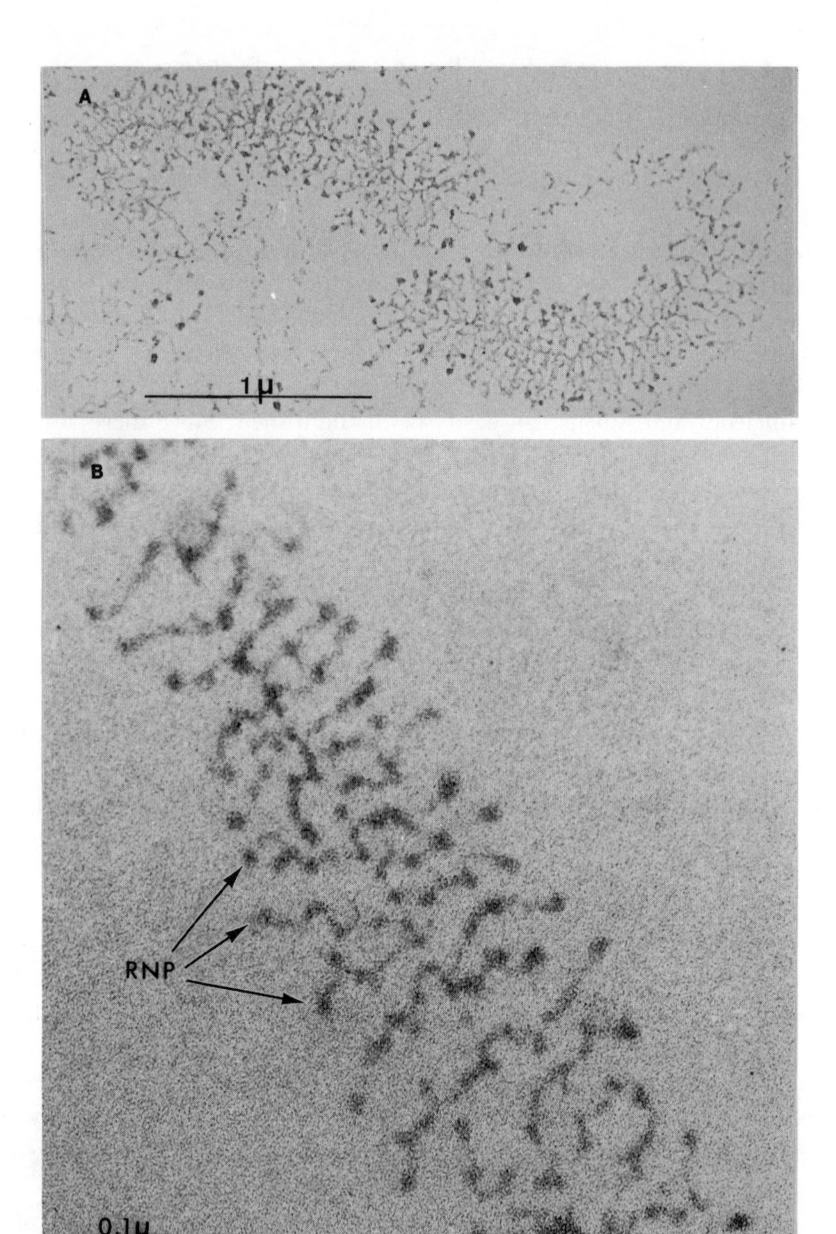

Figure 6. A: Electron micrograph of rRNA genes from the nucleus of a stage 10 nurse cell of D. melanogaster. Non-immune IgGs were added to this preparation. B: Enlargement of RNP fibrils from an rRNA gene to which no antibodies were added. C: Electron micrograph of rRNA genes from stage 10 nurse cells of D. melanogaster. Ferritin-labeled Fabs of anti 50 S (E. coli) were added to this preparation. In all the preparations related to the assembly of ribosomal proteins on rRNA transcription units, several modifications to the procedure described in Figure 1 were made. Ferritin-labeled antibodies or ferritin labeled nonimmune IgGs were added immediately to the preparation upon cell lysis at 0°C. The mixture was centrifuged immediately onto grids. The preparation was stained with 0.5% phosphotungstic acid for about 10 sec, rinsed in 95% ethanol for 5 sec, and air-dried. While this produced very

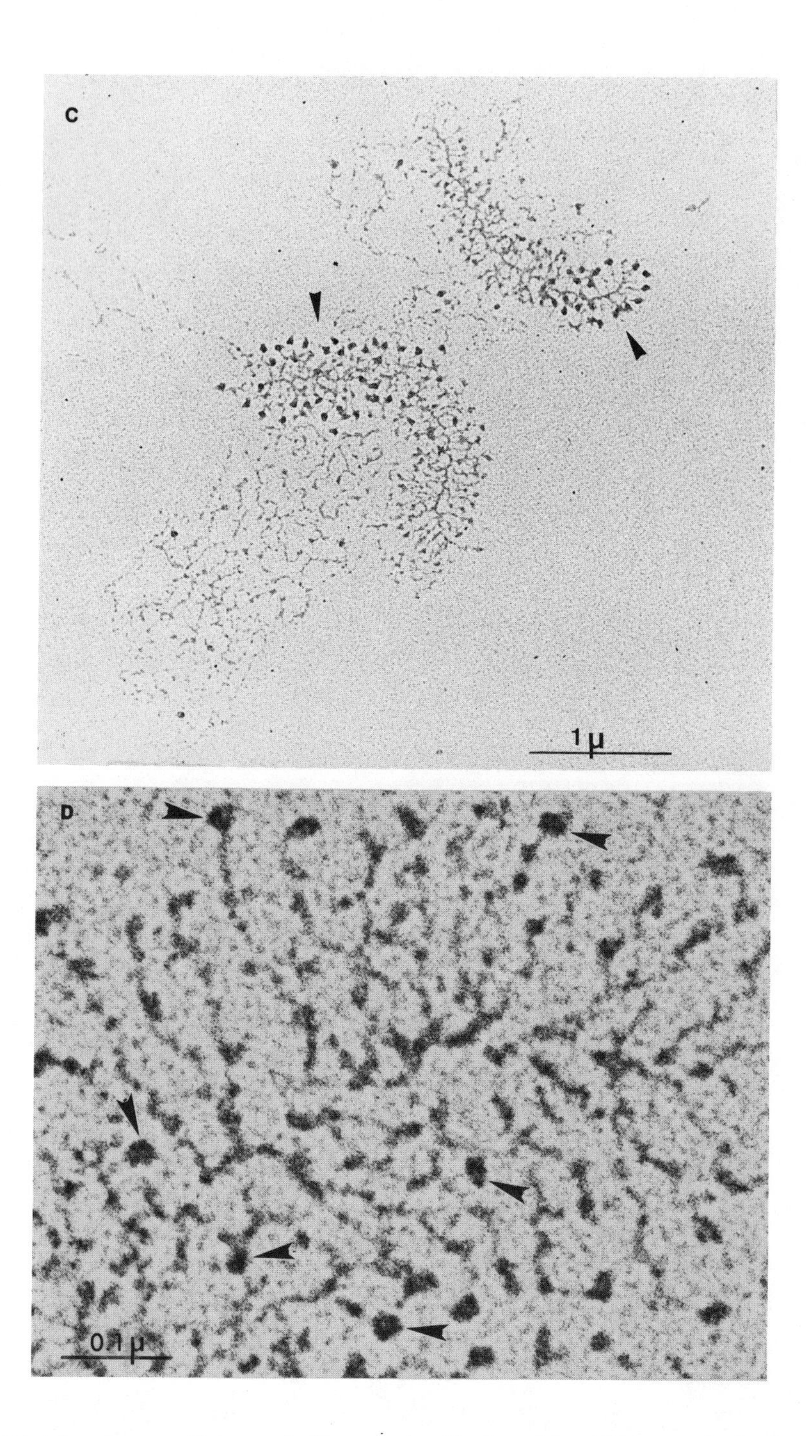

light staining, it was sufficient to facilitate the localization of the rRNA genes in the electron microscope but not enough to interfere with the localization of the electron-dense ferritin. The arrows indicate the position of the ferritin molecules. D: Enlargement of RNP fibrils from a rRNA gene to which ferritin-labeled Fabs of anti 50 S (E. coli) were added. The arrows indicate the positions of some of the electron-dense ferritin molecules on the RNP fibrils.

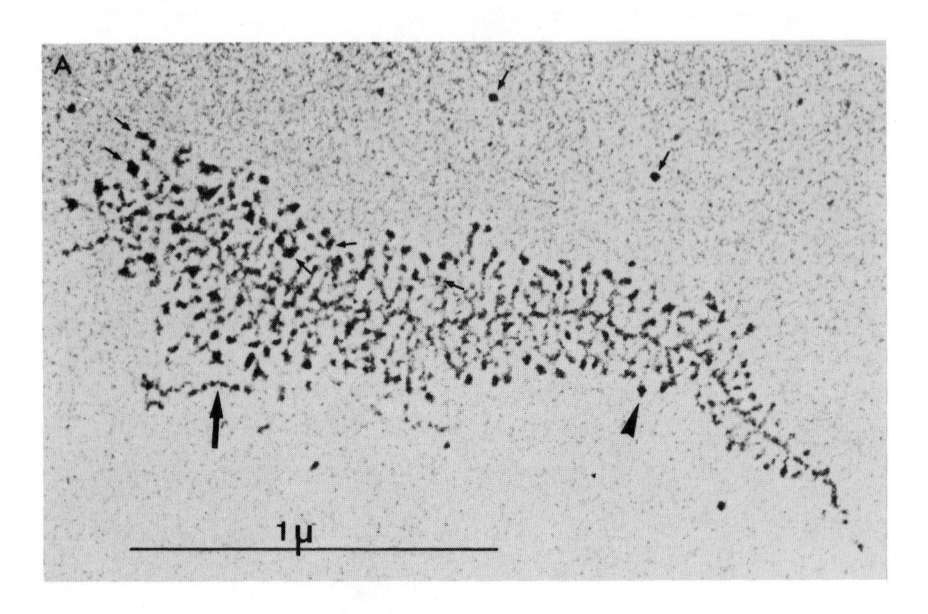
A
1μ

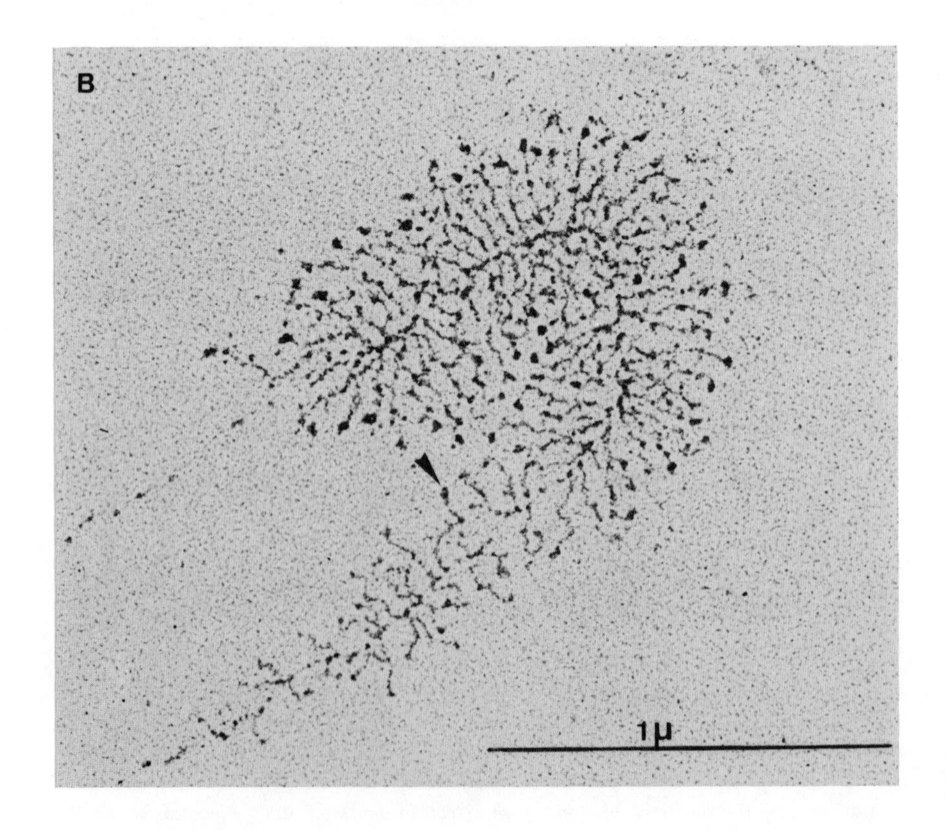
B
1μ

It is important to note that the ferritin-labeled IgG molecules do not uniformly envelop the whole RNP fibril but are restricted to some of the terminal knobs and other specific areas. This feature of the localization is not an artifact created by competition between antibodies to many different proteins for common determinants, since a higher concentration of antibodies and the use of further enriched antibody preparations produce essentially the same results. Such a specialized localization of proteins on the RNP particles may have very interesting implications on the postulated selectivity of cleavage sites during RNA maturation. If the majority of ribosomal proteins from *D. melanogaster* are similar to those in rat liver, we can conclude that the assembly of ribosomal proteins from the large subunit begins fairly early during the transcription process.

Ferritin-labeled Fab molecules specific to the proteins from the 50 S subunit of *E. coli* were also found attached to the terminal knobs and to some specific areas on the RNP fibrils (Figures 6C and 6D). The distributions of the antibodies appear to begin at a mean distance of about 0.3 unit and are found from that position to the end of the transcription unit. This distribution is similar to that observed with anti TP 60 (rat liver) (Figure 7A), suggesting a similar sequence of assembly of the overall proteins in the large subunit, although not necessarily a similar sequence of assembly of individual proteins. The lower density of ferritin-labeled Fabs on the transcription units in the reaction with anti TP 50 (*E. coli*) may reflect a lesser homology between the proteins of the fly and the bacterium.

By comparison, the proteins of the small subunit of *E. coli* (30 S) showed even less homology with *Drosophila* proteins. Few ferritin-labeled IgGs were found complexed to the nascent RNP fibrils (Figure 8A) even when enriched anti TP 30 was used. When complexes were observed, they were frequently located from a mean distance of about 0.64 unit to the end of the transcription unit. Figure 8B shows the localization of ferritin-labeled Fabs to anti L18 (*E. coli*). It appears to be added from a mean

←

Figure 7. A: Electron micrograph of an rRNA gene from a stage 10 nurse cell of D. melanogaster. Ferritin-labeled IgGs of anti TP 60 (rat liver) were added to the preparation. The triangular arrow indicates the approximate position of the starting point of protein assembly. The thin arrows with tails point to the two rows of ferritin-labeled IgGs on the RNP fibrils. Some unbound ferritin-labeled IgGs (arrows) are seen in the background. An RNP fibril that is detached from the DNA template is found adjacent to the transcription unit (thicker arrow with tail). Note that the ferritin-labeled IgGs are distributed only at intervals along the fibril and do not coat it uniformly. B: Ferritin-labeled IgGs of anti 50 S (E. coli) were added to the preparation. The triangular arrow indicates the approximate starting point of protein assembly.

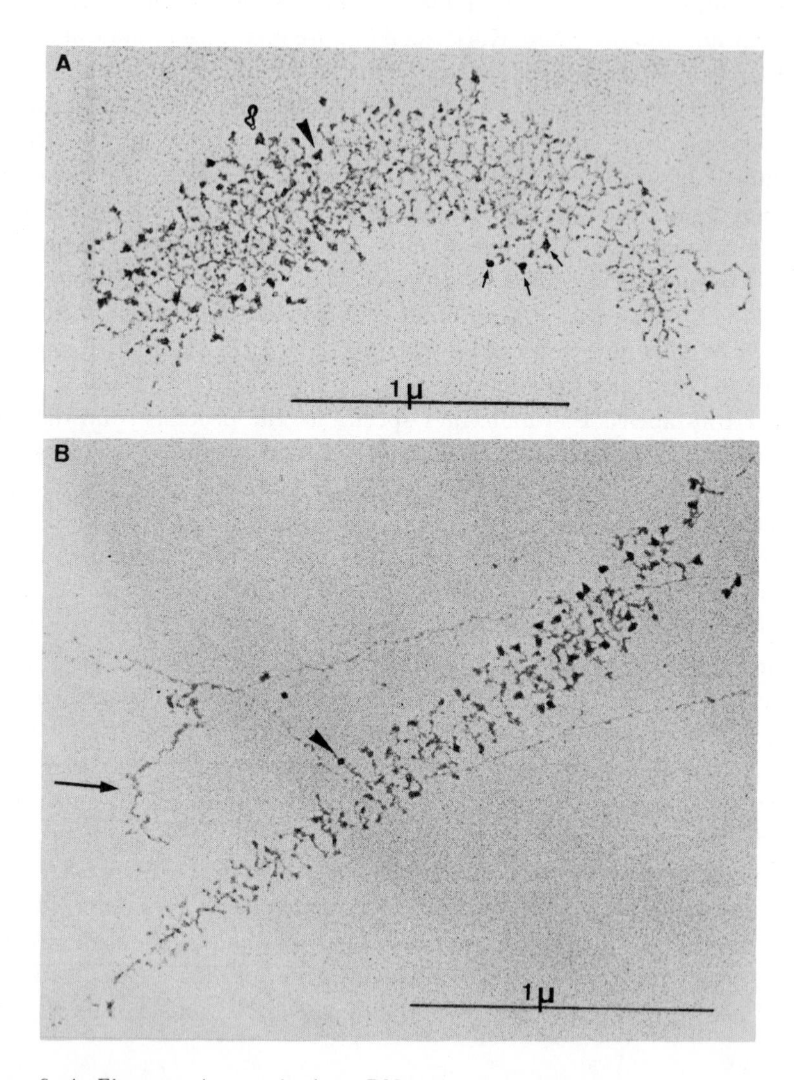

Figure 8. A: Electron micrograph of an rRNA gene from a stage 10 nurse cell of D. melanogaster. Ferritin-labeled IgGs of anti TP 30 (E. coli) were added to the preparation. The triangular arrow indicates the approximate starting point of protein assembly. Beside the rRNA transcription unit are some nascent RNP particles (arrows with tails). B: In this preparation, ferritin-labeled IgGs of anti L18 E. coli were added. The triangular arrow indicates the approximate starting point of protein assembly. Ferritin-labeled IgGs are absent on a non-rRNP fibril (arrow with tail) which is lying adjacent to the rRNA transcription unit.

distance of about 0.55 unit to the 3′ end of the transcription unit. Its localization is compatible with the localization of proteins from the large subunit.

It could be argued that the observed interactions do not reflect the *in vivo* assembly of ribosomal proteins, that instead ribosomal proteins attach specifically or nonspecifically to ribonucleoprotein particles and other cell structures during experimental lysis. This is unlikely for two reasons: The pool size of free ribosomal proteins in prokaryotes and eukaryotes as determined by immunological techniques is very small (0.2%, Wool and Stöffler, 1974). Furthermore, if one argues for nonspecific binding to nascent ribonucleoprotein particles, then nonspecific binding to other cell organelles and structures should also be expected, and this was not observed.

The experimental evidence described above indicates that some proteins are indeed complexed to specific sites on the nascent rRNA precursors during the transcription process, prior to the completion of the synthesis of the rRNA precursors. There may be compartmentalization between the assembly of ribosomal proteins derived from the two subunits, since the evidence indicates that proteins from the large subunit are probably assembled before those from the small subunit. Before the assembly of the small subunit proteins can be ascertained, it should be clarified as to whether anti TP 40 (rat liver) (i.e., antibodies raised against total proteins from the 40 S subunits of rat liver) would give a similar sequence of assembly as anti TP 30 (*E. coli*). However, the degree of homology between anti TP 40 (rat liver) and ribosomal proteins from *Drosophila* would not be expected to be similar.

It has been shown in *E. coli* that protein L18 is essential for the formation of a 5 S RNA–23 S RNA complex. Protein L18 binds adjacent to protein L25 on 5 S RNA. A complex containing 5 S RNA and proteins L18, L25, and L5 has been shown to catalyze the hydrolysis of GTP and ATP (Horne and Erdmann, 1973). In *D. melanogaster* I have found that this protein appears to be assembled about halfway down the transcription unit, a location which is very close to the postulated position of the internal transcribed spacer in *X. laevis* (Wellaeur and Reeder, 1975).

The experimental results presented above demonstrate that antibodies to ribosomal proteins can be localized with a great deal of accuracy and specificity in cell organelles such as ribosomes and nascent ribonucleoprotein particles. The experimental approach used in this study of assembly is novel, and its success is encouraging. Not only are further experiments with the assembly and transport of other nonhistone proteins possible, but

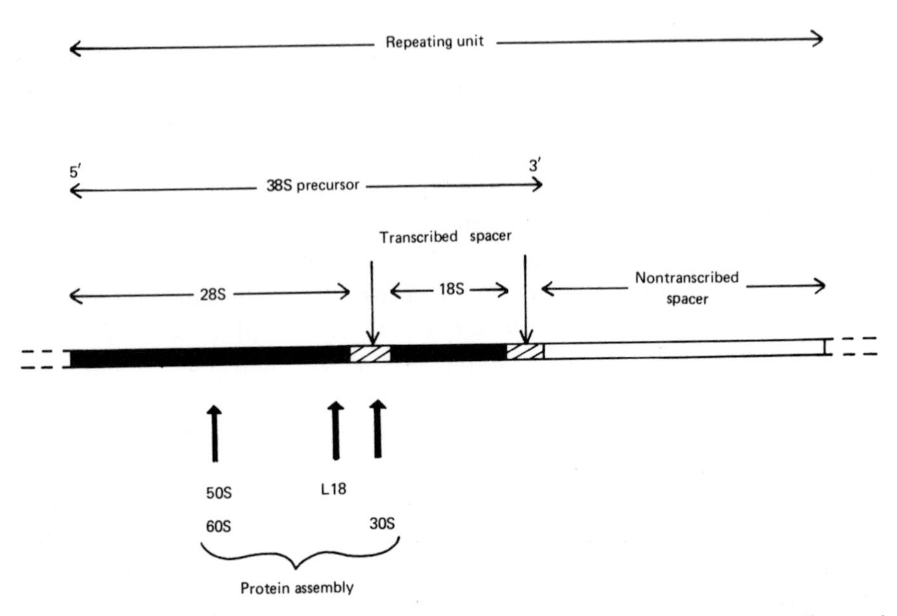

Figure 9. General model of the sequence of assembly of ribosomal proteins on the ribosomal transcription unit in nurse cells of D. melanogaster. The thick arrows indicate the initial point of assembly of the corresponding proteins. The structure of the repeating unit is based on measurements described here and elsewhere (Perry et al., 1970; Wellauer and Dawid, 1974). The locations of the transcribed spacers are hypothetical, as these have not been determined.

the application of this approach to membrane and chromosome structure appears feasible.

Evolution of Ribosomal Proteins

The phylogenetic relationship between organisms is often reflected in the structure of their ribosomal proteins. A comparison of the ribosomes of one bacterial family (Enterobacteriaceae) by the elution profiles of the proteins from chromatography columns by immunological techniques and by two-dimensional polyacrylamide gel electrophoresis patterns (Otaka *et al.*, 1968; Geisser *et al.*, 1973*a*) revealed similar protein structures. Compared with the Enterobacteriaceae, the bacterial family Bacillaceae was immunologically more heterogeneous.

The hypothesis that functionally important proteins are conserved during evolution and thus maintained in distantly related organisms is supported by the finding that *E. coli* proteins L7 and L12, which have important functions in initiation, elongation, and termination (see the review by Moller, 1974), share structural homologies with certain eukaryotic mitochondrial and cytoplasmic proteins. Antisera against *E. coli* L7 and L12 give significant cross-reactions with ribosomal proteins from related bacteria (Geisser *et al.*, 1973*b*) and from yeast and rat liver (Wool and Stöffler, 1974). Anti L7 and L12 are also capable of inhibiting poly(U)-directed phenylalanine synthesis as well as preventing EF–G·GDP·ribosome complex formation within *E. coli* (Highland *et al.*, 1973) and with other bacterial systems (Stöffler, 1974). A number of additional ribosomal proteins are believed to be evolutionarily conserved (Stöffler, private communication).

It was not surprising therefore that immunological similarities were detected between some ribosomal proteins of *E. coli, D. melanogaster,* and rat liver. Since antisera against total ribosomal subunits were used in the preliminary experiments (except in the case of L18), the nature and number of proteins from each subunit that demonstrated homology in protein structure were uncertain. It is clear from the degree of cross-reaction shown in the electron micrographs that, as expected, there is probably less homology between *E. coli* and *Drosophila* ribosomal proteins compared with rat liver and *Drosophila* ribosomal proteins. I have evidence that will not be presented here which indicates that the homology is shared by more than two proteins. The success of these experiments has led me to test the relationship of more single proteins, and these results will be presented elsewhere.

Drosophila mitochondrial ribosomal proteins have also displayed cross-reactions to antisera prepared against *E. coli* ribosomal proteins. This is surprising, since there is no evidence for common antigenic determinants between bacterial ribosomes and chloroplast ribosomes (Gualerzi *et al.*, 1974). If weak homology existed between the two types of ribosomes, their reaction with antisera could have been so weak as to go undetected by current immunological techniques. The immune electron microscopic approach described here for studying protein assembly in *Drosophila* may have some superior advantages. One major advantage is that the specificity and conclusiveness of the immune reaction do not depend on precipitation of the antibody–antigen complex.

How Much of a Band in *D. melanogaster* Is Transcribed?

The question of whether *Drosophila* has very large functional groups has fascinated molecular geneticists. One can estimate the average size of a band in *D. melanogaster* by the following calculation: According to cytochemical (Rudkin, 1965) and kinetic data (Laird, 1971), the haploid genome size is about 10×10^{10} daltons. Since 80% of this is replicated during polytenization (Rudkin, 1969; Gall *et al.,* 1971) and there are about 5000 bands in *D. melanogaster,* we arrive at a figure of about 1.6×10^7 daltons of DNA per haploid chromomere. This value corresponds to 2.4×10^4 base pairs or 7–8 μ of DNA per chromomere. By comparison, an average bacterial gene is about 10^3 base pairs. Therefore, an average size chromomere in *D. melanogaster* is about 20 times larger and has the capacity to code for about 20–30 average cistrons. An attractive hypothesis advanced to try to resolve this dilemma was the possibility that the chromomeres represented repetitive copies of structural cistrons (Callan, 1967). One major problem with the reiterative sequence model is that a quarter to a half of the DNA in polytene chromosomes must be repeated and the number of repeated gene families must correspond to the same number of genes.

Measurements using the technique of nucleic acid hybridization showed this to be unlikely. Less repetitive DNA was found in salivary glands than in adult diploid, nonpolytene tissues (Entingh, 1970; Rae, 1970, 1972; Gall *et al.,* 1971; Laird, 1971). Furthermore, highly repetitive sequences were largely confined to regions of centric heterochromatin which were not polytenized in salivary gland chromosomes and comprised only 5–10% of nuclear DNA from nonpolytene diploid cells. Moderately repetitive sequences accounted for another 5–15% and the remaining 75–80% of the genome in *D. melanogaster* was composed of unique sequences (Laird, 1971; Hennig, 1972; Peacock *et al.,* 1974, Manning *et al.,* 1975).

The average size of messenger RNA in *D. melanogaster* embryos as determined from the size of purified polyribosomes has been estimated to be about 1500 nucleotides or 0.5 μ, with a range from 200 to 4000 nucleotides (Laird *et al.,* 1974; Chooi and Laird, in preparation). An average-sized messenger RNA would be able to code for a protein of about 50,000 daltons. If one chromomere codes for only one protein, then only about 4% of an average band is required to code for messenger RNA. This does not mean, however, that the remaining 96% of the DNA is nonfunctional. An estimate of the size of the precursors would be required for this purpose. This was done in two ways. The first method was biochemical, where size estimates were obtained from poly(A)-containing RNA. In many instances, poly A is added posttranscriptionally to the 3′ end of HnRNA (see Darnell *et al.,* 1973; Weinberg, 1973). The second method involved the electron microscopic visualization of non-rRNA transcription, including HnRNA transcription. This method has several advantages. With biochemical extractions of RNA, it has been observed that RNA can form aggregates even under so-called denaturing conditions (Bramwell, 1972; Macnaughton *et al.*, 1974; McKnight and Schimke, 1974). These observations have raised doubts as to the accuracy of the size determinations attributed to some HnRNA molecules. Such a problem has not been encountered with the type of microscopy used here (see p. 227). Electron microscopic visualization of transcription complexes also permits the enumeration of the number of transcripts made per unit time by each active cistron and measurements of the size of non-rRNA molecules while they are still attached to the DNA template. There is no conclusive evidence to show that cleavage does not take place during transcription. However, the amount is probably minimal compared to that after removal from the DNA template.

Electron Microscopic Morphology of Nonribosomal RNA Transcription Units in *D. melanogaster*

The primary transcription units of non-rRNA in *D. melanogaster* (Figures 10–12) possess different morphologies from those of rRNA (Figures 1, 3, 5–8). Some of the quantitative differences between them are summarized in Table 1, and the two types of genes are illustrated diagrammatically in Figure 10. Some of the distinctive characteristics of the non-rRNA transcription units are catalogued below:

1. The lengths of the matrix units are generally longer.
2. The lengths of the RNP fibrils are generally longer.
3. The steepness of the gradient in length of RNP fibrils is not proportional to the length of the corresponding transcription unit. This suggests interaction with proteins and/or cleavage.
4. The RNP fibrils are spaced farther apart on the transcription unit.
5. The RNP fibrils exhibit a variety of configurations, although they tend to be similar within each matrix. Often there are large loops, knobs, or kinks within the RNP fibrils which may be a reflection of secondary and/or tertiary structure.
6. There are some differences between non-rRNA transcription units observed during oogenesis and those observed during embryogenesis (see p. 255 and Table 1).

Large Functional Groups of *D. melanogaster*

The mean length (4.37 ± 2.26 μm) of non-rRNA transcription units in *D. melanogaster* embryos was used as an estimate of the size of an average functional group. This represented only a minimum estimate,

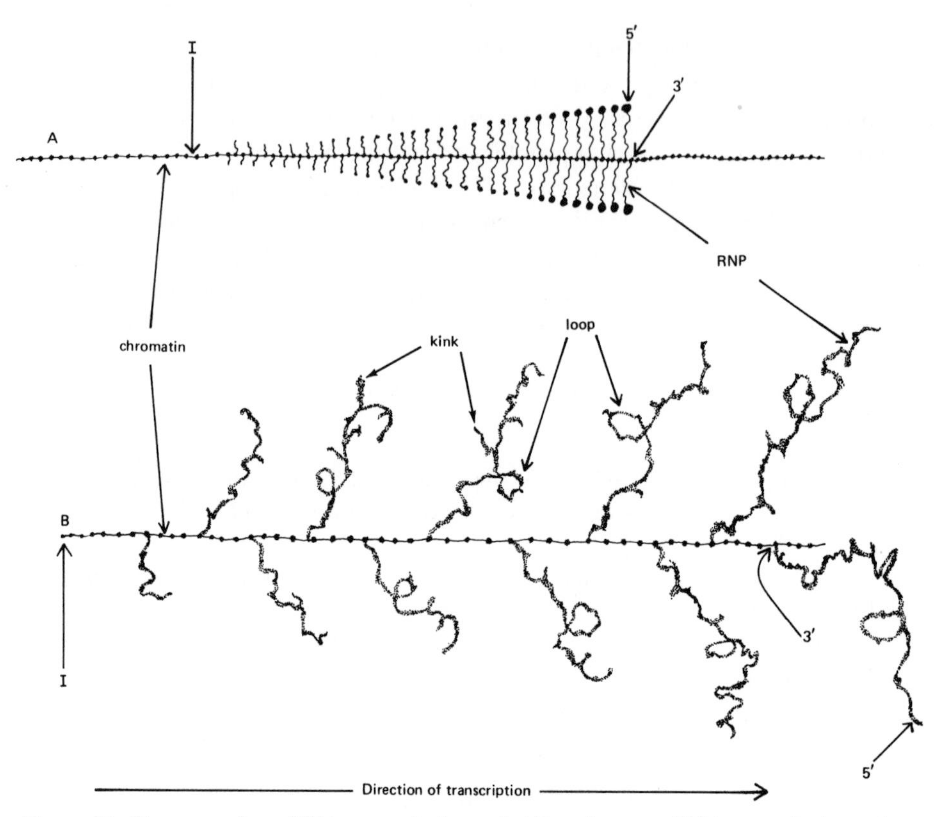

Figure 10. Diagrams of an rRNA transcription unit (A) and a non-rRNA transcription unit (B) in D. melanogaster. Transcription is initiated at I.

TABLE 1. Characteristics of Nonribosomal and Ribosomal Non-rRNA Transcription Units

	Nonribosomal		*Ribosomal*
	Embryo	*Nurse cells*	*nurse cells*
Distance between fibrils (μm)	0.17 ± 0.08	0.50 ± 0.39	0.02 ± 0.01
Apparent length of matrix (μm)	4.37 ± 2.26	1.70 ± 1.24	2.66 ± 0.41
Length of most mature RNP fibril (μm)	0.88 ± 0.28	2.21 ± 1.28	0.28 ± 0.08
Average number of fibrils of matrix	6	2	41
Correction factor	5.7 ± 2.4		10.9 ± 2.1

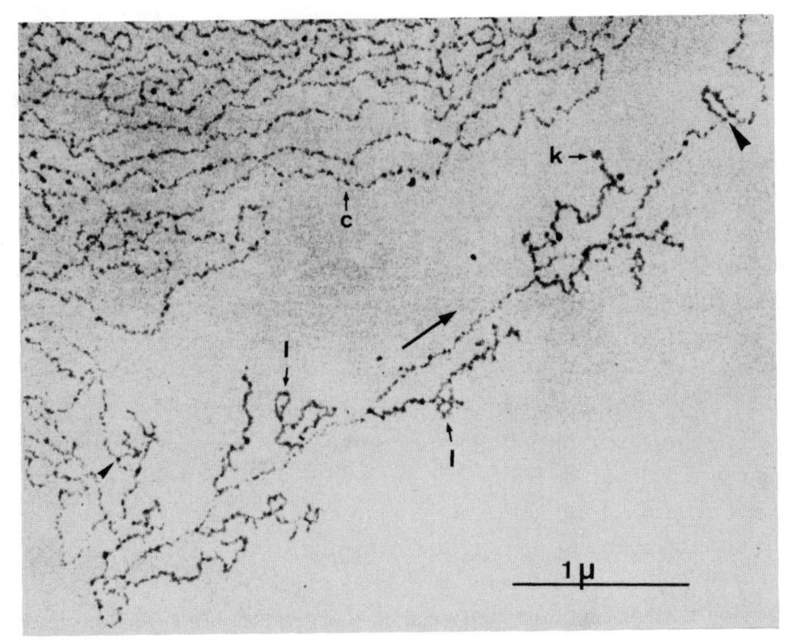

Figure 11. Electron micrograph of a non-rRNA transcription unit from a blastoderm embryo of D. melanogaster. The method of preparation used was as described in Figure 1. The smaller triangular arrow points to the approximate position of the 5′ end of the transcription unit and the larger triangular arrow points to the approximate position of the 3′ end of the transcription unit. Loops (l) and kinks (k) are found on the RNP fibrils. The thinner arrow with a tail indicates the direction of transcription. Adjacent chromatin (c) appears to have a beaded structure.

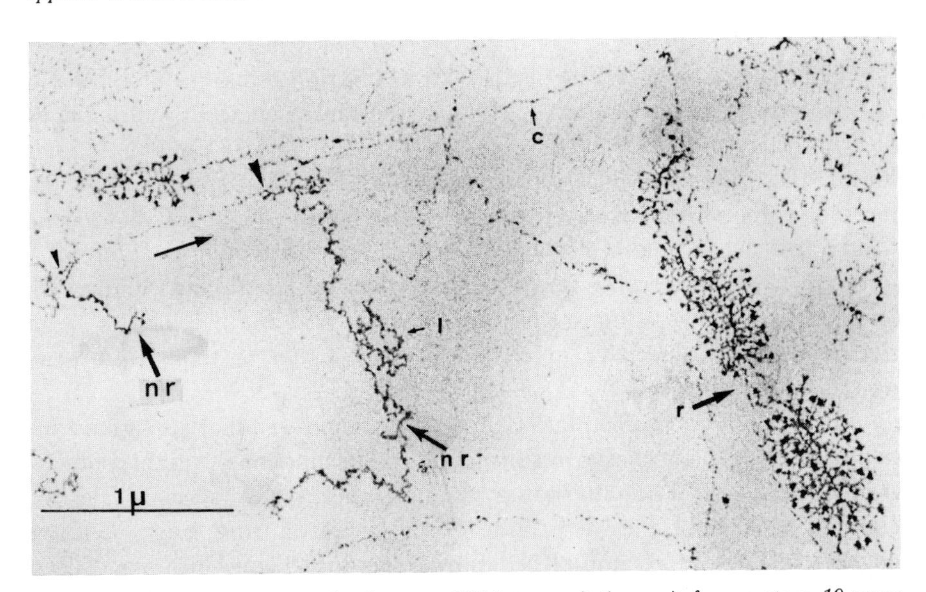

Figure 12. Electron micrograph of a non-rRNA transcription unit from a stage 10 nurse cell of D. melanogaster. The notation used is the same as that described in Figure 11. An rRNA (r) gene in which some of the rRNA fibrils have been lost during preparation is found adjacent to the non-rRNA (nr) transcription unit.

since the accuracy of length measurements was complicated by the observation that, unlike those in the rRNA transcription units, the initiation and termination points on non-rRNA transcription unit were less obvious, since the RNP fibrils were spaced farther apart (see above and Table 1). In *D. melanogaster,* the approximate sizes of the chromomeres range from about 2 to 20 μm, with a mean of about 7 μm. I conclude from these data that perhaps half an average chromomere is transcribed, and in some cases a whole or perhaps more than a whole chromomere (see below) is transcribed as a single unit.

The above estimates are somewhat larger than the biochemical estimates. The mean size of poly(A)-containing RNA extracted from *D. melanogaster* ovaries is about 28 S (1.7 μm), with a range from 18 S (0.7 μm) to 50 S (5.7 μm) (Lamb, private communication). Lengyel and Penman (1975) obtained similar average size estimates (26 S) for HnRNA extracted from tissue culture cells of *D. melanogaster.* Several reasons could account for this discrepancy between EM estimates and biochemical estimates. One reason could be that poly(A) is added posttranscriptionally to molecules that have been cleaved. Another possibility is that endogenous nucleases present in insects are responsible for the rapid breakdown of very high molecular weight RNAs.

Are Termination Signals Bypassed During Oogenesis in *D. melanogaster?*

Some of the non-rRNP fibrils of transcription units in nurse cells of *D. melanogaster* were surprisingly long. the mean length of mature non-rRNP fibrils in stage 10 nurse cell nuclei was 2.21 ± 1.28 μm or 2.5 times longer than those from blastoderm nuclei. If it is assumed that the foreshortening of non-rRNP fibrils is 5.7-fold (see below and Table 1), the "corrected" mean length of non-rRNP in nurse cells would be about 12.5 μm, which is considerably larger than an average-sized band. Figure 12, for example, shows a non-rRNP fibril which is 5.3 μm long. Its large size is striking, especially in comparison with that of the rRNA gene present in the same field.

It is tempting to postulate that perhaps some termination signals are bypassed during oogenesis, resulting in the production of giant polycistronic messages. The alternative is that there is a special selective expression of only the very large chromomeres during stage 9–10 of oogenesis. Very large non-rRNP fibrils (up to 10 μm) also have been reported during spermatogenesis in *D. hydei* (Hennig *et al.,* 1974) and in the mouse (Kierszenbaum and Tres, 1974), and during oogenesis in *Triturus viridescens* (Miller *et al.,* 1970).

Developmental Stage Differences in Nonribosomal RNA Transcription in *D. melanogaster*

In *D. melanogaster* the characteristics of non-rRNA transcription units in the nuclei of stage 9 and 10 nurse cells are somewhat different from those of nuclei of blastoderm embryos. In contrast, the rRNA genes form a unique class of genes which appear homogeneous in length, "foreshortening" ratios, and spacing of adjacent RNP fibrils on a cistron. The unique homogeneity of the spacing of adjacent RNP fibrils on active rRNA cistrons may suggest a constant rate of reinitiation if a constant rate of transcription is assumed. Some of the differences between non-rRNA transcription units at embryogenesis and oogenesis are quantitated and are summarized in Table 1. Figures 11 and 12 show electron micrographs of non-rRNA transcription units taken from the nuclei of blastoderm embryos and of stage 10 nurse cells.

The lengths of the transcription units described here represent minimum estimates. This is because the RNP fibrils are spaced much farther apart on non-rRNA transcription units, and this makes the initiation and termination points difficult to ascertain. The mean length of the transcription units in embryos is $4.4 \pm 2.3\ \mu m$, while that in nurse cells is $1.7 \pm 1.2\ \mu m$. This does not indicate that the mean length of the transcriptional units in embryos is exactly 2.6 times longer than that in nurse cells. The distance between the non-rRNP fibrils in transcription units of nurse cells is about 3 times greater, thereby magnifying the difficulty in determining the positions of the initiation and termination points.

The mean length ($0.88 \pm 0.28\ \mu m$) of the most mature RNP fibrils in non-rRNA transcription units of blastoderm embryos is about 2.5 times shorter than that in nurse cells. These lengths may not be directly comparable (see below). In electron microscopic observations, mature RNP fibrils often appear shorter than expected from the corresponding lengths of their transcription units. This could be attributed to a number of possibilities, such as the formation of secondary structures, interactions with proteins, or cleavage. The magnitudes of these variables have not been determined and thus have not been included in the length estimates of the RNP fibrils. Moreover, the magnitudes of these variables may be gene and stage specific. In the rRNA genes, for example, there is about an elevenfold difference between the length of the mature rRNP fibril and the length of the transcription unit. The magnitude of this difference (referred to here as the *correction factor*) is subject to further small changes depending on the method of preparation of the cell lysates for microscopy. Very low ionic strength buffers and detergents have varying effects on it. However, within a particular set of defined conditions, the correction factor does not vary significantly between different rRNA genes in the same preparation and between different preparations. Workers have frequently used the correction factor derived from rRNA genes for non-rRNA genes. Is such an extrapolation valid? I could not find a single correction factor that would be applicable to all the non-rRNA genes that were examined. In embryos the correction factors have a very broad range, with a mean of 5.7 ± 2.4. In nurse cells the range was even broader. Caution should therefore be used in correcting the observed lengths of non-rRNP fibrils. A simple extrapolation from rRNA genes would, in all likelihood, prove incorrect.

This difference in the correction factor between non-rRNA genes within one developmental stage and between developmental stages may have some interesting biological implications. It could reflect the heterogeneity in the processing of the RNP from different genes, since a large sample of active genes was examined. The differences observed in the correction factor between developmental stages could be attributed in part to the expression of a different spectrum of genes or to differential rates of transcription (see review by Church and Schultz, 1974). The characteristic smaller numbers of non-rRNP fibrils and the larger interfibrillar distances in egg chambers are especially relevant in this context.

Nucleic acid hybridization experiments have also demonstrated that there are considerable differences in complexity in the unique HnRNA or the total RNA sequences between developmental stages of *Dictyostelium* (Firtel, 1972) and the mouse (Grouse *et al.*,1972). These biochemical data are also consistent with cytological evidence obtained from studies of insect polytene chromosomes (Pelling, 1964; 1970; Daneholt, 1975; Daneholt *et al.*, 1969). While some chromosome sites active in transcription (puffs) are common to some tissues and developmental stages, there are some puffs that are stage specific or tissue specific (see chapter 7).

Conclusions

We have estimated that the mean size of messenger RNA in *D. melanogaster* embryos is about 1500 nucleotides or about 10 times smaller than the mean size of non-rRNA precursors. The mean size of non-rRNA precursors in *D. melanogaster* embryos is about 12,000–13,000 nucleotides, a size within the range expected for higher eukaryotic HnRNA (Darnell *et al.*, 1973; Davidson and Britten, 1973; Lewin, 1975*a,b*; Weinberg, 1973). It is uncertain as to what proportion of these non-rRNA transcription units observed represent HnRNA transcription units. It has been estimated that at least 30% of the single copy sequence DNA is active in transcription during *D. melanogaster* embryogenesis (Turner and Laird, 1973). With the onset of blastoderm formation it seems reasonable to assume that a good proportion of the transcription units observed represent HnRNA. While we cannot be certain of how much of the precursor molecule is lost in cleavage with the formation of messenger RNA, the data presented are consistent with the hypothesis that *D. melanogaster* has very large primary transcription units, some of which may approach chromomere size.

The question of whether mRNA is derived from very high molecular weight HnRNAs remains to be resolved. Except in the case of Balbiani ring 2 (see Daneholt, 1972, and Chapter 7), in all the documented cases where precursor–product relationships have been established, mRNAs do not appear to be derived from very high molecular weight precursors. The precursor to globin mRNA (9.5 S), for example, is only 14 S (Macnaughton *et al.*, 1974). The precursor to the ovalbumin mRNA is very similar in size to the final mRNA product (18 S) (McKnight and Schimke, 1974) so that very little cleavage takes place. In *Dictyostelium*, giant RNA molecules have not been detected (Lodish *et al.*, 1973), and only a small amount of cleavage takes place.

The differential expression of some non-rRNA genes and the selective expression of some other non-rRNA genes during development may be controlled at the transcriptional level. Some of these controlling factors are manifested in rates of transcription, rates and patterns of processing, or some combination of these. In the case of the rRNA genes, the control of expression seems to be chromosome specific (see p. 229).

Acknowledgments

Unless otherwise stated, the data on *D. melanogaster* transcription discussed here represent my unpublished research; the morphology of transcription units was done in collaboration with Charles Laird at the University of Washington, and ribosomal protein assembly and evolution were done in collaboration with Hewson Swift and Ira Wool at the University of Chicago and with Georg Stöffler at the Max Planck Institute. I am especially grateful to Hewson Swift for many helpful discussions and particularly for creating a scientific environment that permitted me not only to pursue my scientific interests but also to enjoy them immensely. Many thanks are also due Robert King for his constructive criticisms of the manuscript, which was completed while I was at the Max Planck Institute for Molecular Genetics in Berlin. I am very grateful to Professor H. G. Wittmann for his constant support, advice, and encouragement and for making my stay in his Abteilung such a scientifically rewarding experience. Research carried out at the University of Chicago was supported by NIH grants to Professor Swift (5P30CA 14599-03 and PHF-HD174-09).

Literature Cited

Amaldi, F. and M. Buongiorno-Nardelli, 1971 Molecular hybridization of Chinese hamster 5 S, 4 S and "pulse labelled" RNA in cytological preparations. *Exp. Cell Res.* **65:**329–334.

Avanzi, S., M. Durante, P. G. Cionini, and F. D'Amato, 1972 Cytological localization of ribosomal cistrons in polytene chromosomes of *Phaseolus coccineus. Chromosoma (Berl.)* **39:**191–204.

Beermann, W., 1972 Chromomeres and genes. In *Results and Problems in Cell Differentiation,* edited by W. Beermann, pp. 1–33, Springer-Verlag, Berlin.

Bird, A., E. Rogers, and M. Birnstiel, 1973 Is gene amplification RNA directed? *Nature (London) New Biol.* **242:**226–229.

Birnstiel, M. L., M. Grunstein, J. Speirs, and W. Hennig, 1969 Family of ribosomal genes of *Xenopus laevis. Nature (London)* **223:**1265–1267.

Birnstiel, M. L., M. Chipchase, and J. Speirs, 1971 The ribosomal RNA cistrons. *Prog. Nucleic Acid Res. Mol. Biol.* **11:**351–389.

Birnstiel, M., J. Spiers, I. Purdom, K. Jones, and U. E. Loening, 1968 Properties and composition of the isolated ribosomal DNA satellite of *Xenopus laevis. Nature (London)* **219:**454–463.

Bostock, C., 1971 Repetitious DNA. *Adv. Cell Biol.* **2:**153–223.

Bramwell, M. E., 1972 A comparison of gel electrophoresis and density gradient centrifugation of heterogeneous nuclear RNA. *Biochim. Biophys. Acta* **28:**329–337.

Branlant, C., A. Krol, J. Sriwidada, P. Fellner, and R. Crichton, 1973 The identification of the RNA binding site for a 50 S ribosomal protein by a new technique. *FEBS Lett.* **35:**265–272.

Branlant, C., A. Krol, J. Sriwidada, J. P. Ebel, P. Sloof, and R. A. Garett, 1975 A partial localisation of the binding sites of the 50 S subunit proteins. *FEBS Lett.* **52:**195–201.

Britten, R. J. and E. H. Davidson, 1971 Repetitive and non-repetitive DNA sequences and a speculation on the origins of evolutionary novelty. *Q. Rev. Biol.* **46:**111–133.

Britten, R. J. and D. E. Kohne, 1968 Repeated sequences in DNA. *Science* **161:**529–534.

Brown, D. D. and J. M. Gurdon, 1964 Absence of ribosomal RNA synthesis in the anucleolate mutant of *Xenopus laevis. Proc. Nat. Acad. Sci. USA* **51:**139–142.

Brown, D. D. and I. B. Dawid, 1968 Specific gene amplification in oocytes. Oocyte nuclei contain extrachromosomal replicas of the genes for ribosomal RNA. *Science* **160:**272–280.

Brown, D. D., P. C. Wensinck, and E. Jordon, 1972 A comparison of the ribosomal DNA's of *Xenopus laevis* and *Xenopus mulleri:* The evolution of tandem genes. *J. Mol. Biol.* **63:**53–73.

Brownlee, G. G., E. M. Cartwright, and D. D. Brown, 1974 Sequence studies of the 5 S DNA of *Xenopus laevis. J. Mol. Biol.* **89:**703–718.

Busch, H. and K. Smetana, 1970 *The Nucleolus.* Academic Press, New York.

Callan, H. G., 1967 The organization of genetic units in chromosomes. *J. Cell Sci.* **2:**1–7.

Callan, H. G. and L. Lloyd, 1960 Lampbrush chromosomes of crested newts *Triturus cristatus* (Laurenti). *Philos. Trans. R. Soc. London, Ser.B:* **234:**135–219.

Chooi, W. Y. and C. D. Laird, 1976 DNA and polyribosome-like structures in lysates of mitochondria of *Drosophila melanogaster. J. Mol. Biol.* **100:**493–518.

Church, R. B. and G. A. Schultz, 1974 Differential gene activity in the pre- and post-implantation mammalian embryo. *Curr. Top. Dev. Biol.* **8:**179–203.

Crick, F., 1971 General model for the chromosomes of higher organisms. *Nature (London)* **234:**25–27.

Crippa, M. and G. P. Tocchini-Valentini, 1971 Synthesis of amplified DNA that codes for ribosomal RNA. *Proc. Natl. Acad. Sci. USA* **68:**2769–2773.

Daneholt, B., 1972 Giant RNA transcript in a Balbiani ring, *Nature (London) New Biol.* **240:**229–232.

Daneholt, B., 1975 Transcription in polytene chromosomes. *Cell* **4:**1–9.

Daneholt, B., J. E. Edström, E. Egyházi, B. Lambert, and U. Ringborg, 1969 Chromosomal RNA synthesis in polytene chromosomes of *Chironomus tentans. Chromosoma* **28:**399–417.

Darnell, J. E., W. R. Jelinek, and G. R. Molloy, 1973 Biogenesis of messenger RNA. Genetic regulation in mammalian cells. *Science* **181:**1215–1221.

Davidson, E. H. and R. J. Britten, 1973 Organization, transcription and regulation in the animal genome. *Q. Rev. Biol.* **48:**565–613.

Davidson, E. H., B. R. Hough, C. S. Amenson, and R. J. Britten, 1973 General interspersion of repetitive with non-repetitive sequence elements in the DNA of *Xenopus. J. Mol. Biol.* **77:**1–23.

Davidson, E. H., D. E. Graham, B. R. Neufeld, M. E. Chamberlin, C. S. Amenson, B. R. Hough, and R. J. Britten, 1974 Arrangement and characterization of repetitive sequence elements in animal DNAs. *Cold Spring Harbor Symp. Quant. Biol.* **38:**295–302.

Dawid, I. B., D. D. Brown, and R. H. Reeder, 1970 Composition and structure of chromosomal and amplified ribosomal DNA's of *Xenopus laevis. J. Mol. Biol.* **51:**341–360.

Elsdale, T. R., M. Fishberg, and S. Smith, 1958 A mutation that reduces nucleolar number in *Xenopus laevis. Exp. Cell Res.* **14:**642–643.

Entingh, T. D., 1970 DNA hybridization in the genus *Drosophila. Genetics* **66:**55–68.

Firtel, R. A., 1972 Changes in the expression of single copy DNA during development of the cellular slime mold *Dictyostelium discoideum. J. Mol. Biol.* **66:**363–377.

Ford, P. J. and E. M. Southern, 1973 Different sequences for 5 S RNA in kidney cells and ovaries of *Xenopus laevis. Nature (London) New Biol.* **241:**7–12.

Forsheit, A. B., N. Davidson, and D. D. Brown, 1974 An electron microscope heteroduplex study of the ribosomal DNAs of *Xenopus laevis* and *Xenopus mulleri. J. Mol. Biol.* **90:**301–315.

Gall, J. G., 1969 The genes for ribosomal RNA during oogenesis. *Genetics* **61:**121–132.

Gall, J., E. Cohen, and M. Polan, 1971 Repetitive DNA sequences in *Drosophila. Chromosoma* **33:**319–344.

Geisser, M., G. W. Tischendorf, G. Stöffler, and H. G. Wittmann, 1973*a* Immunological and electrophoretic comparison of ribosomal proteins from eight species belonging to Enterobacteriaceae. *Mol. Gen. Genet.* **127:**111–128.

Geisser, M., G. W. Tischendorf, and G. Stöffler, 1973*b* Comparative immunological and electrophoretic studies on ribosomal proteins of Baccillaceae. *Mol. Gen. Genet.* **127:**129–145.

Georgiev, G. P., 1969 On the structural organization of the operon and the regulation of RNA synthesis in animal cells. *J. Theor. Biol.* **25:**473–490.

Gilbert, W. and D. Dressler, 1968 DNA replication: The rolling circle model. *Cold Spring Harbor Symp. Quant. Biol.* **33:**473–484.

Golberg, R. B., W. R. Crain, J. V. Ruderman, G. P. Moore, T. R. Barnett, R. C. Higgins, R. A. Gelfend, G. A. Galau, R. J. Britten, and E. H. Davidson, 1975 DNA sequence organization in the genomes of five marine invertebrates. *Chromosoma* (*Berl.*) **51:**225–251.

Graham, D. E., B. R. Neufeld, E. H. Davidson, and R. J. Britten, 1974 Interspersion of repetitive and non-repetitive DNA sequences in the sea urchin genome. *Cell* **1:**127–137.

Granboulan, N. and P. Granboulan, 1965 Cytochimie ultrastructurale du nucleole. *Exp. Cell Res.* **38:**604–619.

Gray, P., G. Bellemare, and R. Monier, 1972 Degradation of a specific 5 S RNA–23 S RNA protein complex by pancreatic ribonuclease. *FEBS Lett.* **24:**156–160.

Gray, P. N., G. Bellemare, R. Monier, R. A. Garett, and G. Stöffler, 1973 Identification of the nucleotide sequences involved in the interaction between *Escherichia coli* 5 S RNA and specific 50 S subunit protein. *J. Mol. Biol.* **77:**133–152.

Green, M. and C. G. Kurland, 1973 Molecular interactions of ribosomal components. IV. Co-operative interactions during assembly *in vitro*. *Mol. Biol. Rep.* **1:**105–111.

Grouse, L., M. D. Chilton, and B. J. McCarthy, 1972 Hybridization of ribonucleic acid with unique sequences of mouse deoxyribonucleic acid. *Biochemistry* **11:**798–805.

Gualerzi, G., H. G. Janda, H. Passow, and G. Stöffler, 1974 Studies on the protein moiety of plant ribosomes. *J. Biol. Chem.* **249:**3347–3355.

Hackett, P. B. and W. Sauerbier, 1975 The transcriptional organization of the ribosomal RNA genes in mouse L cells. *J. Mol. Biol.* **91:**235–256.

Hamkalo, B. A., O. L. Miller, and A. H. Bakken, 1974 Ultrastructure of active eukaryotic genomes. *Cold Spring Harbor Symp. Quant. Biol.* **38:**915–919.

Hecht, R. M. and M. L. Birnstiel, 1972 Integrity of the DNA template, a prerequisite for the faithful transcription of *Xenopus* rDNA *in vitro*. *Eur. J. Biochem.* **29:**489–499.

Hedgepeth, J., H. M. Goodman, and H. W. Boyer, 1972 DNA nucleotide sequences restricted by the R1 endonuclease, *Proc. Natl. Acad. Sci. USA* **69:**3448–3452.

Heitz, E., 1933 Cytologische Untersuchungen an Dipteran. III. Die somatische Heteropyknose bei *Drosophila melanogaster* und ihre genetische Bedeutung. *Z. Zellforsch. Mikrosk. Anat.* **20:**237–289.

Henderson, A. and F. Ritossa, 1970 On the inheritance of rDNA of magnified bobbed loci in *D. melanogaster*. *Genetics* **66:**463–473.

Hennig, W., 1972 Highly repetitive DNA sequences in the genome of *Drosophila hydei*. I. Preferential localization in the X chromosome heterochromatin. *J. Mol. Biol.* **71:**407–417.

Hennig, W., G. F. Meyer, I. Hennig, and O. Leoncini, 1974 Structure and function of the Y chromosome of *Drosophila hydei*. *Cold Spring Harbor Symp. Quant. Biol.* **38:**673–683.

Highland, J. H., J. Bodley, J. Gordon, R. Hasenbank, and G. Stöffler, 1973 Identity of the ribosomal proteins involved in the interaction with elongation factor G. *Proc. Natl. Acad. Sci. USA* **70:**147–150.

Hochman, B., 1974 Analysis of a whole chromosome in *Drosophila*. *Cold Spring Harbor Symp. Quant. Biol.* **38:**581–589.

Honjo, T. and R. H. Reeder, 1973 Preferential transcription of *Xenopus laevis* ribosomal RNA in interspecific hybrids between *X. laevis* and *X. mulleri*. *J. Mol. Biol.* **80:**217–228.

Horne, J. R. and V. A. Erdmann, 1973 ATPase and GTPase activities associated with a specific 5 S RNA–protein complex. *Proc. Natl. Acad. Sci. USA* **70:**2870–2873.

Hourcade, D., D. Dressler, and J. Wolfson, 1974 The nucleolus and the rolling circle. *Cold Spring Harbor Symp. Quant. Biol.* **38:**537–550.

Judd, B. H., M. W. Shen and T. C. Kaufmann, 1972 The anatomy and function of a segment of the X chromosome of *Drosophila melanogaster. Genetics* **71:**139–156.

Kaufmann, B. P., 1933 Interchange between X- and Y-chromosomes in attached X-females of *Drosophila melanogaster. Proc. Natl. Acad. Sci. USA* **19:**830–838.

Kaufmann, B. P., 1934 Somatic mitosis in *Drosophila melanogaster. J. Morphol.* **56:**125–155.

King, R. C. 1970 *Ovarian Development in Drosophila melanogaster,* Academic Press, New York.

Kierszenbaum, A. L. and L. L. Tres, 1974 Transcription sites in spread meiotic prophase chromosomes from mouse spermatocytes. *J. Cell Biol.* **63:**923–935.

Kumar, A. and A. R. Subramaniam, 1975 Ribosome assembly in HeLa cells: Labelling pattern of ribosomal proteins by two-dimensional resolution. *J. Mol. Biol.* **94:**409–423.

Kumar, A. and J. R. Warner, 1972 Characterization of ribosomal precursor particles from HeLa cell nucleoli. *J. Mol. Biol.* **63:**233–246.

Laird, C. D., 1971 Chromatid structure: Relationship between DNA content and nucleotide sequence diversity. *Chromosoma* **32:**378–406.

Laird, C. D., W. Y. Chooi, E. H. Cohen, E. Dickson, N. Hutchison, and S. Turner, 1974 Organization and transcription of DNA in chromosomes and mitochondria of *Drosophila Cold Spring Harbor Symp. Quant. Biol.* **38:**311–327.

Lefevre G., Jr., 1974 The relationship between genes and polytene chromosome bands. *Annu. Rev. Genet.* **8:**54–62.

Lengyel, J. and S. Penman, 1975 HnRNA Size and processing as related to different DNA content in two dipterans: *Drosophila* and *Aedes. Cell* **5:**281–290.

Lewin, B., 1975*a* Units of transcription and translation: The relationship between heterogeneous nuclear RNA and messenger RNA. *Cell* **4:**11–20.

Lewin, B., 1975*b* Units of transcription and translation: Sequence components of heterogeneous nuclear RNA and messenger RNA. *Cell* **4:**77–93.

Lindsley, D. L. and E. H. Grell, 1968 *Genetic Variation of Drosophila melanogaster,* Carnegie Institution of Washington, Publication 627, Carnegie Institution, Washington, D.C.

Lodish, H. F., R. A. Firtel, and A. Jacobson, 1974 Transcription and structure of the genome of the cellular slime mold *Dictyostelium discoideum. Cold Spring Harbor Symp. Quant. Biol.* **38:**899–914.

MacGregor, H. C., 1972 The nucleolus and its genes in amphibian oogenesis. *Biol. Rev.* **47:**177–210.

McClintock, B., 1934 The relationship of a particular chromosomal element to the development of the nucleoli in *Zea mays. Z. Zellforsch Mikrosk. Anat.* **21:**294–328.

McKnight, G. S. and R. T. Schimke, 1974 Ovalbumin messenger RNA: Evidence that the initial product of transcription is the same size as polysomal ovalbumin messenger. *Proc. Natl. Acad. Sci. USA* **71:**4327–4331.

Macnaughton, M., K. B. Freeman, and J. O. Bishop, 1974 A precursor to hemoglobin mRNA in nuclei of immature duck red blood cells. *Cell* **1:**117–125.

Maden, B. E. H., 1968 Ribosome formation in animal cells. *Nature (London)* **219:**685–689.

Maden, B. E. H. and K. Tartof, 1974 Nature of the ribosomal RNA transcribed from the X and Y chromosomes of *Drosophila melanogaster*. *J. Mol. Biol.* **90:**51–64.

Mangiarotti, G., E. Terco, A. Ponzetto, and F. Altruda, 1974 Precursor 16 S RNA in active 30 S ribosome. *Nature (London)* **247:**147–148.

Mangiarotti, E., E. Terco, C. Perlo, and F. Altruda, 1975 Role of precursor 16 S RNA in assembly of *E. coli* 30 S ribosomes. *Nature (London)* **253:**569–571.

Manning, J. E., C. W. Schmid, and N. Davidson, 1975 Interspersion of repetitive and non-repetitive DNA sequences in the *Drosophila melanogaster* genome. *Cell* **4:**141–155.

Marvaldi, J., J. Pichon, and G. Marchis-Mouren, 1972 The *in vivo* order of addition of ribosomal proteins in the course of *Escherichia coli* 30 S subunit biogenesis. *Biochim. Biophys. Acta* **269:**173–177.

Miller, O. L., Jr. and A. H. Bakken, 1972 Morphological studies of transcription. *Acta Endocrinol.* **168:**155–177.

Miller, O. L., Jr. and B. R. Beatty, 1969 Visualization of nucleolar genes. *Science* **164:**955–957.

Miller, O. L., Jr., B. A. Hamkalo and C. A. Thomas, Jr., 1970 Visualization of bacterial genes in action. *Science* **169:**392–395.

Mizushima, S. and M. Nomura, 1970 Assembly mapping of 30 S ribosomal proteins of *E. coli. Nature (London)* **226:**1214–1218.

Möller, W., 1974 The ribosomal components involved in EF-G- and EF-TU-dependent GTP hydrolysis. In *Ribosomes,* edited by M. Nomura, P. Lengyel, and A. Tissières, pp. 711–731, Cold Spring Harbor Press, Cold Spring Harbor, N.Y.

Morrow, J. R., S. N. Cohen, A. C. Y. Chang, H. W. Boyer, H. M. Goodman, and R. B. Hellings, 1974 Replication and transcription of eukaryotic DNA in *Escherichia coli. Proc. Natl. Acad. Sci. USA* **71:**1743–1747.

Muller, H. J., 1965 Report of new mutants. *Drosophila Inform. Serv.* **40:**35–36.

Muto, A., C. Ehresmann, P. Fellner, and R. A. Zimmermann, 1974 RNA–protein interactions in the ribosome. I. Characterization and ribonuclease digestion of 16 S RNA–ribosomal protein complexes. *J. Mol. Biol.* **86:**411–432.

Nierhaus, K. H. and F. Dohme, 1974 Total reconstitution of functionally active 50 S ribosomal subunits from *Escherichia coli. Proc. Natl. Acad. Sci. USA* **71:**4713–4717.

Nomura, M., 1972 Assembly of bacterial ribosomes. *Fed. Proc.* **31:**18–20.

Otaka, E., T. Itoh, and S. Osawa, 1968 Ribosomal proteins of bacterial cells: Strain and species specificity. *J. Mol. Biol.* **33:**93–107.

Paul, J., 1972 General theory of chromosome structure and gene activation in eukaryotes. *Nature (London)* **238:**444–446.

Pardue, M. L. and J. G. Gall, 1969 Molecular hybridization of radioactive RNA to the DNA of cytological preparations. *Proc. Natl. Acad. Sci. USA* **64:**600–604.

Pardue, M. L., S. A. Gerbi, R. A. Eckhardt, and J. G. Gall, 1970 Cytological localization of DNA complementary to ribosomal RNA in polytene chromosomes of Diptera. *Chromosoma* **29:**268:290.

Peacock, W. J., D. Brutlag, E. Goldring, R. Appels, C. W. Hinton, and D. L. Linsley, 1974 The organization of highly repeated DNA sequences in *Drosophila melanogaster* chromosomes. *Cold Spring Harbor Symp. Quant. Biol.* **38:**405–416.

Pelling, C. 1964 Ribonukleinsäure-synthese der Riesenchromosomen. *Chromosoma* **15:**71–122.

Pelling, C., 1970 Puff RNA in polytene chromosomes. *Cold Spring Harbor Symp. Quant. Biol.* **35:**521–531.

Perry, R. P., T. Y. Cheng, J. J. Freed, J. R. Greenberg, D. E. Kelley, and K. D. Tartof, 1970 Evolution of the transcription unit of ribosomal RNA. *Proc. Natl. Acad. Sci. USA* **65:**609–616.

Rae, P. M. M., 1970 Chromosomal distribution of rapidly reannealing DNA in *Drosophila melanogaster. Proc. Natl. Acad. Sci. USA* **67:**1018–1025.

Rae, P. M. M. 1972 The distribution of repetitive DNA sequences in chromosomes. *Adv. Cell Mol. Biol.* **2:**109–149.

Reeder, R. H. and D. D. Brown, 1970 Transcription of the ribosomal RNA genes of an amphibian by the RNA polymerase system of a bacterium. *J. Mol. Biol.* **51:**361–377.

Ritossa, F. M., 1968 Unstable redundancy in genes for ribosomal RNA. *Proc. Natl. Acad. Sci. USA* **60:**509–516.

Ritossa, F. M. and S. Spiegelman, 1965 Localization of DNA complementary to ribosomal RNA in the nucleolus organizer region of *Drosophila melanogaster. Proc. Natl. Acad. Sci. USA* **53:**737–745.

Rudkin, G. T., 1965 The relative mutabilities of DNA in regions of the X chromosome of *Drosophila melanogaster. Genetics* **52:**665–681.

Rudkin, G. T., 1969 Nonreplicating DNA in *Drosophila. Genetics* (*Suppl*). **61:**227–238.

Schaup, H. W. and C. G. Kurland, 1972 Molecular interactions of ribosomal components. III. Isolation of the RNA binding site for a ribosomal protein. *Mol. Gen. Genet.* **114:**350–357.

Schaup, H. W., M. L. Sogin, C. Woese, and C. G. Kurland, 1971 Characterization of an RNA "binding site" for a specific ribosomal protein of *Escherichia coli. Mol. Gen. Genet.* **114:**1–8.

Schaup, H. W., M. L. Sogin, C. G. Kurland, and C. Woese, 1973 Localization of a binding site for ribosomal protein S 8 within the 16 S ribosomal ribonucleic acid of *Escherichia coli. J. Bacteriol.* **115:**82–87.

Shepherd, J. and B. E. H. Maden, 1972 Ribosome assembly in HeLa cells. *Nature* (*London*) **236:**211–214.

Smith, G. P., 1974 Unequal crossing over and the evolution of multigene families. *Cold Spring Harbor Symp. Quant. Biol.* **38:**507–513.

Spring, H., M. F. Trendelenburg, U. Scheer, W. W. Franke, and W. Herth, 1974 Structural and biochemical studies of the primary nucleus of two green algal species *Acetabularia mediterranea* and *Acetabularia major. Cytobiologie* **10:**1–65.

Stöffler, G., 1974 Structure and function of the *Escherichia coli* ribosome. Immunochemical analysis. In *Ribosomes,* edited by M. Nomura, P. Lengyel, and A. Tissieres, pp. 615–667, Cold Spring Harbor Press, Cold Spring Harbor, N.Y.

Tartof, K. D., 1971 Increasing the multiplicity of the ribosomal RNA genes in *Drosophila melanogaster. Science* **171:**294–297.

Tartof, K. D., 1973 Regulation of ribosomal RNA gene multiplicity in *Drosophila melanogaster. Genetics* **73:**57–71.

Thomas, C. A. Jr., 1970 The theory of the master gene. In *The Neurosciences: Second Study Program,* edited by F. O. Schmitt, pp. 973–998, Rockefeller University Press, New York.

Thomas, C. A. Jr., 1974 The rolling helix: A model for the eukaryotic gene. *Cold Spring Harbor Symp. Quant. Biol.* **38:**347–352.

Trendelenburg, M. F., 1974 Morphology of ribosomal RNA cistrons in oocytes of the water beetle, *Dytiscus marginalis* L. *Chromosoma* **48:**119–135.

Trendelenburg, M. F., H. Spring, U. Scheer, and W. W. Franke, 1974 Morphology of nucleolar cistrons in a plant cell, *Acetabularia mediterranea. Proc. Natl. Acad. Sci. USA* **71:**3626–3630.

Tsurugi, K., T. Morita, and K. Ogata, 1973 Identification and metabolic relationship between proteins of nucleolar 60 S particles and of ribosomal large subunits of rat liver by means of two-dimensional disc electrophoresis. *Eur. J. Biochem.* **32:**555–562.

Turner, S. H. and C. D. Laird, 1973 Diversity of RNA sequences in *Drosophila melanogaster. Biochem. Genet.* **10:**263–274.

Ungewickell, E., R. A. Garrett, C. Ehresmann, P. Stiegler, and P. Fellner, 1975 An investigation of the 16 S RNA binding sites of ribosomal proteins S 4, S 15 and S 20 from *Escherichia coli. Eur. J. Biochem.* **5:**165–180.

Wabl, M. R., 1974 Electron microscopic localization of two proteins on the surface of the 50 S ribosomal subunit of *Escherichia coli* using specific antibody markers. *J. Mol. Biol.* **84:**241–247.

Wallace, H. and M. L. Birnstiel, 1966 Ribosomal cistrons and the nucleolar organizer. *Biochim. Biophys. Acta* **114:**296–310.

Warner, J. R., 1974 The assembly of ribosomes in eukaryotes. In *Ribosomes,* edited by M. Nomura, P. Lengyel, and A. Tissières, pp. 461–488, Cold Spring Harbor Press, Cold Spring Harbor, N.Y.

Warner, J. R. and R. Soeiro, 1967 Nascent ribosomes from HeLa cells. *Proc. Natl. Acad. Sci. USA* **58:**1984–1990.

Weinberg, R. A., 1973 Nuclear RNA metabolism. *Annu. Rev. Biochem.* **42:**329–354.

Wellauer, P. K. and I. B. Dawid, 1974 Secondary structure maps of ribosomal RNA and DNA. I. Processing of *Xenopus laevis* ribosomal RNA and structure of single stranded ribosomal DNA. *J. Mol. Biol.* **89:**379–397.

Wellauer, P. K., R. H. Reeder, D. Carroll, D. D. Brown, A. Deutsch, T. Higashinakagawa, and I. B. Dawid, 1974 Amplified ribosomal DNA from *Xenopus laevis* has heterogeneous spacer lengths. *Proc. Natl. Acad. Sci. USA* **71:**2823–2827.

Wellauer, P. K. and R. H. Reeder, 1975 A comparison of the structural organization of amplified ribosomal RNA from *Xenopus mulleri* and *Xenopus laevis. J. Mol. Biol.* **94:**151–161.

Wensinck, P. C. and D. D. Brown, 1971 Denaturation map of the ribosomal DNA of *Xenopus laevis. J. Mol. Biol.* **60:**235–247.

Whitehouse, H. L. K., 1973 Hypothesis of post-recombination resynthesis of gene copies. *Nature (London)* **245:**295–298.

Wireman, J. W. and P. S. Sypherd, 1974 *In vitro* assembly of 30 S ribosomal particles from precursor 16 S RNA of *Escherichia coli. Nature (London)* **247:**552–554.

Wittmann, H. G., 1974 Purification and identification of *Escherichia coli* ribosomal proteins. In *Ribosomes,* edited by M. Nomura, P. Lengyel, and A. Tissières, pp. 93–114, Cold Spring Harbor Press, Cold Spring Harbor, N.Y.

Wool, I. G. and G. Stöffler, 1974 Structure and function of eukaryotic ribosomes. In *Ribosomes.* edited by M. Nomura, P. Lengyel, and A. Tissières, pp. 417–460, Cold Spring Harbor Press, Cold Spring Harbor, N.Y.

Zimmermann, R. A., A. Muto, P. Fellner, C. Ehresmann, and C. Branlant, 1972 Localization of ribosomal protein binding sites on 16 S ribosomal RNA. *Proc. Natl. Acad. Sci. USA* **69:**1282–1286.

Zimmermann, R. A., A. Muto and G. A. Mackie, 1974 RNA–Protein interactions in the ribosome. II. Binding of ribosomal proteins to isolated fragments of the 16 S RNA. *J. Mol. Biol.* **86:**433–450.

Zimmermann, R. A., G. A. Mackie, A. Muto, R. A. Garrett, E. Engewickell, C. Ehresmann, P. Stiegler, J. P. Ebel, and P. Fellner, 1975 Localization and characteristics of ribosomal protein binding sites in the 16 S RNA of *Escherichia coli. Nucl. Acids Res.* **2:**279–302.

9

Cytoplasmic Messenger RNA

Anton J. M. Berns and Hans Bloemendal

Introduction

The term *messenger RNA* (mRNA) was first used by Jacob and Monod (1961) in their interpretation of experiments concerning the synthesis of RNA after phage infection and the kinetics of enzyme induction and repression. Their brilliant ideas rapidly found wide acceptance, and studies of cell-free protein synthesis under direction of synthetic and viral mRNAs followed (Singer and Leder, 1966), as well as investigations on cellular messenger RNA (Brammar, 1969). At present, it is well established that in eukaryotes mRNA is also the intermediate carrier of information from DNA to protein. In this connection, it must be kept in mind that a number of oligopeptides can be synthesized *in vivo* without mRNA and the ribosomal machinery (*cf.* Lipmann, 1973).

The sequence of nucleotides, of which four different species are present in RNA, determines the order in which the 20 naturally occurring amino acids are assembled. Each trinucleotide contains the code for an amino acid. This triplet code is recognized by the anticodon of the aminoacyl-tRNA (*cf.* Chapter 10). The stepwise growth of the polypeptide

Anton J. M. Berns and Hans Bloemendal—Department of Biochemistry, University of Nijmegen, Nijmegen, The Netherlands.

chain is paralleled by the concomitant movement of the ribosome along the messenger.

Since the synthesis of each protein requires a specific mRNA, messenger species have to exist in a large variety. The bulk of mRNA, which accounts for about 3–4% of the total RNA present in the cytoplasm, has molecular weights ranging from 2×10^5 to 15×10^5 daltons.

Messenger RNA synthesis and messenger function play an important role in the regulation of various processes in the cell. Therefore, the subsequent events from synthesis to degradation of mRNA in a eukaryotic cell will be described in the text, and are diagrammed in Figure 1. After derepression of a certain gene, DNA is transcribed into RNA by RNA polymerase. The transcriptional product is believed to be processed partially into functional mRNA molecules (see Chapter 7; also Scherrer *et al.*, 1963, 1970; Penman *et al.*, 1963; Scherrer and Marcaud, 1968; Darnell, 1968; Ryskov and Georgiev, 1970; Darnell *et al.*, 1971*a,b*; Stevens and Williamson, 1973*c*; Jacobson *et al.*, 1974; Molloy *et al.*, 1974) and transported to the cytoplasm while coupled to proteins (Penman *et al.*, 1968; Spohr *et al.*, 1970; Georgiev and Samarina, 1971; Williamson *et al.*, 1974; Adesnik *et al.*, 1972; Pederson, 1974; Jelinek *et al.*, 1973). In the cytoplasm, mRNA can be stored (Slater and Spiegelman, 1968; Terman, 1970; Weeks and Marcus, 1971; Gander *et al.*, 1973), degraded (Spohr *et al.*, 1970), or involved in protein biosynthesis. In the last case, a complex is formed between the mRNA, initiation factors, initiator-methionyl-tRNA, and the small ribosomal subunit (Burgess and Mach, 1971; Heywood and Thompson, 1971, Levin *et al.*, 1973; Schreier and Staehelin, 1973*d*; Legon *et al.*, 1973). The 60 S ribosomal subunit joins the complex, and translation starts in the direction of the 5′ to the 3′ end (Salas *et al.*, 1965).

During translation, the messenger moves through the ribosome, each time offering a following trinucleotide to the tRNA recognition site. For each trinucleotide translated, one amino acid is inserted in the polypeptide chain (Crick *et al.*, 1961). After completion of the chain, a signal on the messenger causes the release of the newly synthesized polypeptide (Beaudet and Caskey, 1971; Ilan, 1973) and the dissociation of the messenger–ribosome complex (see Chapter 11). Now the ribosome may enter a new cycle (Kaempfer, 1969; Falvey and Staehelin, 1970; Kurland, 1970; Davis, 1971), and the messenger can be translated again or is degraded by RNAse.

The availability of a variety of purified messenger RNAs has contributed significantly to the elucidation of the events occurring in this complex and highly regulated process. Translation of mRNAs in heterolo-

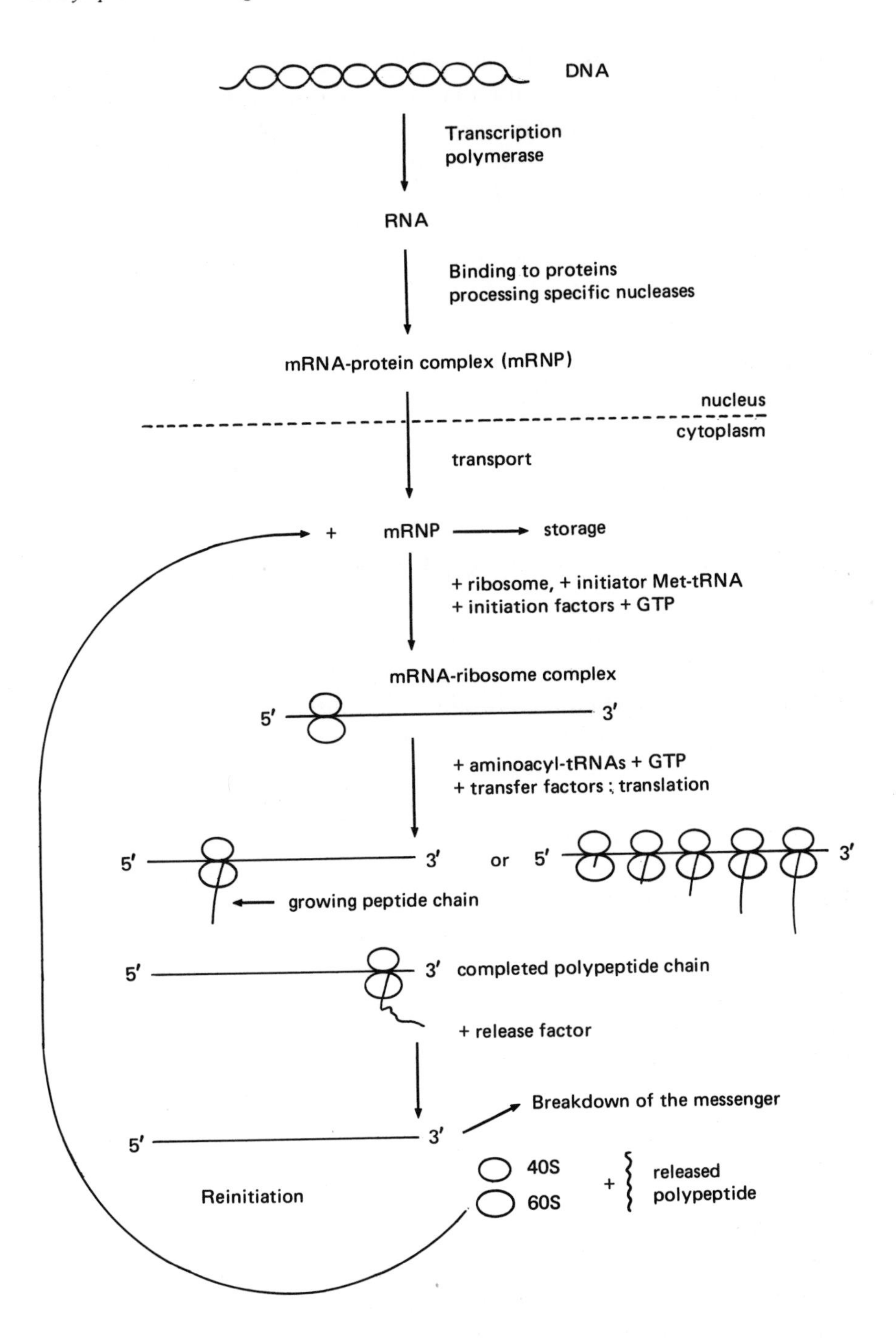

Figure 1. The messenger RNA life cycle.

gous systems has supplied information about the existence of tissue- and species-specific mRNA recognition factors (Fuhr and Natta, 1972; Rourke and Heywood, 1972; Bogdanovsky *et al.*, 1973; Nudel *et al.*, 1973*b*; Wigle and Smith, 1973; Heywood *et al.*, 1974; Jay and Kaempfer, 1974; Nakajima *et al.*, 1974; Berns *et al.*, 1975).

Hybridization experiments with DNA from different tissues have provided information about gene multiplicity (Bishop *et al.*, 1972; Packman *et al.*, 1972; Delovitch and Baglioni, 1973; Sullivan *et al.*, 1973; Faust *et al.*, 1974; P. R. Harrison *et al.*, 1974; Stavnezer *et al.*, 1974). The secondary structure and its influence on initiation and translation can be studied now (Fukami and Imahori, 1971; Jelinek *et al.*, 1974; Favre *et al.*, 1974, 1975). Certainly considerably more progress will be made in the near future which may lead to clearer understanding of the mechanisms underlying the synthesis, processing, transport, and translation of mRNA.

Specific Interactions with Nucleotide Sequences of mRNA

The nucleotide sequence of messenger RNA does not serve exclusively to provide for the information for the amino acid sequence of the encoded polypeptide. It also contains so-called untranslated regions. These sequences may vary in length in different mRNAs, and their extent may be even larger than the coding region (Partington *et al.*, 1973; Berns *et al.*, 1974*a*). The excess of nucleotide sequences is present both at the 5′ and 3′ ends of the mRNA, and these sequences may fulfill a role in ribosome recognition, cellular localization, stability, and regulation of translation. Most mammalian mRNAs contain at their 3′ end a poly(A) track of 50–200 adenine residues added to the mRNA in a posttranscriptional step (Lim and Canellakis, 1970; Edmonds *et al.*, 1971; Slater *et al.* 1972; Adesnik and Darnell, 1972). Also, mRNAs from more primitive organisms such as yeast (McLaughlin *et al.*, 1973) and slime molds (Firtel *et al.*, 1972), mitochondrial mRNAs (Hirsch and Penman, 1974; Ojala and Attardi, 1974), and mRNAs of viral origin (Kates, 1970; Gillespie *et al.*, 1972; Johnston and Bose, 1972; Weinberg *et al.*, 1972; Yogo and Wimmer, 1972; Soria and Huang, 1973) contain poly(A) at their 3′ end. However, it may well be that non-poly(A)-containing mRNAs, except histone mRNA (Adesnik *et al.*, 1972), may have escaped detection, as they might be difficult to recognize, especially if ribosomal RNA contaminants are present. This may be the reason that only recently other populations of non-poly(A)-containing mRNAs have been described (Milcarek and Penman, 1974; Nemer *et al.*, 1974). The role of the extracistronic regions in mRNA is poorly understood. They may play an active role in the cytoplasm or may have been involved in the nuclear processing and transport of the mRNA to the cytoplasm.

Ribosome Recognition

So far, there is no direct evidence to show that in eukaryotic mRNAs extracistronic regions play a role in the recognition of the ribosome. However, the initiation process in eukaryotic systems parallels in many respects the initiation process in prokaryotic systems. This has been shown most convincingly by the proper translation of bacteriophage mRNA in eukaryotic systems (Aviv *et al.,* 1972; Morrison and Lodish, 1973; Schreier *et al.,* 1973). Although the mammalian system also recognizes false initiation codons (Morrison and Lodish, 1974), it may well be that in eukaryotic mRNAs, as in their prokaryotic counterparts (Steitz, 1973), a specific nucleotide sequence is present that is different from the AUG codon which plays an essential role in the formation of the initiation complex. This could be accomplished either by the recognition of a primary structure by base pairing or by interaction of a protein with a defined secondary structure in the mRNA.

Cellular Localization

The bulk of messenger RNA is localized on cytoplasmic polyribosomes, which can be divided roughly into two classes with respect to cellular environment: free polysomes and membrane-bound polysomes (see Chapter 11; also Palade, 1958; Bloemendal *et al.,* 1964, 1974). The membrane-bound polysomes have been suggested to be involved in the synthesis of proteins which have to be exported out of the cell (Stavnezer and Huang, 1971; Milstein *et al.,* 1972; Zauderer *et al.,* 1973; T. M. Harrison *et al.,* 1974*a*), whereas the free polysomes seem to be mainly involved in the synthesis of cellular proteins. Although the situation is certainly more complicated (Rosbash and Penman, 1971; T. M. Harrison *et al.,* 1974*a*), since nonexporting cells also contain large amounts of membrane-bound polysomes (Rosbash and Penman, 1971), it seems reasonable to assume that the mRNA contains the information which specifies its site of translation either on free or membrane-bound polysomes. This site may also be the coding sequence, in which case the binding could be accomplished via the growing polypeptide chain (Sabatini *et al.,* 1972) or by the interaction of noncoding regions directly with the membrane. Studies by Milcarek and Penman (1974) suggest that the poly(A) region is a possible candidate for this function. However, since poly(A)-containing mRNAs are also found on free polysomes, the poly(A) track cannot be responsible exclusively for the confinement of a fraction of the polysomes in membrane structures. Therefore, other sequences have to be involved in this interaction.

Regulation of Protein Synthesis in the Cytoplasm: mRNA Stability and the Role of Poly(A) and Associated Proteins

Messenger RNA stability is one of the tools which a cell can use to promote a higher level of synthesis of a specific class of proteins without the need for regulation at the transcriptional level. Originally, it was thought that poly(A) might play an important role in mRNA stability by acting in the manner of a time clock, as suggested by Sussman (1970). Additional evidence was provided by the fact that histone mRNA, which lacks poly(A) (Adesnik and Darnell, 1972), showed a relatively short half-life (Gallwitz and Mueller, 1969) and that during aging of the mRNA the length of the poly(A) segment was reduced (Brawerman, 1973; Sheines and Darnell, 1973; Gorski *et al.*, 1974). On the other hand, the half-life of the recently described poly(A)-free mRNA in HeLa cells (Milcarek *et al.*, 1975) revealed almost the same value as the poly(A)-containing mRNAs. Furthermore, mRNA from which the poly(A) track is removed enzymatically appears to be translated as efficiently as its poly(A)-containing counterpart (Huez *et al.*, 1974; Sippel *et al.*, 1974; Soreq *et al.*, 1974; Williamson *et al.*, 1974), and only a minor difference in stability is observed. Therefore, although the presence of the poly(A) track may affect the stability of mRNA, it seems not to be a main factor which determines the stability of mRNA.

It has been observed that transport of mRNA is blocked if the addition of poly(A) in the nucleus is prevented by cordycepin (Edmonds and Caramela, 1969; Mendecki *et al.*, 1972; Perry *et al.*, 1973), suggesting that poly(A) may have a transport function (see Chapter 7). However, there is also some evidence that the blocking may be a side effect of the drug, as it also inhibits the entrance of poly(A)-free mRNA in the cytoplasm (Milcarek *et al.*, 1975). Furthermore, the presence of poly(A) in viral mRNAs which replicate exclusively in the cytoplasm (Kates, 1970; Yogo and Wimmer, 1972; Soria and Huang, 1973), the occurrence of poly(A) tracks in mitochondrial RNA (Hirsch and Penman, 1974), and the observed adenylation of mRNA in the cytoplasm (Slater *et al.*, 1973; Wilt, 1973; Diez and Brawerman, 1974) all favor a cytoplasmic function. Altogether, the exact function of poly(A) in mRNA is still rather obscure, and it would not be surprising if it is found to play multiple roles.

In view of the foregoing consideration, it may be assumed that the stability of the messenger is probably more dependent on sequences other than poly(A). Such sequences might, for instance, determine how sensitive the mRNA is for specific nucleases or limit the extent of masking toward nucleases by interaction with specific proteins or other cell components. So the same messenger might well exhibit completely different half-lives in different tissues. In HeLa cells, Singer and Penman, using oligo(dT)-cellulose to estimate labeled mRNA among other labeled RNA species, described two classes of mRNA with half-lives of 7 and 24 hr, respectively (Singer and Penman, 1973). Half-lives as long as cell generation times have also been reported by others (Greenberg, 1972; Murphy and Attardi, 1973). Earlier reports of estimates of half-life, based on the action of actinomycin D to prevent labeling of ribosomal RNA, gave much shorter half-lives. It has been mentioned already that the proteins associated with mRNA may protect the messenger against enzymatic degradation. Two main proteins have been identified (Blobel, 1972), one with a molecular weight of about 78,000 daltons, which is claimed to be associated with the poly(A) part of the mRNA (Blobel, 1973), and one of approximately 52,000 daltons, which is bound to the coding region (Blobel, 1973; Bryan and Hyashi, 1973). Other authors have described different protein species (Lebleu *et al.*, 1971; Gander *et al.*, 1973; Heywood *et al.*, 1974; Kaempfer and Kaufman, 1972; Nakajima *et al.*, 1974; Nudel *et al.*, 1973*a*; Sierra *et al.*, 1974; Stevens, 1974). This regulation could be accomplished by the existence of various initiation factors which recognize only certain classes of mRNA. As IF_3 is a high molecular weight aggregate of several polypeptides (Hirsch *et al.*, 1973), a minor alteration in the composition could result in an IF_3 with different specificity. The "interference" factors described by Lebleu *et al.* (1972) could well belong to such a class of modifying agents.

An RNA species (Fuhr and Natta, 1972; Bogdanovsky *et al.*, 1973) which seems to be a constituent of EIF_3 (Heywood *et al.*, 1974) is a possible candidate for a regulatory function. Also, experiments of Goldstein and Penman (1973) predict the involvement in protein initiation of an RNA species differing from mRNA. Experiments in our laboratory (Berns *et al.*, 1975) showed that the "RNA factor" described by Bogdanovsky *et al.* (1973) had no discriminative features toward different mRNAs, since all the mRNAs studied were affected equally. So far, this "RNA factor" is ill-characterized, and it is even doubtful that it is an RNA species.* Although many workers have accomplished shifts in mRNA translation by adding different factors, only a few provide evidence for an almost obligatory requirement for a different IF. The experiments of Heywood and coworkers

* We demonstrated recently that the factor can fully be replaced by spermine or spermidine (Salden and Bloemendal, 1976).

(Rourke and Heywood, 1972; Heywood *et al.*, 1974) have shown that myosin mRNA translation in a heterologous system requires initiation factors derived from the same type of tissue from which the mRNA is isolated. Another report (Nakijama *et al.*, 1974) claims the modification of EIF_3 in monkey cells after infection with SV40. This observation gives a molecular base to the observed helper function of SV40 after infection of monkey cells with adenovirus. On the other hand, there are numerous reports (see Table 1) which show the translation of mRNA in heterologous systems without any apparent need for special recognition factors. However, in these systems the regulatory functions may be inactivated during the preparation of the extracts or by isolation of the mRNA. So far, the results suggest that if regulation exists at this level it takes place by relatively small shifts in mRNA selection and that an absolute requirement for specific recognition factors is restricted to a minor class of mRNAs.

Reports describing specific interaction which affects translocation of eukaryotic mRNAs are scarce. Weeks and Baxter (1972) and Uenoyama and Ono (1972) described the action of an inhibitor at this level of protein synthesis. There is also some evidence that interferon might act at this stage (Content *et al.*, 1974). It has been suggested earlier that proteins which are found aggregated in fixed ratios (e.g., hemoglobin or antibodies) might interact at the polysomal level. The described interaction of H_2L_2 antibody with heavy-chain mRNA is a well-documented example (Stevens and Williamson, 1973*b*), and the synthesis of L chains could well be regulated at this level. Moreover, the translation of globin mRNA in oocytes seems to be regulated in a translocation event (Lingrel and Woodland, 1974).

mRNA Identification

Template Activity

The ability to direct the synthesis of a specific polypeptide chain is the most valuable functional assay for a particular messenger. The stimulatory effect on amino acid incorporation after addition of RNA fractions to a cell-free system cannot be considered as a reliable criterion for mRNA activity (Chantrenne *et al.*, 1967; Bloemendal, 1972; Williamson, 1972). Addition of mRNA to a reticulocyte lysate may even inhibit amino acid incorporation (Lingrel *et al.*, 1971; Milstein *et al.*, 1972), albeit added mRNA is translated. On the other hand, stimulation of amino acid incorporation may be obtained without the synthesis of any new identifiable polypeptide. Therefore, the only conclusive test for mRNA activity is the

TABLE 1

Type of mRNA	*Assay system*	*Reference*
Albumin	Reticulocytes	Taylor and Schimke, 1973; Shapiro *et al.*, 1974; Shafritz, 1974
Avidin	Reticulocytes	O'Malley *et al.*, 1972
Catalase	Reticulocytes	Uenoyama *et al.*, 1972
Collagen	Ascites	Benveniste *et al.*, 1973
	Reticulocytes	Boedtker *et al.*, 1974
Crystallin (calf)	Reticulocytes	Berns *et al.*, 1972*a*, 1973*a*
	Ascites	Mathews *et al.*, 1972*a*; Chen *et al.*, 1974
	Wheat germ	Berns *et al.*, 1974*a*
	Xenopus oocytes	Berns *et al.*, 1972*b*
Crystallin (chick)	Ascites	Zelenka and Piatigorski, 1974
	Reticulocytes	Williamson *et al.*, 1972
Ferritin	RCFS[a]	Shafritz *et al.*, 1973
Globin (duck)	Reticulocytes	Pemberton *et al.*, 1972
	Xenopus oocytes	Lane *et al.*, 1973
	RCFS	Schreier and Staehelin, 1973*b*
Globin (human)	Ascites	Metafora *et al.*, 1972
Globin (mouse)	Reticulocytes	Lingrel *et al.*, 1971; Lockard and Lingrel, 1969
	Ascites	Metafora *et al.*, 1972; Mathews *et al.*, 1971
Globin (rabbit)	Reticulocytes	Schapira *et al.*, 1968
	Ascites	Mathews *et al.*, 1972*b*; Aviv and Leder, 1972; Housman *et al.*, 1971
	Wheat germ	Roberts and Paterson, 1973; Efron and Marcus, 1973
	Xenopus oocytes	Lane *et al.*, 1971
	Liver	Sampson *et al.*, 1972
	RCFS	Schreier and Staehelin, 1973*a*
	Liver	Prichard *et al.*, 1971
	Lens	Strous *et al.*, 1974*a*
Histone	Reticulocytes	Breindl and Gallwitz, 1973
	Ascites	Jacobs-Lorena *et al.*, 1972
	RCFS	Breindl and Gallwitz, 1974
	HeLa	Gabrielli and Baglioni, 1974
Immunoglobin (light chain)	Reticulocytes	Stavnezer and Huang, 1971
	Ascites	Milstein *et al.*, 1972; Swan *et al.*, 1972; Schechter, 1973
	Xenopus oocytes	Stevens and Williamson, 1972
Immunoglobin (heavy chain)	*Xenopus* oocytes	Stevens and Williamson, 1973*a*

[a] RCFS, reconstituted cell-free system.

TABLE 1. *Continued*

Type of mRNA	*Assay system*	*Reference*
Keratin	Reticulocytes	Partington *et al.*, 1973; Kemp *et al.*, 1974
Leg-hemoglobin	Wheat germ	D. P. S. Verma *et al.*, 1974
Mitochondrial	*E. coli*	Küntzel and Blossey, 1974
Myoglobin	Reticulocytes	Thompson *et al.*, 1973
Myosin	Reticulocytes	Rourke and Heywood, 1972
Ovalbumin	Reticulocytes	Rhoads *et al.*, 1971; Palmiter, 1973; Palacios *et al.*, 1972; Comstock *et al.*, 1972
Preproparathyroid hormone	Wheat germ	Kemper *et al.*, 1974
Promellitin (honeybee)	*Xenopus* oocytes	Kindås-Mügge *et al.*, 1974
Silk fibroin	(Identified by sequence analysis)	Suzuki and Brown, 1972
Virus	Wheat germ	
Virus (adeno)	Ascites	
Virus (Semliki Forest)	Ascites	
Virus (simian)	Hamster ovary	

synthesis of a specific polypeptide in a system in which this protein is normally absent. This limits in fact the application of cell-free synthesis as a simple tool for messenger identification, as it requires a significant amount of at least a partly purified messenger preparation. Furthermore, the failure to demonstrate messenger activity may be caused by the inability of the chosen cell-free system to translate the mRNA under study. This failure may be caused by mRNA discriminating factors which are constituents of the cell-free system or by degradation of the mRNA itself. It would even be more difficult and elaborate to discriminate between these possibilities.

Moreover, the mRNA isolated may code for a polypeptide different from the protein observed *in vivo*. This could be due to unfaithful translation or the synthesis of a precursor which has to undergo posttranslational modification. There are many examples of this "processing" such as the enzymatic trimming of a precursor (Baltimore, 1971*a,b*; Milstein *et al.*, 1972), acetylation (Strous *et al.*, 1974), hydroxylation (Winstead and Wold, 1964), phosphorylation (Jergil, 1972; Louie and Dixon, 1972), and glycosidation (Winterburn and Phelps, 1972).

Although it has been shown, especially with viral mRNAs, that a large precursor polypeptide may be synthesized which is later trimmed down to size (Baltimore, 1971*b*; Milstein *et al.,* 1972; Vogt and Eisenman, 1973), this seems not to be the general mechanism by which eukaryotic proteins are synthesized, as most labeled completed polypeptides do not diminish in size in a "chase." Moreover, incorporation of amino acid analogues does not increase the distribution of polypeptide chain lengths, as might be expected if posttranslational cleavage were involved. On the other hand, a few cases are known of the biosynthesis of relatively small precursor proteins which are cleaved by proteolytic action. Examples are the syntheses of immunoglobulin (Milstein *et al.,* 1972) and mellitin (Kindås-Mügge *et al.,* 1974), which arise from precursor proteins by proteolytic cleavage. Messengers of eukaryotic origin (viral mRNA not included) which have so far been identified more or less thoroughly are summarized in Table 1.

Hybridization

Complementary polynucleotides can form nuclease-resistant duplexes when incubated under appropriate conditions (Britten and Kohne, 1968). The nucleic acid used may be either DNA or RNA, or, alternatively, a mixture of each species, in which case a hybrid duplex results. The kinetics of annealing give a reasonable estimate of the frequency of occurrence of the complementary species. This technique can be utilized to detect complementary sequences to a particular nucleic acid species and to find identical sequences by performing competition hybridization. By use of labeled RNA, the multiplicity of specific genes can be determined (Kedes and Birnstiel, 1971; Bishop *et al.,* 1972; Delovitch and Baglioni, 1973) and the concentration of mRNA sequences estimated (Williamson *et al.,* 1970; Skoultchi and Gross, 1973).

Two main problems in using the mRNA *per se* as species to follow the process of hybridization are its purity and its plus strand character. One of the earlier problems, that of obtaining mRNA labeled with a high specific activity, has been overcome by chemical labeling with radioactive iodine (Prensky *et al.,* 1973; Scherberg and Refetoff, 1973). However, the use of radioactive complementary DNAs offers more advantages. The discovery of the enzyme reverse transcriptase in oncornaviruses made it possible to produce highly labeled DNA complementary to mRNA (see Verma and Baltimore, 1973). Since most eukaryotic mRNAs have been shown to contain at their 3′ end regions rich in poly(A), an oligomer of (dT) can hydrogen-bond to it and serve as a primer to support the synthesis of complementary DNA catalyzed by reverse transcriptase in the presence of actinomycin D. Besides the advantage that a complementary DNA with high specific activity can be produced (approximately 3×10^8 cpm/μg), this method selects only mRNA for transcription as other RNAs are lacking the poly(A) track. This favorable condition reduces the need of purifying the mRNA from ribosomal contaminants. Once a DNA copy (cDNA) has been synthesized, this DNA, unlike I-labeled mRNA, can be used as a probe to detect mRNA-specific sequences in any nucleic acid fraction. Complementary DNA has been synthesized on a variety of mRNAs: globin mRNA (I. Verma *et al.*, 1972; Ross *et al.*, 1972; Kacian *et al.*, 1972; Packman *et al.*, 1972), immunoglobulin (Aviv *et al.*, 1973; Diggelman *et al.*, 1973; Stavnezer *et al.*, 1974), ovalbumin mRNA (Sullivan *et al.*, 1973), crystallin mRNA (Chen *et al.*, 1973; Berns *et al.*, 1973*a*), and slime mold mRNA (I. Verma *et al.*, 1974).

Hybridization with cellular DNA reveals the amount of copies present in the DNA (Packman *et al.*, 1972; Bishop *et al.*, 1972; Sullivan *et al.*, 1973; P. R. Harrison *et al.*, 1974; Stavnezer *et al.*, 1974); with nuclear RNA, it elucidates the characteristic of precursor of mRNA (Imaizumi *et al.*, 1973). Hybridization with mRNA from different developmental stages can show differences in gene expression (I. Verma *et al.*, 1974). By synthesizing RNA on the cDNA, it is possible to sequence the mRNA indirectly (Marotta *et al.*, 1974; Forget *et al.*, 1974; Poon *et al.*, 1974) without the need for extreme purification as in the RNA labeling by iodine (Robertson *et al.*, 1973) or with the aid of polynucleotide kinase (Szekely and Sanger, 1969).

Molecular Weight

Messenger RNAs exhibit a broad range of molecular weights. In general, the distribution of the bulk of cellular RNA is found between 8 and 30 S, corresponding to $1\text{–}15 \times 10^5$ daltons. In general, there exists a relationship between the molecular weight of the mRNA and the size of the protein it codes for. However, some exceptions are reported in which the size of the mRNA exceeds the expected molecular weight considerably (Brownlee *et al.*, 1973; Zelenka and Piatigorski, 1974; Berns *et al.*, 1974*a*; Kemp *et al.*, 1974). In case of the 14 S crystallin mRNA, which codes for the A_2 chain of α-crystallin, a mRNA with a molecular weight of about 510,000 ($\pm$1500 nucleotides) directs the synthesis of a polypeptide of 173 amino acid residues (Ouderaa *et al.*, 1973). No evidence could be provided in this case for the synthesis of a precursor molecule as in the translation of picornaviruses (Baltimore,

1971*a*) where one mRNA codes for a giant polypeptide chain which is later trimmed to size, resulting in several polypeptides of moderate length. Also, the poly(A) track seems to have the normal length (Favre *et al.*, 1974). Therefore, the most likely explanation is that lens 14 S mRNA contains large untranslated regions, the function of which (if any) is unclear so far.

The differences in molecular weights of mRNAs are the basis of one of the most applied techniques used to separate mRNAs from each other, as will be discussed later.

Labeling Kinetics

The synthesis of mRNA as a response to a stimulus is one of the ways in which protein synthesis is regulated and adaptation of the cell to changed conditions is achieved. This phenomenon, which was found to be the main regulatory mechanism in prokaryotes, does not have the same significance in eukaryotic systems. Here, a great deal of regulation seems to take place after transcription, both in the nucleus (see Chapter 7) and in the cytoplasm. In contrast to prokaryotic systems, a great number of eukaryotic mRNAs have a long half-life (Singer and Penman, 1973). As a consequence, synthesis of mRNA occurs less fast than in prokaryotic systems, and mRNA is labeled slowly, although somewhat faster than ribosomal RNA. The messenger is labeled predominantly only with very short pulses. Various drugs (e.g., actinomycin D) can be used to suppress ribosomal RNA synthesis without affecting the labeling of mRNA (Penman *et al.*, 1968). However, for kinetic studies drugs may change the half-lives of mRNAs. Therefore, pulse-chase experiments in the absence of drugs seem to be more valuable for the study of synthesis, transport, and half-life of mRNA (Singer and Penman, 1973). Since poly(A)-containing mRNA can be separated easily from ribosomal RNA by the use of affinity chromatography on oligo(dT)-cellulose or poly(U)-Sepharose, labeling of ribosomal and tRNA does not interfere too seriously in such studies (Singer and Penman, 1973).

Polysomal Location

Since mRNA directs the synthesis of polypeptide chains, it is not surprising that in normally functioning cells mRNA can be found mainly on polyribosomes. The mRNA distribution among the polysomal population is not random; larger mRNAs are located generally on larger polysomes and synthesize larger polypeptides. Some isolation procedures take advantage of this fact by isolating a specific class of polysomes (see Chapter 11) as a first step in mRNA purification (Heywood and Nwagwu, 1969). However, mRNAs are not found exclusively in the polysomal fraction, and there may also be mRNA, complexed with proteins, free in the cytoplasm (Neyfakh, 1971; Edmonds *et al.*, 1971; Tyler and Tyler, 1972; Jacobs-Lorena and Baglioni, 1972).

There are several ways to release mRNA from polysomes: by the use of a chelating agent such as EDTA or pyrophosphate which dissociates the ribosomes and releases the mRNA as a messenger ribonucleoprotein (mRNP) complex (Chantrenne *et al.*, 1967; Labrie, 1969; Darnell, 1968; Penman *et al.*, 1968; Henshaw, 1968; Lebleu *et al.*, 1971; Bloemendal *et al.*, 1972) or by puromycin in the presence of high salt concentrations (Blobel, 1972).

The dissociation of mRNPs from polysomes by chelating agents is especially useful to show that messengerlike RNA is really polysomal bound mRNA (Penman *et al.*, 1968). Since these treatments leave the ribosomal structures virtually intact, this method allows the separation of mRNP from ribosomal subunits, even if these contain partially degraded RNA (as the degraded RNA is kept together in the ribosomal structure). Although the dissociation with chelating agents also releases a 8 S particle composed of 5 S RNA and protein (Blobel, 1971; Lebleu *et al.*, 1971), the procedure is useful as a first purification step of mRNA (Lebleu *et al.*, 1971; Williamson *et al.*, 1972). After fixation with formaldehyde, the released mRNP has a low density in CsCl (± 1.45), indicating a very high protein/RNA ratio (Henshaw, 1968; Perry and Kelley, 1968; Lebleu *et al.*, 1971).

Isolation of mRNA

The difficulties encountered with the isolation of mRNA are threefold: (1) mRNAs are present in only minute quantities as compared with rRNA and tRNA, (2) mRNAs are broken down easily as they are, in general, more accessible to RNAse than ribosomal RNA (Williamson *et al.*, 1969; Shafritz *et al.*, 1973), and (3) often a messenger has to be purified from other mRNA species having similar molecular weights. There are three steps which determine the successful isolation of a specific messenger RNA: (1) the choice of the system from which the mRNA has to be isolated, (2) the purification at the polysomal level, and (3) the purification at the nucleic acid level.

Choice of System

The higher the relative amount of a specific mRNA in a certain system, the easier one can get rid of contaminating mRNAs. This is one of the reasons why a large variety of mRNAs have been isolated from highly differentiated tissues which synthesize a limited amount of proteins in relatively high amounts. Another way to facilitate mRNA isolation at this level is to affect the content of a specific mRNA in a tissue or cell line. This can be achieved in some instances by the administration of hormones before isolating the mRNA (Rhoads *et al.*, 1971; Comstock *et al.*, 1972; O'Malley *et al.*, 1972). For the isolation of histone mRNA it was shown that synchronization of the cells is favorable, since the histone mRNA content is higher at a certain phase in the cell cycle (Jacobs-Lorena *et al.*, 1972; Breindl and Gallwitz, 1973).

Fractionation of Polysomes

Separation on Basis of Density or Size. The isolation of polyribosomes as a first step in messenger purification is almost generally applied as it enables separation from most of the undesired cytoplasmic contaminants. Therefore, cells are ruptured by hypotonic shock, by mechanical means, or by detergents, nuclei are spun down at low speed, and polysomes are isolated from the supernatant by sedimentation (Noll, 1969), by magnesium precipitation (Lee and Brawerman, 1971), or by gel filtration (Uenoyama and Ono, 1972). When membrane-bound ribosomes have to be isolated, cells are disrupted mechanically or by very low detergent concentration in order to keep the endoplasmic reticulum intact. The membrane-bound polysomes can be isolated by fractionated sedimentation or by isopycnic banding in a discontinuous sucrose gradient (Bloemendal *et al.*, 1964, 1974). The isolation of the membrane-bound polysomes may already result in an enrichment of a specific mRNA, as there is evidence that proteins which have to be exported from the cell are mainly synthesized on membrane-bound polysomes (Stavnezer and Huang, 1971; Milstein *et al.*, 1972; Zauderer *et al.*, 1973; T. M. Harrison *et al.*, 1974*a*).

A second way to achieve partial purification of mRNA on the polysomal level is separation of polysomes into different size classes. This method is especially effective if a relatively large-sized messenger has to be isolated. This technique has been successfully applied in the isolation of myosin mRNA (Heywood and Nwagwu, 1969). If a relatively small mRNA is wanted (which is located on small polysomes), contamination with large mRNAs is more likely, but it can be prevented to some extent by using cycloheximide and RNAse inhibitors during the isolation procedure to prevent "runoff" and degradation of the larger polysomes.

Sometimes the size of the polysomes can be altered artificially. This is best illustrated by the elegant procedure to separate α- and β-globin chain mRNA from rabbit reticulocytes (Kazazian and Freedman, 1968; Temple and Housman, 1972). Although there is normally a small difference in the size of α- and β-globin-synthesizing polysomes (three or four ribosomes per polysome and four or five ribosomes per polysome, respectively) caused by the lower efficiency of the α-globin initiation (Lodish and Jacobson, 1972), this difference can be amplified dramatically by incubating the reticulocytes with a high concentration of *o*-methylthreonine. This analogue replaces isoleucine, but further elongation at this residue is blocked. As a consequence, polypeptide synthesis stops as soon as the analogue is incorporated in the growing chain. Isoleucine occurs in the β-chain at position 110 (from the *N*-terminus) and in the α-chain at position 10. Preincubation with *o*-methylthreonine results in β mRNA heavily loaded with ribosomes (seven to ten) and almost unloaded α-globin mRNA (one to three). Isolation of the fast-sedimenting population results in an almost complete purification of β-chain mRNA.

Separation on Base of Information. One of the most promising techniques, which has been applied successfully by several groups of workers (Uenoyama and Ono, 1972; Palacios *et al.*, 1972; Delovitch *et al.*, 1972; Shapiro *et al.*, 1974), takes advantage of the antigenic capacity of growing polypeptide chains. By recognizing growing polypeptide chains, the antibody is able to precipitate a group of polysomes which are synthesizing the antigen. Although this technique seems very simple, a variety of difficulties are encountered in the application. First, the antibodies have to be purified extensively in order to remove even traces of RNAses which otherwise degrade the mRNA during the often long reaction times. Second, a great deal of aspecific precipitation may occur, resulting from interactions of the antibody Fe fragment with ribosomal structures or by trapping of polyribosomes in the precipitate. The aspecific interaction can be reduced by pepsin treatment of the antibody; the $F(ab')_2$ fragment is then the active component (Holme *et al.*, 1971). Schimke and coworkers diminished the aspecific trapping significantly by conducting the adsorption reaction on a solid matrix (Palacios *et al.*, 1972). This group has obtained an even more effective method of polysome–antibody precipitation by applying an indirect precipitation reaction (Shapiro *et al.*, 1974). This method seems the most promising as it is highly specific and precipitates the antigen-synthesizing polysomes almost quantitatively. Moreover, it requires very small amounts of antibody. It seems that this technique will allow the isolation of even minor messenger populations in the near future.

RNA Purification and Fractionation

Deproteinization. In order to obtain nucleic acids either from polysomes or directly from cells or tissues, several deproteinization methods have been described. Polysomal RNA can be obtained by treating the polysomes with sodium dodecylsulfate in the absence or presence of proteolytic enzymes (Lockard and Lingrel, 1969; Wiegers and Hilz, 1972), followed by gradient centrifugation and ethanol (or salt) precipitation.

Often phenol or phenolic "cocktails" are used as deproteinizing agents (Penman, 1966; Kirby, 1968), sometimes in combination with proteolytic enzymes (Mach *et al.*, 1973). The efficiency of extracting poly(A)-containing mRNA during phenolic extractions can be improved by the use of chloroform (Perry *et al.*, 1972) or by the appropriate pH and salt concentration (Lee *et al.*, 1971). By using different pHs during sequential extractions, a significant enrichment of poly(A)-containing mRNA can even be obtained (Brawerman *et al.*, 1971). Also, a method has recently been described in which RNA is purified in one step from protein, DNA, and glycogen (Glišin *et al.*, 1974) by applying CsCl centrifugation.

Separation of RNA from NonRNA Components. After deproteinization the RNA may still be contaminated with DNA, glycogen, and minor amounts of proteins. RNA can be partially purified from the contaminants by salt precipitation (Kirby, 1968; Wiegers and Hilz, 1972) which preferentially insolubilizes single-stranded polynucleotides (Baltimore, 1966). Complete separation can be obtained by CsCl centrifugation (Glišin *et al.*, 1974). DNA and glycogen are banded whereas RNA is pelleted. For analytical purposes, DNA and RNA may both be banded in Cs_2SO_4 (Kacian *et al.*, 1972). DNA may also be removed by DNAse digestion. However, it is often difficult to prevent partial degradation of the RNA during the treatment. Both RNA and DNA may be purified from minor protein contaminations by precipitation with cetylammonium-bromide (Bellamy and Ralph, 1968).

Separation from Other RNA Species. Separation of RNA on the basis of sedimentation velocity with the aid of sucrose density centrifugation is one of the most applied methods in messenger purification. It often enables the separation of the mRNA from ribosomal RNAs (Lockard and

Lingrel, 1969) and from the other mRNAs present in the system (Berns *et al.*, 1973*a*). For relatively small mRNAs (8–12 S), it may be worthwhile to first isolate the messenger ribonucleoprotein. Sodium dodecylsulfate treatment of the mRNP followed by sucrose gradient centrifugation results in mRNA preparations which are almost completely devoid of ribosomal contaminants (Lebleu *et al.*, 1971; Williamson *et al.*, 1972). The use of zonal rotors, which share high resolution power with high capacity, has proven to be especially useful for large-scale preparations of mRNA (Berns *et al.*, 1971; Williamson *et al.*, 1971). More powerful resolution can be achieved by polyacrylamide gel electrophoresis, especially when the electrophoresis is conducted under denaturing conditions (Lanyon *et al.*, 1972; Gould and Hamlyn, 1973; Reynders *et al.*, 1973; Brownlee *et al.*, 1973; Berns *et al.*, 1974*a,b*). Application on a preparative scale of this widely used analytical technique would enable the separation of RNA species with only minor differences in molecular weight (Berns *et al.*, 1974*b*). The occurrence of poly(A) tracks at the 3′ end of the majority of eukaryotic messengers is a fortunate characteristic and the basis for the most simple and powerful technique in mRNA purification. The poly(A) of the mRNA can form hydrogen bonds with oligo(dT) or poly(U) under extremely mild conditions. Using immobilized oligo(dT) or poly(U), separation of poly(A)-containing mRNA from other RNA species becomes extremely simple. After binding of mRNA at relatively high salt concentration and elution of the nonbound RNA, the mRNA is eluted from oligo(dT)-cellulose by low ionic strength buffer at elevated temperatures or by high formamide concentrations. For this type of separation, poly(U)-cellulose filters (Sheldon *et al.*, 1972*a,b*), poly(U)-cellulose (Firtel *et al.*, 1972; Sheldon *et al.*, 1972*b*; Morrison *et al.*, 1972), poly(U)-Sepharose (Adesnik *et al.*, 1972; Lindberg and Persson, 1972), oligo(dT)-cellulose (Aviv and Leder, 1972; Swan *et al.*, 1972; Nakazato and Edmonds, 1972; Nakazato *et al.*, 1974), or poly(dT)-cellulose columns (Piperno *et al.*, 1974) may be used. In order to obtain separation on the basis of the length of the poly(A) track, either stepwise elution can be applied (Morrison *et al.*, 1972) or elution with a salt, formamide, or temperature gradient can be used (Ihle *et al.*, 1974).

Instead of employing immobilized polynucleotides, plain Millipore filters (Lee *et al.*, 1971) or cellulose-Sepharose can be used for absorbing poly(A)-containing RNA (Schutz *et al.*, 1972; Woo *et al.*, 1974). As the binding is not based on hybrid formation but on a poorly understood interaction between lignine and poly(A) segments, this technique has been successfully used to fractionate mRNA with different poly(A) contents (Gorski *et al.*, 1974).

A simplified combination of methods has recently been used in our laboratory to isolate mRNA in a fast and effective procedure. Polysomes were dissolved in 0.5 M NaCl, 0.5% SDS, 0.01 M tris-Cl (pH 7.5) and applied directly on an oligo(dT)-cellulose column at room temperature. The column was washed with starting buffer, followed by a 0.1 M salt wash. This eluted all of the protein and more than 95% of the ribosomal RNAs. The mRNA was eluted in 0.01 M Tris-Cl at 37°C, and collected in small fractions. About 3% of the applied RNA was found in this eluate (260/280 ratio > 2.05). The mRNA-containing fractions were heated at 60°C for 3 min, quenched in ice, and applied directly on a sucrose gradient to separate the mRNA species. Heating before centrifugation appeared to be necessary in order to prevent aggregation.

The selection of a mRNA species may also be accomplished by hybridizing the mRNA to immobilized, complementary nucleic acids. This method has been applied to enrich SV40-specific mRNAs (Weinberg *et al.*, 1972; Prives *et al.*, 1974). This type of annealing permits the isolation of genes, specific mRNAs, and their precursors. This technique allowed the isolation of the genes coding for the 5 S region, the isolation of tumor virus RNA specific sequences (Prives *et al.*, 1974), and the isolation of globin mRNA (Venetianer and Leder, 1974). In the last example, a cDNA was synthesized on globin mRNA using oligo(dT)-cellulose as a primer. In order to use this procedure, it is necessary to first isolate the mRNA in another way. However, once the cellulose-bound cDNA is obtained, it can be used repeatedly for the isolation of any nucleic acid with the complementary sequence. As for the hybridization of unique sequences, relatively high temperatures are required, and during the rather long incubation times special caution is needed to prevent degradation.

Certain proteins may interact directly and specifically with mRNAs. Such interaction has been reported to fulfill a regulatory function (Stevens and Williamson, 1973*b*). If such interaction is strong enough, it can be used as an isolation procedure for the mRNA by using antibodies against the interacting protein moiety (Stevens and Williamson, 1973*a*). Although this procedure is very elegant, its application will be limited.

Assay Systems for Translation

For identification of mRNA, the translation in a heterologous system is the most applied procedure. During the past 5 years, numerous cell-free systems for the assay of heterologous messengers have been described. It would go beyond the scope of this chapter to deal with all of them. We will restrict ourselves by reviewing the systems which have been used most frequently and which show a high efficiency in their translation capacity.

Reticulocyte Lysate

Reticulocytes are immature red blood cells which are synthesizing hemoglobin almost exclusively. Enhanced reticulocyte titers can be obtained by inducing anemia, either by repeated bleeding or by injection with phenylhydrazine for 5 or 6 days. In most animals, the reticulocytes no longer contain nuclei. Nevertheless, they show a high rate of protein synthesis, indicating that they possess a very stable protein-synthesizing machinery. Crude extracts of these cells provide an excellent cell-free system (Adamson *et al.,* 1968). The extract is prepared by lysing the cells with distilled water and sedimenting the cell debris. The supernatant (the reticulocyte lysate) can be kept active when frozen at low temperature (−80°C).

Although very efficient in the translation of exogenous mRNA, the crude extract still exhibits a high level of endogenous protein synthesis which often interferes with the analysis of newly synthesized products. Sometimes direct analysis can be performed by SDS gel electrophoresis followed by autoradiography. However, it often happens that the performance of elaborate purification procedures is necessary in order to identify the product. One of the frequently applied procedures to obtain an enrichment in newly synthesized heterologous polypeptides is immunoprecipitation.

The main advantage of the reticulocyte system is its high efficiency. Incorporation continues for several hours when conducted at reduced temperature and in the presence of hemin (Adamson *et al.,* 1968). The translation of exogenous mRNA cannot be traced by following the amino acid incorporation, since often an inhibition of incorporation is observed after addition of mRNA (Lingrel *et al.,* 1971; Milstein *et al.,* 1972; Berns, 1972). As the inhibition differs depending on the mRNA preparation, contamination with double-stranded RNA (dsRNA) might well be responsible for this effect since it has been shown that dsRNA at extremely low concentrations inhibits protein synthesis in reticulocytes (Robertson and Mathews, 1973).

Reticulocytes have also been the main source for the preparation of ingredients for the purified systems as discussed below. A large variety of mRNAs have been translated in this system (see Table 1). Furthermore, the lysate appears to be able to modify proteins in a posttranslational process. The *N*-terminal methionine of the α-crystallin polypeptide chains was found acetylated (Berns *et al.,* 1972*a*) in this system just as it occurs in the homologous system (Strous *et al.,* 1974*b*).

Ascites Lysate

The ascites lysate system developed by Mathews and Korner (1972) is derived from ascites cells grown in the peritoneal cavity in mice. After lysis of the cells on a Dounce homogenizer in low ionic strength buffer, the extract is spun at 15,000*g* or 30,000*g* and the supernatant (named S 15 or S 30, respectively) is supplemented with ATP, GTP, and an energy-regenerating system and preincubated for 30–45 min at 37°C. This preincubation reduces endogenous protein synthesis considerably. The lysate is gel-filtrated on a Sephadex G25 column in order to remove low molecular weight components and the flowthrough is frozen in fractions at –80°C. This system has a low endogenous synthesis level, and addition of exogenous mRNA results in a significant stimulation of amino acid incorporation, making possible rapid detection of messenger activity. Incorporation in the ascites lysate can be stimulated significantly by the addition of mammalian initiation factors (Metafora *et al.,* 1972; Mathews *et al.,* 1972*b*). Apparently, initiation factors are limited in this system. One of the main advantages besides the stimulation of amino acid incorporation after addition of mRNA is the easy analysis of the products. An endogenous synthesis is low and the total protein concentration only 4–10% of the reticulocyte lysate, the whole incubation reaction can be analyzed directly by gel electrophoresis (which is in general the most suitable analysis technique), by chromatography, and by peptide mapping. With these techniques, the faithful translation of a variety of mRNAs has been demonstrated (see Table 1). Cell-free extracts have been prepared along the same lines from a variety of cell lines (McDowell *et al.,* 1972). Although the activities varied somewhat, the extracts behaved quite similarly as compared to the ascites cell-free system.

Wheat Germ System

The wheat germ cell-free system described by Weeks and Marcus (1971), Roberts and Paterson (1973), and Shih and Kaesberg (1973) has been favored since it was first reported. It combines several attractive features: (1) it can be prepared easily from commercially available wheat germ without the necessity to grow large amounts of cells, and (2) the system does not need a preincubation step, since protein synthesis is at background level. Furthermore, the wheat germ system has reasonably low RNAse activity, and therefore added mRNA stays intact for a considerable time. While the wheat germ system is one of the most convenient ones to assay mRNA, abortive translation seems to occur more frequently than in other cell-free systems.

Purified Cell-Free Systems

In a purified cell-free system the protein-synthesizing machinery is reconstituted by combining ribosomes or ribosomal subunits together with enzyme fractions containing tRNA, tRNA synthetases, translocation factors and release factors, and initiation factors. The most complete purified system studied hitherto has been described by Schreier and Staehelin (1973*c*) and the group at Anderson's laboratory (Shafritz *et al.,* 1973). These systems offer the opportunity of substituting for protein factors with others from different origins, and this allows the requirements for tissue- or species-specific factors in mRNA translation to be studied more precisely.

Xenopus laevis Oocytes

The use of oocytes of frogs for mRNA translation studies has been promoted by Gurdon *et al.,* (1971). Large amounts of oocytes can easily be obtained from *Xenopus laevis*. Each egg (with a volume of about 1 μl) can be supplied with 50 nl of mRNA solution by microinjection. During incubation at 20°C in a balanced medium, mRNA translation in the living oocyte continues for days.

Although only a relatively small stimulation of protein synthesis is observed after injection of mRNA, translation of mRNA is very efficient and can be detected after analysis of the product. As in the reticulocyte system, several purification steps for the identification of newly synthesized polypeptides may be necessary. One of the main advantages of this system is the physiological environment in which translation takes place. This condition enables the detection of regulatory events in the translation process, which might escape observation in the more artificial cell-free systems. Furthermore, the system is extremely sensitive even to the addition of very limited amounts of mRNA. A number of mRNAs have been translated successfully in this system (see Table 1).

Identification of Newly Synthesized Polypeptides

For identification of newly synthesized polypeptides, a number of techniques are available. Proper identification should implicate a variety of characterizations such as immunoprecipitation, column chromatography, polyacrylamide gel elctrophoresis (in the presence of either SDS or urea, at acid or alkaline pH), isoelectric focusing, and tryptic peptide identification by anionic exchange chromatography, or paper (or thin layer) electrophoresis and chromatography. It is often very useful to perform the identification with a double label—for instance, newly synthesized protein labeled with ^{14}C and the carrier protein with ^{3}H. Among the techniques mentioned above, SDS gel electrophoresis is used for almost all of the polypeptides newly synthesized *in vitro*. After electrophoresis the gel is dried down and autoradiographed with X-ray film (Berns *et al.*, 1974). The efficiency of autoradiography can be improved significantly by impregnating the gel with scintillator (dissolved in DMSO) (Bonner and Laskey, 1974).

Concluding Remarks

It is apparent that many questions related to messenger isolation, characterization, and function are still unanswered. The authors have chosen some subjects in this area which seem to them to be of general interest. Certainly a number of important problems do exist which have not been discussed at all or only superficially. Finally, in view of limitation of space and the extent of the subject, we may have omitted reference to groups which have rendered valuable contribution to this still-growing field of eukaryotic messenger investigation. Therefore, we would like to apologize for these unintentional omissions.

Literature Cited

Adamson, S. D., E. Herbert, and W. Godchaux, 1968 Factors affecting the rate of protein synthesis in lysate systems from reticulocytes. *Arch. Biochem. Biophys.* **125:**671–683.

Adesnik, M. and J. E. Darnell, 1972 Biogenesis and characterisation of histone messenger RNA in HeLa cells. *J. Mol. Biol.* **67:**397–406.

Adesnik, M., M. Salditt, W. Thomas, and J. E. Darnell, 1972 Evidence that all messenger RNA molecules (except histone messenger RNA) contain poly(A) sequences and that the poly(A) has a nuclear function. *J. Mol. Biol.* **71:**21–30.

Aviv, J. and P. Leder, 1972 Purification of biologically active globin mRNA by chromatography on oligo-thymidylic acid-cellulose. *Proc. Natl. Acad. Sci. USA* **69:**1408–1412.

Aviv, H., I. Boime, B. Loyd, and P. Leder, 1972 Translation of bacteriophage Q3 messenger RNA in a murine Krebs II ascites tumor cell-free system. *Science* **178:**1293–1295.

Aviv, H., S. Packman, D. Swan, J. Ross, and P. Leder, 1973 *In vitro* synthesis of DNA complementary to mRNA derived from a light chain producing myeloma tumor. *Nature (London) New Biol.* **241:**174–176.

Baltimore, D., 1966 Purification and properties of poliovirus double stranded ribonucleic acid. *J. Mol. Biol.* **18:**421–428.

Baltimore, D., 1971*a* Expression of animal virus genomes. *Bacteriol. Rev.* **35:**235–241.

Baltimore, D., 1971*b* Polio is not dead. *Perspect. Virol.* **7:**1–22.

Beaudet, A. L. and C. T. Caskey, 1971 Mammalian peptide chain termination II codon specificity and GTPase activity of release factor. *Proc. Natl. Acad. Sci. USA* **68:**619–624.

Bellamy, A. R. and R. L. Ralph, 1968 Recovery and purification of nucleic acids by means of cetyl-trimethyl-ammonium-bromide. *Methods Enzymol.* **12:**156–161.

Benveniste, K., J. Wilczek, and R. Stern, 1973 Translation of collagen mRNA from chick embryo calvaria in a cell-free system derived from Krebs II ascites cells. *Nature (London)* **246:**303–305.

Berns, A., 1972 Isolation of calf lens mRNA and its translation in heterologous systems. Ph.D. thesis, Department of Biochemistry, University of Nijmegen, The Netherlands.

Berns, A. and H. Bloemendal, 1974 Translation of mRNA from vertebrate eye lenses. *Methods Enzymol.* **30:**675–694.

Berns, A., R. A. de Abreu, M. van Kraaikamp, E. L. Benedetti, and H. Bloemendal, 1971 Synthesis of lens protein *in vitro.* V. Isolation of messengerlike RNA from lens by high resolution zonal centrifugation. *FEBS Lett.* **18:**159–163.

Berns, A., G. Strous, and H. Bloemendal, 1972*a* Heterologous *in vitro* synthesis of lens α crystallin polypeptide *Nature (London) New Biol.* **236:**7–9.

Berns, A., M. van Kraaikamp, H. Bloemendal, and C. D. Lane, 1972*b* Calf crystallin synthesis in frog cells. The translation of lens cell 14 S RNA in oocytes. *Proc. Natl. Acad. Sci. USA* **69:**1606–1609.

Berns, A., H. Bloemendal, S. Kaufman, and I. Verma, 1973*a* Synthesis of DNA complementary to 14 S calf lens crystallin mRNA by reverse transcriptase. *Biochem. Biophys. Res. Commun.* **52:**1013–1019.

Berns, A., V. Schreurs, M. van Kraaikamp, and H. Bloemendal, 1973*b* Synthesis of lens protein *in vitro:* Translation of calf lens messengers in heterologous systems. *Eur. J. Biochem.* **33:**551–557.

Berns, A., P. Janssen, and H. Bloemendal, 1974*a* The molecular weight of the 14 S calf lens messenger RNA. *Biochem. Biophys. Res. Commun.* **59:**1157–1164.

Berns, A., P. Janssen, and H. Bloemendal, 1974*b* The separation of α and β rabbit globin mRNA by polyacrylamide gel electrophoresis. *FEBS Lett.* **47:**343–347.

Berns, A., M. Salden, H. Bloemendal, D. Bogdanovsky, M. Raymondjean, and G. Schapira, 1975 Non specific stimulation of cell-free protein synthesis by a dialysable factor from reticulocyte initiation factor "iRNA." *Proc. Natl. Acad. Sci. USA* **72:**714–718.

Bishop, J. O., R. Pemberton, and C. Baglioni, 1972 Reiteration frequency of haemoglobin genes in the duck. *Nature (London) New Biol.* **235:**231–234.

Blobel, G., 1971 Isolation of a 5 S RNA–protein complex from mammalian ribosomes. *Proc. Natl. Acad. Sci. USA* **68:**1881–1885.

Blobel, G., 1972 Protein tightly bound to globin mRNA. *Biochem. Biophys. Res. Commun.* **47:**88–95.

Blobel, G., 1973 A protein of molecular weight 78,000 bound to be the polyadenylate region of eukaryotic messenger RNA's. *Proc. Natl. Acad. Sci. USA* **70:**924–928.

Bloemendal, H., 1972 Mammalian messenger RNA. In *Frontiers of Biology,* Vol. 27, pp. 487–514, North-Holland, Amsterdam.

Bloemendal, H., W. S. Bont, and E. L. Benedetti, 1964 Preparation of rat-liver polysomes without the utilization of detergent. *Biochim. Biophys. Acta* **87:**177–180.

Bloemendal, H., A. Berns, G. Strous, M. Mathews, and C. D. Lane, 1972 RNA viruses/ribosomes. *Proc. 8th Meeting Fed. Europ. Biochem. Soc.* **27:**237 –250.

Bloemendal, H., E. L. Benedetti, and W. S. Bont. 1974 Preparation and characterization of free- and membrane-bound polysomes. *Methods Enzymol.* **30:**313–325.

Boedtker, H., R. B. Crkvenjakov, J. A. Last, and P. Doty, 1974 The identification of collagen messenger RNA. *Proc. Natl. Acad. Sci. USA* **71:**4208–4212.

Bogdanovsky, D., W. Hermann, and G. Schapira, 1973 Presence of a new RNA species among the initiation protein factors active in eukaryotes translation. *Biochem. Biophys. Res. Commun.* **54:**25–32.

Bonner, W. M. and R. A. Laskey, 1974 A film detection method for tritium labeled proteins and nucleic acids in polyacrylamide gels. *Eur. J. Biochem.* **46:**83–88.

Brammer, W. J., 1969 The genetic approach to the study of protein biosynthesis. *Protein Synthesis* **2:**1–54.

Brawerman, G., 1973 Alterations in size of the poly(A) segment in newly synthesized messenger RNA of mouse sarcome 180 ascites cells. *Mol. Biol. Rep.* **1:**7–13.

Brawerman, G., J. Mendecki, and S. Y. Lee, 1971 A procedure for the isolation of mammalian messenger ribonucleic acid. *Biochemistry* **11:**637–641.

Breindl, M. and D. Gallwitz, 1973 Identification of histone messenger RNA from HeLa cells: Appearance of histone mRNA in the cytoplasm and its translation in a rabbit reticulocyte cell-free system. *Eur. J. Biochem.* **32:**381–391.

Breindl, M. and D. Gallwitz, 1974 On the translational control of histone synthesis. *Eur. J. Biochem.* **45:**91–97.

Britten, R. J. and D. E. Kohne, 1968 Repeated sequences in DNA. *Science* **161:**529–540.

Brownlee, G. G., E. M. Cartwright, N. J. Cowan, J. M. Jarvis, and C. Milstein, 1973 Purification and sequence of messenger RNA for immunoglobulin light chains. *Nature (London) New Biol.* **244:**236–240.

Bryan, R. N. and M. Hyashi, 1973 Two proteins are bound to most species of polysomal mRNA. *Nature (London) New Biol.* **244:**271–274.

Burgess, A. B. and B. Mach, 1971 Formation of an initiation complex with purified mammalian ribosomal subunits. *Nature (London) New Biol.* **233:**209–210.

Chantrenne, J., A. Burny, and G. Marbaix, 1967 The search for the messenger RNA of hemoglobin. *Prog. Nucleic Acid Res. Mol. Biol.* **7:**173–194.

Chen, J. H., G. C. Lavers, and A. Spector, 1973 Synthesis of complementary DNA from lens mRNA with RNA-dependent DNA polymerase. *Biochem. Biophys. Res. Commun.* **52:**767–773.

Chen, J. H., G. C. Lavers, A. Spector, G. Schutz, and G. Feigelson, 1974 Translation and reverse transcription of lens mRNA isolated by chromatography. *Exp. Eye Res.* **18:**189–199.

Comstock, J. P., G. C. Rosenfeld, B. W. O'Malley, and A. R. Means, 1972 Estrogen-induced changes in translation, and specific mRNA levels during oviduct differentiation. *Proc. Natl. Acad. Sci. USA* **69:**2377–2380.

Content, J., B. Lebleu, A. Zilberstein, H. Berissi, and M. Revel, 1974 Mechanism of the interferon-induced block of mRNA translation in mouse L cells: Reversal of the block by transfer RNA. *FEBS Lett.* **41:**125–130.

Crick, F. H. C., L. Barnett, S. Brenner, and R. J. Watts-Tobin, 1961 General nature of the genetic code for proteins. *Nature* (*London*) **192:**1227–1232.

Darnell, J. E., 1968 Ribonucleic acid from animal cells. *Bacteriol. Rev.* **32:**262–290.

Darnell, J. E., L. Philipson, R. Wall, and M. Adesnik, 1971*a* Poly A sequences: Role in conversion of nuclear RNA into mRNA. *Science* **174:**507–510.

Darnell, J. E., R. Wall, and R. J. Tushinski, 1971*b* An adenylic acid rich sequence in messenger RNA of HeLa cells and its possible relationship to reiterated sites in DNA. *Proc. Natl. Acad. Sci. USA* **68:**1321–1325.

Davis, B. D., 1971 Role of subunits in the ribosomal cycle. *Nature* (*London*) **231:**153–157.

Delovitch, T. L. and C. Baglioni, 1973 Estimation of light chain gene reiteration of mouse immunoglobin by DNA-RNA hybridization. *Proc. Natl. Acad. Sci. USA* **70:**173–178.

Delovitch, T. L., B. K. Davis, G. Holme, and A. H. Schon, 1972 Isolation of messenger-like RNA from immunochemically separated polyribosomes. *J. Mol. Biol.* **69:**373–386.

Diez, J. and G. Brawerman, 1974 Elongation of the polyadenylate segment of mRNA in cytoplasm of mammalian cells. *Proc. Natl. Acad. Sci. USA* **71:**4091–4095.

Diggelman, H., C. H. Faust, and B. March, 1973 Enzymatic synthesis of DNA complementary to purified 14 S mRNA of immunoglobulin light chain. *Proc. Natl. Acad. Sci. USA* **70:**693–696.

Edmonds, M. and M. G. Caramela, 1969 The isolation and characterization of adenosine monophosphate rich polynucleotides synthesized by Ehrlich ascites cells. *J. Biol. Chem.* **244:**1314–1324.

Edmonds, M., M. Vaughan, and N. Nakamoto, 1971 Poly A sequences in the heterogeneous nuclear RNA and rapidly labeled polyribosomal RNA of HeLa cells: Possible evidence for a precursor relationship. *Proc. Natl. Acad. Sci. USA* **68:**1338–1340.

Efron, D. and A. Marcus, 1973 Efficient translation of rabbit globin mRNA in a cell-free system from wheat embryo. *FEBS Lett.* **33:**23–27.

Falvey, A. K. and T. Staehelin, 1970 Structure and function of mammalian ribosomes. II. Exchange of ribosomal subunits at various stages of *in vitro* polypeptide synthesis. *J. Mol. Biol.* **53:**21–34.

Faust, C. H., H. Diggelman, and B. Mach, 1974 Estimation of number of genes coding for the constant part of the 19 kappa chain. *Proc. Natl. Acad. Sci. USA* **71:**2491–2495.

Favre, A., U. Bertazzoni, A. Berns, and H. Bloemendal, 1974 The poly A content and secondary structure of the 14 S calf lens mRNA. *Biochem. Biophys. Res. Commun.* **56:**273–280.

Favre, A., C. Morel, and K. Scherrer, 1975 The secondary structure and poly(A) content of globin messenger as a pure RNA and in polyribosome derived ribonucleoprotein complexes. *Eur. J. Biochem.* **57:**147–157.

Firtel, R. A., A. Jacobson, and H. F. Lodish, 1972 Isolation and hybridization kinetics of messenger RNA from *Dictyostelium discoideum*. *Nature* (*London*) *New Biol.* **239:**225–228.

Forget, B. G., C. A. Marotta, S. M. Weismann, I. M. Verma, R. P. McCaffrey, and D. Baltimore, 1974 Nucleotide sequence of human globin mRNA. *Ann. N.Y. Acad. Sci.* **241:**290–309.

Fuhr, J. E. and C. Natta, 1972 Translational control of β and α globin chain synthesis. *Nature (London) New Biol.* **240**:274–276.

Fukami, H. and K. Imahori, 1971 Control of translation by the conformation of messenger RNA. *Proc. Natl. Acad. Sci. USA* **68**:570–573.

Gabrielli, F. and C. Baglioni, 1974 Translation of histone messenger RNA by homologous cell-free systems from synchronized HeLa cells. *Eur. J. Biochem.* **42**:121–128.

Gallwitz, D. and G. C. Mueller, 1969 Histone synthesis *in vitro* on HeLa cell microsomes: The nature of the coupling to deoxyribonucleic acid synthesis. *J. Biol. Chem.* **244**:5947–5952.

Gander, E., A. G. Steward, C. M. Morel, and K. Scherrer, 1973 Isolation and characterization of ribosome-free cytoplasmic messenger-ribonucleoprotein complexes from avian erythroblasts. *Eur. J. Biochem.* **38**:443–452.

Georgiev, G. P. and O. P. Samarina, 1971 D-RNA containing ribonucleoprotein particles and messenger RNA transport. *Adv. Cell Biol.* **2**:45–110.

Gillespie, D., S. Marshall, and R. C. Gallo, 1972 RNA of RNA tumour viruses contain poly A. *Nature (London) New Biol.* **236**:227–231.

Glišin, V., R. Crkvenjakov, and C. Byus, 1974 Ribonucleic acid isolated by cesium chloride centrifugation. *Biochemistry* **13**:2633–2637.

Goldstein, E. S. and S. Penman, 1973 Regulation of protein synthesis in mammalian cells. V. Further studies on the effect of actinomycin D on translation control in the HeLa cells. *J. Mol. Biol.* **80**:243–254.

Gorski, J., M. R. Morrison, C. G. Merkel, and J. B. Lingrel, 1974 Size heterogeneity of polyadenylate sequences in mouse globin messenger RNA. *J. Mol. Biol.* **86**:363–371.

Gould, H. J. and P. H. Hamlyn, 1973 The molecular weight of rabbit globin messenger RNAs. *FEBS Lett.* **30**:301–304.

Greenberg, J. R., 1972 High stability of messenger RNA in growing cultured cells. *Nature (London)* **240**:102–104.

Gurdon, J. B., C. D. Lane, H. R. Woodland, and G. Marbaix, 1971 Use of frog eggs and oocytes for the study of messenger RNA and its translation in living cells. *Nature (London)* **233**:177–182.

Harrison, P. R., G. D. Binnie, A. Hell, S. Humphries, B. D. Young, and J. Paul, 1974 Kinetic studies of gene frequency. I. Use of a DNA copy of reticulocyte 9 S RNA to estimate globin gene dosage in mouse tissues. *J. Mol. Biol.* **84**:539–554.

Harrison, T. M., G. G. Brownlee, and C. Milstein, 1974*a* Studies on polysome–membrane interactions in mouse myeloma cells. *Eur. J. Biochem.* **47**:613–620.

Harrison, T. M., G. G. Brownlee, and C. Milstein, 1974*b* Preparation of I9 light chain mRNA from microsomes without the use of detergent. *Eur. J. Biochem.* **47**:621–627.

Henshaw, E. C., 1968 Messenger RNA in rat liver polyribosomes: Evidence that it exists as ribonucleoprotein particles. *J. Mol. Biol.* **36**:401–411.

Heywood, S. M. and M. Nwagwu, 1969 Partial characterization of presumptive myosin messenger ribonucleic acid. *Biochemistry* **8**:3839–3845.

Heywood, S. M. and W. C. Thompson, 1971 Studies on the formation of the initiation complex in eukaryotes. *Biochem. Biophys. Res. Commun.* **43**:470–475.

Heywood, S. M., D. S. Kennedy, and A. J. Bester, 1974 Separation of specific initiation factors involved in the translation of myosin and myoglobin messenger RNA's and the isolation of a new RNA involved in translation. *Proc. Natl. Acad. Sci. USA* **71**:2428–2431.

Hirsch, M. and S. Penman, 1974 Post transcriptional addition of poly A to mitochondrial RNA by a cordycepin-insensitive process. *J. Mol. Biol.* **83**:131–142.

Hirsch, C. A., M. A. Cox, W. J. W. van Venrooij, and E. C. Henshaw, 1973 The ribosomal cycle in mammalian protein synthesis. II. Association of the native smaller ribosomal subunit with protein factors. *J. Biol. Chem.* **248:**4377–4385.

Holme, G., S. L. Boyd, and A. H. Sehon, 1971 Precipitation of polyribosomes with pepsin digested antibodies. *Biochem. Biophys. Res. Commun.* **45:**240–245.

Housman, D., R. Pemberton, and R. Taber, 1971 Synthesis of α and β chains of hemoglobin in a cell-free extract of Krebs II ascites cells. *Proc Natl. Acad. Sci. USA* **68:**2716–2719.

Huez, G., G. Marbaix, E. Hubert, M. Leclercq, U. Nudel, H. Soreq, R. Salomom, B. Lebleu, M. Revel, and U. Z. Littauer, 1974 Role of polyadenylate segment in the translation of globin messenger RNA in *Xenopus* oocytes. *Proc. Natl. Acad. Sci. USA* **71:**3143–3146.

Ihle, J. N., K.-L. Lee, and F. T. Kenney, 1974 Fractionation of 34 S ribonucleic acid subunits from oncornaviruses on polyuridylate-Sepharose columns. *J. Biol. Chem.* **249:**38–42.

Ilan, J., 1973 Release factor regulating termination of complete protein in an eukaryotic organism. *J. Mol. Biol.* **77:**437–448.

Imaizumi, T., H. Diggelman, and K. Scherrer, 1973 Demonstration of globin mRNA sequences in giant nuclear precursors of mRNA of avian erythroblasts. *Proc. Natl. Acad. Sci. USA* **70:**1122–1126.

Jacob, F. and J. Monod, 1961 Genetic regulatory mechanisms in the synthesis of proteins. *J. Mol. Biol.* **3:**318–356.

Jacobs-Lorena, M. and C. Baglioni, 1972 Messenger RNA for globin in the postribosomal supernatant of rabbit reticulocytes. *Proc. Natl. Acad. Sci. USA* **69:**1425–1428.

Jacobs-Lorena, M., C. Baglioni, and T. W. Borun, 1972 Translation of messenger RNA for histones from HeLa cells by a cell-free extract from mouse ascites tumor. *Proc. Natl. Acad. Sci. USA* **69:**2095–2099.

Jacobson, A., R. A. Firtel, and H. F. Lodish, 1974 Synthesis of messenger and ribosomal RNA precursors in isolated nuclei of the cellular slime mold *Dictyostelium discoideum. J. Mol. Biol.* **82:**213–230.

Jay, G. and R. Kaempfer, 1974 Sequence of events in initiation for translation: A role for initiator transfer RNA in the recognition of messenger RNA. *Proc. Natl. Acad. Sci. USA* **71:**3199–3203.

Jelinek, W., M. Adesnik, M. Salditt, D. Sheiness, R. Wall, G. Molloy, L. Philipson, and J. E. Darnell, 1973 Further evidence on the nuclear origin and transfer to the cytoplasm of polyadenylic acid sequences in mammalian cell RNA. *J. Mol. Biol.* **75:**515–532.

Jelinek, W., G. Molloy, R. Fernandez-Munoz, M. Salditt, and J. E. Darnell, 1974 Secondary structure in heterogeneous nuclear RNA: Involvement of regions from repeated DNA sites. *J. Mol. Biol.* **82:**361–370.

Jergil, B., 1972 Protein kinase from rainbow trout-testis ribosomes: Partial purification and characterization. *Eur. J. Biochem.* **28:**546–554.

Johnston, R. E. and H. R. Bose, 1972 An adenylate-rich segment in the virion RNA of sindbis virus. *Biochem. Biophys. Res. Commun.* **46:**712–718.

Kacian, D. L., S. Spiegelman, A. Bank, M. Terada, S. Metafora, L. Dow, and P. A. Marks, 1972 *In vitro* synthesis of DNA components of human genes for globins. *Nature (London) New Biol.* **235:**167–169.

Kaempfer, R., 1969 Ribosomal subunit exchange in the cytoplasm of a eukaryote. *Nature (London)* **222:**950–953.

Kaempfer, R. and J. Kaufman, 1972 Translational control hemoglobin synthesis by an initiation factor required for recycling of ribosomes and for their binding to mRNA. *Proc. Natl. Acad. Sci. USA* **69:**3317–3322.

Kates, J., 1970 Transcription of the vaccinia virus genome and the occurrence of poly-adenylic acid sequences in messenger RNA. *Cold Spring Harbor Symp. Quant. Biol.* **35:**743–752.

Kazazian, H. H. and M. L. Freedman, 1968 The characterization of separated α and β chain polyribosomes in rabbit reticulocytes. *J. Biol. Chem.* **243:**6446–6450.

Kedes, L. H. and M. L. Birnstiel, 1971 Reiteration and clustering of DNA sequences complementary to histone messenger RNA. *Nature (London) New Biol.* **230:**165–169.

Kemp, D. J., G. A. Partington, and G. E. Rogers, 1974 Isolation and molecular weight of pure feather keratin mRNA. *Biochem. Biophys. Res. Commun.* **60:**1006–1014.

Kemper, B., J. F. Habener, R. C. Mulligan, J. T. Potts, and A. Rich, 1974 Pre-proparathyroid hormone: A direct translation product of parathyroid messenger RNA. *Proc. Natl. Acad. Sci. USA* **71:**3731–3735.

Kindås-Mügge, I., C. D. Lane, and G. Kreil, 1974 Insect protein synthesis in frog cells. Translation of honey bee promellitin messenger RNA in *Xenopus* oocytes. *J. Mol. Biol.* **87:**451–462.

Kirby, K. S., 1968 Isolation of nucleic acids with phenolic solvents. *Methods Enzymol.* **12:**87–99.

Kozak, M. and D. Nathans, 1972 Translation of the genome of a ribonucleic acid acid bacteriophage. *Bacteriol. Rev.* **36:**109–134.

Küntzel, H. and H. C. Blossey, 1974 Translation products *in vitro* of mitochondrial mRNA from *Neurospora crassa. Eur. J. Biochem.* **47:**165–171.

Kurland, C. G., 1970 Ribosome structure and function emergent. *Science* **169:**1171–1177.

Labrie, F., 1969 Isolation of an RNA with the properties of haemoglobin messenger. *Nature (London)* **221:**1217–1222.

Lane, C. D., G. Marbaix, and J. B. Gurdon, 1971 Rabbit haemoglobin synthesis in frog cells: The translation of reticulocyte 9 S RNA in frog oocytes. *J. Mol. Biol.* **61:**73–91.

Lane, C. D., C. M. Gregory, and C. Morel, 1973 Duck hemoglobin synthesis in frog cells: The translation and assay of reticulocyte 9 S RNA in oocytes of *Xenopus laevis. Eur. J. Biochem.* **34:**219–227.

Lanyon, W. G., J. Paul, and R. Williamson, 1972 Studies on the heterogeneity of mouse globin mRNA. *Eur. J. Biochem.* **31:**38–43.

Lebleu, B., G. Marbaix, G. Huez, J. Temmerman, A. Burny, and H. Chantrenne, 1971 Characterization of the messenger ribonucleoprotein released from reticulocyte polyribosomes by EDTA treatment. *Eur. J. Biochem.* **19:**264–269.

Lebleu, B., U. Nudel, E. Falcoff, C. Prives, and M. Revel, 1972 A comparison of the translation of mengo virus RNA and globin mRNA in Krebs ascites cell-free extracts. *FEBS Lett.* **25:**97–103.

Lee, S. J. and G. Brawerman, 1971 Pulse labeled ribonucleic acid complexes released by dissociation of rat liver polysomes. *Biochemistry* **10:**510–516.

Lee, S. J., J. Mendecki, and G. Brawerman, 1971 A polynucleotide segment rich in adenylic acid in the rapidly-labeled polyribosomal RNA component of mouse sarcoma 180 cells. *Proc. Natl. Acad. Sci. USA* **68:**1331–1335.

Legon, S., C. H. Darnbrough, T. Hunt, and R. J. Jackson, 1973 Initiation of eukaryotic protein synthesis. *Biochem. Soc. Trans.* **1:**553–557.

Levin, D. H., D. Kyner, and G. Aes, 1973 Protein initiation in eukaryotes formation and function of a ternary complex composed of a partially purified ribosomal factor: Methionyl transfer RNA_f and guanosine triphosphate. *Proc. Natl. Acad. Sci. USA* **70:**41–45.

Lim, L. and E. S. Canellakis, 1970 Adenine rich polymer associated with rabbit reticulocyte messenger RNA. *Nature (London)* **227:**710–712.

Lindberg, U. and T. Persson, 1972 Isolation of mRNA from KB-cells by affinity chromatography on polyuridilic acid covalently linked to Sepharose. *Eur. J. Biochem.* **31:**246–254.

Lindberg, U. and B. Sindquist, 1974 Isolation of messenger ribonucleoprotein from mammalian cells. *J. Mol. Biol.* **86:**451–468.

Lingrel, J. B. and H. R. Woodland, 1974 Initiation does not limit the rate of globin synthesis in messenger injected *Xenopus* oocytes. *Eur. J. Biochem.* **47:**47–56.

Lingrel, J. B., R. E. Lockard, R. F. Jones, H. E. Burr, and J. W. Holder, 1971 Biologically active messenger RNA for haemoglobin. *Ser. Haematol.* **4:**37–69.

Lipmann, F., 1973 Nonribosomal polypeptide synthesis on polyenzyme templates. *Acc. Chem. Res.* **6:**361–367.

Lockard, R. E. and J. B. Lingrel, 1969 The synthesis of mouse hemoglobin β chains in a rabbit reticulocyte cell-free system programmed with mouse reticulocyte 9 S RNA. *Biochem. Biophys. Res. Commun.* **37:**204–212.

Lodish, H. F. and M. Jacobson, 1972 Regulating hemoglobin synthesis. *J. Biol. Chem.* **247:**3622–3629.

Louie, A. J. and G. H. Dixon, 1972 Synthesis, acetylation and phosphorylation of histone IV and its binding to DNA during spermatogenesis in trout. *Proc. Natl. Acad. Sci. USA* **69:**1975–1979.

Mach, B., C. Faust, and P. Vassalli, 1973 Purification of 14 S messenger RNA of immunoglobulin light chain, that codes for a possible light chain precursor. *Proc. Natl. Acad. Sci. USA* **70:**451–455.

Marotta, C. A., B. G. Forget, S. M. Weissman, I. M. Verma, R. P. McCaffrey, and D. Baltimore, 1974 Nucleotide sequences of human globin messenger RNA. *Proc. Natl. Acad. Sci. USA* **71:**2300–2304.

Mathews, M. B. and A. Korner, 1972 Mammalian cell-free protein synthesis directed by viral ribonucleic acid. *Eur. J. Biochem.* **17:**328–338.

Mathews, M. B., M. Osborn, and J. B. Lingrel, 1971 Translation of globin messenger RNA in a heterologous cell-free system. *Nature (London) New Biol.* **233:**206–209.

Mathews, M. B., M. Osborn, A. Berns, and H. Bloemendal, 1972*a* Translation of two messenger RNAs from lens in a cell-free system from Krebs II ascites cells. *Nature (London) New Biol.* **236:**5–7.

Mathews, M. B., I. B. Pragnell, M. Osborn, and H. R. V. Arnstein, 1972*b* Stimulation by reticulocyte initiation factors of protein synthesis in a cell-free system from Krebs II ascites cells. *Biochim. Biophys. Acta* **287:**113–123.

McDowell, M. J., W. K. Koflik, L. Villa-Komaroff, and H. F. Lodish, 1972 Translation of reovirus mRNA's synthesized *in vitro* into reovirus polypeptides by several mammalian cell-free extracts. *Proc. Natl. Acad. Sci. USA* **69:**2649–2653.

McLaughlin, C. S., J. R. Warner, J. Edmonds, H. Nakazato, and M. Vaughan, 1973 Polyadenylic acid sequences in yeast messenger ribonucleic acid. *J. Biol. Chem.* **248:**1466–1471.

Mendecki, J., S. Y. Lee, and G. Brawerman, 1972 Characteristics of the poly A segment associated with messenger RNA in mouse sarcome 180 cells. *Biochemistry* **11**:792–798.

Metafora, S., M. Terada, L. W. Dow, P. A. Marks, and A. Banks, 1972 Increased efficiency of exogenous mRNA translation in a Krebs ascites cell lysate. *Proc. Natl. Acad. Sci. USA* **69**:1299–1303.

Milcarek, C., and S. Penman, 1974 Membrane bound polyribosomes in HeLa cells: Association of polyadenylic acid with membranes. *J. Mol. Biol.* **89**:327–338.

Milcarek, C., R. Price, and S. Penman, 1975 The metabolism of a poly(A) minus mRNA fraction in HeLa cells. *Cell* **3**:1–10.

Milstein, C., G. G. Brownlee, T. M. Harrison, and M. B. Mathews, 1972 A possible precursor of immunoglobulin light chains. *Nature* (*London*) *New Biol.* **239**:117–120.

Moar, V. A., J. B. Gurdon, C. D. Lane, and G. Marbaix, 1971 Translational capacity of living frog eggs and oocytes as judged by messenger RNA injection. *J. Mol. Biol.* **61**:93–103.

Molloy, G. R., W. Jelinek, M. Salditt, and J. E. Darnell, 1974 Arrangement of specific oligo nucleotides within poly A terminated Hn RNA molecules. *Cell* **1**:43–54.

Morel, C., E. S. Gander, M. Herzberg, J. Dubochet, and K. Scherrer, 1973 The duck-globin messenger–ribonucleoprotein complex. *Eur. J. Biochem.* **36**:455–464.

Morrison, M. R., J. Gorski, and J. B. Lingrel, 1972 The separation of mouse reticulocyte 9 S RNA into fractions of different biological activity by hybridization to poly(U) cellulose. *Biochem. Biophys. Res. Commun.* **49**:775–781.

Morrison, T. G. and H. F. Lodish, 1973 Translation of bacteriophage Qβ by cytoplasmic extracts of mammalian cells. *Proc. Natl. Acad. Sci. USA* **70**:315–319.

Morrison, T. and H. Lodish, 1974 Recognition of protein synthesis initiation signals on bacteriophage ribonucleic acid by mammalian ribosomes. *J. Biol. Chem.* **249**:5860–5866.

Murphy, W. and G. Attardi, 1973 Stability of cytoplasmic messenger RNA in HeLa cells. *Proc. Natl. Acad. Sci. USA* **70**:115–119.

Nakajima, K., H. Ishitsuka, and K. Oda, 1974 An SV40 induced initiation factor for protein synthesis concerned with the regulation of permissiveness. *Nature* (*London*) **252**:649–653.

Nakazato, H. and M. Edmonds, 1972 The isolation and purification of rapidly labeled polysome bound ribonucleic acid on polythymidylate cellulose. *J. Biol. Chem.* **247**:3365–3367.

Nakazato, H., M. Edmonds, and D. W. Kopp, 1974 Differential metabolism of large and small poly(A) sequences in the heterogeneous nuclear RNA of HeLa cells. *Proc. Natl. Acad. Sci. USA* **71**:200–204.

Nemer, M., M. Graham, and L. M. Dubroff, 1974 Co-existence of non-histone messenger RNA species lacking and containing polyadenylic acid in sea urchin embryos. *J. Mol. Biol.* **89**:435–454.

Neyfakh, A. A., 1971 Steps in realization of genetic information in early development. *Curr. Top. Dev. Biol.* **6**:45–77.

Noll, H., 1969 Polysomes: Analysis of their structure and function. *Tech. Protein Synthesis* **2**:101–179.

Nudel, U., B. Lebleu, T. Sehavi-Willner, and M. Revel, 1973*a* Messenger ribonucleoprotein and initiation factos in rabbit-reticulocyte polyribosomes. *Eur. J. Biochem.* **33**:314–322.

Nudel, U., B. Lebleu, and M. Revel, 1973*b* Discrimination between messenger ribonucleic acids by a mammalian translation initiation factor. *Proc. Natl. Acad. Sci. USA* **70:**2139–2144.

Ojala, D. and G. Attardi, 1974 Expression of mitochondrial genomes in HeLa cells. XIX. Occurrence in mitochondria of polyadenylic sequences, "free" and covalently linked to mitochondrial DNA-coded RNA. *J. Mol. Biol.* **82:**151–174.

O'Malley, B. W., G. C. Rosenfeld, J. P. Comstock, and A. R. Means, 1972 Steroid hormone induction of a specific translatable messenger RNA. *Nature (London) New Biol.* **240:**45–48.

Ouderaa, F. J., W. W. De Jong, and H. Bloemendal, 1973 The amino acid sequence of the αA_2 chain of bovine α crystallin. *Eur. J. Biochem.* **39:**207–222.

Packman, S., H. Aviv, J. Ross, and P. Leder, 1972 A comparison of globin genes in duck reticulocytes and liver cells. *Biochem. Biophys. Res. Commun.* **49:**813–819.

Palacios, R., D. Sullivan, N. M. Summers, M. L. Kiely, and R. T. Schimke, 1972 Purification of ovalbumin messenger RNA by specific immuno adsorption of ovalbumin-synthesizing polysomes and Millipore partition of ribonucleic acid. *J. Biol. Chem.* **248:**540–548.

Palade, G. E., 1958 Microsomes and ribonucleoprotein particles. In *Microsomal Particles and Protein Synthesis,* pp. 36–49, Pergamon Press, New York.

Palmiter, R. D., 1973 Ovalbumin messenger ribonucleic acid translation. *J. Biol. Chem.* **248:**2095–2106.

Partington, G. A., D. J. Kemp, and G. E. Rogers, 1973 Isolation of feather keratin mRNA and its translation in a rabbit reticulocyte cell-free system. *Nature (London) New Biol.* **246:**33–36.

Pederson, T., 1974 Proteins associated with heterogeneous nuclear RNA in eukaryotic cells. *J. Mol. Biol.* **83:**163–183.

Pemberton, R. E., D. Housman, H. F. Lodish, and C. Baglioni, 1972 Isolation of duck haemoglobin messenger RNA and its translation by rabbit reticulocyte cell-free system. *Nature (London)* **235:**99–102.

Penman, S., 1966 RNA metabolism in the HeLa cell nucleus. *J. Mol. Biol.* **17:**117–130.

Penman, S., K. Scherrer, Y. Becker, and J. E. Darnell, 1963 Polyribosomes in normal and poliovirus-infected HeLa cells and their relationship to messenger RNA. *Proc. Natl. Acad. Sci. USA* **49:**654–662.

Penman, S., C. Vesco, and M. Penman, 1968 Location and kinetics of formation of nuclear heterodisperse RNA, cytoplasmic disperse RNA and polysomal-associated messenger RNA in HeLa cells. *J. Mol. Biol.* **34:**49–69.

Perry, R. P. and D. E. Kelley, 1968 Messenger-RNA–protein complexes and newly synthesized ribosomal subunits: Analysis of free particles and components of polyribosomes. *J. Mol. Biol.* **35:**37–59.

Perry, R. P. and D. E. Kelley, 1974 Existence of methylated mRNA in mouse L cells. *Cell* **1:**37–42.

Perry, R. P., J. La Torre, D. E. Kelley, and J. R. Greenberg, 1972 On the lability of poly(A) sequences during extraction of messenger RNA. *Biochim. Biophys. Acta* **262:**220–226.

Perry, R. P., J. R. Greenberg, D. E. Kelley, J. La Torre, and G. Schochetman, 1973 Messenger RNA, its origin and fate in mammalian cells. In *Gene Expression and Its Regulation* (F. T. Kenney *et al.,* eds.), Chap. 12, pp. 149–168, Plenum Press, New York.

Piperno, G., U. Bertazzoni, A. Berns, and H. Bloemendal, 1974 Calf lens crystallin

RNA's contain polynucleotide sequences rich in adenylic acid. *Nucl. Acid Res.* **1**:245–255.

Poon, R., G. V. Paddock, H. Heindell, P. Whitcome, W. Salser, D. Kacian, A. Bank, R. Gambino, and F. Ramirez, 1974 Nucleotide sequence analysis of RNA synthesized from rabbit globin complementary DNA. *Proc. Natl. Acad. Sci. USA* **71**:3502–3596.

Prensky, W., D. M. Steffensen, and W. L. Hughes, 1973 The use of iodinated RNA for gene location. *Proc. Natl. Acad. Sci. USA* **70**:1860–1864.

Prichard, P. M., D. J. Picciano, D. G. Laycock, and W. F. Anderson, 1971 Translation of exogenous messenger RNA for hemoglobin on reticulocyte and liver ribosomes. *Proc. Natl. Acad. Sci. USA* **68**:2752–2756.

Prives, C. L., H. Aviv, B. M. Paterson, B. E. Roberts, S. Rozenblatt, M. Revel, and E. Winicour, 1974 Cell-free translation of messenger RNA of simian virus 40: Synthesis of the major capsid protein. *Proc. Natl. Acad. Sci. USA* **71**:302–306.

Reynders, L., P. Sloof, J. Sival, and P. Borst, 1973 Gel electrophoresis of RNA under denaturing conditions. *Biochim. Biophys. Acta* **324**:320–333.

Rhoads, R. E., G. S. McKnight, and R. T. Schimke, 1971 Synthesis of ovalbumin in a rabbit reticulocyte cell-free system programmed with hen oviduct ribonucleic acid. *J. Biol. Chem.* **246**:7407–7410.

Rhoads, R. E., G. S. McKnight, and R. T. Schimke, 1973 Quantitative measurement of ovalbumin messenger ribonucleic acid activity. *J. Biol. Chem.* **248**:2031–2039.

Roberts, B. E. and B. M. Paterson, 1973 Efficient translation of tobacco mosaic virus RNA and rabbit globin 9 S RNA in a cell-free system from commercial wheat germ. *Proc. Natl. Acad. Sci. USA* **70**:2330–2334.

Robertson, H. D. and M. B. Mathews, 1973 Double stranded RNA as an inhibitor of protein synthesis and as a substrate for a nuclease in extract of a Krebs II ascites cells. *Proc. Natl. Acad. Sci. USA* **70**:225–229.

Robertson, H. D., E. Dickson, P. Model, and W. Prensky, 1973 Application of fingerprinting techniques to iodinated nucleic acids. *Proc. Natl. Acad. Sci. USA* **70**:3260–3264.

Rosbash, M. and S. Penman, 1971 Membrane associated protein synthesis of mammalian cells. *J. Mol. Biol.* **59**:227–241.

Ross, J., H. Aviv, E. Scolnick, and P. Leder, 1972 *In vitro* synthesis of DNA complementary to purified rabbit globin messenger RNA. *Proc. Natl. Acad. Sci. USA* **69**:264–268.

Rourke, A. W. and S. M. Heywood, 1972 Myosin synthesis and specificity of eukaryotic initiation factors. *Biochemistry* **11**:2061–2066.

Ryskov, A. P. and G. P. Georgiev, 1970 Polyphosphate groups at the 5′ ends of nuclear dRNA fractions. *FEBS Lett.* **8**:186–188.

Sabatini, D. D., N. Borgese, M. Adelman, G. Kreibiek, and G. Blobel, 1972 Studies on the membrane-associated protein synthesis apparatus of eukaryotic cells. In *FEBS Symposium,* Vol. 24, pp. 147–171, Academic Press, New York.

Salas, M., M. Smith, W. Stanley, A. Wahba, and S. Ochoa, 1965 Direction of reading of the genetic message. *J. Biol. Chem.* **240**:3988–3995.

Salden, M. and H. Bloemendal, 1976 Polyamines can replace the dialyzable component from crude reticulocyte initiation factors. *Biochem. Biophys. Res. Comm.* In press.

Sampson, J., M. B. Mathews, M. Osborn, and A. F. Borghetti, 1972 Hemoglobin messenger ribonucleic acid: Translation in a cell-free system from rat and mouse liver and Lauschutz ascites cells. *Biochemistry* **11**:3636–3640.

Schapira, G., J. G. Dreyfus, and N. Maleknia, 1968 The ambiguities in the rabbit hemoglobin: Evidence for a messenger RNA translated specifically into hemoglobin. *Biochem. Biophys. Res. Commun.* **32:**558–561.

Schechter, J., 1973 Biologically and chemically pure mRNA coding for mouse 19 light chain prepared with the aid of antibodies and immobilized oligo thymidine. *Proc. Natl. Acad. Sci. USA* **70:**2256–2260.

Scherberg, N. H. and S. Refetoff, 1973 Hybridization of RNA labeled with ^{125}I to high specific activity. *Nature (London) New Biol.* **242:**142–145.

Scherrer, K. and L. Marcaud, 1968 Messenger RNA in avian erythroblasts at the transcriptional and translational levels and the problem of regulation in animal cells. *J. Cell. Physiol.* **72:**181–212.

Scherrer, K., H. Latham, and J. E. Darnell, 1963 Demonstration of an unstable RNA and of a precursor to ribosomal RNA in HeLa cells. *Proc. Natl. Acad. Sci. USA* **49:**240–248.

Scherrer, K., G. Spohr, N. Granboulan, C. Morel, J. Grosclaude, and E. Chezzi, 1970 Nuclear and cytoplasmic messenger like RNA and their relation to the active messenger RNA in polyribosomes of HeLa cells. *Cold Spring Harbor Symp. Quant. Biol.* **35:**539–554.

Schreier, M. H. and T. Staehelin, 1973*a* Translation of rabbit hemoglobin messenger RNA *in vitro* with purified and partially purified components from brain or liver of different species. *Proc. Natl. Acad. Sci. USA* **70:**462–465.

Schreier, M. H. and T. Staehelin, 1973*b* Translation of duck globin messenger RNA in partially purified mammalian cell-free system. *Eur. J. Biochem.* **34:**213–218.

Schreier, M. H. and T. Staehelin, 1973*c* Initiation of mammalian protein synthesis: The importance of ribosome and initiation factor quality for the efficiency of *in vitro* systems. *J. Mol. Biol.* **73:**329–349.

Schreier, M. H. and T. Staehelin, 1973*d* Initiation of eukaryotic protein synthesis (Met-$tRNA_f \cdot 40$ S ribosome) initiation complex catalysed by purified initiation factors in the absence of mRNA. *Nature (London) New Biol.* **242:**35–38.

Schreier, M. H., T. Staehelin, F. R. Gesteland, and P. F. Spahr, 1973 Translation of bacteriophage R17 and QβRNA in a mammalian cell-free system. *J. Mol. Biol.* **75:**575–578.

Schutz, G., M. Beato, and P. Feigelson, 1972 Isolation of eukaryotic messenger RNA in cellulose and its translation *in vitro*. *Biochem. Biophys. Res. Commun.* **49:**680–689.

Shafritz, D. A., 1974 Protein synthesis with messenger RNA fractions from membrane-bound and free liver polysomes. *J. Biol. Chem.* **249:**81–88.

Shafritz, D. A., J. W. Drysdale, and K. J. Isselbacher, 1973 Translation of liver messenger RNA in a messenger RNA-dependent reticulocyte cell-free system. *J. Biol. Chem.* **248:**3220–3227.

Shapiro, D. J., J. M. Taylor, G. S. McKnight, R. Palacios, C. Gonzalez, M. L. Kiely, and R. T. Schimke, 1974 Isolation of hen oviduct ovalbumin and rat liver albumin polysomes by indirect immunoprecipitation. *J. Biol. Chem.* **249:**3665–3671.

Shatkin, A. J., 1974 Methylated messenger RNA synthesis *in vitro* by purified reovirus. *Proc. Natl. Acad. Sci. USA* **71:**3204–3207.

Sheines, D. and J. E. Darnell, 1973 Polyadenylic acid segment in mRNA becomes shorter with age. *Nature (London) New Biol.* **241:**265–269.

Sheldon, R., J. Kates, D. E. Kelley, and R. P. Perry, 1972*a* Polyadenylic acid sequences on 3′ termini of vaccinia messenger ribonucleic acid and mammalian nuclear and messenger ribonucleic acid. *Biochemistry* **11:**3829–3834.

Sheldon, R., C. Jurale, and J. Kates, 1972*b* Detection of polyadenylic acid sequences in viral and eukaryotic RNA. *Proc. Natl. Acad. Sci. USA* **69:**417–421.

Shih, P. S. and P. Kaesberg, 1973 Translation of Brome mosaic viral ribonucleic acid in a cell-free system derived from wheat embryo. *Proc. Natl. Acad. Sci. USA* **70:**1799–1803.

Sierra, J. M., D. Meier, and S. Ochoa, 1974 Effect of development on the translation of messenger RNA in *Artemia salina* embryos. *Proc. Natl. Acad. Sci. USA* **71:**2693–2697.

Singer, M. F. and P. Leder, 1966 Messenger RNA: An evaluation. *Ann. Rev. Biochem.* **35:**195–230.

Singer, R. H. and S. Penman, 1972 Stability of HeLa cell mRNA in actinomycin. *Nature (London)* **240:**100–102.

Singer, R. and S. Penman, 1973 Messenger RNA in HeLa cells kinetics of formation and decay. *J. Mol. Biol.* **78:**321–334.

Sippel, A. E., J. G. Stavrianopoulos, G. Schutz, and P. Feigelson, 1974 Translational properties of rabbit globin messenger RNA after specific removal of poly A with ribonuclease H. *Proc. Natl. Acad. Sci. USA* **71:**4635–4639.

Skoultchi, A. and R. R. Gross, 1973 Maternal messenger RNA: Detection of molecular hybridization. *Proc. Natl. Acad. Sci. USA* **70:**2840–2844.

Slater, D. W. and S. Spiegelman, 1968 Template capabilities and size distribution of echinoid RNA during early development. *Biochim. Biophys. Acta* **166:**82–93.

Slater, D. W., I. Slater, and D. Gillespie, 1972 Post-fertilization synthesis of polyadenylic acid in sea urchin embryos. *Nature (London)* **240:**333–337.

Slater, I., D. Gillespie, and D. W. Slater, 1973 Cytoplasmic adenylation and processing of maternal RNA. *Proc. Natl. Acad. Sci. USA* **70:**406–411.

Soreq, H., U. Nudel, R. Salomon, M. Revel, and U. Z. Littauer, 1974 *In vitro* translation of polyadenylic acid-free rabbit globin messenger RNA. *J. Mol. Biol.* **88:**233–245.

Soria, M. and A. S. Huang, 1973 Association of polyadenylic acid with messenger RNA of vesicular stomatitis virus. *J. Mol. Biol.* **77:**449–455.

Spohr, G., N. Granboulan, C. Morel, and K. Scherrer, 1970 Messenger RNA in HeLa cells: An investigation of free and polyribosome bound cytoplasmic messenger ribonucleoprotein particles by kinetic labeling and electron microscope. *Eur. J. Biochem.* **17:**296–318.

Stavnezer, J. and R. C. C. Huang, 1971 Synthesis of a mouse immunoglobulin light chain in a rabbit reticulocyte cell-free system. *Nature (London) New Biol.* **230:**172–176.

Stavnezer, J., R. C. C. Huang, E. Stavnezer, and J. M. Bishop, 1974 Isolation of mRNA for immunoglobulin kappa chain and enumeration of the genes for the constant region of kappa chain in the mouse. *J. Mol. Biol.* **88:**43–63.

Steitz, J. A., 1973 Specific recognition of non-initiator regions in RNA bacteriophage messengers by ribosomes of *Bacillus stearothermophilus*. *J. Mol. Biol.* **73:**1–16.

Stevens, R. H., 1974 Translational control of antibody synthesis: Control of light chain production. *Eur. J. Biochem.* **42:**553–559.

Stevens, R. H. and A. R. Williamson, 1972 Specific I_9G mRNA molecules from myeloma cells in heterogeneous nuclear and cytoplasmic RNA containing poly A. *Nature (London)* **239:**143–146.

Stevens, R. H. and A. R. Williamson, 1973*a* Isolation of mRNA coding for mouse heavy chain immunoglobulin. *Proc. Natl. Acad. Sci. USA* **70:**1127–1131.

Stevens, R. H. and A. R. Williamson, 1973*b* Translational control of immunoglobulin synthesis. II. Cell-free interaction of myeloma immunoglobulin with messenger RNA. *J. Mol. Biol.* **78:**517–525.

Stevens, R. H. and A. R. Williamson, 1973*c* Isolation of nuclear pre-mRNA which codes for immunoglobulin heavy chain. *Nature (London) New Biol.* **245:**101–104.

Strous, G., H. van Westreenen, J. van der Logt, and H. Bloemendal, 1974*a* Synthesis of lens protein *in vitro*: The lens cell-free system. *Biochim. Biophys. Acta* **353:**89–98.

Strous, G., A. Berns, and H. Bloemendal, 1974*b* *N*-Terminal acetylation of the nascent chains of α crystallin. *Biochem. Biophys. Res. Commun.* **58:**876–884.

Sullivan, D., R. Palacios, J. Stavnezer, J. Taylor, A. J. Faras, M. L. Kiely, N. M. Summers, J. M. Bishop, and R. T. Schimke, 1973 Synthesis of a deoxyribonucleic acid and quantification of ovalbumin genes. *J. Biol. Chem.* **248:**7530–7539.

Sussman, M., 1970 Model for quantitative and qualitative control of mRNA translation in eukaryotes. *Nature (London)* **225:**1245–1246.

Suzuki, Y. and D. D. Brown, 1972 Isolation and identification of the messenger RNA for silk fibroin from *Bombyx mori. J. Mol. Biol.* **63:**409–429.

Swan, D., H. Aviv, and P. Leder, 1972 Purification and properties of biologically active mRNA for a myeloma light chain. *Proc. Natl. Acad. Sci. USA* **69:**1967–1971.

Szekely, M. and F. Sanger, 1969 Use of polynucleotide kinase in fingerprinting non-radioactive nucleic acids. *J. Mol. Biol.* **43:**607–617.

Taylor, J. M. and R. T. Schimke, 1973 Synthesis of rat liver albumin in a rabbit reticulocyte cell-free protein synthesizing system. *J. Biol. Chem.* **248:**7661–7668.

Temple, G. F. and D. E. Housman, 1972 Separation and translation of the mRNA's coding for α and β chains of rabbit globin. *Proc. Natl. Acad. Sci. USA* **69:**1574–1577.

Terman, S. A., 1970 Relative effect of transcription level and translation level control of protein synthesis during early development of the sea urchin. *Proc. Natl. Acad. Sci. USA* **65:**985–992.

Thompson, W. C., E. A. Buzash, and S. M. Heywood, 1973 Translation of myoglobin messenger ribonucleic acid. *Biochemistry* **12:**4559–4565.

Tyler, A. and B. S. Tyler, 1972 Informational molecules and differentiation. In *Cell Differentiation,* edited by O. A. Schjeide and J. de Vellis, pp. 42–118, Van Nostrand Reinhold, New York.

Uenoyama, K. and T. Ono, 1972 Nascent catalase and its messenger RNA on rat liver polyribosomes. *J. Mol. Biol.* **65:**75–89.

Venetianer, P. and P. Leder, 1974 Enzymatic synthesis of solid phase bound DNA sequences corresponding to specific mammalian genes. *Proc. Natl. Acad. Sci. USA* **71:**3892–3895.

Verma, D. P. S., D. T. Nash, and H. M. Schulman, 1974 Isolation and *in vitro* translation of soybean leg-hemoglobin mRNA. *Nature (London)* **251:**74–77.

Verma, I. M. and D. Baltimore, 1973 Purification of the RNA-directed DNA polymerase from avian myeloblastosis virus and its assay with polynucleotide templates. *Methods Enzymol.* **29:**125–130.

Verma, I. M., G. F. Temple, H. Fan, and D. Baltimore, 1972 *In vitro* synthesis of DNA complementary to rabbit reticulocyte 10 S RNA. *Nature (London) New Biol.* **235:**163–167.

Verma, I., R. A. Firtel, H. Lodish, and D. Baltimore, 1974 Synthesis of DNA complementary to cellular slime mold mRNA by reverse transcriptase. *Biochemistry* **13:**3917–3922.

Vogt, V. M. and R. Eisenman, 1973 Identification of a large precursor polypeptide of avian oncornavirus proteins. *Proc. Natl. Acad. Sci. USA* **70:**1734–1738.

Weeks, D. P. and A. Marcus, 1971 Preformed messenger of quiescent wheat embryos. *Biochim. Biophys. Acta* **232:**671–684.

Weeks, D. P. and R. Baxter, 1972 Specific inhibition of peptide chain initiation by MDMP. *Biochemistry* **11:**3060–3064.

Weinberg, R. A., S. O. Warnaar, and E. Winicour, 1972 Isolation and characterization of simian virus 40 ribonucleic acid. *J. Virol.* **10:**193–201.

Wiegers, U. and H. Hilz, 1972 Rapid isolation of undegraded polysomal RNA without phenol. *FEBS Lett.* **23:**77–82.

Wigle, D. T. and A. E. Smith, 1973 Specificity in initiation of protein synthesis in a fractionated mammalian cell-free system. *Nature (London) New Biol.* **242:**136–140.

Williamson, R., 1972 mRNA in animal cells. *J. Med. Genet.* **9:**348–355.

Williamson, R., 1973 The protein moieties of animal messenger ribonucleoproteins. *FEBS Lett.* **37:**1–6.

Williamson, R., G. Lanyon, and J. Paul, 1969 Preferential degradation of "messenger RNA" in reticulocytes by ribonuclease treatment and sonication of polysomes. *Nature (London)* **223:**628–630.

Williamson, R., M. Morrison, and J. Paul, 1970 DNA-RNA hybridization of 9 S messenger RNA for mouse globin. *Biochem. Biophys. Res. Commun.* **40:**740–745.

Williamson, R., M. Morrison, W. G. Lanyon, R. Eason, and J. Paul, 1971 Properties of mouse globin messenger ribonucleic acid and its preparation in milligram quantities. *Biochemistry* **10:**3014–3021.

Williamson, R., R. Clayton, and D. E. S. Truman, 1972 Isolation and identification of chick lens crystallin messenger RNA. *Biophys. Biochem. Res. Commun.* **46:**1936–1943.

Williamson, R., J. Crossley, and S. Humphries, 1974 Translation of mouse globin mRNA from which the poly(adenylic acid) sequence has been removed. *Biochemistry* **13:**703–707.

Wilt, F. H., 1973 Polyadenylation of maternal RNA of sea urchin eggs after fertilization. *Proc. Natl. Acat. Sci. USA* **70:**2345–2349.

Winstead, J. A. and F. Wold, 1964 Studies on rabbit muscle enolase: Chemical evidence for two polypeptide chains in the active enzyme. *Biochemistry* **3:**791–795.

Winterburn, P. J. and C. H. Phelps, 1972 The significance of glycosylated proteins. *Nature (London)* **236:**147–151.

Woo, S. L. C., S. E. Harris, J. M. Rosen, L. Chan, P. J. Sperry, A. R. Means, and B. W. O'Malley, 1974 Use of Sepharose 4B for preparative scale fractionation of eukaryotic messenger RNA's. *Prep. Biochem.* **4:**555–572.

Yogo, Y. and E. Wimmer, 1972 Polyadenylic acid at the 3′ terminus of poliovirus RNA. *Proc. Natl. Acad. Sci. USA* **69:**1877–1882.

Zauderer, M., P. Liberti, and C. Baglioni, 1973 Distribution of histone messenger RNA among free and membrane associated polyribosomes of a mouse myeloma cell line. *J. Mol. Biol.* **79:**577–586.

Zelenka, P. and J. Piatigorski, 1974 Isolation and *in vitro* translation of δ crystallin mRNA from embryonic chick lens fibers. *Proc. Natl. Acad. Sci. USA* **71:**1896–1900.

10

Transfer RNAs

JAMES T. MADISON

This is an attempt to summarize what is known about the nucleotide sequences of transfer ribonucleic acids (tRNAs). Cloverleaf arrangements of the 34 tRNAs whose nucleotide sequences have been determined are shown in Figure 1. Also shown is the tyrosine tRNA precursor. Recent reviews on tRNA include those by Philipps (1969), Zachau (1969), Cramer (1971), Jacobson (1971), Chambers (1971), Gauss *et al.* (1971), Mehler and Chakraburthy (1971), Staehelin (1971), and Venkstern (1973).

In Figure 1, amino acid acceptor activities are shown by the first three letters of the amino acids except for glutamine (Gln), isoleucine (Ile), and tryptophan (Trp). Nucleotides are indicated by capital letters: A, adenylic acid; C, cytidylic acid; G, guanylic acid; U, uridylic acid; D, 5,6,-dihydrouridylic acid; T, ribothymidylic acid; ψ, pseudouridylic acid (5-ribosyluridine-5′-phosphate); I, inosinic acid; X, unknown nucleotide; Y, fluorescent nucleotide whose structure has been determined by Nakanishi *et al.* (1970); Y_w, fluorescent nucleotide found in wheat $tRNA^{Phe}$. Substituents on the nucleotides are shown in lowercase letters (with the site of attachment indicated by the superscript): m, methyl group; m^2G, N^2-methyl G; m^2_2, N^2,N^2-dimethyl G; s, thio; mam, methylaminomethyl; t^6A, *N* [9-(β-D-ribofuranosyl)purin-6-ylcarbamoyl] -L-threonine-5′-phosphate; ms, thiomethyl; i, Δ^2-isopentenyl; ac, acetyl; oac, oxyacetyl; mcm, methylcarboxymethyl; $^+$, unknown modification of a major nucleotide; m

JAMES T. MADISON—U.S. Plant, Soil, and Nutrition Laboratory, U.S. Department of Agriculture, Ithaca, New York.

(without a superscript), 2′-*O*-methyl. A solid line represents a Watson–Crick base pair; ●, a G-U or G-ψ base pair (the "wobble" base pair, Crick, 1966); ····, where alternative structures have been proposed and one of the structures is unable to form a base pair. A p represents the 5′-terminal phosphate group; OH, the 3′-terminal nucleoside. Nucleotides inside parentheses are alternative structures, where (1) two or more related sequences have been found as yeast (*Saccharomyces cerevisiae*) Arg III, *Escherichia coli* Gln I and II, *Staphylococcus epidermidis* Gly IA and IB, bacteriophage T_4 Leu, *E. coli* Met F, rat liver Ser I, II, and III, yeast Ser I and II, *E. coli* Tyr precursor, *E. coli* Val IIA and IIB; or (2) alternative structures have been proposed by different workers, as in the case of yeast Ala, yeast Leu III, *E. coli* Tyr II and yeast Val; or (3) tRNA mutants have been found as *E. coli* Gly (ins), Trp (su^-), and Tyr I (su^+_{III} and su^-_{III}); or finally (4) not all of the nucleotides at a particular location are modified as with yeast Ala, *E. coli* Ile, yeast Lys II, *E. coli* Met, and yeast Trp. For example, about one-half of the yeast $tRNA^{Ala}$ molecules have U in the extra loop while the rest of the molecules have D at this location. (su) are suppressor tRNAs, su^+ are suppressor-positive and su^- are suppressor-negative forms of the tRNA. (ins) is a mutant form of $tRNA^{Gly\ III}$ that contains U in the anticodon. The Roman numerals indicate different tRNAs that accept the same amino acid. Some very closely related forms are further subdivided into forms A and B, as in the case of $tRNA^{Gly\ IA}$ and $tRNA^{Gly\ IB}$ and $tRNA^{Val\ IIA}$ and $tRNA^{Val\ IIB}$ from *E. coli*. $tRNA^{Met_F}$ is capable of accepting a formate group on the amino of methionine, whereas Met_M will not accept formate. In *E. coli* $tRNA^{Tyr\ I}$ the C in the anticodon $^+$(C) converts the molecule to the amber suppressor form (su^+). Those tRNAs that contain the nucleotides in the parentheses with a minus sign –(A) are not amber suppressors even though they have the anticodon C-U-A which should allow it to suppress the amber triplet U-A-G. See text for further discussion. The nucleotides underlined in the *E. coli* $tRNA^{Tyr}_{I}$ were found only in double mutants. Structures have also been determined for tRNAs from *Salmonella typhimurium*, *Torulopsis utilis* (also known as *Candida utilis*), and wheat germ (*Triticum vulgare*).

The various parts of the tRNA molecule will be referred to (starting at the top of the cloverleaf and going clockwise) as the acceptor stem, the T-ψ limb, the extra loop, the anticodon limb, and the dihydrouridine limb; the base-repaired regions will be called "stems" and the single-stranded regions "loops." The "anticodon" is the three nucleotides in the middle of the anticodon loop, e.g., IGC in $tRNA^{Ala}$. In the $tRNA^{Tyr}$ precursor sequence, most of the tRNA molecule has been omitted. The $tRNA^{Tyr}$ is produced when the precursor is "trimmed" at the dashed line. The small circles indicate the rest of the acceptor stem.

There are still questions about several of the sequences shown in Figure 1. Merril (1968) has suggested there should be an additional G in the extra loop of yeast $tRNA^{Ala}$, making the sequence AGGUC. It is likely that Merril is right, but the large pancreatic ribonuclease fragment should be analyzed through three cycles of periodate degradation to pin down the sequence of this oligonucleotide.

Two similar sequences have been reported for the "denaturable" $tRNA^{Leu}$ ($tRNA^{Leu}_{III}$) from yeast. Kowalski *et al.* (1971) found DC in the dihydrouridine loop, an unmodified C in the anticodon, and ψ in the first position of the anticodon stem. In these three positions Chang *et al.* (1971) found CD, m^5C, and C. Although it is possible that alternative forms of $tRNA^{Leu}_{III}$ have been analyzed by the two groups, it is likely that the same tRNA has been studied and that the discrepancies will be resolved by further work.

Quite different structures were originally proposed for *E. coli* $tRNA^{Tyr}$ by Goodman *et al.* (1968) and Doctor *et al.* (1969). But Doctor (private communication) has concluded that the sequence reported by Goodman is more nearly correct. The only remaining discrepancy is the identity of the nucleotide in the mG-G-C(X) sequence, where Doctor *et al.* found some radioactivity in ^{35}S-labeled RNA.

Two sequences have been reported for yeast $tRNA^{Val}$. Bayev *et al.* (1967) originally found the sequence $ACDm^5C$ for the extra loop and the fragment GGGGC at the junction of the T-ψ and acceptor stems. Bonnet *et al.* (1971) found ACm^7GDCm^5C and GGGC at these positions. It is very likely that the correct sequences are GGGC and ACm^7GD, but the location of the m^5C remains in doubt. Both positions for m^5C are known in other tRNAs. The very similar $tRNA^{Val}_{I}$ from *T. utilis* has the sequence m^5CC, while $tRNA^{Phe}$ from yeast has the sequence Cm^5C. A different sequence than that shown has also been proposed for parts of the *E. coli* $tRNA^{Phe}$ (Uziel and Gassen, 1969*a,b*).

The most striking thing about the tRNA nucleotide sequences that have been determined is the fact that they can all be arranged into a cloverleaf. The only variations are the number of base pairs in the dihydrouridine stem (three or four), the number of non-base-paired nucleotides in the dihydrouridine loop (seven to 12), and the size of the extra loop (three to 16 nucleotides). Six tRNAs (yeast Ala, *E. coli* Leu I, Met F and Met M, *Staphylococcus* Gly IB, and wheat Phe) have nucleotides that are unable to form base pairs in parts of the cloverleaf that are normally base paired. *Salmonella* $tRNA^{His}$ (Singer and Smith, 1972; Harada *et al.*, 1972) has eight base pairs in the stem and so has only three (rather than four) unpaired nucleotides at the 3′-terminus. Finally, *Staphylococcus* $tRNA^{Gly}_{I}$ (Roberts, 1972), which is not active in protein synthesis but

(Text continues on p. 318.)

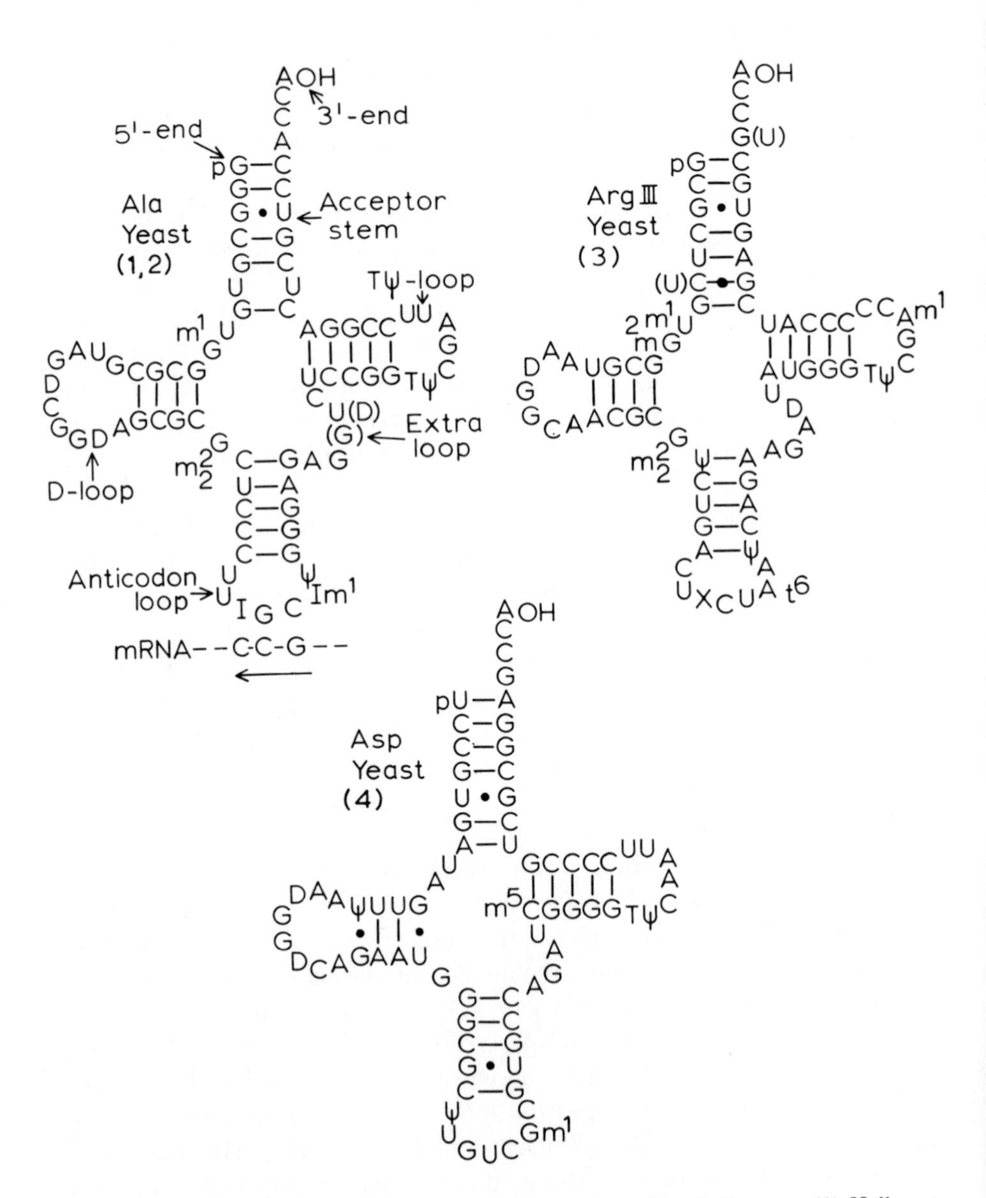

Figure 1. Cloverleaf arrangements of transfer ribonucleic acids. References: (1) Holley et al. (1965); (2) Merril (1968); (3) Kuntzel et al. (1972); (4) Gangloff et al. (1971);

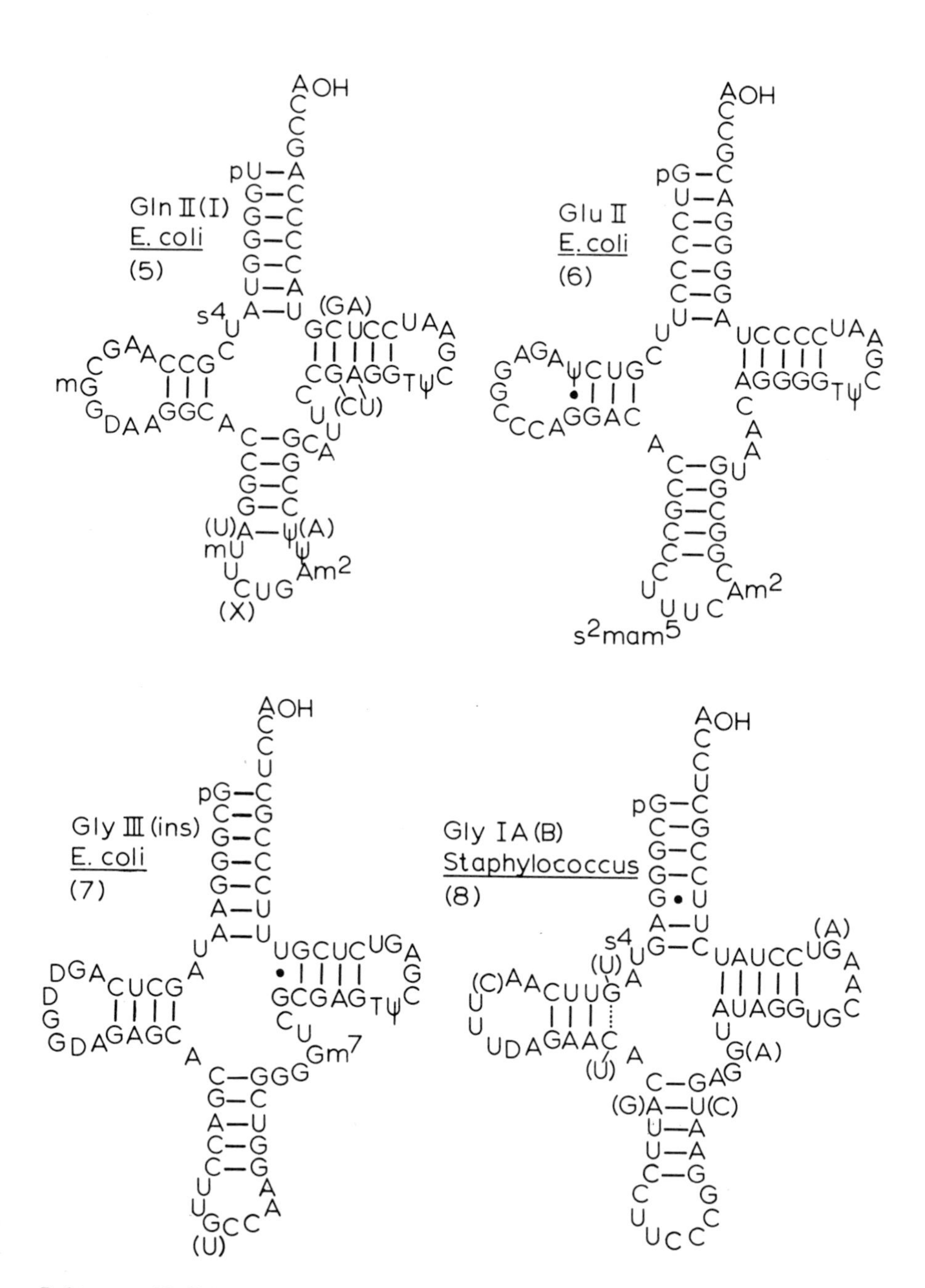

References: (5) Folk and Yaniv (1972); (6) Ohashi et al. (1972); (7) Squires and Carbon (1971); (8) Roberts (1972);

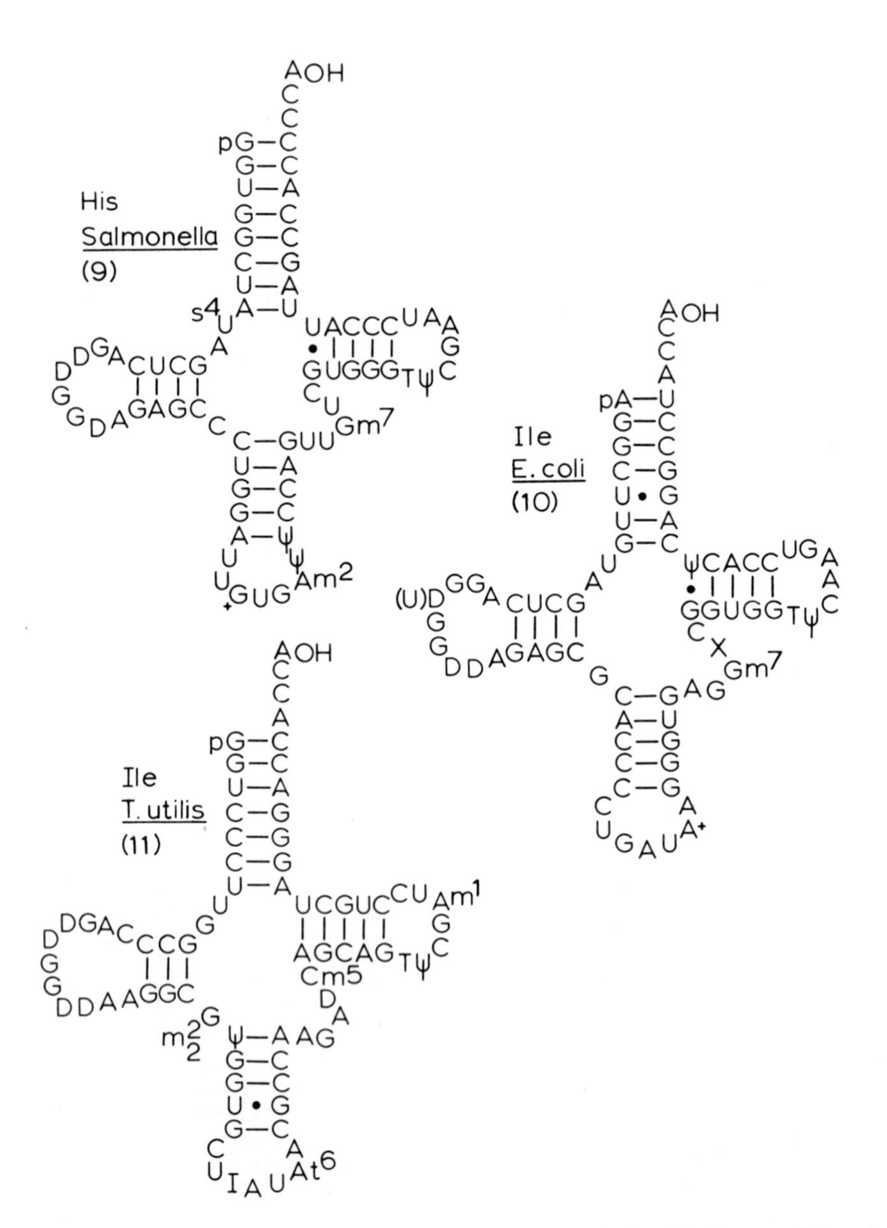

References: (9) Singer and Smith (1972) and Harada et al. (1972); (10) Yarus and Barrell (1971); (11) Takemura et al. (1969a,b);

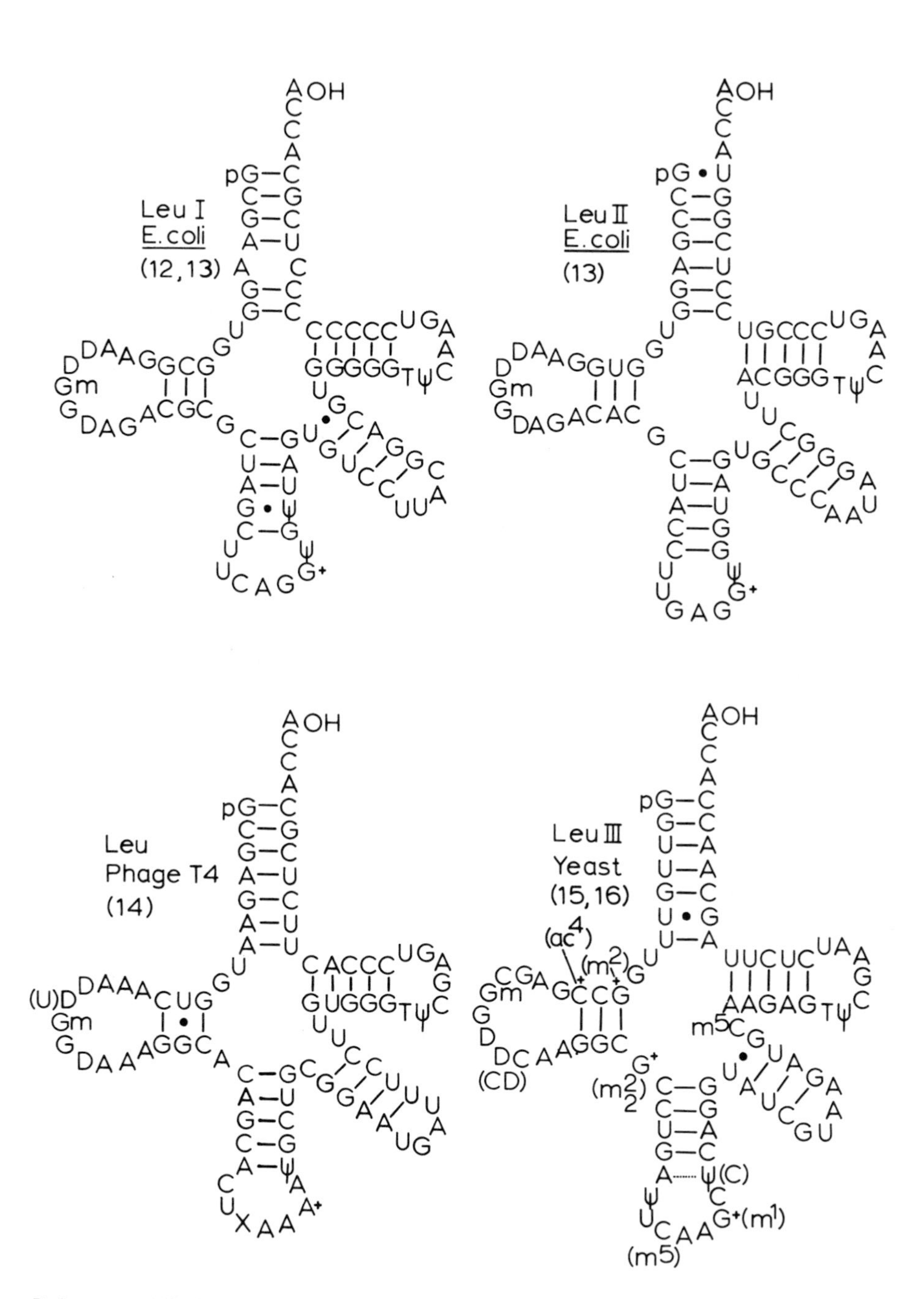

References: (12) Dube et al. (1970); (13) Blank and Söll (1971); (14) Pinkerton et al. (1972); (15) Kowalski et al. (1971); (16) Chang et al. (1971);

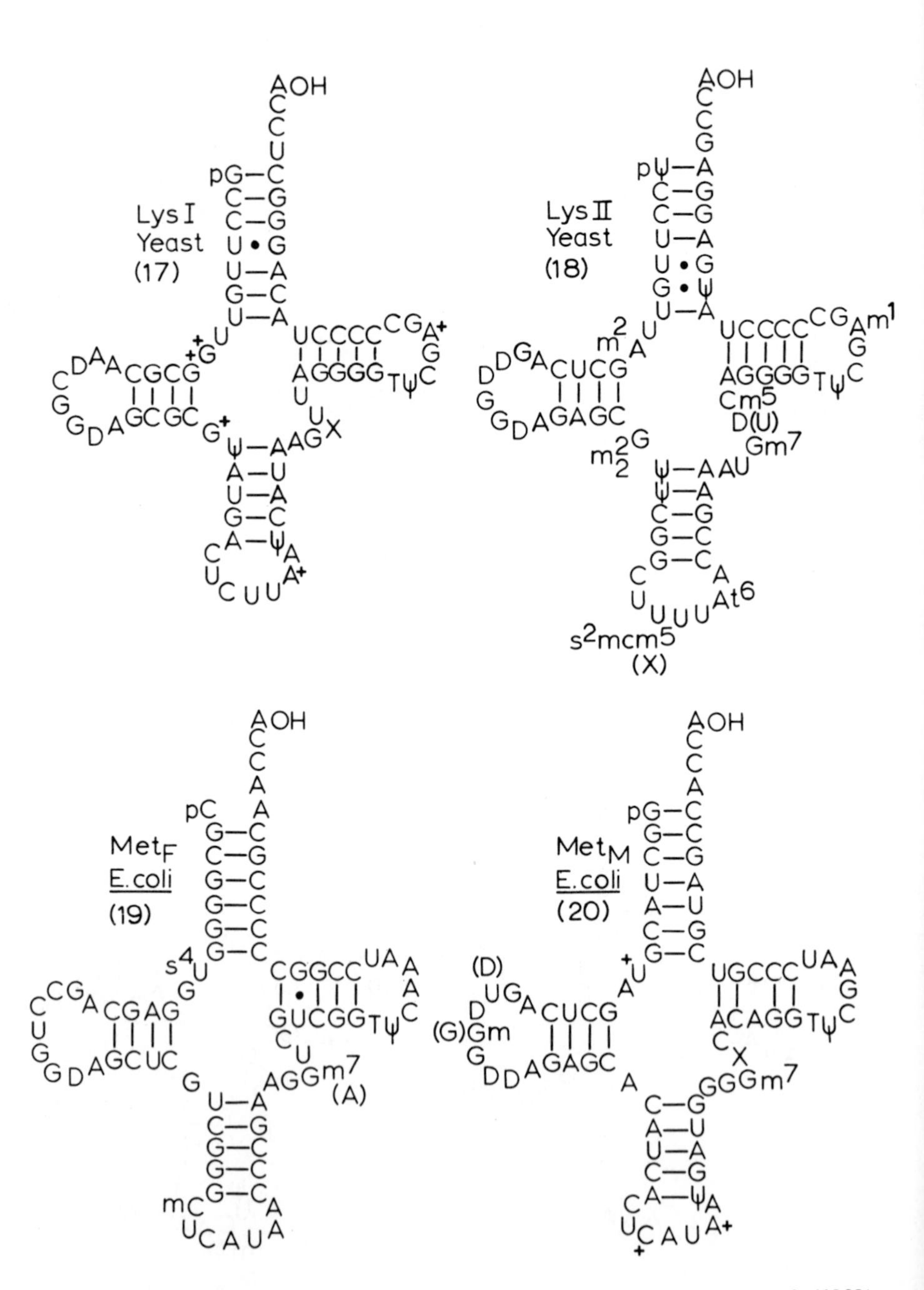

References: (17) Smith et al. (1971); (18) Madison et al. (1972); (19) Dube et al. (1968); (20) Cory et al. (1969) and Cory and Marcker (1970);

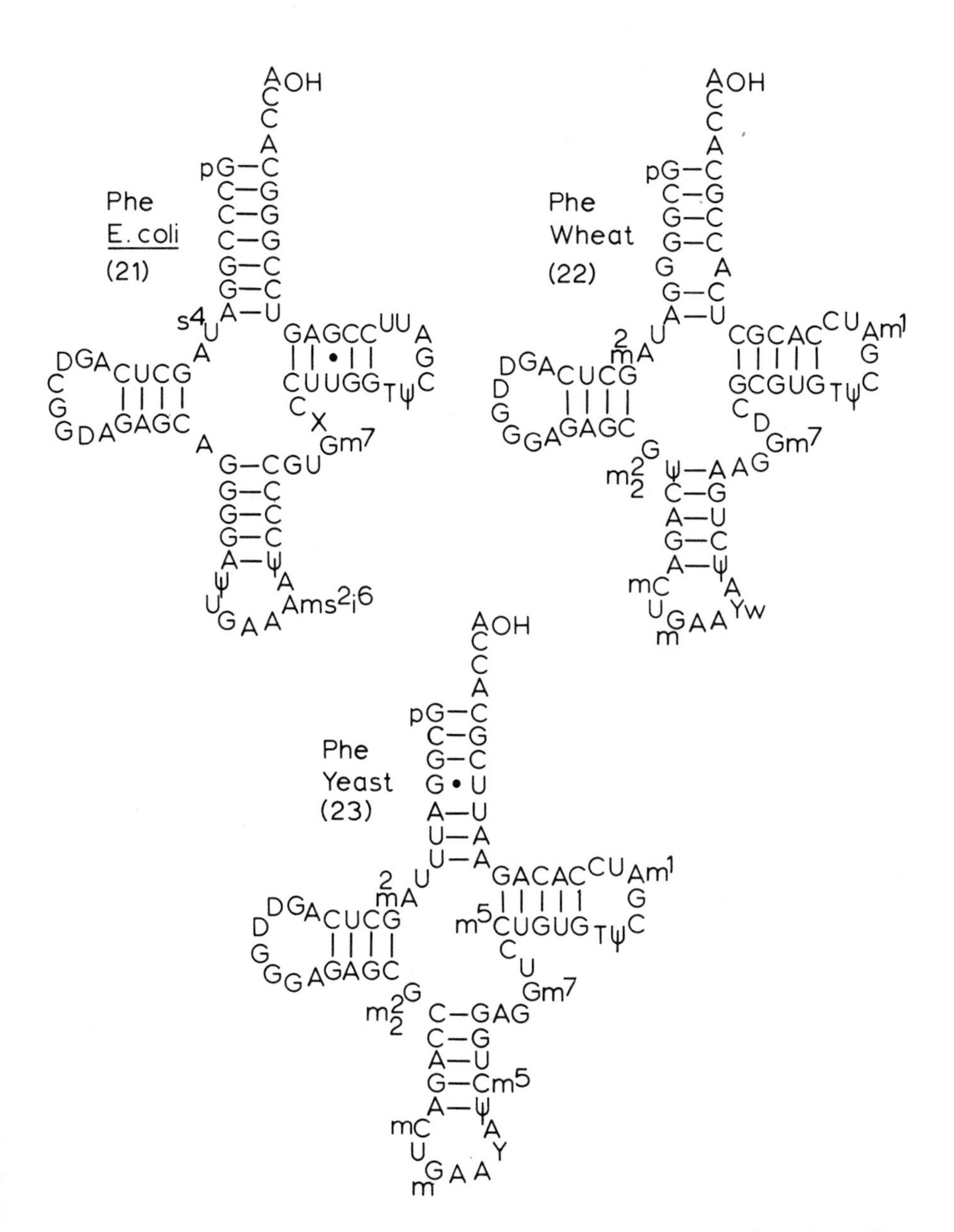

References: (21) Barrell and Sanger (1969); (22) Dudock et al. (1969) and Dudock and Katz (1969); (23) RajBhandary et al. (1967) and RajBhandary and Chang (1968);

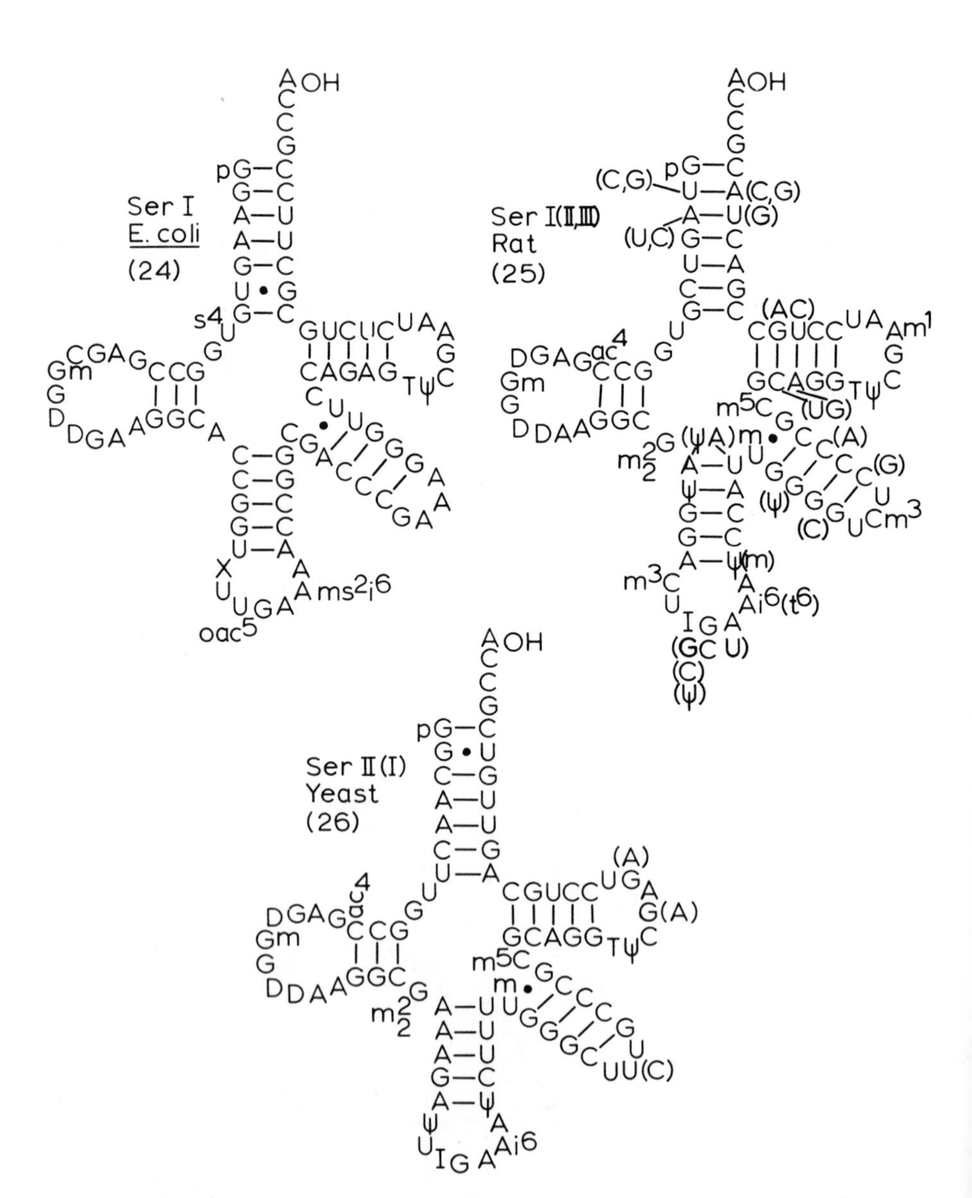

References: (24) Ishikura et al. (1971); (25) Staehelin et al. (1968), Ginsberg et al. (1971), and Staehelin (1971); (26) Zachau et al. (1966a,b);

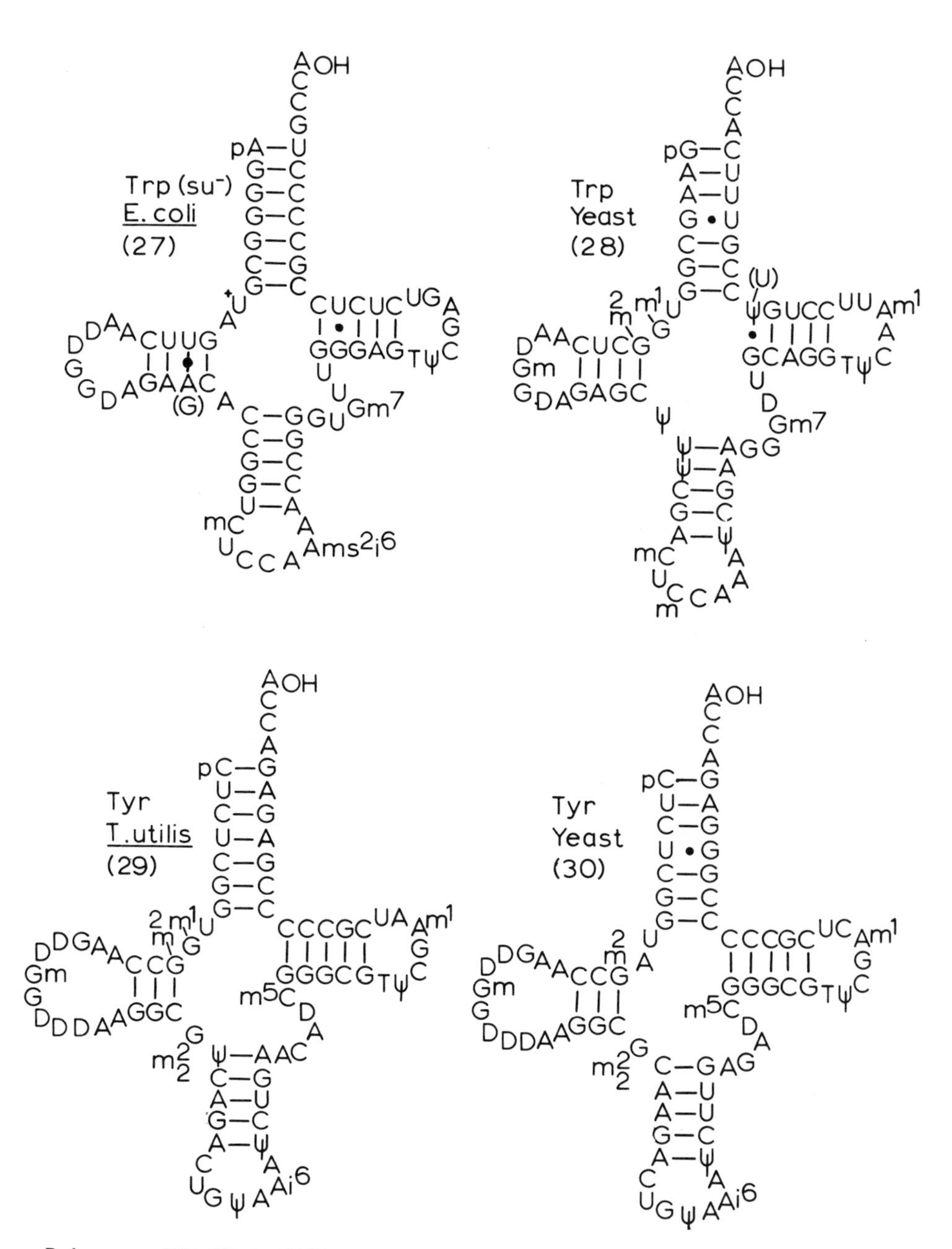

References: (27) Hirsh (1970, 1971); (28) Keith et al. (1971); (29) Hashimoto et al. (1969); (30) Madison et al. (1966) and Madison and Kung (1967);

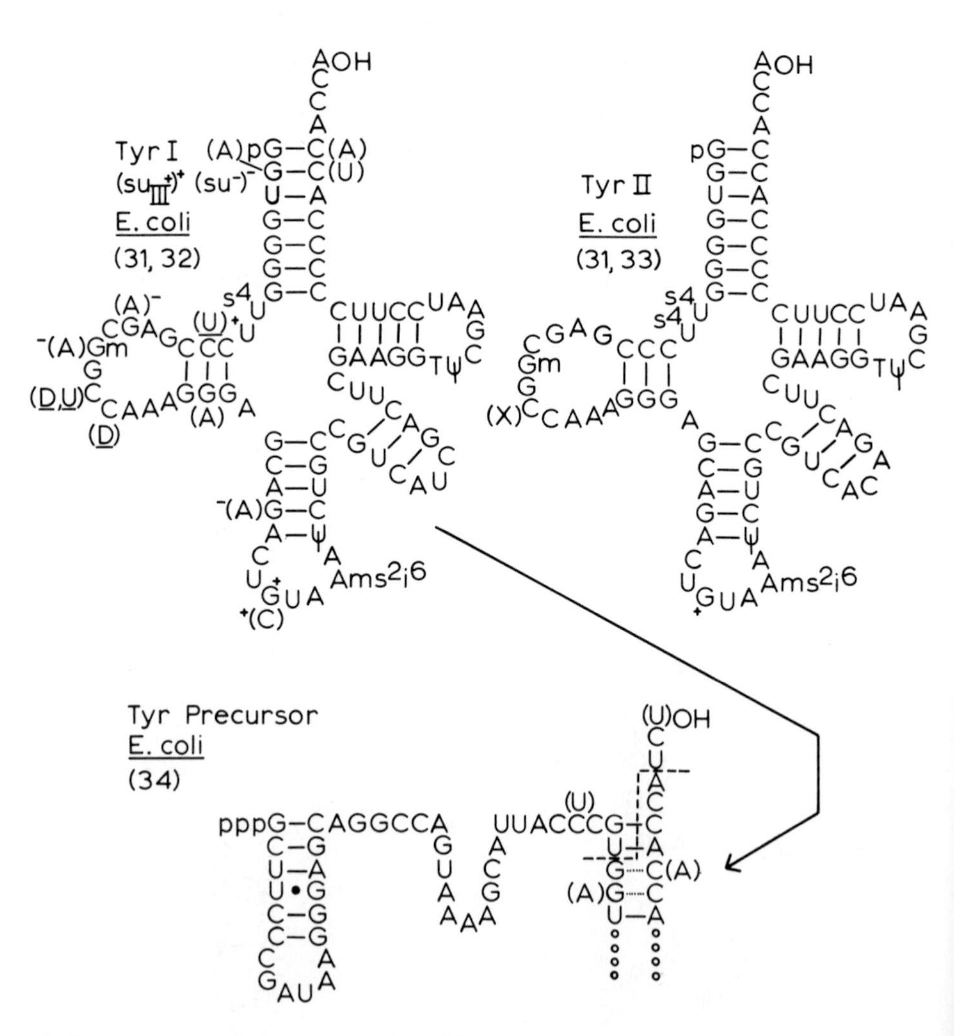

References: (31) Goodman et al. (1968, 1970); (32) Abelson et al. (1969, 1970) and Smith et al. (1970); (33) Doctor et al. (1969); (34) Altman and Smith (1971);

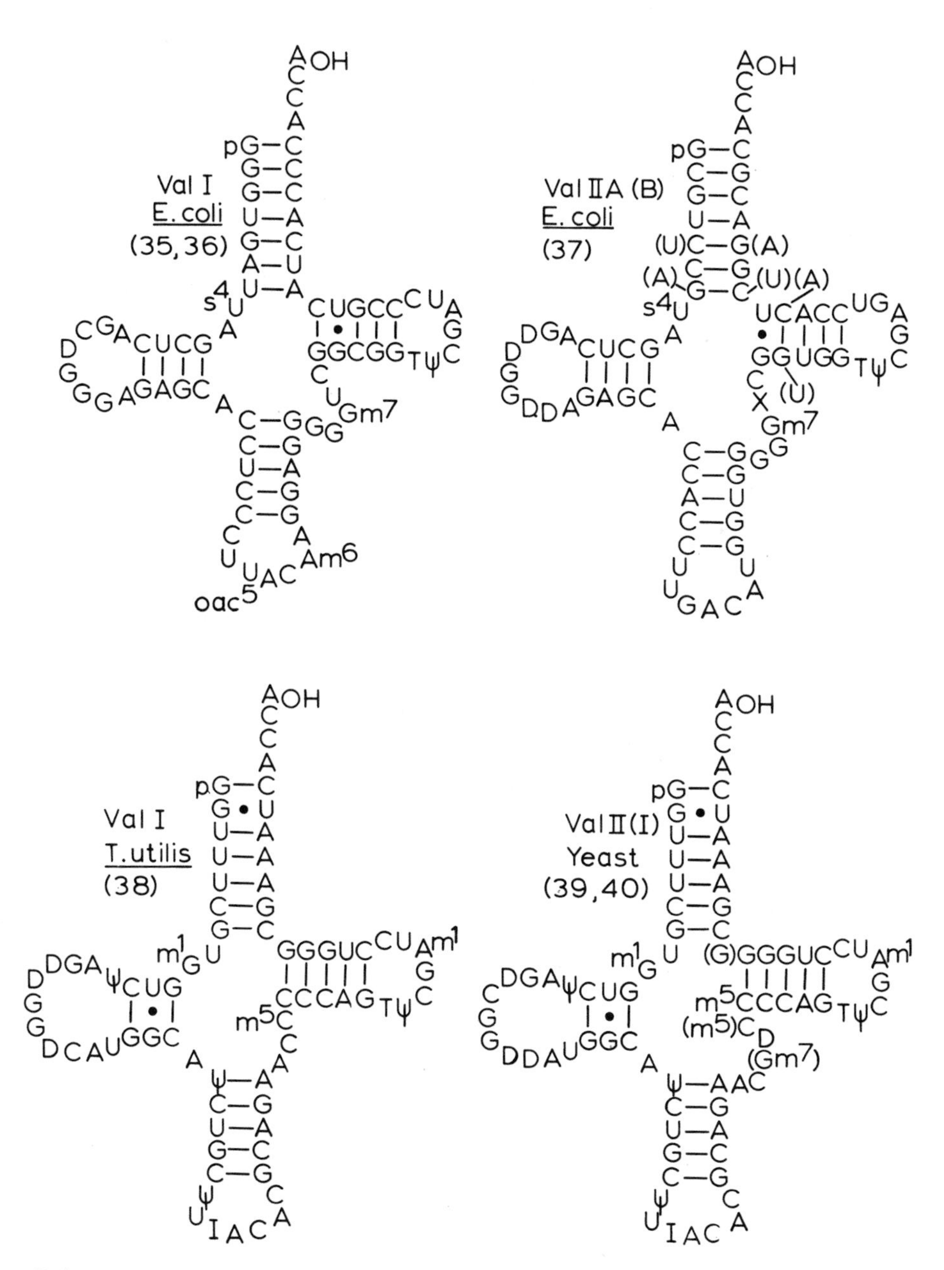

References: (35) Yaniv and Barrell (1969); (36) Harada et al. (1969) and Kimura et al. (1971); (37) Yaniv and Barrell (1971); (38) Takemura et al. (1968) and Mizutani et al. (1968); (39) Bayev et al. (1967); and (40) Bonnet et al. (1971).

which is active in peptidoglycan synthesis, differs from all the other tRNAs in not having the sequences G-G in the dihydrouridine loop and T-ψ in the opposite loop. Even this tRNA still conforms to the cloverleaf arrangement. It is clear, however, that the cloverleaf is folded into a three-dimensional structure (Kim *et al.,* 1973).

In almost every case, the anticodons that have been found explain the coding properties of the tRNA if the "wobble" base pairs are allowed (Crick, 1966). The second and third nucleotides of the anticodon form normal A-U and G-C base pairs, but the first position of the anticodon is allowed to "wobble." So that I in the first position of the anticodon pairs with A, C, and U; U (or ψ) in this position can pair with A and G; G can interact with C and U; while C can form a base pair only with G. Adenosine has not been found in the wobble position; very likely it is converted to I whenever it is in this location. The 2-thiouridine derivatives apparently interact only with A, so that those tRNAs with a derivative of s^2U in the first position of the anticodon—*E. coli* $tRNA_{II}^{Gly}$ (Ohashi *et al.,* 1972), *E. coli* $tRNA_{I}^{Gln}$ (Folk and Yaniv, 1972), and yeast $tRNA_{II}^{Lys}$ (Madison *et al.,* 1972)—should code only for triplets ending with A. The *E. coli* amber suppressor tRNA specific for Tyr (su_{III}^{+}) is explained by the conversion of the anticodon from G^+-U-A to C-U-A, which can base pair with U-A-G, the amber codon (Goodman *et al.,* 1968).

The only cases where the anticodons do not explain the coding properties of tRNAs involved are in the case of *E. coli* $tRNA^{Trp}$ (Hirsh, 1970) where the wild-type tRNA and the U-G-A suppressing tRNA both have the anticodon sequence C-C-A, which should code only with U-G-G. The suppressor strain has an A:U pair in the dihydrouridine stem, while the wild-type tRNA has a G:U base pair in the same location. Why an A:U so far away allows the C in the wobble position to more efficiently code with A is unknown. Yeast $tRNA_{II}^{Lys}$ has a derivative of 2-thiouridine in the wobble position (Madison *et al.,* 1972) so it should code only with A-A-A, but Woodward and Herbert (1971) found that it reacted with both A-A-A and A-A-G. This nucleotide has not been completely identified, so knowledge of its structure might help clarify its coding properties.

Nucleotides in the wobble position are often modified, but the modifications (except for I and s^2U) apparently do not influence the coding properties very much. An exception is the modified G(Q) found in several *E. coli* tRNAs, which binds better with trinucleotides ending with U than it does for those ending in C in the ribosome binding assay (Harada and Nishimura, 1972).

The part of the tRNA that is the aminoacyl-tRNA ligase recognition site has not been identified. It is entirely possible that more than one part of the molecule is involved. It is hard to distinguish between binding sites and specificity sites. Two or more binding sites, neither of which is unique to any one tRNA, could constitute a recognition site. It is entirely possible that different ligases recognize different parts of the different tRNA molecules. Since most isoaccepting tRNAs have the same, or very nearly the same, number of nucleotides, it is probable that overall size and shape are important in the recognition process. While other parts of the molecule are undoubtedly important, several groups (Schulman and Chambers, 1968; Imura *et al.*, 1969; Beltchev and Grunberg-Manago, 1970*a,b*; Mirzabekov *et al.*, 1971; Roe *et al.*, 1972; Schulman, 1972) have evidence that nucleotides in the amino acid acceptor stem are important in the recognition process. Dudock and coworkers (Dudock *et al.*, 1970, 1971; Roe *et al.*, 1971) have found that tRNAs with the sequence $\frac{\text{CUCGA}}{\text{GAGC}}$ in the dihydrouridine stem, m^7G in the extra loop, and A in the fourth position from the 3′-terminus are all charged with phenylalanine by the Phe-tRNA ligase. For a discussion of this problem, see the review by Chambers (1971).

The $tRNA^{Tyr}$ precursor and its mutants that have been sequenced by Altman and Smith (1971) are interesting in many ways. Since many mutants with nucleotide changes in tRNA have decreased amounts of the altered tRNA, it seems likely that the enzymes that "trim" the precursor to form tRNA are very sensitive to changes in the structure of the precursor. However, several mutants of *E. coli* with altered tRNA—$tRNA^{Gly}_{III}$ (Squires and Carbon, 1971), $tRNA^{Trp}$ (Hirsh, 1970, 1971), and $tRNA^{Tyr}_{I}$ (Goodman *et al.*, 1968, 1970; Abelson *et al.*, 1969, 1970; Smith *et al.*, 1970)—have been identified and the sequences of the mutant tRNA determined.

Compilations of tRNA sequences have been assembled by Holmquist *et al.* (1973) and by Barrell and Clark (1974). The structure of yeast $tRNA^{Phe}$ as determined by X-ray diffraction has been reported by Robertus *et al.* (1974) and by Kim *et al.* (1974*a*). The two groups have described slightly different tertiary structures. The general structure of transfer RNA molecules has been discussed by Kim *et al.* (1974*b*).

Singer *et al.* (1972) found that the $tRNA^{His}$ from a mutant of *Salmonella* that was depressed for the enzymes of the histidine pathway did not contain the two pseudouridine residues near the anticodon. This finding strongly suggests that at least these two modified nucleosides are involved in the regulation of protein synthesis.

Literature Cited

Abelson, J., L. Barnett, S. Brenner, M. Gefter, A. Landy, R. Russell, and J. D. Smith, 1969 Mutant tyrosine transfer ribonucleic acids. *FEBS Lett.* **3**:1–4.

Abelson, J. N., M. L. Gefter, L. Barnett, A. Landy, R. L. Russell, and J. D. Smith, 1970 Mutant tyrosine transfer ribonucleic acids. *J. Mol. Biol.* **47**:15–28.

Altman, S. and J. D. Smith, 1971 Tyrosine tRNA precursor molecule: Polynucleotide sequence. *Nature (London) New Biol.* **233**:35–39.

Barrell, B. G. and B. F. C. Clark, 1974 *Handbook of Nucleic Acid Sequences,* Johnson-Bruvvers, Oxford.

Barrell, B. G. and F. Sanger, 1969 The sequence of phenylalanine tRNA from *E. coli. FEBS Lett.* **3**:275–278.

Bayev, A. A., T. V. Venkstern, A. D. Mirzabekov, A. I. Krutilina, L. Li, and V. D. Axelrod, 1967 The primary structure of valine tRNA 1 of baker's yeast. *Mol. Biol. SSSR* **1**:754–766.

Beltchev, B. and M. Grunberg-Manago, 1970*a* Preparation of a pG-fragment from $tRNA^{Phe}_{yeast}$ by chemical scission at the dihydrouracil, and inhibition of $tRNA^{Phe}_{yeast}$ charging by this fragment when combined with the -CCA half of this tRNA. *FEBS Lett.* **12**:24–26.

Beltchev, B. and M. Grunberg-Manago, 1970*b* Competitive inhibition of the acceptor activity of $tRNA^{Tyr\ II}_{E.\ coli}$ by a combination of oligo G and a CCA terminated nineteen residue oligonucleotide of $tRNA^{Tyr\ II}_{E.\ coli}$. *FEBS Lett.* **12**:27–29.

Blank, H. U. and D. Söll, 1971 The nucleotide sequence of two leucine tRNA species from *Escherichia coli* K12. *Biochem. Biophys. Res. Commun.* **43**:1192–1197.

Bonnet, J., J. P. Ebel, and G. Dirheimer, 1971 Primary structure of $tRNA^{Val}_{2}$ from brewer's yeast. *FEBS Lett.* **15**:286–290.

Chambers, R. W., 1971 On the recognition of tRNA by its aminoacyl-tRNA ligase. *Prog. Nucleic Acid Res. Mol. Biol.* **11**:489–525.

Chang, S. H., N. R. Miller, and C. W. Harmon, 1971 Nucleotide sequence of "renaturable" leucine transfer ribonucleic acid. *FEBS Lett.* **17**:265–268.

Cory, S. and K. A. Marcker, 1970 The nucleotide sequence of methionine $tRNA_M$. *Eur. J. Biochem.* **12**:177–194.

Cory, S., K. A. Marcker, S. K. Dube, and B. F. C. Clark, 1969 Primary structure of a methionine transfer RNA from *Escherichia coli. Nature (London)* **220**:1039–1040.

Cramer, F., 1971 Three-dimensional structure of tRNA. *Prog. Nucleic Acid Res. Mol. Biol.* **11**:391–421.

Crick, F. H. C., 1966 Codon–anticodon pairing: The wobble hypothesis. *J. Mol. Biol.* **19**:548–555.

Doctor, B. P., J. E. Loebel, M. A. Sodd, and D. B. Winter, 1969 Nucleotide sequence of *Escherichia coli* tyrosine transfer ribonucleic acid. *Science* **163**:693–695.

Dube, S. K., K. A. Marcker, B. F. C. Clark, and S. Cory, 1968 Nucleotide sequence of *N*-formyl-methionyl-transfer RNA. *Nature (London)* **218**:232–233.

Dube, S. K., K. A. Marcker, and A. Yudelevich, 1970 The nucleotide sequence of a leucine transfer RNA from *E. coli. FEBS Lett.* **9**:168–170.

Dudock, B. S. and G. Katz, 1969 Large oligonucleotide sequences in wheat germ phenylalanine transfer ribonucleic acid. *J. Biol. Chem.* **244**:3069–3074.

Dudock, B. S., G. Katz, E. K. Taylor, and R. W. Holley, 1969 Primary structure of wheat germ phenylalanine transfer RNA. *Proc. Natl. Acad. Sci. USA* **62**:941–945.

Dudock, B. S., C. DiPeri, and M. S. Michael, 1970 On the nature of the yeast phenylalanine transfer ribonucleic acid synthetase recognition site. *J. Biol. Chem.* **245**:2465–2468.

Dudock, B. S., C. DiPeri, K. Scileppi, and R. Reszelbach, 1971 The yeast phenylalanyl-transfer RNA synthetase recognition site: The region adjacent to the dihydrouridine loop. *Proc. Natl. Acad. Sci. USA* **68**:681–684.

Folk, W. R. and M. Yaniv, 1972 Coding properties and nucleotide sequences of *E. coli* glutamine tRNAs. *Nature (London) New Biol.* **273:**165–166.

Gangloff, J., G. Keith, J. P. Ebel, and G. Dirheimer, 1971 Structure of aspartate-tRNA from brewer's yeast. *Nature (London) New Biol.* **230:**125–126.

Gauss, D. H., F. von der Haar, A. Maelicke, and F. Cramer, 1971 Recent results of tRNA research. *Annu. Rev. Biochem.* **40:**1045–1078.

Ginsberg, T., H. Rogg, and M. Staehelin, 1971 Nucleotide sequences of rat liver serine-tRNA. 3. The partial enzymatic digestion of serine-$tRNA_1$ and derivation of its total primary structure. *Eur. J. Biochem.* **21:**249–257.

Goodman, H. M., J. Abelson, A. Landy, S. Brenner, and J. D. Smith, 1968 Amber suppression: A nucleotide change in the anticodon of a tyrosine transfer RNA. *Nature (London)* **217:**1019–1024.

Goodman, H. M., J. N. Abelson, A. Landy, S. Zadrazil, and J. D. Smith, 1970 The nucleoside sequences of tyrosine transfer RNAs of *Escherichia coli*: Sequences of the amber suppressor su^{+}_{III} transfer RNA, the wild type su^{-}_{III} transfer RNA and tyrosine transfer RNAs species I and II. *Eur. J. Biochem.* **13:**461–483.

Harada, F. and S. Nishimura, 1972 Possible anticodon sequences of $tRNA^{His}$, $tRNA^{Asn}$, and $tRNA^{Asp}$ from *Escherichia coli* B: Universal presence of nucleoside Q in the first position of the anticodons of these transfer ribonucleic acids. *Biochemistry* **11:**301–308.

Harada, F., F. Kimura, and S. Nishimura, 1969 Nucleotide sequence of valine tRNA 1 from *E. coli* B. *Biochim. Biophys. Acta* **195:**590–592.

Harada, F., S. Sato, and S. Nishimura, 1972 Unusual CCA-stem structure of *E. coli* B $tRNA_1^{His}$. *FEBS Lett.* **19:**352.

Hashimoto, S., M. Miyazaki, and S. Takemura, 1969 Nucleotide sequence of tyrosine transfer RNA from *Torulopsis utilis. J. Biochem.* **65:**659–661.

Hirsh, D., 1970 Tryptophan tRNA of *Escherichia coli. Nature (London)* **228:**57.

Hirsh, D., 1971 Tryptophan transfer RNA as the UGA suppressor. *J. Mol. Biol.* **58:**439–458.

Holley, R. W., J. Apgar, G. A. Everett, J. T. Madison, M. Marquisee, S. H. Merrill, J. R. Penswick, and A. Zamir, 1965 Structure of a ribonucleic acid. *Science* **147:**1462.

Holmquist, R., T. H. Jukes, and S. Pangburn, 1973 Evolution of transfer RNA. *J. Mol. Biol.* **78:**91–116.

Imura, N., G. B. Weiss, and R. W. Chambers, 1969 Reconstitution of alanine acceptor activity from fragments of yeast $tRNA_{II}^{Ala}$. *Nature (London)* **222:**1147–1148.

Ishikura, H., Y. Yamada, and S. Nishimura, 1971 The nucleotide sequence of a serine tRNA from *Escherichia coli. FEBS Lett.* **16:**68–70.

Jacobson, K. B., 1971 Reaction of aminoacyl-tRNA synthetases with heterologous tRNA's. *Prog. Nucleic Acid Res. Mol. Biol.* **11:**461–488.

Keith, G., A. Roy, J. P. Ebel, and G. Dirheimer, 1971 The nucleotide sequences of two tryptophane-tRNAs from brewer's yeast. *FEBS Lett.* **17:**306–308.

Kim, S. H., G. J. Quigley, F. L. Suddath, A. McPherson, D. Sneeden, J. J. Kim, J. Weinzierl, and A. Rich, 1973 Three-dimensional structure of yeast phenylalanine transfer RNA: Folding of the polynucleotide chain. *Science* **179:**285–288.

Kim, S. H., F. L. Suddath, G. J. Quigley, A. McPherson, J. L. Sussman, A. H. J. Wang, N. C. Seeman, and A. Rich, 1974*a* Three-dimensional tertiary structure of yeast phenylalanine transfer RNA. *Science* **185:**435–440.

Kim, S. H., J. L. Sussman, F. L. Suddath, G. J. Quigley, A. McPherson, A. H. J. Wang, N. C. Seeman, and A. Rich, 1974*b* The general structure of transfer RNA molecules. *Proc. Natl. Acad. Sci. USA* **71:**4970–4974.

Kimura, F., F. Harada, and S. Nishimura, 1971 Primary sequence of $tRNA_{I}^{Val}$ from *Escherichia coli* B. II. Isolation of large fragments by limited digestion with RNases, and overlapping of fragments to deduce the total primary sequence. *Biochemistry* **10:**3277–3283.

Kowalski, S., T. Yamane, and J. R. Fresco, 1971 Nucleotide sequence of the "denaturable" leucine transfer RNA from yeast. *Science* **172:**385–387.

Kuntzel, H., J. Weissenbach, and G. Dirheimer, 1972 The sequences of nucleotides in $tRNA_{III}^{Arg}$ from brewer's yeast. *FEBS Lett.* **25:**189–191.

Madison, J. T., and H. K. Kung, 1967 Large oligonucleotides isolated from yeast tyrosine transfer ribonucleic acid after partial digestion with ribonuclease T_1. *J. Biol. Chem.* **242:**1324–1330.

Madison, J. T., G. A. Everett, and H. Kung, 1966 Nucleotide sequence of a yeast tyrosine transfer RNA. *Science* **153:**531–534.

Madison, J. T., S. J. Boguslawski, and G. H. Teetor, 1972 Nucleotide sequence of a lysine transfer ribonucleic acid from baker's yeast. *Science* **176:**687–689.

Mehler, A. H. and K. Chakraburthy, 1971 Some questions about the structure and activity of aminoacyl-tRNA synthetases. *Adv. Enzymol.* **35:**443–501.

Merril, C. R., 1968 Reinvestigation of the primary structure of yeast alanine tRNA. *Biopolymers* **6:**1727–1735.

Mirzabekov, A. D., D. Lastity, E. S. Levina, and A. A. Bayev, 1971 Localization of two recognition sites in yeast valine tRNA I. *Nature (London) New Biol.* **229:**21–22.

Mizutani, T., M. Miyazaki, and S. Takemura, 1968 The primary structure of valine-I transfer ribonucleic acid from *Torulopsis utilis*. II. Partial digestion with ribonuclease T_1 and derivation of the complete sequence. *J. Biochem.* **64:**839–848.

Nakanishi, K., N. Furutachi, M. Funamizu, D. Grunberger, and I. B. Weinstein, 1970 Structure of the fluorescent Y base from yeast phenylalanine transfer ribonucleic acid. *J. Am. Chem. Soc.* **92:**7617–7619.

Ohashi, Z., F. Harada, and S. Nishimura, 1972 Primary sequence of glutamic acid tRNA II from *Escherichia coli. FEBS Lett.* **20:**239–241.

Philipps, G. R., 1969 Primary structure of transfer RNA. *Nature (London)* **223:**374–377.

Pinkerton, T. C., G. Paddock, and J. Abelson, 1972 Bacteriophage T_4 $tRNA^{Leu}$. *Nature (London) New Biol.* **240:**88–90.

RajBhandary, U. L. and S. H. Chang, 1968 Studies on polynucleotides. LXXXII. Yeast phenylalanine transfer ribonucleic acid: Partial digestion with ribonuclease T_1 and derivation of the total primary structure. *J. Biol. Chem.* **242:**598–608.

RajBhandary, U. L., S. H. Chang, A. Stuart, R. D. Faulkner, R. M. Hoskinson, and H. G. Khorana, 1967 Studies on polynucleotides. LXVIII. The primary structure of yeast phenylalanine transfer RNA. *Proc. Natl. Acad. Sci. USA* **57:**751–758.

Roberts, R. J., 1972 Structures of two glycyl-tRNAs from *Staphylococcus epidermidis. Nature (London) New Biol.* **237:**44–45.

Robertus, J. D., J. E. Ladner, J. T. Finch, D. Rhodes, R. S. Brown, B. F. C. Clark, and A. Klug, 1974 Structure of yeast phenylalanine tRNA at 3 Å resolution. *Nature (London)* **250:**546–551.

Roe, B., M. Sirover, R. Williams, and B. Dudock, 1971 New heterologous mischarging reactions with yeast phenylalanyl transfer RNA synthetase. *Arch. Biochem. Biophys.* **147:**176–177.

Roe, B., M. Sirover, and B. Dudock, 1972 On the nature of the yeast phenylalanine t-RNA synthetase recognition site—The role of three nucleotides in the acceptor stem region. *Fed. Proc.* **31:**450.

Schulman, L. H., 1972 Loss of methionine acceptor activity by modification of a specific guanosine residue in the acceptor stem of *E. coli* $tRNA^{fMet}$. *Fed. Proc.* **31:**449.

Schulman, L. H. and R. W. Chambers, 1968 Transfer RNA. II. A structural basis for alanine acceptor activity. *Proc. Natl. Acad. Sci. USA* **61:**308–315.

Singer, C. E. and G. R. Smith, 1972 Histidine regulation in *Salmonella typhimurium.* XIII. Nucleotide sequence of histidine transfer ribonucleic acid. *J. Biol. Chem.* **247:**2989–3000.

Singer, C. E., G. R. Smith, R. Cortese, and B. N. Ames, 1972 Mutant $tRNA^{His}$ ineffective in repression and lacking two pseudouridine modifications. *Nature (London) New Biol.* **238:**72–74.

Smith, C. J., A. N. Ley, P. D'Obrenan, and S. K. Mitra, 1971 The structure and coding specificity of a lysine transfer ribonucleic acid from the haploid yeast *Saccharomyces cerevisiae* αS288C. *J. Biol. Chem.* **246:**7817–7819.

Smith, J. D., L. Barnett, S. Brenner, and R. L. Russell, 1970 More mutant tyrosine transfer ribonucleic acids. *J. Mol. Biol.* **54:**1–14.

Squires, C. and J. Carbon, 1971 Normal and mutant glycine transfer RNAs. *Nature (London) New Biol.* **233:**274–277.

Staehelin, M., 1971 The primary structure of transfer ribonucleic acid. *Experientia* **27:**1–11.

Staehelin, M., H. Rogg, B. C. Baguley, T. Ginsberg, and W. Wehrli, 1968 Structure of a mammalian serine tRNA. *Nature (London)* **219:**1363–1365.

Takemura, S., T. Mizutani, and M. Miyazaki, 1968 The primary structure of valine-I transfer ribonucleic acid from *Torulopsis utilis. J. Biochem.* **63:**277–278.

Takemura, S., M. Murakami, and M. Miyazaki, 1969*a* Nucleotide sequence of isoleucine transfer RNA from *Torulopsis utilis. J. Biochem.* **65:**489.

Takemura, S., M. Murakami, and M. Miyazaki, 1969*b* The primary structure of isoleucine tRNA from *Torulopsis utilis*: Complete digestion with ribonucleases and construction of model of its secondary structure. *J. Biochem.* **65:**553–565.

Uziel, M. and H. G. Gassen, 1969*a* Phenylalanine transfer ribonucleic acid from *E. coli* B: Isolation and characterization of oligonucleotides from ribonuclease T_1 and ribonuclease A hydrolysates. *Biochemistry* **8:**1643–1656.

Uziel, M. and H. G. Gassen, 1969*b* Structure of $tRNA^{Phe}$. *Fed. Proc.* **28:**409.

Venkstern, T., 1973 *The Primary Structure of tRNA,* Plenum Press, New York.

Woodward, W. R. and E. Herbert, 1971 Coding properties of reticulocyte and yeast isoacceptor lysine transfer ribonucleic acids in hemoglobin synthesis. *Fed. Proc.* **30:**1217.

Yaniv, M. and B. G. Barrell, 1969 Nucleotide sequence of *E. coli* B $tRNA_I^{Val}$. *Nature (London)* **222:**278–279.

Yaniv, M. and B. G. Barrell, 1971 Sequence relationship of three valine acceptor tRNAs from *Escherichia coli. Nature (London) New Biol.* **233:**113–114.

Yarus, M. and B. G. Barrell, 1971 The sequence of nucleotides in $tRNA^{Ile}$ from *E. coli* B. *Biochem. Biophys. Res. Commun.* **43:**729–734.

Zachau, H. G., 1969 Transfer ribonucleic acids. *Angew. Chem. Int. Ed.* **8:**711–726.

Zachau, H. G., D. Dütting, and H. Feldman, 1966*a* Nucleotidsequenzen zweier serinspezifischer Transfer-ribonucleinsauren. *Angew. Chem.* **78:**392–393.

Zachau, H. G., D. Dütting, and H. Feldman, 1966*b* The structures of two serine transfer ribonucleic acids. *Hoppe Seyler's Z. Physiol. Chem.* **347:**212–235.

11

Eukaryotic Ribosomes

MARY G. HAMILTON

This chapter is dedicated to the memory of Dr. Mary L. Petermann, who died on December 13, 1975. She will be remembered for her many pioneering contributions to the study of ribosomes.

Introduction

In order to solve the problem of how the translation of genetic information is regulated in growth and differentiation, it is necessary to know in detail at the molecular level how that information is translated into protein. The key structural element of the cellular apparatus for translation is the ribosome, for it is on the ribosome that the many components of the protein-synthesizing machinery assemble and the peptide bond is formed. Thus the ribosome is an important site for epigenetic control.

A popular and fruitful approach to the study of the mechanisms of protein synthesis has been to study the factor-mediated steps in which the ribosome is considered a necessary black box. A different but complementary view is that from the level of the ribosome itself. This view requires knowledge of the structure of the ribosomal subunits. Compared to the wealth of knowledge concerning the ribosomes of the prototype prokaryote, *Escherichia coli,* surprisingly little is known about eukaryotic ribosomes. While data on the size and composition of the subunits were obtained fairly easily for the bacterial ribosome, these basic data were not obtained for the larger and more complex eukaryotic ribosomes until recently when methods were finally developed for preparing ribosomal subunits that have biological activity.

MARY G. HAMILTON—Sloan-Kettering Institute for Cancer Research, New York City, New York.

Much of the complexity of protein synthesis in prokaryotes is compounded in eukaryotes by features unique to the eukaryotic cell. Among these are cellular organization into nucleus and cytoplasm, segregation of the protein synthetic process in the cytoplasm, remote from the source of genetic information; involvement of the extensive membrane systems of the endoplasmic reticulum, and the relatively long-life and defined functions of a differentiated cell. Moreover, in addition to cytoplasmic ribosomes, eukaryotic cells contain ribosomes within mitochondria and chloroplasts that resemble the prokaryotic ones. Thus in animal cells there are two classes of ribosomes with separate genetic controls, and in plants with chloroplasts there are three. This chapter will deal mainly with the cytoplasmic ribosomes of animal cells, although frequent comparisons with prokaryotic ribosomes will be made. The chapter by Davies in Volume 1 of this series describes prokaryotic ribosomes, and Chapters 13 and 16 in this volume cover chloroplast ribosomes and mitochondrial ribosomes, respectively.

A Brief History of Ribosomology

In the past 10 years the study of protein synthesis and ribosomes has expanded and progressed rapidly. Lately, at least one symposium each year has been devoted to some aspect of ribosome structure and function (Cox and Hadjiolov, 1972; Bloemendal *et al.*, 1973; Smillie, 1973; Nomura *et al.*, 1974; Bielka and Scheler, 1974). The older literature was reviewed comprehensively by Petermann in 1964. Spirin and Gavrilova (1969) have also written a monograph. A monthly bibliography that lists titles of papers on ribosomes is available from the Department of Physiology, the University of Sheffield, Sheffield S102TN, England. The subject of protein synthesis is reviewed every few years in Annual Reviews of Biochemistry. The series *Methods in Enzymology* has devoted volume 12, Parts A and B, Volume 20, Part C, and Volume 30, Part F, to the technology of ribosomes and protein synthesis.

Although ribosomes were first observed in normal tissues of the chick embryo as macromolecular components sedimenting in an ultracentrifuge (Taylor *et al.*, 1942), their study was not pursued until the early 1950s. Then, in rat liver, the "small, dense particles" later named *ribosomes* by Roberts (1958) were described by electron microscopy by Palade, by their physical properties in solution by Petermann, and by their ability to incorporate amino acids into "protein" by Siekevitz (for references, see Petermann, 1964). Palade and Siekevitz (1956) combined biochemical and cytological techniques to prove that the "microsomes" of Claude (1946) consisted of pieces of the rough endoplasmic reticulum. Chao and Schachman (1956) studied the ribosomes of yeast. It was Chao who discovered that Mg ions are required to prevent the dissociation of ribosomes to subunits. Also during this period Ts'o *et al.* (1958) made physical studies of pea ribosomes.

Despite this early work on eukaryotic ribosomes, the prokaryotic ribosomes of *E. coli* have been studied in much greater depth. Practically the whole of present knowledge of the mechanisms of protein synthesis is based on the study of *E. coli*. In a classic work, Tissières *et al.* (1959) studied the physical properties of the *E. coli* ribosome, observed the reversible, Mg^{2+}-dependent dissociation of the 70 S ribosome to 50 S and 30 S subunits, and measured their molecular weights. Ribosomes are named by their sedimentation coefficients. These values have only recently been subject to minor revisions (Hill *et al.*, 1969). Kurland (1960) characterized the two main RNA components, the 23 S and 16 S molecules of the large and small subunits. Waller and Harris (1961) discovered that ribosomes contain many different proteins. Nomura (1970) ingeniously reassembled the small ribosomal subunit from its RNA and proteins and established the order of assembly of the proteins. The study of the genetics of prokaryotic ribosomes has also been fruitful, as detailed by Davies in his review of prokaryotic ribosomes in Volume 1 of this series.

A review by Garrett and Wittmann (1973) presents an excellent summary of the great progress that has been made in elucidating the molecular structure of the *E. coli* ribosome and points the way for studies of animal ribosomes.

Most of the early work on ribosomes was carried out on rat liver, chiefly because rat liver had been the subject of many fundamental biochemical studies. Reticulocytes, induced in rabbits by the administration of phenylhydrazine, have been studied in great detail. This reticulocyte lacks a functioning nucleus, and therefore the polyribosomes can be readily isolated and purified after cell lysis. Since the reticulocyte mainly synthesizes hemoglobin, its messenger RNAs mainly code for globin. Because it contains little ribonuclease, its ribosomes and the other factors of the translation apparatus are very active *in vitro*. It was in reticulocytes that the polysome structure was first observed (Warner *et al.*, 1963).

The Krebs ascites cell is a popular source of ribosomes for heterologous translation of messenger RNAs of many kinds, especially viral ones (Mathews and Korner, 1970). These cells can be grown in quantity by using the mouse peritoneum as an *in vitro* culturing system.

The HeLa cell (see the chapter by Shannon and Macy in Volume 4 of this series) has been extremely useful in elucidating the biosynthesis and processing of ribosomal RNA in eukaryotes (Darnell, 1968). Other cell lines from various mammals have also been used for such studies (Perry and Kelley, 1968). The chief advantage of cultured cells is that radioactive isotopes can be used to follow the biosynthesis of macromolecules. That kind of study is difficult to carry out in an animal's organs such as liver. Fungi and protists have the advantage that they can be manipulated nutritionally and genetically in much the same way as bacteria.

Certain antibiotics act differentially at various steps in protein synthesis in prokaryotes and eukaryotes (Vazquez *et al.*, 1969; Pestka, 1971; Grollman and Huang, 1973). Generally, they inhibit ribosomal functions by binding to specific sites on the ribosome, but, as Pestka observes, because the process of translation is so complex they can produce a variety of effects far removed from the site of interaction. Nevertheless, they provide useful probes into the structural, functional, and, in the case of *E. coli*, genetic aspects of the translation apparatus (see Chapter 5 of Volume 1 of this series).

Protein Synthesis on Polysomes in Eukaryotes: An Overview and Some Definitions

The system that translates the genetic information encoded in the messenger RNA (mRNA) consists of ribosomes, aminoacylated transfer RNAs (aa-tRNAs), protein factors, and low molecular weight cofactors, especially GTP. Since several ribosomes can traverse the same messenger, complexes that contain varying numbers of ribosomes can be isolated from the cytoplasm. These complexes are called *polyribosomes* or *polysomes*. Since it is the messenger RNA that defines polysome structure, polysomes are susceptible to attack by ribonuclease, and as an elongated structure they are also sensitive to shearing forces. A monosome then is a ribosome that is still attached to mRNA and that still binds tRNA, peptidyl-tRNA, and possibly the protein factors. It may be an artifact produced by shear or RNAse action on a polysome, although an intact mRNA bearing a single ribosome is also a monosome. A monomer, however, is defined as a ribosome free of nonribosomal components. Dissociability by dilute KCl has been used to differentiate monomers from monosomes (Tai and Davis, 1972).

All ribosomes are made up of two subunits of unequal size. Small amounts of so-called native subunits are also present in the cytosol. The native small subunit has a higher protein content than one "derived" or dissociated from the mononer structure, and appears to bind initiation factors (Hirsch *et al.*, 1973). Figure 1 is a scheme for the ribosome "cycle" (Ayuso-Parilla *et al.*, 1973). Figure 2 shows the various states of association of rat liver ribosomes as observed in an analytical ultracentrifuge.

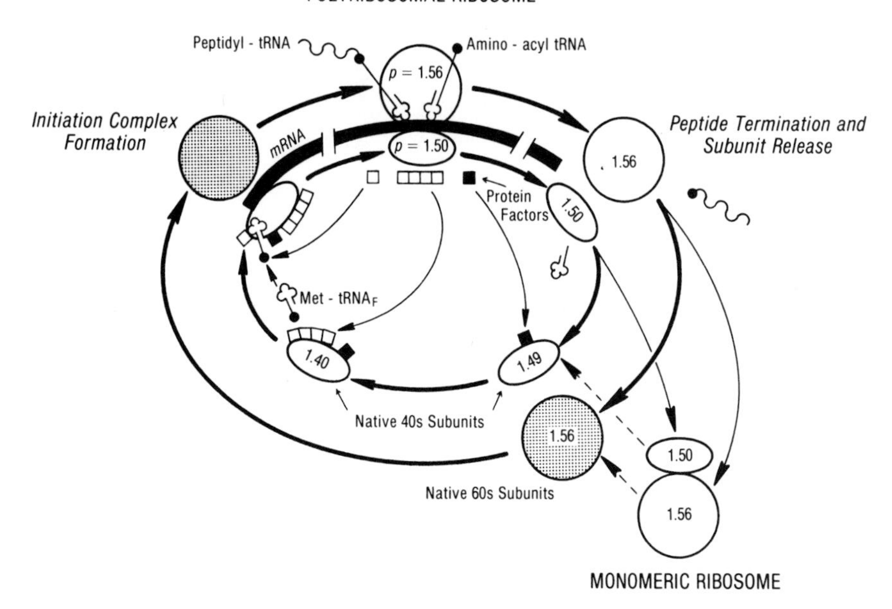

Figure 1. A scheme according to Ayuso-Parilla et al. (1973) for cycling of the ribosomal subunits during protein synthesis. The subunits are labeled with their buoyant densities in CsCl, which are inversely related to their protein contents (Hamilton, 1971). The small subunit binds different amounts of protein at various stages.

Comparative physical data for ribosomes and their subunits from representative bacteria, fungi, and mammals are given in Table 1. Each subunit is a ribonucleoprotein consisting of a high molecular weight RNA and many small proteins. Magnesium ions stabilize the association of the subunits, but other "environmental" factors affect the association. Increases in the concentration of neutral salts, temperature, pH, pressure, or agents that decrease hydrogen bonding favor dissociation to subunits (Petermann *et al.,* 1972*b*). In general, ribosomes of eukaryotes require lower Mg^{2+} concentrations than do those of prokaryotes to remain as monomers. This property led to difficulties in dissociating eukaryotic ribosomes without an irreversible unfolding of their structure. In the case of the functioning ribosomal subunits, other components such as the aa-tRNAs, various factors, and messenger RNA affect the association (see Figure 1).

Free and Membrane-Bound Polysomes

Electron microscopy has revealed that the architecture of the cytoplasm of the differentiated cell differs significantly from that of an em-

TABLE 1. A Comparison of the Sizes and Other Properties of Some Ribosomes and Their Subunits[a]

Source	Sedimentation coefficient[b] $s^0_{20,w}$ (S)	Molecular weight[c] $\times 10^{-6}$	RNA: protein[d]	RNA molecular weight[c] $\times 10^{-6}$	Partial specific volume[e] $\bar{v}$ (ml/g)	Frictional ratio[f] f/f_0	Hydration[g] w (g H_2O/g)	Number of proteins[h]
Escherichia coli								
Monomer	70.5	2.65	2.27		0.606	1.51		
Large subunit	50.2	1.55	2.33	1.05	0.592	1.55	1.4	34
Small subunit	31.8	0.9	2.17	0.56	0.601	1.66	1.3	21
Saccharomyces cerevisiae								
Monomer	82.0	3.60	1.11		0.630	1.48	1.34	
Large subunit	60.7			1.20		1.57	1.82	
Small subunit	37.8			0.65		1.42	0.50	
EDTA-dissociated								
Large subunit	45.1	2.53	1.00		0.635	1.89		
Small subunit	25.6	1.01	1.35		0.620	1.86		

Rat liver cytoplasm								
Monomer	81.0	4.5	1.08		0.631	1.69		
Large subunit	59.1	3.00	1.24	1.7	0.623	1.86	3.3	40
Small subunit	40.9	1.50	0.83	0.62	0.654	1.53	1.4	30
EDTA-dissociated								
Large subunit	49.9	3.12	1.24		0.623	2.26		
Small subunit	28.6	1.18	1.02		0.633	2.01		
Calf-liver mitochondria								
Monomer	56.3	2.83	0.45		0.674	1.58		
Large subunit	44.9	1.65	0.49	0.54	0.671	1.40		44
Small subunit	30.1	1.10	0.49	0.36	0.671	1.59		35

[a] References: *E. coli* ribosomes, Van Holde and Hill (1974); *E. coli* rRNA, Stanley and Bock (1965); *E. coli* ribosomal proteins, Garrett and Wittmann (1973); *S. cerevisiae* ribosomes, Mazelis and Petermann (1973); *S. cerevisiae* rRNA, Bruening and Bock (1967); rat liver ribosomes, Hamilton *et al.* (1971) and Haga *et al.* (1970); rat liver rRNA, Hamilton (unpublished data, 1970); rat liver ribosomal proteins, Lin and Wool (1974); calf liver mitochondria ribosomes, Hamilton and O'Brien (1974); calf liver mitochondria ribosomal proteins, Matthews and O'Brien (1974).

[b] The sedimentation coefficient in H_2O at 20°C by extrapolation to infinite dilution, except for mitochondria where the values were obtained at low concentrations with the UV absorption optical system.

[c] Measured by high-speed equilibrium centrifugation, except for mitochondria RNA where the values were calculated from the molecular weight and composition of the subunit.

[d] Calculated from chemical analyses for all except mitochondria where the protein content was estimated from the buoyant density of HCHO-fixed material.

[e] Calculated from the composition, except for *E. coli* for which $\bar{v}$ was measured.

[f] Calculated from data given here for s, M, and $\bar{v}$.

[g] Calculated from low-angle X-ray scattering for *E. coli*; from shape measurements by electron microscopy and frictional coefficient for yeast and rat liver.

[h] From two-dimensional polyacrylamide gel electrophoresis analysis.

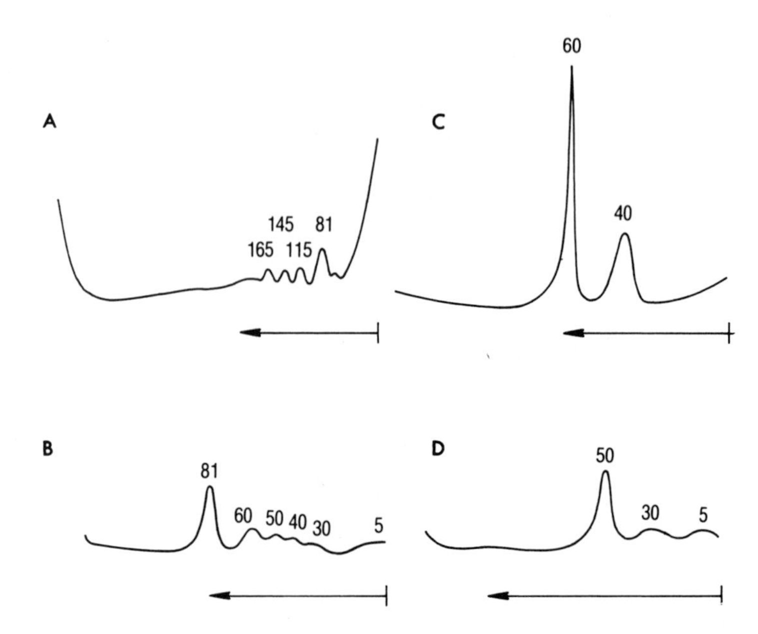

Figure 2. Analytical ultracentrifugal patterns illustrating with schlieren optics various states of association of rat liver ribosomes. The direction of sedimentation is indicated by the arrows. The numbers represent sedimentation coefficients in Svedberg units at infinite dilution. All solutions were buffered at about pH 7, except D. A and C contained 1 mM dithiothreitol. (A) Polysomes in 0.25 M sucrose, 0.03 M KCl, 1 mM $MgCl_2$, after 4 min at 36,000 rpm at 5°C. (B) Intermediate stage of dissociation following dialysis against 0.1 M KCl, 1 mM $MgCl_2$, after 14 min at 37,020 rpm at 26°C. (C) Active subunits produced by dissociation in 0.5 M KCl, 5 mM $MgCl_2$, following incubation with 0.01 M puromycin according to Blobel and Sabatini (1971), after 10 min at 44,770 rpm at 24°C. (D) Inactive subunits obtained by dialysis against 0.1 M $KHCO_3$, pH 8.0, after 18 min at 37,020 rpm at 26°C.

bryonic cell whose main function is self-duplication. A striking feature of many differentiated cells is the array of ribosome-studded membranes, the "rough" surfaced endoplasmic reticulum that fills the cytoplasm of some tissues (e.g., pancreas). The "smooth" surfaced endoplasmic reticulum and the membranes of the Golgi apparatus are free of ribosomes.

In eukaryotic cells two populations of polysomes can be identified, those that are bound to membranes and those that are "free" in the cytosol. It has been postulated that the membrane-bound polysomes synthesize proteins that are to be exported from the cell, whereas free ones synthesize intracellular proteins. Rolleston (1974) has reviewed the large number of studies that have tested this hypothesis. Apparently the evidence for the first part is more convincing than that for the second; i.e., the synthesis of proteins of the cytosol may not occur exclusively on free polysomes. For example, some brain proteins are synthesized on membrane-bound polysomes (Andrews and Tata, 1971), and Faiferman *et al.* (1973) found only membrane-bound polysomes in Ehrlich ascites tumor cells that synthesize no proteins for export.

The nature of the association of ribosomes with the membranes of the endoplasmic reticulum is not completely known. The problem has been attacked by dissociating the polysomes from the membrane. One of the earliest studies (Sabatini *et al.,* 1966) showed that on treatment with the chelating agent sodium ethylenediamine tetraacetate, EDTA, the small subunit was preferentially released, leaving the large subunit still attached to the membrane. In addition, Redman and Sabatini (1966) demonstrated that in liver the polypeptide product was vectorially discharged through the membrane into the cisternum of the endoplasmic reticulum.

Instead of EDTA, Adelman *et al.* (1973*a,b*) used a high concentration of KCl to dissociate the small subunit from the large one and then showed that treatment with puromycin, the antibiotic that causes premature release of the nascent polypeptide, is required to dissociate the large subunit from the membrane, corroborating the idea that the nascent peptide is involved in the binding.

Although structural differences have not been found in the two populations of ribosomes, there may be differences among their proteins. A report of two differences in the one-dimensional polyacrylamide gel electrophoresis patterns of the proteins of free and membrane-bound ribosomes of chick embryo cells (Fridlender and Wettstein, 1970) was corroborated in reticulocytes by two-dimensional analyses (Fehlmann *et al.,* 1975). However, in rat liver, Hanna *et al.* (1973) found that the differences observed by two-dimensional polyacrylamide gel electrophoresis between the two populations disappeared if the free ribosomes were also treated with the detergent sodium deoxycholate that had been used to detach the bound ones from the membranes. These discrepancies may be real or technological. Chick embryo cells and reticulocytes share the property that they have a low ratio of bound to free polysomes, while in rat liver 80% of the polysome population may be membrane bound (Blobel and Potter, 1966).

What mechanism operates to direct certain mRNAs to the membrane while others remain free? The mRNA that specifies a protein to be secreted may locate itself on the membrane either directly or by preferentially seeking a large subunit already bound to a "high-affinity site" in the membrane. These sites have been defined by treatment with agents, EDTA or RNAse, which disrupt the polysome structure and results in a subdivision of bound polysomes into those that are loosely bound and those that are tightly bound to membranes (Rosbash and Penman, 1971). Sequences at the 5′ end of mRNA may serve to both direct and bind the mRNA to the membrane. Milstein *et al.* (1972) found that an extra *N*-terminal peptide sequence was present on immunoglobulin light chains synthesized *in vitro* that was not present in the secreted protein. Milcarek and Penman (1974) suggest that it is the poly(A) tail at the opposite end of mRNA that is attached directly to the membrane. Both findings can be accommodated by a model where the association of mRNA with the membrane is either cyclic in time or physically a circle. Two proteins are found associated with the mRNA bound to polysomes. One of them may be associated with the poly(A) tail (Kwan and Brawerman, 1972; Blobel, 1973). Whether that protein is a membrane protein is not known. These studies may advance more rapidly now that methods have been developed which permit the nondestructive release of bound polysomes (Adelman *et al.*, 1973*b*).

It is also possible that initiation factors or other special translational controls operate to direct mRNA to the correct location. These problems cannot be dissociated from the problem of mRNA transport and polysome formation. If all mRNAs have an obligatory association with membranes, as has been suggested by Shiokawa and Pogo (1974), many conjectures are irrelevant. Figure 3 presents a new model for the flow of ribosomes that accommodates the association of mRNA with membranes during or shortly after it leaves the nucleus (Branes and Pogo, 1975). If mRNA is bound to

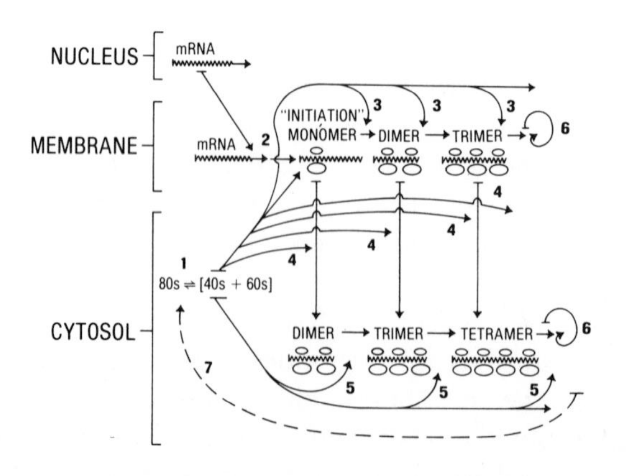

Figure 3. Model proposed by Branes and Pogo (1975) for the formation of polysomes based on their data for yeast cells. (1) Dissociation of the free ribosomes to subunits; (2) coupling of the subunits with the newly transported mRNA and formation of the "initiation monomer" at the level of the "membrane"; (3) site at which the coupling process of the ribosomal subunits occurs on the bound polysomes; (4) this coupling, when the "membrane" releases the mRNA; (5) this coupling in the free polysomes; (6) recycling of free and bound ribosomes; and (7) flowback of ribosomes or ribosomal subunits.

the membrane, how can it travel through the ribosomes? This seeming dilemma may disappear when the fluidity of membrane structure is better understood.

Ribosome Structure

The eukaryotic ribosome consists of two subunits of unequal size and composition (Table 1). The basic structure of each subunit consists of a single strand of RNA of high molecular weight and many proteins of low molecular weight. However, other macromolecules are associated with the subunits either transiently or stoichiometrically (Figure 1). To analyze its structure, the complex can be dismantled in stages proceeding from the polysome, to the ribosome, to its active subunits, to its inactive, unfolded subunits, and finally to rRNAs and proteins. The following scheme illustrates these stages for the rat liver ribosome:

1. Polysomes or ribosomes $\xrightleftharpoons{\text{urea or KCl + puromycin}}$ active subunits + protein factors + mRNA + aa-tRNA + peptidyl-tRNA.
2. Active subunits $\xrightarrow{\text{EDTA}}$ unfolded subunits + SH-proteins + aa-tRNA + mRNA fragments + 5 S RNA–protein.
3. Unfolded subunits $\xrightarrow{\text{1 M LiCl}}$ stripped subunits ("core") + r-proteins ("split proteins").
4. Stripped subunits $\xrightarrow{\text{2 M LiCl}}$ rRNAs + r-proteins

Although step 2 has not yet been reversed, the released material may serve a structural function. Operationally, the proteins released in (2) can be considered as a set distinguishable from either the "factors" released in (1) or the true ribosomal proteins released during the dissociations of (3) and (4). In bacteria, steps (3) and (4) are reversible.

Ribosomes and their subparticles, like viruses, have a size and structure accessible to physical-chemical techniques as well as electron microscopy. In an analytical ultracentrifuge, a solution of ribosomes is paucidisperse (Figure 2A); i.e., several discrete peaks are observed. Moreover, by manipulation of the ionic conditions, both dissociation and association occur (Figure 2B, C, D). For convenience, the various species are named by their approximate sedimentation rates at infinite dilution. Ribosomes exhibit an unusually high concentration dependence in sedimentation, a property attributable to an asymmetrical shape or a high hydration or both. For example, a 1% solution of 80 S ribosomes sediments at 65 S. Since they appear to be roughly spherical in electron microscopy, this property is usually attributed to hydration. Calculating the hydration from the frictional coefficient, assuming that the shape is spherical, gives very high values for the hydration (approximately 2–3 g water per gram of ribonucleoprotein). Table 1 presents such calculations for the cytoplasmic ribosomes of rat liver and yeast. The hydration has been measured directly for *E. coli* ribosomes by low-angle X-ray scattering Eukaryotic ribosomes have sedimentation coefficients of about 80 S, in contrast to the 70 S of prokaryotes and plastids. Mitochondrial ribosomes sediment at 56 S. Plant cytoplasmic ribosomes also sediment at 80 S, although like yeast ribosomes they have a lower molecular weight than animal 80 S ribosomes.

The dissociation pattern of bacterial ribosomes is straightforward. When the Mg^{2+} concentration is reduced, 70 S ribosomes dissociate to 50 S and 30 S subunits. The ratio of the masses of the 50 S and 30 S particles is 2:1. At higher Mg^{2+} concentrations, 100 S dimers of 70 S are observed. When the Mg^{2+} concentration is decreased, rat liver 80 S ribosomes dissociate to four species, 60 S, 50 S, 40 S, and 30 S (Figure 2B). If all Mg^{2+} is complexed with EDTA, only 50 S and 30 S species result (Figure 2D). While the dissociation of the subunits of prokaryotic ribosomes can be reversed by Mg^{2+}, the eukaryotic 50 S and 30 S species aggregate and have no activity in *in vitro* protein synthesis. Only when the subunits are maintained in a compact conformation by Mg^{2+} and the KCl concentration is increased (Martin *et al.*, 1971) or dilute urea is added (Petermann and Pavlovec, 1971) is it possible to dissociate 80 S ribosomes to 60 S and 40 S species (Figure 2C) which can reassociate to 80 S and retain their biological activity.

Improved techniques of electron microscopy have given a better idea of what the ribosome looks like (Nonomura *et al.*, 1971; Kiselev *et al.*, 1974). Instead of roughly spherical objects, the subunits have asymmetries that may be functional. Figure 4 is a schematic view of a rat liver ribosome according to Lake *et al.* (1974).

The following description of the subunits of rat liver is based on data (Table 1) obtained for the rat liver ribosomal subunits (Hamilton and Ruth, 1969; Hamilton *et al.*, 1971; Hamilton, unpublished observations). The large and small subunits have a mass ratio of 2:1 as in bacteria, but the mass ratio of their RNAs is 2.6:1. Therefore, the small subunit has a larger protein to RNA ratio than does the large subunit.

The large subunit of the rat liver ribosome has a molecular weight of 3 million, and it sediments at about 60 S. Its RNA components include three molecular species: (1) 28 S (molecular weight of 1.7 million); (2) "7 S" or "light" RNA (molecular weight of 53,600), probably associated with the 28 S molecule by hydrogen bonding as in reticulocytes (King and

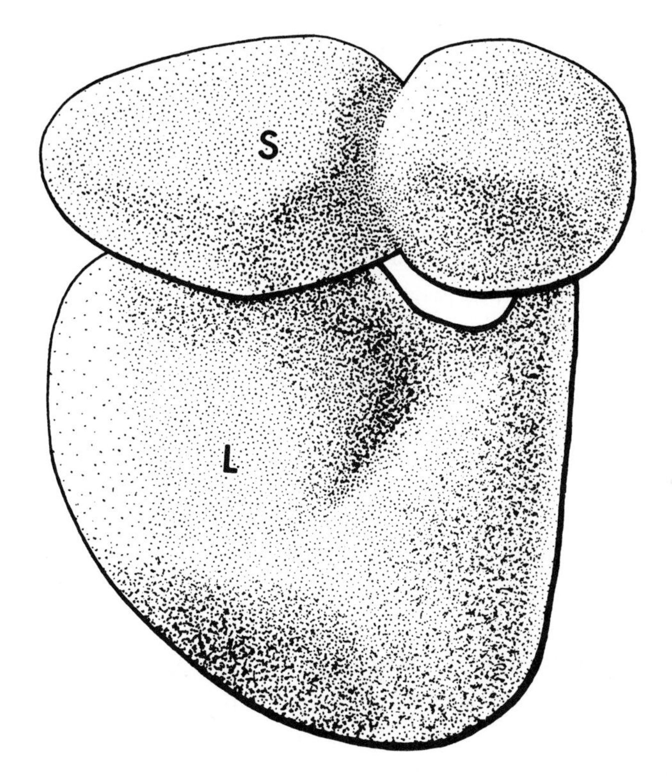

Figure 4. Three-dimensional structure of a rat liver ribosome based on a styrofoam model derived from electron microscopy by Lake et al. (1974). The small subunit (S) has a bipartite structure and is perched asymmetrically on the large subunit (L), forming a channel through which the messenger RNA may pass. Since the nascent peptide is sequestered from enzymatic degradation by the ribosome itself (Malkin and Rich, 1967; Blobel and Sabatini, 1970), there may be a channel in the large subunit through which the peptide is extruded.

Gould, 1970); and (3) 5 S RNA (molecular weight of about 40,000). The large subunit contains 45% protein, about 1.2 million daltons distributed among 40 species with molecular weights ranging from 54,000 to 9000.

The compact, roughly spherical structure can be caused to distend, flatten, and unfold by treatment with EDTA (Haga *et al.*, 1970). The sedimentation coefficient of this unfolded form is 50 S, not 60 S. Its molecular weight and chemical composition are not greatly changed, but it can no longer associate with a small subunit or perform in protein synthesis. The 5 S RNA dissociates from the structure and with it one molecule of protein (Petermann *et al.*, 1972*a*). Grummt *et al.* (1974) have found that the complex has GTPase activity.

The small subunit of the rat liver ribosome has a molecular weight of

1.5 million and sediments of 40 S. It contains one large RNA molecule, 18 S, with a molecular weight of about 0.62 million. Its protein content of 0.825 million daltons consists of 30 species with molecular weights ranging from 39,000 to 11,000. When the Mg^{2+} is chelated by EDTA, the small subunit sediments at 30 S but, unlike the large subunit, undergoes a drastic structural change. With the release of some RNA and protein, its molecular weight decreases by 20% to 1.2 million. The RNA molecules are mRNA fragments and tRNA. The proteins have not yet been identified.

Ribosomal RNAs

Attardi and Amaldi (1970) have reviewed work on ribosomal RNAs. The exact functions of the RNAs in ribosomes are not known. Presumably, sequences on each rRNA help to bind the mRNA and tRNAs by base pairing to the appropriate subunit. The ribosomal RNA seems to be the organizing element for the many proteins and thus the skeleton of the ribosomal subunit. Shine and Dalgarno (1974) have proposed a direct base-pairing role for the 3′-OH end of the 18 S rRNA in termination because it is complementary to the terminator codons. The 5 S RNA molecules that are associated with the large subunit of all ribosomes except those of mitochondria may interact with tRNA (Richter *et al.*, 1973).

Ribosomal RNA molecules are continuous polynucleotide chains which are folded and internally base-paired to varying degrees depending on environmental conditions such as pH, ionic strength, and temperature (Spirin, 1964; Cox, 1966). This "secondary structure" is disrupted by exposure to denaturing conditions and any hidden breaks in the strand can be observed by changes in the sedimentation or electrophoretic pattern. The small RNA called 7 S or light RNA that dissociates from the 28–30 S RNA of reticulocyte ribosomes under these conditions has been fully characterized by King and Gould (1970). Prokaryotes lack this species.

In general, ribosomal RNAs have a high GC content, although the percentages vary from about 60% in man to 40% in *Drosophila melanogaster* (Attardi and Amaldi, 1970). In higher eukaryotes the GC content of the smaller rRNA is less than that of the larger RNA molecule. Besides the four ribonucleotides, adenylic, guanylic, uridylic, and cytidylic acids, ribosomal RNAs contain small amounts of pseudouridylic acid and methylated nucleotides. The methylation occurs both on the bases, mainly on adenine, guanine, and cytidine, and on the 2′-OH group of ribose. Eukaryotic rRNAs seem to have a higher proportion of base methylation than do *E. coli* rRNAs.

The complete base sequence of the *E. coli* 16 S RNA is known, and much of the sequence of its 23 S RNA. The 16 S molecule can be arranged in a series of base-paired loops of varying lengths separated by short single-stranded regions (Fellner, 1974). Since the sequencing methods require large quantities of radioactively-labeled RNA of high specific activity, they have not yet been applied to eukaryotic RNAs except for 5 S RNAs, which have been sequenced for various eukaryotes and prokaryotes (Monier, 1974).

Ribosomal Proteins

One of the most interesting features of ribosomal structure is the large number of unique proteins that each subunit contains. What are their functions? One or two proteins could perform a protective or structural function as in simple viruses. How many different ribosomal proteins are there? What are their sizes? Do they occur in multiple copies? What are their amino acid sequences? What conformations do they assume in solution? Are they chemically modified? What is their arrangement in the subunit?

For eukaryotic ribosomes, very few of these questions have been answered, although for *E. coli* ribosomal proteins even the functions of some individual proteins are known. Garrett and Wittmann (1973) have summarized the present state of this knowledge for the *E. coli* ribosome. Each protein is named by the letter L or S for the subunit of origin and a number depicting its position in the two dimensional polyacrylamide gel electrophoretic system of Kaltschmidt and Wittmann (1970) (see Appendix).

Work on the proteins of eukaryotic ribosomes is just beginning. Unexpectedly, fractionation methods that resolved prokaryotic proteins and led to the separation and complete characterization of all the proteins of the *E. coli* ribosome were not directly applicable. For one thing, the proteins are more numerous and somewhat larger (Table 1, and Bickle and Traut, 1971). Moreover, they are chemically different and therefore their solubilities and other properties differ. A practical matter is the availability of large quantities of ribosomes.

Wool and Stöffler (1974) have reviewed what is known about the proteins of rat liver ribosomes. Ribosomal proteins are basic; i.e., they have isoionic points between pH 8 and 10. An interesting exception is a halophilic bacterium that grows in 1 M NaCl whose ribosomes have acidic proteins (Bayley, 1966). The amino acid compositions of eukaryotic and prokaryotic ribosomal proteins are very similar, with a large number of basic and acidic, probably amidated, residues and small amounts of the sulfur amino acids and tryptophan (Petermann, 1964). Since a main function of the ribosome is to bind the reactants of the translation apparatus, it is not surprising that ribosomes tend to bind extraneous proteins. Purification schemes should therefore include a dilute salt wash.

Probably any simple definition for a ribosomal protein will prove incorrect when more is known about the functions of the individual proteins. An appropriate definition might be a protein that does not leave the subunit at any step in protein synthesis. This definition eliminates the factors known to cycle on and off, such as initiation factors and elongation factors, but would require that all ribosomes be identical. For the *E. coli* ribosome, where every protein has been separated and characterized, it was thought that certain small subunit proteins were "dispensable" at certain stages. Stoichiometry, the size of the subunit and the protein content, seemed to indicate that not all the proteins could fit on the subunit. However, the most recent numerology suggests that there are no "fractional" proteins except for S1, the largest protein. The problem arises from the difficulty in determining the size and composition of the subunit with precision. Since the individual proteins are so small (10,000–30,000 daltons), the numbers of molecules of each kind can change drastically and still fall within the error of the measurements. Held *et al.* (1974) have demonstrated that one of the so-called dispensable proteins, S21, is required for activity.

A unique feature of eukaryotic ribosomal proteins is their modification by phosphorylation. Although *E. coli* proteins can be phosphorylated *in vitro,* they are not phosphorylated *in vivo*. Such a reversible modification might constitute a control mechanism in translation in eukaryotes. *In vivo* experiments of Kabat (1970) suggested a correlation between phosphorylation and the functional state of reticulocyte ribosomes. Although Eil and Wool (1973) could not detect an effect of phosphorylation on several ribosome functions *in vitro,* Gressner and Wool (1974) found that one small subunit protein, but no large subunit protein, was labeled *in vivo*.

Wool and Stöffler (1974) have shown by immunological techniques that two acidic proteins of the rat liver ribosome, L40/L41, are structurally and functionally homologous to the L7/L12 protein pair of the *E. coli* ribosome, although twice the size. In *E. coli* these two proteins are the only acidic ones; they are identical except that one is acetylated at the *N*-terminal. They are involved in the binding of the elongation factors and in the hydrolysis of GTP (Möller, 1974). No homologies were detected among the other proteins, although Wool and Stöffler admit that more sensitive immunochemical techniques may eventually detect some.

Functions of the Ribosome in Protein Synthesis

Pain and Clemens (1973) have summarized current data on the intracellular distribution of factors required for protein synthesis. Since the function of the initiation factors is to cycle on and off the ribosome (Figure 1), it is not surprising to find some fraction of them in the cytosol as well as bound to the polysome. Generally, the elongation factors, EF1 and EF2, are found in the cytosol. The commonly used pH 5 fraction contains tRNAs and their aminoacyl transferases, but there are instances where the latter enzymes cosediment with ribosomes (or subunits). Thus ribosomes are not free of tRNA, IFs, or even fragments of mRNA unless the ribosomes are subjected to extensive "washing" with salt solutions.

Three stages in protein synthesis can be identified: initiation, elongation, and termination. At each stage, specific accessory protein factors named for those steps become associated with the ribosome, aminoacyl tRNA (aa-tRNA) and messenger RNA (mRNA), and then are successively displaced. In *E. coli,* the best-studied system, factors IF1, IF2, and IF3 participate in initiation; EF-Tu and EF-Ts in the binding of aa-tRNA to the ribosome; EF-G in translocation (movement along the mRNA); and RF1 and RF2 in release of the completed polypeptide. Table 2 lists these factors for prokaryotes and eukaryotes. The first step in the synthesis of a protein is the attachment of each amino acid to the terminal adenine of a specific transfer RNA molecule by an aminoacyl transferase to form the aminoacyl-tRNA (aa-tRNA) (Loftfield, 1972).

Chapter 10 in this volume deals with the structure of the transfer RNAs and Chapter 9 with messenger RNA. Lengyel (1974) has reviewed the translation process in general. See also Lucas-Lenard and Lipmann (1971) and Haselkorn and Rothman-Denes (1973) for reviews of protein synthesis. The following discussion focuses on the role of the ribosome in the protein synthetic process.

Ribosomes can function *in vitro*. Given mRNA, amino acids, tRNAs, the aminoacylating enzymes, ATP and an ATP-generating system, GTP, K^+ and Mg^{2+}, and the various accessory protein factors, a faithful copy of the protein specified by the mRNA is synthesized. Since translation can occur in heterologous systems, it has been argued that factors controlling messenger RNA selection do not exist. But since *in vitro* systems are far less efficient than *in vivo* ones and are subject to artifacts from such simple causes as nonoptimal values of Mg^{2+} or monovalent cation concentrations (Anderson, 1974), subtle controls may be difficult to duplicate *in vitro*.

The first step in translation is the formation of an initiation complex. The small subunit associates with a special initiator tRNA, Met-

TABLE 2. *Protein Factors Involved in Protein Synthesis*

E. coli			*Mammals*	
Mol wt	*Factor*	*Function*	*Factor*	*Mol wt*
I. Initiation (after Anderson, 1974)				
9,400	IF1	Ribosomal protein?	IF-M_{2B}	~20,000
80,000	IF2	Binds fMet-$tRNA_F$	IF-M_1	
		GTPase activity	IF-M_{2A}	
21,500	IF3	Binds natural mRNAs	IF-M_3	
		Dissociates monomers	DF	
II. Elongation (from Leder, 1973)				
34,000	EF-Ts	Complexes with and activates EFTu · GDP	EF-1	100,000–300,000
47,000	EF-Tu	Binds aa-tRNA and GTP		
		GTPase activity		
83,000	EF-G	Translocation	EF-2	60,000–65,000
III. Termination (from Caskey, 1973)				
44,000	RF1	Recognizes UAG, UAA	RF	225,000
47,000	RF2	Recognizes UGA, UAA		
	S	Stimulates release	S?	

$tRNA_f^{Met}$, that recognizes an initiator codon, either AUG or GUG. This tRNA, unlike the bacterial one, is not formylated *in vivo,* although it can be formylated *in vitro* by *E. coli* transformylase. Eukaryotes lack this enzyme. The eukaryotic initiator tRNA has a unique sequence in loop IV (the T4 loop, Chapter 10) from which pseudouridine is absent (Petrissant, 1973; Simsek *et al.,* 1973).

These structural differences of the eukaryotic initiator tRNA may be related to a different mode of initiation in eukaryotes. They may ensure binding of the initiator tRNA to the small subunit rather than the large one, since the T4 loop (loop IV) is involved in binding tRNA to the large subunit during elongation (Richter *et al.,* 1973). Binding of the initiator tRNA to the 40 S subunit is independent of the template, suggesting that the tRNA bound to the ribosome may select the initiation region of the mRNA rather than the reverse (Darnborough *et al.,* 1973; Schreier and Staehelin, 1973).

In prokaryotes three protein initiation factors are involved. In eukaryotes the exact number and nature of the factors are not known with certainty; several are found associated with "native" small subunits (Hirsch *et al.,* 1973). Figure 1 shows how the ribosomal subunits may associate and dissociate during protein synthesis (Ayuso-Parilla *et al.*, 1973).

There is evidence that translation may be controlled at the level of initiation by messenger RNA selection, for example. Such control mechanisms are postulated to be implemented by proteins (Revel *et al.*, 1972) or small RNA molecules (Heywood *et al.*, 1974) or other substances such as hemin (Clemens *et al.*, 1974). It has been suggested that association of mRNA with recruitment or protection factors may be a "preinitiation" event and a genetic control mechanism for mRNA selection (Lingrel and Woodland, 1974).

There are two sites for the binding and reaction of aa-tRNAs. These are named A for the acceptor site and D for the donor site. The latter is also called the P site because it binds the growing peptide chain attached to the last preceding tRNA. Peptide bond formation occurs between amino acids attached by the carboxyl terminus to the 3′-terminal adenine of their specific tRNAs. Since the peptide chain grows from the amino terminal to the carboxyl end, a shuttling mechanism operates. Components present in the P site also react with the antibiotic puromycin, which is structurally analogous to aminoacyl tRNA.

The simple two-site model has been criticized, and Pestka (1972), for example, has proposed that there are two P sites on the large subunit. Swan *et al.* (1969) suggest that there is a third site for aa-tRNA on the small subunit. The fact is that more than two aa-tRNA molecules are found bound to ribosomes (Tai and Davis, 1972).

The elongation phase can be resolved into three steps: (1) codon recognition by an aa-tRNA entering the A site (EF1 and GTP participating); (2) peptide bond formation between the peptide in the P site and the new aa-tRNA in the A site by the ribosomal peptidyltransferase; (3) translocation, i.e., release of the uncharged tRNA, movement of the elongated peptidyl-tRNA to the A site, and advancement of the mRNA by precisely one codon (EF2 and GTP participating).

Termination of protein synthesis occurs when one of three terminator codons (UAA, UAG, or UGA) on the mRNA is in the ribosomal A site. In prokaryotes, instead of special tRNAs, two protein "release factors," RF1 and RF2, participate. In eukaryotes, only one protein release factor has been isolated; it responds to all three terminator codons (Caskey, 1973).

The Ribosome, the Organelle for Translation

The ribosome may be considered a multienzyme complex, and some aspects of ribosome function may be viewed as allosterism on a multimolecular level. In addition to binding the reactants in the correct registry, the ribosome must simultaneously move aong the mRNA and shuttle the peptidyl-tRNA from the acceptor site to the donor site after formation of the new peptide linkage. These movements are called *translocation*. Ribosomal subunits respond to changes in the ionic environment by changes in their shape and conformation. We assume that they must be flexible in order to move, perhaps inchworm style (Petermann, 1971*a*), on the mRNA, although other mechanical models such as a rotating sprocket (Leder, 1973) and a locking–unlocking of the subunits (Spirin, 1969) have been discussed. In contrast, others (Woese, 1970; Rich, 1974) would attribute all movement to conformational changes in the tRNAs. Hydrolysis of GTP by ribosomal GTPase may supply the energy for some phase of translocation. That energy apparently is not required for peptide bond formation. The unique role of GTP in these steps has been lucidly described by Lengyel (1974). He speculates that GTP is required for the binding of factors, causes a conformational change in the ribosome, and is cleaved to GDP and inorganic phosphate, whereupon the factor is released. In *E. coli* the four factors involved, IF2, EF-Tu, EF-G, and RF, cannot be attached to the same ribosome simultaneously. That the ribosome can sequester both the product and the mRNA from enzymatic degradation supports the idea that it has an "interior" and is not just a surface on which reactions occur.

A division of labor exists, however, since each subunit performs special functions. The small subunit is involved in initiation, while the main reaction, peptide bond formation, is a function of the large subunit. Although the binding of the initiator tRNA is a small subunit function, the acceptor (A) and donor (D or P) binding sites for the aa-tRNA and the peptidyl-tRNA are mainly on the large subunit. The peptidyltransferase activity is an integral function of large subunit proteins. Under special conditions, in the presence of alcohol, a peptide bond can be formed in the absence of the small subunit, mRNA, elongation factors, and GTP. This is known as the "fragment reaction" (Monro *et al.*, 1969). Moore *et al.* (1975) have identified the large subunit protein, L16, as essential for peptidyl transferase activity in *E. coli*. That protein is one of the 18 proteins that dissociate from the subunit "core" when it is treated with LiCl.

RNA Synthesis and Processing: Ribosome Biosynthesis

Table 3 identifies the RNA species found in a typical animal cell. The discovery of low molecular weight RNAs in both the nucleus (Weinberg, 1973) and cytoplasm (Bogdanovsky *et al.*, 1973; Heywood *et al.*, 1974), with as yet not completely defined functions, shows that the listing is probably far from complete. Chapter 7 of this volume deals with the nonribosomal RNAs of the nucleus.

Most RNAs are modified after transcription. Such posttranscriptional events may offer points in the processing at which genetic controls may be exerted. Except for 5 S RNA, RNA molecules are synthesized as "precursors" which are later cleaved. Further posttranscriptional modification occurs with specific methylation of the RNA (Greenberg and Penman, 1966). In eukaryotes, the methylation pattern of ribosomal RNA differs from that of tRNA or mRNA in that most of the methyl groups are found on the 2′-hydroxyl groups of the ribose instead of on the bases. Methylation occurs on the nascent 45 S precursor RNA and only in the conserved regions. Figure 5 shows the processing patterns for HeLa and yeast RNAs. In the HeLa cell only half the precursor molecule is conserved in the maturation process. Methylation may serve to identify the

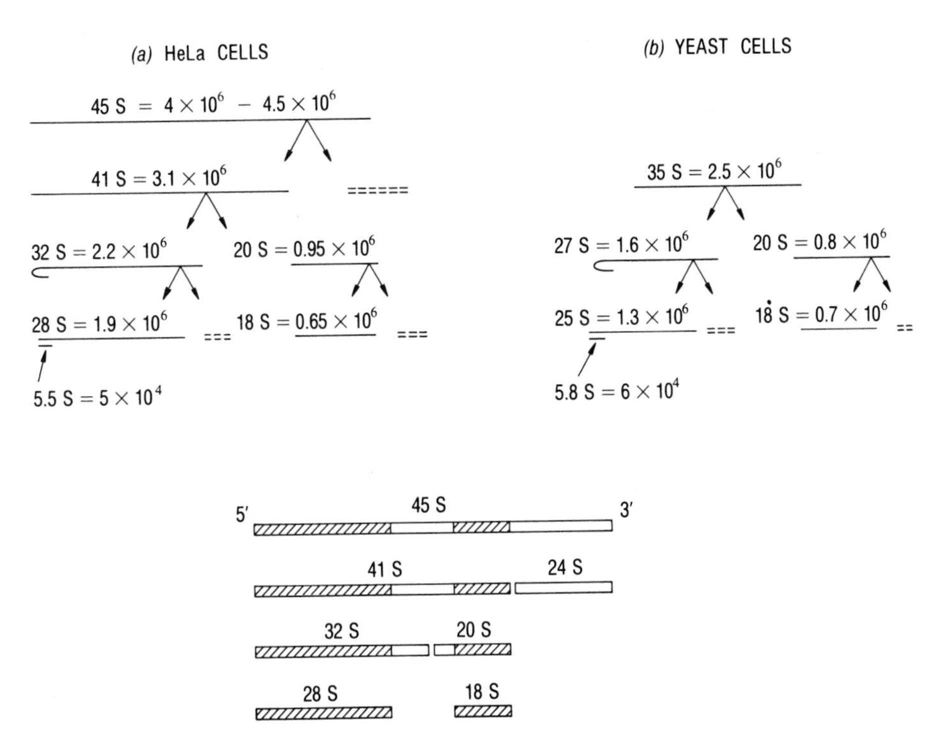

Figure 5. Upper diagrams: Comparison of the formation of ribosomal RNA in the nuclei of (a) HeLa and (b) yeast cells as presented by Warner et al. (1973) from biochemical data of Weinberg and Penman (1970) for HeLa cells and of Udem and Warner (1972) for yeast. The sedimentation rate and molecular sizes are given. Dashed lines indicate the spacer regions. Lower diagram: Processing of HeLa cell ribosomal RNA from secondary structure maps based on electron microscopy by Wellauer and Dawid (1973).

TABLE 3. RNA Species in Animal Cells

	Nucleus[a]		*Cytoplasm*[b]				
	Nucleoplasm	*Nucleolus*	*Cytosol*	*Bound to endoplasmic reticulum or free*			*Bound to polysomes, membranes, and/or protein*
				Ribosomal RNAs			
	Heterogeneous RNA (HnRNA)	*Precursor ribosomal RNA (pre-rRNA)*	*Transfer RNA (tRNA)*	*Large rRNA (light rRNA)*	*Small rRNA*	*5 S RNA*	*Messenger RNA (mRNA)*
Percent of total[c]	≪1	4	12	53	24	~1	~3
Sedimentation rate (S)[d]	~20–70 polydisperse	~45	~4	28–30 (5.8)	16–18	4.5	10–20 polydisperse
Approximate molecular weight (× 10^{-6})	0.5–10	4	0.02	1.7 (0.05)	0.7	0.04	0.03–0.2
Approximate number of nucleotides	2000–50,000		75	~5000 (160)	~2300	120	
Composition	DNA-like	GC-rich	Many unusual bases	GC-rich		GC-rich	DNA-like

Methylation[e]	—	Chiefly on ribose	On bases 7%	Chiefly on ribosome 1.3%	—	0.2%
Structure[f]	Poly(A) at 3′ terminus; oligo(U); repeated sequences	~40% spacer	Cloverleaf compact L	Mainly double-helical hairpins	Cloverleaf	Poly(A) at 3′ terminus (except histone mRNA); unusual methylated sequence at 5′ terminus
Function	10% precursor of mRNA; 90% "turns over"	Precursor of large and small rRNAs	Adapter in protein synthesis	Structural? Binding functions	Interacts with tRNA? Associated with ribosomal GTPase	Genetic information for translation

[a] Precursors of tRNA are also present in the nucleoplasm as well as other low molecular weight species of RNA of unknown function associated with the chromosomes (Weinberg, 1973).
[b] Mitochondria contain ribosomes of small size with small RNAs, but no 5 S RNA (see Table 1). Small RNA species may be involved in the control of translation (Heywood *et al.*, 1974).
[c] After Darnell (1968).
[d] The sedimentation coefficients of single-stranded polynucleotides are extremely sensitive to the ionic composition of the solvent.
[e] Data on methylation from Perry and Kelley (1974).
[f] Structure of tRNA (Rich, 1974); mRNA (Weinberg, 1973; Adesnik *et al.*, 1972; Rottman *et al.*, 1974).

conserved species for the nucleolytic enzymes. Such a mechanism seems highly probable for mRNA processing, where a unique methylated sequence has been found at the 5′ end (Rottman *et al.,* 1974).

Transcription of rRNA genes occurs in the nucleolus, which is also the site of assembly for the ribosomal subunits (Warner *et al.,* 1973). The ribosomal proteins are synthesized in the cytoplasm and are transported across the nuclear membrane into the nucleolus for assembly into particles. To study the maturation of ribosomal subunits, Kumar and Subramanian (1975) compared the labeling patterns of the proteins of the mature 60 S subunit of the cytoplasm with its 55 S nucleolar precursor that contains 32 S RNA (see Figure 5). Their two-dimensional gel electrophoresis patterns showed that the precursor particles contained 65 spots, only 30 of which were found in the mature subunit's population of 40 components. Moreover, the extra 33 proteins of the precursor were acidic and of higher molecular weight than the ribosomal proteins. The authors suggest that limiting amounts of the rapidly labeled, pre-rRNA-specific proteins may have a regulatory role in ribosome maturation.

The maturation of the small ribosomal subunit may be less complicated since newly synthesized 18 S RNA is detectable in the cytoplasm much more quickly than is 28 S RNA (Darnell, 1968). Perhaps it acquires its "extra proteins" in the cytoplasm.

Perry (1973) has discussed the regulation of ribosome content in eukaryotes. Ribosome content depends on the number of genes coding for ribosomal RNA and their rate of transcription. But for such complex structures regulation could occur at many points during synthesis, processing, and assembly. In addition, in the tissues of higher eukaryotes mature ribosomes "turn over" in the cytoplasm (Abelson *et al.,* 1974).

An unusual posttranscriptional event in eukaryotes is a special modification of mRNA that occurs at the level of its precursor, the giant heterogeneous nuclear RNA (HnRNA), and that is the addition of a sequence of several hundred adenine residues at the 3′-OH end (Darnell *et al.,* 1973). This poly(A) tail, present in all mRNAs except histone mRNA, has provided a convenient method for identifying and separating mRNAs (Weinberg, 1973). Most of the HnRNA "turns over" in the nucleus, while the mRNA, representing only a fraction, is transported to the cytoplasm. It was thought that mRNA was transported with the small subunit. Later it was found that mRNA–protein complexes cosediment with the small subunits but can be resolved from them on the basis of their lower buoyant density in CsCl. They contain more protein than do ribosomes (Williamson, 1973). Other data suggest that the situation is more complicated; mRNA–protein complexes may be part of a large network that includes membranes (Faiferman *et al.,* 1971).

Eukaryotic mRNAs differ from prokaryotic ones in several other ways. There is no evidence for polycistronic or polygenic mRNA in eukaryotes, with the exception of virus-specific RNAs that encode polyproteins that are cleaved after synthesis. The mRNAs of eukaryotes seem to have half-lives of hours rather than minutes as in bacteria (Abelson *et al.*, 1974). See Chapter 9 for a discussion of cytoplasmic mRNA.

Evolution

That ribosomes of diverse organisms differ in structure (Table 1) is intriguing since their functions in translation seem not to differ significantly. At least four classes exist if the sizes of the RNA molecules are considered. In addition, whereas the size distributions of the proteins are similar, the amount of protein increases from prokaryote to eukaryote, but not in proportion to the RNA size. One obvious difference is cellular location. In bacteria, ribosomes attach to mRNA as it is synthesized, whereas eukaryotic ribosomes function in the cytoplasm. The mitochondrial ribosomes have the highest percentage of protein and the smallest RNA species. Yeast ribosomes are smaller than rat liver ribosomes. Attempts have been made by Loening (1973) to rationalize the evolutionary status of an organism with the sizes of its ribosomal RNAs and their precursors. The amount of excess or spacer RNA varies considerably in different organisms (see Figure 5). Perry *et al.* (1970) found that in plants and lower animals the precursor molecule is 2.7–2.8 million daltons, only about 25% larger than the mature rRNA products, while in higher animals it is about 4 million daltons and 80% larger. The size difference also correlates with the development of temperature control, with poikilothermal vertebrates resembling nonvertebrates in having smaller precursor RNAs.

Hultin and Sjöqvist (1971) found great similarities in the two-dimensional polyacrylamide gel electrophoretic patterns of the acid-soluble proteins of liver ribosomes from five different vertebrates. One large subunit protein of a homeothermal species was larger than the corresponding protein in a poikilothermal species and correlated with the conformational stability of its subunit at increased temperature and ionic strength.

Delaunay *et al.* (1973) examined the total ribosomal proteins of 13 animal species by two-dimensional polyacrylamide gel electrophoresis. Their data suggest that many proteins may be common to all eukaryotes. The greatest degree of proximity is between mammals and reptiles and the proximity to mammals decreases as organisms become more primitive. In similar studies on plant ribosomes by electrophoretic and immunochemical techniques, Gualerzi *et al.* (1974) established that, while there is a high degree of evolutionary conservation among the proteins of cytoplasmic ribosomes of higher plants, there is little homology between the proteins of the chloroplast ribosomes and the cytoplasmic ribosomes of the same plant. Moreover, no homologies were detected between chloroplast ribosomal proteins and those from ribosomes of mitochondria, bacteria, or blue-green algae. On the other hand, as described earlier in the section on ribosomal protein, Stöffler *et al.* (1974) did find homology for two acidic ribosomal proteins in *E. coli* and rat liver.

Genes for Ribosomal RNA

Since the RNA of ribosomes constitutes 80% of a cell's RNA, one expects the genes specifying those RNAs to be numerous. The genes for ribosomal RNA have been identified and enumerated by RNA:DNA hybridization. Distinct but clustered genes exist for the two high molecular weight components. They are present in multiple copies of the order of hundreds in animal cells, about 10 times less in bacteria, and 10 times more in plant cells (Attardi and Amaldi, 1970). Regulation of the rate of transcription seems to be the general mechanism by which rRNA production is adjusted to the cell type or stage of differentiation.

In eukaryotic cells, the rRNA genes are physically associated with the nucleolar organizer and in amphibia the selective replication of the organizer that occurs at early stages of oogenesis has permitted electron microscopic visualization of rRNA replication (Hamkalo and Miller, 1973).

The ratio of transfer RNA sites to rRNA sites is constant, although the values vary from 2.5 to 10 in different estimates. There is a nonequivalence of gene dosage for 5 S RNA and the high molecular weight species. Although 5 S RNA is associated with the large subunit on a 1-to-1 molar basis, its genes are found in different chromosomes from those which carry the genes for 18 S and 28 S RNA (Steffensen *et al.*, 1974). Moreover, its synthesis is not coordinated with that of 45 S RNA, the precursor of the two large species.

Ribosomal Mutations

Sager has pointed out that a fundamental feature of the genetics of eukaryotes is the interaction of multiple genetic systems, the nuclear and mitochondrial in animals, and a third system in plants, the chloroplast or plastid (see Chapter 12).

So far no genetic analyses comparable to those with prokaryotes have been possible in animal cells because of the lack of mutants required to detect recombination and to map linkages. The progress that is being made with the genetics of the chloroplast ribosome is described in Chapter 13 of this volume. Schlessinger (1974) has discussed these problems.

Only in certain primitive eukaryotes such as yeast, the genetics of which is described by Sherman and Lawrence in Chapter 24 of Volume 1 of this series, has it been possible to take a direct genetic approach to the study of ribosomes. Certain temperature-sensitive mutants have a defect in the initiation stage of protein synthesis (Hartwell and McLaughlin, 1969). Petersen and McLaughlin (1974) have shown that the polysome profiles reflect the nature of the molecular defect in mutants of yeast. One of the few known mutations in a eukaryotic ribosome is resistance to cycloheximide (Cooper *et al.*, 1967). However, resistance occurs naturally in some strains and seems to be a property of the large subunit (Rao and Grollman, 1967). Chain termination mutations have also been observed in yeast cells; the "nonsense" characteristics may correspond to the two bacterial types (Caskey, 1973).

Genetic Aspects of Translation

Leder (1973) has pointed out that in bacterial cells initiation and termination factors occur in much lesser amounts than do ribosomes and binding and translocation factors. In eukaryotes much evidence points to initiation as the step most subject to control. The genes specifying those factors may be subject to independent control. In higher animals and man, studies of the abnormal hemoglobins and the genetics of human disease (Kitchen and Boyer, 1974) have contributed greatly to understanding the control mechanisms of translation, as have studies of protein synthesis in virus-infected cells. Somatic cell hybridization can provide animal cell "mutants" (Marx, 1973; Baumal *et al.*, 1973; Cotton and Milstein, 1973). Cells fuse under the influence of Sendai virus, some chromosomes are discarded, but exactly what happens is not yet completely understood.

Cancer remains the most baffling "mutation" in higher organisms. An early event after administration of a carcinogen to an animal is a decrease in protein synthesis accompanied by "degranulation" of the endoplasmic reticulum, but these early events may be a result of a defect in some step of protein synthesis or the binding of the ribosomes to the membrane rather than a structural change in the ribosome itself. One comparison of the proteins of ribosomes of normal liver and hepatoma cells by two-dimensional gel electrophoresis showed no differences (Delaunay and Schapira, 1972). Probably it is at the control mechanisms of translation that one must look for the effects of genetic changes in such conditions as transformation to the malignant state rather than at the translation machinery itself. As ribosome technology improves and the detailed mechanisms of translation are revealed, the important question of how translation is controlled will be answered.

Appendix: Techniques of Ribosomology

Two areas of ribosomology may be of special interest to geneticists: (1) alterations in control mechanisms mediated through any one of the many factors that participate in the translation process and (2) alterations in the structural components of the ribosomes themselves that affect their function. The technology for the latter studies has improved greatly in recent years. With only a few micrograms of purified ribonucleoprotein, a thorough physical characterization of ribosomes and their subunits can be made (Hamilton, 1974*a*). With a few milligrams, their proteins can be examined by polyacrylamide gel electrophoresis in systems that permit close comparisons (Hamilton 1974*b*). Determining sequences of RNA molecules or fragments may soon be as feasible as physical examination by ultracentrifugation is now.

Cell fractionation schemes are operational and flexible provided that appropriate ionic conditions are maintained. Figures 6A–6F outline typical fractionation procedures and illustrate schematically various other treatments that are described in the following section, which is a compendium of preparative and analytical methods that are particularly useful in the study of ribosomes.

Extinction Coefficients for RNA, Protein, and Ribonucleoprotein. RNA and proteins absorb ultraviolet light. The spectrum of RNA has a maximum at 260 nm and a minimum near 230 nm. The ratio of absorbancies at 260 nm to either the minimum or 280 nm is close to 2. The protein mixture, while not rich in the aromatic amino acids, has a maximum near 270 nm and a minimum near 250 nm. The ribonucleoprotein complexes have a maximum at 260 nm, but the minimum shifts to 235 nm and the ratio of the absorbancies at 260 nm to the minimum or 280 nm is reduced to 1.5–1.8. The first ratio is more sensitive to the amount of protein than is the 260:280 ratio, but it is difficult to quantitate the measurement without accounting for scattering. Approximate extinction coefficients for RNA, protein, and ribosomes are 25, 1, and 13, respectively. These coefficients are defined as A_{260} units/mg/ml in a 1-cm-thick cuvette. They should be determined experimentally from chemical analyses for RNA (by the orcinol colorimetric assay for purine-bound pentose) and for protein (by the Lowry method).

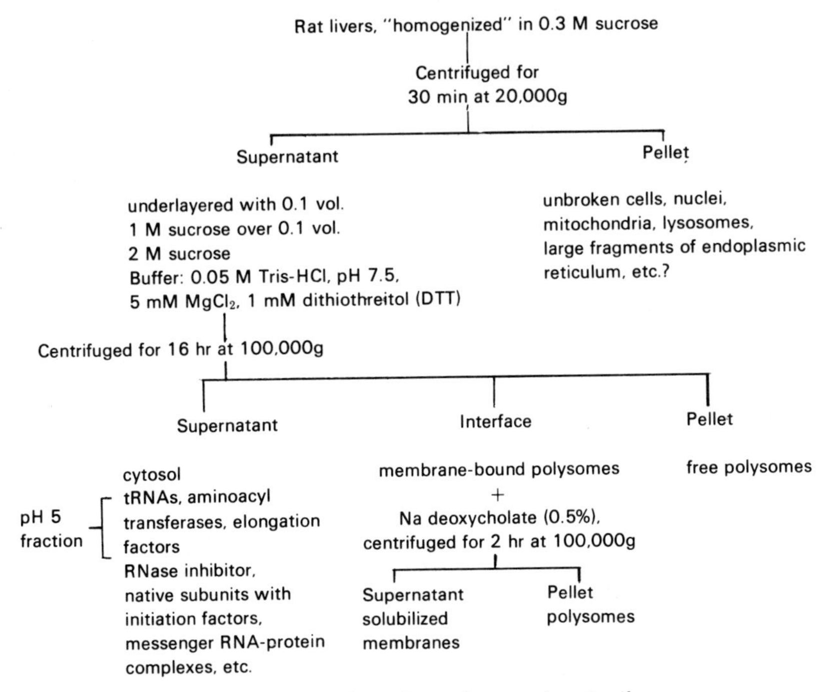

Figure 6A. Cell fractionation scheme for cytoplasmic ribosomes.

Cell Fractionation; Isolation of Free and Bound Polysomes. Many methods have been published for the isolation of ribosomes or polysomes from tissues such as liver. Basically they involve mechanical homogenization with care to avoid the disruption of nuclei, mitochondria, and lysosomes. Bloemendal *et al.* (1973) point out that losses of endoplasmic reticulum inevitably occur in the differential centrifugation methods used to remove the larger structures, and therefore quantitation of the membrane-bound and free polysomes is difficult.

Nevertheless, free and bound polysomes can be separated on the basis of their densities, usually by sedimenting the free species into dense sucrose (2 M) into which the membrane-bound ones do not penetrate because of their lower density. A detergent, sodium deoxycholate, which solubilizes the membranes without affecting the ribosomes, is used to release the "bound" polysomes, which can then be sedimented through 2 M sucrose.

Ribonuclease and Its Inhibitors. Enzymes that degrade RNA are ubiquitous and extremely stable basic proteins that bind to their substrate. While the ribosome will retain its structure despite nicks in its RNA (Cox, 1969), intact RNA cannot be recovered without scrupulous attention to avoiding contamination by extraneous RNAses or attempting to inhibit endogenous ones. The cytosol of rat liver contains an inhibitor of

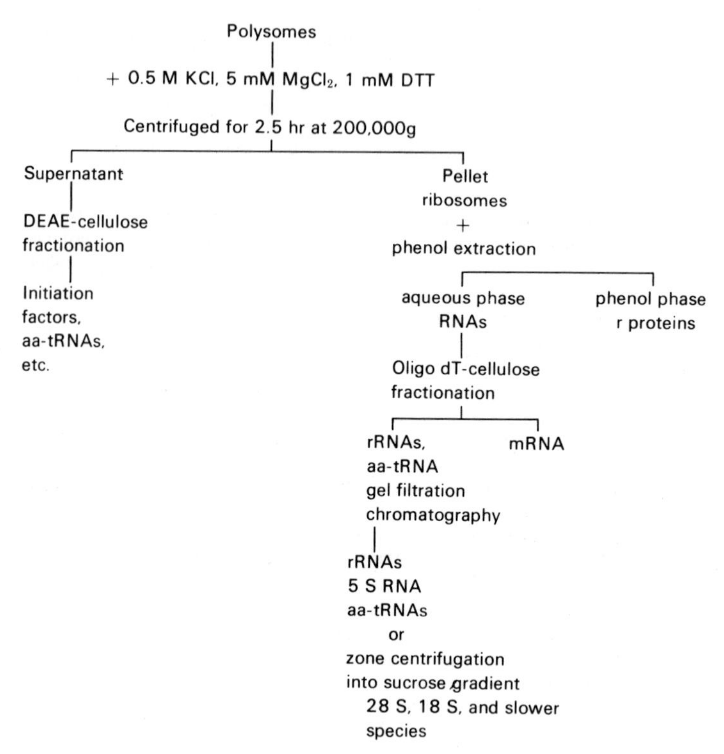

Figure 6B. Extraction of factors for protein synthesis and recovery of mRNA, etc.

RNAse (Roth, 1958) and that fraction can be added back to ribosomes (Blobel and Potter, 1966). The clay bentonite can be used during the purification of rat liver ribosomes provided that it is saturated with Mg^{2+} (Petermann, 1971*b*).

For RNA, other substances such as polyvinyl sulfate and sodium dodecylsulfate have been used. Diethyl pyrocarbonate can be used to decontaminate glassware, etc., but it chemically modifies RNA.

Isolation and Purification of Ribosomes. One method for rat liver ribosomes involves the use of sodium deoxycholate, a pH 8 wash, and $MgCl_2$ precipitation (Petermann, 1971*b*). The resulting fragmented polysomes have endogenous activity in *in vitro* amino acid incorporation.

Dissociation of Ribosomes to Active Subunits and Separation of Subunits. Only in the last few years have methods been developed to obtain active subunits from eukaryotic ribosomes. Treatment with low concentrations of urea after depletion of the Mg^{2+} dissociates rat liver ribosomes (Petermann and Pavlovec, 1971). A widely used method is exposure to a high concentration of KCl after removal of mRNA and nascent peptides by puromycin (Martin *et al.*, 1971; Blobel and Sabatini, 1971).

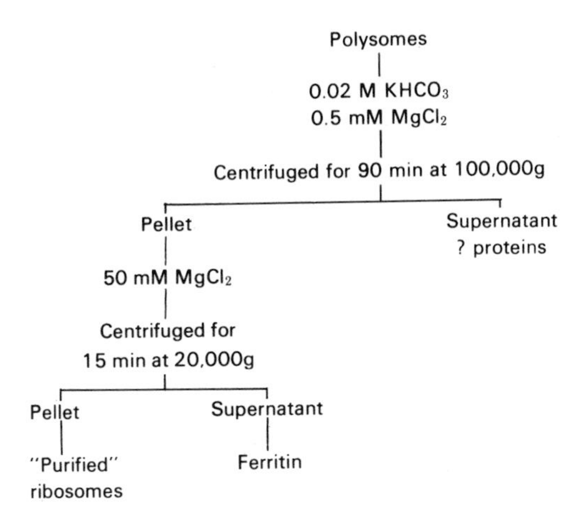

Figure 6C. Purification of rat liver ribosomes by magnesium ion precipitation.

The subunits are separated by zonal centrifugation in a sucrose gradient (Hamilton and Ruth, 1969). The batch-style zonal rotors (Anderson, 1966) can separate several hundred milligrams of ribosomal subunits in an overnight centrifugation. Pooled fractions are pelleted after dilution of the sucrose. Sucrose and reducing agents such as dithiothreitol (DTT) preserve the activity of isolated subunits and ribosomes.

Characterization of Ribosomal Subunits. The purity of the isolated subunits must be assessed before extensive characterization is worth undertaking. Since ribosomal subunits have a tendency to dimerize as well as to unfold, cross-contamination of the subunits is a very common occurrence. One way to detect either dimers or unfolded forms is to deliberately unfold the subunit by treating it with the chelator, EDTA. Then either sedimentation velocity analysis or isodensity equilibrium centrifugation will reveal the presence of the "contaminant" (Hamilton, 1974*a*). Another method requires examination of the RNA species. The ionic detergent sodium dodecylsulfate (SDS) forms complexes with the proteins, releasing the RNA species which can then be identified by their sedimenta-

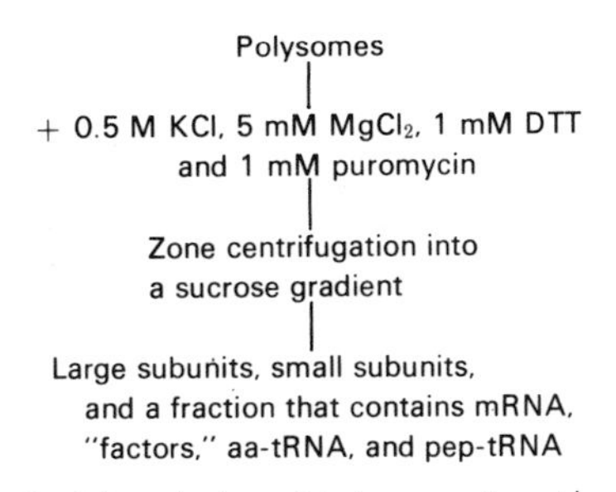

Figure 6D. Dissociation of polysomes to active subunits.

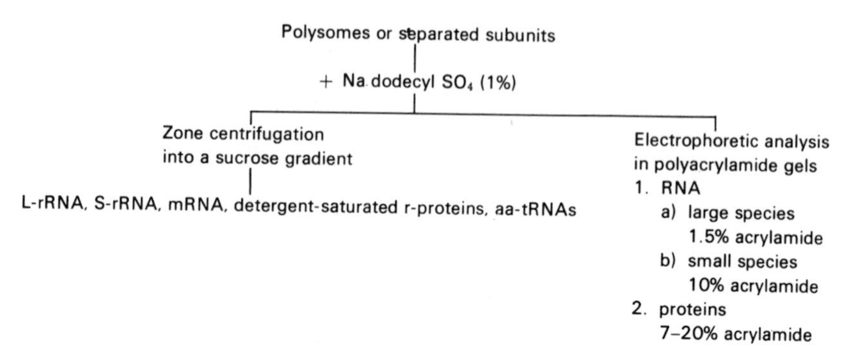

Figure 6E. Detergent treatment of polysomes or isolated subunits for separation of RNA species or polyacrylamide gel electrophoresis.

tion properties or mobilities in sieving polyacrylamide gels as described below.

The properties of interest are the molecular size, the chemical composition, the species of RNA, and the size and number of proteins. A complete characterization as to size and composition can be made in an analytical ultracentrifuge equipped with ultraviolet absorption optics with only a few micrograms of material. The sedimentation coefficient is measured by moving-boundary centrifugation, the molecular weight by high-speed equilibrium centrifugation, and if the subunit is fixed with formaldehyde the buoyant density is measured in CsCl, from which the protein content can be estimated by isodensity equilibrium centrifugation (Hamilton, 1971, 1974*a*).

Extraction of RNA and Protein from Ribosomes. The ionic detergent sodium dodecylsulfate dissociates the RNA and protein of ribosomes, the proteins bind detergent and the larger RNA molecules can be sedimented into sucrose density gradients (Noll and Stutz, 1968). Other extraction methods can be used to obtain pure RNA. In the phenol method of Kirby (1968), RNA remains in the aqueous phase when the ribosome

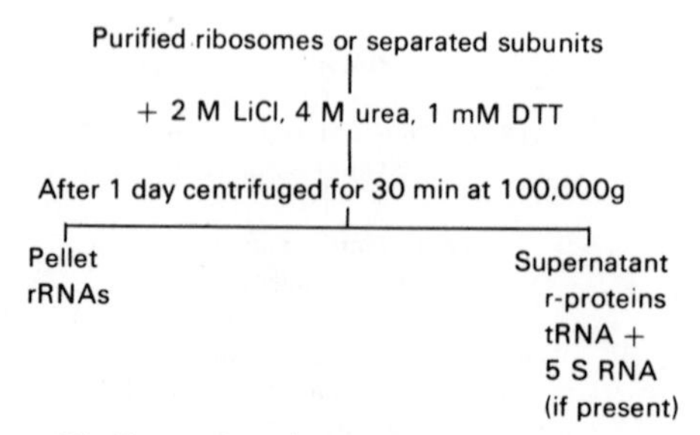

Figure 6F. Extraction of r-proteins for characterization.

solution is extracted by an equal volume of buffer-saturated phenol. The proteins are partitioned into the phenol layer, from which they can be precipitated by the addition of several volumes of acidified methanol. The RNA is recovered from the aqueous phase by the addition of 2 vol of ethanol, to which acidified sodium or potassium acetate has been added.

The proteins can be extracted from ribosomes by a variety of methods. It was Chao (1957) who first noted that strong salts such as 1 M $MgCl_2$ or 2 M NaCl would disrupt ribosomal structure. LiCl is now the preferred agent, usually with urea in a concentration of at least 6 M (Leboy *et al.*, 1964). Lithium salts have a destabilizing or "chaotropic" effect on the structure of water and the conformation of macromolecules (von Hippel and Wang, 1964). The large, single-stranded ribosomal RNAs precipitate, while the proteins and RNAs that have a high proportion of secondary structure, i.e., double-strandedness, remain soluble. Thus transfer RNAs contaminate the protein fraction when intact ribosomes are extracted, but the ribosomal subunits are usually free of tRNA, although native small subunits are not. The ribosomal proteins are insoluble in the absence of urea or other such agents at neutral pH, although they are soluble at very low pH. Other extraction methods have been used, such as the acetic acid extraction method originally applied to viruses (Kurland *et al.*, 1971). Among methods suitable for recovery of both RNA and protein is the use of 6 M guanidine HCl; Cox (1968) recovered RNA by differential precipitation with 0.5 vol of ethanol. The proteins are recovered from such a supernatant (or from 2 M LiCl) by exhaustive dialysis against 0.5 mM dithiothreitol (DTT) before lyophilization.

Identification and Enumeration of RNA and Protein Species by Gel Electrophoretic Techniques. Polyacrylamide gel electrophoresis provides an analytical tool of great resolving power for both RNAs and proteins. For the larger ribosomal RNAs, agarose is frequently added to strengthen the loose gels (less than 2% acrylamide) that are required (Peacock and Dingman, 1968). The RNA species can be stained with methylene blue or scanned directly for absorbancy at 260 nm when purified gel ingredients are used. Usually sodium dodecylsulfate must be removed by presoaking the gel in trichloroacetic acid (which also serves to "fix" the macromolecules) since the detergent is stained by methylene blue as well as by Coomassie blue, a stain used for proteins.

Leboy *et al.* (1964) initiated the use of polyacrylamide gel electrophoresis for the study of ribosomal proteins. The analyses are generally carried out at pH 4.5 in urea, and, while sieving occurs, the net charge affects the migration rate. Shapiro *et al.* (1967) introduced the use of the detergent sodium dodecylsulfate, which imposes a net negative charge by

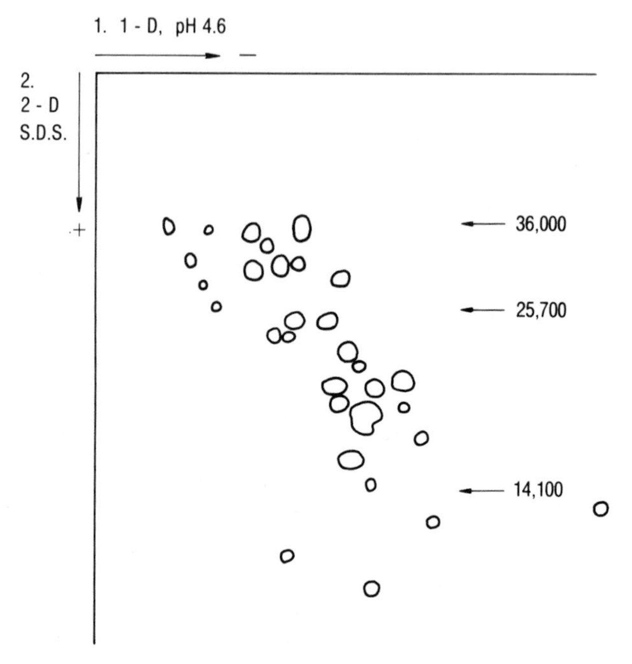

Figure 7. Schematic molecular map of the proteins of the small subunit of the rat liver ribosome. Each spot represents one protein species. (1) After electrophoresis in a cylindrical 7.5% polyacrylamide gel at pH 4.6, the gel cylinder was (2) embedded in a 15% polyacrylamide slab that contained sodium dodecylsulfate and was subjected to an electric field. The numbers at the right indicate the positions of standard proteins that had migrated only in the slab gel.

binding to the proteins; resolution occurs solely on the basis of size. It was a logical development to combine these analyses to obtain a two-dimensional map (Hamilton, 1974*b*). After resolution at pH 4.5, a gel strip is reoriented at 90° to permit migration into a gel slab. With these molecular maps the number of protein species can be counted, and if they are quantitated the number of copies of each molecule can be calculated. A two-dimensional map for the small subunits of rat liver ribosomes is shown in Figure 7. Quantitation is difficult, because in addition to counting the spots, their intensities, i.e., the relative amount of protein in each, must be measured. This has not yet been achieved. Faint spots have generally been ignored. Numerous versions of two-dimensional analysis are possible, since pH, gel concentration, and "additives" can be varied. For example, isoelectric focusing can be used in the first dimension. Another innovation is the use of a concentration gradient of acrylamide to improve resolution in the second dimension.

Kaltschmidt and Wittmann (1970) have varied both pH and gel concentration to display all of the *E. coli* ribosomal proteins in a two-dimensional map. Since each *E. coli* ribosomal protein has been separately characterized, their map now serves as a primary identification for each protein. But for eukaryotic ribosomes the additional molecular size data obtained in the system employing SDS are useful. Sherton and Wool (1972), however, have modified Kaltschmidt and Wittmann's method for rat liver ribosomal proteins. Lin and Wool (1974) have determined the molecular weights of the proteins by a three-dimensional analysis; i.e., they reanalyzed individual spots by a third electrophoretic analysis in one-dimensional gels containing SDS.

Acknowledgment

The author is grateful for support from The National Cancer Institute (Grant CA-08748).

Literature Cited

Abelson, H. T., L. F. Johnson, S. Penman, and H. Green, 1974 Changes in RNA in relation to growth of the fibroblast. II. The lifetime of mRNA, rRNA, and tRNA in resting and growing cells. *Cell* **1:**161–165.

Adelman, M. R., G. Blobel, and D. D. Sabatini, 1973*a* An improved cell fractionation procedure for the preparation of rat liver membrane-bound ribosomes. *J. Cell Biol.* **56:**191–205.

Adelman, M. R., D. D. Sabatini, and G. Blobel, 1973*b* Ribosome–membrane interaction: Nondestructive disassembly of rat liver rough microsomes into ribosomal and membranous components. *J. Cell Biol.* **56:**206–229.

Adesnik, M., M. Salditt, W. Thomas, and J. E. Darnell, 1972 Evidence that all messenger RNA molecules (except histone messenger RNA) contain poly(A) sequences and that the poly(A) has a nuclear function. *J. Mol. Biol.* **71:**21–30.

Anderson, N. G., 1966 Zonal centrifuges and other separation systems. *Science* (*Wash. D.C.*) **154:**103–112.

Anderson, W. F., 1974 Cell-free synthesis of globin chains: An overview. *Ann. N.Y. Acad. Sci.* **241:**142–155.

Andrews, T. M. and J. R. Tata, 1971 Protein synthesis by membrane-bound and free ribosomes of secretory and non-secretory tissues. *Biochem. J.* **121:**683–694.

Attardi, G. and F. Amaldi, 1970 Structure and synthesis of ribosomal RNA. *Annu. Rev. Biochem.* **39:**183–226.

Ayuso-Parilla, M., E. C. Henshaw, and C. A. Hirsch, 1973 The ribosome cycle in mammalian protein synthesis. III. Evidence that the nonribosomal proteins bound to the native smaller subunit are initiation factors. *J. Biol. Chem.* **248:**4386–4393.

Baumal, R., B. K. Birshtein, P. Coffino, and M. D. Scharff, 1973 Mutations in immunoglobulin-producing mouse myeloma cells. *Science* (*Wash. D.C.*) **182:**164–166.

Bayley, S. T., 1966 Composition of ribosomes of an extremely halophilic bacterium. *J. Mol. Biol.* **15:**420–427.

Bickle, T. A. and R. R. Traut, 1971 Differences in the size and number of 80 S and 70 S ribosomal proteins by dodecyl sulfate gel electrophoresis. *J. Biol. Chem.* **246:**6828–6834.

Bielka, H. and W. Scheler, Editors, 1974 *Ribosomes and Biosynthesis of Proteins. Acta Biol. Med. Ger.* **33:**525–980.

Blobel, G., 1973 A protein of molecular weight 78,000 bound to the polyadenylate region of eukaryotic messenger RNAs. *Proc. Natl. Acad. Sci. USA* **70:**924–928.

Blobel, G. and V. R. Potter, 1966 Relation of ribonuclease and ribonuclease inhibitor to the isolation of polysomes from rat liver. *Proc. Natl. Acad. Sci. USA* **55:**1283–1288.

Blobel, G. and D. D. Sabatini, 1970 Controlled proteolysis of nascent polypeptides in rat liver cell fractions. I. Location of the polypeptides within ribosomes. *J. Cell Biol.* **45:**130–145.

Blobel, G. and D. Sabatini, 1971 Dissociation of mammalian polyribosomes into subunits by puromycin. *Proc. Natl. Acad. Sci. USA* **68:**390–394.

Bloemendal, H., E. M. J. Jaspars, A. van Kammen, and R. J. Planta, Editors, 1972 *RNA Viruses: Replication and Structure; Ribosomes: Structure, Function and Biogenesis,* Federation of European Biochemical Societies Symposium, Vol. 27, American Elsevier, New York.

Bloemendal, H., E. L. Benedetti, and W. S. Bont, 1973 Preparation and characterization of free and membrane-bound polysomes. *Methods Enzymol.* **30:** Part F, 313–327.

Bogdanovsky, D., W. Hermann, and G. Schapira, 1973 Presence of a new RNA species among the initiation protein factors active in eukaryotes translation. *Biochem. Biophys. Res. Commun.* **54:**25–32.

Branes, L. and A. O. Pogo, 1975 Biogenesis of polysomes and transport of messenger RNA in yeast. *Eur. J. Biochem.* **54:**317–328.

Bruening, G. E. and R. M. Bock, 1967 Covalent integrity and molecular weights of yeast ribosomal ribonucleic acid components. *Biochim. Biophys. Acta* **149:**377–386.

Caskey, C. T., 1973 Peptide chain termination. *Adv. Protein Chem.* **27:**243–276.

Chao, F.-C., 1957 Dissociation of macromolecular ribonucleoprotein of yeast. *Arch. Biochem. Biophys.* **70:**426–431.

Chao, F.-C., and H. K. Schachman, 1956 The isolation and characterization of a macromolecular ribonucleoprotein from yeast. *Arch. Biochem. Biophys.* **61:**220–230.

Claude, A., 1946 Fractionation of mammalian liver cells by differential centrifugation. I. Problems, methods, and preparation of extract. II. Experimental procedures and results. *J. Exp. Med.* **84:**51–89.

Clemens, M. J., E. C. Henshaw, H. Rahamimoff, and I. M. London, 1974 Met-$tRNA_f^{Met}$binding to 40 S ribosomal subunits: A site for the regulation of initiation of protein synthesis by hemin. *Proc. Natl. Acad. Sci. USA* **71:**2946–2950.

Cooper, D., D. V. Banthorpe, and D. Wilkie, 1967 Modified ribosomes conferring resistance to cycloheximide in mutants of *Saccharomyces cerevisiae J. Mol. Biol.* **26:**347–350.

Cotton, R. G. H. and C. Milstein, 1973 Fusion of two immunoglobulin-producing myeloma cells. *Nature (London)* **244:**42–43.

Cox, R. A., 1966 The secondary structure of ribosomal ribonucleic acid in solution. *Biochem. J.* **98:**841–857.

Cox, R. A., 1968 The use of guanidinium chloride in the isolation of nucleic acids. *Methods Enzymol.* **12:**Part B, 120–129.

Cox, R. A., 1969 The effect of pancreatic ribonuclease on rabbit reticulocyte ribosomes and its interpretation in terms of ribosome structure. *Biochem. J.* **114**:753–767.

Cox, R. A. and A. A. Hadjiolov, editors, 1972 *Functional Units in Protein Biosynthesis,* Federation of European Biochemical Societies Symposium, Vol. 23, Academic Press, New York.

Darnborough, C., S. Legon, T. Hunt, and R. J. Jackson, 1973 Initiation of protein synthesis: Evidence for messenger RNA-independent binding of methionyl-transfer RNA to the 40 S ribosomal subunit. *J. Mol. Biol.* **76**:379–403.

Darnell, J. E., Jr., 1968 Ribonucleic acids from animal cells. *Bacteriol. Rev.* **32**:262–290.

Darnell, J. E., W. R. Jelinek, and G. R. Molloy, 1973 Biogenesis of mRNA: Genetic regulation in mammalian cells. *Science* (*Wash. D.C.*) **181**:1215–1221.

Delaunay, J. and G. Schapira, 1972 Rat liver and hepatoma ribosomal proteins. Two-dimensional polyacrylamide gel electrophoresis. *Biochim. Biophys. Acta* **259**:243–246.

Delaunay, J., F. Creusot, and G. Schapira, 1973 Evolution of ribosomal proteins. *Eur. J. Biochem.* **39**:305–312.

Eil, C. and I. G. Wool, 1973 Function of phosphorylated ribosomes. The activity of ribosomal subunits phosphorylated *in vitro* by protein kinase. *J. Biol. Chem.* **248**:5130–5136.

Faiferman, I., L. Cornudella, and A. O. Pogo, 1971 Messenger RNA particles and their attachment to cytoplasmic membranes in Krebs tumour cells. *Nature* (*London*) *New Biol.* **233**:234–237.

Faiferman, I., A. O. Pogo, J. Schwartz, and M. E. Kaighn, 1973 Isolation and characterization of membrane-bound polysomes from ascites tumor cells. *Biochim. Biophys. Acta* **312**:492–501.

Fehlmann, M., G. Bellemare, and C. Godin, 1975 Free and Membrane-bound ribosomes. II. Two-dimensional gel electrophoresis of proteins from free and membrane-bound rabbit reticulocyte ribosomes. *Biochim. Biophys. Acta* **378**:119–124.

Fellner, P., 1974 Structure of the 16 S and 23 S ribosomal RNAs. In *Ribosomes,* edited by M. Nomura, A. Tissières, and P. Lengyel, pp. 169–191, Cold Spring Harbor Laboratory, Cold Spring Harbor, N.Y.

Fridlender, B. R. and F. O. Wettstein, 1970 Differences in the ribosomal protein of free and membrane bound polysomes of chick embryo cells. *Biochem. Biophys. Res. Commun.* **39**:247–253.

Garrett, R. A. and H. G. Wittmann, 1973 Structure of bacterial ribosomes. *Adv. Protein Chem.* **27**:277–347.

Greenberg, H. and S. Penman, 1966 Methylation and processing of ribosomal RNA in HeLa cells. *J. Mol. Biol.* **21**:527–535.

Gressner, A. M. and I. G. Wool, 1974 The phosphorylation of liver ribosomal proteins *in vivo*. *J. Biol. Chem.* **249**:6917–6925.

Grollman, A. P. and M. T. Huang, 1973 Inhibitors of protein synthesis in eukaryotes: Tools in cell research. *Fed. Proc.* **32**:1673–1678.

Grummt, F., I. Grummt, and V. A. Erdmann, 1974 ATPase and GTPase activities isolated from rat liver ribosomes. *Eur. J. Biochem.* **43**:343–348.

Gualerzi, C., H. G. Janda, H. Passow, and G. Stöffler, 1974 Studies on the protein moiety of plant ribosomes: Enumeration of the proteins of the ribosomal subunits and determination of the degree of evolutionary conservation by electrophoretic and immunochemical methods. *J. Biol. Chem.* **249**:3347–3355.

Haga, J. Y., M. G. Hamilton, and M. L. Petermann, 1970 Electron microscopic observations on the large subunit of the rat liver ribosome. *J. Cell Biol.* **47**:211–221.

Hamilton, M. G., 1971 Isodensity equilibrium centrifugation of ribosomal particles; the calculation of the protein content of ribosomes and other ribonucleoproteins from buoyant density measurements. *Methods Enzymol.* **20:** Part C, 512–521.

Hamilton, M. G., 1974*a* The characterization of ribosomal subunits of eukaryotes by ultracentrifugal techniques. Criteria of purity. *Methods Enzymol.* **30:** Part F, 387–395.

Hamilton, M. G., 1974*b* Estimation of the number and size of ribosomal proteins by two-dimensional gel electrophoresis. *Methods Enzymol.* **30:** Part F, 540–545.

Hamilton, M. G. and T. W. O'Brien, 1974 Ultracentrifugal characterization of the mitochondrial ribosome and subribosomal particles of bovine liver: Molecular size and composition. *Biochemistry* **13:**5400–5403.

Hamilton, M. G., and M. E. Ruth, 1969 The dissociation of rat liver ribosomes by EDTA; molecular weights, chemical composition, and buoyant densities of the subunits. *Biochemistry* **8:**851–856.

Hamilton, M. G., A. Pavlovec, and M. L. Petermann, 1971 The molecular weight, buoyant density, and composition of active subunits of rat liver ribosomes. *Biochemistry* **10:**3424–3427.

Hamkalo, B. A. and O. L. Miller, Jr., 1973 Electronmicroscopy of genetic activity. *Annu. Rev. Biochem.* **42:**379–396.

Hanna, N., G. Bellemare, and C. Godin, 1973 Free and membrane-bound ribosomes. 1. Separation by two-dimensional gel electrophoresis of proteins from rat liver monosomes. *Biochim. Biophys. Acta* **331:**141–145.

Hartwell, L. H. and C. S. McLaughlin, 1969 A mutant of yeast apparently defective in the initiation of protein synthesis. *Proc. Natl. Acad. Sci. USA* **62:**468–474.

Haselkorn, R. and L. B. Rothman-Denes, 1973 Protein synthesis. *Annu. Rev. Biochem.* **43:**397–438.

Held, W. A., M. Nomura, and J. W. B. Hershey, 1974 Ribosomal protein S 21 is required for full activity in the initiation of protein synthesis. *Mol. Gen. Genet.* **128:**11–22.

Heywood, S. M., D. S. Kennedy, and A. J. Bester, 1974 Separation of specific initiation factors involved in the translation of myosin and myoglobin messenger RNAs and the isolation of a new RNA involved in translation. *Proc. Natl. Acad. Sci. USA* **71:**2428–2431.

Hill, W. E., G. P. Rossetti, and K. E. Van Holde, 1969 Physical studies of ribosomes from *Escherichia coli. J. Mol. Biol.* **44:**263–277.

Hirsch, C. A., M. A. Cox, W. J. W. van Venrooij, and E. C. Henshaw, 1973 The ribosome cycle in mammalian protein synthesis. II. Association of the native smaller ribosomal subunit with protein factors. *J. Biol. Chem.* **248:**4377–4385.

Hultin, T. and A. Sjöqvist, 1971 Analogies in protein pattern and conformation among ribosomes from different classes of vertebrates. *Comp. Biochem. Physiol.* **40:**Part B, 1011–1027.

Kabat, D., 1970 Phosphorylation of ribosomal proteins in rabbit reticulocytes. Characterization and regulatory aspects. *Biochemistry* **9:**4160–4175.

Kaltschmidt, E. and H. G. Wittmann, 1970 Ribosomal proteins. VII. Two-dimensional polyacrylamide gel electrophoresis for fingerprinting of ribosomal proteins. *Anal. Biochem.* **36:**401–412.

King, H. W. S. and H. Gould, 1970 Low molecular weight ribonucleic acid in rabbit reticulocyte ribosomes. *J. Mol. Biol.* **51:**687–702.

Kirby, K. S., 1968 Isolation of nucleic acids with phenolic solvents. *Methods Enzymol.* **12:**Part B, 87–99.

Kiselev, N. A., V. Ya. Stel'mashchuk, M. I. Lerman, and O. Yu. Abakumova, 1974 On the structure of liver ribosomes. *J. Mol. Biol.* **86:**577–586.

Kitchen, H. and S. H. Boyer, editors, 1974 Hemoglobin: Comparative molecular biology models for the study of disease. *Ann. N.Y. Acad. Sci.* **241:**1–737.

Kumar, A. and A. R. Subramanian, 1975 Ribosome assembly in HeLa cells: Labeling pattern of ribosomal proteins by 2-dimensional resolution. *J. Mol. Biol.* **94:**409–423.

Kurland, C. G., 1960 Molecular characterization of ribonucleic acid from *Escherichia coli* ribosomes. I. Isolation and molecular weights. *J. Mol. Biol.* **2:**83–91.

Kurland, C. G., S. J. S. Hardy, and G. Mora, 1971 Purification of ribosomal proteins from *Escherichia coli. Methods Enzymol.* **20:**Part C, 381–391.

Kwan, S.-W. and G. Brawerman, 1972 A particle associated with the polyadenylate segment in mammalian messenger RNA. *Proc. Natl. Acad. Sci. USA* **69:**3247–3250.

Lake, J. A., D. D. Sabatini, and Y. Nonomura, 1974 Ribosome structure as studied by electron microscopy. In *Ribosomes,* edited by M. Nomura, A. Tissières, and P. Lengyel, pp. 543–557, Cold Spring Harbor Laboratory, Cold Spring Harbor, N.Y.

Leboy, P. S., E. C. Cox, and J. G. Flaks, 1964 The chromosomal site specifying a ribosomal protein in *Escherichia coli. Proc. Natl. Acad. Sci. USA* **52:**1367–1374.

Leder, P., 1973 The elongation reactions in protein synthesis. *Adv. Protein Chem.* **27:**213–242.

Lengyel, P., 1974 The process of translation: A bird's eye view. In *Ribosomes,* edited by M. Nomura, A. Tissières, and P. Lengyel, pp. 13–52, Cold Spring Harbor Laboratory, Cold Spring Harbor, N.Y.

Lin, A. and I. G. Wool, 1974 The molecular weights of rat liver ribosomal proteins determined by "three-dimensional" polyacrylamide gel electrophoresis. *Mol. Gen. Genet.* **134:**1–6.

Lingrel, J. B. and H. R. Woodland, 1974 Initiation does not limit the rate of globin synthesis in message-injected *Xenopus* oocytes. *Eur. J. Biochem.* **47:**47–56.

Loening, U. E., 1973 Ribosomal ribonucleic acid in evolution. *Biochem. Soc. Symp.* **37:**95–104.

Loftfield, R. B., 1972 The mechanism of aminoacylation of transfer RNA. *Prog. Nucleic Acid Res.* **12:**87–128.

Lucas-Lenard, J. and F. Lipmann, 1971 Protein Biosynthesis. *Annu. Rev. Biochem.* **40:**409–448.

Malkin, L. I. and A. Rich, 1967 Partial resistance of nascent polypeptide chains to proteolytic digestion due to ribosomal shielding. *J. Mol. Biol.* **26:**329–346.

Martin, T. E., I. G. Wool, and J. J. Castles, 1971 Dissociation and reassociation of ribosomes from eukaryotic cells. *Methods Enzymol.* **20:**Part C, 417–433.

Marx, J. L., 1973 Research News: Somatic cell hybrids: impact on mammalian genetics. *Science* (*Wash. D.C.*) **179:**785–787.

Mathews, M. B. and A. Korner, 1970 Mammalian cell-free protein synthesis directed by viral ribonucleic acid. *Eur. J. Biochem.* **17:**328–338.

Matthews, D. E. and T. W. O'Brien, 1974 Proteins of the bovine 55 S mitochondrial ribosome. *Fed. Proc.* **33:**1584.

Mazelis, A. G. and M. L. Petermann, 1973 Physical-chemical properties of stable yeast ribosomes and ribosomal subunits. *Biochim. Biophys. Acta* **312:**111–121.

Milcarek, C. and S. Penman, 1974 Membrane-bound polyribosomes in HeLa cells: Association of polyadenylic acid with membranes. *J. Mol. Biol.* **89:**327–338.

Milstein, C., G. G. Brownlee, T. M. Harrison, and M. B. Mathews, 1972 A possible precursor of immunoglobulin light chains. *Nature* (*London*) *New Biol.* **239:**117–120.

Möller, W., 1974 The ribosomal components involved in EF-G and EF-Tu-dependent

GTP hydrolysis. In *Ribosomes,* edited by M. Nomura, A. Tissières, and P. Lengyel, pp. 711–731, Cold Spring Harbor Laboratory, Cold Spring Harbor, N.Y.

Monier, R., 1974 5 S RNA. In *Ribosomes,* edited by M. Nomura, A. Tissières, and P. Lengyel, pp. 141–168, Cold Spring Harbor Laboratory, Cold Spring Harbor, N.Y.

Monro, R. E., T. Staehelin, M. L. Celma, and D. Vazquez, 1969 The peptidyl transferase activity of ribosomes. *Cold Spring Harbor Symp. Quant. Biol.* **34:**357–368.

Moore, V. G., R. E. Atchison, G. Thomas, M. Moran, and H. F. Noller, 1975 Identification of a ribosomal protein essential for peptidyl transferase activity. *Proc. Natl. Acad. Sci. USA* **72:**844–848.

Noll, H. and E. Stutz, 1968 The use of sodium and lithium dodecyl sulfate in nucleic acid isolation. *Methods Enzymol.* **12:** Part B, 129–155.

Nomura, M., 1970 Bacterial ribosome. *Bacteriol. Rev.* **34:**228–277.

Nomura, M., A. Tissières, and P. Lengyel, editors, 1974 *Ribosomes,* Cold Spring Harbor Monograph Series, Cold Spring Harbor, N.Y.

Nonomura, Y., G. Blobel, and D. Sabatini, 1971 Structure of liver ribosomes studied by negative staining. *J. Mol. Biol.* **60:**303–324.

Pain, V. M. and M. J. Clemens, 1973 The role of soluble protein factors in the translational control of protein synthesis in eukaryotic cells. *FEBS Lett.* **32:**205–212.

Palade, G. E. and P. Siekevitz, 1956 Liver microsomes: An integrated morphological and biochemical study. *J. Biophys. Biochem. Cytol.* **2:**171–200.

Peacock, A. C. and C. W. Dingman, 1968 Molecular weight estimation and separation of ribonucleic acid by electrophoresis in agarose-acrylamide composite gels. *Biochemistry* **7:**668–674.

Perry, R. P., 1973 Regulation of ribosome content in eukaryotes. *Biochem. Soc. Symp.* **37:**105–116.

Perry, R. P. and D. E. Kelley, 1968 Messenger RNA–protein complexes and newly synthesized ribosomal subunits: Analysis of free particles and components of polyribosomes. *J. Mol. Biol.* **35:**37–59.

Perry, R. P. and D. E. Kelley, 1974 Existence of methylated messenger RNA in mouse L cells. *Cell* **1:**37–42.

Perry, R. P., T.-Y. Cheng, J. J. Freed, J. R. Greenberg, D. E. Kelley, and K. D. Tartof, 1970 Evolution of the transcription unit of ribosomal RNA. *Proc. Natl. Acad. Sci. USA* **65:**609–616.

Pestka, S., 1971 Inhibitors of ribosome functions. *Annu. Rev. Microbiol.* **25:**487–562.

Pestka, S., 1972 Studies on transfer ribonucleic acid–ribosome complexes. XIX. Effect of antibiotics on peptidyl puromycin synthesis on polyribosomes from *Escherichia coli*. *J. Biol. Chem.* **247:**4669–4678.

Petermann, M. L., 1964 *The Physical and Chemical Properties of Ribosomes,* Elsevier, Amsterdam.

Petermann, M. L., 1971*a* How does a ribosome translate linear genetic information? *Sub-Cell. Biochem.* **1:**67–73.

Petermann, M. L., 1971*b* The dissociation of rat liver ribosomes to active subunits by urea. *Methods Enzymol.* **20:** Part C, 429–433.

Petermann, M. L. and A. Pavlovec, 1971 Dissociation of rat liver ribosomes to active subunits by urea. *Biochemistry* **10:**2770–2775.

Petermann, M. L., M. G. Hamilton, and A. Pavlovec, 1972*a* A 5 S ribonucleic acid–protein complex extracted from rat liver ribosomes by formamide. *Biochemistry* **11:**2323–2326.

Petermann, M. L., A. Pavlovec, and M. G. Hamilton, 1972*b* Effects of agents that influence hydrogen bonding on the structure of rat liver ribosomes. *Biochemistry* **11:**3925–3933.

Petersen, N. S. and C. S. McLaughlin, 1974 Polysome metabolism in protein synthesis mutants of yeast. *Mol. Gen. Genet.* **129:**189–200.

Petrissant, G., 1973 Evidence for the absence of the G-T-ψ-C sequence from two mammalian initiator transfer RNAs. *Proc. Natl. Acad. Sci. USA* **70:**1046–1049.

Rao, S. S. and A. P. Grollman, 1967 Cycloheximide resistance in yeast: A property of the 60 S ribosomal subunit. *Biochem. Biophys. Res. Commun.* **29:**696–704.

Redman, C. M. and D. D. Sabatini, 1966 Vectorial discharge of peptides released by puromycin from attached ribosomes. *Proc. Natl. Acad. Sci. USA* **56:**608–615.

Revel, M., Y. Pollack, Y. Groner, R. Schleps, H. Inouye, H. Berissi, and H. Zeller, 1972 Protein factors in *Escherichia coli* controlling initiation of mRNA translation. *Fed. Eur. Biochem. Soc. Symp.* **27:**261–280.

Rich, A., 1974 How transfer RNA may move inside the ribosome. In *Ribosomes,* edited by M. Nomura, A. Tissières, and P. Lengyel, pp. 871–884, Cold Spring Harbor Laboratory, N.Y.

Richter, D., V. A. Erdmann, and M. Sprinzl, 1973 Specific recognition of GTψC loop (loop IV) of tRNA by 50 S ribosomal subunits from *E. coli. Nature (London) New Biol.* **246:**132–135.

Roberts, R. B., 1958 Introduction. In *Microsomal Particles and Protein Synthesis,* edited by R. B. Roberts, pp. vii–viii, Pergamon Press, Oxford.

Rolleston, F. S., 1974 Membrane-bound and free ribosomes. *Sub-Cell. Biochem.* **3:**91–117.

Rosbash, M. and S. Penman, 1971 Membrane-associated protein synthesis of mammalian cells. I. The two classes of membrane-associated ribosomes. *J. Mol. Biol.* **59:**227–241.

Roth, J. S., 1958 Ribonuclease. VII. Partial purification and characterization of a ribonuclease inhibitor in rat liver supernatant fraction. *J. Biol. Chem.* **231:**1085–1095.

Rottman, F., A. J. Shatkin, and R. P. Perry, 1974 Sequences containing methylated nucleotides at the 5′ termini of messenger RNAs: Possible implications for processing. *Cell* **3:**197–199.

Sabatini, D. D., Y. Tashiro, and G. E. Palade, 1966 On the attachment of ribosomes to microsomal membranes. *J. Mol. Biol.* **19:**503–524.

Schlessinger, D., 1974 Genetic and antibiotic modification of protein synthesis. *Annu. Rev. Genet.* **8:**135–154.

Schreier, M. H. and T. Staehelin, 1973 Initiation of eukaryotic protein synthesis: Met-$tRNA_f$-40 S ribosome initiation complex catalyzed by purified initiation factors in the absence of mRNA. *Nature (London) New Biol.* **242:**35–38.

Shapiro, A. L., E. Viñuela, and J. V. Maizel, Jr., 1967 Molecular weight estimation of polypeptide chains by electrophoresis in SDS-polyacrylamide gels. *Biochem. Biophys. Res. Commun.* **28:**815–820.

Sherton, C. C. and I. G. Wool, 1972 Determination of the number of proteins in liver ribosomes and ribosomal subunits by two-dimensional polyacrylamide gel electrophoresis. *J. Biol. Chem.* **247:**4460–4467.

Shine, J. and L. Dalgarno, 1974 Identical 3′-terminal octanucleotide sequence in 18 S ribosomal ribonucleic acid from different eukaryotes. *Biochem. J.* **141:**609–615.

Shiokawa, K. and A. O. Pogo, 1974 The role of cytoplasmic membranes in controlling the transport of nuclear messenger RNA and initiation of protein synthesis. *Proc. Natl. Acad. Sci. USA* **71:**2658–2662.

Simsek, M., J. Ziegenmeyer, J. Heckman, and U. L. RajBhandary, 1973 Absence of the sequence G-T-ψ-C-G-(A)- in several eukaryotic cytoplasmic initiator transfer RNAs. *Proc. Natl. Acad. Sci. USA* **70:**1041–1045.

Smillie, R. M. S., editor, 1973 *The Structure and Formation of Eukaryotic Ribosomes,* Biochemical Society Symposium, No. 37, The Biochemical Society, London.

Spirin, A. S., 1964 *Macromolecular Structure of Ribonucleic Acids,* Reinhold, New York.

Spirin, A. S., 1969 A model of the functioning ribosome: Locking and unlocking of the ribosome subparticles. *Cold Spring Harbor Symp. Quant. Biol.* **34:**197–207.

Spirin, A. S. and L. P. Gavrilova, 1969 *The Ribosome,* Springer-Verlag, New York.

Stanley, W. M., Jr. and R. M. Bock, 1965 Isolation and physical properties of the ribosomal ribonucleic acid of *Escherichia coli. Biochemistry* **4:**1302–1311.

Steffensen, D. M., P. Duffey, and W. Prensky 1974 Localization of 5 S ribosomal RNA genes on human chromosome 1. *Nature (London)* **252:**741–743.

Stöffler, G., I. G. Wool, A. Lin, and K.-H. Rak, 1974 The identification of the eukaryotic ribosomal proteins homologous with *Escherichia coli* proteins L7 and L12. *Proc. Natl. Acad. Sci. USA* **71:**4723–4726.

Swan, D., G. Sander, E. Bermek, W. Krämer, T. Kreuzer, C. Arglebe, R. Zöllner, K. Eckert, and H. Matthaei, 1969 On the mechanism of coded binding of aminoacyl-tRNA to ribosomes: Number and properties of sites. *Cold Spring Harbor Symp. Quant. Biol.* **34:**179–196.

Tai, P.-C. and B. D. Davis, 1972 Transfer RNA content of runoff and complexed ribosomes of *Escherichia coli. J. Mol. Biol.* **67:**219–229.

Taylor, A. R., D. G. Sharp, D. Beard, and J. W. Beard, 1942 Isolation and properties of a macromolecular component of normal chick embryo tissue. *J. Infect. Dis.* **71:**115–126.

Tissières, A., J. D. Watson, D. Schlessinger, and B. R. Hollingworth, 1959 Ribonucleoprotein particles from *Escherichia coli. J. Mol. Biol.* **1:**221–233.

Ts'o, P. O. P., J. Bonner, and J. Vinograd, 1958 Structure and properties of microsomal nucleoprotein particles from pea seedlings. *Biochim. Biophys. Acta* **30:**570–582.

Udem, S. A. and J. R. Warner, 1972 Ribosomal RNA synthesis in *Saccharomyces cerevisiae. J. Mol. Biol.* **65:**227–242.

Van Holde, K. E. and W. E. Hill, 1974 General physical properties of ribosomes. In *Ribosomes,* edited by M. Nomura, A. Tissières, and P. Lengyel, pp. 53–91, Cold Spring Harbor Laboratory, Cold Spring Harbor, N.Y.

Vazquez, D., E. Battaner, R. Neth, G. Heller, and R. E. Monro, 1969 The function of 80 S ribosomal subunits and the effects of some antibiotics. *Cold Spring Harbor Symp. Quant. Biol.* **34:**369–375.

von Hippel, P. H. and K.-Y. Wang, 1964 Neutral salts: The generality of their effects on the stability of macromolecular conformations. *Science (Wash. D.C.)* **145:**577–580.

Waller, J.-P. and J. I. Harris, 1961 Studies on the composition of the protein from *Escherichia coli* ribosomes. *Proc. Natl. Acad. Sci. USA* **47:**18–23.

Warner, J. R., P. M. Knopf, and A. Rich, 1963 A multiple ribosomal structure in protein synthesis. *Proc. Natl. Acad. Sci. USA* **49:**122–129.

Warner, J. R., A. Kumar, S. A. Udem, and R. S. Wu, 1973 Ribosomal proteins and the assembly of ribosomes in eukaryotes. *Biochem. Soc. Symp.* **37:**3–22.

Weinberg, R. A., 1973 Nuclear RNA metabolism. *Annu. Rev. Biochem.* **42:**329–354.

Weinberg, R. A. and S. Penman, 1970 Processing of 45 S nucleolar RNA. *J. Mol. Biol.* **47:**169–178.

Wellauer, P. K. and I. B. Dawid, 1973 Secondary structure maps of RNA: Processing of HeLa ribosomal RNA. *Proc. Natl. Acad. Sci. USA* **70:**2827–2831.

Williamson, R., 1973 The protein moieties of animal messenger ribonucleoproteins. *FEBS Lett.* **37:**1–6.

Woese, C., 1970 Molecular mechanics of translation: A reciprocating ratchet mechanism. *Nature* (*London*) **226:**817–820.

Wool, I. G. and G. Stöffler, 1974 Structure and function of eukaryotic ribosomes. In *Ribosomes,* edited by M. Nomura, A. Tissières, and P. Lengyel, pp. 417–460, Cold Spring Harbor Laboratory, Cold Spring Harbor, N.Y.

PART S
CHLOROPLASTS AND MITOCHONDRIA

12

Chloroplast DNA: Physical and Genetic Studies

RUTH SAGER AND GLADYS SCHLANGER

Introduction

This chapter will examine what is known about chloroplast DNA (its physical and chemical properties, functions, and genetic analysis) from studies of algae and higher plants. This subject matter has been brought together for consideration in a book, *Cytoplasmic Genes and Organelles* (Sager, 1972), which was preceded by two excellent symposia (Boardman *et al.*, 1971*b*; Miller, 1970). Specific aspects have been reviewed in three books (Gibbs, 1971; Stewart, 1974; Birky *et al.*, 1975) and in several review articles (Tewari, 1972; Kirk, 1972; Gillham, 1974; Ellis and Hartley, 1974). The aim of this chapter will be twofold: to bring the subject matter up to date and to provide a framework for further research.

Physical and Chemical Properties of Chloroplast DNA

Identification and Base Composition

The likelihood that chloroplasts contained unique sequences of DNA was considered by numerous investigators in the late 1950s. They based

RUTH SAGER—Sidney Farber Cancer Center and Department of Microbiology and Molecular Genetics, Harvard Medical School, Boston, Massachusetts. GLADYS SCHLANGER—Division of Human Genetics, Cornell University Medical College, New York City, New York.

their views primarily on the genetic evidence of cytoplasmic genes affecting chloroplast development (Rhoades, 1955) and on the recognition of DNA as the nuclear genetic material (Watson and Crick, 1953). In 1962, Ris and Plaut (1962) demonstrated the presence of DNA in chloroplasts of *Chlamydomonas* by fluorescence and electron microscopy, but this cytochemical evidence could not distinguish whether the DNA was unique or of nuclear origin. Experimental evidence of uniqueness, double-strandedness, and high molecular weight was provided by the demonstration that DNA extracted from chloroplasts could be distinguished from nuclear DNA by its distinct and characteristic buoyant density in CsCl gradients (Sager and Ishida, 1963; Chun *et al.*, 1963).

Since 1963, CsCl gradient centrifugation has been used routinely to identify and isolate chloroplast DNA from algae and higher plants. A compilation of buoyant density data obtained for nuclear, chloroplast, and mitochondrial DNAs is given in Table 1. Values presented in this table are considered the best estimates, although some others have been reported in the literature. For example, difficulties were encountered in identifying the chloroplast DNA of some higher plants because of the closeness of their buoyant densities to that of the nuclear DNA, and as a result some of the early assignments of higher plant chloroplast DNA densities were erroneous (Kirk, 1971).

The technique of extraction of chloroplast DNA (chlDNA) was improved by Wells and Birnstiel (1969), who were able to treat plant chloroplasts with DNAse under conditions that permitted digestion of contaminating DNA on the outer membrane without degradation of the internal chlDNA. This method had been used with mitochondria, but not successfully with chloroplasts from some species, presumably because of their large size and membrane lability.

Initially, investigators were primarily concerned with distinguishing chlDNA from nuclear DNA, but it soon became apparent that a more pernicious problem was distinction between chlDNA and mitDNA. Both mitochondrial and chloroplast DNAs renature after heat denaturation much more rapidly than do nuclear DNAs. A clear-cut distinction between chlDNA and mitDNA in *Euglena* was facilitated by the use of colorless ("bleached") mutants, which lack chlDNA (Edelman *et al.*, 1966; Ray and Hanawalt, 1965).

The buoyant densities of chlDNAs of higher plants so far reported are all in the range of 1.695–1.697 g/cm^3, and those of the algae *Chlorella, Chlamydomonas,* and *Porphyra* are in the same range. The leucoplast DNA of *Polytoma,* a colorless relative of *Chlamydomonas,* is reported as 1.683 g/cm^3, and has physical properties typical of a DNA with a very low GC content. The buoyant densities of mitDNAs from higher plants are reported as 1.705–1.707 g/cm^3, which is the same range seen in animals. However, a wide range of buoyant densities of mitDNAs have been found in the lower eukaryotes including algae, fungi, and protozoa.

Buoyant density in CsCl is a measure of average base composition within the range from about 35 to 75% GC content (Schildkraut *et al.*, 1962). However, in nuclear DNAs of higher plants, 5-methylcytosine substitutes for some of the cytosine residues, resulting in a lower buoyant density in CsCl, and therefore requiring a correction in the conversion equation (Kirk, 1967). Methylated bases have not been detected in chloroplast DNAs, but the methods used were not sensitive below the 0.5% level.

Physical Size of Chloroplast DNAs

Following the discovery of circularity of some viral and mitochondrial DNAs, it became apparent that size of DNA molecules could be established definitively by measurements in the electron microscope of the contour length of closed circles. After several investigators published negative results, failing to find circles in chloroplast DNA, Manning *et al.* (1971) and Manning and Richards (1972) succeeded in isolating circular chlDNA molecules from *Euglena*. These molecules had an average contour length of 44.5 μm, corresponding to 9.2×10^7 daltons of double-stranded DNA. Subsequently, circular chlDNAs have been recovered from other organisms as well (Table 2).

Manning *et al.* (1972) reported circular chlDNAs from spinach and maize with contour lengths of 44 and 43 μm, respectively, and Kolodner and Tewari (1972) obtained a high yield of 37–42 μm circles from pea chloroplasts.

Kolodner and Tewari (1975*a*) obtained high yields of DNA circles from chloroplasts of spinach, corn, lettuce, and oats. In CsCl–ethidium bromide gradients, typical yields were 10–30% of chlDNA in covalently closed (supercoiled) circles and 40–50% in open circles; of the circular molecules, 1–2% were catenated dimers and 3–4% were circular dimers. Molecular weights of open circles were obtained from both contour lengths (Kolodner and Tewari, 1975*a*) and sedimentation values (Kolodner *et al.*, 1975*a*). The results are given in Table 2.

At the University of Düsseldorf, Herrmann's group succeeded in obtaining yields of up to 80% of chlDNA molecules in circles from *Antirrhinum majus, Beta vulgaris,* and spinach, and fewer circles from *Oenothera hookeri* (Herrmann *et al.*, 1975). They too observed circular dimers (about 15% of the total in spinach chlDNA) and supercoiled

TABLE 1. Buoyant Densities of DNAs of Algae and Higher Plants

Organism	*DNA (g/cm³)*			*Reference*
	Nuclear	*Chloroplast*	*Mitochondria*	
Algae				
Acetabularia mediterranea	1.696	1.702–1.704	1.714–1.715	Green, 1974*a*
				Heilporn and Limbosch, 1971*a*
				Green *et al.*, 1967
				Green and Burton, 1970
Acetabularia (Polyphysa) cliftonii	—	1.706	—	Green, 1973
Chlamydomonas	1.724	1.695	—	Sager and Ishida, 1963
				Sueoka *et al.*, 1967
				Chun *et al.*, 1963
	1.723	—	1.706	Ryan *et al.*, 1973
Chlorella ellipsoidea	1.716	1.695	—	Chun *et al.*, 1963
	1.717	1.692	—	Iwamura and Kuwashima, 1969
Chlorella pyrenoidosa	—	1.687	—	Bayden and Rode, 1973
	1.710	—	—	Rode and Bayen, 1972
Euglena	1.707	1.685	1.690	Brawerman and Eisenstadt, 1964
				Ray and Hanawalt, 1964
				Ray and Hanawalt, 1965
				Edelman *et al.*, 1965
Olisthodiscus luteus	1.700	1.686	—	Cattolico and Gibbs, 1974
Polytoma obtusum	1.710	1.683	1.715	Siu *et al.*, 1975*a*
Porphyra tenera	1.720	1.696	—	Ishida *et al.*, 1969

Plants				
Broad bean (*Vicia faba*)	1.695	1.697	1.705	Wells and Birnstiel, 1969
	1.696	1.696	—	Kung and Williams, 1968
Mung bean (*Phaseolus aureus*)	1.691	—	1.706	Suyama and Bonner, 1966
	1.695	1.697	—	Wells and Ingle, 1970
Daffodil (*Narcissus pseudonarcissus*)	1.698	1.697	1.707	Herrmann, 1972
Lettuce	1.694	1.697	1.706	Wells and Birnstiel, 1969
Oenothera hookeri	1.703	1.697	1.706	Herrmann *et al.*, 1975
Onion (*Allium cepa*)	1.689	—	1.706	Suyama and Bonner, 1955
	1.691	1.696	1.706	Wells and Ingle, 1970
Snapdragon (*Antirrhinum majus*)	1.689	1.698	—	Ruppel, 1967
	1.691	1.697	1.706	Herrmann *et al.*, 1975
Spinach (*Spinacia oleracea*)	1.694	1.696	—	Whitfield and Spencer, 1968
	1.694	1.696	—	Wells and Ingle, 1970
	1.695	1.697	1.706	Herrmann *et al.*, 1975
Sweet pea (*Pisum sativa*)	1.695	1.697	—	Wells and Birnstiel, 1969
	1.694	1.697	1.705	Kolodner and Tewari, 1972
Sweet potato (*Ipomoea batatas*)	1.692	—	1.706	Snyama and Bonner, 1966
Swiss chard (*Beta vulgaris*)	1.689	1.700	—	Kislev *et al.*, 1965
	1.694	1.697	1.705	Wells and Ingle, 1970
	1.695	1.697	1.706	Herrmann *et al.*, 1975
Tobacco (*Nicotiana tabacum*)	1.698	1.698	—	Lyttleton and Petersen, 1964
	1.695	1.697	—	Wells and Ingle, 1970
	1.697	1.697	—	Whitfield and Spencer, 1968
	1.697	1.700	—	Tewari and Wildman, 1970
Turnip (*Brassica rapa*)	1.692	1.695	1.706	Suyama and Bonner, 1966
Wheat (*Triticum vulgare*)	1.702	1.698	—	Wells and Ingle, 1970

TABLE 2. Chloroplast DNA: Molecular Weight of Circular Molecules and Genomic Sizes Estimated from Renaturation Kinetics[a]

Organism	*Molecular weight*	*Reference*	*Estimated genomic size*	*Reference*
Algae				
Acetabularia mediterranea			$\sim 1.1 \times 10^9$	Green *et al.*, 1975*b*
Acetabularia cliftonii			$1.48 \pm 0.07 \times 10^9$	Green *et al.*, 1975*b*
Chlamydomonas			2×10^8[b]	Wells and Sager, 1971
			Fast fraction (10%)	
			1.94×10^8	Bastia *et al.*, 1971*a*
			[9.9×10^7][c]	Kolodner and Tewari, 1972
Chlamydomonas (plasmatic *Sd3* mutant)			1.38×10^8	Metzke and Herrmann, 1975
Chlorella			$2.1–2.3 \times 10^8$ (92%)	Bayen and Rode, 1973
			Fast fraction (8%)	
			[1.2×10^8][d]	Bayen and Rode, 1973
Euglena	9.2×10^7[a]	Manning and Richards, 1972	1.8×10^8	Stutz, 1970
			[9.0×10^7][d]	Stutz and Vandrey, 1971
Polytoma obtusum			$1.7–2.1 \times 10^8$ (63%)	Siu *et al.*, 1975*a*
			Zero-time fraction (27%)	
			Very rapid fraction (10%)	

Plants				
Corn (*Zea mays*)	8.47×10^{7e}	Kolodner and Tewari, 1975*a*		
Lettuce	9.58×10^{7e}	Kolodner and Tewari, 1975*a*	1.2×10^8 (76%)	Wells and Birnstiel, 1969
(*Lactuca*			Fast fraction (24%)	
sativa)	$9.82\ (\pm 0.15) \times 10^{7f}$	Kolodner *et al.*, 1975*a*	$[9.8 \times 10^7]^c$	Kolodner and Tewari, 1972
Oat (*Avena sativa*)	8.57×10^{7e}	Kolodner and Tewari, 1975*a*		
Oenothera hookeri	$\sim 1 \times 10^{8e}$	Herrmann *et al.*, 1975	1.01×10^8	Metzke and Herrmann, 1975
Pea (*Pisum*	8.84×10^{7e}	Kolodner and Tewari, 1975*a*	9.5×10^7	Kolodner and Tewari, 1972
sativum)	8.91×10^{7f}	Kolodner *et al.*, 1975*a*		
Snapdragon *Antirrhinum majus*	$\sim 1 \times 10^{8e}$	Herrmann *et al.*, 1975	1.00×10^8	Metzke and Herrmann, 1975
Spinach	9.43×10^{7e}	Kolodner and Tewari, 1975*a*	$\sim 1.0 \times 10^8$	Herrmann *et al.*, 1974
(*Spinacia oleracea*)	$9.72\ (\pm 0.15) \times 10^{7f}$	Kolodner *et al.*, 1975*a*		
	$\sim 9 \times 10^{7e}$	Manning *et al.*, 1972		
	$\sim 1 \times 10^{8e}$	Herrmann *et al.*, 1975		
Swiss chard (*Beta vulgaris*)	$\sim 1 \times 10^{8e}$	Herrmann *et al.*, 1975	1.12×10^8	Metzke and Herrmann, 1975
Tobacco			1.1×10^8	Tewari and Wildman, 1970
(*Nicotiana tabacum*)			$[9.3 \times 10^7]^c$	Kolodner and Tewari, 1972

[a] Values in brackets and sizes obtained when correction factors are applied to experimental data cited in line above.
[b] See text for discussion.
[c] Genomic size corrected assuming T_4 mol wt = 1.06×10^8.
[d] Genomic size corrected for low G+C content.
[e] Molecular weight obtained from electron microscopy.
[f] Molecular weight obtained from centrifugation studies.

molecules. Herrmann's group attributes their success in obtaining circles in high yield to the recovery of chloroplasts with unbroken envelopes, which were then subjected to both DNAse and phosphodiesterase treatment to remove contaminating nuclear DNA strands prior to sodium sarkosinate lysis. DNA circles were further purified by centrifuging onto a CsCl cushion.

Herrmann's group has also obtained electron microscopic evidence (Herrmann *et al.*, 1974) that aggregates of circular molecules are attached to membranes. From their studies of chlDNA circles, they concluded that the low proportion of covalently closed duplex circles (supertwisted forms) obtained is probably an artifact, resulting from breaks that occurred during membrane release. Only supertwisted forms, not open circles, were often seen attached to small protein particles (Herrmann *et al.*, 1975).

In *Acetabularia,* no large circles have been observed, but linear DNA molecules up to about 8×10^8 daltons were seen by Green and Burton (1970) in Kleinschmidt spreads. Circles of this size would be much more subject to breakage by shear than those of 1×10^8 daltons.

Genomic Size and Organization of Chloroplast DNAs

The nuclear DNAs of eukaryotes contain reiterated sequences of various lengths repeated many times. Both the repeat size and the extent of repetition can be estimated from measurements of the rate of reassociation of denatured DNA under standard conditions (Wetmur and Davidson, 1968; Britten and Kohne, 1968). On this basis, mitDNAs of animal cells, which are about 5 μm in length, appear to have little or no reiteration, and the same holds true for yeast mitDNA, which is about 25 μm long.

Genomic size is the term used to denote the single copy length of DNA, based on reassociation kinetic data. A comparison of genomic and physical sizes of several chlDNAs is given in Table 2. The genomic sizes for major algal and higher plant chlDNAs are all, with the exception of *Acetabularia,* in the range of $1–2 \times 10^8$ daltons, values that agree well (i.e., within a factor of 2) with the contour lengths of circles seen in electron micrographs of chlDNAs from *Euglena* and higher plants. Independent evidence that the chlDNA circles are genetically uniform comes from recent work of Kolodner and Tewari (1975*b*). They generated physical maps of chlDNA from peas and other plants by thermal denaturation and found identical denaturation patterns within each species, but different patterns between species. A partially denatured pea chlDNA molecule is shown in Figure 1.

The techniques employed in most studies summarized in Table 2 were not sensitive enough to detect either a fast-reassociating fraction, or a very slow fraction. Wells and Sager (1971) did find a fast fraction in *Chlamydomonas* chlDNA, comprising about 10% of the total, by taking special precautions to measure very early time points at the onset of reannealing. A similar fraction may be present though undetected in the chlDNAs of other organisms.

The available evidence supports the view that the chlDNAs of *Euglena* and of higher plants are organized simply as many circles of uniform size and informational content. In *Euglena* there are 10–12 chloroplasts per cell (Wolken and Palade, 1953), whereas in higher plants there are at least 200 chloroplasts per cell (e.g., see Ingle *et al.*, 1971). However, both *Euglena* and higher plants contain about 1×10^{-14} g of chlDNA (Smillie and Scott, 1969) representing an estimated 30–60 chlDNA circles per cell.

In the algae listed in Table 2 that contain only one large chloroplast

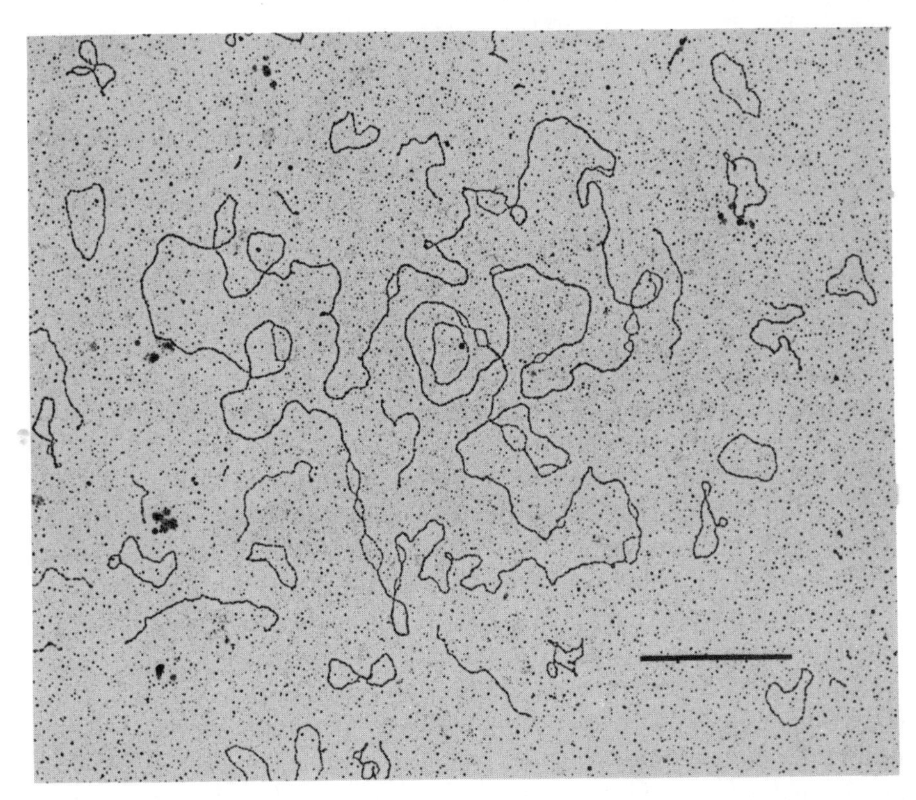

Figure 1. A partially denatured pea chlDNA molecule showing 22 denatured regions. The small circular molecules are single-stranded and double-stranded ϕX DNA. The bar indicates 1 μM. From Kolodner and Tewari (1975b).

per cell, i.e., *Chlamydomonas, Chl'orella,* and *Polytoma,* chlDNA circles have not been found. Is each chloroplast in these species equivalent to the mutliple set of chloroplasts per cell in higher plants? Does each large chloroplast contain a set of 20–60 circles, each with one genome worth of chlDNA?

What other information do we have that might shed light on the organization of chlDNA in *Chlamydomonas* and related forms? Genetic evidence, to be presented below, has shown that the chloroplast genome segregates and recombines as if present in only two copies per cell, not 20 or 40. This discrepancy between the genetic evidence of diploidy and the physical evidence of multiple copies is not unique to *Chlamydomonas.* In all higher eukaryotes, there is vastly more nuclear DNA present than can be accounted for by even the most liberal estimate of the number of genes present (Bishop, 1974).

One interpretation of the organization of chlDNA in *Chlamydomonas* involves a reconsideration of the reassociation kinetic data (Wells and Sager, 1971; Bastia *et al.,* 1971*a*). In these experiments, reannealing could not be carried out meaningfully beyond about 50–60% of the total DNA for technical reasons (e.g., fraying and mismatching of sheared ends), and it was assumed that no additional components were present with reassociation rates slower than the $1–2 \times 10^8$ dalton component. To look for a slower fraction it would be necessary first to fractionate the total DNA by partial reannealing in order to remove most of the 10^8 component, and as yet this experiment has not been done. If a slow component comprised of single-copy DNA were present to the extent of about 20% of the total DNA in *Chlamydomonas,* it would have a genomic content of about 8×10^8 daltons, and would contain the genes detected by mutation and studied by genetic analysis. The $1–2 \times 10^8$ dalton component might contain sets of reiterated genes not detected by genetic methods. Resolving this problem is one of the major current tasks in chloroplast genetics.

In *Acetabularia,* Green and Burton (1970) and Green (1973) describe lengths of chlDNA of up to 200 μm from osmotically lysed or Sarkosyl-lysed isolated chloroplasts. The likelihood that the chlDNA of *Acetabularia* is much longer than that of other species is strengthened by renaturation studies. Heilporn and Limbosch (1971*a*) report incomplete renaturation of chlDNA from *Acetabularia mediterranea,* indicating a large genomic size, and Green *et al.* (1975*b*) report a genomic size of $1.48 \pm 0.07 \times 10^9$ daltons for *A. cliftonii* and approximately 1.1×10^9 daltons for *A. mediterranea.* Green *et al.* (1975*b*) suggest that the large size of *Acetabularia* chlDNA may indicate its closeness to a postulated ancestral chlDNA.

Compositional Heterogeneity of Chloroplast DNA

Buoyant density and base compositional measurements provide average values. When high molecular weight DNA molecules are centrifuged to equilibrium in CsCl gradients, localized regions with a different average base composition may not be seen. Shearing such molecules to smaller pieces sometimes reveals shoulders or even separate small bands. The presence of sequences with a different average base composition from the major component, large enough to be seen by physical methods, is referred to as *compositional heterogeneity*. The concept is operational since it depends on the power of the method used to detect the heterogeneity.

Compositional heterogeneity can be detected by means of thermal denaturation curves. DNA molecules of homogeneous composition denature at a specific temperature determined by their average base composition. Plotting the first derivative of such a denaturation curve shows a Gaussian distribution, with the peak taken as the characteristic T_m of the DNA. Heterogeneous DNAs deviate from this ideal, and derivative plots exhibit shoulders or even secondary peaks. Bernardi *et al.* (1970) used such plots to demonstrate the compositional heterogeneity of mitDNAs from various yeast petite mutants.

By use of this method, compositional heterogeneity has been demonstrated in the chlDNAs of wild-type *Chlamydomonas* (Wells and Sager, 1971; Bastia *et al.*, 1971*a*), in *Chlorella* (Bayen and Rode, 1973), and in *Euglena* (Stutz and Vandrey, 1971). In *Chlamydomonas,* chlDNA from several chloroplast mutants were examined, but no reproducible differences were detected (O'Connor, 1972).

Single bands of DNA obtained in CsCl gradients can be fractionated into components with different average base compositions by recentrifuging in special gradients such as Cs_2SO_4 in the presence of Ag^+ ions. With this method, Vedel *et al.* (1972) found two satellite bands, one heavier and one lighter than the main band in chlDNA of maize, and a heavier satellite in *Chlorella.* The authors suggest that these satellites represent reiterated sequences and that this technique may yield additional satellites by using presheared DNA.

An example of the effect of shearing comes from studies (Siu *et al.*, 1975*a*) of *Polytoma* leucoplast DNA, which gave a symmetrical peak in CsCl with molecules about 10^7 daltons. The average GC content was 16–18% by chemical analysis. When the DNA was sheared to about 3×10^5 daltons, the peak was skewed to the heavy side, and the GC content of fractions across the peak ranged from 50% down to 10%.

Evidence of reiterated sequences in *Chlamydomonas* chlDNA has come from studies by Rochaix (1972), who used the method of Thomas *et al.* (1970). In this method, the DNA is sheared, the exposed ends are nibbled back by a 3′-exonuclease, and the mixture of DNA molecules with single-stranded ends is then reannealed and examined by electron microscopy to determine the extent of circularization. Rochaix found that up to 20% of the DNA fragments were rings or lariats, the majority of which were 0.75–1.0 μm in length. On the basis of these data plus calculations of the length of the reiterated sequences from melting curves, Rochaix concluded that chloroplast DNA has interspersed reiterated sequences arranged similarly to those described in nuclear DNAs from *Xenopus* and sea urchins (Davidson *et al.*, 1973).

Small circular DNA molecules have been reported in DNA extracts of *Euglena* (Nass and Ben-Shaul, 1972; Manning *et al.*, 1971), but it is not established that they are of chloroplast origin. However, in *Acetabularia* Green (1973) reported the recovery of small DNA circles 4.2 μm in length from isolated chloroplasts of *A. cliftonii.*

Ribosomal Cistrons

In *Euglena,* the DNA sequences transcribed as ribosomal RNA have been identified as 1.700 g/cm^3 in CsCl. Manning *et al.* (1971) lysed purified chloroplasts with sodium dodecylsulfate and found three peaks in CsCl gradients: a major peak at 1.685 g/cm^3, and minor components at 1.690 and at 1.699 g/cm^3 comprising less than 20% of the total. Stutz and Vandrey (1971) reported a main band of chlDNA as 1.685 and a pronounced shoulder at 1.701 g/cm^3. After purification of the shoulder by recentrifuging in CsCl, the DNA banded at 1.700 g/cm^3 and hybridized with chloroplast ribosomal RNA. They further showed that the 1.690 g/cm^3 band of Manning *et al.* (1971) could be resolved into DNAs of 1.701 and 1.687 g/cm^3 by shearing and rebanding. Rawson and Haselkorn (1973) found a single peak of chlDNA at 1.686 g/cm^3 which when sheared produced a satellite at 1.700 g/cm^3 that hybridized with chloroplast ribosomal RNA.

The fraction of *Euglena* chlDNA banding at 1.700 g/cm^3 in CsCl was recovered from total chlDNA by Stutz and Bernardi (1972) by fractionation on a hydroxylapatite column. This component eluted at a buffer concentration between the nuclear DNA and the main chloroplast peak, and it was also obtained from purified chloroplasts. Main-band chlDNA (1.685 g/cm^3) and the 1.700 g/cm^3 satellite contained 25% and 40% GC residues, respectively.

Siu *et al.* (1974*b*) sheared the high molecular weight molecules of plastid DNA from *Polytoma* that banded at 1.685 g/cm^3, and recovered a fraction banding at 1.700 g/cm^3 that hybridized with chloroplast ribosomal RNA from *Chlamydomonas*.

The hybridization of chlDNA from higher plants with the corresponding ribosomal RNAs has been described with nucleic acid preparations from tobacco (Tewari and Wildman, 1970), Swiss chard (Ingle *et al.*, 1971), peas (Thomas and Tewari, 1974*a*), beans, lettuce, spinach, corn, and oats (Thomas and Tewari, 1974*b*). However, the base composition of the ribosomal cistrons has not been established in any of these studies.

There is no evidence for the presence of chloroplast rRNA cistrons in algal nuclear DNA. Earlier positive reports for *Euglena* have been refuted (Scott, 1973), but evidence of this sort persists in higher plants. In tobacco (Tewari and Wildman, 1970) and in Swiss chard (Ingle *et al.*, 1971), chloroplast rRNA was found to hybridize with nDNA to the same extent as cytoplasmic rRNA.

However, Ingle *et al.* (1971) found that the cytoplasmic and chloroplast rRNAs hybridized to nuclear DNA of 1.706 g/cm^3 buoyant density, whereas the same rRNAs hybridized to chloroplast DNA of 1.699 g/cm^3. Evidently, more definitive studies are needed to evaluate the sequence similarities between cytoplasmic and chloroplast ribosomal RNAs of higher plants before the possibility that there may be chloroplast rRNA cistrons in nuclear DNA can be seriously considered.

Thomas and Tewari (1974*a,b*) found that base sequences of chloroplast rRNAs in bean, lettuce, spinach, corn, oats, and peas showed a high order of similarity, as evidenced by competition hybridization and thermal stability experiments involving homologous and heterologous chloroplast rRNA–chlDNA combinations.

Number of 23 S and 16 S Ribosomal RNA Cistrons per Chloroplast Genome

In *Euglena*, Rawson and Haselkorn (1973) used a technique involving renografin flotation for isolation of chloroplasts relatively free of nuclear DNA (Brown and Haselkorn, 1972). Taking 9.2×10^7 daltons as the molecular weight of one genome (Manning and Richards, 1972) and 1.1×10^6 and 0.55×10^6 (Loening, 1968) as the values for the two rRNAs, they found 1.9% of double-stranded chlDNA hybridized at saturation. This amount corresponds to one cistron each of 23 S and 16 S RNA per *Euglena* genome.

Using the same chloroplast isolation technique, Vandrey and Stutz (1973) found 2.8% hybridization corresponding to about 1.5 copies of each rRNA species. The authors suggest this higher value may be due to the growth phase of the cells, which were harvested in exponential growth, while those of Rawson and Haselkorn were in stationary phase. Scott (1973) found 6% hybridization, which corresponds to about three cistrons of each rRNA per *Euglena* circle of chlDNA. However, he excluded DNA of buoyant density 1.701 g/cm^3 from his calculations and consequently the 6% value needs to be recomputed. Scott also showed that the two rRNA species hybridize to different sequences in chlDNA.

Bastia *et al.* (1971*b*) report in an abstract that chloroplast rRNAs of *Chlamydomonas* hybridize with DNA to the heavy side of the chlDNA band (1.695 g/cm^3) and that there are approximately three cistrons per 1.94×10^8 daltons of DNA (or one or two cistrons, if the genome size is corrected to about 1×10^8). The leucoplast of the colorless alga *Polytoma* contains DNA and 73 S ribosomes. The ribosomal RNAs hybridize with leucoplast DNA to the extent of one copy per genome. The plastid rRNA does not anneal with nDNA, and cytoplasmic rRNAs hybridize specifically with nuclear DNA (Siu *et al.*, 1975*b*).

In higher plants, Ingle *et al.* (1971) reported that approximately one cistron of each rRNA species hybridized per chloroplast genome of 1.5×10^8 daltons of DNA in Swiss chard. Tewari and Wildman (1970) reported a minimum of one copy per tobacco genome, and Thomas and Tewari (1974*a*) reported two cistrons per genome in peas. In tobacco and peas, the specificity of hybridization was checked by temperature melts. In peas, Thomas and Tewari (1974*a*) further checked their data by hybridizing the rRNA with the circular fraction of chlDNA. They suggest their value of two rRNA genes per "genome" is higher than reported for other plants because their conditions were optimal for hybridization. Chloroplast rRNAs were relatively intact during the procedure, and the loss of filter-bound DNA was less than 5%. They also demonstrated that the cistrons for the two rRNA species are nonoverlapping.

The genomic origin of the tRNAs present in chloroplasts has not been rigorously established in any algal or higher plant species. However, chloroplast-specific tRNAs have been identified in *Euglena*, where at least three are induced by light: phenylalanyl, glutamyl, and isoleucyl tRNAs (Barnett *et al.*, 1969). In a further study from the same laboratory, the light-induced isoleucyl and phenylalanyl tRNAs were shown to be acylated only by the corresponding chloroplast synthestases, whose formation was also light-stimulated (Reger *et al.*, 1970).

Evidence for the transcription of one tRNA on chlDNA comes from a detailed study of leucyl-tRNAs from mature bean leaves (Williams *et al.*, 1973). At least seven distinct isoaccepting species of leucyl-tRNA were separated by reversed-phase chromatography, and three of them were identified tentatively as chloroplast in origin because of their preferential increase during greening. These fractions hybridized 14–21 times as well with chlDNA as with nuclear DNA from bean leaves. The authors do not claim that these tRNAs are transcription products of chloroplast DNA, but in our view that is the most likely interpretation of their findings.

Replication and Repair of Chloroplast DNA

Chloroplast DNA synthesis has been reported in isolated chloroplasts of *Euglena* (Scott *et al.,* 1968), spinach (Spencer *et al.,* 1971), tobacco (Tewari and Wildman, 1967), and *Chlamydomonas* (Ho *et al.,* 1974), and also in enucleated *Acetabularia* (Gibor, 1967) by methods that do not differentiate between repair insertion of nucleotides and replication of whole strands.

Evidence for the compartmentalization of chlDNA replication is the finding that thymidine is incorporated into chlDNA but not into nDNA, probably because only chloroplasts have the enzyme thymidine kinase (Steffensen and Sheridan, 1965; Swinton and Hanawalt, 1972). Keller *et al.* (1973) reported the partial purification of a DNA polymerase isolated from a *Euglena* chloroplast fraction. Chiang and Sueoka (1967), using density transfer (Meselson and Stahl, 1958), showed that *Chlamydomonas* chlDNA replicates semiconservatively in whole cells. Subsequently, Manning and Richards (1972) showed the same for *Euglena* chlDNA, both main band and the 1.700 satellite discussed above.

As to the mode of chlDNA replication, Kolodner and Tewari (1975*c*) observed replicative intermediates in DNA isolated from the chloroplasts of pea leaves. Multiple displacement (D) loops (see Chapter 15) were seen when covalently closed DNA circles were spread with formamide. These loops map in four specific places along the chlDNA molecule. Kolodner and Tewari (1975*d*) suggest that chlDNA replication may involve replicating intermediates of the Cairns type, as well as replicating molecules resulting from the rolling circle mechanism.

ChlDNA molecules isolated from peas were shown to contain RNA at 19 specific sites (Kolodner *et al.,* 1975*b*), and RNA was also identified in chlDNA from lettuce. These findings are in line with reports of short polyribonucleotide sequences in mitochondrial DNA (see Chapter 15) and in viral DNAs (*cf.* Kolodner *et al.,* 1975*b*).

Some studies have been directed toward the control of chlDNA replication. In *Euglena,* Cook (1966) reported that some chlDNA synthesis was triggered by light, out of synchrony with nuclear synthesis. In somatic cells of *Pteridium aquilinum,* a multicellular sexual alga, Sigee (1972) observed, using light and electron microscope autoradiography, that incorporation of label into chloroplast (and mitochondrial) DNA was continuous throughout interphase and showed an increase during the nuclear S phase.

Populations of *Chlamydomonas* can be synchronized by growth in a light–dark cycle. Using a 12-hr light/12-hr dark regime, Chiang and Sueoka (1967) found that chlDNA replicated independently of nDNA: one round of chlDNA replication occurred within 4.5 hr after the onset of the light period, a second round was completed about 2 hr later, and perhaps 20% of the chlDNA again replicated at the end of the light or in the beginning of the dark period. Nuclear DNA replicated twice in the dark, followed by cell division, and each cell gave rise to four progeny. These results show that the replication of nuclear and chlDNAs occurs at different times in the cell cycle, but since the amount of chlDNA per cell is constant in exponentially growing cells the replication of both classes of DNA is probably coupled.

In an attempt to uncouple nuclear and chlDNA replication, Blamire *et al.* (1974) examined the effects of protein synthesis inhibitors on the uptake of [^{3}H]adenine into DNA. They labeled cells *in vivo* in the absence and presence of inhibitors, fractionated the extracted DNA in preparative CsCl gradients, and compared the relative incorporation rates in the nuclear, chloroplast, and mitochondrial DNA regions of the gradients. They found that each inhibitor of chloroplast protein synthesis (streptomycin, neamine, spectinomycin, cleocin, chloramphenicol) and rifampicin, a inhibitor of chloroplast RNA synthesis, inhibited nuclear DNA replication at drug concentrations at which there was little or no inhibition of adenine incorporation into chlDNA. Incorporation into a small DNA peak, probably mitochondrial DNA, was higher in antibiotic-treated cultures than in controls. Cycloheximide, an inhibitor of protein synthesis on cell-sap ribosomes, did not uncouple organelle and nuclear DNA synthesis; rather, it strongly inhibited both. We interpret these results as indirect evidence of regulation by the chloroplast of cellular events directly involving nuclear DNA replication, and thereby cellular growth.

Chiang and Sueoka (1967) observed that replication of chloroplast DNA and that of a shoulder on the nuclear peak, 1.713 g/cm^3 in density, appear to be coordinate. This shoulder was later found to contain cytoplasmic ribosomal RNA cistrons. In *Physarum polycephalum* (Zellweger *et al.,* 1962), a measurable synthesis of nuclear and ribosomal genes occurs during the G_2 phase. Exposure of macroplasmodia to cyloheximide during the S phase strongly suppressed [^{3}H]thymidine labeling of the major nuclear component, but incorporation into mitochondrial and nucleolar DNA during the G_2 phase was only slightly affected by the antibiotic (Werry and Wanka, 1972). These experiments raise the possibility that in some organisms nuclear ribosomal cistron DNA and organelle DNA replications may be coordinate.

Manning and Richards (1972) observed DNA replication in an unsynchronized culture of *Euglena* grown autotrophically. In a density shift experiment, nDNA replicated one time per generation while major band chlDNA (1.685) replicated 1.5 times per cell division. These data, obtained on analytical gradients, were corroborated by findings with radioactive labels in the same experiment. Since the chlDNA : nDNA ratio in log cells remains constant, the authors conclude that one-third of the total chlDNA in each cell generation must be degraded in a process of "turnover."

Swinton and Hanawalt looked for evidence of dark-repair in ultraviolet-irradiated *Chlamydomonas*. Although irradiation induces the formation of pyrimidine dimers in both nuclear and chloroplast DNA, neither excision of dimers (Swinton and Hanawalt, 1973*a*) nor repair replication (Swinton and Hanawalt, 1973*b*) was observed to occur in either nuclear or chloroplast DNA. These negative results should not be considered conclusive, since the high levels of irradiation used gave no more than 1% survival of cells (Swinton and Hanawalt, 1973*b*). Genetic evidence for a dark-repair system in *Chlamydomonas* comes from crosses between irradiated and nonirradiated gametes in which all zygotes are viable and nuclear markers from both parents are distributed to the progeny in normal Mendelian fashion (Sager, 1972).

Induced Loss of Chloroplast DNA

Euglena readily loses its chloroplast-forming ability following any one of numerous treatments, including ultraviolet irradiation, heat shock, and growth in the presence of streptomycin (Schiff and Epstein, 1965), nalidixic acid (Lyman, 1967), or other inhibitors of DNA and protein synthesis (Ebringer, 1972). Chloroplast DNA is absent from certain aplastidic mutants (Edelman *et al.*, 1965; Ray and Hanawalt, 1965) but is present in many mutants producing abnormal plastids (Edelman *et al.*, 1965). Recently, Pienkos *et al.* (1975) found that nalidixic acid did not affect cell division and that chlDNA disappeared at a greater rate than would be expected from dilution at cell division. They suggested that nalidixic acid promoted two events: inhibition of chlDNA replication and the degradation of preexisting plastid DNA. Herschberger *et al.* (1974)

described a mutant of *Euglena* that produced permanently bleached cells at high rate.

In *Chlamydomonas,* yellow and pale green mutants have been described, but no aplastidic mutants have ever been found (Sager, 1959). Flechtner and Sager (1973) showed that ethidium bromide produced a reversible loss of a major part of the chlDNA in *Chlamydomonas.* The synthesis of new chloroplast chlDNA was inhibited and existing DNA was degraded. The effect was reversible only if the drug was removed no later than 12 hr after addition.

The reversibility of the ethidium bromide effect might be a consequence of genetic redundancy (Wells and Sager, 1971; Bastia *et al.,* 1971*a*), if one or a few copies of the chloroplast genome were sequestered in some way and protected from the action of the drug. This possibility could be tested by physical studies (e.g., reannealing kinetics) to find out whether the residual ethidium bromide resistant chlDNA contains a complete copy of the chloroplast genome.

The effects of various treatments on *Euglena* and *Chlamydomonas* may be analogous to the situation with mitDNA in yeast. In some yeasts ethidium bromide induced loss of mitDNA is irreversible (Goldring *et al.,* 1970; Perlman and Mahler, 1971; Nass, 1970), but a reversible decrease in the amount of mitDNA has been described in an obligate aerobe, *Kluyveromyces lactis* (Luha *et al.,* 1971).

Chloroplast DNA in Zygotes of *Chlamydomonas*

When total DNA from vegetative cells of *Chlamydomonas* is centrifuged to equilibrium in CsCl, two major bands are seen: nuclear DNA at 1.723–5 g/cm^3 and chloroplast DNA at 1.695 g/cm^3, comprising about 15% of the total DNA extracted from cells in log-phase growth. A shoulder at 1.713 g/cm^3 representing nuclear cistrons for ribosomal RNA (Bastia *et al.,* 1971*a*) and a very small peak at 1.706 g/cm^3 representing mitochondria DNA (Ryan *et al.,* 1973) are also sometimes seen without further fractionation.

When total DNA is extracted from young zygotes 6–24 hr after mating, the nuclear DNA appears unchanged, but the chloroplast is now 0.005 g/cm^3 lower in buoyant density, banding at 1.689 g/cm^3 (Sager and Lane, 1972) as shown in Figure 2. After replication this DNA returns to its former density.

We have shown that the chloroplast DNA in zygotes banding at 1.689 g/cm^3 originates in the female (mt^+) parent and that the homologous chlDNA from the male (mt^-) parent is degraded soon after mating.

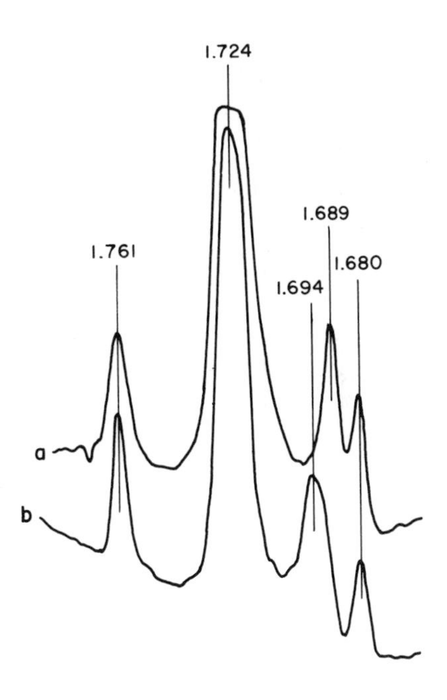

Figure 2. Density shift of chloroplast DNA in zygotes of Chlamydomonas. Microdensitometer tracings of DNAs extracted from (a) mt^+ gametes and (b) zygotes kept in nitrogen-free medium for 24 hr after mating. Nuclear DNA bands at 1.724 gm/cm³. In (b) chloroplast DNA bands at 1.694 g/cm³, but in (a) it has shifted to a new density of 1.689 g/cm³. Density markers are crab poly[d(AT)] (from N. Sueoka) at 1.680 g/cm³ and B. subtilis phage 15 DNA (from J. Marmur) at 1.761 g/cm³.

The different fates of chlDNAs from the two parents were followed by prelabeling the parental DNAs differentially with $^{15}N/^{14}N$ in some experiments (Sager and Lane, 1972) and with [$^{3}H/^{14}C$]adenine in others (Schlanger and Sager, 1974*b*; Schlanger and Sager, 1976). A typical preparative gradient is shown in Figure 3, in which the chlDNA of male origin is no longer present.

What causes the density shift? We infer that it results from covalent addition of some component with a lighter buoyant density in CsCl than chloroplast DNA, covalent because the density shift withstands detergent extraction with sodium laurylsulfate, sarkosyl, Triton X-100, and deoxycholate, as well as enzymatic digestion with pronase, ribonucleases, and amylase. If the density shift results from methylation as in bacterial modification, about 5% methylation would be required (Szybalski and Szybalska, 1971; Kirk, 1967) and should be detectable. Experiments are in progress to look for it.

In the reciprocal crosses between parents grown in ^{15}N-medium and in ^{14}N-medium, a single chloroplast peak was found in DNA from zygotes sampled at 6 hr and at 24 hr after mating. The DNA density was 1.692 g/cm³ in the $^{14}N/^{15}N$ cross and 1.698 g/cm³ in the reciprocal cross. As listed in Table 3, the densities were similar to, but not identical with, those from $^{15}N \times ^{15}N$ and from $^{14}N \times ^{14}N$ crosses. The $^{14}N \times ^{15}N$ average value was about 0.002 g/cm³ heavier than the $^{14}N \times ^{14}N$ average, and, similarly, the

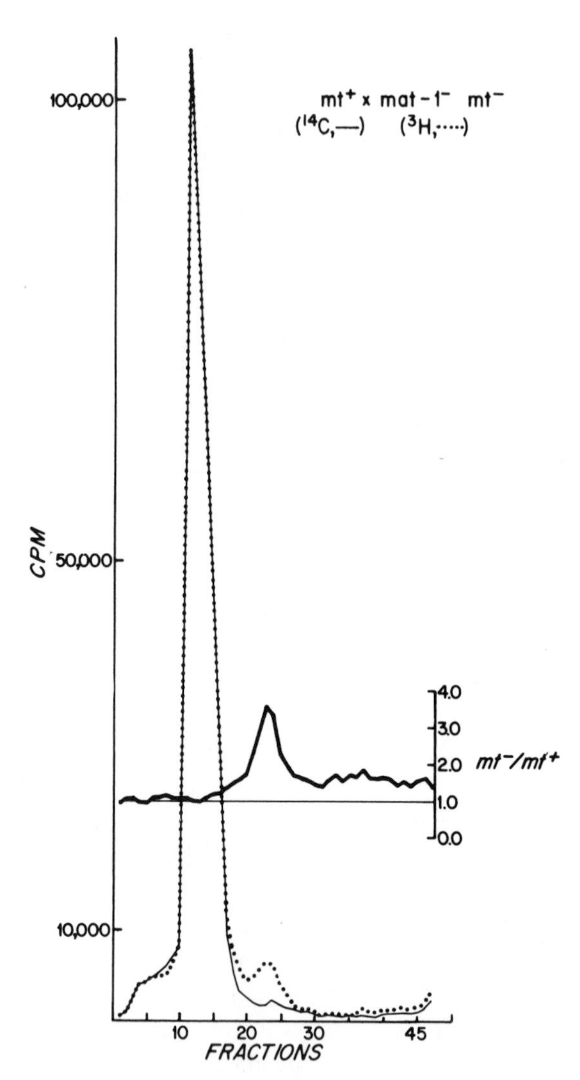

Figure 3. CsCl density gradient of Chlamydomonas zygote DNA from a cross between wild-type and mat-1 gametes. Vegetative cells were labeled differentially with [³H]- and [¹⁴C]adenine, gametes were prepared, and zygotes were harvested approximately 24 hr after the gametes were mixed. From Schlanger and Sager (1976).

$^{15}N \times {}^{14}N$ value was about 0.002 g/cm^3 lighter than the $^{15}N \times {}^{15}N$ value. These small but significant differences could be the result of limited recombination, except that we have genetic evidence indicating that recombination of our markers, situated all around the map, does not occur in unirradiated crosses at a frequency that would be detectable in the DNA. Another possibility, which we favor, is that the 0.002 g/cm^3 density difference reflects either repair or partial replication in which the nucleotides come from a common ^{15}N-^{14}N pool.

Another point to be noted from Table 3 is that the density shift in the $^{15}N \times {}^{15}N$ chloroplast DNA is greater than in the $^{14}N \times {}^{14}N$ DNA and similarly in the pairs of reciprocal crosses. This difference could be the result of a heavy isotope effect, in which the presence of ^{15}N influences the extent of covalent addition of whatever component is responsible for the density shift of maternal chloroplast DNA in zygotes.

Primarily on the basis of these results, we postulated a molecular mechanism involving modification of chlDNA from the female parent and restriction of the homologous chlDNA from the male, to account for the physical and genetic behavior of chlDNA in crosses. As will be detailed in the next section, the linked set of non-Mendelian genes located on chloroplast DNA show maternal inheritance—i.e., the genes from the male parent are lost—and this genetic pattern precisely parallels the physical disappearance of male chlDNA.

Further experimental evidence of the correlated behavior of chlDNA and chloroplast genes has come from studies of crosses in which maternal inheritance was inhibited. As discussed below (p. 392), we have shown that chloroplast genes of male origin can be preserved in zygotes by UV irradiation of the female parent before mating (Sager and Ramanis, 1967) and also by the presence of a mutant gene *mat-1* (Sager and Ramanis, 1974). These findings provided material for a direct test of whether the transmission of chloroplast genes from the mt^- parent following UV irradiation of mt^+ gametes, or in the presence of the *mat*-1 mutant allele, would be paralleled by transmission of chloroplast DNA from the mt^- parent. The results were unambiguous. Chloroplast DNAs of the mt^- (male) origin were recovered from 24-hr zygotes in both classes of experiments: those in which the mt^+ parent had been UV-irradiated before mat-

TABLE 3. Buoyant Densities of Chloroplast DNAs from Chlamydomonas Zygotes[a]

Cross[b]	*Observed density*[c]	*Computed*[d]	*Discrepancy*
$^{14}N \times {}^{14}N$	1.690 ± 0.0009	1.690	0
$^{14}N \times {}^{15}N$	1.692 ± 0.001	1.6935	0.0015
$^{15}N \times {}^{14}N$	1.698 × 0.0009	1.7005	0.0025
$^{15}N \times {}^{15}N$	1.7005	1.704	0.0035
^{14}N vegetative cells	1.695 ± 0.0005		
^{14}N gametes	1.695 ± 0.0005		
^{15}N vegetative cells	1.709 ± 0.0005		
^{15}N gametes	1.709 ± 0.0005		

[a] From Sager and Lane (1972) and Sager (1975).
[b] Female (mt^+) parent given first.
[c] Data from five or more preparations except $^{15}N \times {}^{15}N$ from two preparations.
[d] Computation assumes one round of semiconservative replication: one strand conserved and one strand replicated from equal pools of ^{14}N and ^{15}N precursors in reciprocal crosses; both strands modified (0.005 g/cm^3 lighter).

ing and those involving crosses of $mt^+ \times mat\text{-}1^- \; mt^-$ strains. The genetic behavior was examined in the same experiments: transmission of male chloroplast DNA paralleled transmission of chloroplast genes from the male parent (Schlanger and Sager, 1976).

Genetic Analysis of Chloroplast DNA in *Chlamydomonas*

Background

The first non-Mendelian gene described in *Chlamydomonas* conferred streptomycin resistance (Sager, 1954). The mutation was later shown to map in chloroplast DNA (Sager and Ramanis, 1970; Sager, 1972), and the resistance phenotype was shown to result from mutationally altered chloroplast ribosomes (Schlanger and Sager, 1974*a*) containing a single altered ribosomal protein (Ohta *et al.*, 1975). The principal aim of this section is to trace the thread of genetic analysis that led from the discovery of the streptomycin-resistant mutant strain to the chloroplast genetic map. The methods and results of genetic studies have been summarized in Sager (1972), reviewed (Sager, 1974; Gillham, 1974), and discussed in depth (Sager, 1976).

Pattern of Transmission in Crosses. The first cytoplasmic gene mutation (Sager, 1954) and all mutations subsequently studied that are located in the chloroplast linkage group show the same pattern of transmission in crosses. Chloroplast genes are transmitted from the mating type *plus* (mt^+) parent to all progeny, whereas the alleles carried in the mating type *minus* (mt^-) parent are not transmitted at all (with exceptions to be discussed below). The alleles disappear in the zygote and do not reappear in later generations. This pattern of 4 : 0 segregation resembles the maternal inheritance of chloroplast genes seen in higher plants. By analogy, the mt^+ cells are considered female and the mt^- cells male. The molecular basis for maternal inheritance in *Chlamydomonas* lies in the destruction of chloroplast DNA from the mt^- parent in zygotes soon after their formation (see p. 388). The occurrence of maternal inheritance provides a useful means to distinguish chloroplast genes from nuclear genes that show 2 : 2 Mendelian inheritance. However, maternal inheritance interferes with further genetic analysis, since the male genome is destroyed.

Blocking Maternal Inheritance by UV Irradiation. In an effort to induce biparental inheritance, gametes were pretreated with a variety of agents. The most effective by far was UV irradiation of the mt^+ gametes immediately before mating (Sager and Ramanis, 1967). With an optimal exposure to UV, the frequency of biparental zygotes that transmit the complete genomes of both parents was increased from about one per thousand to 50%, with little or no zygote lethality. Zygote survival resulted from fusion of the irradiated parental cells with unirradiated mates, and was enhanced by photoreactivation following cell fusion. This method has been used routinely in our laboratory to induce biparental inheritance (Sager and Ramanis, 1970; Sager, 1972, 1976). As the UV dose is increased, the frequency of maternal zygotes decreases, the frequency of biparental zygotes increases, and a third class, paternal zygotes, appears. Biparental and paternal zygotes are collectively referred to as exceptional zygotes. The three classes are distinguished as shown in Figure 4.

A nuclear gene mutation, *mat-1*, has been discovered (Sager and Ramanis, 1974) that produces an effect similar to UV irradiation: high frequencies of biparental and paternal zygotes. A second mutation, *mat-2*, has the opposite effect: it enhances the frequency of maternal inheritance,

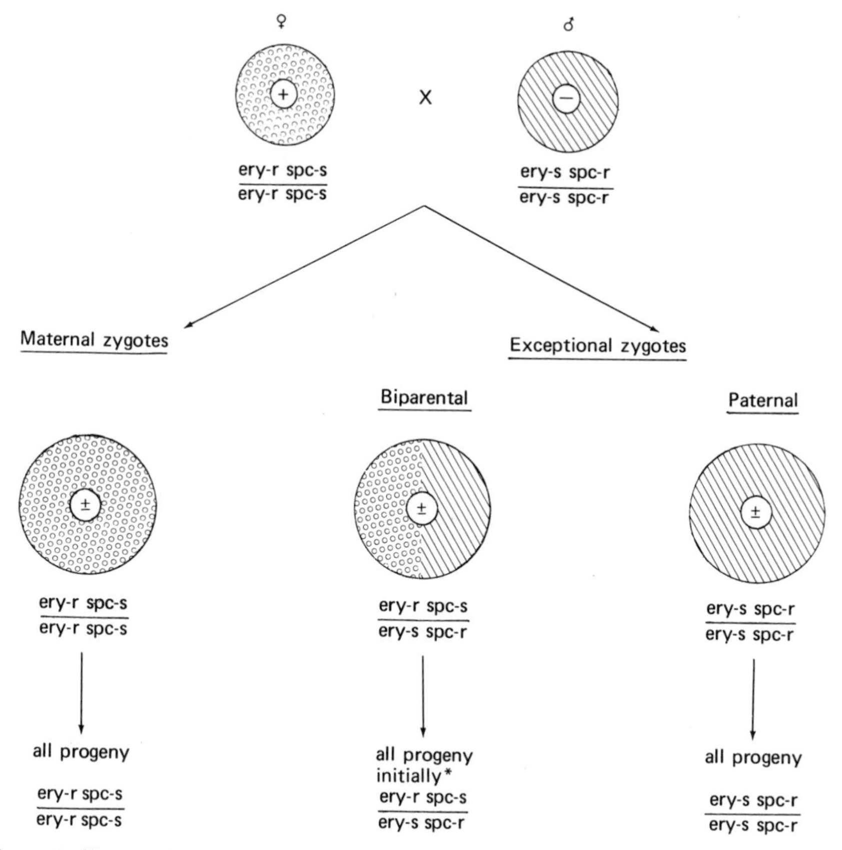

*Figure 4. Classes of zygotes. *Recombination occurs later, during multiplication of progeny clones. From Sager and Ramanis (1974).*

TABLE 4. *Cytoplasmic Gene Mutations in Chlamydomonas*[a]

Gene	*Origin*[b]	*Phenotype*	*Mapped*
ac1	SM induced	Requires acetate (leaky)	Yes
ac2	SM induced	Requires acetate (stringent)	Yes
ac3	SM induced	Requires acetate (stringent)	Yes
ac4	SM induced	Requires acetate (leaky)	Yes
tm1	SM induced	Cannot grow at 35°C	Yes
tm2	SM induced	Conditional: grows at 35°C only in the presence of streptomycin	No
Seven *tm* mutants	SM induced	Cannot grow at 35°C	No
ti1 through *ti5*	NG	Tiny colonies on all media	No
ery1	SM induced	Resistant to 50 μg/ml erythromycin	Yes
kan1	SM induced	Resistant to 100 μg/ml kanamycin	No
spc1	SM induced	Resistant to 50 μg/ml spectinomycin	Yes
spi1 through *5*	SM induced	Resistant to 100 μg/ml spiramycin	Yes
ole1 through *3*	SM induced	Resistant to 50 μg/ml oleandomycin	Yes
car1	SM induced	Resistant to 50 μg/ml carbamycin	Yes
cle1	SM induced	Resistant to 50 μg/ml cleosine	Yes
ery3	SM induced	Resistant to erythromycin, carbamycin, oleandomycin, spiramycin (same concentrations as single mutations above)	No
ery11	SM induced	Same as for *ery3*	No
sm2	SM induced	Resistant to 500 μg/ml SM	Yes
sm3	SM induced	Resistant to 50 μg/ml SM	Yes
sm4	SM induced	SM dependent	Yes
sm5	SM induced	Resistant to 500 μg/ml SM; recombines with *sm2*	Yes
D-371 and *D-310*	Induced by growth of strain *sm4* without SM	Resistant to 500 μg/ml SM	No
Four *D* mutants	Induced by growth of strain *sm4* without SM	Resistant to 500 μg/ml SM; segregate like persistent hets (*sd/sr*)	No
Eleven *D* mutants	Induced by growth of strain *sm4* without SM	Resistant to various low levels of SM: 20, 50, 100 μg/ml; segregate like persistent hets (*sd*/low *sr*)	No

TABLE 4. Continued

Gene	Origin[b]	Phenotype	Mapped
D-769	Induced by growth of strain *sm4* without SM	Conditional *sd*	Yes
Three *D* mutants	Induced by growth of strain *sm4* without SM	Conditional *sd*; segregate like persistent HETS (*sd*/cond. *sd*)	No
UV-16 *UV-17*	UV induced in strain *sm4*	Resistant to 500 μg/ml SM	No
Four *UV* mutants	UV induced in strain *sm4*	Resistant to 20 μg/ml SM	No
Three *UV* mutants	UV induced in strain *sm4*	Resistant to 20 μg/ml SM; segregate like persistent HETS (*sd*/low *sr*)	No
sr-2-1 *sr-2-60* *sr-2-280* *sr-2-218*	Spontaneous mutations selected on SM	Resistant to SM 500 μg/ml	No
kan-1	Spontaneous; selected on kanamycin	Resistant to kanamycin 50 μg/ml	No
ery-2-y	NG	Resistant to erythromycin 100 μg/ml	No
ery-3-6	NG	Resistant to erythromycin 100 μg/ml	No
spr-1-27	NG	Resistant to spectinomycin 100 μg/ml	No
sd-3-18	NG	Dependent on at least 20 μg/ml SM	No
nea-2-1	NG	Resistant to 1 mg/ml neamine	Yes

[a] From Sager (1972).
[b] SM, streptomycin; NG, nitrosoguanidine.

even after UV irradiation. Both mutations are linked to the mating type locus, suggesting that the control of maternal inheritance is a mating type function.

Mutagenesis. Mutant acquisition is a prerequisite for genetic analysis. Since the chloroplast genome is much smaller than the nuclear genome, it is necessary to use a mutagen with specificity for chloroplast DNA; otherwise, most mutants recovered are nuclear. Streptomycin is such a mutagen (Sager, 1962), and most of the mutants used in our studies were selected after growth in the presence of a toxic (but sublethal) concentration of this antibiotic. The mechanism of its mutagenic action and basis of its specificity are unknown. An alternative procedure has been proposed (Lee *et al.*, 1973) using diploids which can be recovered under special circumstances (Ebersold, 1967). Diploids treated with conventional mutagens should give rise primarily to cytoplasmic mutants, since nuclear mutations, if recessive, would not be expressed. No report of the efficacy of this method has yet appeared. The cytoplasmic mutants so far known are listed in Table 4.

Phenotypic Classes of Chloroplast Mutants. Four classes of mutants that show maternal inheritance in crosses (unirradiated) have been reported: acetate-requiring strains that require acetate as a reduced carbon source for growth in the light; temperature-sensitive mutants that grow at 25°C, but not at 35–37°C; mutants giving tiny colonies on growth factor supplemented agar; and the most popular class, those which are drug resistant. The drugs mainly used are those that inhibit protein synthesis on 70 S but not on 80 S ribosomes: streptomycin, neamine, spectinomycin, kanamycin, erythromycin, spiromycin, cleocin, oleandomycin, and carbomycin. Most of the mutants so far studied show little or no cross-resistance to other antibiotics at the cellular or ribosomal level (Schlanger and Sager, 1974*a*). However, some strains have also been examined (Sager, unpublished) that do show cross-resistance, with patterns similar to those that have been described in bacterial mutants (Pestka, 1971).

Methods of Genetic Analysis

Pedigree Analysis. The progeny recovered by pedigree analysis (Sager and Ramanis, 1965, 1967, 1968, 1970; Sager, 1972) represent the first and second doublings of each of the four zoospores originating from single zygotes, as shown in Figure 5. We concentrate on this set because with it we are able to reconstruct the segregational and recombinational events that occurred during meiosis as well as during the first two mitotic doublings of each zoospore clone. Since the frequencies of segregation and recombination are high, a great deal of information can be obtained in this way.

The four products of meiosis (zoospores) are distinguished from each other by the segregation of three pairs of unlinked nuclear genes: *act* (actidione resistance), *ms* (resistance to methionine sulfoximine), and *mt* (mating type), which also identify the zoospore sisters or octospores. The pairs of octospore daughters are recognized by their position on the plates following local respreading (Figure 6), and are checked by replicate plating to classify each colony for all markers segregating in the cross.

Pedigree analysis by this procedure has made it possible for us to determine the extent of segregation and recombination occurring in zygotes and to distinguish it from events occurring in each of the first two mitotic doublings after meiosis. These data have been essential in establishing the patterns of segregation and recombination which are uninterpretable from pooled data or from large zygote colonies sampled 10–20 doublings after meiosis.

A further advantage of the method is that no selection is involved: the entire "family" is recovered on nonselective media, and aliquots of each clone are subsequently classified on selective media. Random lethality or the nonrandom loss of particular progeny or markers is thus immediately detected. Usually a set of 50 zygotes, giving 200 zoospore clones, is sufficient for analysis. For closely linked markers, clones can be scored after many doublings. We also need to know whether the exchange events we have observed in the first two doublings continue at later stages of clonal development, e.g., whether they continue with the same rate and pattern. To answer these questions, we developed a second method, complementary to pedigree analysis.

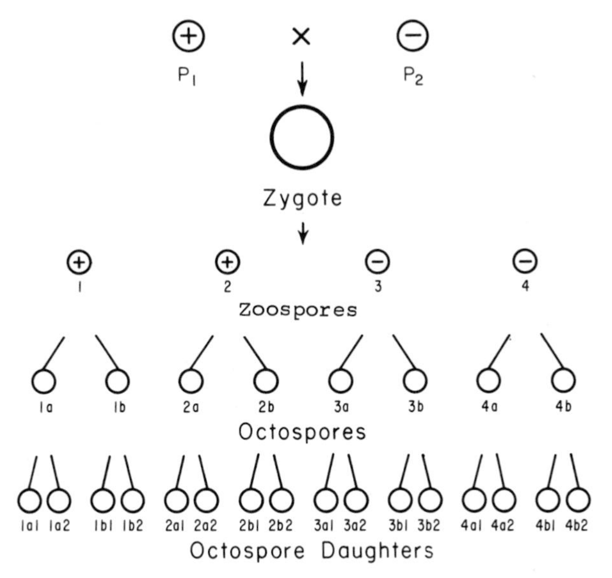

Figure 5. Procedure for pedigree analysis in Chlamydomonas. After germination, zoospores are allowed to undergo one mitotic doubling and then the eight cells (octospores) are transferred to a fresh petri plate and respread. After one further doubling, each pair of octospore daughters is separated and allowed to form colonies. The 16 colonies derived from the first two doublings of each zoospore are then classified for all segregating markers. From Sager (1972).

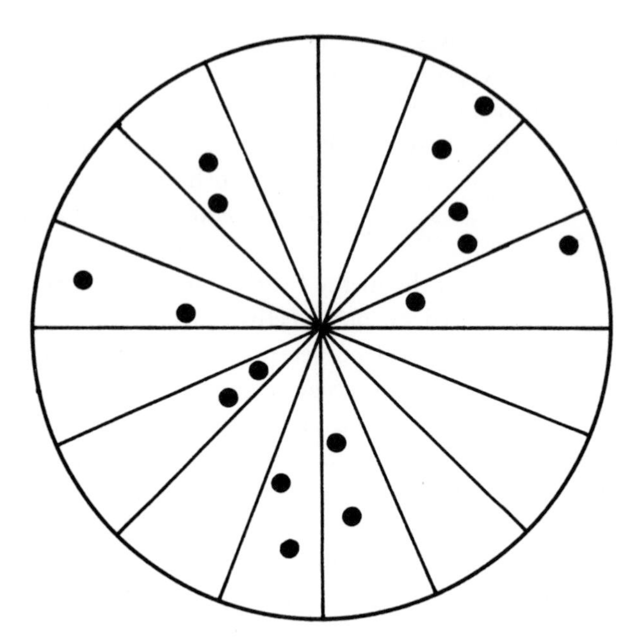

Figure 6. Recovery of octospore daughters for pedigree analysis in Chlamydomonas. A petri plate with 16 pie-shaped sectors marked on the bottom is used. A germinated zygote at the eight-cell stage is placed in the center and the cells are spread with a glass spreader. After one cell division, each sector is restreaked individually to separate octospore daughters. The cells then grow to form colonies which are classified for all markers segregating in the cross.

Liquid Culture Progeny Analysis. Zygotes can be germinated synchronously, and populations of zoospores, recovered just after germination, can be grown up at exponential rate in liquid culture under nonselective conditions. For progeny analysis, aliquots are sampled at 1- to 2-hr intervals, plated for viable counts, and replicated to various media to classify each colony for whether it is a *het* (containing both parental alleles) or a pure type (containing only one of the parental alleles) for each marker involved in the cross. The method lends itself to "global analysis," i.e., to examining the simultaneous behavior of a large number of linked genes. Each gene pair is scored separately for the percent of remaining hets at succeeding times after germination.

By use of this method, it was shown that the rate of segregation, i.e., the rate of disappearance of hets under nonselective conditions, was constant over ten doublings, the longest time during which a measurable fraction of hets could be followed (Sager and Ramanis, 1968; Sager, 1972). More recently, we have used this method for mapping: i.e., to estimate the distance of each segregating gene pair from the centromerelike attachment point (see p. 403).

Sampling from Large Zygote Colonies. Gillham *et al.* (1974) have used a third method of genetic analysis. They start with large zygote colonies containing about 10^6 cells, growing on nonselective media, replicate them to determine which are maternal and which exceptional, and then sample the original colonies from exceptional zygotes to determine the ratios of maternal and paternal alleles of each of the three drug-resistance markers segregating in the cross.

The method is good for distinguishing exceptional from maternal zygotes, with respect to whichever gene is used for identification purposes. The method does not allow distinction between segregation events occurring in zygotes and in postmeiotic divisions, or between type II and type III segregation patterns, nor does it protect the investigator from differential growth rates of progeny clones. Furthermore, the method provides no check on aberrant nuclear events, since the four products of meiosis are not identified.

Results of Genetic Analysis

Our principal aim has been to establish the rules of segregation and recombination of genes in chloroplast DNA as they occur during the meiotic and subsequent mitotic cell divisions. We began with no preconceptions other than those based on the behavior of other genetic systems, and attempted to proceed entirely from the empirical evidence. We have summarized here the sequential steps in the development of our present understanding of the rules, and have referred the interested reader to papers in which these assertions are documented.

1. Chloroplast genes of *Chlamydomonas* show maternal inheritance and can be identified by this transmission pattern (Sager, 1954, 1972).
2. Spontaneous exceptions to the rule of maternal inheritance in which chloroplast genes from the male parent are transmitted to progeny occur at frequencies that are genetically determined, e.g., about 0.1% in our wild-type stocks, up to 20–30% in some of Gillham's stocks (Gillham, 1969) and up to 100% in crosses involving the gene *mat-1* (Sager and Ramanis, 1974).
3. The frequency of exceptional zygotes can be increased to 50% or more by UV irradiation of the female gametes just before mating (Sager and Ramanis, 1967). The yield of exceptional zygotes can also be increased by pretreatment of gametes with drugs that inhibit transcription or protein synthesis in the chloroplast (Sager and Ramanis, 1973).

4. Exceptional zygotes are of two types: biparental zygotes that transmit the chloroplast genome from both parents, and paternal zygotes, in which the maternal complement of chloroplast genes is lost (Figure 4). In our stocks, maternal inheritance affects all chloroplast genes alike: in individual zygotes one parental genome is either all lost or all preserved. However, marker rescue may occur in some strains (e.g., Gillham, 1969; Gillham *et al.,* 1974).
5. Zoospores from biparental zygotes are usually heterozygous for the entire genome, and they are referred to as *cytohets.*
6. In cytohet progeny of biparental zygotes, whether spontaneous or UV induced, two types of exchanges occur: reciprocal (type III) and nonreciprocal (type II). Exchanges occur at a "four-strand stage" when the DNA molecules have replicated and are paired but before the cells have divided. The reciprocal event produces two daughter cells, each carrying one of the parental alleles. The nonreciprocal event, resembling gene conversion, produces one daughter cell which is a pure parental type, and the other which is still a cytohet (Figures 7, 8, and 9). Different genes in the same DNA molecule (i.e., from the same parent) may undergo type II and type III segregation in the same round of replication.
7. Exchanges occur at each cell division beginning with the first postmeiotic doubling and continue indefinitely. We have followed the process in liquid cultures as long as any heterozygous markers remain to detect it (Sager and Ramanis, 1968; Sager *et al.,* 1975).
8. Zoospores from biparental zygotes are genetically diploid for the chloroplast genome, while haploid for the nuclear genome. The diploidy is shown by (a) the occurrence of segregation at the first postmeiotic division, continuing at the same rate in further divisions, and (b) the patterns of segregation, reciprocal and nonreciprocal.

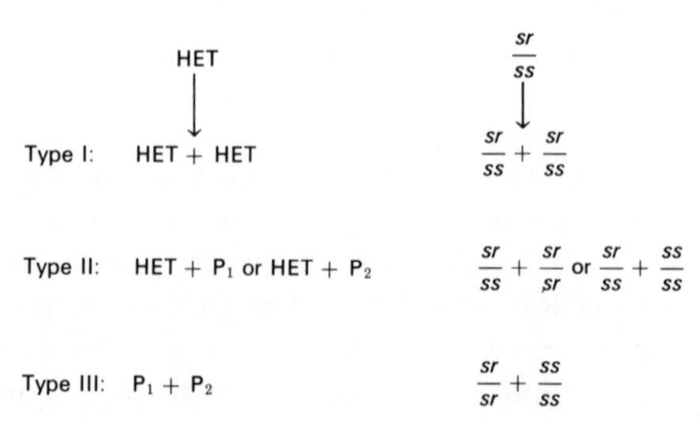

Figure 7. Segregation patterns of cytogene alleles at mitotic cell division.

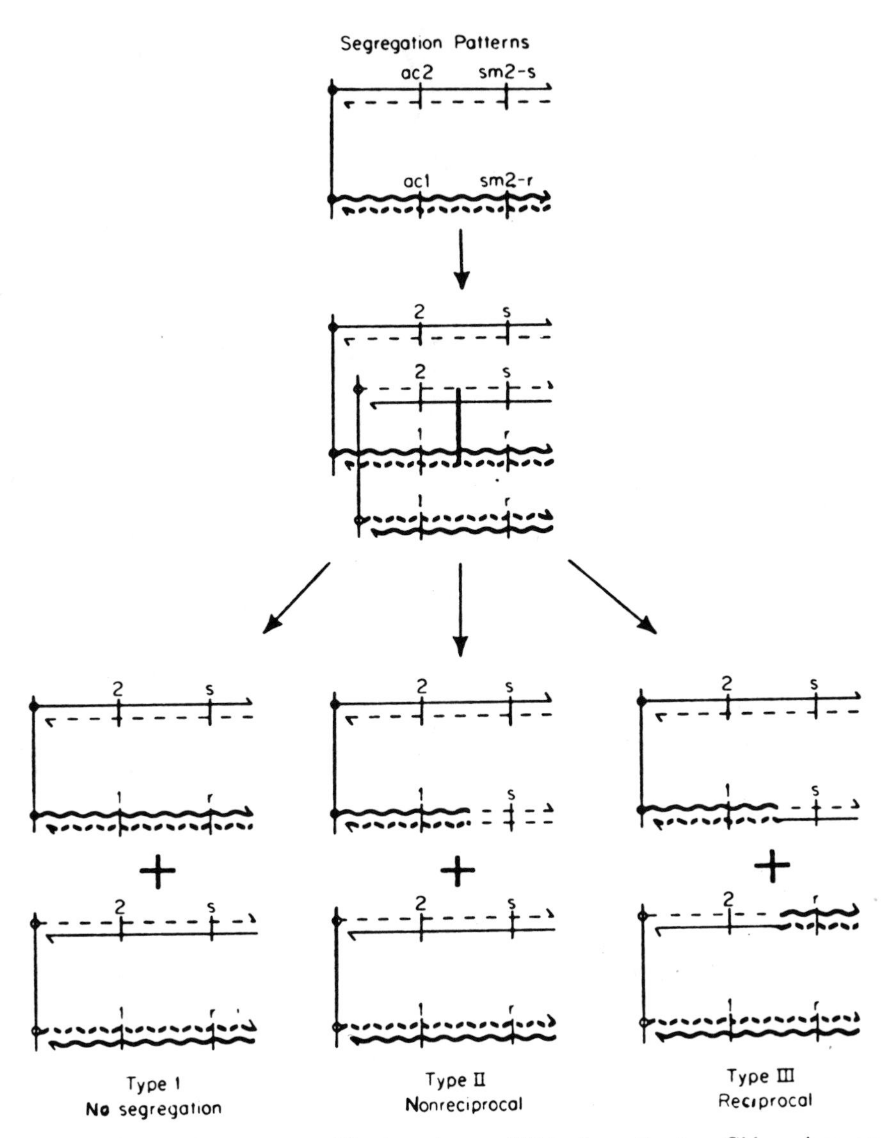

Figure 8. Segregation patterns. The homologous DNAs from the two Chlamydomonas parents are shown as straight (ac2 parent) and wavy (ac1 parent) solid lines. The complementary replicated strands are shown as dashed lines. The homologous DNAs are shown attached to a hypothetical membrane by one of the complementary strands. The model predicts that the "old" attachment points go to one daughter at cell division and the "new" points to the other daughter, thus determining the regularity of distribution following semiconservative replication. Type II segregation is pictured as a double-stranded loss and replacement of a segment, and type III is shown as a double-stranded exchange. This model is consistent with the genetic data but is otherwise speculative. From Sager (1972).

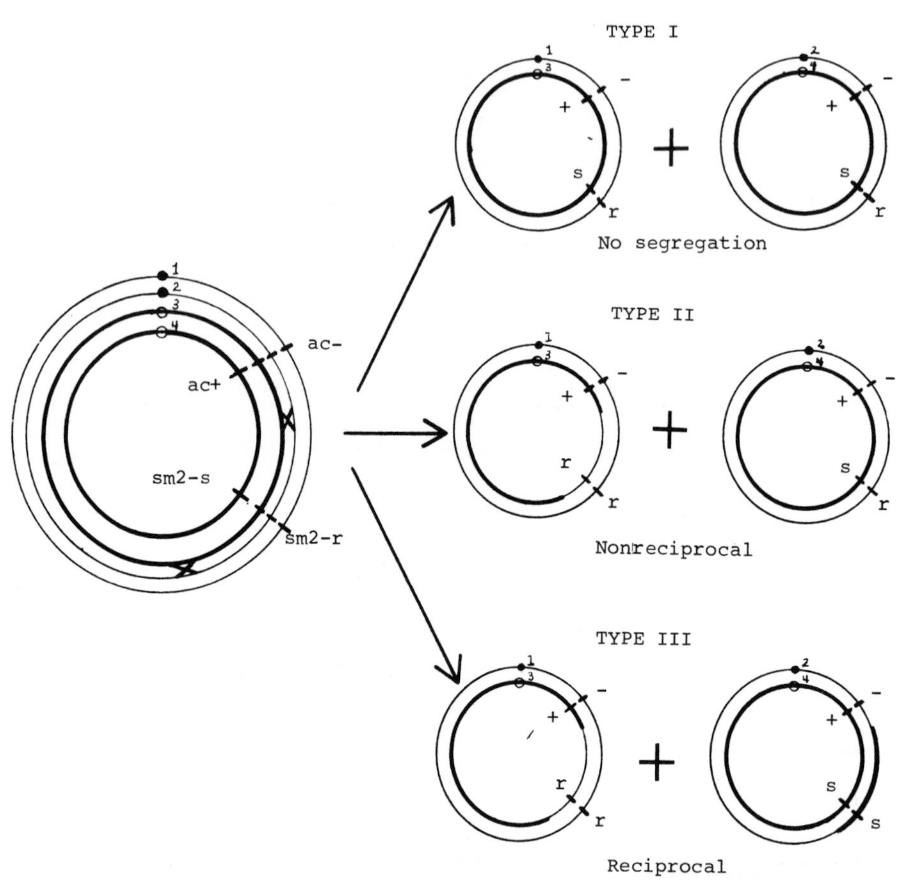

Figure 9. Segregation patterns in circular molecules. The homologous DNAs from the two parents are shown as thick and thin lines, each representing double-stranded molecules with corresponding attachment points. After replication but before cell division, the four molecules are arranged so that exchange events can occur. The attachment points are oriented so that 1 and 2 go to separate cells at cell division, as do 3 and 4. If no exchanges occur, both daughter cells are fully heterozygous (type I). Nonreciprocal (type II) exchange is viewed as the miscopying of a region. Reciprocal (type III) segregation requires an exchange event between the genes and attachment point, as well as an exchange beyond the gene to ensure proper separation of the circles. "X" indicates point of exchange between strands 2 and 3. From Sager (1972).

Mapping Procedures

The first genetic map of chloroplast DNA in *Chlamydomonas* included seven genes in a linear array (Sager and Ramanis, 1970). Gene order and the relative distances between genes were determined with a series of multigenic crosses. Subsequently, when additional genes were mapped, the initial set of seven genes was found to be a sector of a larger, genetically circular map (Sager, 1972, 1976). Our mapping procedures were developed to measure three different but related parameters: (1) the distance from each gene to the centromerelike attachment point, (2) the frequencies of simultaneous segregation of pairs or sets of genes, and (3) intergenic recombination frequencies.

Gene-Attachment Point Distances. As shown in Figure 8, a vegetative cell that is heterozygous for a particular cytogene may give rise to homozygous daughter cells at mitosis, if a reciprocal exchange occurs between the gene and the attachment point (Type III segregation). Alternatively, one homozygote and one het will result from a nonreciprocal event (type II segregation). Extensive studies (Sager, 1972, 1976) have shown that the frequencies of type II events are the same for each gene and therefore do not themselves contribute information for mapping. However, type II events occur with a frequency that is different and characteristic for each gene, as shown in Table 5. The differential frequency of type III events has been measured by two methods. In pedigree analysis, the type III frequencies are determined directly for events occurring in the first two postmeiotic mitoses. The sample size is necessarily small.

We have developed a new method (Singer *et al.*, 1976) based on the relative rates of segregation of different gene pairs in liquid culture (cf. p. 398). The rate of disappearance of hets and the emergence of pure parental types have been expressed as the slope of the het decay curve (Figure 10). The slope values measure the relative distance of each gene from the attachment point (ap). This method does not distinguish between one-armed and two-armed arrangements. To determine whether genes lie on one or two sides of ap, it is necessary to compare these results with data obtained by other methods.

Cosegregation and Recombination Frequencies. The frequency of type II cosegregation of two or more genes in multigenic crosses provides a basis for computing intergene distances. The closer two genes are, the more frequently will they cosegregate. As shown in Figures 7, 8, and 9, cosegregation may result from either a type II or type III exchange

TABLE 5. Frequencies of Type II and Type III Allelic Segregation in Pedigrees of 16 Crosses of Chlamydomonas (from Sager and Ramanis, 1976b)

	ac	*sm2*	*ery*	*spc*	*tm*
Type II	39.1 ± 1.2	39.4 ± 1.3	39.6 ± 1.4	38.4 ± 1.6	43.9 ± 1.2
Type III	5.9 ± 0.56	9.7 ± 1.23	12.3 ± 1.25	11.75 ± 1.20	6.7 ± 0.58

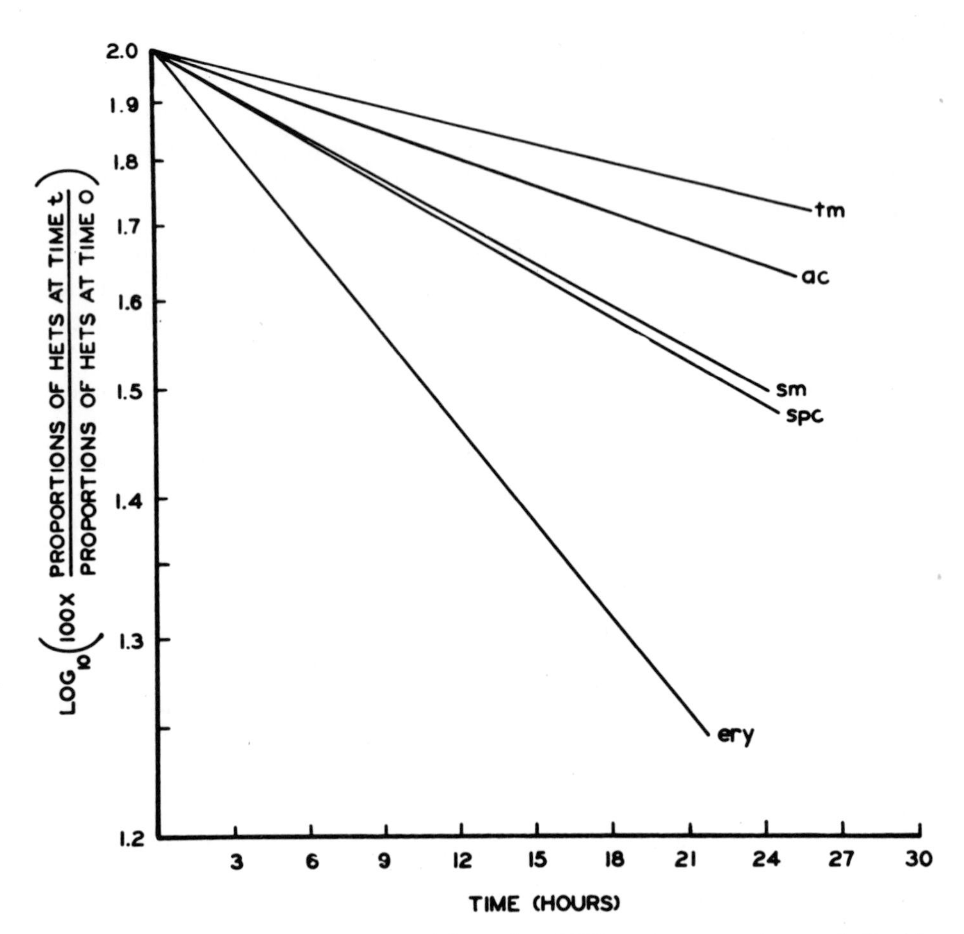

Figure 10. Relative segregation rates of five pairs of Chlamydomonas cytogenes in liquid culture of zoospores. A synchronously germinating population of zygotes in liquid culture was sampled at 1- or 2-hr intervals and each gene pair was classified as heterozygous or segregated to the homozygous state at the time of sampling. The slopes are based on data from 10–15 different crosses in which all five markers were present. From Singer et al. (1976).

event. With pedigree analysis, these two classes of events can be distinguished. In general, type II events, whether involving single genes or clusters of two or more genes, are much more frequent than are type III events. Consequently, both can be utilized for mapping purposes. However, it is the high frequency of type II events which makes mapping of this genome a special problem, since gene conversion events involving several markers obscure the strand exchanges which may have been involved. Furthermore, at each successive doubling, events which occurred earlier may be wiped out by later conversions. For this reason, only very closely linked markers can be mapped by recombination data which were obtained later than the first few doublings of zoospore clones.

By use of pedigree data, a set of markers were mapped by two methods which gave good agreement: (1) the frequency of pairs of markers segregating together in parental configuration, referred to as cosegregation, and (2) the frequency of recombinant pairs (Sager and Ramanis, 1970). The results gave a linear map of genes in a sector from the attachment point, *ap,* to *csd* (conditional streptomycin resistance), and including two closely linked *ac* genes, streptomycin dependence and neamine and erythromycin resistances. More recently, by use of the same methods (Sager and Ramanis, unpublished) these markers have been found to represent a sector of a circular map (Sager, 1972).

The evidence for circularity comes from a comparison of cosegregation and recombination data with the slope data of Figure 10. The cosegregation and recombination data have shown that *ac* and *tm1* are not closely linked, nor are *sm2* and *spc,* but *ac* is linked to *sm2* as previously shown (Sager and Ramanis, 1970), and *spc* lies between *tm1* and *ery*. These data indicate that the most likely gene order is *ap–ac–sm2–ery–spc–tm1*, whereas the slope data indicate the relative gene order to be *ap–tm–ac–(sm2–spc)–ery*. The simplest way to accommodate both sets of data is with a circular map, as shown in Figure 11.

Diploidy of the Chloroplast Genome

The number of genetic copies of the chloroplast genome is of particular importance in the correlation of physical and genetic evidence to give a coherent view of the organization of chloroplast DNA. As stated above (p. 400), our genetic data have provided unambiguous evidence that the chloroplast genome is diploid, at least in gametes and in young zoospore clones. This conclusion has recently been disputed by Gillham *et al.*, (1974). We therefore feel that it is incumbent on us to try to clarify this problem for the reader.

The crux of the experimental problem lies in the ratios of cytoplasmic gene alleles from the two parents recovered in the progeny of biparental zygotes. In our laboratory, the ratios have been close to 1 : 1 for each allelic pair in all crosses (Sager and Ramanis, 1970, 1976*a,b*). However, virtually all of our data come from progeny recovered after UV irradiation of the mt^+ parent, since the spontaneous frequency of biparental zygotes is so low (about 0.1%). Gillham's group have strains which give a high yield of spontaneous biparental zygotes (20–30%), and consequently they can observe allelic ratios of progeny in unirradiated

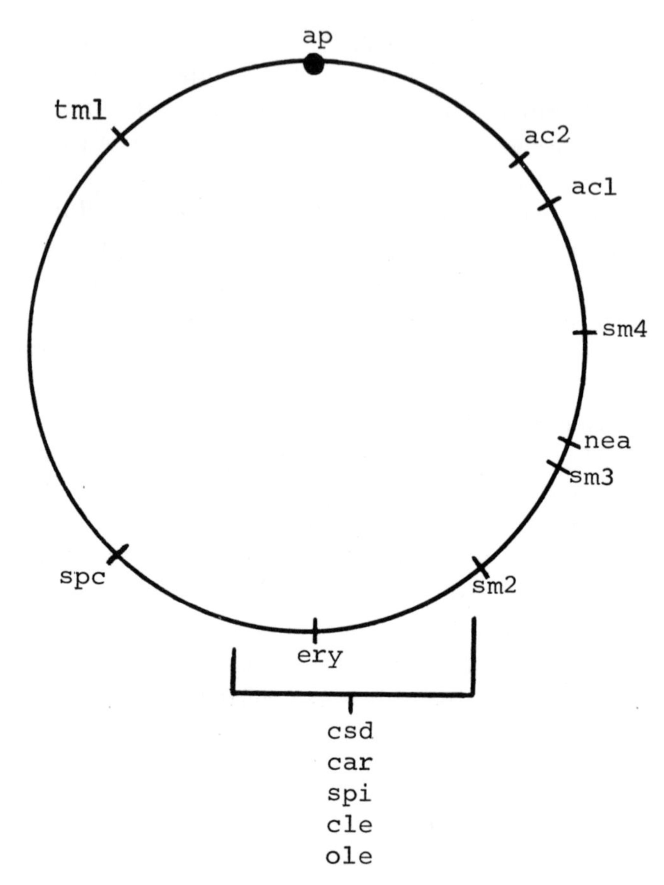

Figure 11. A circular genetic map of Chlamydomonas incorporating data from several crosses. The qualitative evidence of gene order, relative distances, and circularity is well supported by the data but quantitative map distances need further clarification. The bracketed genes all map close to ery and no linear order has been established. Symbols: ap, attachment point; ac2 and ac1, acetate requirement; sm4, streptomycin dependence; nea, neamine resistance; sm3, low-level streptomycin resistance; sm2, high-level streptomycin resistance; ery, erythromycin resistance; csd, conditional streptomycin dependence; car, carbomycin resistance; spi, spiramycin resistance; cle, cleosine resistance; ole, oleandomycin resistance; spc, spectinomycin resistance; tm1, temperature sensitivity. From Sager (1972).

crosses. They have in the past consistently found an excess of alleles from the mt^+ parent (Gillham, 1969), and more recently have shown that UV irradiation of the mt^+ parent in their crosses decreases the relative frequencies of alleles from the mt^+ parent, resulting in a pseudo 1 : 1 ratio, pseudo because UV irradiation, in their view, reduced the number of copies of chlDNA in the irradiated parent. They conclude not only that their strains are multiploid but also that *our* data are artifactual, as a result of our UV irradiation procedure.

We have tested this possibility directly (Sager and Ramanis, 1976*a*) by comparing the allelic ratios as well as other genetic parameters in progeny of a cross involving six chloroplast and three nuclear gene markers, mated after 0-, 15-, 30-, and 50-sec UV irradiation of the mt^+ parent (50 sec is our usual dose). In brief summary, we found that at each UV dose all six pairs of cytogene alleles from each parent were recovered 1 : 1 in progeny sampled from large zygote colonies as well as in pedigrees. The frequency of segregation of each pair of alleles from heterozygous to homozygous state was not UV dose dependent, nor was the frequency of reciprocal or nonreciprocal recombination in zoospore clones. Increasing UV dose did have a small stimulatory effect on exchanges occurring in the zygote before germination. No effect of irradiation was seen on nuclear gene segregation ratios. These results support our previous conclusion that the chloroplast genome behaves as if diploid and that the genetic events are consistent with a model of mitotic recombination at a four-strand stage. Our results provide no support for the multicopy model proposed by Gillham's group.

It seems likely therefore that the differences in experimental results are the consequence of strain differences. However, the basis of the strain differences is less evident. Do the data of Gillham's group support the view that their strains carry multiple copies of the chloroplast genome? We think not, for the following reasons:

1. Gillham *et al.* (1974) have not reported tetrad or pedigree analysis of their crosses. There may be lethal factors present that distort their progeny recovery ratios.
2. The three markers they used are rather closely linked, and may not represent the behavior of the whole genome.
3. Their crosses were done with strains giving a high frequency (20–30%) of spontaneous exceptional zygotes. Thus their strains contain a molecular mechanism that partially inhibits maternal inheritance, analogous to our *mat-1* mutation or to about 30 sec to UV irradiation of our mt^+ parent before mating. If, as we have proposed (Sager, 1972, 1974; Sager and Kitchin, 1975), the enzymatic modification and restriction of chloroplast DNA are at the basis of maternal inheritance (see p. 388), then the results of Gillham *et al.* (1974) can be interpreted in these terms. We suggest, therefore, an imbalance between modification and restriction activity in their zygotes, such that partial degradation and marker rescue occur, giving rise to zygotes carrying some chloroplast genes from the male parent, thus scored as exceptional or biparental, but with an excess of copies *per zygote* from the female parent. A careful pedigree analysis could test this possibility directly.

The Distinction between Chloroplast and Mitochondrial Genes

No mitochondrial genes have yet been clearly identified in *Chlamydomonas,* but the presence of mitochondrial DNA assures the investigator of their existence. Alexander *et al.* (1974) have proposed that minute colonies recovered after acriflavin treatment of wild-type cells are the result of mutations in mitDNA, but the evidence is circumstantial since the mutations are lethal. Some drug-resistant mutants showing non-Mendelian but biparental inheritance (possibly mitochondrial) have been studied (Sager and Ramanis, unpublished), but their location has not been established.

In a series of papers, Arnold, Schimmer, and Behn (Schimmer and Arnold, 1970*a,b,c*; Behn and Arnold, 1972, 1973) have attributed a mitochondrial location to a streptomycin-dependent mutation, *sd3,* obtained from Gillham. Their inference rests primarily on the high reversion rates back to *sd* they observed with streptomycin-sensitive (*ss*) clones recovered from the *sd* mutant. In extensive studies of a similar sort, we (Sager, 1972) found that the apparent "revertants" from *ss* to *sd* were actually segregants, and that the supposed *ss* clones were in fact *ss/sd* heterozygotes. The *sd* gene used in our studies maps in the chloroplast genome. Thus we conclude that the behavior of the *sd3* gene attributed to multiple copies by Schimmer and his colleagues results rather from the segregation of what we have termed *persistent cytohets* (Sager, 1972).

Functions of Chloroplast DNA

The chlDNA circular molecules of higher plants, with a genomic complexity of about 10^8 daltons (p. 378), have a coding potential for about 250 proteins of 30,000 daltons each. If the genomic complexity is higher than 10^8, as suspected in *Chlamydomonas* (p. 380) and *Acetabularia* (p. 380), then the coding potential is correspondingly greater in these algae. The total number of proteins coded by chlDNA is not a question of immediate concern, since very few proteins have yet been identified as products of the chloroplast genome. As discussed above (p. 382), the RNAs of chloroplast ribosomes are transcribed from chlDNA, and it seems likely, although not yet established, that some of the chloroplast-specific tRNAs as well as chloroplast-specific 5 S RNA are also transcription products of chlDNA. Thus one function of chlDNA is to provide cistrons that are transcribed as RNAs involved in the machinery of chloroplast protein synthesis. However, only a small fraction of chlDNA can be accounted for by these cistrons.

In speculations on the proteins coded by chlDNA, the following classes of proteins have been suggested (e.g., Sager, 1972): (1) the fraction I protein, RUdP-carboxylase, which constitutes some 50% of the total protein of chloroplasts; (2) ribosomal proteins; (3) membrane-bound proteins, including some of those that function in photosynthetic electron transport and photophosphorylation; and (4) regulatory proteins that function as trasmitters of signals either within the chloroplast or from the organelle to the cytoplasm or nucleus. Evidence exists that proteins in each of these classes are products of the chloroplast genome, but in no instance is the evidence as complete as it is in classical bacterial systems. Interactions between nuclear and organelle genomes in *Acetabularia* are discussed in Chapter 14.

Fraction I Protein

Recent advances in the study of fraction I protein from *Nicotiana* species have demonstrated that the protein can be readily crystallized (Chan *et al.*, 1972) and that the previously described large and small subunits (Kawashima and Wildmaro, 1972) have different genetic origins (Sakano *et al.*, 1974). Electrofocusing of carboxymethylated fraction I protein from *N. tabacum* resulted in separation of the large and small subunits, and revealed previously unsuspected heterogeneity at the polypeptide level (Kung *et al.*, 1974). The large subunit was resolved into three major peptides and the small subunit into two peptides. The crystallized fraction I protein was then examined by this method in preparations from several species of *Nicotiana* and from pairs of F_1 reciprocal interspecific hybrids.

Considering the large subunit first, a clear difference in position of peptide bands was seen on polyacrylamide gels after electrofocusing of the protein from different species. For example, two of the three bands of *N. glauca* were located at the same position as two of the three from *N. excelsior* and *N. gossei,* but the third was clearly displaced. Similarly, *N. tabacum* had only one band coincident with those of *N. excelsior* and *N. gossei,* and two with *N. glauca*. In proteins of hybrids from reciprocal crosses between *N. tabacum* and *N. gossei,* and between *N. tabacum* and *N. glauca,* the peptide patterns in each instance resembled those of the maternal parent. Similar results were also found in hybrids from crosses between *N. tabacum* and *N. glutinosa,* and between *N. glauca* and *N. langsdorfii*. In each of these studies, the peptides of the small subunit were also examined, and in each instance the isozymes of both parents were present in the hybrids, providing evidence that the small subunit is coded by presumably nuclear genes transmitted by both parents.

This extensive evidence, representing the culmination of many years of pursuit of the fraction I protein by Wildman and his students, clearly demonstrates the maternal inheritance of the genes coding for the large subunit. This is composed of at least three different polypeptides and is therefore the product of at least three different genes. These data are consistent with Mendelian inheritance of genes for the small subunit. In view of the elegance of this work, it is perhaps unkind to point out that the genes themselves have not been identified, or even localized to nuclear and chloroplast DNAs (although any other locations are highly unlikely).

Chloroplast Ribosomal Proteins

The likelihood that chlDNA may code for chloroplast ribosomal proteins was first suggested when chloroplast gene mutations were found to confer resistance to drugs that inhibit protein synthesis of 70 S bacterial ribosomes, and chloroplast ribosomes themselves were shown to resemble bacterial ribosomes (cf. Sager, 1972). This question has been examined extensively with mutants of *Chlamydomonas* and the present evidence will be briefly reviewed.

Streptomycin- and Spectinomycin-Resistant Mutants. Both nuclear and chloroplast gene mutations have been described that confer streptomycin resistance (Sager, 1954; Gillham, 1974), and both classes of mutations have been shown to affect chloroplast ribosome sedimentation values (Gillham *et al.,* 1970). Chloroplast ribosomes isolated from a streptomycin-resistant strain were shown to exhibit resistance to the drug at concentrations as high as 280 μg/ml, in measurements of *in vitro* amino acid incorporation activity in a poly(U)-directed assay. In contrast, similar preparations from wild-type cells were inhibited by streptomycin at concentrations as low as 2 μg/ml (Schlanger and Sager, 1974*a*). In subunit exchange experiments between 30 S and 50 S subunits from sensitive and resistant strains, the resistance to streptomycin was localized to the 30 S subunit. In similar studies with ribosomes from other drug-resistant strains, neamine and spectinomycin resistances were also localized to the 30 S subunit, while resistances to cleocin and carbomycin were each localized to the 50 S subunit (Schlanger and Sager, 1974*a*).

A single protein of the 30 S subunit was identified as altered or missing in preparations made from chloroplast ribosomes of the streptomycin-resistant strain compared with those from wildtype in double-label experiments. Proteins from one strain were labeled with [^{3}H]arginine, and from the other strain with [^{14}C]arginine, and they were next cochromatographed on CM-cellulose. Peak 17 was absent from the mutant profile, and, instead, a pronounced peak was seen in the region of peak 16 in the mutant, but not in the wide-type profile. The molecular weights in both Peak 16 and Peak 17 regions determined by discontinous sodium dodecylsulfate polyacrylamide gel electrophoresis were indistinguishable (about 18,000 daltons). The results show that the chloroplast gene mutation to streptomycin resistance has altered one chloroplast ribosomal protein of the 30 S subunit. We interpret the additional protein in the mutant eluting at Peak 16 as most likely the mutationally altered form of the Peak 17 protein (Ohta *et al.,* 1975).

Similar experiments were carried out with a spectinomycin-resistant strain and with a second streptomycin-resistant strain (both chloroplast mutants), but no differences from wild-type profiles were seen. Burton (1972) has described the loss of spectinomycin-binding capacity by a *spc* mutant, and he has attributed phenotypic resistance to this loss of affinity.

Erythromycin-Resistant Mutants. Mets and Bogorad (1971, 1972) have described several erythromycin-resistant mutants, some of which result from nuclear gene mutations, and one from a chloroplast mutation. The proteins of the large chloroplast ribosomal subunit were examined by two-dimensional gel electrophoresis. A single spot, prominent in the wild type, was missing from the mutant gel. In addition, a series of small new spots appeared in the mutant gel, which the authors suggested are higher molecular weight aggregates of the missing protein. In subsequent studies of the Mendelian mutants (Davidson *et al.,* 1974; Hanson *et al.,* 1974), four *ery* mutants (all resulting from mutations at the same locus, *ery* M1, on linkage group XI) have shown altered mobility of the same protein spot, LC6.

Taken together, these results all show that mutations in the chloroplast genome alter chloroplast ribosomal proteins, but the question of how many of these proteins are coded by the chloroplast genome and how many by the nuclear genome and the question of how the interaction between the genomes operates smoothly in ribosome biogenesis remain problems for the future. Of even more general significance is the question of how this interdigitation in ribosome function is utilized by the organism in the control of its growth and metabolism.

Membrane-Bound Proteins

No membrane-bound proteins have been identified directly as products of the chloroplast genome, but the synthesis of five membrane-bound proteins has been detected in *in vitro* studies of protein synthesis in isolated chloroplasts (*cf.* Ellis and Hartley, 1974). This result confirms previous reports that the synthesis of some membrane-bound proteins can be differentially inhibited by drugs that block protein synthesis on 70 S ribosomes (*cf.* Boulter *et al.*, 1972).

Regulatory Proteins

The need for signals to coordinate DNA replication and protein synthesis in organelle compartments with that of the rest of the cell is obvious, but no specific mechanisms have yet been described. Indirect evidence that a product(s) of chloroplast protein synthesis is involved in the initiation of nuclear DNA replication has been described in *Chlamydomonas* (Blamire *et al.*, 1974) (see p. 386). In *Euglena,* a mutant lacking chloroplast DNA has been described in which the light-induced synthesis of several nuclear-coded enzymes does not occur, although low levels of the enzymes, characteristic of dark-grown cells, are present. The authors (Schmidt and Lyman, 1974) suggest that a product of the chloroplast genome is required to mediate the light-induced signal.

Maternally Inherited Genes with Unknown Functions

In *Chlamydomonas,* numerous mutants have been characterized genetically as non-Mendelian, some of them already mapped in the chloroplast genome, such as *tm1* (conferring temperature sensitivity), but these have not yet been studied biochemically. The list includes several temperature-sensitive mutants, acetate-requiring mutants (nonphotosynthetic), and mutants showing multiple drug resistance, possibly the consequence of altered chloroplast membrane permeability.

In higher plants, the many maternally inherited mutations affecting chloroplast development that have been described are not well characterized at the biochemical level (*cf.* Sager, 1972). Wildman and his students have examined a variegated mutant strain of *Nicotiana tabacum* in which the maternally inherited mutation induces defective development of some but not all chloroplasts. In a comparison of green and colorless leaf sectors, normal fraction I protein, 70 S ribosomes, and normal mitochondrial activity (cytochrome *c* oxidase and catalase) were present in the white sectors, showing that these functions were not directly involved in the mutation (Wildman *et al.*, 1973). A small difference in composition of chlDNA from mutant and wild type was detected in a comparison of melting profiles, indicating that the mutant chlDNA was about 1% higher in GC content than the wild type (Wong-Staal and Wildman, 1973).

Concluding Remarks

This chapter has aimed at an integration of physical, genetic, and biochemical evidence to provide the reader with a picture of current understandings and some future directions of research into the organization and functions of chloroplast DNA. In our view, research in this area holds great promise, with many lines of investigation barely started but clearly available for a multidisciplinary approach. New knowledge about the functions of chloroplast DNA and its integration into cell metabolism is of key importance not only in basic cell biology but also in practical applications of worldwide importance in agriculture. Hardly less wideranging in scope are the uses of chloroplast DNA in medical research as a model system for understanding mitochondrial function, since the two organelle systems bear close similarities, and for some studies the chloroplast rather than the mitochondrion is the system of choice.

General questions that need to be answered in both organelle systems include the following:

1. What are the special functions of organellar DNAs that have been responsible for their continuing presence as unique genomes?
2. Why are organellar genomes transmitted maternally (or uniparentally) in microbial eukaryotes?
3. How is the mutational potential of organellar DNAs regulated to maintain relative if not absolute homogeneity?
4. What is the regulatory organization of organellar DNAs, in comparison with the operons of bacteria and the interspersed reiterated sequences of animal cells?
5. What are the genetic mechanisms that regulate the tight coupling and interdigitation between organellar metabolism and that of the rest of the cell?

Some experimental approaches have been started or have become feasible on each of these questions:

1. Organellar DNAs undoubtedly code for key peptides involved in electron transport and phosphorylation. As with the fraction I protein of chloroplasts, it seems likely that certain peptides of the mitochondrial ATPase and cytochrome oxidase are of organellar genetic origin, while other peptides of the same complex proteins are nuclear-coded and synthesized on cell-sap ribosomes. The identification of more organellar-coded proteins waits on the

combined use of chloroplast mutants and advanced biochemical techniques.

2. The mechanism of maternal inheritance in *Chlamydomonas* and probably in all organellar DNAs depends on the activity of DNA modification and restriction enzymes. The intricate regulation of organellar inheritance shows that this pattern is of key importance to the organism. The mechanism acts to decrease if not eliminate the possibilities for genetic recombination of parental alleles, and the evolutionary importance of uniparental transmission probably lies in minimizing genetic variability. Why this is advantageous is not obvious.
3. Genetic homogeneity of organellar DNAs has not been demonstrated, except insofar as cytogene alleles follow empirically established rules, and insofar as genomic and analytical sizes of organellar DNAs coincide. There is probably little enough heterogeneity, so that a mechanism must be operating to regulate it. Quantitative mutational studies with organellar DNAs need very much to be done, and for this purpose the genetically diploid chloroplast genome of *Chlamydomonas* has much to recommend it.
4. The molecular organization of organellar DNAs is now being investigated with the use of restriction endonucleases, much as has already been done with some viral DNAs (Boyer, 1974). This highly fruitful approach should soon provide vastly more information than is yet available about the presence and arrangement of reiterated sequences, regulatory sites recognized by restriction enzymes, etc. With this powerful tool, as well as nucleic acid hybridization and reannealing kinetic analysis, questions such as the paradox between genomic and analytical size of *Chlamydomonas* chlDNA should be resolvable. The identification of regulatory signals, however, is a more long-range problem, and one which will probably require the combined use of regulatory mutants and of molecular methods, just as in bacterial systems.
5. Perhaps the most difficult problem of all is the identification and characterization of specific genetic mechanisms that regulate interactions between organelle and nucleus. A small hint in this direction came with our finding that protein synthesis in the chloroplast provides a signal for nuclear DNA replication. In our view, direct applications of relevance to agriculture and to medical research could be made of any new understandings in this area of research, embodying as it does the integration of organelle and overall cell function.

Acknowledgments

Research supported by grants to Ruth Sager from the National Institutes of Health (GM-13970) and the American Cancer Society. This chapter was written while both authors were at Hunter College of the City University of New York.

Literature Cited

Alexander, N. J., N. W. Gillham, and J. E. Boynton, 1974 The mitochondrial genome of *Chlamydomonas:* Induction of minute colony mutations by acriflavin and their inheritance. *Mol. Gen. Genet.* **130:**275–290.

Barnett, W. E., C. J. Pennington, Jr., and S. A. Fairfield, 1969 Induction of *Euglena* transfer RNAs by light. *Proc. Natl. Acad. Sci. USA* **63:**1261–1268.

Bastia, D., K.-S. Chiang, and H. Swift, 1971*a* Studies on the ribosomal RNA cistrons of chloroplast and nucleus in *Chlamydomonas reinhardi. Abstracts of the American Society for Cell Biology, 11th Annual Meeting,* p. 25.

Bastia, D., K.-S. Chiang, H. Swift, and P. Siersma, 1971*b* Heterogeneity, complexity and repetition of the chloroplast DNA of *Chlamydomonas reinhardi. Proc. Natl. Acad. Sci. USA* **68:**1157–1161.

Bayen, M. and A. Rode, 1973 Heterogeneity and complexity of *Chlorella* chloroplastic DNA. *Eur. J. Biochem.* **39:**413–420.

Behn, W. and C. G. Arnold, 1972 Zur Lokalization eines nichtmendelnden Gens von *Chlamydomonas reinhardi. Mol. Gen. Genet.* **114:**266–272.

Behn, W. and C. G. Arnold, 1973 Localization of extranuclear genes by investigations of the ultrastructure in *Chlamydomonas reinhardi. Arch. Mikrobiol.* **92:**85–90.

Bernardi, G., M. Faures, G. Piperno, and P. P. Slonimski, 1970 Mitochondrial DNA's from respiratory-sufficient and cytoplasmic respiratory-deficient mutant yeast. *J. Mol. Biol.* **48:**23–42.

Birky, C. W., Jr., P. S. Perlman, and T. J. Byers, editors, 1975 *Genetics and Biogenesis of Mitochondria and Chloroplasts,* Ohio State University Press, Columbus, Ohio.

Bishop, J. O., 1974 The gene numbers game. *Cell* **2:**81–86.

Blamire, J., V. R. Flechtner, and R. Sager, 1974 Regulation of nuclear DNA replication by the chloroplast in *Chlamydomonas. Proc. Natl. Acad. Sci. USA* **71:**2867–2871.

Boardman, N. K., J. M. Anderson, A. Kahn, S. W. Thorne, and T. E. Treffry, 1971*a* Formation of photosynthetic membranes during chloroplast development. In *Autonomy and Biogenesis of Mitochondria and Chloroplasts,* North-Holland, Amsterdam, pp. 70–84.

Boardman, N. K., A. W. Linnane, and R. M. Smillie, editors, 1971*b* *Autonomy and Biogenesis of Mitochondria and Chloroplasts,* North-Holland, Amsterdam, 511 pp.

Boulter, D., R. J. Ellis, and A. Yarwood, 1972 Biochemistry of protein synthesis in plants. *Biol. Rev.* **47:**113–175.

Boyer, H. W., 1974 Restriction and modification of DNA: Enzymes and substrates. *Fed. Proc.* **33:**1125–1127.

Brawerman, G. and J. M. Eisenstadt, 1964 Deoxyribonucleic acid from the chloroplasts of *Euglena gracilis. Biochem. Biophys. Acta* **91:**477–485.

Britten, R. J. and D. E. Kohne, 1968 Repeated sequences in DNA. *Science* **161**:529–540.

Brown, R. D. and R. Haselkorn, 1972 The isolation of *Euglena gracilis* chloroplasts uncontaminated by nuclear DNA. *Biochim. Biophys. Acta* **259**:1–4.

Burton, W. G., 1972 Dehydrospectinomycin binding to chloroplast ribosomes from antibiotic-sensitive and resistant strains of *Chlamydomonas reinhardi. Biochim. Biophys. Acta* **272**:305–311.

Cattolico, R. A. and S. P. Gibbs, 1974 Analysis of chloroplast DNA and chloroplast replication in the synchronously grown wall-less alga *Olisthodiscus luteus. J. Cell Biol.* **63**:52*a*.

Chan, P. H., K. Sakano, S. Singh, and S. G. Wildman, 1972 Crystalline fraction 1 protein: Preparation in large yield. *Science* **176**:1145–1146.

Chiang, K.-S. and N. Sueoka, 1967 Replication of chloroplast DNA in *Chlamydomonas reinhardi* during vegetative cell cycle: Its mode and regulation. *Proc. Natl. Acad. Sci. USA* **57**:1506–1513.

Chun, E. H. L., M. H. Vaughan, and A. Rich, 1963 The isolation and characterization of DNA associated with chloroplast preparations. *J. Mol. Biol.* **7**:130–141.

Cook, J. R., 1966 The synthesis of cytoplasmic DNA in synchronized *Euglena. J. Cell Biol.* **29**:369–373.

Davidson, E. G., D. E. Graham, B. R. Neufeld, M. E. Chamberlin, C. S. Amenson, B. R. Hough, and R. J. Britten, 1973 Arrangement and characterization of repetitive sequence elements in animal DNAs. *Cold Spring Harbor Symp. Quant. Biol.* **38**:295–302.

Davidson, J. N., M. R. Hanson, and L. Bogorad, 1974 An altered chloroplast ribosomal-protein in ery-ML mutants of *Chlamydomonas reinhardi. Mol. Gen. Genet.* **132**:119–129.

Ebersold, W. T., 1967 *Chlamydomonas reinhardi:* Heterozygous diploid strains. *Science* **157**:447–449.

Ebringer, L., 1972 Are plastids derived from prokaryotic micro-organisms? Action of antibiotics on chloroplasts of *Euglena gracilis. J. Gen. Microbiol.* **71**:35–52.

Edelman, M., J. A. Schiff, and H. T. Epstein, 1965 Studies of chloroplast development in *Euglena.* XII. Two types of satellite DNA. *J. Mol. Biol.* **11**:769–774.

Edelman, M., H. T. Epstein, and J. A. Schiff, 1966 Isolation and characterization of DNA from the mitochondrial fraction of *Euglena. J. Mol. Biol.* **17**:463–469.

Ellis, R. J. and M. R. Hartley, 1974 Nucleic acids of chloroplasts. In *Biochemistry of Nucleic Acids,* edited by K. Burton, Vol. 6 of MTP International Review of Science, Biochemistry Series 1, University Park Press, Baltimore.

Flechtner, V. R. and R. Sager, 1973 Ethidium bromide induced selective and reversible loss of chloroplast DNA. *Nature (London) New Biol.* **241**:277–279.

Gibbs, M., editor, 1971 *Structure and Function of Chloroplasts,* Springer-Verlag, New York, 286 pp.

Gibor, A., 1967 DNA synthesis in chloroplasts. In *Biochemistry of Chloroplasts,* Vol. 2, edited by T. W. Goodwin, Academic Press, New York.

Gillham, N. W., 1969 Uniparental inheritance in *Chlamydomonas reinhardi. Am. Nat.* **103**:355–388.

Gillham, N. W., 1974 Genetic analysis of the chloroplast and mitochondrial genomes. *Ann. Rev. Genet.* **8**:347–391.

Gillham, N. W., J. E. Boynton, and B. Burkholder, 1970 Mutations altering chloroplast ribosomes phenotype in *Chlamydomonas.* I. Non-Mendelian mutations. *Proc. Natl. Acad. Sci. USA* **67**:1026–1033.

Gillham, N. W., J. E. Boynton, and R. W. Lee, 1974 Segregation and recombination of non-Mendelian genes in *Chlamydomonas.* 13th International Congress of Genetics Symposium *Genetics* **78:**439–457.

Goldring, E. S., L. I. Grossman, D. Krupnick, D. R. Cryer, and J. Marmur, 1970 The petite mutation in yeast: Loss of mitochondrial deoxyribonucleic acid during induction of peptides with ethidium bromide. *J. Mol. Biol.* **52:**323–335.

Green, B. R., 1973 Small circular DNA molecules associated with *Acetabularia* chloroplasts. *Proc. Can. Fed. Biol. Soc.* **17:**20.

Green, B. R., 1974*a* Nucleic acids and their metabolism. In *Algal Physiology and Biochemistry,* edited by W. D. P. Stewart, Blackwell, Oxford, pp. 281–313.

Green, B. R., 1974*b* Small circular DNA molecules associated with *Acetabularia* chloroplasts. *Can. Fed. Biol. Stud.* **17:**20.

Green, B. R. and H. Burton, 1970 *Acetabularia* chloroplast DNA: Electron microscope visualization. *Science* **168:**981–982.

Green, B. R., V. Heilporn, S. Limbosch, M. Boloukhere, and J. Brachet, 1967 The cytoplasmic DNA's of *Acetabularia mediterranea. Proc. Natl. Acad. Sci. USA* **58:**1351–1358.

Green, B. R., H. Burton, V. Heilporn, and S. Limbosch, 1975*a* The cytoplasmic DNA's of *Acetabularia mediterranea:* Their structure and biological properties. In *The Biology of Acetabularia,* edited by J. Brachet and S. Bonotto, Academic Press, New York, pp. 35–60.

Green, B. R., U. Padmanabhan, and B. L. Muir, 1975*b* The kinetic complexity of *Acetabularia* chloroplast DNA. In *Proceedings of the 12th International Botanical Congress,* Leningrad.

Hanson, M. R., J. N. Davidson, and L. J. Mets, 1974 Characterization of chloroplast and cytoplasmic ribosomal proteins of *Chlamydomonas reinhardi* by two-dimensional gel electrophoresis. *Mol. Gen. Genet.* **132:**105–118.

Heilporn, V. and S. Limbosch, 1971*a* Recherches sur les acides désoxyribonucléiques d'*Acetabularia mediterranea. Eur. J. Biochem.* **22:**573–579.

Heilporn, V. and S. Limbosch, 1971*b* Les effets du bromure d'ethidium sur *Acetabularia mediterranea. Biochim. Biophys. Acta* **240:**94–108.

Herrmann, R. G., 1972 Do chromoplasts contain DNA? II. The isolation and characterization of DNA from chromoplasts, chloroplasts, mitochondria and nuclei of *Narcissus. Protoplasma* **74:**7–17.

Herrmann, R. G., H. J. Bohnert, and K. V. Kowallik, 1974 Arrangement of circular DNA in chloroplasts. *Hoppe-Seylers Z. Physiol. Chem.* **355:**1205.

Herrmann, R. G., H.-J. Bohnert, K. V. Kowallik, and J. M. Schmitt, 1975 Size, conformation and purity of chloroplast DNA of some higher plants. *Biochim. Biophys. Acta* **378:**305–317.

Herschberger, C. L., D. Morgan, F. A. Weaver, and P. Pienkos, 1974 An unusual mutant affecting the frequency of organelle mutation. *J. Bacteriol.* **118:**434–441.

Ho, C., L. Lipsich, G. Fisher, and S. J. Keller, 1974 DNA replication in isolated chloroplasts of *Chlamydomonas reinhardi. J. Cell Biol.* **68:**104*a*.

Ingle, J., R. Wells, J. V. Possingham, and C. J. Leaver, 1971 The origins of chloroplast ribosomal-RNA. In *Autonomy and Biogenesis of Mitochondria and Chloroplasts,* edited by N. K. Boardman, A. W. Linnane, and R. M. Smillie, North-Holland, Amsterdam, pp. 393–401.

Ishida, M. R., K. Tadatoshi, T. Matsurbara, F. Hayashi, and E. Yokomura, 1969 Characterization of satellite DNA from the cells of *Porphyra tenera. Res. Reactor Inst. Kyoto Univ.* **2:**73–75.

Iwamura, T. and S. Kuwashima, 1969 Two DNA species in chloroplasts of *Chlorella*. *Biochim. Biophys. Acta* **174:**330–339.

Kawashima, N. and S. G. Wildman, 1972 Studies on fraction 1 protein. IV. Mode of inheritance of primary structure in relation to whether chloroplast or nuclear DNA contains the code for a chloroplast protein. *Biochim. Biophys. Acta* **262:**42–49.

Keller, S. J., S. A. Biedenbach, and R. R. Meyer, 1973 Partial purification of a chloroplast DNA polymerase from *Euglena gracilis*. *Biochem. Biophys. Res. Commun.* **50:**620–628.

Kirk, J. T. O., 1967 Effect of methylation of cytosine residues on the buoyant density of DNA in caesium chloride solution. *J. Mol. Biol.* **28:**171–172.

Kirk, J. T. O., 1971 Will the real chloroplast DNA please stand up? In *Autonomy and Biogenesis of Mitochondria and Chloroplasts,* edited by N. K. Boardman, A. W. Linnane, and R. M. Smillie, North-Holland, Amsterdam, pp. 267–276.

Kirk, J. T. O., 1972 The genetic control of plastid formation: Recent advances and strategies for the future. *Sub-cell. Biochem.* **1:**333–361.

Kislev, N., H. Swift, and L. Bogorad, 1965 Nuclei acids of chloroplasts and mitochondria in Swiss chard. *J. Cell Biol.* **25:**327–344.

Kolodner, R. and K. K. Tewari, 1972 Molecular size and conformation of chloroplast deoxyribonucleic acid from pea leaves. *J. Biol. Chem.* **247:**6355–6364.

Kolodner, R. and K. K. Tewari, 1975*a* The molecular size and conformation of chloroplast DNA of higher plants. *Biochim. Biophys. Acta* **402:**372–390.

Kolodner, R. and K. K. Tewari, 1975*b* Denaturation mapping studies on the circular chloroplast deoxyribonucleic acid from pea leaves. *J. Biol. Chem.* **250:**4888–4895.

Kolodner, R. and K. K. Tewari, 1975*c* The presence of displacement loops in the covalently closed circular chloroplast deoxyribonucleic acid from higher plants. *J. Biol. Chem.* **250:**8840–8847.

Kolodner, R. and K. K. Tewari, 1975*d* Chloroplast DNA from higher plants replicates by both the Cairns and the rolling circle mechanism. *Nature* (*London*) **256:**708–711.

Kolodner, R., K. K. Tewari, and R. C. Warner, 1975*a* Physical studies on the size and conformation of the chloroplast DNA from higher plants. *Biochim. Biophys. Acta* (in press).

Kolodner, R., R. C. Warner, and K. K. Tewari, 1975*b* The presence of covalently linked ribonucleotides in the closed circular deoxyribonucleic acid from higher plants. *J. Biol. Chem.* **250:**7020–7026.

Kung, S. D. and J. P. Williams, 1968 Isolation of chloroplasts free from nuclear DNA contamination. *Biochim. Biophys. Acta* **169:**265–268.

Kung, S. D., K. Sakano, and S. G. Wildman, 1974 Multiple peptide composition of the large and small subunits of *Nicotiana tabacum:* Fraction 1 protein ascertained by fingerprinting and electrofocusing. *Biochim. Biophys. Acta* **1:**365–138.

Lee, R. W., N. W. Gillham, K. P. VanWinkle, and J. E. Boynton, 1973 Preferential recovery of uniparental streptomycin resistant mutants from diploid *Chlamydomonas reinhardi*. *Mol. Gen. Genet.* **121:**109–116.

Loening, U. E., 1968 Molecular weights of ribosomal RNA in relation to evolution. *J. Mol. Biol.* **38:**355–365.

Luha, A. A., L. E. Sarcoe, and P. A. Whittaker, 1971 Biosynthesis of yeast mitochondria: Drug effects on the petite-negative yeast *Kluyveromyces lactis*. *Biochem. Biophys. Res. Commun.* **44:**396–402.

Lyman, H., 1967 Specific inhibition of chloroplast replication in *Euglena gracilis* by nalidinic acid. *J. Cell Biol.* **35:**726–730.

Lyttleton, J. W. and G. B. Petersen, 1964 The isolation of deoxyribonucleic acid from plant tissues. *Biochim. Biophys. Acta* **80:**391–398.

Manning, J. E. and O. C. Richards, 1972 Synthesis and turnover of *Euglena gracilis* nuclear and chloroplast deoxyribonucleic acid. *Biochemistry* **11:**2036–2043.

Manning, J. E., D. R. Wolstenholme, R. S. Ryan, J. A. Hunter, and O. C. Richards, 1971 Circular chloroplast DNA from *Euglena gracilis. Proc. Natl. Acad. Sci. USA* **68:**1169–1173.

Manning, J. E., D. R. Wolstenholme, and O. C. Richards, 1972 Circular DNA molecules associated with chloroplasts of spinach, *Spinacia oleracea. J. Cell Biol.* **53:**594–601.

Meselson, M. and F. W. Stahl, 1958 The replication of DNA in *Escherichia coli. Proc. Natl. Acad. Sci. USA* **44:**671–682.

Mets, L. J. and L. Bogorad, 1971 Mendelian and uniparental alteration in erythromycin binding by plastid ribosomes. *Science* **174:**707–709.

Mets, L. and L. Bogorad, 1972 Altered chloroplast ribosomal proteins associated with erythromycin-resistant mutants in two genetic systems of *Chlamydomonas reinhardi. Proc. Natl. Acad. Sci. USA* **69:**3779–3783.

Metzke, J. and R. G. Herrmann, 1975 *Z. Naturforsch. Teil C* (in press).

Miller, P. L., editor, 1970 *Control of Organelle Development,* Symposium of the Society for Experiments Biology, Vol. 24, Cambridge University Press, London, 524 pp.

Nass, M. M. K., 1970 Abnormal DNA patterns in animal mitochondria: Ethidium bromide-induced breakdown of closed circular DNA and conditions leading to oligomer accumulation. *Proc. Natl. Acad. Sci. USA* **67:**1926–1933.

Nass, M. M. K. and Y. Ben-Shaul, 1972 A novel closed circular duplex DNA in bleached mutant and green strains of *Euglena gracilis. Biochim. Biophys. Acta* **272:**130–136.

O'Connor, B., 1972 Preparation and comparison of chloroplast DNA's of wild type and mutant *Chlamydomonas reinhardi.* M.A. thesis, Department of Biological Sciences, Hunter College, New York.

Ohta, N., M. Inouye, and R. Sager, 1975 Identification of a chloroplast ribosomal protein adhered by a chloroplast gene mutation in *Chlamydomonas. J. Biol. Chem.* **250:**3655–3659.

Perlman, P. S. and H. R. Mahler, 1971 Molecular consequences of ethidium bromide mutagenesis. *Nature (London) New Biol.* **231:**12–16.

Pestka, S., 1971 Inhibitors of ribosome functions. *Ann. Rev. Microbiol.* **25:**488–562.

Pienkos, P., A. Walfield, and C. L. Herschberger, 1975 Effect of nalidixic acid on *Euglena gracilis:* Induced loss of chloroplast deoxyribonucleic acid. *Arch. Biochem. Biophys.* (in press).

Rawson, J. R. and R. Haselkorn, 1973 Chloroplast ribosomal RNA genes in the chloroplast DNA of *Euglena gracilis. J. Mol. Biol.* **77:**125–132.

Ray, D. S. and P. C. Hanawalt, 1964 Properties of the satellite DNA associated with the chloroplasts of *Euglena gracilis. J. Mol. Biol.* **9:**812–824.

Ray, D. S. and P. C. Hanawalt, 1965 Satellite DNA components in *Euglena gracilis* cells lacking chloroplasts. *J. Mol. Biol.* **11:**760–768.

Reger, B. J., S. A. Fairfield, J. L. Epler, and W. E. Barnett, 1970 Identification and origin of some chloroplast aminoacyl-tRNA synthetases and tRNAs. *Proc. Natl. Acad. Sci. USA* **67:**1207–1213.

Rhoades, M. M., 1955 Interaction of genic and non-genic hereditary units and the physiology and non-genic inheritance. In *Encyclopedia of Plant Physiology,* 1, edited by W. Ruhland, Springer-Verlag, Berlin, pp. 19–57.

Ris, H. and W Plaut, 1962 Ultrastructure of DNA-containing areas in the chloroplast of *Chlamydomonas*. *J. Cell Biol.* **13:**383–391.

Rochaix, J. D., 1972 Cyclization of DNA fragments of *Chlamydomonas reinhardi*. *Nature (London) New Biol.* **238:**76–78.

Rode, A. and M. Bayen, 1972 Action de la 8-azagnanine sur la division cellulaire de *Chlorella pyrenoidosa*. I. Inhibition selective de la synthese de l'ADN nucleaire. *Planta* **102:**237–246.

Ruppel, H. G., 1967 Nucleinsaüren in Chloroplasten. I. Charakterisierung der DNS und RNS von *Antirrhinum majors*. *Z. Naturforsch.* **228:**1068–1076.

Ryan, R. S., D. Grant, K.-S. Chiang, and H. Swift, 1973 Isolation of mitochondria and characterization of the mitochondrial DNA of *Chlamydomonas reinhardi*. *J. Cell Biol.* **59:**297*a*.

Sager, R., 1954 Mendelian and non-Mendelian inheritance of streptomycin resistance in *Chlamydomonas reinhardi*. *Proc. Natl. Acad. Sci. USA* **40:**356–363.

Sager, R., 1959 The architecture of the chloroplast in relation to its photosynthetic activities. *Brookhaven Symp. Biol.* **11:**101–117.

Sager, R., 1962 Streptomycin as a mutagen for non-chromosomal genes. *Proc. Natl. Acad. Sci. USA* **48:**2018–2026.

Sager, R., 1972 *Cytoplasmic Genes and Organelles,* Academic Press, New York, 405 pp.

Sager, R., 1974 Nuclear and cytoplasmic inheritance in green algae. In *Algal Physiology and Biochemistry,* edited by W. D. P. Stewart, Blackwell, Oxford, pp. 314–345.

Sager, R., 1975 Patterns of inheritance of organelle genomes: Molecular basis and evolutionary significance. In *Genetics and Biogenesis of Mitochondria and Chloroplasts,* edited by C. W. Birky, Jr., P. S. Perlman, and T. J. Byers, Ohio State University Press, Columbus, Ohio.

Sager, R., 1976 Genetic analysis of chloroplast DNA in *Chlamydomonas*. *Adv. Genet.* **19:** in press.

Sager, R. and M. R. Ishida, 1963 Chloroplast DNA in *Chlamydomonas*. *Proc. Natl. Acad. Sci. USA* **50:**725–730.

Sager, R. and R. Kitchin, 1975 Selective silencing of eukaryotic DNA. *Science* **189:**426–433.

Sager, R. and D. Lane, 1972 Molecular basis of maternal inheritance. *Proc. Natl. Acad. Sci. USA* **69:**2410–2414.

Sager, R. and Z. Ramanis, 1965 Recombination of non-chromosomal genes in *Chlamydomonas*. *Proc. Natl. Acad. Sci. USA* **53:**1053–1061.

Sager, R. and Z. Ramanis, 1967 Biparental inheritance of non-chromosomal genes induced by ultraviolet irradiation. *Proc. Natl. Acad. Sci. USA* **58:**931–937.

Sager, R. and Z. Ramanis, 1968 The pattern of segregation of cytoplasmic genes in *Chlamydomonas*. *Proc. Natl. Acad.* **61:**324–331.

Sager, R. and Z. Ramanis, 1970 A genetic map of non-Mendelian genes in *Chlamydomonas*. *Proc. Natl. Acad. Sci. USA* **65:**593–600.

Sager, R. and Z. Ramanis, 1973 The mechanism of maternal inheritance in *Chlamydomonas:* Biochemical and genetic studies. *Theor. Appl. Genet.* **43:**101–108.

Sager, R. and Z. Ramanis, 1974 Mutations that alter the transmission of chloroplast genes in *Chlamydomonas*. *Proc. Natl. Acad. Sci. USA* **71:**4698–4702.

Sager, R. and Z. Ramanis, 1976*a* Chloroplast genetics of *Chlamydomonas:* I. Allelic segregation ratios. *Genetics* (in press).

Sager, R. and Z. Ramanis, 1976*b* Chloroplast genetics of *Chlamydomonas:* II. Mapping by cosegregation frequency analysis. *Genetics* (in press).

Sakano, K., S. D. Kung, and S. G. Wildman, 1974 Identification of several chloroplast DNA genes which code for the large subunit of *Nicotiana* fraction 1 proteins. *Mol. Gen. Genet.* **130:**91–97.

Schiff, J. A. and H. T. Epstein, 1965 The continuity of the chloroplast in *Euglena.* In *Reproduction: Molecular, Subcellular and Cellular,* edited by M. Locke, *Symp. Soc. Dev. Biol.* **24:**131.

Schildkraut, C. L., J. Marmur, and P. Doty, 1962 Determination of the base composition of deoxyribonucleic acid from its buoyant density in CsCl. *J. Mol. Biol.* **4:**430–433.

Schimmer, O. and C. G. Arnold, 1970*a* Untersuchungen über Reversions- und Segregationsverhalten eines ausserkaryotischen Gens von *Chlamydomonas reinhardi* zur Bestimmung des Esträgers. *Mol. Gen. Genet.* **107:**281–290.

Schimmer, O. and C. G. Arnold, 1970*b* Über die Zahl der Kopien eines ausserkaryotischen Gens bei *Chlamydomonas reinhardi. Mol. Gen. Genet.* **107:**366–371.

Schimmer, O. and C. G. Arnold, 1970*c* Hin- und Rücksegregation eines ausserkaryotischen Gens bei *Chlamydomonas reinhardi. Mol. Gen. Genet.* **108:**33–40.

Schlanger, G. and R. Sager, 1974*a* Localization of five antibiotic resistances at the subunit level in chloroplast ribosomes of *Chlamydomonas. Proc. Natl. Acad. Sci. USA* **71:**1715–1719.

Schlanger, G. and R. Sager, 1974*b* Correlation of chloroplast DNA and cytoplasmic inheritance in *Chlamydomonas* zygotes. *J. Cell Biol.* **63:301***a*.

Schlanger, G. and R. Sager, 1976 Correlation of chloroplast DNA and cytoplasmic inheritance in *Chlamydomonas.* (manuscript in preparation).

Schmidt, G. and H. Lyman, 1974 Photocontrol of chloroplast enzyme synthesis in mutant and wild-type *Euglena. J. Cell Biol.* **63:**302*a*.

Scott, N. S., 1973 Ribosomal RNA cistrons in *Euglena gracilis. J. Mol. Biol.* **81:**327–336.

Scott, N. S., V. C. Shah, and R. Smillie, 1968 Synthesis of chloroplast DNA in isolated chloroplast. *J. Mol. Biol.* **38:**151–152.

Sigee, D. C., 1972 Pattern of cytoplasmic DNA synthesis in somatic cells of *Peteridium aquilinum. Exp. Cell Res.* **73:**481–486.

Singer, B., R. Sager, and Z. Ramanis, 1976 Chloroplast genetics of *Chlamydomonas:* III. Closing the circle. *Genetics* (in press).

Siu, C.-H., K.-S. Chiang, and H. Swift, 1975*a* Characterization of cytoplasmic and nuclear genomes in the colorless alga *Polytoma.* V. Molecular structure and heterogeneity of leucoplast DNA. *J. Mol. Biol.* **98:**369–391.

Siu, C.-H., K.-S. Chiang, and H. Swift, 1975*b* Characterization of cytoplasmic and nuclear genomes in the colorless alga *Polytoma.* III. Ribosomal RNA cistrons of the nucleus and leucoplast. *J. Cell Biol.* (in press).

Smillie, R. M. and N. S. Scott, 1969 *Organelle Biosynthesis: The Chloroplast,* Progress in Molecular and Subcellular biology, Vol. 1, Springer-Verlag, New York, pp. 136–202.

Spencer, D., P. R. Whitfield, W. Bottomley, and A. M. Wheeler, 1971 The nature of the proteins and nucleic acids synthesized by isolated chloroplasts. In *Autonomy and Biogenesis of Mitochondria and Chloroplasts,* edited by (N. K. Boardman, A. W. Linnane, and R. M. Smillie, North-Holland, Amsterdam, pp. 372–382.

Steffensen, D. M. and W. F. Sheridan, 1965 Incorporation of ^{3}H-thymidine into chloroplast DNA of marine algae. *J. Cell Biol.* **25:**619–626.

Stewart, W. D. P., editor, 1974 *Algal Physiology and Biochemistry,* Vol. 10, Blackwell, Oxford, 989 pp.

Stutz, E., 1970 The kinetic complexity of *Euglena gracilis* chloroplast DNA. *FEBS Lett.* **8**:25–28.

Stutz, E. and G. Bernardi, 1972 Hydroxyapatite chromatography of deoxyribonucleic acids from *Euglena gracilis*. *Biochimie* **54**:1013–1021.

Stutz, E. and J. P. Vandrey, 1971 Ribosomal DNA satellite of *Euglena gracilis* chloroplast DNA. *FEBS Lett.* **17**:277–280.

Sueoka, N., K. S. Chiang, and J. R. Kates, 1967 Deoxyribonucleic acid replication in meiosis of *Chlamydomonas reinhardi*. *J. Mol. Biol.* **25**:47–66.

Suyama, Y. and W. D. Bonner, Jr., 1966 DNA from plant mitochondria. *Plant Physiol.* **41**:383–388.

Swinton, D. C. and P. C. Hanawalt, 1972 *In vivo* specific labeling of *Chlamydomonas* chloroplast DNA. *J. Cell Biol.* **54**:592–597.

Swinton, D. C. and P. C. Hanawalt, 1973*a* The fate of pyrimidine dimers in ultraviolet-irradiated *Chlamydomonas*. *Photochem. Photobiol.* **17**:361–375.

Swinton, D. C. and P. C. Hanawalt, 1973*b* Absence of ultraviolet-stimulated repair replication in the nuclear and chloroplast genomes of *Chlamydomonas reinhardi*. *Biochim. Biophys. Acta* **294**:385–395.

Szybalski, W. and E. H. Szybalska, 1971 Equilibrium density gradient centrifugation. In *Procedures in Nucleic Acid Research,* edited by G. L. Cantoni and D. R. Davies, Harper and Row, New York, pp. 311–354.

Tewari, K. K., 1972 Genetic autonomy of extranuclear organelles. *Ann. Rev. Plant Physiol.* **22**:141–168.

Tewari, K. K. and S. G. Wildman, 1967 DNA polymerase in isolated tobacco chloroplasts and nature of the polymerized product. *Proc. Natl. Acad. Sci. USA* **58**:689–696.

Tewari, K. K. and S. G. Wildman, 1970 Information content in the chloroplast DNA. In *Control of Organelle Development,* Cambridge University Press, Cambridge, pp. 147–179.

Thomas, C. A., B. A. Hamkalo, D. N. Misra, and C. S. Lee, 1970 Cyclization of eucaryotic deoxyribonucleic acid fragments. *J. Mol. Biol.* **51**:621–632.

Thomas, J. R. and K. I. Tewari, 1974*a* Ribosomal-RNA genes in the chloroplasts DNA of pea leaves. *Biochim. Biophys. Acta* **361**:73–83.

Thomas, J. R. and K. K. Tewari, 1974*b* Conservation of 70 S ribosomal RNA genes in the chloroplast DNAs of higher plants. *Proc. Natl. Acad. Sci. USA* **71**:3147–3151.

Vandrey, J. P. and Stutz, E., 1973 Evidence for a novel DNA component in chloroplasts of *Euglena gracilis*. *FEBS Lett.* **37**:174–177.

Vedel, F., F. Quetier, M. Bayen, A. Rode, and J. Dalmon, 1972 Intramolecular heterogeneity of mitochondrial and chloroplast DNA. *Biochem. Biophys. Res. Commun.* **46**:972–978.

Watson, J. D. and F. H. C. Crick, 1953 Molecular structure of nucleic acids. A structure for deoxyribose nucleic acid. *Nature* (*London*) **171**:737–738.

Wells, R. and M. Birnstiel, 1969 Kinetic complexity of chloroplastal deoxyribonucleic acid and mitochondrial deoxyribonucleic acid from higher plants. *Biochem. J.* **112**:777–786.

Wells, R. and J. Ingle, 1970 The constancy of the buoyant density of chloroplast and mitochondrial DNA's in a range of the higher plants. *Plant Physiol.* **46**:178–179.

Wells, R. and R. Sager, 1971 Denaturation and renaturation kinetics of chloroplast DNA from *Chlamydomonas reinhardi*. *J. Mol. Biol.* **58**:611–622.

Werry, P. A. T. J. and F. Wanka, 1972 The effect of cycloheximide on the synthesis of major satellite DNA components in *Physarum polycephalum*. *Biochim. Biophys. Acta* **287**:232–235.

Wetmur, J. G. and N. Davidson, 1968 Kinetics of renaturation of DNA. *J. Mol. Biol.* **31**:349–370.

Whitfield, P. R. and D. Spencer, 1968 Buoyant density of tobacco and spinach chloroplast DNA. *Biochim. Biophys. Acta* **157**:333–343.

Wildman, S. G., C. Lu-Liao, and F. Wong-Staal, 1973 Material inheritance, cytology and macromolecular composition of defective chloroplasts in a variegated mutant of *Nicotiana tabacum*. *Planta* **113**:293–312.

Williams, G. R., A. S. Williams, and S. A. George, 1973 Hybridization of leucyl-transfer ribonucleic acid isoacceptors from green leaves with nuclear and chloroplast deoxyribonucleic acid. *Proc. Natl. Acad. Sci. USA* **70**:3498–3501.

Wolken, J. J. and G. E. Palade, 1953 An electron microscope study of two flagellates: Chloroplast structure and variation. *Ann. N.Y. Acad. Sci.* **56**:873–889.

Wong-Staal, F. and S. G. Wildman, 1973 Identification of a mutation in chloroplast DNA correlated with formation of defective chloroplasts in a variegated mutant of *Nicotiana tabacum*. *Planta* **113**:313–326.

Zellweger, A., U. Ryser, and R. Brown, 1962 Ribosomal genes of *Physarum:* Their isolation and replication in the mitotic cycle. *J. Mol. Biol.* **64**:681–691.

13

Chloroplast Ribosomes

ERHARD STUTZ AND ARMINIO BOSCHETTI

Introduction

Chloroplasts have a certain degree of genetic independence, since they possess their own set of double-stranded DNA which is replicated, transcribed, and translated by chloroplast-specific enzyme systems. The ribosomes occurring within the chloroplasts (hereafter referred to as *chloroplast ribosomes*) play an important role in chloroplast genome expression, and since their discovery by Lyttleton (1960, 1962) an impressive number of papers have appeared dealing with numerous aspects of the structure and function of these ribosomes. The earlier work in this field was mainly concerned with the identification of chloroplast ribosomes and their separation from the cell-sap ribosomes. It was, and to a certain degree still is, difficult to obtain preparative amounts of pure chloroplast ribosomes (or their subunits) possessing acceptable polymerizing activity. It is not surprising, therefore, that our knowledge of chloroplast ribosome functions (such as the mechanisms of the initiation, elongation, and termination of polypeptides) and our ability to reconstitute ribosomes from their components are modest compared to the corresponding advances with bacterial or animal ribosomes. Today, research on chloroplast ribosomes centers around such areas as (1) the detailed structural analyses of the

ERHARD STUTZ—Laboratory of Plant Biochemistry, University of Neuchâtel, Neuchâtel, Switzerland. ARMINIO BOSCHETTI—Institute for Biochemistry, University of Berne, Berne, Switzerland.

subunits (and their protein and RNA components) and of the special requirements of translation, (2) the functioning of chloroplast ribosomes, (3) the genetic aspects and loci of synthesis and assembly of chloroplast ribosomes, and (4) thé phylogenetic relationship of chloroplast ribosomes with ribosomes from other sources (i.e., from prokaryotes, mitochondria, and the cell sap of eukaryotic cells).

Common to these questions is the fundamental problem of gene expression and its control in the green eukaryotic cell. The study of chloroplast ribosomes can contribute novel elements to our understanding of the interplay between the nuclear and organellar genomes, since chloroplasts are themselves a product of both genomes and they serve as an instrument to express particular genes of one and possibly both genomes (see Chapter 14).

In the following account, we shall review the newer literature on chloroplast ribosomes stressing the abovementioned topics. Earlier work which has been competently reviewed (Smillie and Scott, 1969; Woodcock and Bogorad, 1971; Kirk, 1970, 1972; Boulter *et al.*, 1972; Sager, 1972; Ellis *et al.*, 1973) will be discussed only briefly.

General Properties of Chloroplast Ribosomes

Depending on the source of ribosomes and the aim of the study, different isolation and purification procedures are recommended. Some newer procedures for the isolation of chloroplast ribosomes, polysomes, or subunits have been reported, i.e., for *Euglena gracilis* (Avadhani and Buetow, 1972; Schwartzbach *et al.*, 1974), for *Chlamydomonas reinhardi* (Chua *et al.*, 1973*a*), for *Acetabularia mediterranea* (Kloppstech and Schweiger, 1973*a*), for *Spinacia oleracea* (Grivell and Groot, 1972), and for *Pisum sativum* (Tao and Jagendorf, 1973).

The number of chloroplast ribosomes per cell and the amount relative to the cytoplasmic ribosomes have been estimated in several instances. The ratio is not constant, but depends on growth conditions and the developmental stage of the cell. As an example, we take the values found in *Chlamydomonas reinhardi* grown either mixotrophically or heterotrophically. The analyses were done either by electron microscopy or by determination of the respective ribonucleic acids (Bourque *et al.*, 1971). The number of chloroplast ribosomes per cell was 0.53×10^5 (mixotroph) and 0.57×10^5 (heterotroph). The number of cell-sap ribosomes per cell was 0.98×10^5 (mixotroph) and 1.43×10^5 (heterotroph). Therefore, chloroplast ribosomes amount to 35% and 29% of the total ribosome content of mixotrophically and heterotrophically grown cells, respectively. Similar analyses with higher plants gave ratios of chloroplast ribosomes to cell-sap ribosomes between 1 : 3 and 1 : 2, i.e., for *Pisum sativum* and *Phaseolus vulgaris* (Bruskov and Odintsova, 1968), for *Vicia faba* (Dyer *et al.*, 1971), and for *Hordeum vulgare* (Sprey, 1972).

Chloroplast ribosomes can occur free or membrane bound (Chen and Wildman, 1970; Phillipovitch *et al.*, 1973; Chua *et al.*, 1973*b*; Tao and Jagendorf, 1973; Margulies and Michaels, 1974). The ratio of free ribosomes to membrane-bound ribosomes in cells of the pea leaf was found to be in the range of 4 : 1. According to Ellis (1976), there seem to be functional differences between the two kinds of chloroplast ribosomes.

All chloroplast ribosomes or ribosomal subunits so far characterized are smaller than the cell-sap ribosomes or subunits, but similar in size to bacterial ribosomes or subunits. The overall dimensions were found to be $268 \pm 24 \times 219 \pm 20$ Å (tobacco, negative staining, Miller *et al.*, 1966) or $266 \pm 32 \times 208 \pm 33$ Å (*Chlamydomonas reinhardi,* thin sections of whole cell, Ohad *et al.*, 1967). The buoyant density of chloroplast ribosomes (from species of *Chlorella, Chenopodium,* and *Pisum*) was given as 1.568 g/cm^3, corresponding to 47% protein. Under similar conditions, the buoyant densities of ribosomes from prokaryotes or blue-green algae are in the range of 1.610–1.640 g/cm^3 (Yurina and Odintsova, 1974). The ribosomal protein amounts to a total of 0.82×10^6 daltons for chloroplasts and 2.8×10^6 daltons for cell-sap ribosomes. Functional chloroplast ribosomes are bipartite: each of the two subunits contains one ribosomal RNA (rRNA) molecule with a molecular weight in the range of 0.56×10^6 for the small-subunit RNA or 1.1×10^6 for the large-subunit RNA. It is convenient to characterize ribosomes, ribosomal subunits, and rRNA in accordance with their S values. In numerous cases S values of both chloroplast and cell-sap ribosomes and subunits have been determined, and the results have been tabulated in several instances (Smillie and Scott, 1969; Boulter *et al.*, 1972; Sager, 1972). There is agreement that chloroplast ribosomes belong to the 70 S class, as do bacterial ribosomes, and cell-sap ribosomes belong to the 80 S class, as do cell-sap ribosomes from animal cells.

Chloroplast ribosomes change their shape and therefore their sedimentation characteristics as a function of the absolute and relative concentrations of monovalent and divalent ions (Hoober and Blobel, 1969; Dyer and Koller, 1971). Moreover, chloroplast ribosomes, as compared to cell-sap ribosomes, require a higher Mg^{2+} concentration in order to remain functionally associated. Chua *et al.* (1973*a*) compared the shape changes (S values) and association–dissociation behaviors of *Chlamy-*

domonas ribosomes. They found that cell-sap ribosomal subunits which were dissociated into subunits in a buffer containing 500 mM K^+ and 25 mM Mg^{2+} would reassociate to 80 S ribosomes in a milieu of 25 mM K^+ and 25 mM Mg^{2+}, while chloroplast ribosomal subunits obtained under identical conditions would not reassociate. This difference in the ionic requirements for structural and functional integrity can be exploited in *in vitro* studies to enhance or inhibit either of the protein-synthesizing systems. One should keep in mind, however, that depending on their functional stage in the ribosome cycle, subunits will undergo association–dissociation reactions at different ionic conditions.

Chloroplast ribosomes specifically bind to those antibiotics which are known to inhibit prokaryotic protein synthesis, but they do not bind to antibiotics known to inhibit protein synthesis on 80 S ribosomes. A list of antibiotics which bind to chloroplast ribosomes *in vitro* is given in Table 1.

Structure of Chloroplast Ribosomes

Ribosomal RNA

The isolation of intact chloroplast rRNA in good yield is still a difficult task, for two reasons: (1) 80 S cell-sap ribosomes with 25 S and 17 S RNA components very often strongly adhere to chloroplasts and contaminate subsequent chloroplast nucleic acid preparations, and (2) the chloroplast 23 S RNA is easily degraded and therefore may be partly or totally lost. This may have been the reason why 23 S and 16 S rRNAs of chloroplast were described 5 years after the discovery of the chloroplast

TABLE 1. In Vitro Binding of Antibiotics to Chloroplast Ribosomes

Antibiotic	*Source of chloroplast ribosomes*	*Reference*
Chloramphenicol	*Pisum sativum*	Anderson and Smillie, 1966
	Triticum vulgare	
	Euglena gracilis	
	Chlamydomonas reinhardi	Hoober and Blobel, 1969
Erythromycin	*Chlamydomonas reinhardi*	Mets and Bogorad, 1971
Dihydrospectinomycin	*Chlamydomonas reinhardi*	Burton, 1972
Dihydrostreptomycin	*Chlamydomonas reinhardi*	Boschetti and Bogdanov, 1973*b*
	Euglena gracilis	Schwartzbach and Schiff, 1974

ribosomes (Loening and Ingle, 1967; Stutz and Noll, 1967). Today, it is accepted that all chloroplast ribosomes have an equimolar ratio of the large-subunit rRNA (23 S, approximate molecular weight 1.1×10^6) to the small-subunit rRNA (16 S, approximate molecular weight 0.56×10^6), very similar to bacterial ribosomes (Loening, 1968).

The large subunit of chloroplast ribosomes contains, in addition to the 23 S RNA, a 5 S RNA which is slightly different from 5 S RNA of the cell-sap ribosomes (Dyer and Leech, 1968). A hydrogen-bound 5.8 S RNA, sometimes referred to as 7 S or 25 SA, was not found in chloroplast ribosomes, but it is part of the large ribosomal subunits of the cell-sap ribosomes of several species of lower and higher plants (Payne and Dyer, 1972*a*).

Plant cells are rich in nucleases which rapidly degrade rRNA. The 23 S rRNA is particularly vulnerable. Ingle and collaborators (Ingle *et al.*, 1970; Leaver and Ingle, 1971) showed that the 23 S RNA from radish, spinach, broad bean, pea, maize, and *Chlamydomonas* cells broke apart during isolation, yielding upon gel electrophoresis a specific pattern of RNA fragments, while the 16 S RNA remained intact. The use of Mg^{2+} or Ca^{2+} in the isolation buffer stabilized the 23 S rRNA. These observations suggest that "hidden breaks" exist in the RNA molecule which may or may not become apparent, depending on the isolation conditions. In recent similar studies, attempts were made to decide whether the 23 S rRNA is already nicked *in vivo*, i.e., in a particular loop region, or alternatively is degraded by tenaciously bound ribonucleases being released or activated during extraction (Atchison *et al.*, 1973; Grierson, 1974; Munsche and Wollgiehn, 1974). The question was not unequivocally settled, and its answer bears importance for further work on rRNA biosynthesis and degradation during aging.

The base composition of chloroplast rRNA has been analyzed in many instances. For several species of higher plants, Rossi and Gualerzi (1970) made a comparative study and found good base compositional coincidence among various chloroplast rRNAs and a small but significant difference with respect to the rRNA of the corresponding cell-sap ribosomes. In Table 2, the base composition of 23 S and 16 S rRNA of *Spinacia oleracea* and the 5 S rRNA of *Vicia faba* are compared with the base compositions of the respective rRNAs of *Euglena gracilis, Anacystis nidulans,* and *Escherichia coli.* It is evident that higher plant chloroplast 23 S and 16 S rRNAs are similar to prokaryotic rRNAs, but are quite unlike *Euglena gracilis* chloroplast and mitochondrial rRNAs. There also exists a remarkable difference between *Vicia faba* and *Escherichia coli* 5 S rRNAs.

It has been recognized for some time that the 23 S and 16 S rRNAs

TABLE 2. Base Composition of rRNA from Plant Cell Organelles, Plant Cell Sap, and Prokaryotes

		Moles percent					
Source	*RNA class*	*A*	*U*	*G*	*C*	*G + C*	*Reference*
Spinacia oleracea							
Cell sap	25 S and 17 S	21.6	23.3	31.7	23.5	55.2	Rossi and Gualerzi, 1970
Chloroplast	23 S	24.5	19.0	34.8	21.7	56.5	
	16 S	23.0	22.4	30.6	24.0	54.6	
Euglena gracilis							
Cell sap	25 S and 20 S	21.6	22.3	32.0	24.1	56.1	Rawson and Stutz, 1969
Chloroplast	23 S and 16 S	26.9	26.0	28.2	18.6	46.8	Crouse *et al.*, 1974
Mitochondria	23 S and 16 S	40.0	30.2	15.8	14.0	29.8	Avadhani and Buetow, 1972
Vicia faba							
Cell sap	5 S	24.1	23.5	28.6	23.8	52.4	Payne and Dyer, 1971
Chloroplast	5 S	26.5	22.2	28.7	22.6	51.3	
Anacystis nidulans	23 S	29.4	17.7	32.0	20.8	52.8	Payne and Dyer, 1972*b*
	16 S	28.7	18.6	32.7	20.0	52.7	
Escherichia coli	23 S	25.5	21.0	32.5	21.0	53.5	Stanley and Bock, 1965
	16 S	24.2	21.3	32.1	22.3	54.4	
	5 S	20.8	17.0	33.4	28.8	62.2	Payne and Dyer, 1971

are transcripts of the chloroplast genome. Evidence comes from rRNA : DNA hybridization experiments where it was shown that 4–8% of the chloroplast genome codes for rRNA (Scott, 1973; Thomas and Tewari, 1974; for reviews of the earlier literature, see Kirk, 1971, and Tewari, 1971). Enrichment of DNA segments carrying the rDNA was achieved with chloroplast DNA from *Euglena gracilis* (Vandrey and Stutz, 1973). Direct evidence for the synthesis of chloroplast rRNA within the chloroplast comes from the *in vitro* studies with *Euglena gracilis*. Carritt and Eisenstadt (1973*a*) reported the incorporation of [^{3}H]UTP into 23 S and 16 S rRNA.

By analogy with other cellular systems, precursor RNA molecules are expected to occur, but their detection in chloroplasts is difficult. The precursor molecules of plant cell-sap rRNA have been known for some time, but reports giving evidence for the presence of precursor RNA in chloroplasts have appeared only recently. In all cases there seem to be two different precursor molecules. One of them has a molecular weight of approximately 1.2×10^6 and matures to the 23 S rRNA of 1.1×10^6 molecular weight; the other precursor of approximately 0.64×10^6 molecular weight yields the 16 S rRNA of 0.56×10^6 molecular weight. These molecules occur in *Euglena gracilis* (Carritt and Eisenstadt, 1973*b*; Heizmann, 1974*a*), *Chlamydomonas reinhardi* (Miller and McMahon, 1974), and spinach (Detchon and Possingham, 1973; Hartley and Ellis, 1973). Galling (1974) reported the finding of two rRNA molecules in *Chlorella pyrenoidosa* which were rapidly labeled and had slightly higher molecular weights than the 23 S and 16 S RNA. Contrary to these reports, Grierson and Loening (1974) were not able to detect the 1.2×10^6 and 0.65×10^6 precursor molecules in *Phaseolus aureus*. A large common precursor RNA is likely to occur (R. G. Herrmann, private communication).

The resemblance of chloroplast rRNA with prokaryotic rRNA does not include base sequences, although enzymatic degradation patterns are similar (Zablen *et al.*, 1975). Scott *et al.* (1971) showed that *Escherichia coli* rRNA did not compete with *Euglena* chloroplast rRNA in chloroplast RNA : DNA hybridization experiments.

In summary, we can say that with respect to the mode of biosynthesis, molecular weight, and general base composition, chloroplast rRNA resembles prokaryotic rRNA. However, specific base sequences differ considerably.

Ribosomal Proteins

The three standard procedures for the isolation of ribosomal proteins are discussed in detail by Wittmann (1972): (1) treatment of the ribosomal subunits with LiCl and urea, (2) treatment with 60% acetic acid, or (3) treatment with ribonucleases.

None of the chloroplast ribosomal proteins has been obtained in sufficient amounts to allow for the detailed analyses of its amino acid sequences or for producing specific antibodies. All the work done so far has been analytical (i.e., one- or two-dimensional gel electrophoresis).

Lyttleton (1968) found 15 bands of basic proteins when analyzing spinach chloroplast ribosomes in one-dimensional gels (6 M urea, pH 4.5). Odintsova and Yurina (1969) distinguished 23 bands in gels (6 M urea, pH 3.5) from pea and bean chloroplast ribosomes. Both groups compared the chloroplast ribosomal protein patterns with cell-sap ribosomal protein patterns and found them to be different in all cases. Similar comparative studies with similar results were made with ribosomes from *Chlamydomonas reinhardi* (Hoober and Blobel, 1969), *Zea mays* and *Phaseolus aureus* (Vasconcelos and Bogorad, 1971), and *Euglena gracilis* (Freyssinet and Schiff, 1974).

With the introduction of two-dimensional gel electrophoresis, more detailed analysis became possible. Jones *et al.*, (1972) distinguished 75 spots (of which 26 migrated toward the anode at pH 8.6) when analyzing chloroplast ribosomes from *Triticum vulgare*. Gualerzi *et al.* (1974) separated spinach chloroplast ribosomal proteins and resolved 40 spots, (11 or 12 of these proteins traveled toward the anode at pH 8.3). Hanson *et al.* (1974) analyzed separately the chloroplast ribosomal subunits from *Chlamydomonas reinhardi*. The small subunit contained 22 proteins with molecular weights between 14,000 and 37,000; the large subunit contained 26 proteins with molecular weights between 13,500 and 40,000. Brügger and Boschetti (1975), also working with *Chlamydomonas reinhardi*, resolved 25 spots (small subunit) and 35 spots (large subunit). The discrepancy in the number of proteins reported by the two groups may be due to different isolation and analysis conditions. For comparative reasons, it may be said that the small and large ribosomal subunits from *Escherichia coli* have 21 and 34 proteins, respectively (Kaltschmidt and Wittmann, 1970).

The chloroplast ribosomal protein patterns so far mentioned were obtained from wild-type algae. Mutants have been described in *Chlamydomonas* which show resistance to antibiotics that poison chloroplast ribosomes. In a few cases, the ribosomal protein patterns were analyzed and the following deviations from the wild-type situation were found:

1. A uniparental, erythromycin-resistant mutant, *eryU1a*, has one protein from the large subunit altered (Mets and Bogorad, 1972).
2. The Mendelian mutants *eryM1a, eryM1b, eryM1c, eryM1d,* and

eryM2d have a ribosomal protein of the large subunit altered which is not identical to the mutated protein of the uniparental mutant, *eryU1a* (Davidson *et al.,* 1974).

3. A uniparental (*sr35*) and a Mendelian (*sr3*) streptomycin-resistant mutant differ from each other and from the wild-type strain by one protein (spot) of the small and the large subunit (Brügger and Boschetti, 1975). This somewhat surprising result, namely the alteration of a protein in both subunits, could be explained by a mutation of a ribosomal protein which can be attached to either of the two subunits (interphase protein).

Whether chloroplast ribosomal proteins are synthesized inside of the chloroplast envelope on 70 S ribosomes or outside on the 80 S cytoplasmic ribosomes has not been unequivocally determined. Experimental results obtained so far are best explained by the assumption that some, probably a minority, of chloroplast ribosomal proteins are translated on 70 S ribosomes, while the majority are imported from the cell sap. Ellis and Hartley (1971) reported the inhibition of chloroplast ribosome formation in the presence of lincomycin, a known inhibitor of prokaryotic protein synthesis. Schlanger *et al.* (1972) and Schlanger and Sager (1974) working with a series of non-Mendelian, uniparental antibiotic-resistant *Chlamydomonas* mutants provided strong, albeit indirect, evidence that uniparental genes which are most likely situated on the chloroplast DNA confer drug resistance by coding for an altered chloroplast ribosomal protein. Such proteins would be synthesized within the chloroplast, provided that chloroplast mRNA is exclusively translated on 70 S chloroplast ribosomes. On the other hand, results from amino acid incorporation studies on the *arg1* mutant of *Chlamydomonas reinhardi* by Honeycutt and Margulies (1973) indicate that chloroplast ribosomal proteins are mainly synthesized on 80 S ribosomes. Boschetti *et al.* (1973) came to a similar conclusion after working with the *Chlamydomonas* mutant, *sr3,* where the quantity of chloroplast ribosomes was not affected by streptomycin. For *Chlamydomonas,* a series of Mendelian mutants with changed chloroplast ribosomes have been described (see Table 3). Assuming that transcripts of nuclear origin are translated on 80 S ribosomes, this finding would suggest an outside origin of chloroplast ribosomal proteins. Bourque and Wildman (1973) describe two tobacco varieties with ribosomes that give different electrophoretic protein patterns. Crosses between the two varieties allowed Bourque and Wildman to conclude that the respective chloroplast ribosomal proteins follow the rules of Mendelian genetics, again suggesting that the proteins are translated on 80 S ribosomes. Kloppstech and Schweiger (1973*b*) exchanged nuclei of dif-

TABLE 3. Mutants with Altered Chloroplast Ribosomes or Lacking Chloroplast Ribosomes

Plant	*Mutant*	*Type of inheritance*	*Modification of the chloroplast ribosome*	*Reference*
Chlamydomonas reinhardi	*ery-U1a*	U	No binding of erythromycin	Mets and Bogorad, 1971
	ery-M1a	M		
	ery-M1b	M		
	ery-M1c	M		
	ery-M1d	M		
	ery-M2d	M		
	ery-U1a	U	Resistance to erythromycin in poly(C)-directed incorporation of proline	Conde *et al.*, 1975
	ery-U-37	U		
	ery-U1a	U	One large-subunit protein with marked tendency to aggregate	Mets and Bogorad, 1972
	ery-M2d	M	One large-subunit protein with altered electrophoretic properties	
	ery-M1a	M	Protein No. 6 of large subunits with altered net charges at pH 5	Davidson *et al.*, 1974
	ery-M1b	M		
	ery-M1c	M		
	ery-M1d	M		
	ery-M1b	M	Protein No. 6 of large subunit, reduced 30% in molecular weight	Davidson *et al.*, 1974
	spr-U-1-27-3 identical to *sp-1-73* and *sp-2-73*	U	Spectinomycin resistant, reduced S value (66 S), no binding of dihydrospectinomycin, partially resistant in poly(U)-directed phenylalanine incorporation	Gillham *et al.*, 1970; Burton, 1972; Conde *et al.*, 1973

spr-U-1-6-2	U	Very resistant to poly(U)-directed phenylalanine incorporation	Conde *et al.*, 1975
sr3	M	Altered binding characteristic for dihydrostreptomycin	Boschetti and Bogdanov, 1973*a,b;* Boschetti *et al.*, 1973
sr35	U	One protein of small and large subunits, subunit association, no binding of dihydro-streptomycin	Brügger and Boschetti, 1975
car	U	Resistance to carbomycin in poly(U)-directed amino acid polymerization, alteration of large subunit	Schlanger *et al.*, 1972
cle	U	Resistance to cleocin in poly(U)-directed amino acid polymerization, alteration of large subunit	Schlanger and Sager, 1974
nea	U	Resistance to neamine in poly(U)-directed amino acid polymerization, alteration of small subunit	
spc	U	Resistance to spectinomycin in poly(U)-directed amino acid polymerization, alteration of small subunit	
sm2	U	Resistance to streptomycin in poly(U)-directed amino acid polymerization, one altered protein of small subunit	Ohta *et al.*, 1975

TABLE 3. *Continued*

Plant	*Mutant*	*Type of inheritance*	*Modification of the chloroplast ribosome*	*Reference*
Chlamydomonas reinhardi (contd)	*sr-2-60* (a) *sr-U-2-23* (b)	U U	Fully (a) and partially (b) resistant to streptomycin induced incorporation of isoleucine in poly(U)-directed polymerization (misreading)	Conde *et al.*, 1975
	sr-2-60 *sr-2-281*	U U	Streptomycin resistant, reduced S value (66 S)	Gillham *et al.*, 1970
	nr-2-1- spr-1-1 (double mutant)	U	Neamine and spectinomycin resistant, reduced S value (66 S)	
	ac20	M	Reduced number of ribosomes, impaired rRNA formation	Goodenough and Levine, 1970; Bourque *et al.*, 1971; Harris *et al.*, 1974
	cr1 *cr2* *cr3*	M M M	Reduced number of ribosomes, accumulation of large subunits, defect in small subunit synthesis	Boynton *et al.*, 1970; Boynton *et al.*, 1972; Harris *et al.*, 1974
	cr4	M	Reduced number of 70 S ribosomes, impaired rRNA formation suggested	
Hordeum vulgare	—	—	No 70 S ribosomes in white parts of leaves	Sprey, 1972
Pelargonium zonale	—	U	No 70 S ribosomes in white parts of leaves	Börner *et al.*, 1973
Gossypium hirsutum	—	U	No 70 S ribosomes in white parts of leaves	Katterman and Endrizzi, 1973

ferent Mendelian mutants of *Acetabularia* and found that some plastidal ribosomal proteins are synthesized on 80 S ribosomes (see Chapter 14).

Function of Chloroplast Ribosomes

Boulter *et al.* (1972) have thoroughly reviewed the literature on protein synthesis by plants. Two aspects will be considered here: (1) the requirements for translation and (2) the nature of the proteins synthesized within the chloroplast. A thorough investigation of point (1) would require *in vitro* studies with highly purified ribosomal subunits, natural messenger RNA molecules, and specific translation factors. While such an intact chloroplast system has not yet been formulated, it has been shown that purified chloroplast ribosomes in combination with (a) a synthetic mRNA, i.e., poly(U), (b) postribosomal supernatants containing the necessary components for translation (of either prokaryotic, eukaryotic, or chloroplast origin), and (c) an energy source are able to perform marginal protein synthesis. The conclusion from many such studies was that chloroplast ribosomes of higher plants and algae resemble prokaryotic ribosomes in their translational requirements. More recent work by Grivell and Groot (1972) with very active chloroplast ribosomes confirmed these earlier findings.

Chloroplasts were shown to contain special tRNAs, some synthetases, *N*-formyl-methionyl-tRNA, and active transformylases (Schwartz *et al.*, 1967; Burkhard *et al.*, 1969, 1970, 1973; Leis and Keller, 1970, 1971; Reger *et al.*, 1970; Merrick and Dure, 1971; Kislev *et al.*, 1972; Parthier, 1973; Hecker *et al.*, 1974). Ciferri and Tiboni (1973) report the presence of a chloroplast-specific elongation factor (EF-G) in *Chlorella vulgaris*. This is the only case where a chloroplast-specific translational factor has been somewhat characterized.

Some interesting results with respect to point (2) were obtained on the characterization of chloroplast proteins synthesized on 70 S ribosomes. The molecular weight of chloroplast DNA from lower and higher plants is in the range of 1×10^8 daltons, sufficient to code for more than 100 proteins. However, it seems that only a limited number of genes are transcribed and translated. *In vitro* experiments utilizing cell-free systems containing intact or broken chloroplasts and *in vivo* experiments utilizing whole plants, algae, or plant organs or tissues have been done where the incorporation of labeled amino acids into discrete chloroplast proteins has been measured. The full *in vitro* translational capacity is conserved only when chloroplasts with intact outer membranes are used. Intact chloroplasts show light-dependent protein synthesis. Earlier work (see review by Boynton *et al.,* 1972) does not meet such rigorous standards, and some conflicting results as to the kind of proteins synthesized within the chloroplasts have been reported. More recently, excellent isolation procedures have been described (Ramirez *et al.,* 1968) yielding chloroplasts with intact outer membranes. Using such chloroplast preparations, Blair and Ellis (1973) and Eaglesham and Ellis (1974) have characterized by gel electrophoresis and tryptic peptide mapping proteins made *in vitro*. They conclude that the large subunit of fraction 1 protein is the sole soluble protein synthesized on 70 S ribosomes. The small subunit of fraction 1 protein was shown by Gray and Kekwick (1974) to be synthesized on 80 S ribosomes. In addition, Eaglesham and Ellis (1974) found that five thylakoid membrane proteins were labeled under incubation conditions where only 70 S ribosomes were active. According to Chan and Wildman (1972) and Sakano *et al.* (1974), the large-subunit protein of fraction 1 is also a gene product of the chloroplast DNA. Genetic analyses were done with suitable mutants of *Zea mays,* and analyses of the gene products were made by tryptic peptide mapping. The gene loci are not yet known for the five thylakoid membrane proteins. However, Joy and Ellis (1975) have shown that two proteins of the chloroplast envelope out of at least 25 are synthesized on 70 S ribosomes. According to Ellis (private communication), the membrane-bound polysomes seem to synthesize the membrane proteins while the soluble polysomes specialize in the synthesis of fraction 1 large subunit protein.

A particularly interesting problem is the translational role of 70 S ribosomes in connection with light-induced development of etioplasts to functional chloroplasts. Usually such problems are studied *in vivo,* but Siddell and Ellis (1975) carried out amino acid incorporation experiments *in vitro* with etioplasts and plastids isolated from peas after 24, 48, and 96 hr of greening in continuous white light. The proteins were analyzed on gels, and in all cases six were labeled. Again the large subunit of fraction 1 protein and five membrane proteins were radioactive. The tentative conclusion was that protein synthesis within developing plastids is qualitatively identical to that in mature chloroplasts. Hartley *et al.* (1975) have extended these *in vitro* studies in a very promising direction by incubating chloroplast RNA with *Escherichia coli* ribosomes and a 30,000*g* supernatant. They found the translational product to be primarily fraction 1 large subunit protein as characterized by gel electrophoresis. This result indicates that some of the translational control mechanisms were operative in this heterologous polymerizing system and that these methods can be used for further characterization of chloroplast mRNA. Should it turn out that the non-membrane-bound chloroplast polysomes all contain the mRNA for the fraction 1 large subunit, these chloroplasts would be an excellent source for a defined mRNA (Ellis, 1976).

For *in vivo* studies, several approaches are currently in use: (1) the incorporation of labeled precursors into chloroplast proteins in the presence or absence of selective inhibitors of 70 S or 80 S ribosomes, (2) the utilization of uniparental or Mendelian mutants with impaired chloroplasts having altered chloroplast ribosomes, and (3) the utilization of antibiotic-resistant mutants. Table 3 lists mutants which are potentially useful in studies of chloroplast ribosome function. However, cautious interpretations of results obtained by either of the three approaches is recommended, especially since the 70 S ribosomal proteins are themselves, partly at least, translated on 80 S ribosomes, enhancing thereby the possibility of misinterpretations. A case in point may be mentioned. Machold and Aurich (1972) suggested on the basis of inhibitor studies that the major proteins of photosystem I are synthesized on 70 S ribosomes, leaving some possibility that the minor proteins might be translated in the cytoplasm, while Ellis (1975) working with *Pisum sativum* came to a different conclusion. According to the data of Ellis, photosystem I proteins are synthesized on 80 S ribosomes, but their incorporation into the developing thylakoid membrane requires an immobilizing protein produced by 70 S ribosomes. In Table 4 we summarize a list of chloroplast proteins that are thought to be translational products of the 70 S ribosomes. More ambiguous results have been omitted.

It is beyond the scope of this chapter to discuss the physiological control of protein synthesis. Controlling factors such as light (its intensity, quality, and periodicity), temperature, and nutrition have been studied extensively (see the review by Boulter *et al.,* 1972). We will mention just a few experiments concerning the control of chloroplast rRNA synthesis. Cattolico *et al.* (1973) working with *Chlamydomonas* reported that the amount of cytoplasmic rRNA per cell increased linearly during the entire G_1 phase, whereas chloroplast rRNA accumulated only through 70% of G_1. The amount of cytoplasmic rRNA per cell remained constant during nuclear DNA synthesis, whereas a gradual loss of chloroplast rRNA per cell was noted at this time period. Heizmann (1974*b*), working with *Euglena gracilis*, and Detchon and Possingham (1973), working with spinach, gave evidence that chloroplast rRNA synthesis was light stimulated and dark inhibited, while the synthesis of cell-sap rRNA was not tightly controlled by light. The phytochrome system seems to be involved in the control of chloroplast ribosome synthesis and function (Scott *et al.,* 1971; Pine and Klein, 1972).

TABLE 4. Tentative List of Chloroplast Proteins Synthesized on 70 S Chloroplast Ribosomes

Protein	*Source*	*Reference*
Fraction I	*Hordeum vulgare*	Criddle *et al.*, 1970
	Zea mays	Graham *et al.*, 1970; Chan and Wildman, 1972; Sakano *et al.*, 1974
	Pisum sativum	Ellis and Hartley, 1971; Blair and Ellis, 1973
	Phaseolus vulgaris	Ireland and Bradbeer, 1971
	Euglena gracilis	Schiff, 1971
	Chlamydomonas reinhardi	Armstrong *et al.*, 1971
	Triticum vulgare	Gooding *et al.*, 1973
Cytochrome 552	*Chlamydomonas reinhardi*	Hoober, 1970
	Euglena gracilis	Smillie and Scott, 1969; Schiff, 1971
Cytochrome 561	*Euglena gracilis*	Smillie and Scott, 1969; Smillie *et al.*, 1967
Thylakoid membrane	*Pisum sativum*	Eaglesham and Ellis, 1974; Bottomley *et al.*, 1974
	Chlamydomonas reinhardi	Jennings and Ohad, 1973; Hoober and Stegeman, 1973; Michaels and Margulies, 1975
	Acetabularia mediterranea	Apel and Schweiger, 1973
Outer membrane	*Pisum sativum*	Joy and Ellis, 1975
Photosystem I	*Vicia faba*	Machold and Aurich, 1972

Homologies between Chloroplast and Other Ribosomes

Considerable comparative research has been done in order to measure homologies (in terms of structure and function) and thereby evolutionary relationship between ribosomes.

Homologies on the rRNA level involve similarities in molecular weights, base compositions, and, most importantly, base sequences. Base sequence homologies can be tested by proper hybridization reactions between RNA and DNA (Bendich and McCarthy, 1970). Bigott and Carr (1972) hybridized rRNA from various prokaryotes with chloroplast DNA of *Euglena gracilis*. According to these results, the relationship in terms of rRNA sequences is closest to the blue-green algae, followed by photosynthetic bacteria and heterotrophic bacteria. The least homology was found between chloroplast and cell-sap rRNA.

Similarities between ribosomal proteins from chloroplasts and from prokaryotic and eukaryotic cells were studied using gel electrophoretic and

TABLE 5. *Synopsis of Qualitative Immunological Cross-reactions Between Ribosomal Proteins of Different Origin*[a]

	Antibodies										
	Beans		*Spinach*		*Escherichia coli*			*Bacillus stearothermophilus*			*Bacillus subtilis*
Antigen and (source)	*70 S*	*80 S*	*70 S*	*80 S*	*70 S*	*50 S*	*30 S*	*70 S*	*50 S*	*30 S*	*70 S*
70 S (beans)	++[a]	+[c]	±[d]	–[e]	–	–	–	–	–	–	–
80 S (beans)	±	++	–	++	0[f]	0	0	0	0	0	0
70 S (spinach)	±	–	++	+	–	–	–	–	–	–	–
80 S (spinach)	–	++	+	++	0	0	0	0	0	0	0
70 S mitochondrial (*Neurospora crassa*)	–	0	–	0	–	–	–	–	–	–	–
70 S (*Anacystis nidulans*)	–	0	–	0	±	±	–	–	–	–	–

[a] From Gualerzi *et al.* (1974).
[b] Strong cross-reaction.
[c] Weak cross-reaction.
[d] Very weak or uncertain cross-reaction.
[e] No cross-reaction.
[f] Experiment not performed.

immunological techniques (Gualerzi and Cammarano, 1970; Wittmann *et al.*, 1970; Vasconcelos and Bogorad, 1971). These earlier results have been verified and extended by Gualerzi *et al.* (1974) (Table 5). In summary, they found that chloroplast ribosomal proteins from different higher plant species show only limited intragroup resemblances, certainly less than is the case for the 80 S ribosomes. Furthermore, chloroplast ribosomal proteins show no serologically detectable relationship to bacterial ribosomes.

Relationships may be measured also by functional tests. Lee and Evans (1971) mixed ribosomal subunits from *Euglena* chloroplasts with those from *Escherichia coli.* A combination of 30 S subunits from chloroplast plus 50 S subunits from *Escherichia coli* was functional along with the *Escherichia coli* polymerizing enzymes and poly(U) as template. The reverse combination was nonfunctional. Grivell and Walg (1972) combined ribosomal subunits from spinach chloroplasts with those from either yeast mitochondria or *Escherichia coli.* Chloroplast subunits formed active hybrids with the *Escherichia coli* subunits but not with the mitochondrial subunits. Avadhani and Buetow (1974) compared the functional compatibility of chloroplast, mitochondrial, and cell-sap ribosomes from *Euglena gracilis.* The tRNA and supernatant enzymes from chloroplasts can partly sustain the protein-synthesizing activity of mitochondrial and cytoplasmic ribosomes, but both types of ribosomes are highly specific for the homologous, salt-extractable ribosomal fractions (i.e., initiation factors).

Future work along this line will progress only if better-characterized chloroplast translational factors become available. Such systems will be of interest in the study of questions of evolution and in the eventual definition in precise terms of the chloroplast translational system.

Acknowledgments

We are very grateful to Dr. R. Braun for critically reading the manuscript, and we also thank Miss F. Prieur for her secretarial help. The authors' research is supported in part by Fonds National Suisse de la Recherche Scientifique. One of us (E. S.) receives additional support from Fonds Dr. L. Sauberli.

Literature Cited

Anderson, L. A. and R. M. Smillie, 1966 Binding of chloramphenicol by ribosomes from chloroplasts. *Biochem. Biophys. Res. Commun.* **23:**535–539.

Apel, K. and H. G. Schweiger, 1973 Site of synthesis of chloroplast-membrane proteins: Evidence for three types of ribosomes engaged in chloroplast-protein synthesis. *Eur. J. Biochem.* **38:**373–383.

Armstrong, J. J., S. J. Surzycki, B. Moll, and R. P. Levine, 1971 Genetic transcription and translation specifying chloroplast components in *Chlamydomonas reinhardi. Biochemistry* **10:**692–701.

Atchison, B. B., D. P. Bourque, and S. G. Wildman, 1973 Preservation of 23 S chloroplast RNA as single chain of nucleotides. *Biochim. Biophys. Acta* **331:**382–389.

Avadhani, N. G. and D. E. Buetow, 1972 Isolation of active polyribosome from the cytoplasm, mitochondria and chloroplasts of *Euglena gracilis. Biochem. J.* **128:**353–365.

Avadhani, N. G. and D. E. Buetow, 1974 Mitochondrial and cytoplasmic ribosomes: Distinguishing characteristics and a requirement for the homologous ribosomal salt extractable fraction for protein synthesis. *Biochem. J.* **140:**73–78.

Bendich, A. J. and B. J. McCarthy, 1970 Ribosomal RNA homologies among distantly related organisms. *Proc. Natl. Acad. Sci. USA* **65:**349–356.

Bigott, G. H. and N. G. Carr, 1972 Homology between nucleic acids of blue-green algae and chloroplast of *Euglena gracilis. Science* **175:**1257–1259.

Blair, G. E. and R. J. Ellis, 1973 Protein synthesis in chloroplasts. I. Light-driven synthesis of the large subunit of fraction I protein by isolated chloroplasts. *Biochim. Biophys. Acta* **319:**223–234.

Börner, T., F. Herrmann, and R. Hagemann, 1973 Plastid ribosome deficient mutants of *Pelargonium zonale. FEBS Lett.* **37:**117–119.

Boschetti, A. and S. Bogdanov, 1973*a* Different effects of streptomycin on the ribosomes from sensitive and resistant mutants of *Chlamydomonas reinhardi. Eur. J. Biochem.* **35:**482–488.

Boschetti, A. and S. Bogdanov, 1973*b* Binding of dihydrostreptomycin to ribosomes and ribosomal subunits from streptomycin resistant mutants of *Chlamydomonas reinhardi. FEBS Lett.* **38:**19–22.

Boschetti, A., S. Bogdanov, M. Brügger, and E. Frei, 1973 Zur Streptomycin induzierten Bildung von 70 S-Monosomen und von Oligomeren in *Chlamydomonas reinhardi. FEBS Lett.* **37:**59–63.

Bottomley, W., D. Spencer, and P. R. Whitfeld, 1974 Protein synthesis in isolated spinach chloroplasts: Comparison of light driven and ATP driven synthesis. *Arch. Biochem. Biophys.* **164:**106–117.

Boulter, D., R. J. Ellis, and A. Yarwood, 1972 Biochemistry of protein synthesis in plants. *Biol. Rev.* **47:**113–175.

Bourque, D. P. and S. G. Wildman, 1973 Evidence that nuclear genes code for several chloroplast ribosomal proteins. *Biochem. Biophys. Res. Commun.* **50:**532–537.

Bourque, D. P., J. E. Boynton, and N. W. Gillham, 1971 Studies on the structure and cellular location of various ribosome and ribosomal RNA species in the green alga *Chlamydomonas reinhardi. J. Cell Sci.* **8:**153–183.

Boynton, J. E., N. W. Gillham, and B. Burkholder, 1970 Mutations altering the chloroplast phenotype in *Chlamydomonas.* II. A new Mendelian mutation. *Proc. Natl. Acad. Sci. USA* **67:**1505–1512.

Boynton, J. E., N. W. Gillham, and J. E. Chabot, 1972 Chloroplast ribosome deficient mutants in the green alga *Chlamydomonas reinhardi* and the question of chloroplast ribosome function. *J. Cell Sci.* **10:**267–305.

Brügger, M. and A. Boschetti, 1975 Two-dimensional gel electrophoresis of ribosomal proteins from streptomycin-sensitive and streptomycin-resistant mutants of *Chlamydomonas reinhardi. Eur. J. Biochem.* **58:**603–610.

Bruskov, V. I. and M. S. Odintsova, 1968 Comparative electron microscopic study of chloroplast and cytoplasmic ribosomes. *J. Mol. Biol.* **32:**471–473.

Burkhard, G., B. Eclancher, and J. H. Weil, 1969 Presence of *N*-formyl-methionyl-transfer RNA in bean chloroplasts. *FEBS Lett.* **4:**285–287.

Burkhard, G., P. Guillemont, and J. H. Weil, 1970 Comparative studies of the tRNAs and aminoacyl-tRNA synthetases from the cytoplasm and the chloroplasts of *Phaseolus vulgaris. Biochim. Biophys. Acta* **224:**184–198.

Burkhard, G., P. Guillemont, A. Steinmetz, and J. H. Weil, 1973 Transfer ribonucleic acid and transfer ribonucleic acid-recognizing enzymes in bean cytoplasm, chloroplasts, etioplasts and mitochondria. *Biochem. Soc. Symp.* **38:**43–56.

Burton, W. G., 1972 Dihydrospectinomycin binding to chloroplast ribosomes from antibiotic sensitive and resistant strains of *Chlamydomonas reinhardi. Biochim. Biophys. Acta* **272:**305–311.

Carritt, B. and J. M. Eisenstadt, 1973*a* Synthesis *in vitro* of high molecular weight RNA by isolated *Euglena* chloroplasts. *Eur. J. Biochem.* **36:**482–488.

Carritt, B. and J. M. Eisenstadt, 1973*b* RNA synthesis in isolated chloroplast: Characterization of the newly synthesized RNA. *FEBS Lett.* **36:**116–120.

Cattolico, R. A., J. W. Senner, and R. F. Jones, 1973 Changes in cytoplasmic and chloroplast rRNA during the cell cycle of *Chlamydomonas reinhardi. Arch. Biochem. Biophys.* **156:**58–65.

Chan, P. H. and S. G. Wildman, 1972 Chloroplast DNA codes for the primary structure of the large subunit of fraction I protein. *Biochim. Biophys. Acta* **277:**677–680.

Chen, J. L. and S. G. Wildman, 1970 "Free" and membrane-bound ribosomes, and nature of products formed by isolated tobacco chloroplasts incubated for protein synthesis. *Biochim. Biophys. Acta* **209:**207–219.

Chua, N. M., G. Blobel, and P. Siekewitz, 1973*a* Isolation of cytoplasmic and chloroplast ribosomes and their dissociation into active subunits from *Chlamydomonas reinhardi. J. Cell Biol.* **57:**798–814.

Chua, N. M., G. Blobel, P. Siekewitz, and G. E. Palade, 1973*b* Attachment of chloroplast polysomes to thylakoid membranes in *Chlamydomonas reinhardi. Proc. Natl. Acad. Sci. USA* **70:**1554–1558.

Ciferri, O. and O. Tiboni, 1973 Elongation factor for chloroplast and mitochondrial protein synthesis in *Chlorella vulgaris. Nature (London) New Biol.* **245:**209–211.

Conde, M. F., J. E. Boynton, N. W. Gillham, E. H. Harris, C. L. Tingle, and W. L. Wang, 1975 Chloroplast genes in *Chlamydomonas* affecting organelle ribosomes. *Mol. Gen. Genet.* **140:**183–220.

Criddle, R. S., B. Dau, G. E. Kleinkopf, and R. C. Huffaker, 1970 Differential synthesis of ribulose-diphosphate carboxylase subunits. *Biochem. Biophys. Res. Commun.* **41:**621–627.

Crouse, E., J. P. Vandrey, and E. Stutz, 1974 Comparative analyses of chloroplast and mitochondrial nucleic acids from *Euglena gracilis.* In *Proceedings of the IIIrd International Congress on Photosynthesis Research,* Rehovot, Israel, edited by M. Avron, Elsevier, Amsterdam, pp. 1775–1786.

Davidson, J. N., M. R. Hanson, and L. Bogorad, 1974 An altered chloroplast ribosomal protein in *ery*-M1 mutants of *Chlamydomonas reinhardi. Mol. Gen. Genet.* **132:**119–124.

Detchon, P. and J. V. Possingham, 1973 Chloroplast ribosomal ribonucleic acid synthesis in cultured spinach leaf tissue. *Biochem. J.* **136:**829–836.

Dyer, T. A. and B. Koller, 1971 Chloroplast and cytoplasmic ribosomes of leaves. In *Proceedings of the IInd International Congress on Photosynthesis Research,* Vol. 3, Dr. W. Junk N. V., The Hague, pp. 2537–2544.

Dyer, T. A. and R. M. Leech, 1968 Chloroplast and cytoplasmic low-molecular-weight ribonucleic acid components of the leaf of *Vicia faba. Biochem. J.* **106:**689–698.

Dyer, T. A., R. H. Miller, and A. D. Greenwood, 1971 Leaf nucleic acids. *J. Exp. Bot.* **22:**125–136.

Eaglesham, A. R. J. and R. J. Ellis, 1974 Protein synthesis in chloroplasts. II. Light-driven synthesis of membrane proteins by isolated pea chloroplasts. *Biochem. Biophys. Acta* **335:**396–407.

Ellis, R. J., 1975 Inhibition of chloroplast protein synthesis by lincomycin and 2-(4-methyl-2,6-dinitroanilino)-*N*-methylpropionamide. *Phytochemistry* **14:**89–93.

Ellis, R. J., 1976 The search for plant messenger RNA. In *Perspectives in Experimental Biology,* Vol. 2, edited by N. Sunderland, Pergamon Press, New York, pp. 283–298.

Ellis, R. J. and M. R. Hartley, 1971 Sites of synthesis of chloroplast proteins. *Nature* (*London*) **233:**193–196.

Ellis, R. J., G. E. Blair, and M. R. Hartley, 1973 The nature and function of chloroplast protein synthesis. *Biochem. Soc. Symp.* **38:**137–162.

Freyssinet, G. and J. A. Schiff, 1974 The chloroplast and cytoplasmic ribosomes of *Euglena.* II. Characterization of ribosomal proteins. *Plant Physiol.* **53:**543–554.

Galling, G., 1974 Vorstufen der plastidiären ribosomalen RNA und ihre Reifung bei *Chlorella. Planta* **118:**283–295.

Gillham, N. W., J. E. Boynton, and B. Burkholder, 1970 Mutations altering the chloroplast ribosome phenotype in *Chlamydomonas.* I. Non-Mendelian mutations. *Proc. Natl. Acad. Sci. USA* **67:**1026–1033.

Goodenough, U. W. and R. P. Levine, 1970 Chloroplast structure and function in *ac-20,* a mutant strain if *Chlamydomonas reinhardi.* III. Chloroplast ribosomes and membrane organization. *J. Cell Biol.* **44:**547–562.

Gooding, L. R., H. Roy, and A. T. Jagendorf, 1973 Immunological identification of nascent subunits of wheat ribulose-diphosphate carboxylase on ribosomes of both chloroplast and cytoplasmic origin. *Arch. Biochem. Biophys.* **159:**324–335.

Graham, D., M. D. Hatch, C. R. Slack, and R. M. Smillie, 1970 Light-induced formation of enzymes of the C_4-dicarboxylic acid pathway of photosynthesis in detached leaves. *Phytochemistry* **9:**521–532.

Gray, J. C. and R. G. O. Kekwick, 1974 The synthesis of the small subunit of ribulose-1,5-bisphosphate carboxylase in the French bean *Phaseolus vulgaris. Eur. J. Biochem.* **44:**491–500.

Grierson, D., 1974 Characterization of ribonucleic acid components from leaves of *Phaseolus aureus. Eur. J. Biochem.* **44:**509–515.

Grierson, D. and U. Loening, 1974 Ribosomal RNA precursors and the synthesis of chloroplast and cytoplasmic ribosomal ribonucleic acid in leaves of *Phaseolus aureus. Eur. J. Biochem.* **44:**501–507.

Grivell, L. A. and G. S. P. Groot, 1972 Spinach chloroplast ribosomes active in protein synthesis. *FEBS Lett.* **25:**21–24.

Grivell, L. A. and H. L. Walg, 1972 Subunit homology between *Escherichia coli,* mitochondrial and chloroplast ribosomes. *Biochem. Biophys. Res. Commun.* **49:**1452–1458.

Gualerzi, C. and T. Cammarano, 1970 Species specificity of ribosomal proteins from chloroplasts and higher plants. *Biochim. Biophys. Acta* **199:**203–213.

Gualerzi, C., H. G. Janda, H. Passow, and G. Stöffler, 1974 Studies on the protein moiety of plant ribosomes: Enumeration of the proteins of the ribosomal subunits and determination of the degree of evolutionary conservation by electrophoretic and immunochemical methods. *J. Biol. Chem.* **249:**3347–3355.

Hanson, M. R., J. N. Davidson, L. J. Mets, and L. Bogorad, 1974 Characterization of chloroplast and cytoplasmic ribosomal proteins of *Chlamydomonas reinhardi* by two-dimensional gel-electrophoresis. *Mol. Gen. Genet.* **132:**105–118.

Harris, E. H., J. E. Boynton, and N. W. Gillham, 1974 Chloroplast ribosome biogenesis in *Chlamydomonas:* Selection and characterization of mutants blocked in ribosome formation. *J. Cell Biol.* **63:**160–179.

Hartley, M. R. and R. J. Ellis, 1973 Ribonucleic acid synthesis in chloroplasts. *Biochem. J.* **134:**249–262.

Hartley, M. R., A. Wheeler, and R. J. Ellis, 1975 Protein synthesis in chloroplasts. V. Translation of messenger RNA for the large subunit of fraction I protein in a heterologous cell-free system. *J. Mol. Biol.* **91:**67–77.

Hecker, L. I., J. Egan, R. J. Reynolds, C. E. Nix, J. A. Schiff, and W. E. Barnett, 1974 The sites of transcription and translation for *Euglena* chloroplastic aminoacyl-tRNA synthetases. *Proc. Natl. Acad. Sci. USA* **71:**1910–1914.

Heizmann, P., 1974*a* Maturation of chloroplast rRNA in *Euglena gracilis. Biochem. Biophys. Res. Commun.* **56:**112–118.

Heizmann, P., 1974*b* La synthèse des RNA ribosomiques au cours de l'éclairement d'Euglènes étiolées. *Biochim. Biophys. Acta* **353:**301–312.

Honeycutt, R. C. and M. M. Margulies, 1973 Protein synthesis in *Chlamydomonas reinhardi*: Evidence for synthesis of proteins of chloroplast ribosomes on cytoplasmic ribosomes. *J. Biol. Chem.* **248:**6145–6153.

Hoober, J. K., 1970 Sites for synthesis of chloroplast membrane polypeptides in *Chlamydomonas reinhardi* y-1. *J. Biol. Chem.* **245:**4327–4334.

Hoober, J. K. and G. Blobel, 1969 Characterization of the chloroplastic and cytoplasmic ribosomes of *Chlamydomonas reinhardi. J. Mol. Biol.* **41:**121–138.

Hoober, J. K. and W. J. Stegeman, 1973 Control of the synthesis of a major polypeptide of chloroplast membranes in *Chlamydomonas reinhardi. J. Cell Biol.* **56:**1–12.

Ingle, J., R. Wells, J. V. Possingham, C. J. Leaver, and U. E. Loening, 1970 Properties of chloroplast ribosomal-RNA. In *Control of Organelle Development,* edited by P. L. Miller, Cambridge University Press, Cambridge, pp. 303–326.

Ireland, H. M. M. and J. W. Bradbeer, 1971 Plastid development in primary leaves of *Phaseolus vulgaris*: The effects of D-threo and L-threo chloramphenicol on the light-induced formation of enzymes of the photosynthetic carbon pathway. *Planta* **96:** 254–261.

Jennings, R. C. and I. Ohad, 1973 Biogenesis of chloroplast membranes. XII. The influence of chloramphenicol on chlorophyll fluorescence yield and chlorophyll organization in greening cells of a mutant of *Chlamydomonas reinhardi* y-1. *Plant Sci. Lett.* **1:**3–9.

Jones, B. L., N. Nagablushan, A. Gulyas, and S. Zalik, 1972 Two-dimensional acrylamide gel electrophoresis of wheat leaf cytoplasmic and chloroplast ribosomal proteins. *FEBS Lett.* **23:**167–170.

Joy, K. W. and R. J. Ellis, 1975 Protein synthesis in chloroplasts. IV. Polypeptides of the chloroplast envelope. *Biochim. Biophys. Acta* **378:**143–151.

Kaltschmidt, E. and H. G. Wittmann, 1970 Ribosomal proteins. XII. Number of proteins in small and large ribosomal subunits of *Escherichia coli* as determined by two-dimensional gel electrophoresis. *Proc. Natl. Acad. Sci. USA* **67:**1276–1282.

Katterman, F. R. H. and J. E. Endrizzi, 1973 Studies on the 70 S ribosomal content of a plastid mutant in *Gossypium hirsutum. Plant Physiol.* **51:**1138–1139.

Kirk, J. T. O., 1970 Biochemical aspects of chloroplast development. *Ann. Rev. Plant Physiol.***21:**11–42.

Kirk, J. T. O., 1971 Will the real chloroplast DNA please stand up? In *Autonomy and Biogenesis of Mitochondria and Chloroplasts,* edited by N. K. Boardman, A. W. Linnane, and R. M. Smillie, North-Holland, Amsterdam, pp. 267–276.

Kirk, J. T. O., 1972 The genetic control of plastid formation: Recent advances and strategies for the future. *Sub-Cell. Biochem.* **1:**333–361.

Kislev, N., M. I. Selsky, C. Norton, and J. M. Eisenstadt, 1972 tRNA and tRNA-aminoacyl synthetases of chloroplasts, mitochondria and cytoplasm from *E. gracilis. Biochem. Biophys. Acta* **287:**256–269.

Kloppstech, K. and H. G. Schweiger, 1973*a* Nuclear genome codes for chloroplast ribosomal proteins in *Acetabularia.* I. Isolation and characterization of chloroplast ribosomal particles. *Exp. Cell Res.* **80:**63–68.

Kloppstech, K. and H. G. Schweiger, 1973*b* Nuclear genome codes for chloroplast ribosomal proteins in *Acetabularia.* II. Nuclear transplantation experiments. *Exp. Cell Res.* **80:**69–78.

Leaver, C. J. and J. Ingle, 1971 The molecular integrity of chloroplast ribosomal ribonucleic acid. *Biochem. J.* **123:**235–243.

Lee, S. G. and W. R. Evans, 1971 Hybrid ribosome formation from *Escherichia coli* and chloroplast ribosome subunits. *Science* **173:**241–242.

Leis, J. P. and E. B. Keller, 1970 Protein chain initiating methionine tRNA in chloroplast and cytoplasm of wheat leaves. *Proc. Natl. Acad. Sci. USA* **67:**1593–1599.

Leis, J. P. and E. B. Keller, 1971 *N*-Formylmethionyl-tRNA of wheat chloroplasts: Its synthesis by wheat transformylase. *Biochemistry* **10:**889–894.

Loening, U. E., 1968 Molecular weights of ribosomal RNA in relation to evolution. *J. Mol. Biol.* **38:**355–365.

Loening, U. E. and J. Ingle, 1967 Diversity of RNA components in green plant tissues. *Nature* (*London*) **215:**363–367.

Lyttleton, J. W., 1960 Nucleoproteins of white clover. *Biochem. J.* **74:**82–90.

Lyttleton, J. W., 1962 Isolation of ribosomes from spinach chloroplasts. *Exp. Cell Res.* **26:**312–317.

Lyttleton, J. W., 1968 Proteins constituents of plant ribosomes. *Biochim. Biophys. Acta* **154:**145–149.

Machold, O. and O. Aurich, 1972 Sites of synthesis of chloroplast lamellar proteins in *Vicia faba. Biochim. Biophys. Acta* **281:**103–112.

Margulies, M. M. and A. Michaels, 1974 Ribosomes bound to chloroplast membranes in *Chlamydomonas reinhardi. J. Cell Biol.* **60:**65–77.

Merrick, W. C. and L. S. Dure, III, 1971 Specific transformylation of one methionyl tRNA from cotton seedling chloroplasts by endogeneous and *E. coli* transformylases. *Proc. Natl. Acad. Sci. USA* **68:**641–644.

Mets, L. J. and L. Bogorad, 1971 Mendelian and uniparental alteration in erythromycin binding by plastid ribosomes. *Science* **174:**707–709.

Mets, L. J. and L. Bogorad, 1972 Altered chloroplast ribosomal proteins associated with erythromycin-resistant mutants in two genetic systems of *Chlamydomonas reinhardi*. *Proc. Natl. Acad. Sci. USA* **69:**3779–3783.

Michaels, A. and M. M. Margulies, 1975 Amino acid incorporation into proteins by ribosomes bound to chloroplast thylakoid membranes: Formation of discrete products. *Biochim. Biophys. Acta* **390:**352–362.

Miller, A., U. Karlsson, and N. K. Boardman, 1966 Electron microscopy of ribosomes isolated from tobacco leaves. *J. Mol. Biol.* **17:**487–489.

Miller, M. J. and D. McMahon, 1974 Synthesis and maturation of chloroplast and cytoplasmic ribosomal RNA in *Chlamydomonas reinhardi*. *Biochim. Biophys. Acta* **366:**35–44.

Munsche, D. and R. Wollgiehn, 1974 Altersabhängige Labilität der ribosomalen RNA aus Chloroplasten von *Nicotiana rustica*. *Biochim. Biophys. Acta* **340:**437–445.

Odintsova, M. S. and N. P. Yurina, 1969 Proteins of chloroplast and cytoplasmic ribosomes, *J. Mol. Biol.* **40:**503–506.

Ohad, I., P. Siekevitz, and G. E. Palade, 1967 Biogenesis of chloroplast membranes. I. Plastid dedifferentiation in a dark-grown algal mutant (*Chlamydomonas reinhardi*). *J. Cell Biol.* **35:**521–552.

Ohta, N., R. Sager, and M. Inouye, 1975 Identification of a chloroplast ribosomal protein altered by a chloroplast mutation in *Chlamydomonas*. *J. Biol. Chem.* **250:**3655–3659.

Parthier, B., 1973 Cytoplasmic site of synthesis of chloroplast aminoacyl-tRNA synthetases in *Euglena gracilis*. *FEBS Lett.* **38:**70–74.

Payne, P. I. and T. A. Dyer, 1971 Characterization of cytoplasmic and chloroplast 5 S ribosomal ribonucleic acid from broad-bean leaves. *Biochem. J.* **124:**83–89.

Payne, P. I. and T. A. Dyer, 1972*a* Plant 5.8 S RNA is a component of 80 S but not 70 S ribosomes. *Nature (London) New Biol.* **235:**145–147.

Payne, P. I. and T. A. Dyer, 1972*b* Characterization of the ribosomal ribonucleic acids of blue-green algae. *Arch. Mikrobiol.* **87:**29–40.

Phillipovitch, I. I., I. N. Bezsmertuaya, and A. I. Oparin, 1973 On the localization of polyribosomes in the system of chloroplast lamellae. *Exp. Cell Res.* **79:**159–168.

Pigott, G. H. and N. G. Carr, 1972 Homology between nucleic acids of blue-green algae and chloroplasts of *Euglena gracilis*. *Science* **175:**1259–1261.

Pine, K. and A. O. Klein, 1972 Regulation of polysome formation in etiolated bean leaves by light. *Dev. Biol.* **28:**280–289.

Ramirez, J. M., F. F. del Campo, and D. I. Arnon, 1968 Photosynthetic phosphorylation as energy source for protein synthesis and carbon dioxide assimilation by chloroplasts. *Proc. Natl. Acad. Sci. USA* **59:**606–611.

Rawson, J. R. and E. Stutz, 1969 Isolation and characterization of *Euglena gracilis* cytoplasmic and chloroplast ribosomes and their RNA components. *Biochim. Biophys. Acta* **190:**368–380.

Reger, B. J., S. A. Fairfield, J. L. Epler, and W. E. Barnett, 1970 Identification and origin of some chloroplast aminoacyl-tRNA synthetases and tRNAs. *Proc. Natl. Acad. Sci. USA* **67:**1207–1213.

Rossi, L. and C. Gualerzi, 1970 Non-random differences in the base composition of chloroplast and cytoplasmic ribosomal RNA from some higher plants. *Life Sci.* **9:**1401–1407.

Sager, R., 1972 *Cytoplasmic Genes and Organelles,* Academic Press, New York.

Sakano, K., S. D. Kung, and S. G. Wildman, 1974 Identification of several chloroplast DNA genes which code for the large subunit of *Nicotiana* fraction I protein. *Mol. Gen. Genet.* **130:**91–97.

Schiff, J. A., 1971 Developmental interactions among cellular compartments in Euglena. In *Autonomy and Biogenesis of Mitochondria and Chloroplasts,* edited by N. K. Boardman, A. W. Linnane, and R. M. Smillie, North-Holland, Amsterdam, pp. 98–118.

Schlanger, G. and R. Sager, 1974 Localization of five antibiotic resistances at the subunit level in chloroplast ribosomes of *Chlamydomonas*. *Proc. Natl. Acad. Sci. USA* **71:**1715–1719.

Schlanger, G., R. Sager, and Z. Ramanis, 1972 Mutation of a cytoplasmic gene in *Chlamydomonas* alters chloroplast ribosome function. *Proc. Natl. Acad. Sci. USA* **69:**3551–3555.

Schwartz, J. H., R. Meyer, J. M. Eisenstadt, and G. Brawerman, 1967 Involvement of *N*-formylmethionine in initiation of protein synthesis in cell-free extracts of *Euglena gracilis*. *J. Mol. Biol.* **25:**571–574.

Schwartzbach, S. D. and J. A. Schiff, 1974 Chloroplast and cytoplasmic ribosomes of *Euglena*: Selective binding of dihydrostreptomycin to chloroplast ribosomes. *J. Bacteriol.* **120:**334–341.

Schwartzbach, S. D., G. Freyssinet, and J. A. Schiff, 1974 The chloroplast and cytoplasmic ribosomes of *Euglena*. I. Stability of chloroplast ribosomes prepared by an improved procedure. *Plant Physiol.* **53:**533–542.

Scott, N. S., 1973 Ribosomal RNA cistrons in *Euglena gracilis*. *J. Mol. Biol.* **81:** 327–336.

Scott, N. S., R. Munns, D. Graham, and R. M. Smillie, 1971 Origin and synthesis of chloroplast ribosomal RNA and photoregulation during chloroplast biogenesis. In *Autonomy and Biogenesis of Mitochondria and Chloroplasts,* edited by N. K. Boardman, A. W. Linnane, and R. M. Smillie, North-Holland, Amsterdam, pp. 383–392.

Siddell, S. G. and R. J. Ellis, 1975 Protein synthesis in chloroplasts: Characteristics and products of *in vitro* protein synthesis in etioplasts and developing chloroplasts from pea leaves. *Biochem. J.* **146:**675–685.

Smillie, R. M. and N. S. Scott, 1969 Organelle biosynthesis: The chloroplast. *Prog. Mol. Subcell. Biol.* **1:**136–202.

Smillie, R. M., D. Graham, M. R. Dwyer, A. Grieve, and N. F. Tobin, 1967 Evidence for the synthesis *in vivo* of proteins of the Calvin cycle and of the photosynthetic electron-transfer pathway on chloroplast ribosomes. *Biochem. Biophys. Res. Commun.* **28:**604–610.

Sprey, B., 1972 Ribosomale RNA und Thylakoidmembranen in Plastiden von Chlorophylldefektmutanten der Gerste. *Z. Pflanzenphysiol.* **67:**223–243.

Stanley, W. M and R. M. Bock, 1965 Isolation and physical properties of the ribosomal ribonucleic acid of *E. coli*. *Biochemistry* **4:**1302–1311.

Stutz, E. and H. Noll, 1967 Characterization of cytoplasmic and chloroplast polysomes in plants: Evidence for 3 classes of ribosomal RNA in nature. *Proc. Natl. Acad. Sci. USA* **57:**774–781.

Tao, K. J. and A. T. Jagendorf, 1973 The ratio of free to membrane-bound chloroplast ribosomes. *Biochim. Biophys. Acta* **324:**518–532.

Tewari, K. K., 1971 Genetic autonomy of extranuclear organelles. *Ann. Rev. Plant Physiol.* **22:**141–168.

Thomas, J. R. and K. K. Tewari, 1974 Ribosomal-RNA genes of the chloroplast DNA of pea leaves. *Biochim. Biophys. Acta* **361:**73–83.

Vandrey, J. P. and E. Stutz, 1973 Evidence for a novel DNA component in chloroplasts of *Euglena gracilis*. *FEBS Lett.* **37:**174–177.

Vasconcelos, A. C. L. and L. Bogorad, 1971 Proteins of cytoplasmic, chloroplast and mitochondrial ribosomes of some plants. *Biochim. Biophys. Acta* **228:**492–502.

Wittmann, H. G., 1972 Ribosomal proteins. XXXII. Comparison of several extraction methods for proteins from *Escherichia coli* ribosomes. *Biochimie* **54:**167–175.

Wittmann, H. G., G. Stöffler, E. Kaltschmidt, V. Rudloff, M. G. Janda, M. Dzionara, D. Donner, K. Nierhaus, M. Cech, I. Hindenach, and B. Wittmann, 1970 Protein chemical and serological studies on ribosomes of bacteria, yeast and plants. *FEBS Symp.* **21:**33–46.

Woodcock, C. L. F. and L. Bogorad, 1971 Nucleic acids and information processing in chloroplasts. In *Structure and Function of Chloroplasts,* edited by M. Gibbs, Springer-Verlag, New York, pp. 89–128.

Yurina, N. P. and M. S. Odintsova, 1974 Buoyant density of chloroplast ribosomes in CsCl. *Plant Sci. Lett.* **3:**229–234.

Zablen, L. B., M. S. Kissil, C. R. Woese, and D. E. Buetow, 1975 Phylogenetic origin of the chloroplast and prokaryotic nature of its ribosomal RNA. *Proc. Natl. Acad. Sci. USA.* **72:**2418–2422.

14

Nucleocytoplasmic Interaction in *Acetabularia*

Hans-Georg Schweiger

Introduction

The eukaryotic cell is characterized by a clear separation of the nucleus from the rest of the cell, namely, the cytoplasm. This compartmentalization is achieved by the nuclear envelope, which in the electron microscope appears as a double unit membrane structure. Both of the compartments are involved in the expression of the nuclear genome.

In brief, the nucleus contains the nuclear genome and the complete system for the replication and transcription of nuclear DNA, whereas the system for translating mRNA transcribed in the nucleus into specific proteins is localized in the cytoplasm.

The intracellular separation of replication and transcription from translation directs attention to the transport of materials between the nucleus and the cytoplasm. For example, three classes of RNA molecules must be transported from the nucleus to the cytoplasm for the ultimate expression of the nuclear genome: the various RNAs which contribute to the structure of the 80 S cytosol ribosomes, the mRNAs which will determine the proteins to be synthesized, and the tRNAs required to

---

Hans-Georg Schweiger—Max-Planck-Institut für Zellbiologie, Wilhelmshaven, Federal Republic of Germany.

translate the mRNA nucleotide code into the correct sequence of amino acids. In addition, several classes of cytoplasmic substances must migrate into the nucleus, since the nucleus has to be provided with precursors for DNA and RNA. To date, relatively little is known about other cytoplasmic materials which may enter the nucleus and regulate its activity.

This two-step system of gene expression is less simple than it appears at first glance. First, in a eukaryotic cell, genetic information is contained not only in the nucleus but also in the mitochondria and in the chloroplasts, when the latter are present. Second, translation may take place not only on the 80 S ribosomes of the cytosol but also on 70 S ribosomes which are localized in the organelles (for a detailed discussion, see Chapter 13). Third, gene expression is subjected to temporal organization; i.e., the degree, as well as the type of expression, varies in time as a consequence of a sophisticated type of regulation.

The possibilities for complex interactions between the different genomes and the two types of ribosomes and the existence of this temporal regulation provoke a number of questions. For example, from which genome does the genetic information for a distinct protein originate? On which type of ribosome is such a protein synthesized? These questions are of special relevance for organelle proteins. If, for example, an organelle protein is synthesized on 80 S ribosomes, then one may ask how this protein finds its way into the organelle while other proteins do not enter.

Furthermore, the question arises of where the regulation underlying differentiation takes place. Does this regulation take place solely within the nucleus at the level of transcription of the nuclear genome or does it occur in the cytoplasm? If the latter, is it related to translation or might it instead involve the transcription of the organelle genome? Because of the network of nucleocytoplasmic interrelations, it is quite possible that the regulation of nuclear transcription is mediated by a cytoplasmic factor which could be either specific for a few genes or act more broadly on a large part of the nuclear genome. Finally, it should be kept in mind that either the nuclear envelope or the perinuclear region might regulate the release of information into the cytoplasm. Since little is known about this possibility at present, it will not be discussed further in this chapter.

The Organism

Questions such as those listed above can be profitably studied in an organism whose size, morphology, and biology permit the removal, isolation, and implantation of a nucleus (Hämmerling, 1953, 1963; Schweiger, 1969; Puiseux-Dao, 1970). The unicellular uninucleate green alga *Acetabularia* is such an organism (Figure 1). The cell consists of the

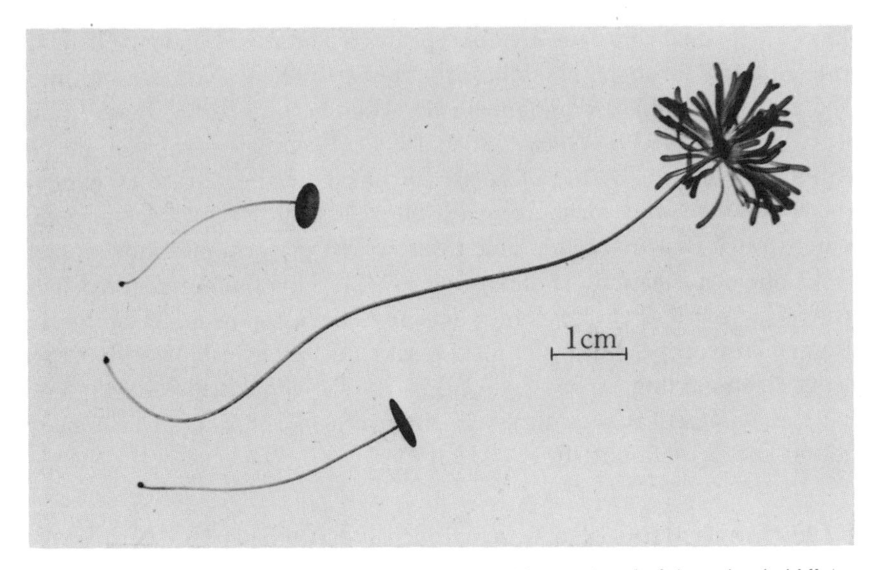

Figure 1. Mature cells of A. mediterranea (top and bottom) and of A. major (middle).

rhizoid, the stalk, and, during the final stages of its vegetative phase, the cap. The maximum length of the cell is 3–100 mm or even more depending on the species (Berger *et al.*, 1974; Schweiger *et al.*, 1974). This giant cell is uninucleate throughout its vegetative phase, and its primary nucleus is in one of the branches of the rhizoid (Figure 2). During the generative

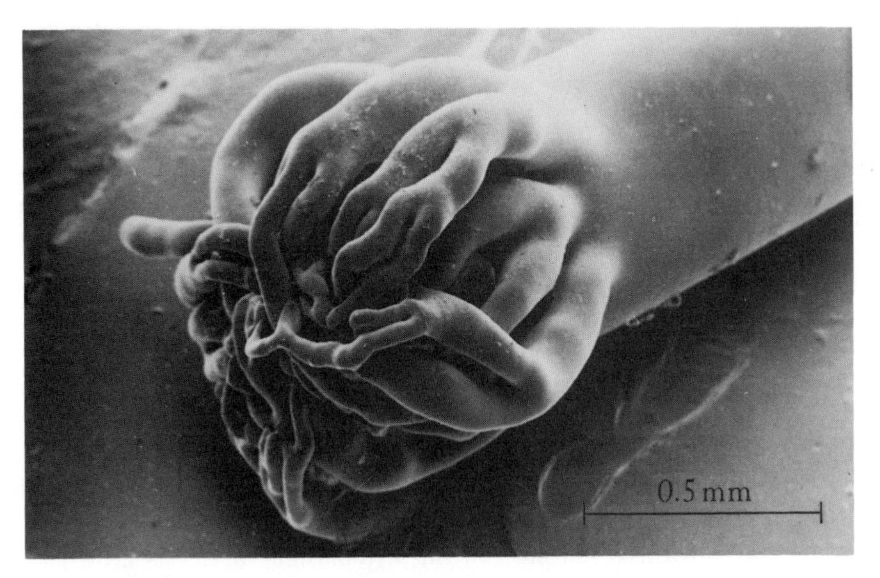

Figure 2. Scanning electron micrograph of a rhizoid from A. major. One of the branches of the rhizoid contains the nucleus. By courtesy of Dr. S. Berger.

phase, the primary nucleus divides repeatedly and the secondary nuclei migrate upward through the stalk into the cap, where cysts are organized around the individual secondary nuclei (Hämmerling, 1963; Berger *et al.*, 1975; Schulze, 1939; Werz, 1968, 1974). Eventually, gametes develop within each cyst. The life cycle is concluded by the release of gametes, which copulate and form zygotes. The classical work of Hämmerling demonstrated that this cell is able to grow and perform morphogenesis for weeks after enucleation (Hämmerling, 1932). This finding resulted in his postulating the existence of "morphogenetische Substanzen." Hämmerling also demonstrated by means of interspecific nuclear transplantation experiments (Hämmerling, 1955; Schweiger, 1969; Sandakhchiev *et al.*, 1973) that the hypothetical substances which govern morphogenesis originate in the cell nucleus (Hämmerling, 1932, 1934).

Demonstration of a Cytoplasmic Effect on the Nucleus

Evidence for a cytoplasmic effect on nuclear activity might be sought by examining the fate of nuclei implanted into enucleated cells which are in a physiological state different from that of the cell from which the nucleus originated. Such experiments were performed by Hämmerling (1958), who showed that the onset of cyst formation can be accelerated by grafting the rhizoid of a young cell onto an old fragment (Hämmerling, 1939). The reciprocal experiment, an old rhizoid grafted onto a young anucleate fragment, had the opposite effect. In a similar way, cyst formation could be repressed by removing part of the cap (Hämmerling, 1953).

This finding prompted us to develop a test system which would demonstrate a cytoplasmic effect on the nucleus and which might even be used for the eventual identification of the active cytoplasmic factors. This test system is based on the observation that the nuclei and the perinuclear regions of "young" and "old" cells exhibit substantial ultrastructural differences (Berger and Schweiger, 1973; Franke *et al.*, 1974; Berger *et al.*, 1975) (Figure 3). While an old cell has a length of more than 40 mm and has already formed a maximum-diameter cap, a young cell has a length of about 10 mm and no cap. Old and young nuclei differ mainly in their nucleolar ultrastructure. In young nuclei, the multiple nucleoli contain a large number of small subunits (Hämmerling, 1931; Schulze, 1939; Berger and Schweiger, 1973; Franke *et al.*, 1974). In old nuclei, only one nucleolus is present, and this is spherical. The perinuclear regions differ in that in a young cell the nuclear envelope is covered by a thin (10 nm) cytoplasmic layer while in an old cell this cytoplasmic layer has grown to a thickness of 300–1000 nm (Berger and Schweiger, 1973, 1975; Berger, 1975).

The influence of the cytoplasm on the nucleus was tested by implanting a young nucleus into an old cytoplasm and *vice versa* (Berger and Schweiger, 1975). Control experiments had shown that the implantation of a young nucleus into a young cytoplasm and of an old nucleus into an old cytoplasm did not result in a significant change of the characteristic ultrastructure. However, when a nucleus was implanted into a cell fragment which differed from it in age, it assumed an ultrastructural appearance consistent with the age of the cytoplasm in less than 10 days after implantation.

This type of experiment indicates that the cytoplasm is able to induce either aging or rejuvenation of the nucleus, depending on its developmental stage. With this system in hand, work can now proceed toward the identification of the active cytoplasmic factor.

Coding Problems

The existence of multiple and diverse genomes in the cell immediately raises the question of where the information for different specific proteins orginates (Figure 4). This question has been investigated for the enzyme malic dehydrogenase (MDH), which exhibits a rather high activity in *Acetabularia* (Schweiger *et al.,* 1967). During growth, the activity of MDH increases in nucleate as well as in anucleate cells.

In order to determine whether the increase in MDH activity is due to *de novo* synthesis of enzyme protein or to an activation of a preexisting protein (Schael and Schweiger, unpublished; Schweiger *et al.,* 1972) (Figure 5a,b), intact and enucleated cells were cultured in a medium in which 50% of the water had been replaced by deuterium, and the buoyant densities of thin MDH fractions were determined by CsCl density gradient centrifugation. Following growth in deuterated medium, the buoyant density of the enzyme was significantly greater than that from cells cultured in normal medium (Figure 5a,b). With appropriate control experiments, it could be shown that the replacement of H atoms by D atoms in the MDH molecule was not due to a physicochemical exchange between H and D in preexisting proteins but rather to a polymerization of the MDH molecule from amino acids which had been synthesized *de novo*. Since a similar density shift was observed in enucleated cells, it may be concluded that MDH is synthesized in *Acetabularia* even in the absence of the nucleus.

In order to find out which genome, nuclear or organelle, codes for the MDH, a nuclear exchange experiment was performed (Schweiger *et al.,* 1967) (Figure 6). *Acetabularia* cells from two species whose MDH isozyme patterns can easily be distinguished were enucleated, and the nuclei were isolated and reciprocally implanted. In this way, hybrids consist-

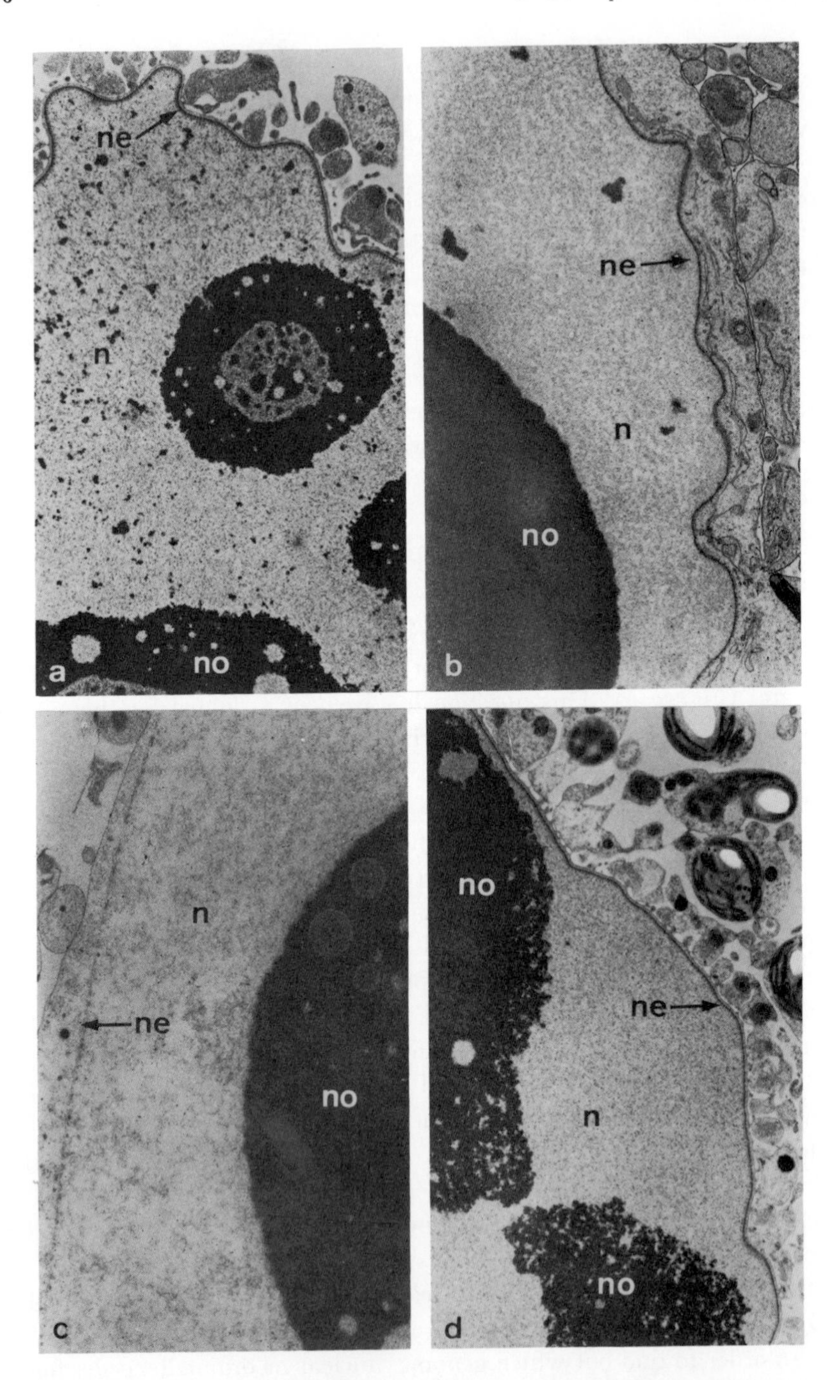

Figure 3. Ultrastructure of nuclei of A. mediterranea with their perinuclear region. (a) Young nucleus. Note the thin cytoplasmic layer covering the nuclear envelope and the multiple-subunit structure of the nucleolar material with the peripheral granular and the central fibrous zones. (b) Old nucleus. Note the thick perinuclear cytoplasmic layer contain-

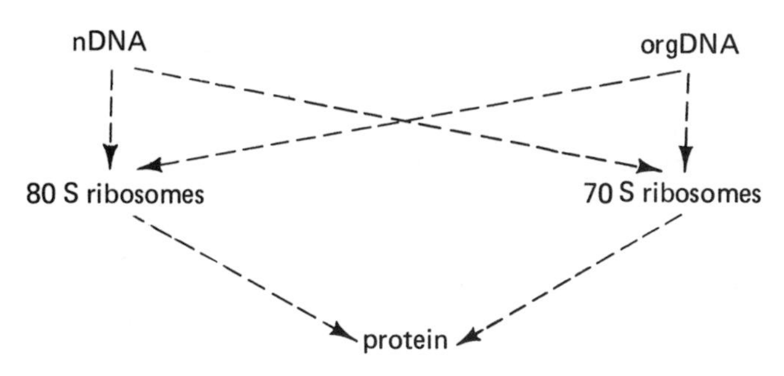

Figure 4. Scheme of theoretically possible pathways of gene expression in a eukaryotic cell.

ing of a nucleus from *A. mediterranea* and a cytoplasm from *A. cliftonii* and vice versa were obtained. Four weeks after the nuclear exchange, the isozyme patterns were established again and compared with the original species patterns of the cytoplasm. The patterns of the hybrid cells were found to correspond with those of the species from which the nuclei had originated. This result suggests that in *Acetabularia* the isozymes of MDH are coded by the nuclear genome.

A negative effect of the implanted nucleus was also observed. The isozymes of the cytoplasmic host species disappeared in hybrid cells. None of the isozymes disappears in anucleate cells, and the ratio of the different isoenzymes remains roughly constant despite the fact, as noted earlier, that MDH activity also increases during the growth of enucleated cell fragments, and this increase is due to *de novo* synthesis of the enzyme protein. At present, it is not clear why the cytoplasmic host isozymes disappear under the influence of the implanted heterologous nucleus. Possibly, mRNA is more stable in the absence of the nucleus and is subject to a high turnover in the presence of a nucleus.

That MDH is localized in the chloroplasts was shown by the following methods: cytochemical tests (Schweiger *et al.*, 1967), density gradient centrifugation (Schweiger *et al.*, 1967), and demonstration of MDH activity in partially purified chloroplast fractions (Schael, unpublished results) and in a single chloroplast (Sandakhchiev, unpublished results). From these experiments, it follows that the chloroplast protein MDH is coded by the nuclear genome.

ing membranous structures and the single spherical, condensed nucleolus. (c) Young nucleus 6 days after implantation into an old cytoplasm. Note the symptoms of aging. (d) Old nucleus 6 days after implantation into a young cytoplasm. Note the symptoms of rejuvenation. Abbreviations: n, nucleus; no, nucleolus; ne, nuclear envelope. Magnifications: (a) 2400×, (b) 5400×, (c) 6300×, (d) 2400×. By courtesy of Dr. S. Berger.

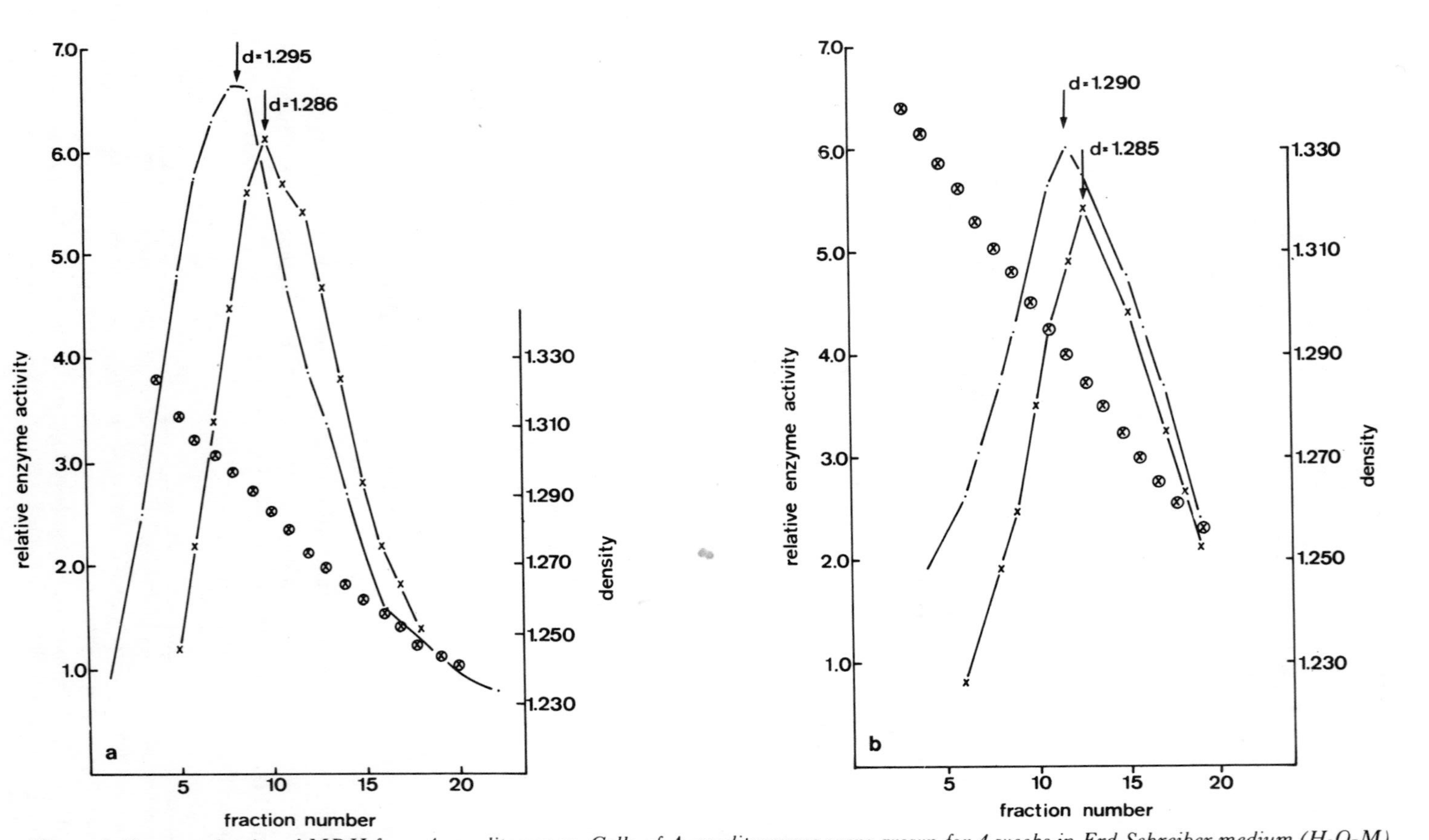

Figure 5. Buoyant density of MDH from A. mediterranea. Cells of A. mediterranea were grown for 4 weeks in Erd-Schreiber medium (H_2O-M) and in a medium in which 50% of the water was replaced by deuterium (D_2O-M). The cells were then washed and homogenized, following which the homogenate was subjected to CsCl density gradient centrifugation for 48 hr. Finally, MDH activity (×, H_2O-M; ·, D_2O-M) and CsCl density (⊗) were estimated for each fraction. Note the increase in buoyant density of MDH in nucleate (a) and anucleate (b) cells following growth in D_2O-

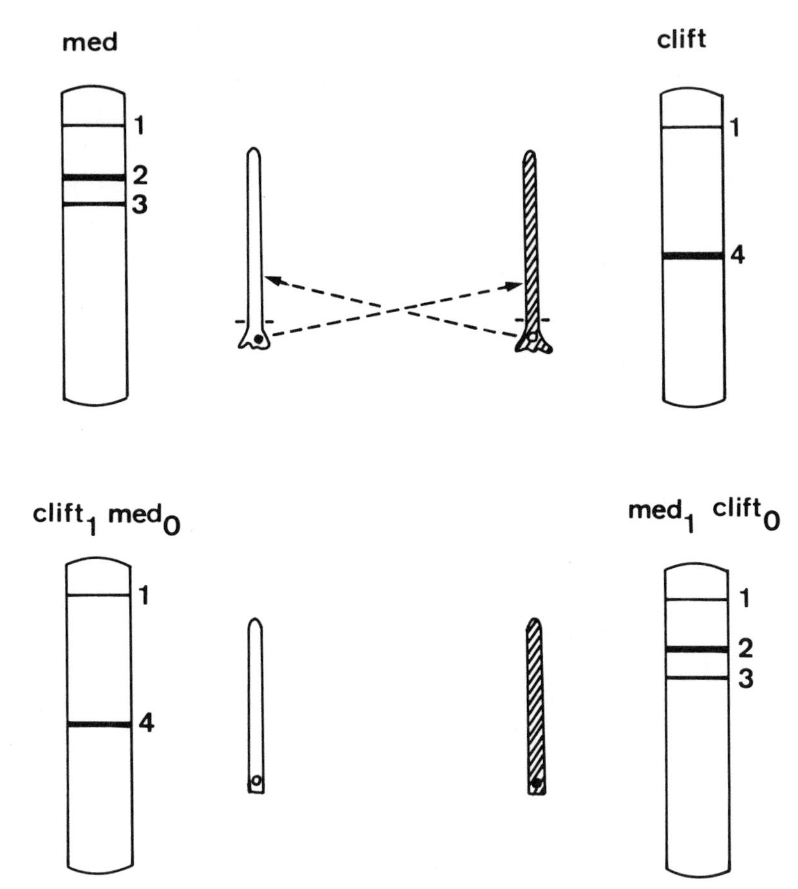

Figure 6. Schematic presentation of the effect of nuclear exchange between A. mediterranea (med) and A. cliftonii (clift) on the MDH isozyme pattern. Single cells of med and clift were homogenized, the homogenates were subjected to polyacrylamide gel electrophoresis, and the gels were stained for MDH activity. In the upper part of the figure are shown the isozyme patterns characteristic of med and clift. In the lower part of the figure are shown the patterns of hybrid cells examined 4 weeks after reciprocal nuclear exchange. Note in the hybrid cell consisting of a clift nucleus and a med cytoplasm ($clift_1$ med_0) that band 4, typical of clift, replaced bands 2 and 3, which are characteristic for med. In the case of the med_1 $clift_0$ hybrid, band 4 had disappeared and bands 2 and 3 were present. Band 1 represents the starting line on each gel.

Nuclear exchange experiments have proven to be a useful tool for gaining more insight into the problem of the origin of the genetic information for distinct cytoplasmic proteins, particularly those of chloroplasts. Lactic dehydrogenase (Reuter and Schweiger, 1969) and a number of chloroplast membrane proteins have been shown to be coded for by the nuclear genome (Apel and Schweiger, 1972). The genetic information for all those chloroplast membrane proteins for which specific differences between two species have been found originates in the nuclear genome.

A problem of special interest is the source of information for the proteins of the chloroplast 70 S ribosomes (for a detailed discussion and for references, see Chapter 13). Without any doubt, these proteins are integral constituents of the chloroplasts. It is well established that the two ribonucleic acids of the chloroplast ribosomes, the 23 S and 16 S RNA, are coded by the genome of the organelle (Schweiger and Berger, 1964; Berger, 1967; Tewari and Wildman, 1970). Surprisingly, it could be shown by two types of experiments that at least some of the proteins of the 70 S chloroplast ribosomes are coded by the nuclear genome. These experiments are based on the observation that the electrophoretic protein patterns of the chloroplast ribosomes from *A. mediterranea, A. crenulata,* and *A. cliftonii* are distinguished by a number of bands (Kloppstech and Schweiger, 1973*b*).

In the first experiment, hybrids of the type *A. mediterranea* nucleus plus *A. crenulata* cytoplasm were prepared and kept in culture until cysts were formed. The F_1 generation, which again should be hybrids with respect to the nuclear and the chloroplast genomes, was analyzed for the protein pattern of the 44 S subunit of the 70 S chloroplast ribosomes. In the second experiment, rhizoids from *A. cliftonii* were grafted to stalks from *A. mediterranea,* and after 6 weeks the protein pattern of the 44 S subunits was analyzed again (Figure 7). Under the influence of the heterologous nucleus, the ribosomal protein pattern had been replaced by the one which is specific for the nuclear species. In both cases, the results suggested that the species-specific ribosomal proteins are coded in the nucleus (Kloppstech and Schweiger, 1973*b*). However, it must be emphasized that the nuclear coding could be shown only for those proteins which are

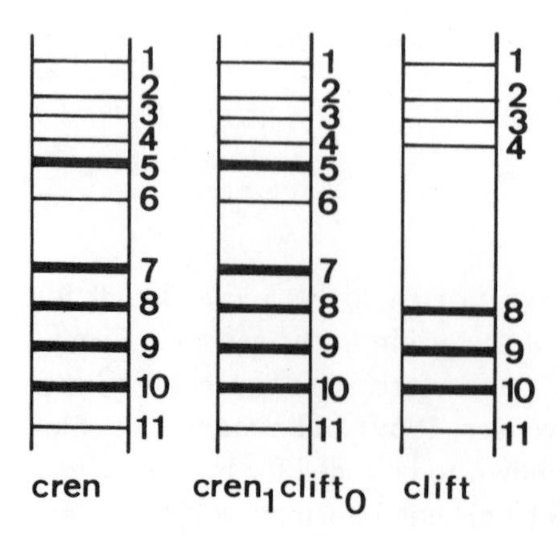

Figure 7. Schematic presentation of the effect of a heterologous nucleus on the protein pattern of chloroplast ribosomes. Proteins of the 44 S subunit of chloroplast 70 S ribosomes from A. crenulata (cren) and A. cliftonii (clift) were subjected to polyacrylamide gel electrophoresis. The patterns obtained from cells of these species (cren and clift) were compared with the pattern obtained 6 weeks after the clift nucleus had been replaced by a cren nucleus ($cren_1$ $clift_0$)*. Note that bands 5, 6, and 7, characteristic for cren and missing in clift chloroplast ribosomes, were present in the ribosomes isolated from hybrid cell chloroplasts.*

species specific. Thus it might be that besides nuclear-coded chloroplast ribosomal proteins there are others which are coded by the organelle genome. Experiments using uniparentally inherited antibiotic resistance mutants in *Chlamydomonas* indicate that in such mutants one protein of the 30 S ribosomal subunit of chloroplasts is altered (Mets and Bogorad, 1972; Schlanger and Sager, 1974; Ohta *et al.*, 1974; Bogorad *et al.*, 1975). The conclusion from the experiments with *Acetabularia* is that the information for the structure of chloroplast ribosomes comes from two sources. The ribonucleic acids are transcription products of the organelle DNA, and at least some of the ribosomal proteins are coded by the nuclear genome. This finding underlines the complex interactions of the nucleus and cytoplasm in cell development.

Site of Translation

Gene expression in a eukaryotic cell is so complex not only because the cell makes use of different types of genomes but also because translation may be performed on at least two types of ribosomes (Figure 4), the 80 S cytosol ribosomes and the 70 S organelle ribosomes. In *Acetabularia* cells the occurrence of both types of ribosomes has been established (Kloppstech and Schweiger, 1973*a,c*). To date, there is no evidence that, in the living cell, a given protein is translated on the 70 S as well as on the 80 S ribosomes. Specific inhibitors of protein synthesis have proven to be most useful in recognizing the site of translation of a distinct protein. Most frequently, cycloheximide has been used as a specific inhibitor of protein synthesis on 80 S ribosomes (Siegel and Sisler, 1965) and chloramphenicol as an inhibitor of translation on 70 S ribosomes (Pestka, 1971).

How good is the evidence for the specificity of their effects? Where structural proteins are involved, caution has to be exercised that the effect observed, i.e., the absence of a distinct protein, is not the result of an inhibition of another protein which is essential for its binding to the membrane. Such indirect effects could be detected by analyzing the kinetics of the inhibitory effect and making a comparison of the effects of inhibitors of 70 S ribosomes and of 80 S ribosomes.

It could be shown by such inhibitor studies that at least some of the 70 S chloroplast ribosomal proteins are translated on 80 S ribosomes (Kloppstech and Schweiger, 1974). This is concluded from the observation that cycloheximide (0.5 μg/ml) greatly inhibits incorporation of labeled amino acids into the large subunit of the 70 S ribosomes (Table 1). Chloramphenicol (10 μg/ml), on the other hand, stimulates the incorporation of amino acids into the organelle ribosomal subunit. The stimulating ef-

TABLE 1. Effect of Cycloheximide and Chloramphenicol on the Incorporation of Labeled Amino Acids into the 44 S Ribosomal Subunit from Chloroplasts of Acetabularia mediterranea

Conditions	*Specific activity cpm/1 unit OD at 260 nm* ($\times 10^{-3}$)	*Change (%)*	*Change (%) on the basis of the specific activity of total chloroplast RNA*
Control	5.25		
Cycloheximide	1.09	−79	−61
Chloramphenicol	7.59	+45	+95

fect of chloramphenicol might reflect a greater availability of amino acids for cytosol (80 S) transplantation due to the antibiotic's inhibition of peptide synthesis on 70 S ribosomes. Thus it appears that the genetic information for some of the 70 S ribosomal proteins originates in the nuclear genome and is translated on 80 S ribosomes of the cytosol, and that the completed proteins migrate into the organelle. It would be of some importance to find out whether all proteins which are coded by the nuclear genome are translated on 80 S ribosomes and vice versa. Furthermore, the question is raised of whether a similar relationship exists between the organelle genome and the 70 S organelle ribosomes.

The 80 S ribosomes of the cytosol and the 70 S chloroplast ribosomes form polysomes in *Acetabularia* (Kloppstech and Schweiger, 1973*a,c*). This was shown by establishing the sedimentation pattern of ribosome fractions in a sucrose gradient before and after treatment with low concentrations of RNase (Figure 8A,B).

The site of translation of the structural proteins of the chloroplast membranes has been studied in *Acetabularia* (Apel and Schweiger, 1973). When the EDTA-insoluble fraction of the chloroplast membranes was solubilized in a phenol–formic acid–water mixture and subjected to gel electrophoresis, three major fractions were obtained which could be clearly distinguished by their incorporation patterns (Figure 9A). Under *in vivo* conditions, all three fractions were labeled (Figure 9B). Isolated chloroplasts were no longer able to incorporate radioactive amino acids into the first of the three fractions (Figure 9B). This result suggested that the absence of labeling in the first membrane protein fraction was due to the loss of 80 S cytosol ribosomes during the isolation procedure.

The labeling of one of the remaining two fractions which were labeled *in vitro* was preferentially sensitive to chloramphenicol (Figure 9C), while incorporation into the other one was inhibited by cycloheximide (Apel and

Schweiger, 1973) (Figure 9D). The sensitivity of one of the fractions against cycloheximide is a striking observation, for it indicates that the chloroplasts contain 80 S ribosomes. However, these 80 S ribosomes of the chloroplasts seem to be different from the 80 S ribosomes of the cytosol in that the two types appear to be involved in the synthesis of different proteins. The existence of 80 S ribosomes in the chloroplast fraction was substantiated by the isolation of 26 S and 18 S RNA from this fraction.

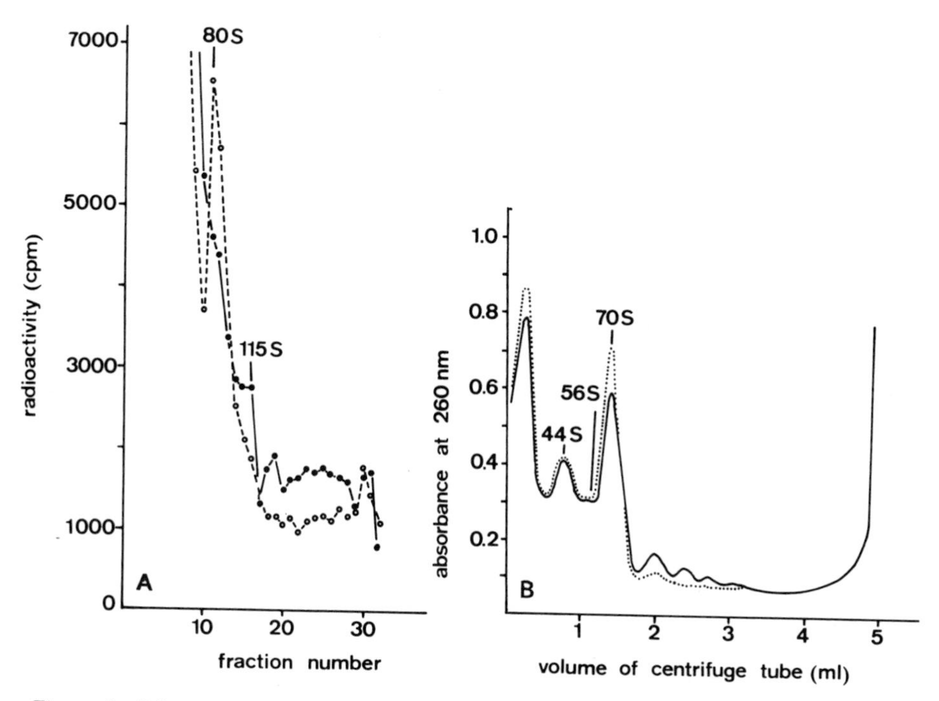

Figure 8. Ribosomes and polysomes in A. mediterranea. (A) 80 S cytosol ribosomes. Acetabularia cells were labeled with [^{3}H]uridine for 22 hr, homogenized, and centrifuged at 15,000g for 10 min. The supernatant was then subjected to sucrose density gradient centrifugation (●). The sedimentation pattern reveals labeled 115 S and larger particulate materials (polysomes), but no discrete peak of 80 S particles. When part of the homogenate was treated with 0.1 μg/ml RNAse for 10 min at 0°C before centrifugation, only a sharp peak of 80 S particles was observed (○). The loss of large particles and their replacement by 80 S material following RNAse treatment would be expected as a result of the degradation of the mRNA strands binding the ribosomes together to form polysomes. (B) 70 S chloroplast ribosomes. Ribosomes were prepared by centrifugation from chloroplasts isolated from Acetabularia cells and subjected to sucrose density gradient centrifugation (——). As in A, part of the ribosomal material was treated with RNAse before centrifugation (····). After centrifugation, the absorbance of each fraction was automatically recorded at 260 nm. Note the absence of large particles and a discrete 70 S peak in the RNAse-treated ribosomal fraction. This 70 S peak is not apparent in untreated material.

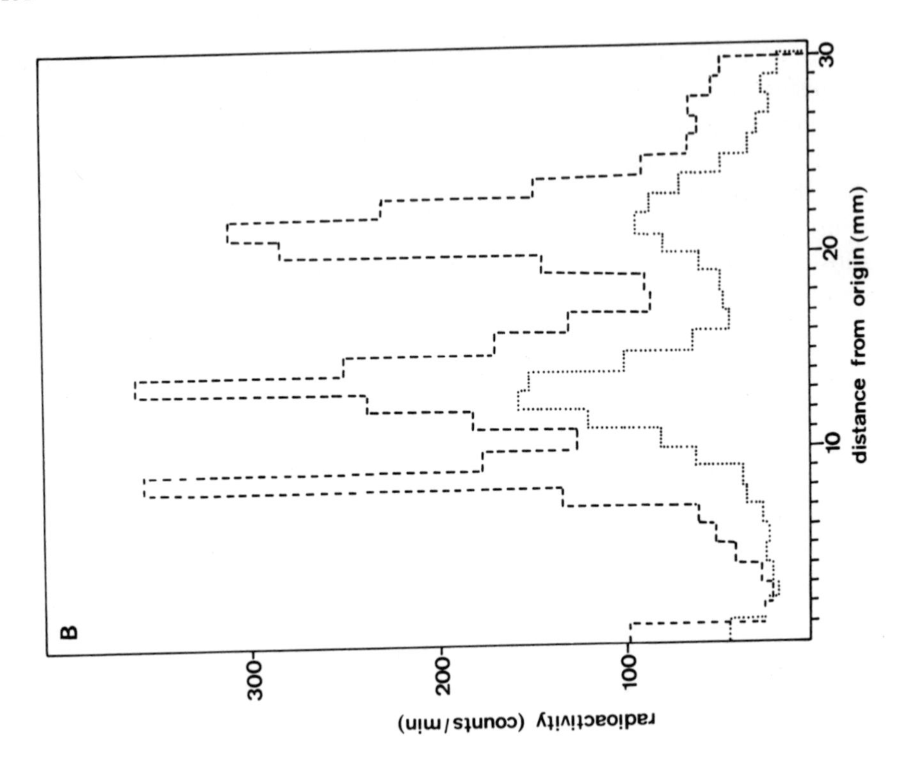
B
300
200
100
radioactivity (counts/min)
10
20
30
distance from origin (mm)

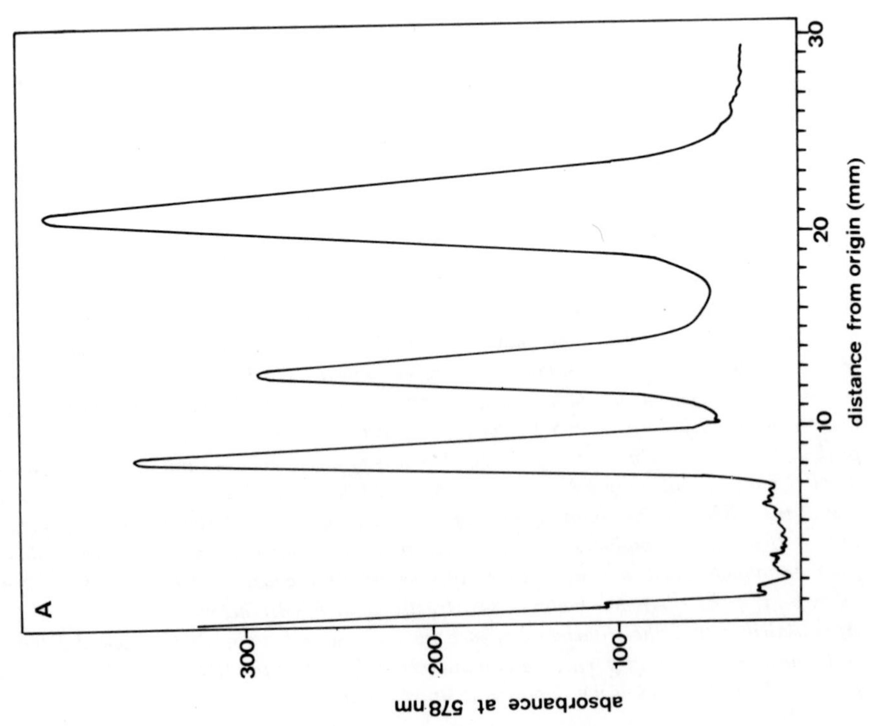
A
300
200
100
absorbance at 578 nm
10
20
30
distance from origin (mm)

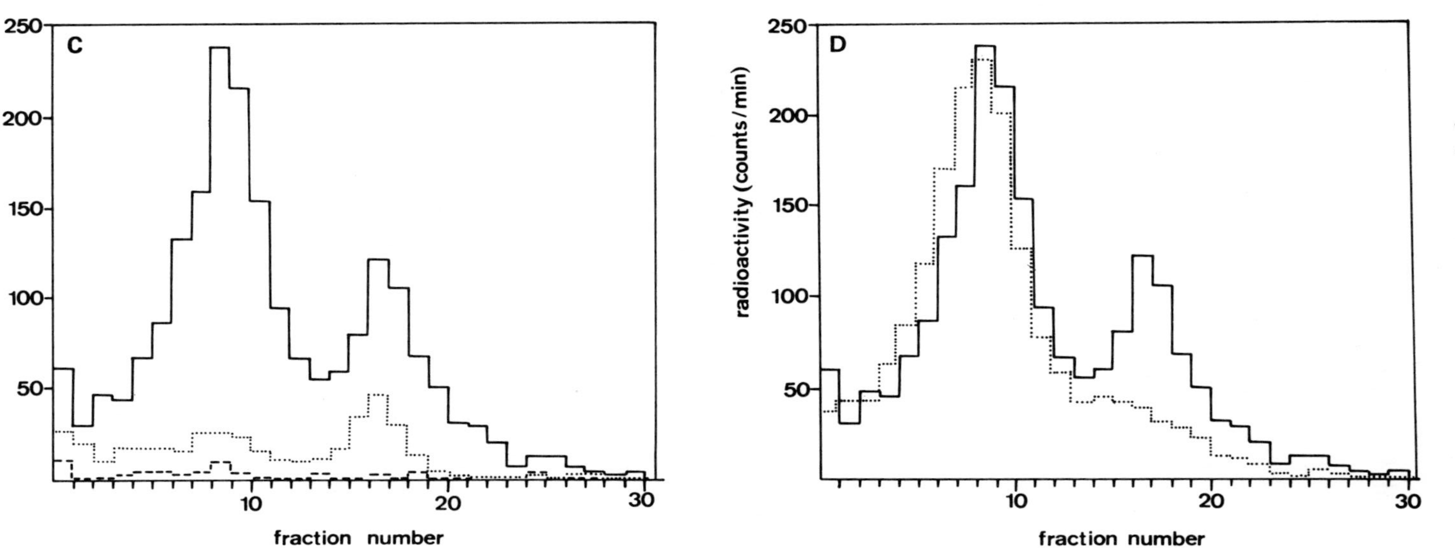

Figure 9. Insoluble chloroplast membrane proteins from Acetabularia after extraction with a solution containing EDTA (1 mM). The proteins were solubilized in a phenol–formic acid–water (2 : 1 : 1) mixture and subjected to polyacrylamide gel electrophoresis. (A) Staining with amido black. (B) Incorporation of ^{14}C-labeled amino acids in vivo (– – – –) and in vitro (····). (C) Inhibitory action of chloramphenicol (CAL) on the incorporation of ^{14}C-labeled amino acids in vitro. ——, – CAL; ····, + CAL (50 μg/ml); – – – –, + CAL + cycloheximide (1 μg/ml). (D) Inhibitory action of cycloheximide (CX) on the incorporation of ^{14}C-labeled amino acids in vitro. ——, – CX; ····, + CX (1 μg/ml).

Site of Regulation

The activities of a number of enzymes vary during the life cycle of *Acetabularia*. There is good evidence that some of the observed changes are due to a regulation of the synthesis of enzyme proteins, and this might occur at the level of either transcription or translation. In *Acetabularia* it has been shown by enucleation experiments that phosphatases (Spencer and Harris, 1964; Triplett *et al.,* 1965) and a number of enzymes which are involved in carbohydrate metabolism (Zetsche, 1966, 1968) undergo typical developmental changes in activity in the absence of the nucleus.

More insight into the mechanisms underlying such regulation of enzyme activity was obtained in investigations of the enzyme thymidine kinase. The activity of this enzyme is closely correlated with nuclear division. This conclusion is based on the observation that, after an increase in the early vegetative phase of the cell, enzyme activity remains constant or decreases slightly. It then steeply increases at the time corresponding to end of the vegetative phase, i.e., when the cap has reached a maximum diameter (Bannwarth and Schweiger, 1974) (Figure 10). This increase is correlated with the time at which nuclear division starts (Figure 11). During the period of nuclear division, the number of nuclei is increased more than a millionfold. Since the degree of ploidy of the primary nucleus seems to be relatively low, substantial amounts of DNA must be synthesized in preparation for these nuclear divisions. Experiments with *Acetabularia* cells which were enucleated about 2 weeks prior to the end of the vegetative phase showed that thymidine kinase activity increases in a manner similar to that observed in nucleate cells (Bannwarth and Schweiger, 1974) (Figure 12).

Puromycin inhibited the increase in thymidine kinase activity (Bannwarth and Schweiger, 1974). Since mixing of homogenates from cells of different developmental stages always resulted in an additive enzyme activity, the explanation that the increase in activity was due to the increased synthesis of an activator could be excluded (Bannwarth and Schweiger, unpublished results). The conclusion to be drawn from these experiments is that the regulation of thymidine kinase is due to regulation of the synthesis of the enzyme protein and, furthermore, that this regulation of enzyme synthesis is not due to an activation of the nuclear genome.

A more detailed study using a number of different inhibitors of protein synthesis revealed that the increase in thymidine kinase activity is not sensitive to cycloheximide but is affected by chloramphenicol and puromycin (Table 2). These results strongly indicate that the thymidine kinase is synthesized on the 70 S ribosomes of the organelle. The question

TABLE 2. *Effect of Inhibitors on the Increase of Thymidine Kinase Activity in Anucleate Cells Acetabularia mediterranea*

Conditions	*Concentration (mg/liter)*	*Increase (%)*	*Inhibition (%)*
Control	—	100	0
Chloramphenicol	10	10	90
Chloramphenicol	100	0	100
Cycloheximide	0.1	94	6
Cycloheximide	2	100	0
Puromycin	30	13	87

of whether the thymidine kinase is coded by the nuclear or the organellar genome remains to be answered.

Circadian Rhythm and Gene Expression

Acetabularia is well suited for investigations of circadian rhythmicity (Schweiger, 1971), since it exhibits a stable endogenous rhythm of photosynthetic O_2 evolution which can be recorded continuously in a single cell or in cell fragments over a period of weeks (Sweeney and Haxo, 1961; Schweiger *et al.*, 1964*a*, Vanden Driessche, 1966; Terborgh and McLeod, 1967; Mergenhagen and Schweiger, 1973, 1975*a*; Karakashian, unpublished results). Paradoxically, the cell is able to retain its rhythm after removal of the nucleus, but an implanted nucleus can shift the phase if it comes from a cell whose phase was different from that of the host cytoplasm (Schweiger *et al.*, 1964*b*) (Figure 13). From these results, it has been concluded that although the nucleus is not necessary for maintaining the circadian oscillation it is able to exert an influence on the phase.

The circadian rhythm of O_2 evolution is a true cytoplasmic function which in an anucleate cell is autonomous. In the presence of a nucleus, this rhythm might be nucleus dependent. This ambivalency resembles the situation found in studies of the MDH isozymes. In that case, enucleation affected neither the isozyme pattern nor the synthesis of the enzyme, but, after implantation of a heterologous nucleus, the typical isozyme pattern of the host cytoplasm disappeared. In that context, I suggested that the underlying regulation takes place at the level of mRNA. A similar interpretation might account for the behavior of the circadian rhythm in *Acetabularia*.

In order to examine this hypothesis, the influence of different inhibi-

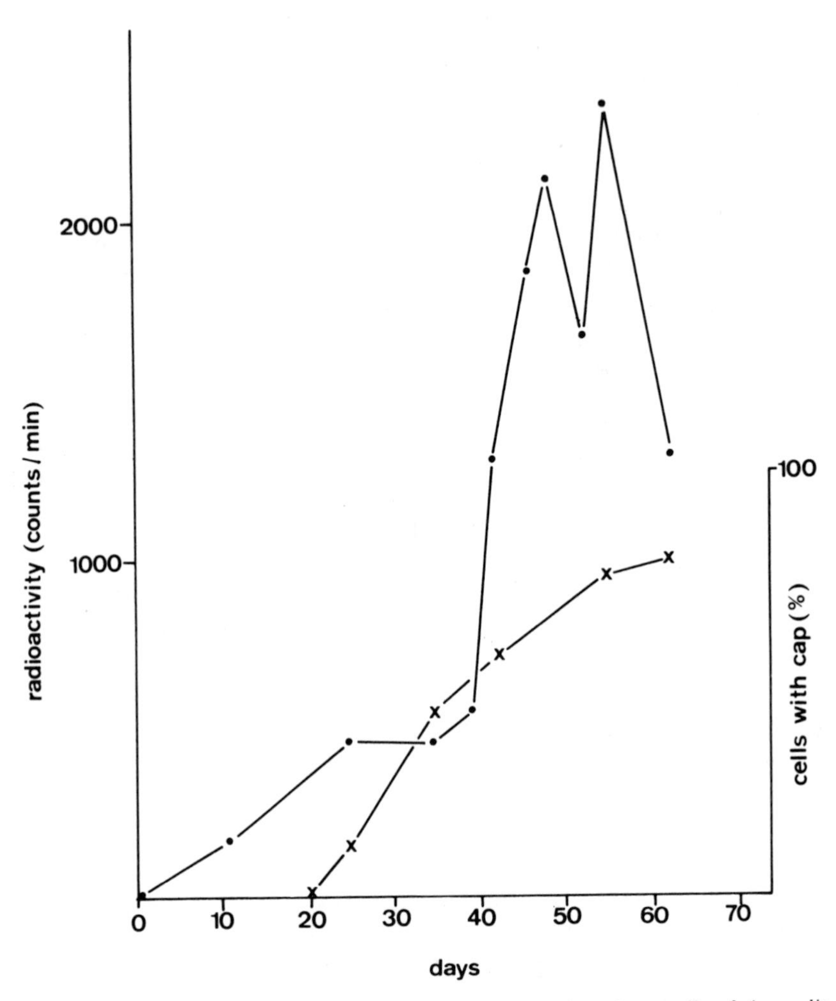

Figure 10. Thymidine kinase activity during development of nucleate cells of A. mediterranea. Acetabularia cells were incubated for 10 min in the presence of [^{3}H]thymidine and homogenized in 0.6 N $HClO_4$. The neutralized supernatant was subjected to anion exchange chromatography and the radioactivity incorporated into deoxythymidylate (TMP) was estimated. ●, *Enzyme activity (measured as radioactivity into TMP);* ×, *percent cells with caps.*

tors on the rhythm was studied (Mergenhagen and Schweiger, 1975*b*). The addition of rifampicin, chloramphenicol, and actinomycin D to anucleate cells did not affect the rhythm. Its expression was, however, blocked by puromycin and cycloheximide (Table 3, Figure 14). These observations suggest that translation on 80 S ribosomes plays an essential part in the expression of the circadian rhythm of O_2 evolution in *Acetabularia*. The role of mRNA in the circadian system in *Acetabularia* remains unclear. It has been observed that prolonged exposure to actinomycin D (10–14 days)

TABLE 3. *Effect of Antibiotics on the Photosynthesis Rhythm*[a]

	Nucleate	Anucleate
Actinomycin D	+	−
Rifampicin	−	−
Chloramphenicol	−	−
Puromycin	+	+
Cycloheximide	+	+

[a] Symbols: +, O_2 evolution blocked; −, no effect.

eventually results in the disappearance of circadian changes in O_2 evolution in nucleate cells (Vanden Driessche, 1966). Conceivably, actinomycin D interferes with the replacement of certain classes of mRNA needed for the normal expression of the photosynthesis rhythm, classes which must be replaced in nucleate cells but which are extremely long lived in enucleated cells.

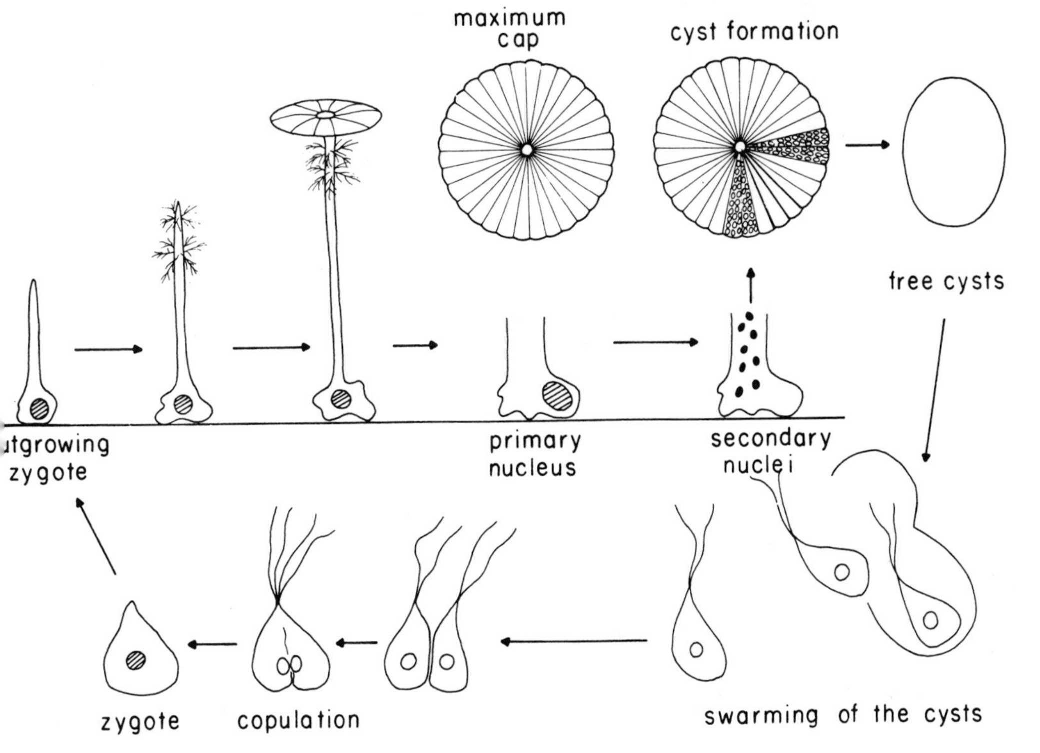

Figure 11. Life cycle of Acetabularia. The gamete is haploid and the zygote diploid. The exact stage at which meiosis occurs is still unknown.

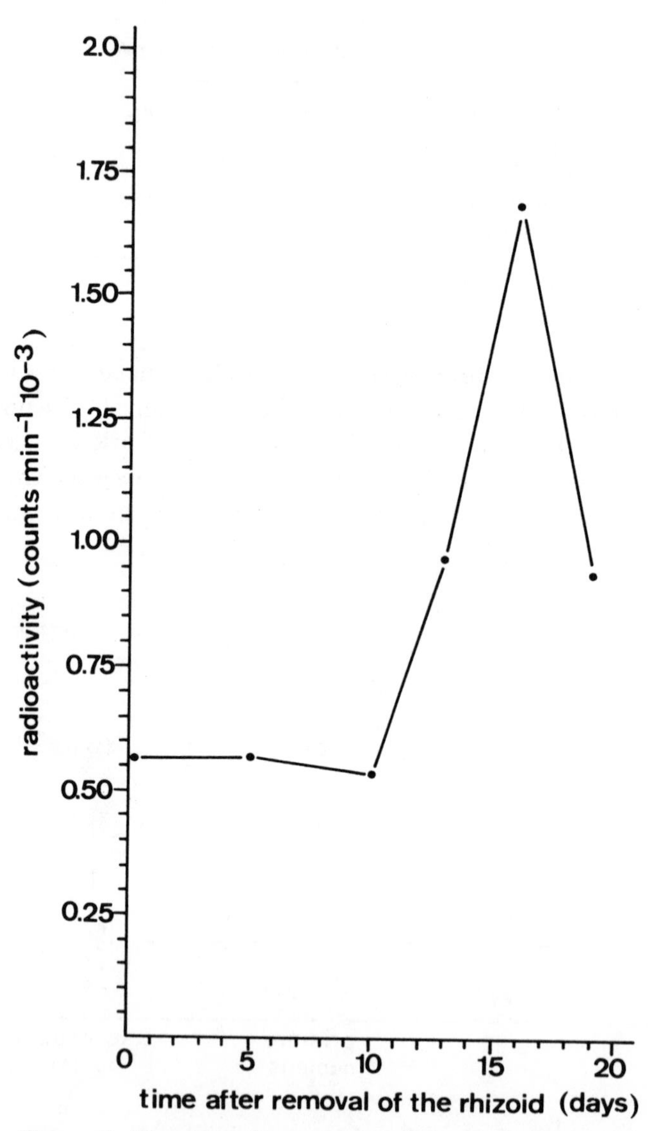

Figure 12. Thymidine kinase activity during development of anucleate cells of A. mediterranea. The conditions were similar to those described in Figure 10 with the exception that cells were homogenized before exposure to [^{3}H]thymidine. Enzyme activity was measured as radioactivity incorporated into TMP.

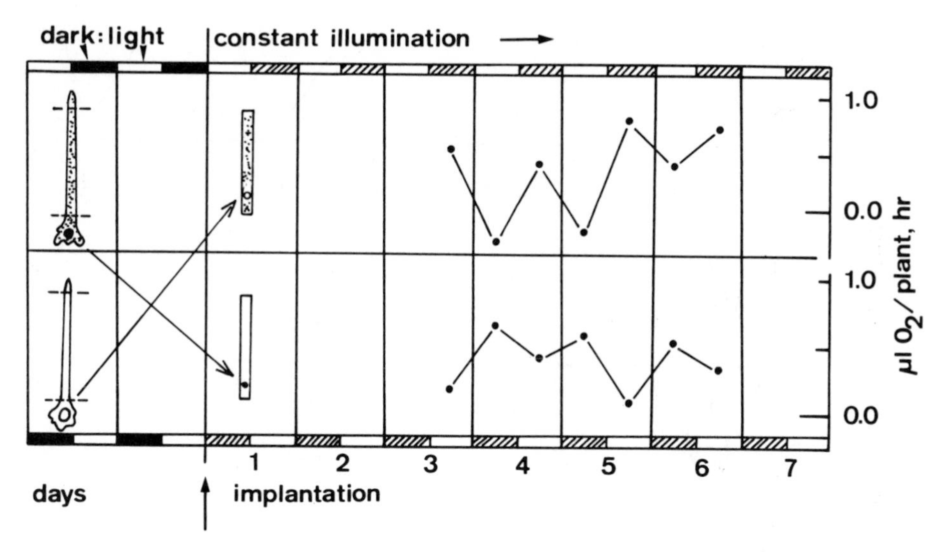

Figure 13. Effect of nuclear exchange on the phase of the endogenous photosynthesis rhythm in Acetabularia. Normally O_2 evolution is highest during the light period. Once the cycle is set, it continues under constant illumination with a period length of roughly 24 hr. Before the operation, the two cells were entrained by an exogenous dark–light cycle so that the phases were different by 180°. Note that 3 days after the operation the phase had changed in both.

Concluding Remarks

This survey of the interrelations between the cell nucleus and cytoplasm with respect to the regulation of gene expression has revealed a rather complex situation. The two sources of genetic information, namely, the nucleus and the organelle genomes, contribute in different degrees to the total set of cellular proteins. Of special interest are the observed contributions of the two types of genomes to the specification of various organellar proteins. In this connection, nuclear exchange experiments have proven to be a useful tool. The complexity of nucleocytoplasmic interactions has also become evident from investigations of the site of translation of distinct proteins. In order to more fully understand the manner in which the various components cooperate in gene expression, it will be important to know whether mRNA which originates in the nucleus can be translated only on cytosol 80 S ribosomes and whether 70 S ribosomes are the only site of translation of organellar information. Cytoplasmic regulatory processes appear to be most directly involved in the expression of the circadian rhythm in *Acetabularia*. Further applications of a cell biological approach may also help to elucidate the molecular mechanism of this complex phenomenon.

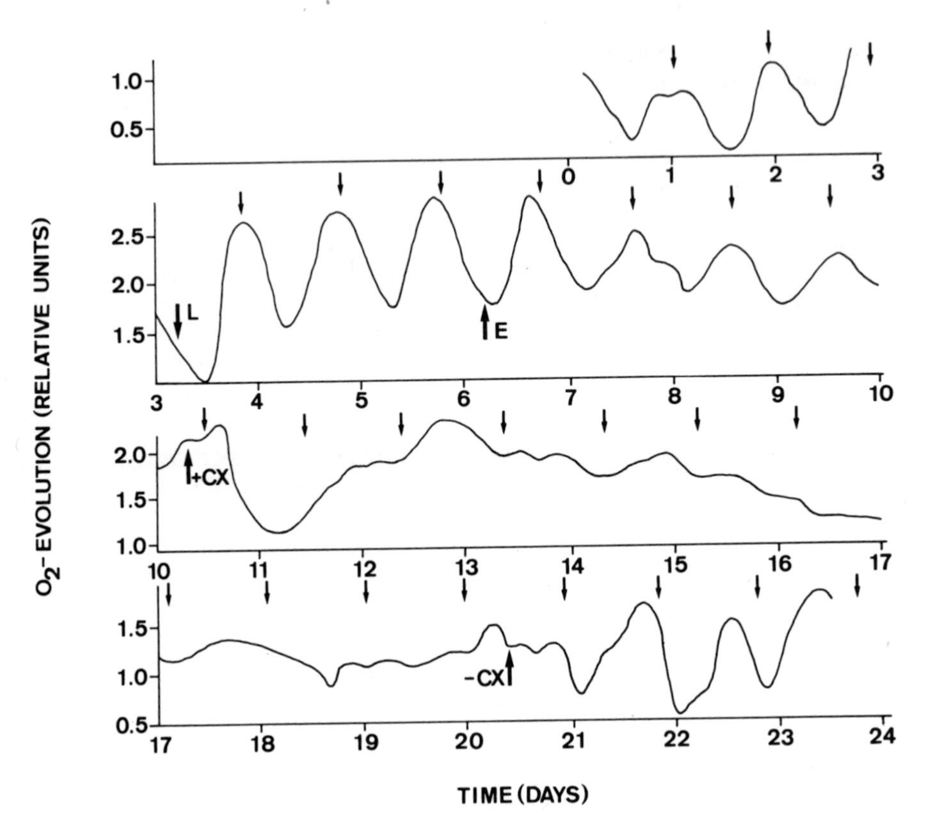

Figure 14. Effect of cycloheximide on the photosynthesis rhythm in anucleate cells of Acetabularia. +CX, cycloheximide added; −CX, cycloheximide removed; small arrows, circadian time (noon); L, ligating; E, enucleation by removing the rhizoid.

Acknowledgments

The author acknowledges the helpful criticism of Dr. Marlene Karakashian during the preparation of this chapter. This contribution is partially based on lectures which were given at the Annual Meeting of the American Society for Cell Biology in Miami Beach in November 1973 and at the Institute of Cytology and Genetics of the Soviet Academy of Sciences in Akademgorodok/Novosibirsk and the Institute of Plant Physiology and Biochemistry of the Soviet Academy of Sciences in Irkutsk in October and November 1974.

Literature Cited

Apel, K. and H. G. Schweiger, 1972 Nuclear dependency of chloroplast proteins in *Acetabularia. Eur. J. Biochem.* **25:**229–238.

Apel, K. and H. G. Schweiger, 1973 Sites of synthesis of chloroplast-membrane proteins. *Eur. J. Biochem.* **38:**373–383.

Bannwarth, H. and H. G. Schweiger, 1974 Regulation of thymidine phosphorylation in nucleate and anucleate cells of *Acetabularia. Proc. R. Soc. London Ser. B* **188:**203–219.

Berger, S., 1967 RNA-synthesis in *Acetabularia.* II. RNA-synthesis in isolated chloroplasts. *Protoplasma* **64:**13–25.

Berger, S., 1975 Ultrastructure of nuclei after implantation. *Protoplasma* **83:**171.

Berger, S. and H. G. Schweiger, 1973 Fine structural changes of the cell nucleus after implantation into a cytoplasm of a different developmental stage. In *Second European Symposium on the Cell Cycle,* Publications of the University of Innsbruck, pp. 49–50.

Berger, S. and H. G. Schweiger, 1975 Cytoplasmic induction of changes in the ultrastructure of the *Acetabularia* nucleus and perinuclear cytoplasm. *J. Cell Sci.* **17:**517–529.

Berger, S., L. Sandakchiev, and H. G. Schweiger, 1974 Fine structural and biochemical markers of *Dasycladaceae. J. Microsc.* **19:**89–104.

Berger, S., W. Herth, W. W. Franke, H. Falk, H. Spring, and H. G. Schweiger, 1975 Morphology of the nucleo-cytoplasmic interactions during the development of *Acetabularia* cells. II. The generative phase. *Protoplasma* **84:**223–256.

Bogorad, L., J. N. Davidson, M. R. Hanson, and L. J. Mets, 1975 Genes for proteins of chloroplast ribosomes and the evolution of eucaryotic genomes. In *Molecular Biology of Nucleocytoplasmic Relationships,* edited by S. Puiseux-Dao, North-Holland, Amsterdam, pp. 111–127.

Franke, W. W., S. Berger, H. Falk, H. Spring, U. Scheer, W. Herth, M. F. Trendelenburg, and H. G. Schweiger, 1974 Morphology of the nucleo-cytoplasmic interactions during the development of *Acetabularia* cells. I. The vegetative phase. *Protoplasma* **82:**249–282.

Hämmerling, J., 1931 Entwicklung und Formbildungsvermögen von *Acetabularia mediterranea.* I. Die normale Entwicklung. *Biol. Zentralbl.* **51:**633–647.

Hämmerling, J., 1932 Entwicklung und Formbildungsvermögen von *Acetabularia mediterranea.* II. Das Formbildungsvermögen kernhaltiger und kernloser Teilstücke. *Biol. Zentralbl.* **52:**42–61.

Hämmerling, J., 1934 Über formbildende Substanzen bei *Acetabularia mediterranea,* ihre räumliche und zeitliche Verteilung. *Arch. Entw. Mech.* **131:**1–81.

Hämmerling, J., 1939 Über die Bedingungen der Kernteilung und der Zystenbildung bei *Acetabularia mediterranea. Biol. Zentralbl.* **59:**158–193.

Hämmerling, J., 1953 Nucleo-cytoplasmic relationships in the development of *Acetabularia. Int. Rev. Cytol.* **2:**475–498.

Hämmerling, J., 1955 Neuere Versuche über Polarität und Differenzierung bei *Acetabularia. Biol. Zentralbl.* **74:**545–554.

Hämmerling, J., 1958 Uber die wechselseitige Abhängigkeit von Zelle und Kern. *Z. Naturforsch.* **13b:**440–448.

Hämmerling, J., 1963 Nucleo-cytoplasmic interactions in *Acetabularia* and other cells. *Ann. Rev. Plant Physiol.* **14:**65–92.

Kloppstech, K. and H. G. Schweiger, 1973*a* Nuclear genome codes for chloroplast ribosomal proteins in *Acetabularia*. I. Isolation and characterization of chloroplast ribosomal particles. *Exp. Cell Res.* **80**:63–68.

Kloppstech, K. and H. G. Schweiger, 1973*b* Nuclear genome codes for chloroplast ribosomal proteins in *Acetabularia*. II. Nuclear transplantation experiments. *Exp. Cell Res.* **80**:69–78.

Kloppstech, K. and H. G. Schweiger, 1973*c* 80-S ribosomes from *Acetabularia*. *Biochim. Biophys. Acta* **324**:365–374.

Kloppstech, K. and H. G. Schweiger, 1974 The site of synthesis of chloroplast ribosomal proteins. *Plant. Sci. Lett.* **2**:101–105.

Mergenhagen, D. and H. G. Schweiger, 1973 A method for recording the oxygen production of a single cell for a prolonged period. *Exp. Cell Res.* **81**:360–364.

Mergenhagen, D. and H. G. Schweiger, 1975*a* Circadian rhythm of oxygen evolution in cell fragments of *Acetabularia mediterranea*. *Exp. Cell Res.* **92**:127–130.

Mergenhagen, D. and H. G. Schweiger, 1975*b* The effect of different inhibitors of transcription and translation on the expression and control of circadian rhythm in individual cells of *Acetabularia*. *Exp. Cell Res.* **94**:321–326.

Mets, L. and L. Bogorad, 1972 Altered chloroplast ribosomal proteins associated with erythromycin-resistant mutants in two genetic systems of *Chlamydomonas reinhardi*. *Proc. Natl. Acad. Sci. USA* **69**:3779–3783.

Ohta, N., M. Inouye, and R. Sager, 1974 A chloroplast ribosomal protein coded by a chloroplast gene in *Chlamydomonas*. *Fed. Proc.* **33**:1584.

Pestka, S., 1971 Inhibitors of ribosome functions. *Ann. Rev. Biochem.* **40**:697–710.

Puiseux-Dao, S., 1970 *Acetabularia and Cell Biology*, Logos Press, London, pp. 1–162.

Reuter, W. and H. G. Schweiger, 1969 Kernkontrollierte Lactatdehydrogenase in *Acetabularia*. *Protoplasma* **68**:357–368.

Sandakhchiev, L., R. Niemann, and H. G. Schweiger, 1973 Kinetics of changes of malic dehydrogenase isozyme pattern in different regions of *Acetabularia* hybrids. *Protoplasma* **76**:403–415.

Schlanger, G. and R. Sager, 1974 Localization of five antibiotic resistances at the subunit level in chloroplast ribosomes of *Chlamydomonas*. *Proc. Natl. Acad. Sci. USA* **71**:1715–1719.

Schulze, K. L., 1939 Cytologische Untersuchungen an *Acetabularia mediterranea* und *Acetabularia wettsteinii*. *Arch. Protistenk.* **92**:179–223.

Schweiger, E., H. G. Wallraff, and H. G. Schweiger, 1964*a* Über tagesperiodische Schwankungen der Sauerstoffbilanz kernhaltiger und kernloser *Acetabularia mediterranea*. *Z. Naturforsch.* **19b**:499–505.

Schweiger, E., H. G. Wallraff, and H. G. Schweiger, 1964*b* Endogenous circadian rhythm in cytoplasm of *Acetabularia:* Influence of the nucleus. *Science* **146**:658–659.

Schweiger, H. G., 1969 Cell biology of *Acetabularia*. *Curr. Top. Microbiol. Immunol.* **50**:1–36.

Schweiger, H. G., 1972 Circadian rhythms: Subcellular and biochemical aspects. In *Proceedings of the International Symposium on Circadian Rhythmicity,* Centre for Agricultural Publishing and Documentation, Wageningen, pp. 157–174.

Schweiger, H. G. and S. Berger, 1964 DNA-dependent RNA synthesis in chloroplasts of *Acetabularia*. *Biochim. Biophys. Acta* **87**:533–535.

Schweiger, H. G., R. W. P. Master, and G. Werz, 1967 Nuclear control of a cytoplasmic enzyme in *Acetabularia*. *Nature* (*London*) **216**:554–557.

Schweiger, H. G., K. Apel, and K. Kloppstech, 1972 Source of genetic information of chloroplast proteins in *Acetabularia*. *Adv. Biosci.* **8**:249–262.

Schweiger, H. G., S. Berger, K. Kloppstech, K. Apel, and M. Schweiger, 1974 Some fine structural and biochemical features of *Acetabularia major* (Chlorophyta, Dasycladaceae) grown in the laboratory. *Phycologia* **13:**11–20.

Siegel, M. R. and H. D. Sisler, 1965 Site of action of cycloheximide in cells of *Saccharomyces pastorianus.* III. Further studies on the mechanism of action and the mechanism of resistance in *Saccharomyces* species. *Biochim. Biophys. Acta* **103:**558–567.

Spencer, T. and H. Harris, 1964 Regulation of enzyme synthesis in an enucleate cell. *Biochem. J.* **91:**282–286.

Sweeney, B. M. and F. T. Haxo, 1961 Persistence of a photosynthetic rhythm in enucleated *Acetabularia. Science* **134:**1361–1363.

Terborgh, J. and G. C. McLeod, 1967 The photosynthetic rhythm of *Acetabularia crenulata.* I. Continuous measurements of oxygen exchange in alternating light-dark regimes and in constant light of different intensities. *Biol. Bull.* **133:**659–669.

Tewari, K. K. and S. G. Wildman, 1970 Information content in the chloroplast DNA. *Symp. Soc. Exp. Biol.* **24:**147–179.

Triplett, E. L., A. Steens-Lievens, and E. Baltus, 1965 Rates of synthesis of acid phosphatases in nucleate and enucleate *Acetabularia* fragments. *Exp. Cell Res.* **38:**366–378.

Vanden Driessche, T., 1966 The role of the nucleus in the circadian rhythms of *Acetabularia mediterranea. Biochim. Biophys. Acta* **126:**456–470.

Werz, G., 1968 Plasmatische Formbildung als Voraussetzung für die Zellwandbildung bei der Morphogenese von *Acetabularia. Protoplasma* **65:**81–96.

Werz, G., 1974 Fine-structural aspects of morphogenesis in *Acetabularia. Int. Rev. Cytol.* **38:**319–367.

Zetsche, K., 1966 Regulation der UDP-Glucose 4-Epimerase Synthese in kernhaltigen und kernlosen *Acetabularien. Biochim. Biophys. Acta* **124:**332–338.

Zetsche, K., 1968 Regulation der UDPG-Pyrophosphorylase-Aktivität in *Acetabularia.* I. Morphogenese und UDPG-Pyrophosphorylase-Synthese in kernhaltigen und kernlosen Zellen. *Z. Naturforsch.* **23b:**369–376.

15

Mitochondrial DNA

Margit M. K. Nass

Introduction

Research on mitochondrial DNA has evolved progressively and rapidly since this DNA species was first clearly identified as an integral and probably universal component of mitochondria (Nass and S. Nass, 1963; S. Nass and Nass, 1963; Nass *et al.,* 1965). It is now known that mitochondria contain specific DNA molecules capable of replication and that mitochondria have their own equipment to transcribe this DNA into RNA and translate RNA into proteins in association with mitochondria-specific ribosomes and transfer RNA. This mitochondrial genetic apparatus has many similarities with microbial genetic systems. The biogenesis and function of the mitochondria depend on the joint operation of the nuclear-cytoplasmic and the mitochondrial genetic system.

A copious number of review articles have dealt with the earlier data on cytochemical, genetic, and physicochemical properties of mtDNA: Wilkie (1964), Granick and Gibor (1967), Borst *et al.* (1967*a*), Nass (1967), Roodyn and Wilkie (1968), Slater *et al.* (1968), Borst and Kroon (1969), Nass (1969*a*), Swift and Wolstenholme (1969), S. Nass (1969), Ashwell and Work (1970), Wolstenholme *et al.* (1970), Rabinowitz and Swift (1970), Paoletti and Riou (1970), Schatz (1970), Wunderlich (1971), and Nass (1971*a,b*). Several articles cover more recent developments:

Margit M. K. Nass—Department of Therapeutic Research, University of Pennsylvania, School of Medicine, Philadelphia, Pennsylvania.

Borst (1972), Linnane *et al.* (1972), Borst and Flavell (1972), Sager (1972), Paoletti and Riou (1973), Kroon and Saccone (1974), Kasamatsu and Vinograd (1974), and Nass (1974*a*).

In this chapter, special attention will be focused on the most recent advances. Some of the earlier work will be cited briefly, and the reader will be directed to the most pertinent papers and reviews. It is hoped that this article will also serve to stimulate new ideas for future experimentation.

Structure and Properties of Mitochondrial DNA

General Properties

Mitochondrial DNA (mtDNA) molecules are found in all aerobically respiring eukaryotic organisms. These molecules are double-stranded and range in size between 1 and 30 μm in different cell types. mtDNA is usually circular and is covalently closed except in some stages of replication. There are also exceptions where only linear molecules could be isolated. mtDNA in each cell type differs from the respective nuclear DNA complement in base composition, in coding sequences for RNA and protein species, and in its DNA replication and transcription mechanism. In addition, the evidence indicates that mtDNA can mutate, can recombine, and is inherited by a maternal mechanism. mtDNA may therefore be considered to serve a genuine genetic function in the cell.

Organization of mtDNA within the Organelle

Electron microscopic studies of mitochondria from a wide range of animal and plant cells (Nass and S. Nass, 1963; Nass *et al.*, 1965; Kislev *et al.*, 1965, Schuster, 1965; Bisalputra and Bisalputra, 1967; Yokomura, 1968) have shown that mitochondrial DNA is localized in one to several electron-lucid regions or "nucleoids" within the matrix of the mitochondrion. In serial sections these regions can be seen to appear and disappear in various locations of the mitochondria (Nass, 1969*b*). These DNA regions are most conspicuous (and usually have the highest DNA content) in embryonic cells, plant cells, unicellular organisms, tissue culture cells, and many tumor cells, but are difficult to see in dense, cristae-rich mitochondria of highly differentiated tissues such as liver, kidney, and muscle (*cf.* Nass *et al.*, 1965). In the latter cell types, mitochondrial DNA can usually be visualized if the tissue is explanted to form primary cell cultures or if certain mitochondrial functions are altered by treatments with drugs. For example, the administration of acriflavin to rats was found to cause the appearance of electron-lucid regions containing DNA fibers in myocardial mitochondria (Lagueus *et al.,* 1972). The visualization of this DNA may be the result of a condensation of preexistent DNA molecules unable to replicate in the presence of acriflavin.

The morphology of DNA in ultrathin sections of mitochondria of plant and animal cells is in the form of condensed clumps with fine fibrils branching out if osmium tetroxide is used as a fixative; the fibers form a fine network of 20-Å-thick units if uranyl acetate is used in post-fixation (Nass and S. Nass, 1963). Exceptional spiral configurations of DNA-associated inclusions have been reported in mitochondria of several unicellular organisms (Tripoldi *et al.,* 1972; Schuster, 1965). The DNA-containing region of the mitochondrion-associated kinetoplast of trypanosomes has a very complex arrangement and large accumulation of organelle DNA (Simpson, 1972; Burton and Dusanic, 1968; Renger and Wolstenholme, 1972). The mtDNA fibrils in all cases tested can be removed by treatment with DNAse. All evidence thus far suggests that no histone or basic protein is associated with mtDNA.

A high proportion of DNA molecules is associated with the inner mitochondrial membranes in several regions of a given mitochondrion (Nass *et al.,* 1965; Nass, 1969*b*). Brief treatment with proteolytic enzymes dissociates the molecules from the membrane (Nass, 1969*b*). In agreement, Van Tuyle and Kalf (1972) described properties of an isolated membrane–DNA–RNA complex. A three-dimensional model to visualize the possible intramitochondrial organization of mtDNA has been presented (Nass, 1969*a*, 1974*a*).

DNA Content and Genomes per Mitochondrion and per Cell

The content of mitochondrial DNA per organelle and per cell is a function of the size of the mitochondrial genome or nonrepeating DNA unit and the size and number of mitochondria per cell. Both of these parameters in turn are a function of the nuclear genetic type of a given cell as well as of various environmental effects on the cell.

Experimentally, mitochondrial DNA content has been quantitated by three general approaches: (1) by relating mtDNA content to nuclear DNA content, (2) by relating mtDNA content to mitochondrial numbers per cell either by electron microscopic techniques or by chemical analyses of isolated mitochondria relative to counts of isolated mitochondria, and (3) by relating mtDNA content to the protein mass of isolated mitochondria (*cf.* Nass, 1969*a*). Each technique has its obvious limitations, and preferably more than one approach should be used to achieve reliable results. The yields of mtDNA in cultured L cells comprise about 0.1% of total cellular DNA (Nass, 1969*b*). The most common range in animal cells is 0.1–1%. The quantity of mtDNA relative to nuclear DNA is considerably greater in yeast, where 12–13% of total DNA was reported for diploid and haploid populations (Grimes *et al.*, 1974). The quantity of mtDNA may actually exceed that of nuclear DNA in amphibian eggs (Dawid, 1966).

The number of mitochondria per cell may vary considerably in different organisms, ranging from one mitochondrion per cell in *Micromonas,* a small pigmented flagellate (Manton, 1959), to 250 $\pm$ 70 mitochondria per L cell (mouse fibroblast) (Nass, 1969*a,b*), to 2000 mitochondria per rat liver cell (Allard *et al.*, 1952). The mean number of mitochondria in yeast was estimated as 22 in diploids and ten in haploids (Grimes *et al.*, 1974). There are additional factors that contribute to varying numbers of mitochondria per cell in a given cell type and that therefore complicate estimates of mitochondrial numbers: mitochondria may be highly pleomorphic, are often branched, and are known to fuse and dissociate during the cell cycle (e.g., in *Euglena,* Calvayrac *et al.*, 1972) and under different conditions of growth (e.g., in yeast, Stevens, 1974). Hoffman and Avers (1973) reported on the basis of serial sections of entire yeast cells that all separate mitochondrial profiles seen were cross-sections through a single branching mitochondrial structure. In mammalian somatic cells there is thus far no evidence to indicate that the mitochondria form a single connected reticulum.

Since the total number of mitochondria per cell may vary drastically, it is not surprising that the total content of mtDNA per cell also varies in different cell types. In L cells, an average quantity of 9×10^{-17} g mtDNA or an average molecular weight of 53×10^{6} daltons of mtDNA per organelle has been calculated (Nass, 1969*b*). This corresponds to an average of five DNA molecules of molecular weight 10×10^{6} daltons per mitochondrion, or 1250 mtDNA genomes per L cell. A similar mtDNA content for L cells and a 4 times higher DNA content for HeLa mitochondria were also reported (Bogenhagen and Clayton, 1974). Electron microscopic analyses showed that the total DNA content per L cell mitochondrion generally varied from two to eight molecules per organelle and that these may be mixtures of monomeric and dimeric or oligomeric molecules, located within one or more nucleoids (Nass, 1969*b*). Only one to two DNA molecules per mitochondrion were reported in sea urchin (Pikó *et al.*, 1967). In yeast, values of about 83 mtDNA genomes (50×10^{6} daltons each) per diploid cell and 47 per haploid cell were found (Grimes *et al.*, 1974).

There appears to be a direct proportionality between DNA content and size of the mitochondrion. Long filamentous and branching types of organelles tend to have a larger number of DNA-containing regions than short forms. Giant mitochondria that develop during the cycle in human uterus mucosa showed an increase in the number of DNA filaments (Merker *et al.,* 1968). Bahr (1971) has suggested from quantitative studies of rat liver mitochondria that there is a constant amount of DNA per unit mitochondrial dry mass in a given cell type. This relationship seems to be valid in many other cases.

mtDNA content has been expressed most frequently per milligram of mitochondrial protein. In general, mtDNA of embryonic cells and cultured mammalian cells contains at least twice as much DNA per milligram of mitochondrial protein ($\geqq 1$ μg DNA/mg protein) than mtDNA of more differentiated tissues such as rat liver. A range of 0.9–1.8 μg/mg has been reported for mtDNA of embryonic and placental tissues of various rodents, and even higher values were reported for some tumor cells (*cf.* Table 1 in Nass, 1969*a*). The obvious difficulties in this type of analysis are contamination of mitochondrial fractions with nuclear DNA and variable protein content of mitochondria in different cell types. The possible pitfalls in expressing DNA values relative to mitochondrial protein, as opposed to per organelle and/or per cell, have been discussed (Nass, 1969*a*).

Sizes, Circular Forms, and Linear Forms of mtDNA

Mitochondrial DNAs of most animal cells occur as small circular duplex molecules with a contour length of 4.4–5.7 μm, corresponding to a molecular weight of $9\text{–}11 \times 10^6$ daltons (see Table 1).

In contrast to the relatively narrow size range of mtDNAs from a wide variety of animal species, a spectrum of sizes and structures has been reported for mtDNAs of some lower organisms and plants. The kinetoplast of trypanosomes contains large networks of circular molecules ranging in various genera from 0.3 to 0.8 μm and containing longer linear filaments as well (Simpson, 1972; Renger and Wolstenholme 1972; Laurent and Steinert, 1970). Only linear DNA was found in mitochondria of *Tetrahymena* (15–17.5 μm) and *Paramecium* (Suyama and Miura, 1968; Flavell and Follett, 1970; Arnberg *et al.,* 1972), and in *Physarum polycephalum* (Sonenshein and Holt, 1968; Evans and Suskind, 1971), while 12.8 μm and a few 16.2 μm circles were seen in mtDNA of *Acanthamoeba castellanii* (Bohnert, 1973; Hettiarchchy and Jones, 1974). The intact form of mtDNA in *Saccharomyces cerevisiae* and *S. carlsbergensis* appears to be a 25 μm circle (Hollenberg *et al.,* 1970) and in *Klebsiella lactis,* an

TABLE 1. Sizes of Circular mtDNAs in Animal Cells

Organism	*Contour length (μm)*	*Reference*
Mammals		
Mouse L cells	4.7	Nass, 1966
Human HeLa cells	4.8	Radloff *et al.*, 1967
Human liver	5.1	Nass, 1969*c*
Birds		
Chicken liver	5.3	Borst *et al.*, 1967*b*
Chicken embryo fibroblasts	5.0	Nass, 1973*b*
Reptiles		
Viper cells	5.4–5.6	Nass, unpublished
Turtle	5.3	Wolstenholme *et al.*, 1970
Amphibians		
Anurans	5.4–5.6	Wolstenholme and Dawid, 1968
Urodeles	4.7–4.8	
Fishes		
Carp	5.4	Van Bruggen *et al.*, 1968
Echinoderms		
Sea urchin	4.4	Pikó *et al.*, 1967
Nematodes		
Ascaris lumbricoides	4.8	Carter *et al.*, 1972
Insects		
Drosophila eggs	5.3	Polan *et al.*, 1973
House fly	5.2	Van Bruggen *et al.*, 1968
Rhynchiosciara hollaenderi spermatocytes	9 μm (dimers?)	Handel *et al.*, 1973

11.4 μm circle (Sanders *et al.*, 1974). Smaller circles (Avers *et al.*, 1968) and usually a high percentage of assorted linear molecules have also been reported. In *Euglena gracilis,* 1 μm linear and 1 μm circular DNA (Nass *et al.*, 1974) and 1–19 μm linear molecules (Manning *et al.*, 1971) were reported. In various strains of *Neurospora,* 19–20 μm circles (Clayton and Brambl, 1972), 0.5–19 μm circles (Agsteribbe *et al.*, 1972), and 26 μm linear molecules (Schäfer *et al.*, 1971) were found. In mtDNA of higher plants, heterogeneous linear molecules (Wolstenholme and Gross, 1968) and 30 μm circular forms (Kolodner and Tewari, 1972) were detected.

The isolation of mtDNA in intact form is relatively uncomplicated for most animal cells, but has met with various degrees of difficulties in plants, protists (especially in *Euglena*), and fungi (yeast and *Neurospora*). Breakdown of mtDNA in these organisms may be facilitated by increased endogenous nuclease activity but also by the possible mode of association or attachment of the DNA molecules to mitochondrial membranes (*cf.* Nass *et al.*, 1974).

Many of the small differences in sizes reported for the circular mitochondrial DNAs that fall into the "5 μm" category are not merely due to experimental variation caused by different methods and investigators, but seem to be true size differences, as shown by mixing experiments with isolated mtDNA from different species (Nass, 1969*c*; Wolstenholme and Dawid, 1968). The significance of such minor differences in terms of informational content is thus far not known.

Physical Properties and Superhelix Density of Closed Circular DNA

The structure and physicochemical properties of mtDNA have been studied extensively and have been found, in the case of the circular 5 μm-type DNA, to be similar to the properties of circular DNAs of polyoma and SV40 viruses. The covalently closed twisted or supercoiled form of mtDNA (component I) is the predominant form isolated in the lower band of cesium chloride–ethidium bromide gradients (Radloff *et al.*, 1967). The open or loosely twisted forms have one or more strand scissions and are found in intermediate and upper bands of CsCl-EB gradients. Assorted portions of these molecules may have D-loop and expanded D-loop strands attached (see later section). The closed circles can be distinguished from open or nicked circles on the basis of their higher electrophoretic mobility in agarose gels (Aaij and Borst, 1972), their higher sedimentation coefficients at neutral and even more so at alkaline pH, their higher buoyant density in alkaline cesium chloride, their elevated melting temperature (T_m), their restricted uptake of the intercalating dyes ethidium bromide and propidium diiodide at high concentrations of dye, and their possession of superhelical turns (*cf.* Borst and Kroon, 1969; Nass, 1969*a*; Borst 1972; see also p. 492).

The superhelix density has been defined as the number of superhelical turns per ten base pairs; it is considered to be a measure of the winding deficiency of the closed circular molecule relative to the winding that is required in the relaxed molecule (*cf.* Schmir *et al.*, 1974). An interrelation has been established between the duplex winding number, β, which can be influenced by temperature, ionic strength, and intercalating dyes, and the topological winding number, α, which can be varied experimentally. Both parameters may modify the superhelix density. By the relatively crude but remarkably reproducible method of electron microscopic examination of supercoils or twists in mtDNA preparations, the number of tertiary turns per DNA molecule was found to be 33–34 for mouse and human mtDNA

(Nass, 1969*c*) and also for chick liver mtDNA (Ruttenberg *et al.*, 1968), corresponding to about 3.6 turns per 1×10^6 molecular weight. The degree of supercoiling of mtDNA was found to increase by treatment of cells with the intercalating drug, ethidium bromide (Smith *et al.*, 1971).

Various mtDNAs and other closed circular DNAs may differ in their superhelix densities and can be distinguished in cesium chloride–propidium diiodide gradients (Hudson *et al.*, 1969). A method for determining superhelix density by viscometric titration in the presence of ethidium bromide (Révet *et al.*, 1971) has also been applied to mouse fibroblast (LA9) mitochondrial DNA by Schmir *et al.* (1974), who also presented evidence that the supercoils are negative in naturally closed circular DNAs.

Information Content (kinetic complexity)

A measure of the informational content of many DNAs has been obtained by studying the renaturation rates of denatured DNA. The resultant kinetic complexity is taken to represent the genetic complexity or nonrepeated base sequence or genome size of the respective DNA. In the limited number of cases of mitochondrial DNAs analyzed, renaturation experiments have shown (with some exceptions noted below) that the "genetic content" of mtDNAs is roughly equivalent to molecular size as obtained by electron microscopy. Molecular weights from renaturation experiments have been reported as 9.9 and 11×10^6 daltons for mtDNA of rat and guinea pig, respectively (Borst and Flavell, 1972), 30×10^6 for *Tetrahymena* (Flavell and Jones, 1970), 35×10^6 for *Paramecium* (Flavell and Jones, 1971), $49–50 \times 10^6$ for *Saccharomyces* (Hollenberg *et al.*, 1970; Christiansen *et al.*, 1974), 74×10^6 for the pea (Kolodner and Tewari, 1972), about 100×10^6 for potato tubers (Vedel and Quetier, 1974), and greater than 100×10^6 for lettuce (Wells and Birnstiel, 1969). Abnormalities and special problems in the kinetic analyses have been discussed for yeast (Christiansen *et al.*, 1971, 1974), *Neurospora* (Wood and Luck, 1969), and *Euglena* (Nass *et al.*, 1974). In *Euglena,* the molecular weight as determined by sedimentation analysis of isolated mtDNA is about 3×10^6 (Ray and Hanawalt, 1965), and the molecular weight as determined by electron microscopy of mtDNA isolated from *Euglena* in the presence of various nuclease inhibitors is about 2.1×10^6 (Nass *et al.*, 1974). Both linear and circular forms of DNA of molecular weight $1.9–2.1 \times 10^6$ were detected in mtDNA fractions (Nass *et al.*, 1974). Kinetic experiments have suggested a complexity that is considerably greater than 3×10^6 (Nass, unpublished), and a value of about 40×10^6 was reported by

Talen *et al.* (1974). Further experiments are required to determine whether the small *Euglena* mtDNA molecules isolated by various methods consist perhaps of classes of molecules of different complexities. It was shown (Nass *et al.*, 1974) from electron microscopic data that *Euglena* mtDNA molecules *in situ* are sharply folded and associated at one or more points with a membrane site where enzymes active in nucleic acid metabolism may be located. Such DNA molecules might well be prone to breakage by mechanical or enzymatic removal of the DNA during isolation.

The kinetic complexities of mtDNA from cytoplasmic petite mutants of yeast were found to be reduced from 1/3 down to 1/500 as compared with the complexity of wild-type yeast (Michel *et al.*, 1974). The mtDNAs of nine petite strains of yeast analyzed by electron microscopy and by kinetic studies (Locker *et al.*, 1974*a,b*) were found to contain circular molecules at frequencies of 2–20% of total mtDNA. The monomer size was found to correlate well with the kinetic complexity.

Dimeric and Oligomeric Forms of mtDNA in Normal and Tumor Cells

The circular mtDNA of animal cells has been shown to contain a small percentage of double- and multiple-length DNA forms whose origin and biological significance are not yet clearly understood. The catenated dimer (trimer and higher oligomers), consisting of two (or more) interlocked monomers, has been identified as a minor component in all cell types examined (e.g., Nass, 1968, 1969*a,b,c,d,* 1974*a*; Clayton *et al.*, 1968). The other form of dimer, a unicircular molecule, is less common. It has been identified in human leukemic leukocytes, where its level correlated with the severity of chronic granulocytic leukemia (Clayton and Vinograd, 1967, 1969); in mouse L cells, where its level could be regulated by growth conditions and the use of metabolic inhibitors (Nass, 1969*d*, 1970); in several human solid tumors (Smith and Vinograd, 1973); and in thyroid cells of presumably nonmalignant origin (Paoletti *et al.*, 1972).

The sedimentation velocity properties of covalently closed and nicked (open) circular forms of oligomeric or complex mtDNA have been examined, with slightly varying results. For HeLa cell mitochondria, values of 63 S were reported for the triply closed trimer, which is converted to a 45 S triply open trimer by DNAse treatment. The catenated dimer sedimented at 51 S in its doubly closed superhelical form, at 42 S in its singly open form, and at 36 S in its doubly open form (Brown and Vinograd, 1971). Values of 52 S were found for the covalently closed unicircular dimer, 37 S for the covalently closed monomer, and 26 S for the nicked monomer (Hudson and Vinograd, 1969). In preparative sucrose gradients of mouse L cell mitochondrial DNA, values of 65 S, 54 S, and 39 S were obtained for closed trimers, dimers, and monomers, respectively (Nass 1969*d*, 1970).

The catenated forms probably consist of interlinked identical monomer forms. Pulse-labeling experiments in L cells suggested that monomers are precursors of dimers of mtDNA (Nass, 1969*d*). Reannealing experiments and electron microscopy of monomeric and unicircular dimeric mtDNA have revealed that these dimers consist of two monomer genomes which appear to be connected in a head-to-tail rather than a head-to-head structure (Clayton *et al.*, 1970).

The distribution of monomeric and complex forms of mtDNA in a given DNA preparation is usually determined by scoring individual molecules in the electron microscope. DNA from lower bands (covalently closed DNA) or lower bands plus intermediate bands of cesium chloride–ethidium bromide gradients has been most commonly examined. The DNA is spread by various modifications of the protein-monolayer technique. The occurrence of complex forms of mtDNA relative to total isolated mtDNA is quantitated either by recording numbers in percent or by converting these values to weight percent. Some values in weight percent have also been obtained by measuring optical density of dimer and monomer fractions obtained by band sedimentation velocity analysis in cesium chloride (e.g., Clayton and Vinograd, 1967) and sucrose (e.g., Nass, 1969*d*, 1970, 1973*b*). The advantage of electron microscopy is obviously that most molecules can be identified unequivocally and that very low frequencies can easily be determined. On the other hand, some skill, practice, and lack of bias are required of the investigator who does the scoring.

The relative frequencies of circular catenated higher oligomers in mtDNA preparations may range between 0.1 and 4% in various organs of adult animals (Clayton *et al.*, 1968; Wolstenholme *et al.*, 1973). The frequencies of total catenated dimers and oligomers may range from 4.3% in adult mouse liver (Nass, 1970; Wolstenholme *et al.*, 1973), to 6% in chick embryo fibroblasts (Nass, 1973*b*), to 10% in cultured baby hamster kidney cells (Nass, 1970), to 15% in Rous virus infected chick embryo fibroblasts (Nass, 1973*b*), and to 31% in cycloheximide-treated chick embryo fibroblasts (Nass, 1973*b*). The relative frequencies of unicircular plus catenated dimers may range from 10% in mouse L cells (Nass, 1970) to about 30–40% in leukocytes from certain cases of chronic leukemia (Clayton *et al.*, 1968, 1970) to 60–80% in L cells grown into stationary phase or treated with cycloheximide (Nass, 1969*d*, 1970, 1974*a*). An ideally controlled cell system of chick embryo fibroblasts made malignant by infection with temperature-sensitive mutants of oncogenic viruses was recently examined (Nass, 1973*b*, 1974*a,b*). The experiments showed a clear-cut correlation of an elevated level of catenated dimeric and oligomeric forms of mtDNA with the phenotypic manifestation of transformation. Temperature shifts from 41° to 36°C and *vice versa* resulted in reversals of the respective levels of complex DNA, rates of deoxyglucose transport, and morphology.

Detailed tabulations and discussions of the relative frequencies of dimeric and oligomeric mtDNA of normal and tumor cells under various conditions can be found in the following papers: Nass (1969*a*, 1970, 1974*a*), Wolstenholme *et al.* (1973), Smith and Vinograd (1973), and Paoletti and Riou (1970, 1973). In general, these studies have shown that (1) the unicircular dimer is not detectable in all types of malignant cells; (2) unicircular dimers, although common in malignant cells, are not unique to them, since very occasional unicircular dimers have been observed in seemingly nonmalignant cells; and (3) catenated dimers and unicircular dimers may exist in high frequencies in many malignant cells. However, the increase is quite variable in different types of tumors, and, moreover, environmental factors (e.g., culture conditions) may cause increases in dimer and oligomer frequencies regardless of whether the cell line is derived from normal or malignant cells (Nass, 1974*a*).

No simple correlations or answers have yet been found concerning the biological significance and mechanism of oligomer accumulation. Obvious possibilities are that these molecules arise during DNA replication, either as normal intermediates which can accumulate under certain imbalanced conditions or as abnormal intermediates which occur due to errors in the DNA replication mechanism. Perhaps complex forms may also conceivably arise as a by-product of recombination (Hudson and Vinograd, 1967; Flory and Vinograd, 1973), or a combination of replication and recombination events may occur. Changes in the replication mechanism leading to oligomer accumulation may occur at the level of the mitochondrial membrane (Nass, 1973*b*, 1974*a*). The mitochondrial and other cellular membranes show distinct structural differences in normal and malignant cells (Soslau *et al.*, 1974). Since the evidence suggests the existence of a mitochondrial DNA–inner membrane complex (as discussed earlier), a conformational and/or functional alteration of the membrane may affect the attachment site of the DNA. Consequently, the activity or content of enzymes that are involved in DNA replication, e.g., a "nickase" that cleaves newly replicated DNA structures into monomeric molecules, may be reduced. In addition, the synthesis of these enzymes or protein factors, which probably occurs outside mitochondria on cytoplasmic ribosomes, may become a limiting factor. For example, amino acid starvation, and cycloheximide, which inhibits primarily cytoplasmic protein synthesis, may lead to the accumulation of oligomeric DNA. Chloramphenicol, which inhibits mitochondrial protein synthesis, has little effect (Nass, 1970, 1973*b*, 1974*a*).

Base Composition, "Spacers," and Nearest-Neighbor Frequencies

The average base composition is commonly expressed as mole percentage of guanine plus cytosine (% G+C) and is usually derived by formula (Nass, 1969*c*; Borst and Kroon, 1969; Simpson, 1972) from buoyant density analyses in cesium chloride or from melting temperature analyses in saline–citrate solutions. In addition, direct base analysis by chromatographic separations of individual bases has been employed in a few cases (e.g., see Nass, 1969*c*, where all three methods have been compared for mouse L cells). The technical difficulty in direct base analysis is obviously that large quantities and/or very highly labeled mtDNA must be available, whereas less than 1 μg quantities of unlabeled DNA can be analyzed for buoyant densities in the analytical ultracentrifuge. The latter method is therefore most prevalent in the literature. Various detailed tables and lists of buoyant densities of mtDNA and nuclear DNA of different uni- and multicellular organisms and plants can be found in the following review papers: Borst and Kroon (1969), S. Nass (1969), Ashwell and Work (1970), Wolstenholme *et al.* (1970) and Swift and Wolstenholme (1969). In addition, the following papers have information on DNAs of insects: Polan *et al.* (1973), Travaglini and Schultz (1972) Travaglini *et al.* (1972), and Tanguay and Chaudhary (1972); DNAs of crustaceans: Skinner and Kerr (1971); and DNAs of trypanosomes: Simpson (1972).

Average base compositions of nuclear DNAs of closely related organisms tend to be very similar. However, although mitochondrial DNAs of many related organisms also tend to have similar average base compositions, there is no clear relation of these values to the phylogenetic position of the organism. Moreover, the buoyant densities of mtDNA may be identical to, similar to, higher than, or lower than those of nuclear DNA of the same cell type, again with no clear relation of these differences to evolutionary classification (see Table 2). During evolution, the two DNA systems thus have not undergone isogenous, parallel, or predictable changes but rather seem to have evolved independently.

Additional satellite peaks of undetermined origin have been reported in yeast, *Neurospora, Euglena,* crustaceans, trypanosomes, and insects.

Some of these satellites consist of dAT polymers with the very low buoyant density of 1.677 g/cm^3. Such satellites occur in several crab species (Skinner and Kerr, 1971) and in the fruit fly (Fansler *et al.*, 1970). The mitochondrial DNAs from several "low-density" petite mutants of yeast also have a very low G+C content (about 4–6%, as compared with about 18% for wild-type yeast). Several unusual properties of this DNA will be discussed in a later section.

It is clear that base sequences rather than the average base composition must be investigated to obtain more insight into the structure and evolution of mtDNA and nuclear DNA, as done with yeast DNA. The structure of yeast mtDNA has been particularly interesting, and originally puzzling, since anomalous physical and chromatographic properties were observed (e.g., Bernardi *et al.*, 1972). Studies to date have shown a compositional heterogeneity of yeast mtDNA (Bernardi *et al.*, 1972; Piperno

TABLE 2. Distribution of Buoyant Densities of mtDNA and Nuclear DNA in Eukaryotes[a]

Classification	Buoyant density (ρ) (g/cm^3)	
	mtDNA	*nucDNA*
ρ of mtDNA similar to or 0.002 unit above or below ρ of nucDNA		
Most mammals (except man)	1.698–1.704	1.698–1.704
Turtle	1.701	1.702
Anurans	1.702	1.700–1.702
ρ of mtDNA > ρ of nucDNA		
Man (Chang, HeLa cells)	1.706–1.707	1.700
Birds	1.707–1.711	1.698–1.701
Fish	1.703–1.708	1.697–1.699
Sea urchin	1.704	1.694
Paramecium	1.702	1.689
Some higher plants	1.705–1.706	1.692–1.698
ρ of mtDNA < ρ of nucDNA		
Salamander	1.695	1.704–1.707
Some crustaceans	1.688	1.698–1.701
Drosophila	1.680–1.685	1.699–1.702
Wild-type yeasts	1.683–1.691	1.698–1.709
Neurospora	1.692–1.702	1.712–1.713
Tetrahymena	1.683–1.686	1.683–1.692
Euglena	1.690–1.692	1.707
Physarum	1.686	1.700
Hemoflagellates	1.688–1.702 (kinetoplast)	1.703–1.721

[a] Data are compiled from various review articles and original references as cited in text.

et al., 1972; Carnevali and Leoni, 1972; Ehrlich *et al.*, 1972) when it is degraded by spleen acid DNAse, which prefers G+C-rich sequences. The conclusions were drawn that wild-type and petite mutants of yeast contain alternating and nonalternating d(AT) sequences. There seem to be long G+C-rich and A+T-rich stretches (molecular weight 10^5–10^6 daltons) that are intermingled, of which the A+T-rich sequences may be responsible for the atypical physical properties. Similar conclusions on the organization of A+T-rich and G+C-rich sequences were drawn from an analysis of pyrimidine tracts in depurinated mtDNA (Ehrlich *et al.*, 1972). In further studies, Prunell and Bernardi (1974) used a micrococcal nuclease to degrade mtDNA under conditions where A+T-rich stretches, presumed to be "spacers" (which are not transcribed), are degraded, but G+C-rich sequences (presumed to be genes) are little affected. It was concluded that the mtDNA genome of yeast consists of about equal amounts of genes and spacers. The spacers are very homogeneous, with a G+C content below 5%; the genes appear very heterogeneous in base composition, with a G+C content of 25–50%. It will be of interest to see to what extent low-melting regions that may correspond to spacers exist in other organisms, especially those that have the smaller 5 μm genome.

The nearest-neighbor frequencies for dinucleotides have been analyzed in a few cases to determine a possible relation of the patterns between diverse mtDNAs and their relation to the patterns of nuclear DNAs or microbial DNAs. The method of examining RNA transcripts of mtDNA copied with a bacterial RNA polymerase was used in most cases (Cummins *et al.*, 1967; Grossman *et al.*, 1971; Antonoglou and Georgatsos, 1972; Antonoglou *et al.*, 1972). This method, however, is not fully satisfactory (*cf.* Borst and Flavell, 1972) since the RNA transcripts copied *in vitro* may not be random. A different method of analyzing doublets with DNA polymerase was applied to mtDNAs of *Tetrahymena*, mouse, and guinea pig (Borst and Flavell, 1972). These patterns were similar for mitochondria but differed from those of nuclei and prokaryotes. In general, differences between the doublet patterns of mtDNA and nuclear DNA were noted by all the investigators. It is too early to say, however, whether mtDNAs of animals and lower eukaryotes, such as *Tetrahymena*, have a common origin.

Methylation

The occurrence and functional significance of methylated bases have been studied primarily in DNAs of bacteria and viruses, and to some extent in nuclear DNA of animal and plant cells (*cf.* Nass, 1973*a*). Relatively few reports have dealt with the problem of whether mitochondrial DNA is methylated. A low DNA methylase activity was found in mitochondrial extracts from rat liver (Sheid *et al.*, 1968) and loach embryos (Vanyushin *et al.*, 1971). It was not known, however, whether incomplete separation from other cell fractions was responsible for this activity. Incorporation of ^{3}H from [*methyl*-^{3}H]methionine into 5-methylcytosine of mitochondrial DNA of *Physarum polycephalum* was reported (Evans and Evans, 1970). Preferential labeling in 7-methylguanine of mitochondrial as compared with nuclear DNA was found after injection of rodents with *N*-nitroso[^{14}C]dimethylamine (Wunderlich *et al.*, 1971, 1972).

The methylation of mitochondrial DNA relative to nuclear DNA of several mammalian cell lines was found to be significantly lower, using four independent methods (Nass, 1973*a*): (1) *in vivo* transfer of the methyl group from [*methyl*-^{3}H]methionine to cytosine of DNA, (2) *in vivo* incorporation of [^{32}P]orthophosphate into DNA, and a combination of (1) and (2), (3) *in vivo* incorporation of [^{3}H]deoxycytidine into DNA, and (4) *in vitro* methylation of isolated DNAs with ^{3}H-labeled *S*-adenosylmethionine as methyl donor and DNA methylase preparations from L cell nuclei. The DNA bases were separated by two-dimensional thin-layer chromatography, using 5-methylcytosine, 5-hydroxymethylcytosine, 6-methylaminopurine, and, in some cases, 7-methylguanine as markers. The level of 5-methylcytosine in mtDNA as compared to nuclear DNA was estimated as one-fourth to one-fourteenth in various cell lines and by different methods. The estimated *maximum* content of 5-methylcytosine per circular DNA molecule was equal to or less than 12 methylcytosine residues for mouse L cells and slightly higher for control and virus-transformed baby hamster kidney cells. Since these values are at the limit of detectability with present methods, a multiple approach for analyzing levels of methylation was found to be advisable (*cf.* Nass, 1973*a*).

A lower methyl content of mtDNA (less than 0.1 mole %) as compared with nuclear DNA (about 2–4 mole %) was also reported for *Paramecium* (Cummings *et al.*, 1974) and for frog ovaries and HeLa cells (Dawid, 1974). In the latter case, although the detection limit of the method was 15–30 residues of 5-methylcytosine per molecule of mtDNA, the author claimed absence of this base. Vanyushin and Kirnos (1974) reported a higher 5-methylcytosine content in mtDNA of beef heart as compared with nuclear DNA. It is possible that different mtDNAs vary with respect to methylation. Analysis of a wide range of mtDNAs and the development of more sensitive methods for detecting methylated bases in mtDNA than currently available should be of benefit in future studies.

The demonstration that mtDNA in many cases is less methylated than nuclear DNA raises, of course, the question of whether mtDNA is methylated at all. A direct approach is to look for an enzyme with DNA methylase activity in mitochondria. Such an activity was indeed associated with purified whole mitochondria and inner membrane-matrix preparations (Nass, 1973*a*). The mitochondrial enzyme was greatly inhibited by mercaptoethanol and dithiothreitol at concentrations that are normally and commonly used in enzyme purification, whereas the enzyme from nuclei was slightly stimulated. Further studies on DNA methylation in mitochondria may give clues as to the biological significance of DNA methylation in these organelles. If methylation is indeed an essential function in mitochondria, likely possibilities are (1) modification, similar to that found in microbial DNA, which may protect mtDNA from degradation by specific endonucleases, if present; (2) involvement in the initiation of DNA replication and/or transcription, e.g., by control of endonuclease action or by other mechanisms.

Properties of mtDNA under Denaturing Conditions

The properties of the covalently closed component I differ from those of the nicked component II because of the topological restraint to unwinding of the double-stranded helix, imposed by the covalently closed structure. Consequently, component I, as compared to II or linear DNA, greatly resists denaturation or strand separation by heat and alkali; it renatures more easily after denaturation because the two strands remain aligned and "zip" together as hydrogen bonds re-form (Nass, 1969*a,c*; Borst and Kroon, 1969).

The hydrodynamically more compact component I, as compared to the relaxed forms, has a higher sedimentation velocity at neutral pH ($S_{20,w}$ ~39 S) and more so at alkaline pH (~87 S). The relaxed forms sediment at about 24 S (linear) and 27 S (nicked circular) at neutral pH, and at 28 S (linear single-stranded) and 32 S (circular single-stranded) at pH above 12.5. Furthermore, component I also has an elevated pH_m (midpoint of pH melting curve in alkaline CsCl) relative to the pH_m of DNA with single-strand scissions. Similarly, component I has an elevated T_m (midpoint of melting temperature) relative to the T_m of DNA with single-strand scissions. DNA I also has a restricted uptake of ethidium bromide. (Examples, tabulations, further discussion, and references can be found in Borst and Kroon, 1969, and Nass, 1969*a,c*). Further, more recent, studies on alkali-denatured closed circular mitochondrial and other DNAs have shown that on renaturation this two-stranded DNA species cobands with native DNA I in CsCl–ethidium bromide gradients, but bands more densely in CsCl–propidium diiodide gradients (Grossman *et al.*, 1974). The latter gradient allows preparative isolation of denatured DNA I, native DNA I, single-stranded DNA, and DNA II from a mixture.

The following discussion deals with the relaxed or nicked circular species (DNA II) of mtDNA. Complete strand separation is achieved in alkaline CsCl equilibrium gradients (above pH 12.5), and the complementary strands of most mtDNAs studied separate into two bands. These bands have a density difference that may vary in different organisms from only 5–6 mg/cm^3 for *Tetrahymena,* sea urchin, and a yeast petite mutant mtDNA (Borst and Flavell, 1972) to 27–42 mg/cm^3 for mtDNAs of most animals studied. For example, rat liver mtDNA has buoyant densities of 1.772 and 1.740 g/cm^3 for heavy and light complementary strands, respectively (Nass and Buck, 1970). We have identified by electron microscopy about 20% intact circular and linear single-stranded DNA molecules in heavy and light fractions.

Since not all mtDNAs show a good strand separation in alkali and since fragmentation occurs in many molecules, strand separation has also been achieved by interaction of the denatured (by heat or alkali) DNA with various polyribonucleotides, followed by equilibrium centrifugation in neutral CsCl (Borst, 1972). Poly(I,G) was used to separate the strands of chick mtDNA and also to prevent aggregation of the strands (Borst and Ruttenberg, 1969). The L strand tends to preferentially bind poly(I,G), poly(I), and poly(U). Poly(C) was shown to shift the H strand of rat mtDNA to a higher density by about 20 mg/cm^3 (Borst and Ruttenberg, 1972). Strand separation of *Tetrahymena* mtDNA in neutral CsCl was done by differential binding of poly(U) or poly(U,G) to the heat-denatured DNA (Schutgens *et al.,* 1973). Plant mitochondrial DNA (from potato tubers) could be resolved into two strands by interaction with poly(G), poly(U), and poly(I,G) (Quetier and Vedel, 1973).

Besides being an aid for the preparative isolation of complementary strands, polyribonucleotides have been used to study the clustering of d(G)-, d(T)-, d(A)-, and d(T)-rich residues on complementary strands (see references cited above). The separated strands have been most useful in genetic mapping experiments of transcription products (see section on transcription).

"Denaturation mapping" is a technique originally developed by Inman (1966) to study the base sequence homogeneity of circular λ bacteriophage DNA by electron microscopy. If the DNA molecules have identical base sequences, the helix random coil transition of DNA during heating in the presence of formaldehyde is expected to occur at specific sites of the DNA molecule; if these regions are adenine–thymine rich, they preferentially melt at the lower temperatures of the heat denaturation curves. By this method, specific sites have been localized in a number of bacteriophage DNAs. This method was first applied to mitochondrial DNA of mouse L cells and ascites tumor cells (Našs, 1968), which showed one to three and up to 12 distinct regions of strand separation depending on whether samples were taken in the early phase of the melting curve (below T_m 50) or above. Shrinkage of the circular molecules by 15–30% was observed during denaturation. This method has also been applied to rat liver mtDNA where specific positions around the molecule were located (Wolstenholme *et al.*, 1972) and to yeast petite mtDNA (Locker *et al.*, 1974*b*).

Ribonucleotides in Mitochondrial DNA

It was originally observed that mitochondrial closed circular duplex DNAs from amphibian eggs (Dawid and Wolstenholme, 1967) and from mouse L cells (Nass, 1969*c*) suffer chain scissions at high alkaline pH, and in band sedimentation studies the newly nicked species were seen to lag behind the intact species (Nass, 1969*c*). The alkali lability, also observed in chick, rat, and other cell types, has been attributed to artifacts in the method of preparation (Borst and Kroon, 1969; Borst, 1972). Alternatively, the susceptibility to alkali is also consistent with the occurrence of covalently linked ribonucleotides in mtDNA.

Several reports appeared that ribonucleases A, T_1, and/or H nick or relax closed mtDNA from rat ascites hepatoma, mouse, HeLa, African green monkey, and rat liver cells (Miyaki *et al.*, 1973; Wong-Staal, *et al.*, 1973; Grossman *et al.*, 1973; Porcher and Koch, 1973; Lonsdale and Jones, 1974). The reported sensitivities vary with the enzyme and cell type used. Grossman *et al.* (1973) report their results to be consistent with the presence of about ten ribonucleotides in the major fraction of mtDNA. Some recent studies dealt with the possibility that short polyribonucleotides, as in some microorganisms, serve as primers for DNA synthesis in mitochondria. The isolated progeny H strand from D-loop DNA (see p. 503) has been examined by testing the alkali lability of a terminal label attached with polynucleotide kinase (Kasamatsu and Vinograd, 1974). Future studies on the ribonucleotides in mtDNA are necessary to elucidate their structure and localization, as well as their origin and significance.

Base Sequence Homology of mtDNA from Different Organisms

Comparisons of base sequence homologies between closely related and more distantly related mtDNAs are of special interest with respect to (1) the number and structure of basic genes on mtDNA for messenger, ribosomal, and transfer RNAs, (2) the evolutionary changes in these and other genes as compared to nontranscribing "spacer regions," and (3) the occurrence and type of possible base alterations, insertions, and deletions in cytoplasmic mutants, tumor and other pathological cells, virus-transformed cells, irradiated cells, and cells subjected to other types of environmental stress.

The relatively few studies that have been done on this subject utilized the following techniques: (1) DNA:DNA hybridization or coannealing techniques in which DNA is analyzed either on nitrocellulose filters (Borst and Flavell, 1972) or by ultracentrifugation in buoyant cesium chloride (Brown and Hallberg, 1972; Dawid and Wolstenholme, 1968; Dawid and Brown, 1970; Dawid, 1972; Flavell and Jones, 1971) or by electron microscopy (Dawid and Wolstenholme, 1968; Nass, 1974*b*) (see Figure 1); (2) DNA hybridization with complementary RNA (cRNA) synthesized *in vitro* with *Escherichia coli* RNA polymerase on respective mtDNA templates (Dawid and Brown, 1970; Dawid, 1972; Nass, unpublished), and (3) comparison of mtDNAs by restriction endonuclease mapping (DeFilippes and Nass, 1975). So far only qualitative or semiquantitative results have been obtained.

In comparisons of closely related mtDNAs, two *Xenopus* species (*X. laevis* and *X. mulleri*) were found to have 30% nonhomologous sequences and various regions of mismatching in the remaining parts of the DNA molecule, including 6% mismatching in regions believed to contain genes for rRNA and tRNA (Dawid, 1972). Dawid concludes that 80% of the sequences of these mtDNAs have evolved so rapidly that they may be the equivalent of "spacer" regions found in some nuclear DNAs. Brown and Hallberg (1972) observed less than 1% mismatching in man–chimpanzee heteroduplexes but 30% mismatching in human–African green monkey heteroduplexes.

In more distantly related mtDNAs, Borst and Flavell (1972) observed greater homology between rat–mouse than between chick–rat mtDNA hybrids. About 50% sequence homology is apparent in rat mtDNA–mouse cRNA hybrids (Nass and Buck, unpublished experiments). Some common base sequences are detectable in mtDNA of the chick, *Xenopus,* and *Urechis,* an echiuroid worm (Dawid and Brown, 1970). The extent of mismatching, which undoubtedly exists, is not known. No detectable sequence homology was reported between mtDNAs of *Xenopus* and yeast (Dawid and Wolstenholme, 1968). Very little homology seemed to exist between mtDNAs of *Paramecium* and *Tetrahymena* (Flavell and Jones, 1971). The genes for mt rRNA and tRNAs must therefore be very different in these organisms.

Figure 1 shows an electron micrograph of a heteroduplex molecule of chick and mouse mtDNA. The advantage of electron microscopy over other techniques is that sequences of true homology can be distinguished from sequences of mismatching down to stretches of about 100 base sequences. In methods using hybridization with cRNA, the fidelity of copying by RNA polymerase is a potential problem. Ideally, of course, a combination of techniques is advisable. Sequence homology studies have also been done between mtDNAs of wild-type yeast and various petite mutant derivatives. These will be discussed later.

mtDNA Sequences in Nuclear DNA?

The possibility that nuclear DNA has integrated one or more "master copies" of mtDNA has been raised (Wilkie, 1964), extensively discussed (Borst *et al.*, 1967*a*; Borst and Kroon, 1969; Borst, 1972; Borst and Flavell, 1972), and subjected to various experimental tests (e.g., Dawid and Wolstenholme, 1968; Fukuhara, 1970; Kung *et al.*, 1972; Tabak *et al.*, 1973; Flavell and Trampé, 1973; Dawid and Blackler, 1972). The question of a nuclear master copy of mtDNA remained unresolved, although its likelihood has somewhat decreased. To overcome obvious technical difficulties in hybridization experiments due to cross-contamination, the sensitivity of detection of mtDNA sequences in nuclear DNA was increased by hybridization of nuclear DNA with complementary RNA of high specific activity that was synthesized on mtDNA. The limit of detection was set at 1–3 mtDNA copies per haploid nuclear DNA in chick (Tabak *et al.*, 1973) and 0.04 copies per haploid nuclear genome in *Tetrahymena* (Flavell and Trampé 1973). Dawid and Blackler (1972) claim that all 14–16 copies of mtDNA found in nuclear DNA result from cytoplasmic contamination. Further experimental evidence is obviously needed to resolve this problem and to assess the situation in different organisms.

Restriction Endonuclease Mapping of mtDNA

Bacterial restriction deoxyribonucleases cleave various foreign bacterial and viral DNAs at specific sites into smaller double-stranded fragments (*cf.* Meselson *et al.*, 1972). Nathans and his associates have used the *Hemophilus influenzae* restriction enzyme, which produces 11 fragments from SV40 DNA, to map transcription sites on the DNA (Khoury *et al.*, 1973). The restriction enzyme from *Hemophilus parainfluenzae*, Hpa I, produces three fragments from SV40 DNA (Sack and Nathans, 1973).

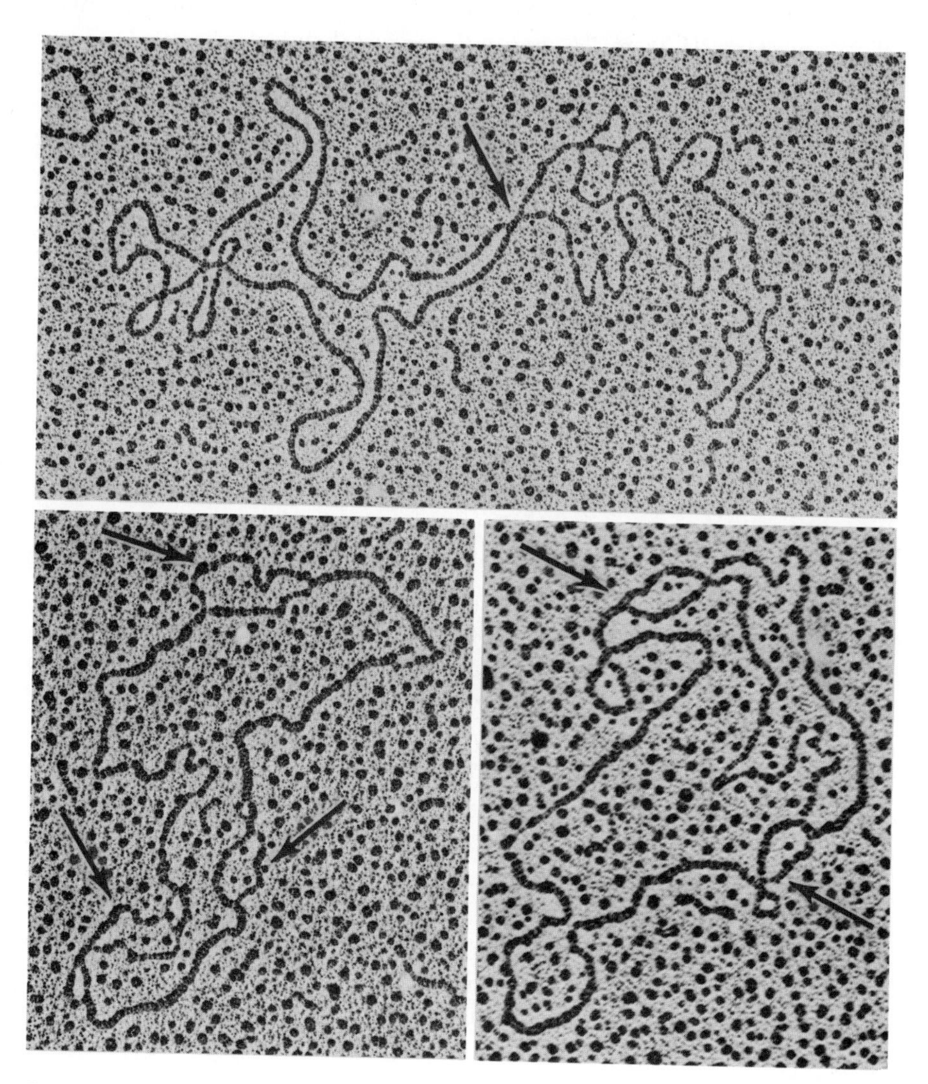

Figure 1. Top: Homoduplex molecule of circular mtDNA in which one complementary strand from a catenated dimer and one from a monomer of chick embryo fibroblasts are renatured in 50% formamide. Left of arrow, double-stranded submolecule; right of arrow, single-stranded submolecule. Bottom: Heteroduplex molecules in which one complementary strand of mtDNA from chicken embryo fibroblasts (5.0 μm) and one strand from mouse L cells (4.7 μm) are renatured to form molecular hybrids. Right frame, partial renaturation (forked single-stranded region at right of two arrows); left frame, complete renaturation showing three major regions of nonhomology at arrows. The upper region shows one single strand to be longer than the other; the longer strand is therefore probably of chicken origin. 56,500×.

The restriction endonucleases Eco RI from *E. coli* (Morrow and Berg, 1972) and Hpa II from *H. parainfluenzae* (Sharp *et al.*, 1973) cleave SV40 DNA once at unique sites. By analysis of overlapping sets of fragments produced by two different restriction enzymes, the order of the enzyme cleavage sites on the DNA molecules of ϕX174 (Lee and Sinsheimer, 1974) and SV40 (Danna *et al.*, 1973) has been mapped.

Studies in our laboratory have shown that the two restriction enzymes from *Hemophilus parainfluenzae,* Hpa I and Hpa II, cleave covalently closed circular mitochondrial DNA into specific, double-stranded, linear fragments which are unique for a given organism (DeFilippes and Nass, 1975). Five or more mtDNA fragments were produced by the Hpa I enzyme in different organisms. The fragments were separated by agarose gel electrophoresis and visualized by staining with ethidium bromide, by autoradiography of ^{32}P-labeled material, or by electron microscopy. A comparison of mtDNA digests from organs (liver and brain of African green monkey) and from cultured cells and virus-transformed cells of human, mouse, hamster, and chicken origin revealed major differences between organisms, but similar patterns for cell types of the same organism. The Eco RI endonuclease was recently used for HeLa mtDNA and mouse L cell mtDNA (Brown and Vinograd, 1974; Robberson *et al.*, 1974), producing three and two fragments, respectively, in these DNAs.

The restriction fragments may be used for localizing initiation sites for DNA synthesis and transcription. The cleavage patterns may also be of use for determining the position of ribonucleotides or methylated bases, for analyzing base sequence homologies, and for isolating specific genes. Isolated genes can be identified by hybridizing with specific mitochondrial RNA species, and the expression of these genes may be studied by combining the mtDNA fragments with plasmid DNA. The plasmids may serve to introduce this foreign DNA into bacteria where its expression can be analyzed. The restriction enzyme mapping technique is a valuable tool for analyzing the structure and function of the mitochondrial genome.

Synthesis and Replication of Mitochondrial DNA

Properties and Rates of mtDNA Synthesis *in Vivo*

Autoradiographic methods have been used in a number of cases to study the incorporation of [^{3}H]thymidine into mitochondrial DNA, e.g., in *Tetrahymena* (Stone and Miller, 1965; Parsons, 1965; Parsons and Rustad, 1968) and in cultured mouse L cells (Nass, 1967). It is obvious that autoradiographic methods that involve ultrathin sections cannot be quantitative, since the mtDNA concentration is very low and not every section of a mitochondrion includes a DNA region. Many biochemical studies have followed the incorporation of deoxyribonucleosides into mtDNA *in vivo* (for references, see Nass, 1969*a*, 1974*a*). In most cases, the incorporation rates into mtDNA differed from corresponding rates into nuclear DNA. An increase of mtDNA synthesis over nuclear DNA synthesis *in vivo* was observed under diverse conditions, e.g., in tissue culture cells upon change of medium (Georgatsos *et al.*, 1972), in rats subjected to undernourishment (Dallman, 1971), and in regenerating rat liver immediately after surgery (Baugnet-Mahieu and Goutier 1972). Specific activities of mtDNA were also greater in certain hepatomas than in liver. Chang *et al.* (1968*a*) reported that a correlation seems to exist between specific activity ratios of tumor-to-host liver mtDNA and growth rates.

In general, the available data are not clear-cut enough to allow generalizations. A survey of the literature indicates that the incorporation of [^{3}H]thymidine into mtDNA may be greater or less than the incorporation into nuclear DNA, depending on a variety of factors, e.g., tissue of origin, growth rates, differences in numbers of mitochondria from normal and certain tumor cells, permeability of membranes, precursor pool sizes, and type of isotope used. In the last case, it is of interest that the ratio of mtDNA to nuclear DNA specific activities in two transplantable rat hepatomas and host livers was 10–100 times greater following [^{3}H]cytidine administration than the ratio following [^{3}H]thymidine administration (Chang *et al.*, 1968*b*). Various specific activities of mtDNA and nuclear DNA following labeling of control and virus-transformed cells with ^{32}P, [^{3}H]thymidine, and [^{3}H]cytidine are summarized and discussed in Nass, 1973*a*.

Relatively little is known about the effects of hormones on mtDNA synthesis. In the case of testosterone administration to rats (Doeg *et al.*, 1972) and cortisone treatment of rats (Kimberg and Loeb, 1972), mitochondrial DNA synthesis was not differentially affected as compared with nuclear DNA synthesis.

There have been reports that mitochondrial DNA synthesis is stimulated by infection of African green monkey cells with SV40 virus (Levine, 1971), 3T3 cells with polyoma virus (Vesco and Basilico, 1971), and HeLa cells with herpes simplex virus (Radsak and Freise, 1972). Experiments with Rous virus infected cells have shown that the specific activities of thymidine-labeled mtDNA are several times greater in the virus-transformed cells than in uninfected control cells or in cells that are infected with a temperature-sensitive mutant virus but grown at the non-permissive temperature at which the cells are phenotypically normal (Bosmann *et al.*, 1974; Nass, 1974*b*).

mtDNA Synthesis During the Cell Cycle

The methods most commonly used for studying organelle DNA synthesis during the mitotic cycle are autoradiographic techniques, either with the light microscope (Parsons and Rustad, 1968) or the electron microscope (Nass, 1967; Charret and André, 1968; Sigee, 1972), as well as diverse biochemical methods using isotope incorporation and density gradient centrifugation (e.g., Koch and Stokstad, 1967; Braun and Evans, 1969; Williamson and Moustacchi, 1971; Bosmann, 1971*a*; Pica-Mattoccia and Attardi, 1972; Volpe and Eremenko, 1973; Küenzi and Roth, 1974; Wells, 1974). Most studies of the temporal relations of mtDNA and nuclear DNA synthesis during the mitotic cycle indicate that the two DNA synthetic systems are not in phase. Although nuclear DNA synthesis can almost always be ascribed to the S phase, there is little agreement in the attempts to correlate mtDNA synthesis with specific phases of the cell cycle. The inconclusive and conflicting results are in part due to (1) the probable existence of different regulatory mechanisms in various organisms, (2) fluctuations in the size of DNA precursor pools, (3) uncertain identification of mitochondrial and additional "satellite" DNAs in a given cell, (4) the type of synchronization method used (e.g., selective detachment of mitotic cells, cold shock, thymidine block, colcemid block, or isoleucine deprivation), and (5) side effects due to additional treatments with drugs, e.g., camptothecin and ethidium bromide. The small amount of mtDNA present relative to nuclear DNA, especially in higher animal cells, is obviously also a potential problem.

The reported DNA synthesis patterns fall into three general classes: (1) A *periodicity* of mtDNA and nuclear DNA synthesis has been reported in many mammalian cells, such as cultured human Chang liver cells (Koch and Stokstad, 1967), mouse L cells (Nass, 1967), mouse lymphoma cells (Bosmann, 1971*a*), and human HeLa cells (Pica-Mattoccia and Attardi, 1972; Volpe and Eremenko, 1973). The maximum rate of mtDNA synthesis in the above cases tends to be in G_2 phase, with some synthesis in S phase. A periodic synthesis of mtDNA and nuclear DNA synthesis was also reported in two species of *Saccharomyces* (Smith *et al.*, 1968; Wells, 1974). (2) *Simultaneous* synthesis of mtDNA and nuclear DNA during the cell cycle was reported for *Euglena gracilis* (Calvayrac *et al.*, 1972). In this case, the replication of mtDNA was also correlated with changes in mitochondrial morphology, ranging from giant reticular forms to small individual mitochondria when DNA replication was completed. (3) In contrast, there are other reports where mtDNA synthesis occurred *continuously* throughout the cell cycle, although not necessarily at a constant rate in each phase. Examples are *Tetrahymena pyriformis* (Charret and André, 1968; Parsons and Rustad, 1968), *Physarum polycephalum* (Braun and Evans, 1969), *Saccharomyces cerevisiae* (Williamson and Moustacchi, 1971; Küenzi and Roth, 1974), *Pteridium aquilinum* (Sigee, 1972), and mouse fibroblasts (Madreiter *et al.,* 1972).

All the mitochondria of *Tetrahymena* seem to incorporate tritiated thymidine and increase their DNA content during each generation (Parsons and Rustad, 1968), which is compatible with the idea that the synthesis of one or more DNA molecules within a given mitochondrion may be discontinuous, but synthesis may occur in some mitochondria at any time during the mitotic cycle. Synchronization of mtDNA synthesis may be achieved after addition of isoleucine to isoleucine-deprived cells (Ley and Murphy, 1973). The mechanism by which the timing of mtDNA and nuclear DNA synthesis is regulated so that the relative amounts of both types of DNA are constant in each generation remains to be determined.

DNA Synthesis in Isolated Mitochondria

Isolated mitochondria from diverse cell types are capable of incorporating into their DNA radioactively labeled deoxyribonucleosides (e.g., Mitra and Bernstein, 1970) and deoxyribonucleoside triphosphates (e.g., Parsons and Simpson, 1973; Wattiaux-de Coninck *et al.,* 1973; Karol and Simpson, 1968). Several review papers summarize findings of the earlier literature: Borst and Kroon (1969), Nass (1969*a*), and Borst (1972). Rat and chick liver mitochondria have been most extensively studied. The rate of incorporation is usually only a fraction of that observed *in vivo* and may range from 1 to 1200 pmoles of nucleotide incorporated per milligram protein per hour at 37°C in liver and yeast mitochondria under various incubation conditions. Approximately 1% of total mtDNA may be synthesized in 2 hr in rat liver mitochondria (Parsons and Simpson, 1973). Sulfhydryl compounds that are routinely used in many assay mixtures were recently found to be inhibitory to the incorporation process in rat liver mitochondria (D'Agostino *et al.,* 1975).

In general, the incorporation (depending on internal precursor pools) at least partially requires the presence of all four deoxyribonucleoside triphosphates. The incorporation is insensitive to deoxyribonuclease in isolated liver mitochondria if they remain intact during the incubation, but it is sensitive to deoxyribonuclease in isolated mitochondria of various cultured mammalian cells (e.g., L cells, BHK cells, chick embryo fibroblasts) because these organelles tend to lose their structural integrity rapidly during incubation at 37°C (Nass, personal observations). The incorporation is inhibited by uncoupling agents and inhibitors of mitochondrial electron transport and by ethidium bromide, acriflavin, actinomycin D, nogalomycin, and phleomycin at suitable concentrations. The labeled product appears to have an ordered structure and is equivalent to mitochondrial DNA. For example, the buoyant density of the labeled product (both native and after denaturation–renaturation) is identical with the density of mitochondrial DNA (Borst and Kroon, 1969; Wintersberger, 1968; Parsons and Simpson, 1973). The label is not of bacterial or nuclear origin. The labeling seems to be internal in the DNA molecule, covalently bound to neighboring nucleotides, and not of a homopolymer or alternating copolymer nature, as shown by end-group and nearest-neighbor frequency analysis (Parsons and Simpson, 1973). The bulk of the label incorporated into isolated chick liver mitochondria was reported to be in the D-loop portion of covalently closed and nicked circular DNA (Ter Schegget and Borst, 1971; Borst and Flavell, 1972). However, a substantial portion of radioactivity is also covalently incorporated into the closed mtDNA circle itself in rat and chick liver (Parsons and Simpson, 1973; Nass, 1969*a*). The sum total of data suggests that mtDNA synthesis *in vitro* represents, to a large extent, an ordered replication process, although the additional elements of repair and recombination cannot be excluded in any of the studies.

Partially Purified mtDNA Polymerase(s)

Mitochondria-specific DNA polymerases have been isolated and partially purified to various extents from mitochondria of rat liver (Kalf and Ch'ih, 1968; Meyer and Simpson, 1970; Probst and Meyer, 1973), rat hepatoma (Tschiersch and Graffi, 1970; Hunter *et al.,* 1973), yeast (Iwashima and Rabinowitz, 1969; Wintersberger and Wintersberger, 1970), HeLa cells (Fry and Weissbach, 1973; Tibbetts and Vinograd, 1973*a,b*), and chick embryo brain (Soriano and Croiselle, 1973). The properties of mtDNA polymerases generally differ from those of nuclear and other cellular DNA polymerases. The mtDNA polymerase is more sensitive to the mutagenic dye ethidium bromide than the nuclear DNA polymerase (Meyer and Simpson, 1970). mtDNA polymerase of rat liver was found to have a higher molecular weight (150,000) than nuclear DNA polymerase (35,000) (Probst and Meyer, 1973). The various studies also indicate that mtDNA polymerases may differ from each other and from nuclear DNA polymerases in primer–template specificity (e.g., native, denatured, foreign, or synthetic templates) and in incubation conditions that lead to maximum activity (e.g., ionic strength, divalent cations). It is therefore not surprising that conflicting results exist in the literature. For example, preference for denatured over native templates (e.g., Meyer and Simpson, 1970) or *vice versa* (e.g., Tibbetts and Vinograd, 1973*a,b*) was reported.

There is also some indication that RNA or RNA-like synthetic templates, e.g., poly(rA) $d(pT)_{10}$, are utilized by mtDNA polymerase preparations or isolated mitochondria (Bosmann, 1971*b*; Reitz *et al.*, 1974; Hunter *et al.*, 1973; Soslau and Nass, unpublished). It is not known at this time whether the DNA-dependent and RNA-dependent DNA polymerase activities are the properties of one or two or more enzymes.

Partially purified DNA polymerase from HeLa mitochondria was found to contain a low level of endonuclease activity, which causes single-strand scissions in the template DNA, with a nicking bias of H over L strands (Tibbetts and Vinograd, 1973*a,b*). Using either SV40 DNA or HeLa mtDNA as template, the product DNA was found, by density shift experiments, to be synthesized in a covalent extension of template DNA strands at the nicked sites. The authors believe that the observed preferential synthesis of H strand DNA is at least in part due to the asymmetrical endonuclease action.

It appears in general that the properties *in vivo* of mtDNA polymerases are not necessarily identical in different organisms. The existence of two or more functionally distinct DNA polymerases in mammalian mitochondria is a further possibility. It is clear that the properties of a partially purified mtDNA polymerase, in a given organism, may also vary depending on the diverse methods used for rupturing mitochondria and depending on the type and extent of purification chosen for the enzyme.

The intramitochondrial location of mtDNA polymerase appears to be the inner side of the inner mitochondrial membrane (Wattiaux-de Coninck *et al.*, 1973). This enzyme may be synthesized extramitochondrially and transported into the organelle (Ch'ih and Kalf, 1969), as are numerous other mitochondrial enzymes (Borst, 1972; Nass, 1974*a*).

D-Loop DNA and Replicative Intermediates of mtDNA

A rapidly increasing literature exists on the structure and role of various forms of mtDNA molecules which, on the basis of electron microscopic observations and pulse-chase labeling experiments, are interpreted to be replicative intermediates which are ordered to fit various replication schemes. Summaries of this topic and diagrammatic representations (not necessarily in complete agreement) can be found in the following papers:

Kasamatsu and Vinograd (1974), Koike and Wolstenholme (1974), Wolstenholme *et al.* (1974), Kasamatsu *et al.*, 1971; Robberson *et al.*, 1972*b*; Borst and Flavell, 1972; Berk and Clayton, 1974).

The original observation was the detection of "D-loop" DNA (circular DNA molecule containing a displacement loop or D loop) found *in vivo* in mtDNA from mouse cells grown in culture (Kasamatsu *et al.*, 1971; Robberson *et al.*, 1972*b*).

A similar structure, I*, was found *in vitro* in the product DNA synthesized in isolated chick liver mitochondria (Ter Schegget and Borst, 1971; Ter Schegget *et al.*, 1971). D-loop DNA was also observed in mtDNA of *Xenopus oocytes* (Hallberg, 1974) and other cell types. To summarize from various studies cited above, D-loop DNA is believed to be formed when replication begins, with the initiation of H-strand synthesis occurring on the parental L strand at a unique site of the mitochondrial genome. The identity of the strands has been determined by hybridization experiments. H-strand synthesis results in the displacement loop. The first stretch of the daughter H strand (sometimes also termed E strand on the initiation sequence) lengthens to 450 ± 50 nucleotides and its synthesis may then temporarily cease, awaiting a further signal for expansion. D-loop molecules may accumulate at this step to a frequency of up to 65% in mouse L cells, less than that in most other cells examined. The daughter H strand piece is hydrogen-bonded to the parental circular L strand, leaving the parental duplex molecule with its displacement loop covalently closed and therefore isolable in the lower bands of cesium chloride–ethidium bromide gradients. The daughter H-strand fragment may be released by partial denaturation to form a 7 S, single-stranded fragment which appears to have a high turnover rate (Robberson and Clayton, 1973). Further replication is presumed to involve nicking and closing events, and it proceeds unidirectionally with further displacement of the parental H strand (Robberson and Clayton, 1972; *cf.* Kasamatsu and Vinograd, 1974). Various intermediates formed in this expansion synthesis are "expanded D-loop molecules" (predominant in intermediate regions between lower and upper bands of CsCl–EB gradients) and expanded D-loop molecules in which daughter L-strand synthesis has begun to various extents on the displaced parental H-strand loop. Because of the asynchrony of H- and L-strand replication, two circular daughter molecules, α and β, separate, which have variously completed H- or L-strand portions, respectively. These molecules are termed "gapped circular molecules" because of the still missing strand portions which are subsequently filled in and closed by ligase. Large gapped molecules are predominant in intermediate bands, small gapped molecules in upper bands of CsCl–EB gradients. The replication process is believed to require 120 min in cultured mouse cells and to occur once per cell doubling (Berk and Clayton, 1974). Somewhat different results were obtained in a study of replicative intermediates in mtDNA from Novikoff rat ascites hepatomas and Chang rat solid hepatomas (Koike and Wolstenholme, 1974; Wolstenholme *et al.*, 1974). Replicative intermediates consisted of total double-stranded forms and of molecules that had single-stranded daughter segments of various lengths. Various replication intermediates have also been studied in chick and rat mtDNA synthesized *in vitro* by incubating isolated mitochondria with radioactive DNA precursors (e.g., Ter Schegget *et al.*, 1971; Flavell *et al.*, 1972; Gause *et al.*, 1973; Arnberg *et al.*, 1973; Koike *et al.*, 1974).

The *in vivo* and *in vitro* studies seem to agree in many respects, but differ in details and interpretation. Nevertheless, the work *in toto* supports unidirectionality, asymmetry, and discontinuity in the replication mechanism of circular mtDNA. It is possible that the control of initiation and extent of asymmetry in the replicating circular forms may vary in different organisms and under different experimental conditions. In general, the mechanism in animal cells seems to be of a modified Cairns type rather than a rolling circle type (Kasamatsu and Vinograd, 1974).

In *Tetrahymena* the mtDNA is a linear molecule of 15 μm. Therefore, the basic features of mtDNA replication in this protozoan may differ markedly from those of circular animal mtDNA (Arnberg *et al.*, 1972, 1974; Clegg *et al.*, 1974). Various types of branched linear forms were seen which contained short single-stranded sections in one or both forks of the otherwise duplex branch. The authors suggest that DNA synthesis starts near the middle and proceeds bidirectionally to the ends of the linear molecule. For details on the replication of kinetoplast DNA of trypanosomes, see Simpson (1972) and Kasamatsu and Vinograd (1974). An increased proportion of replicating double-branched circular DNA molecules was observed after treatment of *Trypanosoma cruzi* with Berenil, a trypanocide (Brack *et al.*, 1972).

The mode of replication of circular oligomers and the question of whether they arise originally by errors in replication or are due to recombination is not yet resolved (Kasamatsu and Vinograd, 1974; Nass, 1974*a*; Flory and Vinograd, 1973). Displacement synthesis seems to take place in circular dimers (Kasamatsu and Vinograd, 1974; Robberson and Clayton, 1972). For catenated dimers and oligomers, evidence is lacking as to how the interlocked submolecules separate. A hypothetical scheme of linearization and recombination has been proposed (Hudson and Vinograd, 1967).

The results from density shift experiments *in vivo* have been consistent with the idea that replication of mtDNA, e.g., in rat liver (Gross and Rabinowitz, 1969) and in *Neurospora* (Reich and Luck, 1966), occurs by a semiconservative mechanism, as has been shown with various nuclear and microbial DNAs. Each strand of the duplex DNA replicates to form a parent–progeny hybrid in the first generation. For mtDNAs, however, the data have not been as clear-cut as for nuclear DNAs. In density labeling experiments with yeast it was concluded that a primary semiconservative mtDNA replication step is followed by a recombinational process in which extensive partial exchanges occur between homologous strands of sister duplexes (Williamson and Fennell, 1974).

Inhibitors of Mitochondrial or Nuclear DNA Synthesis

Ethidium Bromide. The phenanthridine dye ethidium bromide (EB) is of particular interest because the mitochondria are a prime target for this drug (*cf.* Nass, 1974*a*, for review of literature). EB intercalates between the bases of double-stranded DNA and exhibits differential binding affinities for covalently closed circular and nicked circular or linear duplex DNA, which forms the basis for the widely used isolation technique of species of mtDNA in cesium chloride–ethidium bromide gradients (Radloff *et al.*, 1967).

The most striking effect of EB is the quantitative conversion of yeast cells to petite mutants with reduced cytochrome content and altered mtDNA (see later section). In mammalian cells, the effects of EB are usually somewhat less drastic and permanent than in yeast. The transcription of mtDNA is inhibited (e.g., Fukuhara and Kujawa 1970; Zylber *et al.*, 1969) and the synthesis of mitochondrial DNA is differentially inhibited over that of nuclear DNA both *in vivo* (Nass, 1970, 1972; Kato *et al.*, 1972; Koch, 1972; Volpe and Eremenko, 1973) and *in vitro,* using partially purified mitochondrial and nuclear DNA polymerases (Meyer and Simpson, 1970). The synthesis of nuclear DNA is either unaffected or slightly stimulated (Nass, 1970). Concentrations of 0.1–2 μg EB/ml are usually effective in mammalian cells. Higher concentrations, up to 10 μg/ml, are generally needed to achieve a significant inhibition of mtDNA in lower organisms, e.g., *Physarum polycephalum* (Horwitz and Holt, 1971) and *Euglena gracilis* (Nass and Ben-Shaul, 1973). In *Tetrahymena pyriformis,* an accumulation of replicative intermediates of mtDNA was reported after release from EB treatment (Upholt and Borst, 1974). The physical structure of covalently closed mtDNA is drastically altered upon EB treatment; there are strand breakage and some rejoining of strands, manifested in both a greater proportion of nicked circles and linear DNA and some closed circular DNA having a different superhelix density as compared with control cases (Nass, 1972, 1974*a*; Smith *et al.*, 1971).

9β-D-Arabinofuranosyl Adenine (ara-a). An interesting tool in combination with the use of EB is the potential use of ara-a, which has been shown to have the phenotypically opposite effect of EB, namely, to selectively inhibit nuclear DNA synthesis but not circular mitochondrial DNA synthesis, at an optimal concentration of about $5–6 \times 10^{-4}$ M (Shipman *et al.*, 1972; Nass, 1973*b*).

Cycloheximide, Puromycin, Camptothecin, Hydroxyurea. Other agents that seem to have an inhibitory effect on nuclear DNA synthesis but not, or less so, on mtDNA synthesis are cycloheximide and puromycin. These have been tested in chick embryo and mouse cells (Nass, 1970, 1973*b*), in thymidine kinase deficient HeLa cells (Kit and Minekawa, 1972), and in *Physarum polycephalum* (Werry and Wanks, 1972). Camptothecin, although its specificity is rather broad, has also been used to inhibit nuclear but presumably not mtDNA synthesis (Bosmann, 1970, 1971*a*). Similarly, labeling of cytoplasmic (presumably mitochondrial) DNA in HeLa cells was reported to be relatively resistant to hydroxyurea (Vesco and Penman, 1969).

It must be borne in mind that none of these drugs is entirely specific and that secondary effects occur; the usefulness and effective dosages of these compounds may vary considerably in different cell types and must be tested accordingly.

Turnover and Repair of mtDNA

A few studies have been done on turnover of mtDNA in different organisms and tissues (*cf.* Borst and Kroon, 1969; Nass, 1969*a*). Among the obvious difficulties in such studies are pool effects, purity of mtDNA and nuclear DNA, possible recycling effects of radioactive components, and the contribution of replicative processes and repair activities. Gross *et al.* (1969) have therefore used the terms "apparent turnover" and "apparent half-life." There is thus far no concrete evidence for or against the notion that certain mitochondrial protein, lipid, and nucleic acid components of inner membrane, outer membrane, or matrix compartments turn over as a unit. The apparent turnover of mtDNA in various tissues of the rat was found to differ greatly (Gross *et al.*, 1969). The half-life of mtDNA in heart, which has a high aerobic metabolism, was fastest (6–7 days), and the half-life was progressively longer in liver (9.4 days), kidney (10.4 days), and brain (31 days). Similarly, a half-life of 7–10 days was reported earlier for rat liver and kidney mtDNA (Neubert *et al.*, 1968). In both studies, the turnover of nuclear DNA was found to be much longer than that of mtDNA. In logarithmic cultures of *Euglena gracilis*, mtDNA was estimated to turn over with a half-life of 1.8 cell generations (Richards and Ryan, 1974). A half-life of 7.9 hr was reported for the 7 S heavy strand initiation sequence of D-loop DNA from mouse mitochondria (Berk and Clayton, 1974).

Relatively little is known about repair activities in mtDNA (see also Borst and Flavell, 1972). Indirectly, several studies present data that the incorporation of isotopic material into mtDNA is not due to repair (e.g., Karol and Simpson, 1968; Gross and Rabinowitz, 1969). In *Tetrahymena*, an increase in mtDNA polymerase upon UV irradiation was reported which may be related to repair activity (Keiding and Westergaard, 1971). In mitochondria of mouse and human cells grown in culture, the absence of a pyrimidine dimer excision mechanism for repair of UV-irradiated mtDNA was reported (Clayton *et al.*, 1974). In all cases where the repair of damaged mtDNA and its apparent need to the cell are studied (e.g., after UV irradiation, EB treatment, exposure to various drugs, and laser beam treatment, *cf.* Nass, 1974*a*), the possibility must be considered that some undamaged mitochondria undoubtedly remain in a cell that may normally contain from several hundred to several thousand mitochondria. Survival of the cell may be maintained by intact mitochondria outgrowing the damaged ones.

Genetic Properties and Inheritance of mtDNA

Transcription and Gene Products

A mitochondrial RNA polymerase has been purified from mitochondria of *Neurospora* (Küntzel and Schäfer, 1971), *Xenopus laevis* (Wu and Dawid, 1972, 1974), and yeast (e.g., Tsai *et al.*, 1971). The enzyme from *Neurospora* mitochondria appears to consist of a single polypeptide chain (mol wt 64000), and it prefers homologous mtDNA over calf thymus DNA. The enzyme from *Xenopus* shows strand selectivity *in vitro*. With closed circular mtDNA, the H strand is preferred about 3 : 1; with nicked circular mtDNA, the H strand is preferred 20 : 1; with denatured mtDNA, the L strand is preferred 5 : 1.

Hybridization studies have shown that mtDNAs from vertebrates and from lower eukaryotes have complementary base sequences to species of mitochondrial ribosomal RNA (Aloni and Attardi, 1971*a*; Chi and Suyama, 1970; Robberson *et al.*, 1972*a*; Dawid, 1972; Wu *et al.*, 1972;

Schäfer and Küntzel, 1972; Reijnders *et al.*, 1972), mitochondrial transfer RNA (Nass and Buck, 1970; Cohen *et al.*, 1972; Casey *et al.*, 1974*a,b*), mitochondrial *N*-formylmethionyl-tRNA (Halbreich and Rabinowitz, 1971), and mitochondrial messenger-type RNA components (Ojala and Attardi, 1974; Avadhani *et al.*, 1974). The last in some cases were found to contain polyadenylic acid sequences (Hirsch and Penman, 1973, 1974; Ojala and Attardi, 1974; Gaitskhoki *et al.*, 1973).

It appears that in the few cases studied the majority of RNA products that are stable enough to be identified are transcribed from the H strand of mtDNA. Much of the L-strand product identified by pulse-labeling seems to be rapidly eliminated (Aloni and Attardi, 1971*b*).

In the case of ribosomal RNA, the hybridization results (cited above) have indicated that both the 5-μm-type DNA molecule of animal mitochondria and the larger-size mtDNA from lower eukaryotes (yeast) contain one gene each for the two respective ribosomal RNAs. Ribosomal RNA is transcribed mainly from the "heavy" or H strand of mtDNA from higher eukaryotes (Wu *et al.*, 1972; Schutgens *et al.*, 1973). Wu *et al.* (1972) identified by electron microscopy 12 S and 16 S ribosomal RNA-DNA duplex regions on the H strand of mtDNA.

In the case of tRNA, both strands of mtDNA from higher eukaryotes are transcribed. Mitochondrial leucyl- and phenylalanyl-tRNA hybridized exclusively with the H strand, whereas tyrosyl- and seryl-tRNA hybridized exclusively with the L strand of mtDNA (Nass and Buck, 1970), demonstrating "symmetrical transcription" in the case of tRNA. In yeast, 14 aminoacyl tRNAs were shown to hybridize with wild-type mtDNA: leucyl-, isoleucyl-, valyl-, alanyl-, phenylalanyl-, glycyl-, and formylmethionyl-tRNAs (Halbreich and Rabinowitz, 1971; Cohen *et al.*, 1972), and tyrosyl-, aspartyl-, glutamyl-, prolyl-, lysyl-, histidyl-, and seryl-tRNAs (Casey *et al.*, 1974*a*; Carnevali *et al.*, 1973). The localization of different tRNA genes on complementary strands of mtDNA in yeast is unknown because in these organisms it is difficult to separate the complementary strands.

Hybridization saturation experiments with 4 S RNA (which has been assumed but not proven to contain total mt tRNA) have suggested that only 12–15 tRNA cistrons are specified by mtDNA in HeLa and amphibian cells (Wu *et al.*, 1972; Dawid, 1972). By localizing electron-opaque ferritin granules, Wu *et al.* (1972) detected nine binding sites for 4 S RNA on the H strand and three on the L strand of HeLa mtDNA. In wild-type yeast, at least 20 tRNA cistrons are suggested (Reijnders and Borst, 1972; Casey *et al.*, 1974*a*). DNA of the latter is also about 5 times more complex than mtDNA of higher eukaryotes. It is not yet known whether the "missing" tRNA species in higher eukaryotes have remained undetected for unknown technical reasons, e.g., are formed by secondary modification of existing primary mt tRNAs, or whether some nuclear-coded tRNA species are imported from the cytoplasm (see Borst and Flavell, 1972, and Nass, 1974*a*, for further discussions).

Messenger-type poly(A)-containing RNA species, like tRNA, seem to be transcribed from both strands of mtDNA. A 7 S RNA component was found to be complementary mainly to the L strand of HeLa mtDNA, whereas seven other components hybridized to the H strand (Ojala and Attardi, 1974). The authors estimated that 70% of the single-strand informational content of HeLa mtDNA is accounted for by the two mitochondrial rRNA species (16 S and 12 S, mol wt 5.4 and 3.5×10^5, respectively), by twelve 4 S RNA species (assumed mol wt 2.6×10^4 each), and by eight poly(A)-containing RNA components (one 7 S, mol wt 9×10^4, and seven larger components, mol wt $2.6–5.3 \times 10^5$).

In addition to mtRNA species, mtDNA appears to specify a number of hydrophobic proteins that are part of the inner mitochondrial membrane (Schatz and Mason, 1974). There may also be some spacer regions.

Cytoplasmic Mutations That Affect mtDNA and Properties of Mutant mtDNAs in LowerEukaryotes

Heritable alterations in mitochondria have been found mainly in eukaryotic microorganisms (e.g., yeast, *Neurospora, Chlamydomonas*). The mutation(s) is usually expressed phenotypically either as drug resistance (which may result from point mutations) or as a respiratory deficiency in which cytochromes a, a_3, b, and c_1 may be completely or partially lost. In the latter case, especially in yeast, extensive defects may be found in mtDNA.

Since the early studies of Ephrussi and collaborators, who discovered the cytoplasmic peptite mutation in yeast (e.g., Ephrussi, 1953), hundreds of papers have appeared on the genetic analysis and molecular consequences of cytoplasmic mutations in mitochondria. A detailed coverage of this subject is beyond the scope of this chapter, and the reader is referred to the following sources: Ephrussi (1953), Wilkie (1964), Wagner (1969), Sager (1972), Linnane *et al.* (1972), Borst (1972), Wilkie (1972), Wilkie and Thomas (1973), Birky (1973), Flury *et al.* (1974), Nass (1974*a*), and Sherman and Lawrence (1974).

Yeast. By far the most extensive studies have been done with yeast. Cytoplasmic mutations with a respiratory-deficient phenotype may occur spontaneously or are induced by ultraviolet light (e.g., Chanet *et al.*,

1973), by various drugs (*cf.* Sager, 1972), or by DNA-intercalating dyes, such as the acridines and, most extensively studied, ethidium bromide (Slonimski *et al.,* 1968; Perlman and Mahler, 1971). Ethidium bromide inhibits mtDNA synthesis (as discussed earlier), and in various ethidium-induced mutants of yeast, major mtDNA alterations may occur ranging from abnormal base sequences (see below) to a reduced size (Goldring *et al.,* 1971) or apparent absence (Goldring *et al.,* 1970; Nagley and Linnane, 1972) of mtDNA. In extensive studies of the mechanism of mutagenesis by ethidium bromide, Mahler and his collaborators have implicated a complex between mtDNA and inner membrane as target for EB mutagenesis (Mahler and Perlman, 1972) and provided evidence for the formation of a novel, stable, probably covalently linked DNA–ethidium bromide intermediate (Mahler and Bastos, 1974; Bastos and Mahler, 1974).

A number of mitochondrial marker genes conferring resistance to various drugs (e.g., chloramphenicol, erythromycin, paromomycin, spiramycin, oligomycin, or ethidium bromide) have been identified and selected for in several yeast strains (Linnane *et al.,* 1972; Sager, 1972; Wilkie, 1972; Wilkie and Thomas, 1973; Gouhier and Movnolov, 1973). The drug-resistance point mutations have become most useful as markers for the mapping of alterations or loss of sequences on mtDNA of petite mutants. Nagley *et al.* (1973) reported that ethidium bromide treatment of petite clones carrying mitochondrial erythromycin resistance genes leads to the actual elimination of these genes. Casey *et al.* (1974*b*) performed hybridization saturation studies of mtDNA with 11 mt tRNAs from several petite clones carrying chloramphenicol and erythromycin resistance markers. They found that in different petite strains mt tRNA cistrons were lost, retained, or amplified. Some preliminary data on ordering several tRNA genes with respect to the drug resistance markers were obtained. Fukuhara *et al.* (1974) found that a genetic deletion of the chloramphenicol (C^R) or erythromycin (E^R) resistance marker corresponded to the physical loss of specific base sequences in mtDNA. A large fraction of the sequences coding for ribosomal mtRNA appears to be located between the two drug resistance loci. The structural and physical characteristics of the diverse forms and heterogeneous base sequences of mtDNAs isolated from various *petite* strains have been analyzed extensively (e.g., Mol *et al.,* 1974; Sanders *et al.,* 1973; Lazowska *et al.,* 1974; Locker *et al.,* 1974*a,b;* Michel *et al.,* 1974).

Neurospora and Chlamydomonas. Various respiration-deficient mutant strains (poky) of *Neurospora* have been studied with respect to the inheritance of the cytoplasmic character (e.g., Diacumakos *et al.,* 1965; *cf.*

Sager, 1972), the buoyant density and inheritance of mtDNA (e.g., Reich and Luck, 1966), and defective production of mitochondrial ribosomal and tRNA species (Rifkin and Luck, 1971; Brambl and Woodward, 1972). A model on the control of the mitochondrial genetic apparatus in *Neurospora* was formulated (Barath and Küntzel, 1972). In *Chlamydomonas,* the apparent mitochondrial localization and cytoplasmic inheritance of respiratory-deficient mutations (e.g., Alexander *et al.,* 1974) and of drug dependence genes (e.g., Behn and Arnold, 1973) have been investigated. The mitochondrial genome in these organisms is far less characterized than in yeast. For further review of the earlier literature, see Sager (1972).

Kinetoplast of Trypanosomes. The cytoplasmic mutagen ethidium bromide may cause the loss of the mitochondrion-associated kinetoplast of trypanosomes, as was shown earlier using acridine dyes (*cf.* Simpson, 1972). This treatment may lead to alterations and the actual loss of the circular kinetoplast DNA (e.g., Riou and Delain, 1969; Riou, 1968; Renger and Wolstenholme, 1972; Simpson, 1972).

Interspecific Hybrids, Propagation, and Inheritance of mtDNA in Animal Cells

A genetic approach to the structure and function of mtDNA in animal cells, as opposed to similar studies in eukaryotic microorganisms, has met with considerable difficulties because it requires the isolation and propagation analysis of cytoplasmic mutants that are respiratory deficient or have other stable mitochondrial alterations. Such mutants are obviously rarely viable in obligative aerobic cells. A number of studies were directed to develop cell lines with drug resistance markers that are (as in yeast) encoded by mtDNA. Chloramphenicol (CAP) resistant mutants of human HeLa cells were isolated, and CAP resistance was shown to be expressed at the level of mitochondrial protein synthesis (Spolsky and Eisenstadt, 1972). Similar mutants were obtained from mouse A9 cells. Bunn *et al.* (1974) have convincingly shown the cytoplasmic inheritance of CAP resistance in mutants isolated from mouse A9 cells, by fusion and growth selection experiments with enucleated CAP-resistant fragments and nucleated CAP-sensitive cells, and *vice versa.* Further studies are needed to show whether the actual cytoplasmic site is the mitochondria and whether mtDNA is altered. Ethidium bromide resistant hamster cells that have mitochondrial alterations have also been isolated (Klietmann *et al.,* 1973), but the mechanism or location of this resistance is not yet known.

Interspecific somatic cell hybridization has become a useful tool in studying (1) the genetic control of mammalian mitochondrial enzymes (e.g., Van Heyningen *et al.*, 1973; Jeffreys and Craig, 1974; Kit and Leung, 1974) and (2) the propagation of mitochondrial DNA (Clayton *et al.*, 1971; Attardi and Attardi, 1972; Coon *et al.*, 1973; Dawid *et al.*, 1974). It is essential in such experiments that cells are generated which contain various identifiable proportions and different combinations of mitochondrial and chromosomal genomes from different cell types. Also, of course, mtDNAs must be distinguishable either by direct buoyant density analysis (e.g., Clayton *et al.*, 1971) or by comparing ^{3}H-labeled and ^{32}P-labeled complementary RNA transcripts of respective mtDNAs (Coon *et al.*, 1973; Dawid *et al.*, 1974). In hybrid cells derived from Sendai virus induced fusion between established human and mouse cell lines, human mtDNA was lost along with the preferentially eliminated human chromosomes (Clayton *et al.*, 1971; Attardi and Attardi, 1972). However, when hybrid cells were grown from the fusion products of freshly dissociated mouse or rat embryo cells and of human cell lines VA2 or D98/AH2, the chromosomes of either parent segregated, and both rat and human mtDNA molecules persisted and replicated in hybrid cells for at least 40 cell doublings (Coon *et al.*, 1973). Subcloning experiments showed that true hybrid cells were formed rather than a mixture of cells where each contains only one type of mtDNA. A positive correlation undoubtedly exists between the segregation of chromosomes and mtDNA in hybrid cells. The proportions of nuclear DNA in hybrid progeny cells appear to be more balanced than those of mtDNA (Coon *et al.*, 1973).

The maternal mode or cytoplasmic inheritance of mtDNA in animal cells was first shown in interspecific crosses of amphibians (Dawid and Blackler, 1972). Earlier work demonstrated this mode of inheritance in fungi, e.g., *Neurospora* (Reich and Luck, 1966) and yeast (see previous section). In animal cells, egg mitochondria are generally much more numerous than sperm mitochondria. The relative contribution of egg and sperm to the zygote of animal cells is partly dependent, of course, on the ultimate fate of sperm mitochondria, which are located within the sperm midpiece. The midpiece in some organisms does not enter the egg or it enters and degenerates later (Ursprung and Schabtach, 1965; Szollosi, 1965). Mitochondrial DNA has been detected in some sperm mitochondria (e.g., Nass *et al.*, 1965; Handel *et al.*, 1973) but its actual fate at the molecular level remains to be investigated.

Recombination of mtDNA

There is ample genetic evidence that recombination occurs between genetic factors controlling antibiotic resistance in yeast (Linnane *et al.*,

1972; Wilkie, 1972; Coen *et al.*, 1970; Kleese *et al.*, 1972; Howell *et al.*, 1973). The coincident loss of a resistance gene with the induction of the *petite* mutation indicated that these mutations are genetically linked to the ρ factor (mtDNA) and that mtDNA is therefore the probable molecular site for recombinational events. The polarity of mitochondrial gene recombination and the transmission frequency of the resistance factors may vary markedly among recombinant classes and progeny cells, respectively (Wilkie and Thomas, 1972; Coen *et al.*, 1970; Howell *et al.*, 1973). An enhancement of mitochondrial recombination was found in ethidium bromide resistant yeast cells (Gouhier-Monnerot, 1974). Deutsch *et al.* (1974) propose that recombination plays an important role in the induction of petite mutations. They postulate that during the petite mutation process mtDNA molecules which are hit by the mutagen and mutated during the first step in this process recombine with those which remain unaffected; this recombination, if achieved by unequal crossing over or nonreciprocal recombination, may lead to the spreading of the initial errors among almost all the mtDNA molecules of the cell.

Further discussion on the genetic analysis of mitochondrial recombination is found in the following review articles and papers: Coen *et al.*, (1970), Linnane *et al.*, (1972), Sager (1972), Wilkie (1972), and Deutsch *et al.*, (1974).

In more direct physical studies of mtDNA molecules, it was shown that petite mtDNA molecules of yeast may recombine either with grande (wild-type) or other petite mtDNA molecules (Coen *et al.*, 1970; Michaelis *et al.*, 1973). In animal cells, Hudson and Vinograd (1967) hypothesized that catenated oligomers arise by recombination, and Dawid *et al.*, (1974) reported apparent recombinant mtDNA molecules in rodent–human hybrid cells, as evidenced by combined buoyant density and hybridization studies.

Animal mtDNA in Pathological Conditions: Tumors and Virus-Infected Cells

Abnormalities in the structure and function of mitochondria have been reported in many diseases and in some cases following therapy (for review, see Nass, 1974*a*). Although some of these changes may be related to defects in either one or both of the cellular genetic systems, many are undoubtedly due to perturbance of the supply of essential nutrients and to degenerative processes.

There are mainly three types of changes found thus far in mtDNA of certain tumor and oncogenic virus-infected cells: (1) Structural alterations that are manifested as a quantitative shift toward higher oligomer content without necessarily involving changes in base sequence (see the section on dimeric forms; Nass, 1974*a*; Paoletti and Riou, 1970, 1973). (2) Structural changes that possibly involve an alteration of base sequence, as either modification, deletion, or insertion of a sequence. The evidence is still circumstantial, scarce, and preliminary at the moment. For example, a difference in nearest-neighbor frequencies in mtDNA of tumor as compared to normal cells was reported (Antonoglou *et al.*, 1972); the contour lengths of mtDNA from Rous sarcoma virus transformed cells appeared consistently more heterogenous than the contour lengths of control cells (Nass, 1973*b*, 1974*b*). Heteroduplex and restriction enzyme analyses promise to give more conclusive information on this point. (3) Changes in the rates of mtDNA synthesis may occur in tumor cells. In the case of virus infection by oncogenic DNA and RNA viruses, mtDNA synthesis is significantly stimulated over that of control cells (see the section on DNA synthesis *in vivo*). Pertinent to this problem are the increasing number of reports on the association and development of Rous sarcoma type viruses in mitochondria (Nass, 1974*a,b*; Borst, 1972). It remains to be determined whether mitochondria are essential or coincidental to the development of certain viruses or viruslike particles. It also must be shown whether changes in mitochondria or mtDNA are directly and significantly related to malignancy or are merely secondary effects.

Evolutionary Aspects of mtDNA

The evolutionary origin of mtDNA is obviously linked to the general question of mitochondrial evolution. Arguments for an endosymbiotic origin of these organelles and, alternatively, for the evolvement of this organelle more directly from a particularly advanced prokaryotic cell have been advanced (e.g., Margulis, 1974; Raff and Mahler, 1972; Meyer, 1973, Nass, 1971*b*). From our present position, this type of evolutionary problem can be speculated on but cannot possibly be solved. In a comparison of the remarkable similarities in structure and some biochemical mechanisms of mitochondria and present-day bacteria, it was pointed out (Nass, 1971*b*) that the DNA of mitochondria resembles most strikingly the DNA of bacterial plasmids (bacterial extrachromosomal elements or episomes). The size range and structure are similar, oligomeric forms are similar, and in both cases multiple-size DNA can be experimentally built up by interfering with protein synthesis (Nass, 1969*c*, 1970). Schemes suggesting the evolutionary origin of mtDNA as a plasmid element have been elaborated (Raff and Mahler, 1972; Meyer, 1973).

Concluding Remarks

The study of the genetics, molecular organization, and function of mitochondrial DNA has rapidly moved into the foreground of current research. It is apparent that the choice of organism or cell type as a model for studies of mtDNA depends on the individual aspect to be examined and that the numerous new techniques and experimental approaches now available must be adapted to each cell type. Studies in the next few years will undoubtedly deepen our understanding of mitochondrial function and dysfunction and allow genetic and analytical dissection, mapping, and genetic engineering analyses of the mitochondrial genome in a wide variety of organisms.

Acknowledgments

This work was supported by Grants 5-R01-CA 13814 from the Cancer Institute, National Institutes of Health, and NP-93A from the American Cancer Society.

Literature Cited

Aaij, C. and P. Borst, 1972 The gel electrophoresis of DNA. *Biochim. Biophys. Acta* **269**:192–200.

Agsteribbe, E., A. M. Kroon, and E. F. J. Van Bruggen, 1972 Circular DNA from mitochondria of *Neurospora crassa. Biochim. Biophys. Acta* **269**:299–303.

Alexander, N. J., N. W. Gillham, and J. E. Boynton, 1974 The mitochondrial genome of *Chlamydomonas*: Induction of minute colony mutations by acriflavin and their inheritance. *Mol. Gen. Genet.* **130**:275–290.

Allard, C., G. de Lamirande, and A. Cantero, 1952 Mitochondrial population of mammalian cells. II. Variation in the mitochondrial population of the average rat liver cell during regeneration: Use of the mitochondrion as a unit of measurement. *Cancer Res.* **12**:580–583.

Aloni, Y. and G. Attardi, 1971*a* Expression of the mitochondrial genome in HeLa cells. IV. Titration of mitochondrial genes for 16 S, 12 S, and 4 S RNA. *J. Mol. Biol.* **55**:271–276.

Aloni, Y, and G. Attardi, 1971*b* Symmetrical *in vivo* transcription of mitochondrial DNA in HeLa cells. *Proc. Natl. Acad. Sci. USA* **68**:1757–1961.

Antonoglou, O. and J. G. Georgatsos, 1972 Nearest neighbor frequencies of mitochondrial DNA in mouse liver. *Biochemistry* **11**:618–621.

Antonoglou, O., A. Symeonidis, and J. G. Georgatsos, 1972 Nearest neighbor frequencies of mitochondrial DNA in spontaneous mouse hepatoma. *Eur. J. Cancer* **8**:629–631.

Arnberg, A. C., E. F. J. Van Bruggen, R. B. H. Schutgens, R. A. Flavell, and P. Borst, 1972 Multiple D-loops in *Tetrahymena* mitochondrial DNA. *Biochim. Biophys. Acta* **272**:487–493.

Arnberg, A. C., E. F. J. Van Bruggen, R. A. Flavell, and P. Borst, 1973 DNA synthesis by isolated mitochondria. V. Electron microscopy of replicative intermediates. *Biochim. Biophys. Acta* **308:**276–284.

Arnberg, A. C., E. F. J. Van Bruggen, R. A. Clegg, W. B. Upholt, and P. Borst, 1974 An Analysis by electron microscopy of intermediates in the replication of linear *Tetrahymena* mitochondrial DNA. *Biochim. Biophys. Acta* **361:**266–276.

Ashwell, M. and T. S. Work, 1970 The biogenesis of mitochondria. *Ann. Rev. Biochem.* **39:**251–290.

Attardi, B. and G. Attardi, 1972 Fate of mitochondrial DNA in human–mouse somatic cell hybrids. *Proc. Natl. Acad. Sci. USA* **69:**129–133.

Avadhani, N. G., F. S. Lewis, and R. J. Rutman, 1974 Messenger ribonucleic acid metabolism in mammalian mitochondria: Quantitative aspects of structural information coded by the mitochondrial genome. *Biochemistry* **13:**4638–4645.

Avers, C. J., F. E. Billheimer, H. P. Hoffmann, and R. M. Pauli, 1968 Circularity of yeast mitochondrial DNA. *Proc. Natl. Acad. Sci. USA* **61:**90–97.

Bahr, G. F., 1971 A unit mitochondrion: DNA content and response to X-irradiation. In *Advances in Cell and Molecular Biology,* Vol. I, Academic Press, New York, pp. 267–292.

Barath, Z. and H. Küntzel, 1972 Cooperation of mitochondrial and nuclear genes specifying the mitochondrial genetic appartus in *Neurospora crassa. Proc. Natl. Acad. Sci. USA* **69:**1371–1374.

Bastos, R. N. and H. R. Mahler, 1974 Molecular mechanisms of mitchondrial genetic activity: Effects of ethidium bromide on the deoxyribonucleic acid and energetics of isolated mitochondria. *J. Biol. Chem.* **249:**6617–6627.

Baugnet-Mahieu, L. and R. Goutier, 1972 Reutilization of labelled-thymidine and -iododeoxyuridine for nuclear and mitochondrial DNA synthesis in regenerating rat liver. *Arch. Int. Physiol. Biochim.* **80:**319–330.

Behn, W. and C. G. Arnold, 1973 Localization of extranuclear genes by investigations of the ultrastructure in *Chlamydomonas reinhardi. Arch. Mikrobiol.* **92:**85–90.

Berk, A. J. and D. A. Clayton, 1974 Mechanism of mitochondrial DNA replication in mouse L-cells: Asynchronous replication of strands, segregation of circular daughter molecules, aspects of topology and turnover of an initiation sequence. *J. Mol. Biol.* **86:**801–824.

Bernardi, G., G. Piperno, and G. Fonti, 1972 The mitochondrial genome of wild-type yeast cells. I. Preparation and heterogeneity of mitochondrial DNA. *J. Mol. Biol.* **65:**173–189.

Birky, C. W., Jr., 1973 On the origin of mitochondrial mutants: Evidence for intracellular selection of mitochondria in the origin of antibiotic-resistant cells in yeast. *Genetic* **74:**421–432.

Bisalputra, T. and A. A. Bisalputra, 1967 Chloroplast and mitochondrial DNA in a brown alga *Egregia menziesii. J. Cell Biol.* **33:**511–520.

Bogenhagen, D. and D. A. Clayton, 1974 The number of mitochondrial deoxyribonucleic acid genomes in mouse L and human HeLa cells. *J. Biol. Chem.* **249:**7991–7995.

Bohnert, H. J., 1973 Circular mitochondrial DNA from *Acanthamoeba castellanii* (Neff-strain). *Biochim. Biophys. Acta* **324:**199–205.

Borst, P., 1972 Mitochondrial nucleic acids. *Ann. Rev. Biochem.* **41:**333–376.

Borst, P. and R. A. Flavell, 1972 Mitochondrial DNA: Structure, genes, replication. In *Mitochondria/Biomembranes,* North-Holland, Amsterdam, pp. 1–19.

Borst, P. and M. Kroon, 1969 Mitochondrial DNA: Physicochemical properties, replication and genetic function. *Int. Rev. Cytol.* **26:**107–190.

Borst, P. and G. J. C. M. Ruttenberg, 1969 Mitochondrial DNA. IV. Interaction of ribopolynucleotides with the complementary strands of chick-liver mitochondrial DNA. *Biochim. Biophys, Acta* **190:**391–405.

Borst, P. and G. J. C. M. Ruttenberg, 1972 The binding of polyribonucleotides to the complementary strands of mitochondrial DNA. *Biochim. Biophys. Acta* **259:**313–320.

Borst, P., A. M. Kroon, and G. J. C. M. Ruttenberg, 1967*a* Mitochondrial DNA and other forms of cytoplasmic DNA. In *Genetic Elements, Properties and Function,* edited by D. Shugar, Academic Press, New York, pp. 81–116.

Borst, P., E. F. J. Van Bruggen, G. J. C. M. Rutttenberg, and A. M. Kroon, 1967*b* Mitochondrial DNA. II. Sedimentation analysis and electron microscopy of mitochondrial DNA from chick liver. *Biochim. Biophys. Acta* **149:**156–172.

Bosmann, H. B., 1970 Camptothecin inhibits macromolecular synthesis in mammalian cells but not in isolated mitochondria or *E. coli. Biochem. Biophys. Res. Commun.* **41:**1412–1420.

Bosmann, H. B., 1971*a* Mitochondrial biochemical events in a synchronized mammalian cell population. *J. Biol. Chem.* **246:**3817–3823.

Bosmann, H. B., 1971*b* Mitochondrial autonomy: Synthesis of DNA from RNA templates in isolated mammalian mitochondria. *FEBS Lett.* **19:**27–29.

Bosmann, H. B., M. W. Meyers, and H. R. Morgan, 1974 Synthesis of DNA, RNA, protein and glycoprotein in mitochondria of cells transformed with Rous sarcoma viruses. *Biochem. Biophys. Res. Commun.***56:**75–83.

Brack, C., E. Delain, and G. Riou, 1972 Replicating, covalently closed, circular DNA from kinetoplasts of *Trypanosoma cruzi. Proc. Natl. Acad. Sci. USA* **69:**1642–1646.

Brambl, R. M. and D. O. Woodward, 1972 Altered species of mitochondrial transfer RNA associated with the *mi-1* cytoplasmic mutation in *Neurospora crassa. Nature (London) New Biol.* **238:**198–200.

Braun, R. and T. E. Evans, 1969 Replication of nuclear satellite and mitochondrial DNA in the mitotic cycle of *Physarum. Biochim. Biophys. Acta* **182:**511–522.

Brown, I. H. and J. Vinograd, 1971 Sedimentation velocity properties of catenated DNA molecules from HeLa cell mitochondria. *Biopolymers* **10:**2015–2028.

Brown, W. M. and R. L. Hallberg, 1972 Relatedness of mitochondrial DNA in primates. *Fed. Proc.* **31:**426 (Abst. No. 1173).

Brown, W. M. and J. Vinograd, 1974 Restriction endonuclease cleavage maps of animal mitochondrial DNAs. *Proc. Natl. Acad. Sci. USA* **71:**4617–4621.

Bunn, C. L., D. C. Wallace, and J. M. Eisenstadt, 1974 Cytoplasmic inheritance of chloramphenicol resistance in mouse tissue culture cells. *Proc. Natl. Acad. Sci. USA* **71:**1681–1685.

Burton, P. R. and D. G. Dusanic, 1968 Fine structure and replication of the kinetoplast of *Trypanosoma lewisi. J. Cell. Biol.* **39:**318–331.

Calvayrac, R., R. A. Butow, and M. Lefort-Tran, 1972 Cyclic replication of DNA and changes in mitochondrial morphology during the cell cycle of *Euglena gracilis* (Z). *Exp. Cell Res.* **71:**422–432.

Carnevali, F. and L. Leoni, 1972 Intramolecular heterogeneity of yeast mitochondrial DNA. *Biochem. Biophys. Res. Commun.* **47:**1322–1331.

Carnevali, F., C. Falcone, L. Frontali, L. Leoni, G. Macino, and C. Palleschi, 1973 Informational content of mitochondrial DNA from a "low density" petite mutant of yeast. *Biochem. Biophys. Res. Commun.* **51:**651–658.

Carter, C. E., J. R. Wells, and A. J. Macinnis, 1972 DNA from anaerobic adult *Ascaris lumbricoides* and *Hymenolepis diminuta* mitochondria isolated by zonal centrifugation. *Biochim. Biophys. Acta* **262:**135–144.

Casey, J. W., H.-J. Hse, G. S. Getz, and M. Rabinowitz, 1974*a* Transfer RNA genes in mitochondrial DNA of *grande* (wild-type) yeast. *J. Mol. Biol.* **88:**735–747.

Casey, J. W., H.-J. Hsu, M. Rabinowitz, G. S. Getz, and H. Fukuhara, 1974*b* Transfer RNA genes in the mitochondrial DNA of cytoplasmic *petite* mutants of *Saccharomyces cerevisiae. J. Mol. Biol.* **88:**717–733.

Chanet, R., D. H. Williamson, and E. Moustacchi, 1973 Cyclic variations in killing and "petite" mutagenesis induced by ultra violet light in synchronized yeast strains. *Biochim. Biophys. Acta* **324:**290–299.

Chang, L. O., H. P. Morris, and W. B. Looney, 1968*a* Rates of incorporation of tritium-labeled thymidine into the mitochondrial and nuclear DNA of normal rat liver, regenerating liver, and four hepatomas with different growth rates and their host livers. *Br. J. Cancer* **22:**860–866.

Chang, L. O., H. P. Morris, and W. B. Looney, 1968*b* Comparative incorporation of tritiated thymidine and cytidine into mitochondrial and nuclear DNA and RNA of two transplantable hepatomas (3924A and H-35tc2) and host livers. *Cancer Res.* **28:**2164–2167.

Charret, R. and J. André, 1968 La synthèse de l'ADN mitochondrial chez *Tetrahymena pyriformis*: Etude radioautographique quantitative au microscope électronique. *J. Cell Biol.* **39:**369–381.

Chi, J. C. H. and Y. Suyama, 1970 Comparative studies on mitochondrial and cytoplasmic ribosomes of *Tetrahymena pyriformis. J. Mol. Biol.* **53:**531–556.

Ch'ih, J. J. and G. F. Kalf, 1969 Studies on the biosynthesis of the DNA polymerase of rat liver mitochondria. *Arch. Biochem. Biophys.* **133:**38–45.

Christiansen, C., A. Leth Bak, A. Stenderup, and C. Christiansen, 1971 Repetitive DNA in yeasts. *Nature (London) New Biol.* **231:**176–177.

Christiansen, C., G. Christiansen, and A. Leth Bak, 1974 Heterogeneity of mitochondrial DNA from *Saccharomyces carlsbergensis:* Renaturation and sedimentation studies. *J. Mol. Biol.* **84:**65–82.

Clayton, D. A. and R. M. Brambl, 1972 Detection of circular DNA from mitochondria of *Neurospora crassa. Biochem. Biophys. Res. Commun.* **46:**1477–1482.

Clayton, D. A. and J. Vinograd, 1967 Circular dimer and catenate forms of mitochondrial DNA in human leukemic leucocytes. *Nature (London)* **216:**652–657.

Clayton, D. A. and J. Vinograd, 1969 Complex mitochondrial DNA in leukemic and normal human myeloid cells. *Proc. Natl. Acad. Sci. USA* **62:**1077–1084.

Clayton, D. A., C. A. Smith, J. M. Jordan, M. Teplitz, and J. Vinograd, 1968 Occurence of complex mitochondrial DNA in normal tissues. *Nature (London)* **220:**976–979.

Clayton, D. A., R. W. Davis, and J. Vinograd, 1970 Homology and structural relationships between the dimeric and monomeric circular forms of mitochondrial DNA from human leukemic leucocytes. *J. Mol. Biol.* **47:**137–153.

Clayton, D. A., R. L. Teplitz, M. Nabholz, H. Dovey, and W. Bodmer, 1971 Mitochondrial DNA of human–mouse cell hybrids. *Nature (London)* **234:**560–562.

Clayton, D. A., J. N. Doda, and E. C. Friedberg, 1974 The absence of a pryimidine dimer repair mechanism in mammalian mitochondria. *Proc. Natl. Acad. Sci. USA* **71:**2777–2781.

Clegg, R. A., P. Borst, and P. J. Weijers, 1974 Intermediates in the replication of the mitochondrial DNA of *Tetrahymena pyriformis*. *Biochim. Biophys. Acta* **361:**277–287.

Coen, D., J. Deutsch, P. Netter, E. Petrochilo, and P. P. Slonimski, 1970 Mitochondrial genetics. I. Methodology and Phenomenology. *Symp. Soc. Exp. Biol.* **24:**449–496.

Cohen, M., J. Casey, J. Rabinowitz, and G. S. Getz, 1972 Hybridization of mitochondrial transfer RNA and mitochondrial DNA in petite mutants of yeast. *J. Mol. Biol.* **63:**441–451.

Coon, H. G., I. Horak, and I. B. Dawid, 1973 Propagation of both parental mitochondrial DNAs in rat–human and mouse–human hybrid cells. *J. Mol. Biol.* **81:**285–298.

Cummings, D. J., A. Tait, and J. M. Goddard, 1974 Methylated bases in DNA from *Paramecium aurelia*. *Biochim. Biophys. Acta* **374:**1–11.

Cummins, J. E., H. P. Rusch, and T. E. Evans, 1967 Nearest neighbor frequencies and the phylogenetic origin of mitochondrial DNA in *Physarum polycephalum*. *J. Mol. Biol.* **23:**281–284.

D'Agostino, M. A., K. M. Lowry, and G. F. Kalf, 1975 DNA biosynthesis in rat liver mitochondria: Inhibition by sulfhydryl compounds and stimulation by cytoplasmic proteins. *Arch. Biochem. Biophys.* **166:**400–416.

Dallman, P. R., 1971 Malnutrition: Incorporation of thymidine-^{3}H into nuclear and mitochondrial DNA. *J. Cell Biol.* **51:**549–553.

Danna, K. J., G. Sack, and D. Nathans, 1973 Studies of simian virus 40 DNA. VII. A cleavage map of the SV40 genome. *J. Mol. Biol.* **78:**363–376.

Dawid, I. B., 1966 Evidence for the mitochondrial origin of frog egg cytoplasmic DNA. *Proc. Natl. Acad. Sci. USA* **56:**269–276.

Dawid, I. B., 1972 Mitochondrial RNA in *Xenopus laevis*. I. The expression of the mitochondrial genome. *J. Mol. Biol.* **63:**201–216.

Dawid, I. B., 1974 5-Methylcytidylic acid: Absence from mitochondrial DNA of frogs and HeLa cells. *Science* **184:**80–81.

Dawid, I. B. and A. W. Blackler, 1972 Maternal and cytoplasmic inheritance of mitochondrial DNA in *Xenopus*. *Devl. Biol.* **29:**152–161.

Dawid, I. B. and D. D. Brown, 1970 The mitochondrial and ribosomal DNA components of oocytes of *Urechis caupo*. Devl. Biol. **22:**1–14.

Dawid, I. B. and D. R. Wolstenholme, 1967 Ultracentrifuge and electron microscope studies on the structure of mitochondrial DNA. *J. Mol. Biol.* **28:**233–245.

Dawid, I. B. and D. R. Wolstenholme, 1968 Renaturation and hybridization studies of mitochondrial DNA. *Biophys. J.* **8:**65–81.

Dawid, I. B., I. Horak, and H. G. Coon, 1974 Propagation and recombination of parental mt DNAs in hybrid cells. In *The Biogenesis of Mitochondria,* edited by A. M. Kroon and C. Saccone, Academic Press, New York, pp. 255–262.

DeFilippes, F. M. and M. M. K. Nass, 1975 Specific cleavage of mitochondrial DNA by the restriction enzymes Eco RI, Hpa I and Hpa II (manuscript submitted).

Deutsch, J., B. Dujon, P. Netter, E. Petrochilo, P. P. Slonimski, M. Bolotin-Fukuhara, and D. Coen, 1974 Mitochondrial genetics. VI. The petite mutation in *Saccharomyces cerevisiae: Interrelations between the loss of the* ρ^+ factor and the loss of the drug resistance mitochondrial genetic markers. *Genetics* **76:**195–219.

Diacumakos, E. G., L. Garnjobst, and E. L. Tatum, 1965 A cytoplasmic character in *Neurospora crassa*: The role of nuclei and mitochondria. *J. Cell Biol.* **26:**427–443.

Doeg, K. A., L. L. Polomski, and L. H. Doeg, 1972 Androgen control of mitochondrial and nuclear DNA synthesis in male sex accessory tissue of castrate rats. *Endocrinology* **90:**1633–1638.

Ehrlich, S. D., J. P. Thiery, and G. Bernardi, 1972 The mitochondrial genome of wild-type yeast cells. III. The pyrimidine tracts of mitochondrial DNA. *J. Mol. Biol.* **65:**207–212.

Ephrussi, B., 1953 *Nucleo-cytoplasmic Relations in Micro-organisms,* Clarendon Press, Oxford.

Evans, H. H. and T. E. Evans, 1970 Methylation of the deoxyribonucleic acid of *Physarum polycephalum* at various periods during the mitotic cycle. *J. Biol. Chem.* **245:**6436–6441.

Evans, T. E. and D. Suskind, 1971 Characterization of the mitochondrial DNA of the slime mold *Physarum polycephalum. Biochim. Biophys. Acta* **228:**350–364.

Fansler, B. S., E. C. Travaghini, L. A. Loeb, and J. Schulz, 1970 Structure of *Drosophila melanogaster* dAT replicated in an *in vitro* system. *Biochem. Biophys. Res. Commun.* **40:**1266–1272.

Flavell, R. A. and E. A. C. Follett, 1970 Size and configuration of *Tetrahymena* mitochondrial deoxyribonucleic acid. *Biochem. J.* **119:**61P.

Flavell, R. A. and I. G. Jones, 1970 Mitochondrial deoxyribonucleic acid from *Tetrahymena pyriformis* and its kinetic complexity. *Biochem. J.* **116:**811–817.

Flavell, R. A. and I. G. Jones, 1971 *Paramecium* mitochondrial DNA: Renaturation and hybridization studies. *Biochim. Biophys. Acta* **232:**255–260.

Flavell, R. A. and P. O. Trampé, 1973 The absence of an integrated copy of mitochondrial DNA in the nuclear genome of *Tetrahymena pyriformis. Biochim. Biophys. Acta* **308:**101–105.

Flavell, R. A., P. Borst, and J. Ter Schegget, 1972 DNA synthesis by isolated mitochondria. IV. Isolation of an intermediate containing newly synthesized DNA in full-length light strands. *Biochim. Biophys. Acta* **272:**341–349.

Flory, P. J. and J. Vinograd, 1973 5-Bromo-deoxyuridine labelling of monomeric and catenated circular mitochondrial DNA in HeLa cells. *J. Mol. Biol.* **74:**81–94.

Flury, U., H. R. Mahler, and F. Feldman, 1974 A novel respiration-deficient mutant of *Saccharomyces cerevisiae.* I. Preliminary characterization of phenotype and mitochondrial inheritance. *J. Biol. Chem.* **249:**6130–6137.

Fry, M. and A. Weissbach, 1973 A new deoxyribonucleic acid dependent deoxyribonucleic acid polymerase from HeLa mitochondria. *Biochemistry* **12:**3602–3608.

Fukuhara, H., 1970 Transcriptional origin of RNA in a mitochondrial fraction of yeast and its bearing on the problem of sequence homology between mitochondrial and nuclear DNA. *Mol. Gen. Genet.* **107:**58–70.

Fukuhara, H. and C. Kujawa, 1970 Selective inhibition of the *in vivo* transcription of mitochondrial DNA by ethidium bromide and by acriflavin. *Biochem. Biophys. Res. Commun.* **41:**1002–1008.

Fukuhara, H., G. Faye, F. Michel, J. Lazowska, J. Deutsch, M. Bolotin-Fukuhara, and P. P. Slonimski, 1974 Physical and genetic organization of petite and grande yeast mitochondrial DNA. I. Studies by RNA-DNA hybridization. *Mol. Gen. Genet.* **130:**215–238.

Gaitskhoki, V. S., O. I. Kisselev, and N. A. Klimov, 1973 Poly(A)-containing ribonucleic acid in mitochondria from rat liver and Krebs II ascitic carcinoma cells. *FEBS Lett.* **37:**260–263.

Gause, G. G., S. M. Dolgilevich, and V. S. Mikhailov, 1973 Heterogeneous rapidly labeled DNA with the properties of replicating form in isolated rat liver mitochondria. *Biochim. Biophys. Acta* **312:**179–191.

Georgatsos, J. G., J. Taylor-Papadimitriou, and T. Karemfyllis, 1972 Incorporation of labeled precursors into nuclear and mitochondrial DNA of cells grown in culture (36210). *Proc. Soc. Exp. Biol. Med.* **139:**663–666.

Goldring, E. S., L. I. Grossman, D. Krupnick, D. R. Cryer, and J. Marmur, 1970 The petite mutation in yeast: Loss of mitochondrial deoxyribonucleic acid during induction of petites with ethidium bromide. *J. Mol. Biol.* **52:**323–335.

Goldring, E. S., L. I. Grossman, and J. Marmur, 1971 Petite mutation in yeast. II. Isolation of mutants containing mitochondrial deoxyribonucleic acid of reduced size. *J. Bacteriol.* **107:**377–381.

Gouhier, M. and J. C. Movnolov, 1973 Yeast mutants resistant to ethidium bromide. *Mol. Gen. Genet.* **122:**149–164.

Gouhier-Monnerot, M., 1974 Ethidium bromide resistance and enhancement of mitochondrial recombination. *Mol. Gen. Genet.* **130:**65–79.

Granick, S. and A. Gibor, 1967 The DNA of chloroplasts, mitochondria and centrioles. In *Progress in Nucleic Acid Research and Molecular Biology,* Vol. 6, edited by J. N. Davidson and W. E. Cohn, Academic Press, New York, pp. 143–186.

Grimes, G. W., H. R. Mahler, and P. S. Perlman, 1974 Nuclear gene dosage effects on mitochondrial mass and DNA. *J. Cell Biol.* **61:**565–574.

Gross, N. J. and M. Rabinowitz, 1969 Synthesis of new strands of mitochondrial and nuclear deoxyribonucleic acid by semiconservative replication. *J. Biol. Chem.* **244:**1563–1566.

Gross, N. J., G. S. Getz, and M. Rabinowitz, 1969 Apparent turnover of mitochondrial deoxyribonucleic acid and mitochondrial phospholipids in the tissues of the rat. *J. Biol. Chem.* **244:**1552–1562.

Grossman, L. I., D. R. Cryer, E. S. Goldring, and J. Marmur, 1971 The petite mutation in yeast. III. Nearest-neighbor analysis of mitochondrial DNA from normal and mutant cells. *J. Mol. Biol.* **62:**565–575.

Grossman, L. I., R. Watson, and J. Vinograd, 1973 The presence of ribonucleotides in mature closed-circular mitochondrial DNA. *Proc. Natl. Acad. Sci. USA* **70:**3339–3343.

Grossman, L. I., R. Watson, and J. Vinograd, 1974 Restricted uptake of eithidium bromide and propidium diiodide by denatured closed circular DNA in buoyant cesium chloride. *J. Mol. Biol.* **86:**271–283.

Halbreich, A. and M. Rabinowitz, 1971 Isolation of *Saccharomyces cerevisiae* mitochondrial formyltetrahydrofolic acid: Methionyl tRNA transformylase and the hybridization of mitochondrial fMet-tRNA with mitochondrial DNA. *Proc. Natl. Acad. Sci. USA* **68:**294–298.

Hallberg, R. L., 1974 Mitochondrial DNA in *Xenopus laevis* oocytes. I. Displacement loop occurrence. *Dev. Biol.* **38:**346–355.

Handel, M. A., J. Papaconstantinou, D. P. Allison, E. M. Julku, and E. T. Chin, 1973 Synthesis of mitochondrial DNA in spermatocytes of *Rhynchiosciara hollaenderi. Dev. Biol.* **35:**240–249.

Hettiarchchy, N. Ș. and I. G. Jones, 1974 Isolation and characterization of mitochondrial deoxyribonucleic acid of *Acanthamoeba castellanii. Biochem. J.* **141:**159–164.

Hirsch, M. and S. Penman, 1973 Mitochondrial polyadenylic acid-containing RNA: Localization and characterization. *J. Mol. Biol.* **80:**379–391.

Hirsch, M. and S. Penman, 1974 Post-transcriptional addition of polyadenylic acid to mitochondrial RNA by a cordycepin-insensitive process. *J. Mol. Biol.* **83:**131–142.

Hoffman, H. P. and C. J. Avers, 1973 Mitochondrion of yeast: Ultrastructural evidence for one giant, branched organelle per cell. *Science* **181:**749–751.

Hollenberg, C. P., P. Borst, and E. F. J. Van Bruggen, 1970 Mitochondrial DNA. V. A 25-μ closed circular duplex DNA molecule in wild-type yeast mitochondria: Structure and genetic complexity. *Biochim. Biophys. Acta* **209:**1–15.

Horwitz, H. B. and C. E. Holt, 1971 Specific inhibition by ethidium bromide of mitochondrial DNA synthesis in *Physarum polycephalum. J. Cell Biol.* **49:**546–553.

Howell, N., M. K. Trembach, A. W. Linnane, and H. B. Lukins, 1973 Biogenesis of mitochondria. 30. An analysis of polarity of mitochondrial gene recombination and transmission. *Mol. Gen. Genet.* **122:**37–51.

Hudson, B. and J. Vinograd, 1967 Catenated circular DNA molecules in HeLa cell mitochondria. *Nature (London)* **216:**647–652.

Hudson, B. and J. Vinograd, 1969 Sedimentation velocity properties of complex mitochondrial DNA. *Nature (London)* **221:**332–337.

Hudson, B., W. B. Upholt, J. Devinny, and J. Vinograd, 1969 The use of an ethidium analogue in the dye-buoyant density procedure for the isolation of closed circular DNA: The variation of the superhelix density of mitochondrial DNA. *Proc. Natl. Acad. Sci. USA* **62:**813–820.

Hunter, G. R., G. F. Kalf, and H. P. Morris, 1973 Partial characterization of the DNA-dependent DNA polymerases of rat liver and hepatoma. *Cancer Res.* **33:**987–992.

Inman, R. B., 1966 A denaturation map of the λ phage DNA molecule determined by electron microscopy. *J. Mol. Biol.* **18:**464–476.

Iwashima, A. and M. Rabinowitz, 1969 Partial purification of mitochondrial and supernatant DNA polymerase from *Saccharomyces cerevisiae. Biochim. Biophys. Acta* **178:**283–293.

Jeffreys, A. and I. Craig, 1974 Differences in the products of mitochondrial protein synthesis *in vivo* in human and mouse cells and their potential use as markers for the mitochondrial genome in human–mouse somatic cell hybrids. *Biochem. J.* **144:**161–164.

Kalf, G. and J. J. Ch'ih, 1968 Purification and properties of deoxyribonucleic acid polymerase from rat liver mitochondria. *J. Biol. Chem.* **243:**4904–4916.

Karol, M. and M. V. Simpson, 1968 DNA biosynthesis by isolated mitochondria: A replicative rather than a repair process. *Science* **162:**470–472.

Kasamatsu, H. and J. Vinograd, 1974 Replication of circular DNA in eukaryotic cells. *Ann. Rev. Biochem.* **43:**695–719.

Kasamatsu, H., D. L. Robberson, and J. Vinograd, 1971 A novel closed circular mitochondrial DNA with properties of a replicating intermediate. *Proc. Natl. Acad. Sci. USA* **68:**2252–2257.

Kato, K., K. D. Radsak, and H. Koprowski, 1972 Differential effect of ethidium bromide and cytosine arabinoside on mitochondrial and nuclear DNA synthesis in HeLa cells. *Z. Naturforsch.* **27b:**989–991.

Keiding, J. and O. Westergaard, 1971 Induction of DNA polymerase activity in irradiated Tetrahymena cells. *Exp. Cell Res.* **64:**317–322.

Khoury, G., M. Martin, T. N. H. Lee, K. Danna, and D. Nathens, 1973 A map of simian virus 40 transcription sites expressed in productively infected cells. *J. Mol. Biol.* **78:**377–389.

Kimberg, D. V. and J. N. Loeb, 1972 Effects of cortisone administration on rat liver mitochondria: Support for the concept of mitochondrial fusion. *J. Cell Biol.* **55:**635–643.

Kislev, N., H. Swift, and L. Bogorod, 1965 Nucleic acids of chloroplasts and mitochondria in Swiss chard. *J. Cell Biol.* **25:**327–344.

Kit, S. and W.-C. Leung, 1974 Genetic control of mitochondrial thymidine kinase in human–mouse and monkey–mouse somatic cell hybrids. *J. Cell Biol.* **61:**35–44.

Kit, S. and Y. Minekawa, 1972 Mitochondrial thymidine-deoxyuridine-phosphorylating activity and the replication of mitochondrial DNA. *Cancer Res.* **32:**2277–2288.

Kleese, R. A., R. C. Grotbeck, and J. R. Snyder, 1972 Recombination among three mitochondrial genes in yeast (*Saccharomyces cerevisiae*). *J. Bacteriol.* **112:**1023–1025.

Klietmann, W., N. Sato, and M. M. K. Nass, 1973 Establishment and characterization of ethidium bromide resistance in simian virus 40-transformed hamster cells: Effects on mitochondria *in vivo*. *J. Cell Biol.* **58:**11–26.

Koch, J., 1972 The cytoplasmic DNAs of cultured human cells: Effects of ethidium bromide on their replication and maintenance. *Eur. J. Biochem.* **30:**53–59.

Koch, J. and E. L. R. Stokstad, 1967 Incorporation of (^{3}H) thymidine into nuclear and mitochondrial DNA in synchronized mammalian cells. *Eur. J. Biochem.* **3:**1–6.

Koike, K. and D. R. Wolstenholme, 1974 Evidence for discontinuous replication of circular mitochondrial DNA molecules from Novikoff rat ascites hepatoma cells. *J. Cell Biol.* **61:**14–25.

Koike, K., M. Kobayashi, and T. Fujisawa, 1974 Novel properties of two classes of nascent mitochondrial DNA formed *in vitro*. *Biochim. Biophys. Acta* **361:**144–154.

Kolodner, R. and K. K. Tewari, 1972 Physicochemical characterization of mitochondrial DNA from pea leaves. *Proc. Natl. Acad. Sci. USA* **69:**1830–1834.

Kroon, A. M. and C. Saccone, editors, 1974 *The Biogenesis of Mitochondria, Transcriptional, Translational and Biochemical Aspects.* Proceedings of the International Conference on the Biogenesis of Mitochondria, Bari, Italy, 1973, Academic Press, New York.

Küenzi, M. T. and R. Roth, 1974 Timing of mitochondrial DNA synthesis during meiosis in *Saccharomyces cerevisiae*. *Exp. Cell Res.* **85:**377–382.

Kung, S. D., M. A. Moscarello, and J. P. Williams, 1972 Studies with chloroplast and mitochondrial DNA. I. Evidence of sequence homology between chloroplast and nuclear DNA (broad beam) and between mitochondrial and nuclear DNA (rat liver). *Biophys. J.* **12:**474–583.

Küntzel, H. and K. P. Schäfer, 1971 Mitochondrial RNA polymerase from *Neurospora crassa*. *Nature* (*London*) *New Biol.* **231:**265–269.

Lagueus, R., P. C. Meckert, and A. Segal, 1972 Effect of acriflavine on the fine structure of the heart muscle cell mitochondria of normal and exercised rats. *J. Mol. Cell Cardiol.* **4:**185–193.

Laurent, M. and M. Steinert, 1970 Electron microscopy of kinetoplastic DNA from *Trypanosoma mega*. *Proc. Natl. Acad. Sci. USA* **66:**419–424.

Lazowska, J., F. Michel, G. Faye, H. Fukuhara, and P. P. Slonimski, 1974 Physical and genetic organization of *petite* and *grande* yeast mitocondrial DNA. II. DNA-DNA hybridization studies and buoyant density determinations. *J. Mol. Biol.* **85:**393–410.

Lee, A. S. and R. L. Sinsheimer, 1974 A cleavage map of bacteriophage ϕX174 genome. *Proc. Natl. Acad. Sci. USA* **71:**2882–2886.

Levine, A. J., 1971 Induction of mitochondrial DNA synthesis in monkey cells infected by simian virus 40 and (or) treated with calf serum. *Proc. Natl. Acad. Sci. USA* **68:**717–720.

Ley, K. D. and M. M. Murphy, 1973 Synchronization of mitochondrial DNA synthesis in Chinese hamster cells (line CHO) deprived of isoleucine. *J. Cell Biol.* **58:**340–345.

Linnane, A. W., J. M. Haslam, H. B. Lukins, and P. Nagley, 1972 The biogenesis of mitochondria in microorganisms. *Ann. Rev. Microbiol.* **26:**163–198.

Locker, J., M. Rabinowitz, and G. S. Getz, 1974*a* Electron microscopic and renaturation kinetic analysis of mitochondrial DNA of cytoplasmic petite mutants of *Saccharomyces cerevisiae*. *J. Mol. Biol.* **88:**489–502.

Locker, J., M. Rabinowitz, and G. S. Getz, 1974*b* Tandem inverted repeats in mitochondrial DNA of petite mutants of *Saccharomyces cerevisiae*. *Proc. Natl. Acad. Sci. USA* **71:**1366–1370.

Lonsdale, D. M. and I. G. Jones, 1974 Ribonuclease-sensitivity of covalently closed rat liver mitochondrial deoxyribonucleic acid. *Biochem. J.* **141:**155–158.

Madreiter, H., C. Mittermayer, and R. Osieka, 1972 ^{3}H-Thymidine incorporation into mitochondria of synchronized mouse fibroblasts. *Beitr. Pathol.* **145:**249–255.

Mahler, H. R. and R. N. Bastos, 1974 A novel reaction of mitochondrial DNA with ethidium bromide. *FEBS Lett.* **39:**27–34.

Mahler, H. R. and P. S. Perlman, 1972 Mitochondrial membranes and mutagenesis by ethidium bromide. *J. Supramol. Struct.* **1:**105–124.

Manning, J. E., D. R. Wolstenholme, R. S. Ryan, J. A. Hunter, and O. C. Richards, 1971 Circular chloroplast DNA from *Euglena gracilis*. *Proc. Natl. Acad. Sci. USA* **68:**1169–1173.

Manton, I., 1959 Electron microscopical observations on a very small flagellate: The problem of *Chromulina pusilla* Butcher. *J. Mar. Biol. Assoc. U.K.* **38:**319–333.

Margulis, L., 1974 The classification and evolution of prokaryotes and eukaryotes. In *Handbook of Genetics*, Vol. 1, edited by R. C. King, Plenum Press, New York, pp. 1–41.

Merker, H. J., R. Herbst, and K. Kloss, 1968 Elektronenmikroskopische Untersuchungen an den Mitochondrien des menschlichen Uterusepithels während der Sekretionsphase. *Z. Zellforsch.* **86:**139–152.

Meselson, M., R. Yuan, and J. Heywood, 1972 Restriction and modification of DNA. *Annu. Rev. Biochem.* **41:**447–500.

Meyer, R. R., 1973 On the evolutionary origin of mitochondrial DNA. *J. Theor. Biol.* **38:**647–663.

Meyer, R. R. and M. V. Simpson, 1970 Deoxyribonucleic acid biosynthesis of mitochondria: Purification and general properties of rat liver mitochondrial deoxyribonucleic acid polymerase. *J. Biol. Chem.* **245:**3426–3435.

Michaelis, G. E., E. Petrochilo, and P. P. Slonimski, 1973 Mitochondrial genetics. III. Recombined molecules of mitochondrial DNA obtained from crosses between cytoplasmic petite mutants of *S. cerevisiae:* Physical and genetic characterization. *Mol. Gen. Genet.* **123:**51–65.

Michel, F., J. Lazowska, G. Faye, H. Fukuhara, and P. P. Slonimski, 1974 Physical and genetic organization of *petite* and *grande* yeast mitochondrial DNA. III. High resolution melting and reassociation studies. *J. Mol. Biol.* **85:**441–431.

Mitra, R. S. and I. A. Bernstein, 1970 Thymidine incorporation into deoxyribonucleic acid by isolated rat liver mitochondria. *J. Biol. Chem.* **245:**1255–1260.

Miyaki, M., K. Koide, and T. Ono, 1973 RNase and alkali sensitivity of closed circular mitochondrial DNA of rat ascites hepatoma cells. *Biochim. Biophys. Res. Commun.* **50:**252–258.

Mol, J. N. M., P. Borst, F. G. Grosveld, and J. H. Spencer, 1974 The size of the repeating unit of the repetitive mitochondrial DNA from a "low-density" petite mutant of yeast. *Biochim. Biophys. Acta* **374**–115–128.

Morrow, J. F. and P. Berg, 1972 Cleavage of simian virus 40 DNA at a unique site by a bacterial restriction enzyme. *Proc. Natl. Acad. Sci.* **69:**3365–3369.

Nagley, P. and A. W. Linnane, 1972 Biogenesis of mitochondria. XXI. Studies on the nature of the mitochondrial genome in yeast: The degenerative effects of ethidium bromide on mitochondrial genetic information in a respiratory competent strain. *J. Mol. Biol.* **66:**181–193.

Nagley, P., E. B. Gingold, H. B. Lukins, and A. W. Linnane, 1973 Biogenesis of mitochondria. XXV. Studies on the mitochondrial genomes of petite mutants of yeast using ethidium bromide as a probe. *J. Mol. Biol.* **78:**335–350.

Nass, M. M. K., 1966 The circularity of mitochondrial DNA. *Proc. Natl. Acad. Sci. USA* **56:**1215–1222.

Nass, M. M. K., 1967 Circularity and other properties of mitochondrial DNA in animal cells. In *Organizational Biosynthesis,* edited by H. J. Vogel, J. O. Lampen, and V. Bryson, Academic Press, New York, pp. 503–522.

Nass, M. M. K., 1968 Properties of organelle-associated and isolated mitochondrial DNA. In *Biochemical Aspects of the Biogenesis of Mitochondria,* edited by E. C. Slater, J. M. Tager, S. Papa, and E. Quagliariello, Adriatica Editrice, Bari, pp. 27–50.

Nass, M. M. K., 1969*a* Mitochondrial DNA: Advances, problems and goals. *Science* **165:**25–35.

Nass, M. M. K., 1969*b* Mitochondrial DNA. I. Intramitochondrial distribution and structural relations of single- and double-length circular DNA. *J. Mol. Biol.* **42:**521–528.

Nass, M. M. K., 1969*c* Mitochondrial DNA. II. Structure and physicochemical properties of isolated DNA. *J. Mol. Biol.* **42:**529–545.

Nass, M. M. K., 1969*d* Reversible generation of circular dimer and higher multiple forms of mitochondrial DNA. *Nature* (*London*) **223:**1124–1129.

Nass, M. M. K., 1970 Abnormal DNA patterns in animal mitochondria: Ethidium bromide-induced breakdown of closed circular DNA and conditions leading to oligomer accumulation. *Proc. Natl. Acad. Sci. USA* **67:**1926–1933.

Nass, M. M. K., 1971*a* Properties and biological significance of mitochondrial DNA. In *Biological Ultrastructure: The Origin of Cell Organelles,* edited by P. Harris, Oregon State University Press, pp. 41–63.

Nass, M. M. K., 1971*b* Origin of mitochondria: Are they descendants of ancestral bacteria? *Triangle* **10:**29–36.

Nass, M. M. K., 1972 Differential effects of ethidium bromide on mitochondrial and nuclear DNA synthesis *in vivo* in cultured mammalian cells. *Exp. Cell Res.* **72:**211–222.

Nass, M. M. K., 1973*a* Differential methylation of mitochondrial and nuclear DNA in cultured mouse, hamster and virus-transformed hamster cells: *In vivo* and *in vitro* methylation. *J. Mol. Biol.* **80:**155–175.

Nass, M. M. K., 1973*b* Temperature-dependent formation of dimers and oligomers of mitochondrial DNA in cells transformed by a thermosensitive mutant of Rous sarcoma virus. *Proc. Natl. Acad. Sci. USA* **70:**3739–3743.

Nass, M. M. K., 1974*a* Structure, synthesis and transcription of mitochondrial DNA in normal, malignant, and drug-treated cells. In *Hormones and Cancer,* edited by K. W. McKerns, Academic Press, New York, pp. 261–307.

Nass, M. M. K., 1974*b* Studies on the synthesis and structure of mitochondrial DNA in cells infected by Rous sarcoma viruses and on the occurrence of intramitochondrial virus-like particles in certain RSV-induced tumor cells. In *International Symposium on the Genetics of Mitochondria,* Leningrad, USSR, *Mol. Cell. Biochem.* (in press).

Nass, M. M. K. and Y. Ben-Shaul, 1973 Effects of ethidium bromide on growth, chlorophyll synthesis, ultrastructure and mitochondrial DNA in green and bleached mutant *Euglena gracilis*. *J. Cell Sci.* **13:**567–590.

Nass, M. M. K. and C. A. Buck, 1970 Studies on mitochondrial tRNA from animal cells. II. Hybridization of aminoacyl-tRNA from rat liver mitochondria with heavy and light complementary strands of mitochondrial DNA. *J. Mol. Biol.* **54:**187–198.

Nass, M. M. K. and S. Nass, 1963 Intramitochondrial fibers with DNA characteristics. I. Fixation and electron staining reactions. *J. Cell Biol.* **19:**593–611.

Nass, M. M. K., S. Nass, and B. A. Afzelius, 1965 The general occurrence of mitochondrial DNA. *Exp. Cell Res.* **37:**516–539.

Nass, M. M. K., L. Schori, Y. Ben-Shaul, and M. Edelman, 1974 Size and configuration of mitochondrial DNA in *Euglena gracilis*. *Biochim. Biophys. Acta* **374:**283–291.

Nass, S., 1969 The significance of the structural and functional similarities of bacteria and mitochondria. *Int. Rev. Cytol.* **25:**55–129.

Nass, S. and M. M. K. Nass, 1963 Intramitochondrial fibers with DNA characteristics. II. Enzymatic and other hydrolytic treatments. *J. Cell Biol.* **19:**613–629.

Neubert, D., E. Oberdisse, and R. Bass, 1968 Biosynthesis and degradation of mammalian mitochondrial DNA. In *Biochemical Aspects of the Biogenesis of Mitochondria,* edited by E. C. Slater, J. M. Tager, S. Papa, and E. Quagliariello, Adriatica Editrice, Bari, Italy, pp. 103–122.

Ojala, D. and G. Attardi, 1974 Identification and partial characterization of multiple discrete polyadenylic acid-containing RNA components coded for by HeLa cell mitochondrial DNA. *J. Mol. Biol.* **88:**205–219.

Paoletti, C. and G. Riou, 1970 Le DNA mitochondrial des cellules malignes. *Bull. Cancer* **57:**301–334.

Paoletti, C. A. and G. Riou, 1973 The mitochondrial DNA of malignant cells. In *Progress in Molecular and Subcellular Biology,* Vol. 3, Springer-Verlag, New York, pp. 203–248.

Paoletti, C., G. Riou, and J. Pairault, 1972 Circular oligomers in mitochondrial DNA of human and beef monmalignant thyroid glands. *Proc. Natl. Acad. Sci. USA* **69:**847–850.

Parsons, J. A., 1965 Mitochondrial incorporation of tritiated thymidine in *Tetrahymena pyriformis*. *J. Cell Biol.* **25:**641–646.

Parsons, J. A. and R. C. Rustad, 1968 The distribution of DNA among dividing mitochondria of *Tetrahymena pyriformis*. *J. Cell Biol.* **37:**683–693.

Parsons, P. and M. V. Simpson, 1973 Deoxyribonucleic acid biosynthesis in mitochondria: Studies on the incorporation of labeled precursors into mitochondrial deoxyribonucleic acid. *J. Biol. Chem.* **248:**1912–1919.

Perlman, P. S. and H. R. Mahler, 1971 Molecular consequences of ethidium bromide mutagenesis. *Nature (London) New Biol.* **231:**12–16.

Pica-Mattoccia, L. and G. Attardi, 1972 Expression of the mitochondrial genome in HeLa cells. IX. Replication of mitochondrial DNA in relationship to the cell cycle in HeLa cells. *J. Mol. Biol.* **64:**465–484.

Pikó, L., A. Tyler, and J. Vinograd, 1967 Amount, location, priming capacity, circularity and other properties of cytoplasmic DNA in sea urchin eggs. *Biol. Bull.* **132:**68–90.

Piperno, G., G. Fonti, and G. Bernardi, 1972 The mitochondrial genome of wild-type yeast cells. II. Investigations on the compositional heterogeneity of mitochondrial DNA. *J. Mol. Biol.* **65:**191–205.

Polan, M. L., S. Friedman, J. G. Gall, and W. Gehring, 1973 Isolation and characterization of mitochondrial DNA from *Drosophila melanogaster*. *J. Cell Biol.* **56**:580–589.

Porcher, H. H. and J. Koch, 1973 The anatomy of the mitochondrial DNA: The localization of the heat-induced and RNAse-induced scissions in the phosphatediester backbones. *Eur. J. Biochem.* **40**:329–336.

Probst, G. S. and R. R. Meyer, 1973 Subcellular localization of high and low molecular weight DNA polymerases of rat liver. *Biochem. Biophys. Res. Commun.* **50**:111–117.

Prunell, A. and G. Bernardi, 1974 The mitochondrial genome of wild-type yeast cells. IV. Genes and spacers. *J. Mol. Biol.* **86**:825–841.

Quetier, F. and F. Vedel, 1973 Interaction of polyribonucleotides with plant mitochondrial DNA. *Biochem. Biophys. Res. Commun.* **54**:1326–1334.

Rabinowitz, M. and H. Swift, 1970 Mitochondrial nucleic acids and their relation to the biogenesis of mitochondria. *Physiol. Rev.* **50**:376–427.

Radloff, R., W. Bauer, and J. Vinograd, 1967 A dye-buoyant density method for the detection and isolation of closed circular DNA: The closed circular DNA in HeLa cells. *Proc. Natl. Acad. Sci. USA* **57**:1514–1521.

Radsak, K. D. and H. W. Freise, 1972 Stimulation of mitochondrial DNA synthesis in HeLa cells by herpes simplex virus (1). *Life Sci. Part II* **11**:717–724.

Raff, R. A. and H. R. Mahler, 1972 The non-symbiotic origin of mitochondria. *Science* **177**:575–582.

Ray, D. S. and P. C. Hanawalt, 1965 Satellite DNA components in *Euglena gracilis* cells lacking chloroplasts. *J. Mol. Biol.* **11**:760–768.

Reich, E. and D. J. L. Luck, 1966 Replication and inheritance of mitochondrial DNA. *Proc. Natl. Acad. Sci. USA* **55**:1600–1608.

Reijnders, L. and P. Borst, 1972 The number of 4-S RNA genes on yeast mitochondrial DNA. *Biochem. Biophys. Res. Commun.* **47**:126–133.

Reijnders, L., C. M. Kleisen, L. A. Grivell, and P. Borst, 1972 Hybridization studies with yeast mitochondrial RNA's. *Biochim. Biophys. Acta* **272**:396–407.

Reitz, M. S., Jr., R. G. Smith, E. A. Roseberry, and R. C. Gallo, 1974 DNA-directed and RNA-primed DNA synthesis in microsomal and mitochondrial fractions of normal human lymphocytes. *Biochem. Biophys. Res. Commun.* **57**:934–948.

Renger, H. R. and D. R. Wolstenholme, 1972 The form and structure of kinetoplast DNA of *Crithidia*. *J. Cell Biol.* **54**:346–364.

Révet, B. M. J., M. Schmir, and J. Vinograd, 1971 Direct determination of the superhelix density of closed circular DNA by viscometric titration. *Nature (London) New Biol.* **229**:10–13.

Richards, O. C. and R. S. Ryan, 1974 Synthesis and turnover of *Euglena gracilis* mitochondrial DNA. *J. Mol. Biol.* **82**:57–75.

Rifkin, M. R. and D. J. L. Luck, 1971 Defective production of mitochondrial ribosomes in the *poky* mutant of *Neurospora crassa*. *Proc. Natl. Acad. Sci. USA* **68**:287–290.

Riou, G., 1968 Disposition de l'ADN du kinetoplaste de *Trypanosoma cruzi* cultivé au présence de bromure d'ethidium. *C. R. Acad. Sci.* **266**:250–252.

Riou, G. and E. Delain, 1969 Abnormal circular DNA molecules induced by ethidium bromide in the kinetoplast of *Trypanosoma cruzi*. *Proc. Natl. Acad. Sci. USA* **64**:618–625.

Robberson, D. L. and D. A. Clayton, 1972 Replication of mitochondrial DNA in mouse L cells and their thymidine kinase derivatives: Displacement replication on a covalently closed circular template. *Proc. Natl. Acad. Sci.* **69**:3810–3814.

Robberson, D. L. and D. A. Clayton, 1973 Pulse-labeled components in the replication of mitochondrial deoxyribonucleic acid. *J. Biol. Chem.* **248:**4512–4514.

Robberson, D., Y. Aloni, G. Attardi, and N. Davidson, 1972*a* Expression of the mitochondrial genome in HeLa cells. VIII. The relative position of ribosomal RNA genes in mitochondrial DNA. *J. Mol. Biol.* **64:**313–317.

Robberson, D. L., H. Kasamatsu, and J. Vinograd, 1972*b* Replication of mitochondrial DNA: Circular replicative intermediates in mouse L cells. *Proc. Natl. Acad. Sci. USA* **69:**737–741.

Robberson, D. L., D. A. Clayton, and J. F. Morrow, 1974 Cleavage of replicating forms of mitochondrial DNA by EcoR1 endonuclease. *Proc. Natl. Acad. Sci. USA* **71:**4447–4451.

Roodyn, D. B. and D. Wilkie, 1968 *The Biogenesis of Mitochondria,* Methuen, London.

Ruttenberg, G. J. C. M., E. M. Smith, P. Borst, and E. F. J. Van Bruggen, 1968 The number of superhelical turns in mitochondrial DNA. *Biochim. Biophys. Acta* **157:**429–432.

Sack, G. H., Jr. and D. Nathans, 1973 Studies of SV40 DNA. VI. Cleavage of SV40 DNA by restriction endonuclease from *Hemophilus parainfluenzae. Virology* **51:**517–520.

Sager, R., 1972 *Cytoplasmic Genes and Organelles,* Academic Press, New York.

Sanders, J. P. M., R. A. Flavell, P. Borst, and J. N. M. Mol, 1973 Nature of the base sequence conserved in the mitochondrial DNA of a low-density petite. *Biochem. Biophys. Acta* **312:**441–457.

Sanders, J. P. M., P. J. Weijders, G. S. P. Groot, and P. Borst, 1974 Properties of mitochondrial DNA from *Kluyveromyces lactis. Biochim. Biophys. Acta* **374:**136–144.

Schäfer, K. P. and H. Küntzel, H., 1972 Mitochondrial genes in *Neurospora*: A single cistron for ribosomal RNA. *Biochem. Biophys. Res. Commun.* **46:**1312–1319.

Schäfer, K. P., G. Bugge, M. Grandi, and H. Küntzel, 1971 Transcription of mitochondrial DNA *in vitro* from *Neurospora crassa. Eur. J. Biochem.* **21:**478–488.

Schatz, G., 1970 Biogenesis of mitochondria. In *Membranes of Mitochondria and Chloroplasts,* edited by E. Racker, Van Nostrand Reinhold, New York, pp. 251–314.

Schatz, G. and T. L. Mason, 1974 The biosynthesis of mitochondrial proteins. *Annu. Rev. Biochem.* **43:**51–87.

Schmir, M., B. M. J. Révet, and J. Vinograd, 1974 Dependence of the sedimentation coefficient of denatured closed circular DNA in alkali on the degree of strand interwinding: The absolute sense of supercoils. *J. Mol. Biol.* **83:**35–45.

Schuster, F. L., 1965 A deoxyribose nucleic acid component in mitochondria of *Didymium nigripes,* a slime mold. *Exp. Cell Res.* **39:**329–45.

Schutgens, R. B. H., L. Reijnders, S. P. Hoekstra, and P. Borst, 1973 Transcription of *Tetrahymena* mitochondrial DNA *in vivo. Biochim. Biophys. Acta* **308:**372–380.

Sharp, P. A., B. Sugden, and J. Sambrook, 1973 Detection of two restriction endonuclease activities in *Haemophilus parainfluenzae* using analytical agarose–ethidium bromide electrophoresis. *Biochemistry* **12:**3055–3063.

Sheid, B., P. R. Srinivasan, and E. Borek, 1968 Deoxyribonucleic acid methylase of mammalian tissues. *Biochemistry* **7:**280–285.

Sherman, F. and C. W. Lawrence, 1974 Saccharomyces. In *Handbook of Genetics,* Vol. 1, edited by R. C. King, Plenum Press, New York, pp. 359–393.

Shipman, C., Jr., S. H. Smith, and J. C. Drach, 1972 Selective inhibition of nuclear DNA synthesis by 9-β-D-arabinofuranosyl adenine in rat cells transformed by Rous sarcoma virus. *Proc. Natl. Acad. Sci. USA* **69:**1753–1757.

Sigee, D. C., 1972 Pattern of cytoplasmic DNA synthesis in somatic cells of *Pteridium aquilinum*. *Exp. Cell Res.* **73**:481–486.

Simpson, L., 1972 The kinetoplast of the hemoflagellates. *Int. Rev. Cytol.* **32**:139–207.

Skinner, D. M. and M. S. Kerr, 1971 Characterization of mitochondrial and nuclear satellite deoxyribonucleic acids of five species of Crustacea. *Biochemistry* **10**:1864–1872.

Slater, E. C., J. M. Tager, S. Papa, and E. Quagliariello, editors, 1968 *Biochemical Aspects of the Biogenesis of Mitochondria,* Adriatica Editrice, Bari, Italy.

Slonimski, P. P., G. Perrodin, and J. H. Croft, 1968 Ethidium bromide induced mutation of yeast mitochondria: complete transformation of cells into respiratory deficient non-chromosomal "petites." *Biochem. Biophys. Res. Commun.* **30**:232–239.

Smith, C. A. and J. Vinograd, 1973 Complex mitochondrial DNA in human tumors. *Cancer Res.* **33**:1065–1070.

Smith, C. A., J. M. Jordan, and J. Vinograd, 1971 *In vivo* effects of intercalating drugs on the superhelix density of mitochondrial DNA isolated from human and mouse cells in culture. *J. Mol. Biol.* **59**:255–272.

Smith, D., P. Tauro, E. Schweitzer, and H. O. Halvorson, 1968 The replication of mitochondrial DNA during the cell cycle in *Saccharomyces lactis*. *Proc Natl. Acad. Sci. USA* **60**:936–942.

Sonenshein, G. E. and C. H. Holt, 1968 Molecular weight of mitochondrial DNA in *Physarum polycephalum*. *Biochim. Biophys. Res. Commun.* **33**:361–367.

Soriano, L. and Y. Croiselle, 1973 Primer-template specificity of DNA polymerases bound to mitochondria and nuclei. *FEBS Lett.* **31**:143–148.

Soslau, G., J. P. Fuhrer, M. M. K. Nass, and L. Warren, 1974 The effect of ethidium bromide on the membrane glycopeptides in control and virus-transformed cells. *J. Biol. Chem.* **249**:3014–3020.

Spolsky, C. M. and J. M. Eisenstadt, 1972 Chloramphenicol-resistant mutants of human HeLa cells. *FEBS Lett.* **25**:319–324.

Stevens, B. J., 1974 Variation in mitochondrial numbers and volume in yeast according to growth conditions. *J. Cell Biol.* **63**:Abst. No. 671.

Stone, G. E. and O. L. Miller, 1965 A stable mitochondrial DNA in *Tetrahymena pyriformis*. *J. Exp. Zool.* **159**:33–38.

Suyama, Y. and K. Miura, 1968 Size and structural variations of mitochondrial DNA. *Proc. Natl. Acad. Sci.* **60**:235–242.

Swift, H. and D. R. Wolstenholme, 1969 Mitochondria and chloroplasts: Nucleic acids and the problem of biogenesis (genetics and biology). In *Handbook of Molecular Cytology,* edited by A. Lima-de-Faria, Wiley, New York, pp. 972–1046.

Szollosi, D. G., 1965 The fate of sperm middlepiece mitochondria in the rat egg. *J. Exp. Zool.* **159**:367–378.

Tabak, H. F., P. Borst, and A. J. H. Tabak, 1973 Search for mitochondrial DNA sequences in chick nuclear DNA. *Biochim. Biophys. Acta* **294**–184–191.

Talen, J. L., J. P. M. Sanders, and R. A. Flavell, 1974 Genetic complexity of mitochondrial DNA from *Euglena gracilis*. *Biochim. Biophys. Acta.* **374**:129–135.

Tanguay, R. and K. D. Chaudhary, 1972 Studies on mitochondria. II. Mitochondrial DNA of thoracic muscles of *Schistocerca gregaria*. *J. Cell Biol.* **54**:295–301.

Ter Schegget, J. and P. Borst, 1971 DNA synthesis by isolated mitochondria. II. Detection of product DNA hydrogen-bonded to closed duplex circles. *Biochim. Biophys. Acta* **246**:249–257.

Ter Schegget, J., R. A. Flavell, and P. Borst, 1971 DNA synthesis by isolated mitochondria. III. Characterization of D-loop DNA, a novel intermediate in mtDNA synthesis. *Biochim. Biophys. Acta* **254:**1–14.

Tibbetts, C. J. B. and J. Vinograd, 1973*a* Properties and mode of action of a partially purified deoxyribonucleic acid polymerase from the mitochondria of HeLa cells. *J. Biol. Chem.* **248:**3367–3379.

Tibbetts, C. J. B. and J. Vinograd, 1973*b* Synthesis of mitochondrial deoxyribonucleic acid with a partially purified deoxyribonucleic acid polymerase from the mitochondria of HeLa cells. *J. Biol. Chem.* **248:**3380–3385.

Travaglini, E. C. and J. Schultz, 1972 Circular DNA molecules in the genus *Drosophila. Genetics* **72:**441–450.

Travaglini, E. C., J. Petrovic, and J. Schulz, 1972 Characterization of the DNA in *Drosophila melanogaster. Genetics* **72:**419–430.

Tripoldi, G., P. Pizzolongo, and M. Giannattasio, 1972 A DNase-sensitive twisted structure in the mitochondrial matrix of *Polysiphonia* (*Rhodophyta*). *J. Cell Biol.* **55:**530–532.

Tsai, M. J., G. Michaelis, and R. S. Criddle, 1971 DNA-dependent RNA polymerase from yeast mitochondria. *Proc. Natl. Acad. Sci. USA* **68:**473–477.

Tschiersch, B. and A. Graffi, 1970 Untersuchungen über die DNA-Polymerase der Mitochondrien aus Tumor- und Normalgeweben. *Arch. Geschwulstforsch.* **35:**217–226.

Upholt, W. B. and P. Borst, 1974 Accumulation of replicative intermediates of mitochondrial DNA in *Tetrahymena pyriformis* grown in ethidium bromide. *J. Cell Biol.* **61:**383–397.

Ursprung, H. and E. Schabtach, 1965 Fertilization in tunicates: Loss of the paternal mitochondrion prior to sperm entry. *J. Exp. Zool.* **159:**379–384.

Van Bruggen, E. F. J., C. M. Runner, P. Borst, G. J. C. M. Ruttenberg, A. M. Kroon, and F. M. A. H. Schuurmans Stekhoven, 1968 Mitochondrial DNA. III. Electron microscopy of DNA released from mitochondria by osmotic shock. *Biochim. Biophys. Acta* **161:**402–414.

Van Heyningen, V., I. Craig, and W. Bodmer, 1973 Genetic control of mitochondrial enzymes in human–mouse somatic cell hybrids. *Nature* (*London*) **242:**509–512.

Van Tuyle, G. C. and G. F. Kalf, 1972 Isolation of a membrane–DNA–RNA complex from rat liver mitochondria. *Arch. Biochem. Biophys.* **149:**425–434.

Vanyushin, B. F. and M. D. Kirnos, 1974 The nucleotide composition and pyrimidine clusters in DNA from beef heart mitochondria. *Fed. Eur. Biochem. Soc.* **39:**195–199.

Vanyushin, B. F., G. I. Kiryanor, I. B. Kudryashora, and A. N. Belozersky, 1971 DNA methylase in loach embryos (*Misgurnus fossilis*). *FEBS Lett.* **15:**313–316.

Vedel, F. and F. Quetier, 1974 Physico-chemical characterization of mitochondrial DNA from potato tubers. *Biochim. Biophys. Acta* **340:**374–387.

Vesco, C. and C. Basilico, 1971 Induction of mitochondrial DNA synthesis by polyoma virus. *Nature* (*London*) **229:**336–338.

Vesco, C. and S. Penman, 1969 Purified cytoplasmic DNA from HeLa cells: Resistance to inhibition by hydroxyurea. *Biochem. Biophys. Res. Commun.* **35:**249–257.

Volpe, P. and T. Eremenko, 1973 Nuclear and cytoplasmic DNA synthesis during the mitotic cycle of HeLa cells. *Eur. J. Biochem.* **32:**227–232.

Wagner, R. P., 1969 Genetics and phenogenetics of mitochondria. *Science* **163:**1026–1031.

Wattiaux-de Coninck, S., F. Dubois, and R. Wattiaux, 1973 Submitochondrial localization of DNA polymerase in rat liver tissue. *FEBS Lett.* **29:**159–163.

Wells, J. R., 1974 Mitochondrial DNA synthesis during the cell cycle of *Saccharomyces cerevisiae. Exp. Cell Res.* **85:**278–286.

Wells, R. and M. Birnstiel, 1969 Kinetic complexity of chloroplastal deoxyribonucleic acid and mitochondrial deoxyribonucleic acid from higher plants. *Biochem. J.* **112:**777–786.

Werry, P. A. T. J. and F. Wanks, 1972 The effect of cycloheximide on the synthesis of major and satellite DNA components in *Physarum polycephalum. Biochim. Biophys. Acta* **287:**232–235.

Wilkie, D., 1964 *The Cytoplasm in Heredity,* Methuen, London, Wiley, New York.

Wilkie, D., 1972 Genetic aspects of mitochondria. In *Mitochondria/Biomembranes,* North-Holland, Amsterdam.

Wilkie, D. and D. Y. Thomas, 1973 Mitochondrial genetic analysis by zygote cell lineages in *Saccharomyces cerevisiae. Genetics* **73:**367–377.

Williamson, D. H. and D. J. Fennell, 1974 Apparent dispersive replication of yeast mitochondrial DNA as revealed by density labelling experiments. *Mol. Gen. Genet.* **131:**193–207.

Williamson, D. H. and E. Moustacchi, 1971 The synthesis of mitochondrial DNA during the cell cycle in the yeast *Saccharomyces cerevisiae. Biochem. Biophys. Res. Commun.* **42:**195–201.

Wintersberger, E., 1968 Synthesis of DNA in isolated yeast mitochondria. In *Biochemical Aspects of the Biogenesis of Mitochondria,* edited by E. C. Slater, J. M. Tager, S. Papa, and E. Quagliariello, Adriatica Editrice, Bari, pp. 189–201.

Wintersberger, U. and E. Wintersberger, 1970 Studies on deoxyribonucleic acid polymerases from yeast. 2. Partial purification and characterization of mitochondrial DNA polymerase from wild-type and respiration-deficient yeast cells. *Eur. J. Biochem.* **13:**20–27.

Wolstenholme, D. R. and I. B. Dawid, 1968 A size difference between mitochondrial DNA molecules of urodele and anuran amphibia. *J. Cell Biol.* **39:**222–228.

Wolstenholme, D. R. and N. J. Gross, 1968 The form and size of mitochondrial DNA of the red bean, *Phaseolus vulgaris. Proc. Natl. Acad. Sci. USA* **61:**245–252.

Wolstenholme, D. R., K. Koike, and H. Renger, 1970 Form and structure of mitochondrial DNA. In *Oncology,* Vol. I., edited by R. C. Clark, R. W. Cumley, J. E. McCoy, and M. M. Copeland, pp. 627–648. Year Book Medical Publishers, Chicago,

Wolstenholme, D. R., R. G. Kirschner, and N. J. Gross, 1972 Heat denaturation studies of rat liver mitochondrial DNA: A denaturation map and changes in molecular configurations. *J. Cell Biol.* **53:**393–406.

Wolstenholme, D. R., J. D. McLaren, K. Koike, and E. L. Jacobson, 1973 Catenated oligomeric circular DNA molecules from mitochondria of malignant and normal mouse and rat tissues. *J. Cell Biol.* **56:**247–255.

Wolstenholme, D. R., K. Koike, and P. Cochran-Fouts, 1974 Replication of mitochondrial DNA: Replicative forms of molecules from rat tissues and evidence for discontinuous replication. *Cold Spring Harbor Symp. Quant. Biol.* **38:**267–280.

Wong-Staal, F., J. Mendelsohn, and M. Goulian, 1973 Ribonucleotides in closed circular mitochondrial DNA from HeLa cells. *Biochem. Biophys. Res. Commun.* **53:**140–148.

Wood, D. D. and D. J. L. Luck, 1969 Hybridization of mitochondrial ribosomal RNA. *J. Mol. Biol.* **41:**211–224.

Wu, G. J. and I. B. Dawid, 1972 Purification and properties of mitochondrial deoxyribonucleic acid dependent ribonucleic acid polymerase from ovaries of *Xenopus laevis. Biochemistry* **11:**3589–3595.

Wu, G. J. and I. B. Dawid, 1974 *In vitro* transcription of *Xenopus* mitochondrial deoxyribonucleic acid by homologous mitochondrial ribonucleic acid polymerase. *J. Biol. Chem.* **249**:4412–4419.

Wu, M., N. Davidson, G. Attardi, and Y. Aloni, 1972 Expression of the mitochondrial genome in HeLa cells. XIV. The relative positions of the 4 S RNA genes and of the ribosomal RNA genes in mitochondrial DNA. *J. Mol. Biol.* **71**:81–93.

Wunderlich, V., 1971 Mitochondriale DNS. *Pharmazie* **5**:257–271.

Wunderlich, V., I. Tetzlaff, and A. Graffi, 1971/1972 Studies on nitrosodimethylamine: Preferential methylation of mitochondrial DNA in rats and hamsters. *Chem.-Biol. Interact.* **4**:81–89.

Yokomura, D., 1968 An electron microscopic study of DNA-like fibrils in plant mitochondria. *Cytologia* **32**:378–389.

Zylber, E., C. Vesco, and S. Penman, 1969 Selective inhibition of the synthesis of mitochondria-associated RNA by ethidium bromide. *J. Mol. Biol.* **44**:195–204.

16

Mitochondrial Ribosomes

THOMAS W. O'BRIEN AND DAVID E. MATTHEWS

Introduction

Mitochondria are multifunctional organelles found in all eukaryotic cells, their primary function being the aerobic production of ATP. In addition to their important role in cellular energy metabolism, mitochondria make an essential contribution to their own biogenesis. For this purpose, they contain a large complement of biosynthetic enzymes and other macromolecules, distinct from their analogues in the nucleus and extramitochondrial cytoplasm. These components of the mitochondrial biogenetic system include DNA, DNA and RNA polymerases, messenger RNA, ribosomes, translation factors, transfer RNA species, and aminoacyl-tRNA synthetases. Mitochondrial DNA codes for the ribosomal RNA and at least some of the tRNA and mRNA species found in mitochondria, and these macromolecules are indispensable in the biogenesis of functionally active mitochondria. On the other hand, most of the (equally essential) protein components of the mitochondrial biogenetic system—ribosomal proteins, factors, and enzymes—appear to be coded by nuclear DNA and synthesized on cytoplasmic ribosomes. Thus the mitochondrial and nuclear-cytoplasmic macromolecule-synthesizing systems must cooperate as intimately in the production of the mitochondrial

---

THOMAS W. O'BRIEN AND DAVID E. MATTHEWS—Department of Biochemistry, University of Florida, Gainesville, Florida.

biogenetic system itself as they do in the synthesis of the enzymes of oxidative phosphorylation.

The ribosomes in a given organism's mitochondria are generally distinguishable from its cytoplasmic ribosomes on the basis of several functional or physical-chemical criteria. Indeed, especially in their functional properties, mitochondrial ribosomes have been found to be more like bacterial ribosomes than cytoplasmic ones. Some structural similarities between mitochondrial and Moneran ribosomes were noted early (Küntzel and Noll, 1967) and complemented reports of other biochemical homologies between organelles and prokaryotes that had already aroused considerable interest in the question of the evolutionary origin of mitochondria and chloroplasts. More recent comparisons of the structural parameters of bacterial and mitochondrial ribosomes have shown some similarities but also a surprising number of differences, both between the two groups and among mitochondrial ribosomes from different species. Thus it is true that the mitochondrial ribosomes of most organisms studied to date sediment more slowly than the corresponding cytoplasmic ribosomes, and some of them have sedimentation coefficients close to that of prokaryotic ribosomes (70 S). However, mitoribosomes from various species range in sedimentation rates from 55 S to 80 S, a much wider variation than is found among bacterial, eukaryotic cytoplasmic, or chloroplast ribosomes obtained from different organisms. Indeed, when all ribosomal attributes are considered, it seems that only in mitochondria do so many different kinds of ribosomes occur.

The major interspecies differences among mitochondrial ribosomes appear to be structural rather than functional. Mitoribosomes from most sources respond in a similar way to antibiotic inhibitors, monovalent and divalent cations, and protein factors obtained from prokaryotic or eukaryotic protein synthetic systems, and their responses are generally like those of bacterial ribosomes. Hence it is the physical and chemical properties of mitoribosomes which will be emphasized in this chapter. We will attempt to describe and relate these characteristics systematically for mitoribosomes from the large variety of different organisms studied so far, and to compare these properties with those of cytoplasmic and prokaryotic ribosomes. Several excellent reviews have been published on such related topics as the proteins which are synthesized on mitochondrial ribosomes (Schatz and Mason, 1974) and the nucleic acids of mitochondria (Borst, 1972), as well as general aspects of protein synthesis in mitochondria (Dawid, 1972*b*; Kroon *et al.*, 1972) and the overall problem of mitochondrial biogenesis (Mahler, 1973). The properties of nonmitochondrial ribosomes, which will be mentioned here only by way of comparison, have been reviewed more thoroughly elsewhere in this series (J. Davies, "Bacterial Ribosomes," Volume 1; M. G. Hamilton, "Eukaryotic Ribosomes," this volume; E. Stutz and A. Boschetti, "Chloroplast Ribosomes," this volume).

Criteria for Identification of Mitochondrial Ribosomes

The first indication that mitochondria might contain ribosomes was furnished by McLean *et al.* (1958), who demonstrated that isolated rat liver mitochondria could incorporate amino acids into high molecular weight material *in vitro*. The finding that this incorporation could be inhibited by chloramphenicol (Rendi, 1959; Mager, 1960), an inhibitor of protein synthesis on bacterial ribosomes, suggested that the mitochondrial enzyme responsible for the incorporation was also a ribosome, and foreshadowed the later discovery of other properties in common between mitochondrial and prokaryotic ribosomes. Similarly, the demonstration of puromycin inhibition of mitochondrial protein synthesis suggested that both ribosomes and aminoacyl-tRNA were involved (Kroon, 1963; Kalf, 1963). Cytochemical evidence was provided by the observation of small ribonuclease-sensitive particles within mitochondria in thin sections of vertebrate tissue (Andre and Marinozzi, 1965).

In spite of such evidence for the existence of ribosomes within mitochondria and in spite of repeated unsuccessful attempts to isolate them (begun by Rendi in 1959), the true mitochondrial ribosome was not isolated until nearly 10 years after the discovery of mitochondrial protein synthesis (O'Brien and Kalf, 1967*a*; Küntzel and Noll, 1967). The major obstacles to the isolation of intramitochondrial ribosomes were the large quantities of membrane-bound extramitochondrial ribosomes found in mammalian liver (the experimental material used in most of the early studies) and the lack of suitable criteria for distinguishing the mitochondrial ribosomes from the cytoplasmic contaminants. The former problem has been alleviated to some extent by the use of the detergent digitonin to solubilize the endoplasmic reticulum and mitochondrial outer membrane while leaving the mitochondrial inner membrane and its contents intact. As a bonus, many degradative enzymes are simultaneously removed from the mitochondrial fraction, since lysosomal membranes are also disrupted by digitonin. This method has been used with great success in the preparation of animal mitoribosomes (DeVries and Kroon, 1974), but not yet for other organisms.

Several useful criteria have also been developed for the identification of authentic intramitochondrial ribosomes. The first mitochondrial ribosomes to be isolated were identified by a functional criterion, their rapid accumulation of radioactivity when isolated rat liver mitochondria were incubated with labeled amino acids under conditions supporting mitochondrial protein synthesis. It had been shown that extramitochondrial 80 S ribosomes were inactive under these conditions, and in fact the only particles which became labeled had a sedimentation coefficient of 55 S (O'Brien and Kalf, 1967*a,b*).

A second criterion for identifying mitochondrial ribosomes is the antibiotic susceptibility of their protein synthetic activity *in vitro*. To date, all mitochondrial ribosomes have been found to be sensitive to inhibition by chloramphenicol and insensitive to cycloheximide or anisomycin, while 80 S cytoplasmic ribosomes universally show the opposite pattern of susceptibility. Unfortunately, no antibiotics are available that discriminate all known mitochondrial ribosomes from those of bacteria, which are a possible contaminant of mitochondrial preparations. Physical parameters have also been useful in discriminating mitoribosomes. Any of the conveniently measurable properties may be used, including sedimentation coefficients and buoyant densities of the monoribosomes and their subunits, and the size and base composition of the ribosomal RNA molecules. A valuable control for contamination is a mixing experiment, in which known amounts of the likely contaminants are added to the crude homogenate and these contaminants are shown to be removed from the mitoribosomes by the purification scheme. A very good criterion is the ability of the ribosomal RNA to hybridize specifically to mitochondrial DNA.

In general, it is desirable to use as many of these criteria as possible, particularly in cases where only small differences are seen between the putative mitoribosome and the most likely contaminants. In such cases, it may be feared that the "mitoribosomes" are merely a discrete subclass of cytoribosomes, artifactually produced by enzymatic degradation or their own instability under the particular mechanical and chemical conditions used in their isolation. Artifacts of ribosome isolation certainly can occur: mitoribosomes of *Saccharomyces cerevisiae* have been obtained by various methods as 72 S, 75 S, or 80 S particles, and *Neurospora crassa* mitoribosomes as 73 S or 80 S. Fortunately, in these instances there is other evidence that the structures in question—whatever their sedimentation rate—are indeed mitochondrial ribosomes, for without such additional criteria the situation would be hopelessly confused in these two systems (which coincidentally are among the most useful ones for genetic studies of mitochondrial biogenesis).

Distinguishing Physical and Chemical Properties of Mitochondrial Ribosomes

Mitoribosomes are structurally different both from nonmitochondrial ribosomes and among themselves. Here we shall summarize the properties of the nonmitochondrial ribosomes which have been studied, to provide a perspective for later comparison of mitochondrial ribosomes. This task is simplified by the fact that the characteristics of nonmitochondrial ribosomes are not as divergent as might be thought. They all fit reasonably well into two large categories within which the members seem to share more similarities than differences. Table 1 shows the relative homogeneity of the properties of cytoplasmic ribosomes, whatever eukaryotic organism they are obtained from—protists, fungi, plants, or animals. Despite small differences among the cytoplasmic ribosomes from these four taxonomic kingdoms, all of them appear to be members of a single structural class typified by a sedimentation coefficient of 80 S, a buoyant density of 1.57 g/ cm^3, and rRNA molecules of 0.7 and 1.2 million daltons containing 50% G+C.

The second category of nonmitochondrial ribosomes is also relatively uniform in physical and chemical properties. Described in Table 2 are several prokaryotic ribosomes—from *Escherichia coli,* a mycoplasma, and a blue-green alga—and those of various chloroplasts. The prokaryotic particles can all be adequately described by the values 70 S, 1.64 g/cm^3, $(0.56 + 1.10) \times 10^6$ daltons, and 50% for the structural parameters tabulated. Chloroplast ribosomes are similar in all respects except that some of them appear to be significantly lower in buoyant density. Perhaps it should be mentioned that this homogeneity of gross structural properties almost certainly conceals a great deal of diversity in fine structure. Differences in such properties as ribosomal protein electrophoretic mobilities and immunological identities are the rule even for prokaryotic ribosomes related to each other more closely than are the entries of Table 2 (Davies, Volume 1 of this series; Wittmann *et al.,* 1970).

In contrast to the rather simple classification scheme possible for ribosomes from all other sources, mitochondrial ribosomes do not seem to fall into one or even a few structural categories. Reference to Table 3 shows the degree of diversity found in mitochondrial ribosomes from different species. Many of the individual ribosome species in this table are as distinct in physical and chemical properties from each other as the 80 S (Table 1) and 70 S (Table 2) classes are. Furthermore, among the protists and fungi no two genera have yet been shown to contain similar mitoribosomes, so it seems likely that many more structurally different

TABLE 1. Properties of Cytoplasmic Ribosomes of Eukaryotes[a]

Organism	*Sedimentation coefficient*	*Buoyant density in CsCl (g/cm³)*	*Molecular weight of rRNA (× 10⁶ daltons)*	*Base composition of rRNA (% G+C)*
Protists				
Euglena	87 S[1]	[1.53][2b]	(0.85 + 1.20)[1c]	54[3]
Tetrahymena	80 S[4]	1.56[4]	0.52 + 1.18[5]	47[4]
Chlamydomonas	83 S[6]	[1.52][7]	(0.70 + 1.28)[6]	
Fungi				
Neurospora	77 S[8]	1.58[9]	0.67 + 1.28[10]	51[8]
Saccharomyces	80 S[11]	1.55[12]	0.72 + 1.21[5]	53[13]
Plants				
Pea	80 S[14]	1.58[9]	0.7 + 1.3[15]	52[16]
Animals				
Toad	87 S[17]	1.59[18]	0.7 + 1.5[15]	59[19]
Rat	83 S[20]	1.55[21]	0.66 + 1.48[5]	64[22]

[a] References: [1] Avadhani and Buetow (1972). [2] Avadhani and Buetow (1974). [3] Krawiec and Eisenstadt (1970). [4] Chi and Suyama (1970). [5] Reijnders *et al.* (1973). [6] Bourque *et al.* (1971). [7] Sager and Hamilton (1967). [8] Küntzel and Noll (1967). [9] Cammarano *et al.* (1973). [10] Neupert *et al.* (1969). [11] Cooper and Avers (1974). [12] Grivell *et al.* (1971). [13] Morimoto and Halvorson (1971). [14] Ts'o *et al.* (1958). [15] Loening *et al.* (1969). [16] Bonner and Varner (1965). [17] Swanson and Dawid (1970). [18] Leister and Dawid (1974). [19] Dawid *et al.* (1970). [20] Hamilton and Petermann (1959). [21] DeVries and Kroon (1974). [22] Kirby (1965).

[b] [ ] Calculated from the data published, in the cited reference, by the formula buoyant density = 1.89/[1 + 0.0040 (% protein)] (Hamilton, 1971).

[c] ( ) The data were published as ultracentrifugal or electrophoretic S values relative to *Escherichia coli* rRNA standards taken as 16 S and 23 S. To simplify the tabulation, we have converted these S values to molecular weights according to the formula $M = KS^{\alpha}$; the constants K and α were evaluated from the values 0.56×10^6 daltons and 1.10×10^6 daltons for 16 S and 23 S RNA.

TABLE 2. *Properties of Moneran and Chloroplast Ribosomes*[a]

Source	*Sedimentation coefficient*	*Buoyant density in CsCl (g/cm³)*	*Molecular weight of rRNA (× 10⁶ daltons)*	*Base composition of rRNA (% G+C)*
Bacteria				
Escherichia coli	69 S[1]	1.64[2]	0.56 + 1.10[3]	52[4]
Mycoplasma hominis	71 S[5]	[1.63][5b]	(0.56 + 1.01)[5b]	46[5]
Blue-green algae				
Anabaena	72 S[6]	1.63[7]	0.55 + 1.07[8]	
Chloroplasts				
Euglena	69 S[9]	[1.67][10]	(0.63 + 1.01)[9]	52[10]
Chlorella	67 S[7]	1.57[7]	0.56 + 1.10[8]	
Pea	70 S[7]	1.58[7]	0.56 + 1.10[8]	
Spinach	66 S[11]		0.56 + 1.05[12]	54[11]

[a] References: [1] Tissières *et al.* (1959). [2] Sacchi *et al.* (1973). [3] Kurland (1960). [4] Morimoto and Halvorson (1971). [5] Johnson and Horowitz (1971). [6] Taylor and Storck (1964). [7] Yurina and Odintsova (1974). [8] Loening (1968). [9] Avadhani and Buetow (1972). [10] Rawson and Stutz (1969). [11] Lyttleton (1962). [12] Hartley and Ellis (1973).

[b] Parenthesized and bracketed entries as in Table 1.

TABLE 3. Properties of Mitochondrial Ribosomes[a]

Source	*Sedimentation coefficient*	*Buoyant density in CsCl (g/cm³)*	*Molecular weight of rRNA (× 10⁶ daltons)*	*Base composition of rRNA (% G+C)*
Protists				
Euglena	71 S[1]	[1.61][2b]	(0.56 + 0.93)[1b]	27[3]
Tetrahymena	80 S[4]	1.46[4]	0.47 + 0.90[5]	29[4]
Fungi				
Neurospora	73 S[6], 80 S[7]	1.52[8]	0.72 + 1.28[9]	38[6]
Saccharomyces	72 S[10], 75 S[11], 80 S[12]	1.64[13]	0.70 + 1.30[5]	30[14]
Candida utilis	72 S[15]	1.48[15]	0.71 + 1.21[15]	34[15]
Aspergillus			0.66 + 1.27[16]	32[17]
Plants				
Maize	77 S[18]	1.56[19]	0.76 + 1.25[20]	
Bean	78 S[21]		0.78 + 1.15[21]	
Animals				
Locust	60 S[22]		0.28 + 0.52[23]	
Shrimp			0.35 + 0.50[24]	43[24]
Toad	60 S[25]	1.45[26]	0.32 + 0.58[26]	41[27]
Rat	55 S[28]	1.43[29]	0.30 + 0.50[30]	47[31]
HeLa cells	55 S[32]	1.40[32]	0.35 + 0.54[33]	53[34]

[a] References: [1] Avadhani and Buetow (1972). [2] Avadhani and Buetow (1974). [3] Krawiec and Eisenstadt (1970). [4] Chi and Suyama (1970). [5] Reijnders *et al.* (1973). [6] Küntzel and Noll (1967). [7] Agsteribbe *et al.* (1974). [8] Agsteribbe and Kroon (private communication). [9] Neupert *et al.* (1969). [10] Schmitt (1970). [11] Cooper and Avers (1974). [12] Morimoto *et al.* (1971). [13] Grivell *et al.* (1971). [14] Morimoto and Halvorson (1971). [15] Vignais *et al.* (1972). [16] Verma *et al.* (1970). [17] Edelman *et al.* (1970). [18] Pring (1974). [19] D. R. Pring, W. R. Clark, M. J. Critoph, and T. W. O'Brien (unpublished). [20] Pring and Thornbury (1975). [21] Leaver and Harmey (1973). [22] Kleinow *et al.* (1971). [23] Kleinow (1974). [24] Schmitt *et al.* (1974). [25] Swanson and Dawid (1970). [26] Leister and Dawid (1974). [27] Dawid and Chase (1972). [28] O'Brien and Kalf (1967*b*). [29] DeVries and Kroon (1974). [30] Sacchi *et al.* (1973). [31] Bartoov *et al.* (1970). [32] Perlman and Penman (1970*b*). [33] Robberson *et al.* (1971). [34] Vesco and Penman (1969).

[b] Parenthesized and bracketed entries as in Table 1.

ribosomes will be found as other species are investigated. On the other hand, considerable homology is seen among the mitoribosomes of several species of higher animals, from locust to man. Only one of the many mitochondrial ribosomes characterized to date, that of *Euglena,* appears to be similar to bacterial ribosomes in most of the tabulated characteristics. We shall now turn to a more detailed discussion of the properties of mitochondrial ribosomes from the four eukaryotic kingdoms.

Protists

The best-characterized protist mitochondrial ribosomes are those of *Euglena gracilis* and *Tetrahymena pyriformis.* The particle from *Euglena* is in some respects the simplest of mitoribosomes, since it is quite similar to prokaryotic ribosomes in most of its properties. The 71 S monoribosome form is composed of a 32 S and a 50 S subunit, containing rRNA molecules of 16 S and 21 S, respectively (Avadhani and Buetow, 1972). The monoribosomes of *E. coli,* for comparison, are 70 S particles made up of 30 S and 50 S subunits containing 16 S and 23 S rRNA. Also like bacterial ribosomes, *Euglena* mitoribosomes dissociate into their subunits at relatively high magnesium concentrations, while the cytoribosomes require a lower [Mg^{2+}] for dissociation (Avadhani and Buetow, 1974). The *Euglena* mitoribosome contains 43% protein and 57% RNA (Avadhani and Buetow, 1974), a composition similar to the value of 37% protein found for the bacterial ribosome (Tissières *et al.,* 1959). Since the weight percent of RNA in the *Euglena* mitoribosome is known and the molecular weights of the rRNA molecules can be estimated from their S values (see Table 1 notes), an estimate can be made for the total particle weight of the ribosome. The answer is 2.7×10^6 daltons, as compared with 2.65×10^6 daltons determined for the ribosomes of *E. coli* (Hill *et al.,* 1969). By these criteria, then, *Euglena* mitoribosomes indeed resemble bacterial ribosomes. The major structural difference between these two ribosomes is in the base composition of the rRNAs. The mole percentage of guanine plus cytosine in the *Euglena* mitoribosome, 27%, is among the lowest values yet found in any ribosome.

Tetrahymena mitoribosomes are quite exceptional particles in many respects, being dissimilar in structure not only from most other protist mitoribosomes but also from any other ribosome characterized to date. The monoribosome is 80 S, as is the cytoplasmic ribosome of this species, and it does not dissociate to subunits even in solutions containing no magnesium. If EDTA is added, however, the particle does dissociate, and both of the subunits produced have sedimentation coefficients of 55 S; this ribosome and the mitoribosome of *Paramecium,* which is its only structural cousin, are the only reported instances in which the subribosomal particles cannot be separated by rate zonal centrifugation (Chi and Suyama, 1970; Curgy *et al.,* 1974; Tait, 1972). Furthermore, the subunits cannot be resolved by electrophoresing them in polyacrylamide gels, nor can they be distinguished on the basis of their shape or size in the electron microscope (Curgy *et al.,* 1974). Fortunately, the subunits can be separated by isopycnic centrifugation, since one of them is considerably more dense (1.52 g/cm^3) than the other (1.46 g/cm^3) (Chi and Suyama, 1970).

The ribosomal RNAs of this ribosome are on the small side, 0.47 + 0.90 million daltons, and may be viewed as intermediate in this respect between those of bacterial ribosomes and the very small molecules of higher animal mitoribosomes (while fungal mitoribosomes have taken the opposite tack, toward larger rRNAs). However, the low buoyant density of the particle (1.46 g/cm^3 for the monoribosome) indicates that quite a large amount of protein is associated with and organized by these relatively small rRNA molecules—another property we shall see carried to an extreme by higher animal mitoribosomes. An estimate of the particle weight of *Tetrahymena* mitochondrial ribosomes based on their RNA content and the molecular weights of the RNAs is 3.2 million daltons (Chi and Suyama, 1970), quite a bit larger than the value of 2.7×10^6 daltons for the ribosome of *E. coli.* This large molecular weight is indirectly confirmed by the observation that these particles appear even larger by electron microscopy than the cytoplasmic ribosomes of this species: 370 × 240 Å vs. 275 × 230 Å (Stevens *et al.,* 1974; Curgy *et al.,* 1974). Furthermore, the mitoribosome of *Tetrahymena* has a lower electrophoretic mobility in polyacrylamide gels than its cytoribosome (Curgy *et al.,* 1974), although probably both the large size and the low RNA content of the mitochondrial particle are responsible for this effect. *Tetrahymena* mitochondrial rRNA is nearly as low in G+C content (29%) as that of *Euglena* mitoribosomes. Thus in this respect there is no apparent evolutionary continuity from bacterial ribosomes (52% G+C) through *Tetrahymena* mitoribosomes to higher animal mitoribosomes (45% G+C).

The proteins of *Tetrahymena* mitoribosomes have been resolved electrophoretically into about 27 components and have been shown to differ in electrophoretic mobility from the proteins of the cytoribosome in this species (Chi and Suyama, 1970). It has been suggested that some of these mitoribosomal proteins may themselves be synthesized on mitochondrial ribosomes rather than in the cytoplasm, since their synthesis can be inhibited by chloramphenicol (Millis and Suyama, 1972). With regard to this question, some results in *Paramecium,* whose mitoribosomes may be closely related in structure to those of *Tetrahymena,* are of interest. Beale *et al.* (1972) have shown an altered pattern of mitoribosomal proteins associated with a cytoplasmically inherited mutation which confers erythromycin resistance. However, at present the conclusion that some of the proteins of *Paramecium* mitochondrial ribosomes are coded by the mitochondrial DNA is still open to question, since the component directly affected by the mutation may be the ribosomal RNA (Tait, 1972).

There appears to be at least a third structural kind of protist mitoribosome. Mitochondria from the trypanosome *Crithidia luciliae* yield a ribosome of only 60 S, dissociable at low magnesium concentration (3 mM) to a 32 S and a 45 S subunit (Laub-Kupersztejn and Thirion, 1974). On the other hand, *Trypanosoma brucei* mitochondria have been reported to contain 70 S ribosomes (Stuart *et al.,* 1973). Other properties of these interesting particles have not yet been reported, but the trypanosome with its large mitochondrion may prove to be a very rich source of mitoribosomes.

Fungi

The mitochondrial ribosomes of various fungal species appear to be quite divergent in physical and chemical properties. Although it seems likely that much of this diversity is real, the picture is unfortunately clouded by conflicting reports of the characteristics of some individual species. Kroon *et al.* (1972) have commented that "the variability of the physicochemical characteristics attributed to mitochondrial ribosomes from different organisms and to mitochondrial ribosomes from the same organisms by different groups is almost infinite," and it may be added that the variability in properties of mitoribosomes from the same organisms reported by the *same* groups is also remarkable. In all of these respects, it has been the mitoribosomes of fungi that have been the most troublesome.

An example is the case of *Neurospora crassa.* The mitoribosome of this species was one of the first to be characterized, and its reported properties were the basis of a prediction that mitochondrial ribosomes, like those of chloroplasts, would prove to be similar to bacterial ribosomes in structure (Küntzel and Noll, 1967). Not only have most mitoribosomes failed to fit this mold, but even the one from *Neurospora* has proven less prokaryotic in nature than was originally supposed. The monoribosomes of *Neurospora* mitochondria have long been thought to have a sedimentation coefficient of 73 S (Küntzel and Noll, 1967). However, ribosomes from mitochondria isolated in the presence of magnesium, rather than EDTA, and lysed in the presence of heparin to inhibit ribonuclease are

80 S particles (Datema *et al.,* 1974). They are distinguished from a possible altered form of the 77 S cytoplasmic ribosomes of this species by several criteria, including their antibiotic sensitivities. Since they can be converted to 73 S particles by aging or by incubation with a high-speed supernatant of lysed mitochondria, it now appears likely that these 80 S ribosomes are the true *Neurospora* mitoribosome. It is not yet clear, however, whether the major difference between the 80 S and 73 S particles lies in their composition or their conformation. Datema *et al.* (1974) found no difference in the functional activity of the two types of particles, their response to antibiotics, or the sedimentation coefficients of the subunits or ribosomal RNAs. Since polysomes are also found in the preparations that yield 80 S monosomes but not in those that give the 73 S particles (Agsteribbe *et al.,* 1974), it may be that the difference in sedimentation coefficient is related to the presence or absence of bound mRNA.

The *Neurospora* mitoribosome has a buoyant density of 1.52 g/cm^3 (Agsteribbe and Kroon, private communication). Thus it has a slightly higher protein content than most cytoplasmic ribosomes; the buoyant density of the *Neurospora* cytoplasmic ribosome, for example, is 1.58 g/cm^3 (Cammarano *et al.,* 1973). Like bacterial ribosomes, the mitoribosome can be dissociated into its subunits at a relatively high concentration of magnesium, while the cytoribosome requires a much lower concentration. Furthermore, the mitochondrial monoribosome is dissociated by a protein dissociation factor isolated from *E. coli* ribosomes, whereas the cytoribosome is not (Agsteribbe and Kroon, 1973).

The mitoribosomal subunits have sedimentation coefficients of 50–52 S and 37–39 S (Küntzel, 1969*a*; Datema *et al.,* 1974), the small subunit being distinctly larger by this measure than that of bacterial ribosomes. The molecular weights of the rRNAs in these particles have been reported as 1.28 and 0.72 million daltons (Neupert *et al.,* 1969). These molecules are therefore quite large, within the range of sizes reported for cytoplasmic ribosomal RNA. However, it is difficult at present to quote a really reliable molecular weight estimate for the mitoribosomal RNA of *Neurospora.* The problem is that this RNA, like that of other fungi and of protists, has a low G+C content—about 36% in this case (Küntzel and Noll, 1967; Rifkin *et al.,* 1967). Consequently, it contains less secondary structure at room temperature than does RNA of higher G+C content, such as *E. coli* RNA. For this reason, comparison of these rRNAs with those of *E. coli* by sedimentation velocity or electrophoresis experiments do not yield accurate molecular weight values. This difficulty can be overcome by performing the determinations under conditions which denature all the double-stranded regions of the RNA (Reijnders *et al.,* 1973). Reliable values have been obtained by this means for the low-G+C mitochondrial rRNAs of *Tetrahymena* and *Saccharomyces,* but not yet for *Neurospora*.

The proteins of *Neurospora* mitoribosomes have been shown to differ from those of the cytoribosomes by electrophoresis (Gualerzi, 1969), chromatography on carboxymethylcellulose (Küntzel, 1969*b*), and immunological methods (Hallermayer and Neupert, 1974). Lizardi and Luck (1972) were able to resolve the mitoribosomal proteins into 53 components by a combination of isoelectric focusing and electrophoresis. Furthermore, these authors showed that all of the components were labeled if they were obtained from *Neurospora* grown in the presence of radioactive amino acids and chloramphenicol, whereas cycloheximide and anisomycin inhibited the labeling. It may be concluded that in *Neurospora* all of the mitoribosomal proteins are synthesized on ribosomes in the extramitochondrial cytoplasm.

The confusion with regard to the properties of *Saccharomyces* mitoribosomes is even greater than is the case for *Neurospora*. Every laboratory which studies these ribosomes seems to obtain a different result, like the blind men characterizing the elephant. The situation is not apt to be resolved until one group tries all the different methods of preparation, obtains all the different particles that have been reported, and determines the structural and functional differences among them. This tactic has shed more light on the similar problems in *Neurospora* (Datema *et al.*, 1974) (see p. 546) and *Candida utilis* (Vignais *et al.*, 1972; see p. 549) than any degree of characterization of one of the particles alone could have.

The first mitochondrial ribosome to be obtained from *Saccharomyces cerevisiae* was an 80 S particle (Schmitt, 1969). It was distinguished from the cytoplasmic ribosome, which is also 80 S in this species, only by the relative ease with which it could be dissociated into its subunits. Unfortunately, these subunits had sedimentation coefficients of 60 S and 38 S, again like those of the cytoplasmic ribosome. On the basis of this evidence, it was possible that the "mitoribosomes" obtained here were merely a subclass of cytoplasmic ribosomes, associated with mitochondria and having somewhat different dissociation properties. Such an interpretation is not unreasonable, since Keyhani (1973) and Kellems *et al.* (1974) have observed a class of cytoplasmic ribosomes which are at least cytologically distinct: thin sections of yeast cells show large numbers of ribosomes specifically associated with the outer surface of the mitochondrial outer membranes. In fact, Schmitt himself adopted this interpretation in a follow-up study (Schmitt, 1970). He found that mitochondria washed in a medium containing EDTA yielded no 80 S particles. Rather, these mitochondrial ribosomes were 72 S, and dissociated to subunits of 50 S and 38 S. Furthermore, the antibiotic sensitivities of protein synthesis in the two mitochondrial preparations indicated that the EDTA-washed mitochondria were free of functional cytoplasmic ribosomes, whereas 60% of the activity in the earlier preparation could be attributed to cytoribosome contamination.

Such results must cast a shadow of doubt over any other reports of 80 S mitoribosomes from *S. cerevisiae*. Even so, the claim of Morimoto *et al.* (1971) cannot be dismissed lightly. Their 80 S particle also dissociated to subunits of 60 S and 37 S to 40 S (depending on conditions). But the differences between this ribosome and the free 80 S cytoribosome were greater than one would naively expect for two subclasses of cytoribosomes. The ribosomal RNAs and many of the ribosomal proteins of these two particles were distinguishable electrophoretically. The base composition of the rRNA of the mitochondrial 80 S ribosome was 32% G+C (like that of other fungal and protist mitoribosomes), while a value of 45% G+C was obtained for the cytoribosome. Furthermore, the mitoribosomal RNA hybridized specifically to the mitochondrial DNA. Yu *et al.* (1972*b*) likewise obtained only 80 S ribosomes from their preparation of *S. cerevisiae* mitochondria, and approximately 50% of these ribosomes were distinguished from free cytoplasmic ribosomes by the size and low G+C content of their rRNA.

A third variety of *S. cerevisiae* mitoribosome was described by Stegeman *et al.* (1970). This one is 75 S, and is further differentiated from the free cytoplasmic ribosome by the size of its rRNA. Polysomes containing the 75 S particle as a monomer were found in this preparation, whereas none of the other reports mentioned above had obtained such an indication of the structural integrity of their particles. Also relevant was the observation by electron microscopy that the intramitochondrial ribosomes appeared slightly smaller than the cytoplasmic ones in thin sections of whole cells. A similar mitoribosome was obtained from the closely related species *Saccharomyces carlsbergensis* (Grivell *et al.*, 1971). This 74 S particle, composed of 50 S and 37 S subunits, was shown to differ from the cytoplasmic ribosome in buoyant density and in rRNA size. Its functional integrity was demonstrated by its high activity in poly(U)-directed phenylalanine incorporation. Good evidence for its mitochondrial origin was provided by the sensitivity of its activity *in vitro* to chloramphenicol but not to cycloheximide. However, in spite of such evidence it is premature to conclude that the physical properties of this particle are those of the native *Saccharomyces* mitoribosome, especially in view of the inability of Datema *et al.* (1974) to find differences in the level of activity or in antibiotic susceptibility between the 73 S and 80 S forms of the mitoribosome from *Neurospora*.

For these reasons, a physical description of the *Saccharomyces* mitoribosome at this time can only be provisional. Its sedimentation coefficient is 72 S–80 S, and those of its subunits are 50 S–60 S and 37 S–40 S. The mitoribosomal RNA of *S. cerevisiae* has been variously reported to electrophorese either more rapidly or more slowly than that of the cytoribosomes. However, the mit rRNAs of *S. carlsbergensis* have been studied under fully denaturing conditions and shown to have molecular weights of 0.7 and 1.3 million daltons (Reijnders *et al.*, 1973). This rRNA has a low G+C content, in the neighborhood of 30% (Grivell *et al.*, 1971). The buoyant density of the 74 S form of the ribosome is 1.64 g/cm^3 (Grivell *et al.*, 1971), corresponding to an RNA content of about 63% like that of bacterial ribosomes. However, the RNA content of the 80 S form, determined by chemical methods, is only 47% (Morimoto and Halvorson, 1971). The mitoribosomal proteins can be differentiated from those of the cytoplasmic ribosomes or bacterial ribosomes by electrophoresis or ion-exchange chromatography (Morimoto and Halvorson, 1971; Schmitt, 1971, 1972). The site of synthesis of these proteins is thought to be extramitochondrial since they are not made in the presence of cycloheximide (Schmitt, 1972). Several mutations resulting in mitochondrial ribosomes which are resistant to various antibiotics have been found to map in the mitochondrial DNA, but no differences have been found between the mutant and wild-type ribosomal proteins by electrophoresis (Grivell *et al.*, 1973).

Fewer groups are engaged in the study of mitoribosomes from other fungal species; consequently, the data are somewhat sparser but so are the contradictions. The particle from *Candida utilis* has been exceptionally well characterized. It was originally reported to have a sedimentation coefficient of 77–80 S (Vignais *et al.*, 1969). As was the case in *S. cerevisiae*, this value was later revised to 72 S (Vignais *et al.*, 1972), but not for the same reason. The "80 S mitoribosomes" from *C. utilis* were shown by mixing experiments not to be contaminated by the 78 S cytoribosomes; rather, they appeared by electron microscopy to be composed of dimers of the 50 S large mitoribosomal subunits. Furthermore, it was shown that isolated 50 S subunits could be induced to dimerize to the 80 S particles by treatment with the high magnesium concentrations which were used in the earlier study. The 72 S particle can be dissociated into a 50 S and a 36 S subunit by treatment with 10^{-4}M $MgCl_2$, whereas the cytoribosome of this species is not dissociated until the magnesium concentration is lowered to 10^{-6} M. The molecular weights of the ribosomal RNAs were estimated at 0.71 and 1.21 million daltons, based on their sedimentation rates in sucrose density gradients in the presence of EDTA, values distinctly smaller than those of the cytoribosomal RNA. The mitochondrial rRNA has a base composition of 33% G+C, as compared to 50% for the cytoplasmic rRNA. The buoyant density of the mitoribosome is 1.48 g/cm^3, implying a higher protein content than was found for the cytoribosome of this species (1.53 g/cm^3) or the mitoribosomes of *Neurospora* or *Saccharomyces*. In the electron microscope, the size of the mitoribosome (265 × 210 × 200 Å) is roughly the same as that of the cytoribosome (260 × 225 × 205 Å).

Aspergillus nidulans mitochondrial ribosomes have been characterized only partially thus far. It is noteworthy that a molecular weight estimate for their rRNA, free of artifacts due to secondary structure, is available. Verma *et al.* (1970) have measured the lengths of these molecules in the electron microscope and calculated molecular weights of 0.66 and 1.27 million daltons. Their base composition is 32% G+C, and the mitoribosomes themselves, as well as their subunits, have sedimentation coefficients similar to those of the corresponding particles from *E. coli* (Edelman *et al.*, 1970).

70 S mitoribosomes were obtained from the yeast *Candida parapsilosis* (Yu *et al.*, 1972*a*) under the same conditions which yielded 80 S particles from *S. cerevisiae* (see p. 548). The subunits were 50 S and 30 S, and the G+C content of the rRNA was 34%. A similar base composition (33% G+C) was reported for the filamentous fungus *Trichoderma viride* (Edelman *et al.*, 1971).

In view of the limited number of fungal species whose mitochondrial ribosomes have been characterized and the diverse results that have been obtained, any description of these particles at the present time can have but little generality for fungal mitoribosomes as a group. However, two points of similarity exist among all species studied thus far. First, the ribosomal RNA molecules are quite large in these species. Although there is reason to doubt the accuracy of many of the molecular weight estimates, as discussed above (see p. 546) the values obtained for *S. carlsbergensis* and *A. nidulans* are probably reliable. Practically all reports agree that the fungal mitoribosomal RNAs are distinctly larger than those of bacteria, and thus much larger than those of protist or animal mitochondria. Second, these rRNAs are uniformly low in guanine and cytosine content. In general, the mitoribosomes of fungi appear to be intermediate in this respect between the very low values found for protist mitoribosomes (< 30% G+C) and the relatively higher values of animal mitoribosomes (40–50% G+C).

Plants

Study of the mitochondrial ribosomes of plants has been hindered by the difficulty of obtaining large quantities of intact purified mitochondria from these organisms. As a result, the particles which have been isolated have been incompletely characterized, and rigorous criteria for their mitochondrial origin have not always been applied. Furthermore, quite disparate properties have been reported for mitoribosomes of various plant species by various workers.

For example, the mitoribosome of *Phaseolus aureus* (mung bean) hypocotyls has been investigated by two groups. Vasconcelos and Bogorad (1971) reported that a particle of about 70 S could be obtained from a mitochondrial fraction of uncertain purity, but not enough data were presented to evaluate this claim. Leaver and Harmey (1973) obtained 77–78 S ribosomes from the mitochondria of this species, as well as from turnip, cauliflower, and French bean. The mitoribosome of *P. aureus* was further distinguished from the 80 S cytoribosome only by a small difference in the sizes of the rRNAs (0.78 and 1.15 million daltons for the mitochondrial particle vs. 0.70 and 1.30 million daltons for the cytoplasmic one) and the absence of a 5.8 S component in the mitochondrial large rRNA.

Similarly, maize mitochondrial ribosomes of 66 S (Wilson *et al.*, 1968) and 78 S (Pring, 1974) have been reported. Unfortunately, the 66 S particle was not characterized further. The 78 S mitoribosome dissociates to 60 S and 44 S subunits, relative to assumed values of 60 S and 40 S for the maize cytoribosome subunits (Pring, 1974) and the mitochondrial rRNAs appear slightly larger than those of the cytoribosome (0.76 and 1.25 million daltons vs. 0.67 and 1.19 million daltons) (Pring and Thornbury, 1975). The dissociation of the mitoribosome to its subunits occurs at relatively high magnesium concentrations, under which conditions the cytoplasmic monoribosome remains intact (Pring, 1974). The buoyant densities of the mitochondrial and cytoplasmic particles are both 1.56 g/cm^3 (D. R. Pring, W. R. Clark, M. J. Critoph, and T. W. O'Brien, unpublished).

A different result has been obtained by Quetier and Vedel (1974). Mitochondria from cultured cells of *Parthenocissus tricuspidata* (Virginia creeper) were shown to contain major RNA species of 0.42 and 0.84 million daltons. Except for animal mitochondrial rRNAs, these are the smallest rRNA molecules yet reported—much smaller than the other plant rRNAs mentioned above. Fortunately, all of the rRNA molecular weights which have been reported for plant mitochondria have been determined under the denaturing conditions of Reijnders *et al.* (1973). Furthermore, the investigators all agree that electrophoretic conditions do not seriously affect the values obtained, so the discrepancy in these results is probably not due to electrophoretic artifacts.

In order to resolve these conflicting reports, the most pressing need is more criteria for the intramitochondrial origin of these particles. Where there is contradiction, such criteria as the purity of the mitochondrial fraction or small structural differences between the "mitochondrial" ribosome and its most likely contaminants will not suffice. Our conclusions about the actual nature of mitochondrial ribosomes from this kingdom of organisms must await data on the antibiotic susceptibility of their functional activity, the base composition of their rRNA, and the homology of their rRNA to mitochondrial DNA. [Preliminary results indicate that the "78 S mitoribosome" of maize is inhibited by anisomycin and not by chloramphenicol, like the cytoribosome and unlike any other known mitoribosome (Pring, Denslow, and O'Brien, unpublished).]

Animals

In contrast to the fascinating and perplexing diversity of mitochondrial ribosomes from protist, fungal, and plant species, metazoan mitoribosomes appear strikingly uniform in structure. Although most of the animal species investigated have been mammalian, the mitoribosomes of the toad *Xenopus* have also been thoroughly characterized, and enough data have been presented for several invertebrates to justify a tentative conclusion that the mitoribosomes of all multicellular animals may be quite similar. Furthermore, this relatively homogeneous group of ribosomes is distinctly different in structural properties from any other ribosomes yet described, with the possible exception of the mitoribosome of the protist *Crithidia* (see p. 545).

The first difference to be noticed was the low sedimentation coefficient of animal mitoribosomes. Values within the range 54 S to 61 S have been obtained for these particles from rat (O'Brien and Kalf, 1967*b*), HeLa cells (Perlman and Penman, 1970*b*), rabbit, pig, cow (O'Brien, 1971), hamster (Coote *et al.*, 1971), chicken (Rabbitts and Work, 1971), toad (Swanson and Dawid, 1970), shark (O'Brien, 1972), and locust (Kleinow *et al.*, 1971). The early inference drawn from these low sedimentation coefficients was that animal mitoribosomes were smaller (lower in molecular weight) than other ribosomes (Borst and Grivell, 1971), and this inference was reinforced by observations of the unusually small rRNA in these particles (see p. 554). An alternate possibility, that the 55 S particle is actually a subunit of the functional monoribosome, has been excluded by the dissociation of the 55 S structure into two subunits and the demonstration that either the 55 S particle or a mixture of both subunits is competent for poly(U)-dependent phenylalanine incorporation (Leister and Dawid, 1974).

There is a third interpretation of the low sedimentation coefficient of animal mitoribosomes, because in point of fact these 55 S particles are not exceptionally small either in molecular weight or in physical dimensions. The particle weight of the bovine mitoribosome, determined by high-speed equilibrium centrifugation, is 2.8 million daltons (Hamilton and O'Brien, 1974)—slightly greater than the value reported for the ribosome of *E. coli* (Hill *et al.*, 1969). DeVries and Kroon (1974) have presented evidence that rat mitoribosomes are even larger in volume than the *E. coli* particles, although smaller than rat cytoribosomes. These investigators electrophoresed ribosomes into gels composed of a gradient of polyacrylamide concentrations until the particles could make no further progress through the decreasing pore size of the gel matrix. The mitoribosomes penetrated farther into the gels than cytoribosomes but not as far as bacterial ribosomes. The physical dimensions of ribosomes can also be determined by electron microscopic measurements. The results confirm the conclusion that mitoribosomes are smaller than cytoribosomes in rat (O'Brien and Kalf, 1967*b*; Aaij *et al.*, 1972) and locust (Kleinow *et al.*, 1974), but direct comparisons of mitochondrial and bacterial ribosomes (under the same conditions of fixation and staining) have not been performed.

How can two particles of the same molecular weight sediment respectively at 55 S and 70 S? The significant difference appears to be the much lower buoyant density of the animal mitochondrial ribosome. Buoyant density values from 1.40 to 1.46 g/cm^3 have been found for mitoribosomes of HeLa cells (Perlman and Penman, 1970*b*; Wengler et al., 1972), rat (Sacchi *et al.*, 1973; DeVries and Kroon, 1974), cow (Hamilton and O'Brien, 1974), and toad (Leister and Dawid, 1974). As mentioned above, the buoyant density of a ribosome is considered to be a measure of the relative proportions of RNA and protein in the particle (Hamilton, 1971); from this relationship an RNA content of about 30% can be calculated for animal mitoribosomes, as contrasted with 63% for bacterial ribosomes (Tissières *et al.*, 1959). From the RNA content and the sum of the molecular weights of the rRNA molecules, the particle weight of the ribosome may be calculated. By this means, molecular weight estimates equal to or greater than those for bacterial ribosomes were obtained for mitoribosomes from rat (Sacchi *et al.*, 1973; DeVries and Kroon, 1974), cow (O'Brien *et al.*, 1974), and toad (Leister and Dawid, 1974) even before the molecular weight was determined directly by sedimentation equilibrium (Hamilton and O'Brien, 1974).

Doubts have been raised about the validity of buoyant density values as measures of the protein content of ribosomes (McConkey, 1974). Others have suggested that the low buoyant densities of animal and *Tetrahymena* mitoribosomes might be due to membrane fragments adhering to these particles (specifically, due to membrane lipids, which are assumed to be absent in the calculation of protein content from buoyant density) (Borst and Grivell, 1971). Several lines of evidence indicate that these factors do not represent significant objections to the description of animal mitoribosomes presented above. Determinations of the protein content by either chemical analysis or ultraviolet absorption spectra of the ribosomes are in agreement with estimates from buoyant density, yielding values of 70–80% protein in rat (O'Brien and Kalf, 1967*b*) and toad (Leister and Dawid, 1974). No phospholipids were detectable in rat mitoribosomes (DeVries and Kroon, 1974). The molecular weight estimate for bovine mitoribosomes based on their buoyant density and the size of their rRNA agrees quite closely with the molecular weight determined by sedimentation equilibrium (Hamilton and O'Brien, 1974). Finally, similar values for total protein content have been obtained directly by summing the molecular weights of the individual proteins of mitochondrial ribosomes from *Xenopus* (Leister and Dawid, 1974) and cow (Matthews and O'Brien, unpublished).

Besides their unusual sedimentation behavior and high protein content, another unusual characteristic of animal mitoribosomes is the small size of their rRNA molecules. Although all reports of centrifugal and electrophoretic analysis of these components agree that they are smaller than those of bacterial ribosomes, none of these determinations has been performed under the denaturing conditions of Reijnders *et al.* (1973). However, values of 0.35 and 0.54 million daltons have been obtained from HeLa mitoribosomal RNA by electron microscopic length measurements (Robberson *et al.*, 1971), and these numbers are in good agreement with those found by other methods for rat (Sacchi *et al.*, 1973), toad (Dawid and Chase, 1972; Leister and Dawid, 1974), shrimp (Schmitt *et al.*, 1974), and locust (Kleinow, 1974*b*). Thus animal mitoribosomes contain scarcely more than half as much RNA as any nonmitochondrial ribosome known. To a first approximation, these particles may be pictured as *E. coli* ribosomes modified by converting half of the RNA into an equal mass of protein. Such a picture raises questions about the structural and functional roles of that portion of RNA which is replaceable by protein. It may be that this material is only "filler" to make up the physical bulk required of an enzyme whose many cofactors and substrates are almost all macromolecules, or the answer may be more interesting. Unfortunately, very little is known about the function of the high molecular weight ribosomal RNA molecules in other ribosomes, so the immediate prospects for resolving such questions are poor.

The base composition of the rRNA of animal mitoribosomes is distinctly higher in guanine and cytosine than that of protist or fungal mitoribosomes, though still lower than that of animal cytoplasmic ribosomes. G+C contents of 40–47% have been found for mitochondrial rRNA of HeLa (Vesco and Penman, 1969), rat (Bartoov *et al.*, 1970), toad (Dawid and Chase, 1972), and shrimp (Schmitt *et al.*, 1974). The degree of methylation is also lower for animal mitoribosomal RNAs than for either cytoplasmic or prokaryotic rRNAs (Dubin, 1974).

Like prokaryotic ribosomes and mitoribosomes of most other species, the 55 S mitochondrial ribosome is relatively unstable to decreasing concentrations of divalent cations, as compared with eukaryotic cytoplasmic ribosomes. Under these conditions, it dissociates to subunits which sediment at approximately 30 S and 40 S in sucrose gradients (O'Brien, 1971; Greco *et al.*, 1973; Kleinow, 1974*a*). The molecular weights of the subribosomal particles obtained from bovine liver mitoribosomes are 1.1 and 1.7 million daltons, respectively (Hamilton and O'Brien, 1974).

The large protein complement of animal mitoribosomes has been analyzed by two-dimensional electrophoresis and found to be more complex than that of any other ribosome yet described. Eighty-four different proteins were resolved in electropherograms of *Xenopus* mitoribosomal proteins, 40 of them localized in the large subunit and 44 in the small subunit (Leister and Dawid, 1974). Similarly, bovine liver mitoribosomes appear to contain as many as 53 and 41 proteins in the large and small subunits, respectively (Matthews and O'Brien, unpublished). Animal mitoribosomes thus contain considerably more individual proteins than *E. coli* ribosomes (55) or *Neurospora* mitoribosomes (53; see p. 547) or even eukaryotic cytoplasmic ribosomes (about 70; Wool and Stöffler, 1974). But, as mentioned above, when the molecular weight of each of these mitoribosomal proteins is considered (and unit stoichiometry is assumed for each), the total quantity of protein is just what would be predicted from the difference between the mass of the particle and that of the RNAs. Although it is not possible at present to demonstrate conclusively that all of these proteins are required for the functional activity of animal mitochondrial ribosomes, preliminary experiments do suggest that this is the case (O'Brien, Denslow, and Matthews, unpublished).

Although mitochondrial ribosomes from all animal species examined thus far appear quite similar in their physical and chemical properties, it may be anticipated that they will differ in their detailed structure. In fact, detectable nonhomology has already been found between the base sequences of mitoribosomal RNA from two species of toad, *Xenopus laevis* and *X. mulleri* (Dawid, 1972*a*). Electrophoretic differences have also been demonstrated in several of the mitoribosomal proteins of these two species, and have been exploited to study the inheritance of these proteins in sexual hybrids (Leister and Dawid, 1975). An interesting extension of this work would be the use of mitoribosomal proteins as markers in somatic cell hybrids of more distantly related species. Since the mitochondrial DNA contributed by each parental line can be distinguished in the hybrid (Coon *et al.*, 1973), such experiments would provide an opportunity to study the interactions of the mitochondrial (for mitochondrial rRNA) and nuclear-cytoplasmic (for mitoribosomal proteins) biogenetic systems from different species in the synthesis of an enzyme complex (mitoribosomes) which requires both.

Phylogenetic Relationships in Mitochondrial Ribosome Structure

Some of the physical and chemical properties of mitochondrial ribosomes have been investigated in enough species that we may begin to draw a few conclusions about the variation in these properties among mitoribosomes of different organisms and the differences between the ribosomes of mitochondria and those of prokaryotes, eukaryotic cytoplasm, and chloroplasts.

The most obvious generalization is that mitochondrial ribosomes show more diversity in all of their structural properties than do any other kinds of ribosomes. Cytoplasmic ribosomes of all four eukaryotic kingdoms are relatively similar with respect to sedimentation coefficient, buoyant density, rRNA size, and guanine plus cytosine content. Prokaryotic ribosomes display even more uniformity in these characteristics, while chloroplast ribosomes, whether obtained from protists or from higher plants, show remarkable homologies not only among themselves but to a large extent between themselves and prokaryotic ribosomes. Mitochondrial ribosomes, on the other hand, can vary in sedimentation coefficient even within a kingdom: such differences do exist between the protists *Euglena* and *Tetrahymena,* and almost certainly exist among the fungi. Similarly, differences in the buoyant densities of mitoribosomes are found within both the protist and the fungal groups. Mitoribosomal RNA molecular weights and G+C contents, in contrast, appear to vary significantly between kingdoms but not within them.

The correlation between these last two structural properties of mitochondrial ribosomes and the taxonomic kingdoms in which they are found is shown in Figure 1. Plant mitoribosomes unfortunately cannot be included in this comparison because of the uncertainties about their properties (see p. 551) and the fact that their base composition has not yet been reported. Also plotted in Figure 1 are the data for chloroplast and prokaryotic ribosomes, which cluster together as expected, and for four kingdoms of cytoplasmic ribosomes. The latter group also forms a cluster, with the conspicuous exception of the cytoribosome from *Tetrahymena*. Both of these groups are dissimilar from any of the three mitoribosome kingdoms plotted, and these are in turn distinct from each other. Thus the two parameters, rRNA size and G+C content, discriminate ribosomes satisfactorily along the lines of phylogeny and intracellular location.

Sedimentation coefficients and buoyant densities, on the other hand, do not group mitoribosomes from the same kingdom together, nor do they differentiate them from the various nonmitochondrial ribosomes. This observation suggests that these two structural properties have been less conserved than rRNA size and G+C content in the course of ribosome evolution. It is noteworthy that buoyant density is the only one of these four structural characteristics which discriminates chloroplast from prokaryotic ribosomes (see Chapter 13). However, some of the intrakingdom variability reported for sedimentation coefficients and buoyant densities may be due to the possibly greater sensitivity of these parameters to differing conditions used in the preparation of the mitoribosomes.

The data pose several interesting problems for speculation and analysis. Is there some identifiable selective constraint opposing the evolutionary divergence of the size and base composition of ribosomal RNA? The G+C content does have known implications with regard to the thermal stability of the intramolecular double-stranded regions in ribosomal RNA; Freeman *et al.* (1973) have discussed the fact that all known mitoribosomes have lower G+C contents than the corresponding cytoribosomes, and some possible factors which may counteract the lower stability of the mitoribosomal RNA *in vivo*. The finding that ribosomes within mitochondria have diverged more widely during evolutionary history than have their extramitochondrial counterparts a few Angstroms away also demands further investigation, and we shall return to this point later (see p. 568).

Functional Properties of Mitoribosomes

Despite their wide-ranging physical-chemical properties, mitoribosomes share many properties among themselves and with nonmitochondrial ribosomes. All are ribonucleoprotein complexes composed of two subribosomal particles. The overall process of mitochondrial protein synthesis involves ribosomal interactions with mRNA, tRNA, and soluble factors reminiscent of those of nonmitochondrial systems. Further-

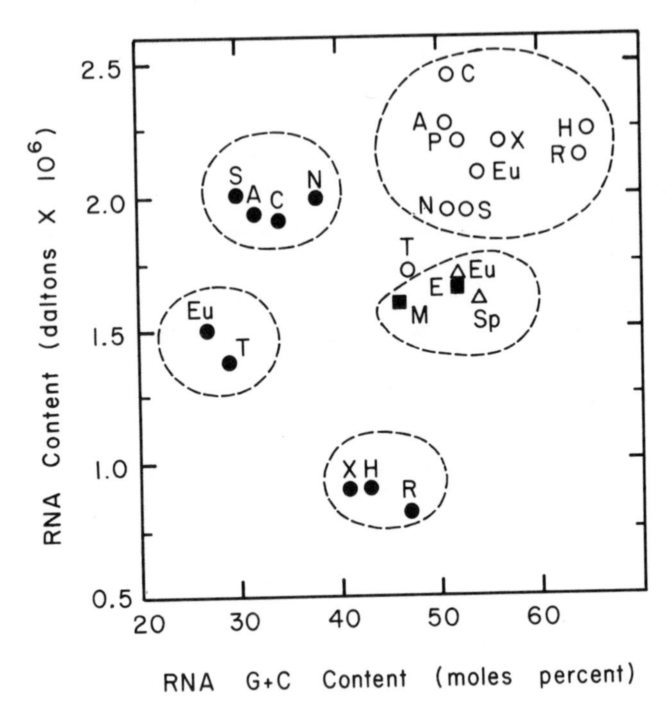

Figure 1.

●, *Mitochondrial Ribosomes*

Fungi

S, Saccharomyces (Reijnders et al., 1973; Morimoto and Halvorson, 1971)

A, Aspergillus (Verma et al., 1970; Edelman et al., 1970)

C, Candida utilis (Vignais et al., 1972)

N, Neurospora (Neupert et al., 1969; Küntzel and Noll, 1967)

Protists

Eu, Euglena (Krawiec and Eisenstadt, 1970; calculated from Avadhani and Buetow, 1972)

T, Tetrahymena (Reijnders et al., 1973; Chi and Suyama, 1970)

Animals

X, Xenopus (Leister and Dawid, 1974; Dawid and Chase, 1972)

H, HeLa (Robberson et al., 1971; Vesco and Penman, 1969)

R, rat (Sacchi et al., 1973; Bartoov et al., 1970)

■, *Prokaryotic Ribosomes*

E, Escherichia coli (Kurland, 1960; Morimoto and Halvorson, 1971)

M, Mycoplasma hominis (Johnson and Horowitz, 1971)

△, *Chloroplast Ribosomes*

Eu, Euglena (Rawson and Stutz, 1969; calculated from Avadhani and Buetow, 1972)

Sp, spinach (Hartley and Ellis, 1973; Lyttleton, 1962)

○, *Cytoplasmic Ribosomes*

Fungi

S, Saccharomyces (Reijnders et al., 1973; Morimoto and Halvorson, 1971)

A, Aspergillus (Verma et al., 1970; Edelman et al., 1970)

C, Candida utilis (Vignais et al., 1972)

N, Neurospora (Neupert et al., 1969; Küntzel and Noll, 1967) more, subunit-specific functions of bacterial and eukaryotic ribosomes appear localized to analogous subunits of mitoribosomes in all cases where such information is available (O'Brien *et al.*, 1974). As in the other ribosomal systems, much of our knowledge about the functional properties of mitoribosomes derives from the use of inhibitors of protein synthesis.

Antibiotic Susceptibility

General information on antibiotics and inhibitors of protein synthesis can be found in reviews by Pestka (1971) and Vazquez (1974). The effects of specific inhibitors of mitochondrial protein synthesis are reviewed in Linnane and Haslam (1970), Ashwell and Work (1970*a*), Mahler (1973), Kroon and DeVries (1971), and Kroon and Arendzen (1972). In addition to the direct action of antibiotics on mitoribosomes, the last two works also consider the consequences of the limited permeability of cellular and mitochondrial membranes to some antibiotics.

The use of antibiotics has contributed to our understanding of mitochondrial protein synthesis from two main perspectives. On the one hand, to gain insight into the mechanism of mitochondrial protein synthesis, inferences may be drawn by analogy to antibiotic action in other ribosomal systems. In this fashion, for example, aminocyl-tRNA was implicated in mitochondrial protein synthesis by the inhibitory action of puromycin on mitochondrial amino acid incorporation, well before mitochondrial ribosomes had even been discovered (Kroon, 1963; Kalf, 1963). On the other hand, the relation of mitochondrial to nonmitochondrial ribosomes can be inferred from their differential response to selected inhibitors. Thus mitochondrial ribosomes appear to be of the prokaryotic type, as they are generally inhibited by antibacterial antibiotics and largely unaffected by inhibitors that act specifically on eukaryotic cytoplasmic ribosomes. Recognition of the differential antibiotic susceptibility of mi-

---

Protists

Eu, Euglena (Krawiec and Eisenstadt, 1970; calculated from Avadhani and Buetow, 1972)

T, Tetrahymena (Reijnders et al., 1973; Chi and Suyama, 1970)

Animals

X, Xenopus (Loening et al., 1969; Dawid et al., 1970)

H, HeLa (Darnell, 1968; calculated from Vesco and Penman, 1969)

R, rat (Reijnders et al., 1973; Kirby, 1965)

Plants

P, pea (Loening et al., 1969; Bonner and Varner, 1965)

toribosomes has provided the investigator with a valuable research tool, enabling him to study the products of mitochondrial or cytoplasmic protein synthesis separately *in vivo*. To inhibit mitoribosome function selectively, chloramphenicol (Lamb *et al.,* 1968; Freeman, 1970) and ethidium bromide (Perlman and Penman, 1970*a*; Kroon and DeVries, 1971) are often used. Inhibitors used to suppress the extramitochondrial protein synthetic system selectively include cycloheximide (Lamb *et al.,* 1968; Pestka, 1971), anisomycin (Pestka, 1971; DeVries *et al.,* 1971), pederine (Brega and Vesco, 1971), and emetine (Perlman and Penman, 1970*a*; Ojala and Attardi, 1972). Although emetine can also affect mitochondrial protein synthesis (Lietman, 1971; Chakrabarti *et al.,* 1972; Kroon and Arendzen, 1972), it is often used in place of cycloheximide since some cells are relatively impermeable to the latter antibiotic.

Phylogenetic Differences. The possibility that phylogenetic differences exist among mitoribosomes in their response to antibiotics was suggested by Firkin and Linnane (1969). They noted that amino acid incorporation by mitochondria isolated from rats, rabbits, and cats was unaffected by amounts of erythromycin and lincomycin that were strongly inhibitory to isolated yeast (*S. cerevisiae*) mitochondria. Both of these antibiotics inhibit protein synthesis by specific interaction with the large subunit of bacterial ribosomes (Pestka, 1971). Extension of the above observation to include some aminoglycoside antibiotics which act on the small subunit of bacterial ribosomes, paramomycin and the neomycins (Davey *et al.,* 1970), made it seem even more likely that the differential antibiotic sensitivity reflected fundamental differences in the ribosomes of the two kinds of mitochondria.

Especially in view of the gross physical differences between fungal and mammalian mitoribosomes, such differences in antibiotic susceptibility are not unexpected. Yet an alternative explanation for the insensitivity of mammalian mitochondrial protein synthesis to these antibiotics was advanced. Kroon and DeVries asserted that the antibiotic resistance of mammalian mitochondria resides not in the mitoribosomes *per se* but in the relative impermeability of the mammalian mitochondrial membrane to some antibiotics (Kroon, 1969; Kroon and DeVries, 1970, 1971). Their arguments were based on studies in which the permeability of mitochondria to the antibiotics in question was increased by hypotonic shock, detergent treatment, or sonication. These methods give variable results (Towers *et al.,* 1972; but see Kroon *et al.,* 1974) and clear resolution of the issue had to await the development of mitochondria-free systems.

In fact, these differences in antibiotic sensitivity proved to be less pronounced when isolated ribosomes were tested. For example, the poly(U)-dependent synthesis of polyphenylalanine by isolated rat mitoribosomes is inhibited by erythromycin (Ibrahim and Beattie, 1973; Greco *et al.,* 1974) at the same concentrations as are effective on isolated yeast mitoribosomes (Ibrahim *et al.,* 1974), confirming the contention that the resistance of mammalian mitochondria to erythromycin resides in their membrane (DeVries *et al.,* 1973). On the other hand, the substantial resistance of protein synthesis by mammalian mitochondria to lincomycin (Towers *et al.,* 1972) appears to be a true property of the isolated mitoribosome when assayed by the fragment reaction for peptidyltransferase activity (Denslow and O'Brien, 1974). It remains to be seen whether isolated mammalian mitoribosomes are resistant to the small-subunit-specific antibiotics paramomycin and the neomycins.

In view of the limited number of species in which the antibiotic susceptibility of isolated mitoribosomes has been tested, it is premature to generalize about differences in antibiotic susceptibility among ribosomes. This point is especially emphasized by the finding that the mitoribosomes isolated from *Neurospora* and rat liver are similar in their response to various macrolide antibiotics such as erythromycin, and dissimilar from the mitoribosomes of another ascomycete, *S. cerevisiae* (DeVries *et al.,* 1973).

Special Cases. *Fusidic Acid:* Fusidic acid inhibits protein synthesis in both eukaryotic and prokaryotic systems by a direct interaction with the soluble translocase factor EF-2, but only when this factor is bound to the ribosome and complexed with GTP or GDP (Vazquez, 1974; Bodley *et al.,* 1970). The inhibitory action of fusidic acid is thought to result from the stabilization of this complex to prevent the cycling of EF-2 during the elongation phase of protein synthesis.

It is of considerable interest that levels of fusidic acid adequate to abolish protein synthesis in a cell-free system from *E. coli* are without effect on such systems from *Neurospora* mitochondria or cytoplasm (Grandi *et al.,* 1971). The mitoribosomal system is inhibited by fusidic acid, but only at hundredfold higher levels of the antibiotic. This result does not apply to mitochondrial systems in general, however, because mitochondria-free systems from yeast mitochondria appear nearly as susceptible to fusidic acid as the *E. coli* system (Richter *et al.,* 1971). The differential susceptibility of protein-synthesizing systems to fusidic acid probably reflects differences in the elongation factor more than in the respective ribosomes. Nevertheless, on the basis of the differential susceptibility noted here for the two fungal mitoribosome systems, fusidic acid seems to be a promising probe for other mitoribosome systems.

Thiostrepton: Thiostrepton, another inhibitor of protein elongation, is specific for prokaryotic ribosomes (Vazquez, 1974). Accordingly, it has been shown to inhibit yeast mitoribosomes (Richter *et al.,* 1971). The molecular basis for thiostrepton inactivation of bacterial ribosomes has recently been elucidated (Highland *et al.*, 1975). This inhibitor binds irreversibly to one of the proteins, L11, near the peptidyltransferase center on the large subunit of *E. coli* ribosomes, preventing further interaction of the ribosome with initiation and elongation factors (Highland *et al.*, 1975; Mazumder, 1973). In view of the defined mechanism of action and the prokaryotic specificity of thiostrepton, this inhibitor should be tested in other mitoribosomal systems. Thiostrepton may turn out to be a useful probe for fine structural homology among mitoribosomes.

Ricin: Ricin, a plant toxin, is a potent inhibitor of protein synthesis occurring on eukaryotic cytoplasmic ribosomes. It is a glycoprotein which irreversibly inactivates the large subunit of the ribosomes by an enzymatic process (Olsnes *et al.*, 1974; Sperti *et al.*, 1973) possibly analogous to the inactivation of *E. coli* ribosomes by colicin E3 (Garrett and Wittmann, 1973). Bacterial ribosomes are not affected by the toxin; unsurprisingly, neither are mitoribosomes from rat liver (Greco *et al.*, 1974).

Ethidium Bromide: Ethidium bromide (EB) is a phenanthridine dye which has some most intriguing effects on the synthesis of macromolecules in mitochondria. In addition to its clinical use as a trypanocidal agent, it has been widely employed to study various aspects of mitochondrial biogenesis (Mahler, 1973; Schatz and Mason, 1974; see also Chapter 15). In this capacity, its popularity derives in part from its diverse actions as a mitochondrial mutagen and a potent inhibitor of mitochondrial RNA synthesis, DNA synthesis, and protein synthesis. It is, moreover, a selective agent. Low levels of the drug elicit these multiple effects on mitochondrial processes without appreciable effects on the nuclear-cytoplasmic system. Ethidium bromide binds to mitochondrial DNA and intercalates between adjacent base pairs, but the precise mechanisms by which it exerts its diverse effects are not well understood. The mutagenic effects of EB, for example, probably involve more than simple intercalation and DNA helix distortion. In fact, Mahler and Bastos (1974) have shown that some of the drug actually becomes covalently attached to the DNA, and they have implicated this reaction in the mutagenic process.

It has been known for some time that ethidium bromide can also inhibit mitochondrial protein synthesis (Kroon *et al.*, 1968; Kroon and DeVries, 1969), but this inhibition was at first considered to be an indirect effect of its inhibitory action on mitochondrial RNA synthesis (Zylber *et al.*, 1969; Knight, 1969; Penman *et al.*, 1970; Kroon and DeVries, 1971). However, mitochondrial protein synthesis is inhibited more rapidly by EB than by other inhibitors of mitochondrial RNA synthesis (Perlman and Penman, 1970*a*; Lederman and Attardi, 1973), suggesting a direct action of the drug on mitoribosomes. To test this possibility, Grivell and Metz (1973) used mitochondria isolated from *Xenopus laevis,* which can import and translate exogenous poly(U) templates (Swanson, 1971), thus providing a convenient protein-synthesizing system which is independent of mitochondrial RNA synthesis. Because EB (2–5 μg/ml) inhibits the translation but not the uptake of poly(U) by *Xenopus* mitochondria, it appears that this drug acts directly on some components of the mitochondrial translation system (Grivell and Metz, 1973).

Difficulties have been encountered, however, in attempts to demonstrate EB sensitivity in isolated mitochondrial ribosomes. Phenylalanine incorporation in the presence of poly(U), by either purified bovine liver mitoribosomes (Denslow and O'Brien, unpublished) or a crude preparation from Ehrlich ascites cells (a low-speed supernatant of lysed mitochondria) (Avadhani and Rutman, 1975), is inhibited only by concentrations of EB tenfold greater than those required for a comparable degree of inhibition *in vivo* or in intact mitochondria. The peptidyltransferase activity of isolated *Neurospora* mitoribosomes has been found to be sensitive to EB, but again only at these higher concentrations (DeVries, 1973). Avadhani and Rutman (1975) have shown that the incorporation of labeled poly(U) into polyribosomes is inhibited by EB at 10 μg/ml, leading these authors to conclude that low levels of the antibiotic inhibit specifically at the initiation step of mitochondrial protein synthesis. This observation may provide a clue to the action of EB *in vivo,* but at present it is difficult to reconcile the high sensitivity of initiation with the lower sensitivity of total amino acid incorporation. Thus, while there is good evidence that EB affects mitochondrial protein synthesis directly, its mechanism of action remains unclear. It may be that intact mitochondria are capable of concentrating this antibiotic, as is the case for carbomycin (Kroon *et al.,* 1974). The availability of radioactively labeled EB (Loomeijer and Kroon, 1972; Bastos and Mahler, 1974) should permit a test of this possibility, as well as the identification of the specific component(s) of the mitochondrial translation apparatus which interact directly with the drug to produce the observed inhibition.

The importance of EB as an inhibitor of mitochondrial protein synthesis lies in the fact that it is at present the only likely candidate for an agent which can distinguish isolated mitoribosomes from all other ribosomes. Even at concentrations as high as 100 μg/ml, the drug has no detectable effect on poly(U)-programmed phenylalanine incorporation by either cytoplasmic ribosomes (Avadhani and Rutman, 1975) or *E. coli* ribosomes *in vitro* (Kroon and DeVries, 1971; Denslow and O'Brien, unpublished). The possibility that EB inhibits some bacterial ribosomes cannot be excluded, however, for there is indirect evidence that the translation apparatus in some strains of *E. coli* is susceptible to EB *in vivo* (Tonnesen and Friesen, 1973).

Are all mitoribosomes susceptible to ethidium bromide? Unqualified answers to this question require that the drug be tested on isolated mitoribosomes to rule out possible secondary effects of EB on protein synthesis (inhibition of mRNA synthesis, interference with ribosome–membrane interaction, inhibited amino acid transport, etc.). To date, such experiments have been done only with mitoribosomes isolated from *Neurospora* (DeVries, 1973), mouse (Ehrlich ascites) cells (Avadhani and Rutman, 1975), and bovine liver (Denslow and O'Brien, unpublished). Available evidence strongly suggests that EB also acts directly on mitoribosomes of the toad, as mentioned above (Grivell and Metz, 1973), as well as those of HeLa cells (Lederman and Attardi, 1973). Thus a significant degree of EB susceptibility is a general property of all mitoribosomes tested to date.

Therapeutic Implications. Recognition of the differential antibiotic susceptibility of mitoribosomes provides a rationale for interpreting the medical side effects of some antibacterial antibiotics. The best-documented cases in point are those for chloramphenicol (Freeman, 1970; Kroon and Arendzen, 1972) and tetracycline (Kroon and DeVries, 1970; Gijzel *et al.,* 1972; Oerter *et al.,* 1974) and their analogues. When administered in therapeutic amounts to man and animals, these antibiotics interfere with mitochondrial protein synthesis. Unable to make those few polypeptides essential for proper assembly, integration, and functioning of components of their electron transfer system (Schatz and Mason, 1974), the mitochondria lose some of their capacity for ATP generation. The resultant effects on energy metabolism and other mitochondrial processes are most noticeable in cells and tissues that are growing or dividing rapidly such as the gastrointestinal lining and the bone marrow, where it is necessary to renew mitochondrial components at elevated rates. The reversible immunosuppression and bone marrow depression that accompany chloramphenicol and thiamphenicol administration, for example, are now thought to be direct consequences of impaired mitochondrial protein synthesis (Nijhof and Kroon, 1974; Oerter *et al.,* 1974; Yunis *et al.,* 1974). In contrast, there are other antibacterial antibiotics (erythromycin, for example) with which serious mitochondrial side effects are not noticed. As discussed above (see p. 560), permeability barriers probably exist in animal cells for many of these "safer" antibiotics, either at the cell membrane or at the level of the mitochondrial membrane (DeVries *et al.,* 1973; Kroon and Arendzen, 1972), so that ribosomes inside mitochondria never become exposed to inhibitory levels of the antibiotics.

5 S Ribosomal RNA

There is one structural component of bacterial ribosomes for which a specific function has been assigned and which appears to be present also in eukaryotic cytoplasmic and chloroplast ribosomes. An RNA molecule sedimenting at about 5 S is found in each of these ribosomes, specifically associated with the large subunit, and in amounts of one molecule per ribosome. In *E. coli,* this component has been implicated as the one which interacts directly with the ribosome-binding tetranucleotide sequence in transfer RNA (Erdmann *et al.,* 1973), and it seems likely that the similar molecules in other ribosomes perform the same function. Consequently, it is of interest to ask whether mitochondrial ribosomes also contain 5 S rRNA.

The answer appears to be that mitoribosomes are as variable in this respect as they are in most of their other properties. 5 S RNA has been found in mitoribosomes of *Tetrahymena* (Chi and Suyama, 1970) and in the 78 S mitoribosomes of higher plants (Leaver and Harmey, 1973). It was also reported to be present in *Euglena* mitoribosomes (Avadhani and Buetow, 1972), but further investigation showed that in fact it is absent from these particles (Avadhani and Buetow, 1974). *Neurospora* mitoribosomes do not appear to contain 5 S RNA, at least in the 73 S particles (Lizardi and Luck, 1971); it would be interesting to confirm this observation for the 80 S ribosomes obtained from these mitochondria by Datema *et al.* (1974) (see p. 546). Mitochondrial ribosomes of animals likewise seem to lack 5 S RNA, but they do contain a smaller RNA molecule which may be functionally equivalent. As isolated from whole hamster-cell mitochondria, this molecule is distinguishable from 4 S transfer RNA by its greater electrophoretic mobility (about 3 S), its lack of methylated nucleotides, and its 1 : 1 stoichiometry with the large ribosomal RNAs (Dubin *et al.,* 1974). A similar component has been obtained directly from isolated mitochondrial ribosomes of HeLa cells, where it was shown to be localized in the large subunit in stoichiometric amounts (Gray and Attardi, 1973).

The presence of a conserved tetranucleotide sequence in the ribosome binding loop of both eukaryotic cytoplasmic and bacterial tRNAs suggests that the functional interaction of this sequence with a complementary one in 5 S RNA (Erdmann *et al.,* 1973; Richter *et al.,* 1973) is a central mechanism for the binding of tRNA molecules to ribosomes. Mitoribosomes can also utilize bacterial (Greco *et al.,* 1973) and eukaryotic cytoplasmic and chloroplast (Avadhani and Buetow, 1974) tRNAs. Although the sequence of none of the mitochondrial tRNAs is presently known, they very likely bind to mitoribosomes by the same mechanism used by other ribosomes. How can this mode of tRNA binding be accomplished by those mitoribosomes which appear to lack a separate 5 S RNA molecule? Perhaps the critical sequence for binding resides in the large rRNA molecule itself in such ribosomes, and not in a separate smaller molecule. A means of testing this possibility, as well as the functional involvement of the putative 5 S RNA analogues in other mitoribosomes, could be the tetranucleotide probe TΨCG, which competes with tRNA for the tRNA binding sites on bacterial ribosomes (Richter *et al.,* 1973).

Protein Synthesis Factors

The protein synthesis factors that have been studied in mitochondrial systems include the initiation factors (IF) and elongation factors (EF). Ribosomes and subribosomal particles prepared in media of low to intermediate ionic strength (0.1–0.3 M) contain variable amounts of bound factors which can be washed from them by exposure to buffers of higher ionic strength (0.5–1.0 M). These factors and others present in the supernatant fraction normally cycle on and off the ribosome during the process of protein synthesis. In bacteria, IF-3 and IF-2 serve to bind mRNA and initiator tRNA (fmet-$tRNA^{f}_{met}$), respectively, to the small subribosomal particle in the initiation steps of protein synthesis. IF-1 promotes the catalytic use of IF-2. In addition to its mRNA binding role, IF-3 binds to free small subribosomal particles, promoting the dissociation of single ribosomes. Because of this latter action, IF-3 has also been identified as a ribosome "dissociation factor" (DF). The elongation factor EF-1 (EF-T) is required to bring aminoacyl-tRNAs to the ribosome during protein elongation, and EF-2 (EF-G) is needed to move mRNA and the tRNA-bound growing peptide chain from one site on the ribosome to another in the process of translocation (Lengyel, 1974). On the basis of limited studies on mitochondrial systems, the protein synthesis factors in mitochondria and bacteria appear to be analogous, able to function interchangeably with either kind of ribosome.

Initiation Factors. Specific proteins were implicated in the initiation of protein synthesis on isolated *Neurospora* mitoribosomes by Sala and Küntzel (1970). To assay chain initiation, they measured the factor-dependent formation of fmet-puromycin when ribosomes were incubated with fmet-tRNA, puromycin, and the initiation codon trinucleotide ApUpG, in the presence of factors which can be washed from mitochondrial and bacterial ribosomes by exposure to 1 M NH_4Cl. These investigators showed that unfractionated initiation factors from either source can restore the activity of both kinds of salt-washed ribosomes. The homology between the initiation factors of bacteria and *Neurospora* mitochondria also includes the ribosome dissociation factor (IF-3). This factor can effect the dissociation of both the 73 S form (Agsteribbe and Kroon, 1973) and the 80 S form (Datema *et al.*, 1974) of *Neurospora* mitoribosomes, but not *Neurospora* cytoribosomes.

The involvement of mitochondrial initiation factors in the translation of natural mRNAs by isolated mitoribosomes has been demonstrated in two other mitochondrial systems, yeast and *Euglena*. The translation of R17 bacteriophage RNA by mitoribosomes isolated from *S. cerevisiae* requires factors contained in the 1 M ammonium chloride wash of the mitoribosomes (Scragg *et al.*, 1971). Similarly, mitoribosomes isolated from *Euglena* depend on added initiation factors to translate *Euglena* mRNA (Avadhani and Buetow, 1974). It is of considerable interest that initiation factors from *Euglena* chloroplasts can substitute for the mitochondrial factors in this system, while cytoplasmic initiation factors cannot.

Elongation Factors. Most of our knowledge of mitochondrial elongation factors derives from studies in the ascomycete systems *Neurospora* and yeast. The crude mitochondrial elongation factors in both of these systems have been resolved into fractions that are functionally interchangeable with the bacterial factors EF-T and EF-G, respectively (Grandi and Küntzel, 1970; Richter and Lipmann, 1970; Morimoto *et al.*, 1971; Scragg, 1971). The ready interchangeability of elongation factors between the two kinds of ribosomes attests an additional homology between bacterial and mitochondrial ribosomes: mitoribosomal binding sites preferentially recognize elongation factors from bacterial systems over those from cytoplasmic systems. Comparison of mitochondrial and cytoplasmic elongation factors from yeast has revealed substantial differences between them. Analogous factors manifest different chromatographic behavior on hydroxylapatite and DEAE-cellulose, antisera directed against the cytoplasmic factors are not inhibitory to the mitochondrial factors, and mitochondrial EF-G—unlike the cytoplasmic factor—is not affected by diphtheria toxin (Richter and Lipmann, 1970; Scragg, 1971). Despite these differences, both mitochondrial and cytoplasmic elongation factors are encoded in the nuclear genome and both are synthesized on cytoplasmic ribosomes. The evidence for the extramitochondrial origin of the mitochondrial factors comes from the observation that these factors are present in yeast petite mutants lacking mitochondrial DNA and mitochondrial ribosomes (Richter, 1971;Scragg, 1971; Parisi and Cella, 1971).

Comparative studies of the elongation factor specificity of mitoribosomes have also been performed in two protists, *Tetrahymena* and *Euglena*. *Tetrahymena* mitoribosomes show a specific requirement for homologous elongation factors, like the ascomycete mitoribosomes above (Allen and Suyama, 1972). In contrast, *Euglena* mitoribosomes are able to use cytoplasmic elongation factors, although not as well as elongation factors from mitochondria or chloroplasts (Avadhani and Buetow, 1974).

These homologies between bacterial and mitochondrial elongation factors extend even to the animal systems. Swanson (1973) has shown that *Xenopus* mitoribosomes will function with elongation factors from mitochondria or *E. coli*, but not from the cytoplasm of *Xenopus* or rat. The ability of mitoribosomes from mammals to function with cytoplasmic factors has not been tested, but it is known that these mitoribosomes can use bacterial factors (Ibrahim *et al.*, 1974; Greco *et al.*, 1973).

Evolution of Ribosomes

The major conclusions to be drawn from the data reviewed above are that some properties of ribosomes vary more widely among different species of organisms than other properties and that some kinds of ribosomes show greater differences with respect to these properties than other ribosomes. The first observation is not surprising, since it is intuitively probable that some properties are more essential for the functional activity of ribosomes than others. Furthermore, this differential variability has been known for some time from the study of nonmitochondrial ribosomes. But findings on the more widely divergent ribosomes from mitochondria reveal variability in some characteristics that are highly conserved in other ribosomes and permit a ranking of the most commonly measured characteristics in order of increasing variability.

1. Mechanism of action. All ribosomes operate by a mechanism requiring two ribosomal subunits, messenger RNA, aminoacyl-transfer RNA, and initiation and elongation factors. The similarity in detail among different ribosomes with respect to these complex interactions forces the conclusion that all of them derive from a common ancestor.

2. Antibiotic sensitivity and factor specificity. These two properties have been very highly conserved in the course of evolution. The evidence suggests that only one major change has ever occurred in each of these two properties, that these changes occurred at nearly the same point in evolutionary history, and that these changes were approximately simultaneous with the event which created eukaryotic cells. This timing is indicated by the lack of any extant species which appear to be descendants of an intermediate organism in which only one or two of the changes had occurred—for example, a eukaryote with chloramphenicol-sensitive, cycloheximide-resistant cytoplasmic ribosomes capable of utilizing bacterial initiation and elongation factors. Major changes in factor specificity and susceptibility to most antibiotics did not occur at all in the organellar ribosomes of eukaryotes, either at the time of their origin or in the billion years since.

The observation of factor interchangeability among ribosomes from widely divergent organisms is relatively easy to rationalize as an example of the evolutionary conservation of functional sites. But the high degree to which specific antibiotic sensitivities have been conserved, even among ribosomes of grossly different structure, is not as easy to explain at our present level of understanding of the mechanism of protein synthesis. There is certainly no selective advantage associated directly with antibiotic susceptibility. One antibiotic, puromycin, has been found to be an obvious structural analogue of one of the components of the protein synthetic machinery. Although this is not the case for chloramphenicol, the large number of different kinds of ribosomes which are sensitive to this inhibitor indicates that it is likewise directed to a functional site at which mutations can only rarely lead to ecologically competitive phenotypes. This conclusion is supported by the failure of attempts to isolate mutants of bacteria with chloramphenicol-resistant ribosomes (Benveniste and Davies, 1973). However, yeast mutants have been obtained with mitochondrial ribosomes which are resistant to this drug and show no impairment of function *in vitro* or *in vivo* (Grivell *et al.*, 1973). It is possible that the isolation of such mutants in mitochondria, but not in prokaryotes, is a further expression of the same unknown mechanisms which have permitted greater evolutionary divergence among mitochondrial ribosomes than among prokaryotic ribosomes with respect to other, less conserved, properties. Certainly the availability of chloramphenicol-resistant ribosomes, from whatever source, will be of value in mechanistic studies of the functionally important site which is altered in the mutant. The possibility that mitochondria may prove to be a richer source of other such mutations than bacteria should not be ignored.

3. *Ribosomal RNA size and G+C content.* As discussed above (see p. 556), there is a sharp distinction between the rRNA molecular weights found in eukaryotic ribosomes and those of ribosomes from prokaryotes and chloroplasts, with relatively little variation within each of the two groups. Much greater variation is observed within the group of mitochondrial ribosomes in this property, as well as in the G+C content of the rRNA, but mitoribosomes of a given kingdom appear to be similar to each other in both of these respects.

4. *Sedimentation coefficient and buoyant density.* These structural properties have diverged widely within kingdoms of organisms in the case of mitochondrial ribosomes, but much less widely, even among different kingdoms, for other kinds of ribosomes. The actual rate of evolution of

these mitoribosome characters is difficult to assess for several reasons. Relatively few species related to each other at intermediate taxonomic levels such as order and family have been compared. Even when such comparisons are made they may be difficult to interpret because the actual phylogenetic relationships among the lower organisms, including protists and fungi, are uncertain (Margulis, Volume 1), particularly for the intermediate taxa. These mitoribosome properties might prove to be useful themselves as characters for taxonomic classification. However, the sedimentation coefficient is an indirect parameter affected by several different structural properties such as molecular weight, buoyant density, and conformation, and its measurement may be subject to unknown experimental artifacts in some cases (see p. 548).

5. *Primary sequence of ribosomal proteins and ribosomal RNA.* Some differences in ribosomal proteins have been observed among eukaryotic cytoplasmic ribosomes of organisms within the same phylum by electrophoretic methods (Delaunay *et al.,* 1973), and even within the same order by sensitive immunological techniques (Delaunay and Schapira, 1974). Such differences are also found in bacterial ribosomes at the family level, using either technique (Garrett and Wittmann, 1973). In view of the greater variability of mitochondrial ribosomes in most of their other properties, it is perhaps not surprising that some electrophoretically different proteins can be found in mitoribosomes from two species of the same genus (Leister and Dawid, 1975). Evolutionary variation in rRNA base sequence can be detected by hybridizing rRNA from one species with denatured DNA from another. By this means, Dawid (1972*a*) detected differences between the mitochondrial rRNAs of two species of *Xenopus*. No differences were found between *S. cerevisiae* and *S. carlsbergensis* in this respect (Groot *et al.,* 1975). However, the mitochondrial rRNA of *C. utilis* shows considerable nonhomology with that of *S. carlsbergensis;* like most of the other properties of ribosomes, the base sequence differs more between these two organisms in the mitochondrial rRNA than in the cytoplasmic rRNA (Groot *et al.,* 1975).

At present, we can only conjecture about the molecular mechanisms which have promoted the relatively rapid evolutionary divergence of mitochondrial ribosomes or (more likely) restricted the divergence of other kinds of ribosomes. Perhaps the best basis for speculations on this subject would be a comparison of mitochondrial mutational and genetic processes with those of chloroplasts. Such processes are probably more alike in these two organelles than they are like the ones which have guided the evolution of nuclear or prokaryotic genes. Thus explanations for the variability of mitochondrial ribosomes must take into account the much lower degree of variability in chloroplast ribosomes.

Acknowledgments

We are very grateful to Miss Lynn Erb and Miss Sharon Bryant for their secretarial help. This work was supported by a grant from the National Institutes of Health (Grant GM-15438-07). D. E. M. was supported by a National Science Foundation Graduate Fellowship.

Literature Cited

Aaij, C., N. Nanninga, and P. Borst, 1972 The structure of ribosome-like particles from rat liver mitochondria. *Biochim. Biophys. Acta* **277:**140–148.

Agsteribbe, E. and A. M. Kroon, 1973 Dissociation of mitochondrial ribosomes of *Neurospora crassa* by a bacterial dissociation factor. *Biochem. Biophys. Res. Commun.* **51:**8–13.

Agsteribbe, E., R. Datema, and A. Kroon, 1974 Mitochondrial polysomes from *Neurospora crassa.* In *The Biogenesis of Mitochondria,* edited by A. Kroon and C. Saccone, Academic Press, New York, pp. 305–314.

Allen, N. E. and Y. Suyama, 1972 Protein synthesis *in vitro* with *Tetrahymena* mitochondrial ribosomes. *Biochim. Biophys. Acta* **259:**369–377.

Andre, J. and V. Marinozzi, 1965 Présence, dans les mitochondries, de particules ressemblant aux ribosomes. *J. Microsc.* **4:**615–626.

Ashwell, M. and T. S. Work, 1970*a* The biogenesis of mitochondria. In *Annual Review of Biochemistry,* edited by E. E. Snell, Annual Reviews, Palo Alto, Calif., pp. 251–290.

Ashwell, M. and T. S. Work, 1970*b* The functional characterization of ribosomes from rat liver mitochondria. *Biochem. Biophys. Res. Commun.* **39:**204–211.

Avadhani, N. G. and D. E. Buetow, 1972 Isolation of active polyribosomes from the cytoplasm, mitochondria and chloroplasts of *Euglena gracilis. Biochem. J.* **128:**353–365.

Avadhani, N. G. and D. E. Buetow, 1974 Mitochondrial and cytoplasmic ribosomes: Distinguishing characteristics and a requirement for the homologous ribosomal salt-extractable fraction for protein synthesis. *Biochem. J.* **140:**73–78.

Avadhani, N. G. and R. J. Rutman, 1975 Differential effects of ethidium bromide on cytoplasmic and mitochondrial protein synthesis. *FEBS Lett.* **50:**303–305.

Bartoov, B., R. S. Mitra, and K. B. Freeman, 1970 Ribosomal-type ribonucleic acid from rodent mitochondria. *Biochem. J.* **120:**455–466.

Bastos, R. N. and H. R. Mahler, 1974 A synthesis of labeled ethidium bromide. *Arch. Biochem. Biophys.* **160:**643–646.

Beale, G. H., J. K. C. Knowles, and A. Tait, 1972 Mitochondrial genetics in Paramecium. *Nature* (*London*) **235:**396–397.

Benveniste, R. and J. Davies, 1973 Mechanisms of antibiotic resistance in bacteria. *Ann. Rev. Biochem.* **42:**471–506.

Bodley, J. W., F. J. Zieve, L. Lin, and S. T. Zieve, 1970 Studies on translocation. III. Conditions necessary for the formation and detection of a stable ribosome–G factor–guanosine diphosphate complex in the presence of fusidic acid. *J. Biol. Chem.* **245:**5656–5661.

Bonner, J. D. and J. E. Varner, 1965 *Plant Biochemistry,* Academic Press, New York.

Borst, P., 1972 Mitochondrial nucleic acids. *Annu. Rev. Biochem.* **41:**333–376.

Borst, P. and L. Grivell, 1971 Mitochondrial ribosomes. *FEBS Lett.* **13:**73–88.

Bourque, D. P., J. E. Boynton, and N. W. Gillham, 1971 Studies on the structure and cellular location of various ribosomes and ribosomal RNA species in the green alga *Chlamydomonas reinhardi. J. Cell Sci.* **8:**153–183.

Brega, A. and C. Vesco, 1971 Ribonucleoprotein particles involved in HeLa mitochondrial protein synthesis. *Nature (London) New Biol.* **229:**136–139.

Cammarano, P., A. Felsani, A. Romeo, and F. M. Alberghina, 1973 Particle weights of active ribosomal subunits from *Neurospora crassa. Biochim. Biophys. Acta* **308:**404–411.

Chakrabarti, S., D. K. Dube, and S. C. Roy, 1972 Effects of emetine and cycloheximide on mitochondrial protein synthesis in different systems. *Biochem. J.* **128:**461–462.

Chi, J. C. H. and Y. Suyama, 1970 Comparative studies on mitochondrial and cytoplasmic ribosomes of *Tetrahymena pyriformis. J. Mol. Biol.* **53:**531–556.

Coon, H. G., I. Horak, and I. B. Dawid, 1973 Propagation of both parental mitochondrial DNA's in rat–human and mouse–human hybrid cells. *J. Mol. Biol.* **81:**285–298.

Cooper, C. and C. Avers, 1974 Evidence of involvement of mitochondrial polysomes and messenger RNA in synthesis of organelle proteins. In *The Biogenesis of Mitochondria,* edited by A. Kroon and C. Saccone, Academic Press, New York, pp. 289–303.

Coote, J. L., T. H. Rabbitts, and T. S. Work, 1971 The mitochondrial ribosomes of baby hamster kidney cells. *Biochem. J.* **123:**279–281.

Curgy, J. J., G. Ledoigt, B. J. Stevens, and J. Andre, 1974 Mitochondrial and cytoplasmic ribosomes from *Tetrahymena pyriformis:* Correlative analysis by gel electrophoresis and electron microscopy. *J. Cell Biol.* **60:**628–640.

Darnell, J. E., 1968 Ribonucleic acid from animal cells. *Bacteriol. Rev.* **32:**262–290.

Datema, R., E. Agsteribbe, and A. M. Kroon, 1974 The mitochondrial ribosomes of *Neurospora crassa.* I. On the occurrence of 80 S ribosomes. *Biochim. Biophys. Acta* **335:**386–395.

Davey, P. J., J. M. Haslam, and A. W. Linnane, 1970 Biogenesis of mitochondria. 12. The effects of aminoglycoside antibiotics on the mitochondrial and cytoplasmic protein-synthesizing systems of *Saccharomyces cerevisiae. Arch. Biochem. Biophys.* **136:**54–64.

Dawid, I. B., 1972*a* Evolution of mitochondrial DNA sequences in *Xenopus. Dev. Biol.* **29:**139–151.

Dawid, I. B., 1972*b* Mitochondrial protein synthesis. In *Mitochondria/Biomembranes,* edited by S. G. van den Bergh, P. Borst, L. L. M. van Deenen, J. C. Riemersma, E. C. Slater, and J. M. Tager, North-Holland, Amsterdam, pp. 35–51.

Dawid, I. B. and J. W. Chase, 1972 Mitochondrial RNA in *Xenopus laevis.* II. Molecular weights and other physical properties of mitochondrial ribosomal and 4 S RNA. *J. Mol. Biol.* **63:**217–231.

Dawid, I. B., D. D. Brown, and R. H. Reeder, 1970 Composition and structure of chromosomal and amplified ribosomal DNA's of *Xenopus laevis. J. Mol. Biol.* **51:**341–360.

Delaunay, J. and G. Schapira, 1974 Phylogenic distance between prokaryotes and eukaryotes as evaluated by ribosomal proteins. *FEBS Lett.* **40:**97–100.

Delaunay, J., F. Creusot, and G. Schapira, 1973 Evolution of ribosomal proteins. *Eur. J. Biochem.* **39:**305–312.

Denslow, N. D. and T. W. O'Brien, 1974 Susceptibility of 55 S mitochondrial ribosomes to antibiotics inhibitory to prokaryotic ribosomes, lincomycin, chloramphenicol and PA114A. *Biochem. Biophys. Res. Commun.* **57:**9–16.

DeVries, H., 1973 The protein synthetic system of rat liver mitochondria: Its characterization and its response to inhibitors. Thesis, University of Groningen, The Netherlands.

DeVries, H. and A. Kroon, 1974 Physicochemical and functional characterization of the 55 S ribosomes from rat liver mitochondria. In *The Biogenesis of Mitochondria,* edited by A. Kroon and C. Saccone, Academic Press, New York, pp. 357–365.

DeVries, H., E. Agsteribbe, and A. M. Kroon, 1971 The "fragment reaction": A tool for the discrimination between cytoplasmic and mitochondrial ribosomes. *Biochim. Biophys. Acta* **246:**111–122.

DeVries, H., A. J. Arendzen, and A. M. Kroon, 1973 The interference of the macrolide antibiotics with mitochondrial protein synthesis. *Biochim Biophys. Acta* **331:**264–275.

Dubin, D. T., 1974 Methylated nucleotide content of mitochondrial ribosomal RNA from hamster cells. *J. Mol. Biol.* **84:**257–273.

Dubin, D. T., T. H. Jones, and G. R. Cleaves, 1974 An unmethylated "3 S" RNA in hamster mitochondria: A 5 S RNA equivalent? *Biochem. Biophys. Res. Commun.* **56:**401–406.

Edelman, M., I. M. Verma, and U. Z. Littauer, 1970 Mitochondrial ribosomal RNA from *Aspergillus nidulans:* Characterization of a novel molecular species. *J. Mol. Biol.* **49:**67–83.

Edelman, M., I. M. Verma, R. Herzog, E. Galun, and U. Z. Littauer, 1971 Physicochemical properties of mitochondrial ribosomal RNA from fungi. *Eur. J. Biochem.* **19:**372–378.

Erdmann, V. A., M. Sprinzl, and O. Pongs, 1973 The involvement of 5 S RNA in the binding of tRNA to ribosomes. *Biochem. Biophys. Res. Commun.* **54:**942–948.

Firkin, F. C. and A. W. Linnane, 1969 Phylogenetic differences in the sensitivity of mitochondrial protein synthesizing systems to antibiotics. *FEBS Lett.* **2:**330–332.

Freeman, K. B., 1970 Inhibition of mitochondrial and bacterial protein synthesis by chloramphenicol. *Can. J. Biochem.* **48:**479–485.

Freeman, K. B., R. S. Mitra, and B. Bartoov, 1973 Characteristics of the base composition of mitochondrial ribosomal RNA. *Sub-Cell. Biochem.* **2:**183–192.

Garrett, R. A. and H. G. Wittmann, 1973 Structure of bacterial ribosomes. *Adv. Protein Synthesis* **27:**277–347.

Gijzel, W. P., M. Strating, and A. M. Kroon, 1972 The biogenesis of mitochondria during proliferation and maturation of the intestinal epithelium of the rat: Effects of oxytetracycline. *Cell Diff.* **1:**191–198.

Grandi, M. and H. Küntzel, 1970 Mitochondrial peptide chain elongation factors from *Neurospora crassa. FEBS Lett.* **10:**25–28.

Grandi, M., A. Helms, and H. Küntzel, 1971 Fusidic acid resistance of mitochondrial G factor from *Neurospora crassa. Biochem. Biophys. Res. Commun.* **44:**864–871.

Gray, P. H. and G. Attardi, 1973 An attempt to identify a presumptive 5 S RNA-equivalent RNA species in mitochondrial ribosomes. *J. Cell Biol.* **59:**120*a*.

Greco, M., P. Cantatore, G. Pepe, and C. Saccone, 1973 Isolation and characterization of rat liver mitochondrial ribosomes highly active in poly(U)-directed polyphenylalanine synthesis. *Eur. J. Biochem.* **37:**171–177.

Greco, M., G. Pepe, and C. Saccone, 1974 Characterization of the monomer form of rat liver mitochondrial ribosome and its activity in poly U-directed polyphenylalanine synthesis. In *The Biogenesis of Mitochondria,* edited by A. Kroon and C. Saccone, Academic Press, New York, pp. 367–376.

Grivell, L. A. and V. Metz, 1973 Inhibition by ethidium bromide of mitochondrial protein synthesis programmed by imported poly(U). *Biochem. Biophys. Res. Commun.* **55:**125–131.

Grivell, L. A., L. Reijnders, and P. Borst, 1971 Isolation of yeast mitochondrial ribosomes highly active in protein synthesis. *Biochim. Biophys. Acta* **247:**91–103.

Grivell, L. A., P. Netter, P. Borst, and P. P. Slonimski, 1973 Mitochondrial antibiotic resistance in yeast ribosomal mutants resistant to chloramphenicol, erythromycin, and spiramycin. *Biochim. Biophys. Acta* **312:**358–367.

Groot, G. S. P., R. A. Flavell, and J. P. M. Sanders, 1975 Sequence homology of nuclear and mitochondrial DNAs of different yeasts. *Biochim. Biophys. Acta* **378:**186–194.

Gualerzi, C., 1969 Electrophoretic comparison of cytoplasmic and mitochondrial ribosomal proteins from *Neurospora crassa. Ital. J. Biochem.* **18:**418–425.

Hallermayer, G. and W. Neupert, 1974 Immunological difference of mitochondrial and cytoplasmic ribosomes of *Neurospora crassa. FEBS Lett.* **41:**264–268.

Hamilton, M. G., 1971 Isodensity equilibrium centrifugation of ribosomal particles; the calculation of the protein content of ribosomes and other ribonucleoproteins from buoyant density measurements. *Methods Enzymol.* **20:** Part C 512–521.

Hamilton, M. G. and T. W. O'Brien, 1974 Ultracentrifugal characterization of the mitochondrial ribosome and subribosomal particles of bovine liver: Molecular size and composition. *Biochemistry* **13:**5400–5403.

Hamilton, M. G. and M. L. Petermann, 1959 Ultracentrifugal studies on ribonucleoprotein from rat liver microsomes. *J. Biol. Chem.* **234:**1441–1446.

Hartley, M. R. and R. J. Ellis, 1973 Ribonucleic acid synthesis in chloroplasts. *Biochem. J.* **134:**249–262.

Highland, J. H., G. A. Howard, E. Ochsner, G. Stoffler, R. Hasenbank and J. Gordon, 1975 Identification of a ribosomal protein necessary for thiostrepton binding to *Escherichia coli* ribosomes. *J. Biol. Chem.* **250:**1141–1145.

Hill, W. E., G. P. Rossetti, and K. E. Van Holde, 1969 Physical studies of ribosomes from *Escherichia coli. J. Mol. Biol.* **44:**263–277.

Ibrahim, N. G. and D. S. Beattie, 1973 Protein synthesis on ribosomes isolated from rat liver mitochondria: Sensitivity to erythromycin. *FEBS Lett.* **36:**102–104.

Ibrahim, N. G., J. P. Burke, and D. S. Beattie, 1974 The sensitivity of rat liver and yeast mitochondrial ribosomes to inhibitors of protein synthesis. *J. Biol. Chem.* **249:**6806–6811.

Johnson, J. D. and J. Horowitz, 1971 Characterization of ribosomes and RNAs from *Mycoplasma hominis. Biochem. Biophys. Acta* **247:**262–279.

Kalf, G., 1963 The incorporation of leucine-1-C^{14} into the protein of rat heart sarcosomes: An investigation of optimal conditions. *Arch. Biochem. Biophys.* **101:**350–359.

Kellems, R. E., V. F. Allison, and R. A. Butow, 1974 Cytoplasmic type 80 S ribosomes associated with yeast mitochondria. II. Evidence for the association of cytoplasmic ribosomes with the outer mitochondrial membrane *in situ. J. Biol. Chem.* **249:**3297–3303.

Keyhani, E., 1973 Ribosomal granules associated with outer mitochondrial membrane in aerobic yeast cells. *J. Cell Biol.* **58:**480–484.

Kirby, K. S., 1965 Isolation and characterization of ribosomal ribonucleic acid. *Biochem. J.* **96:**266–269.

Kleinow, W., 1974*a* Influence of cations on the dissociation of mitochondrial and cytoplasmic ribosomes from *Locusta migratoria. Hoppe-Seylers Z. Physiol. Chem.* **355:**1027–1034.

Kleinow, W., 1974*b* RNA from mitochondrial ribosomes of *Locusta migratoria.* In *The Biogenesis of Mitochondria,* edited by A. M. Kroon and C. Saccone, Academic Press, New York, pp. 337–346.

Kleinow, W., W. Neupert, and T. Bucher, 1971 Small sized ribosomes from mitochondria of *Locusta migratoria. FEBS Lett.* **12:**129–133.

Kleinow, W., W. Neupert, and F. Miller, 1974 Electron microscope study of mitochondrial 60 S and cytoplasmic 80 S ribosomes from *Locusta migratoria. J. Cell Biol.* **62:**860–875.

Knight, E., Jr., 1969 Mitochondria-associated ribonucleic acid of the HeLa cell: Effect of ethidium bromide on the synthesis of ribosomal and 4 S ribonucleic acid. *Biochemistry* **8:**5089–5093.

Krawiec, S. and J. M. Eisenstadt, 1970 Ribonucleic acids from the mitochondria of bleached *Euglena gracilis. Biochim. Biophys. Acta* **217:**132–141.

Kroon, A. M., 1963 Inhibitors of mitochondrial protein synthesis. *Biochim. Biophys. Acta* **76:**165–167.

Kroon, A. M., 1969 On the effects of antibiotics and intercalating dyes on mitochondrial biosynthesis. In *Inhibitors: Tools in Cell Research,* edited by T. H. Bucher and H. Sies, Springer-Verlag, New York, pp. 159–166.

Kroon, A. M. and A. J. Arendzen, 1972 The inhibition of mitochondrial biogenesis by antibiotics. In *Mitochondria/Biomembranes,* edited by S. G. van den Bergh, P. Borst, L. L. M. van Deenen, J. C. Riemersma, E. C. Slater, and J. M. Tager, North-Holland, Amsterdam, pp. 71–83.

Kroon, A. M. and H. DeVries, 1969 The effect of chloramphenicol on the biogenesis of mitochondria of rat liver *in vivo. FEBS Lett.* **3:**208–210.

Kroon, A. M. and H. DeVries, 1970 Antibiotics: A tool in the search for the degree of autonomy of mitochondria in higher animals. In *Control of Organelle Development,* Academic Press, New York, pp. 181–199.

Kroon, A. M. and H. DeVries, 1971 Mitochondriogenesis in animal cells: Studies with different inhibitors. In *Autonomy and Biogenesis of Mitochondria and Chloroplasts,* edited by N. K. Boardman, A. W. Linnane, and R. M. Smillie, American Elsevier, New York, pp. 318–327.

Kroon, A. M., M. J. Botman, and C. Saccone, 1968 Practical procedures for the isolation of mitochondrial preparations suitable for the study of mitochondrial macromolecules and with minimal contamination by other cell fractions or bacteria. In *Biochemical Aspects of the Biogenesis of Mitochondria,* edited by E. C. Slater, J. M. Tager, S. Papa, and E. Quagliariello, Adriatica Editrice, Bari, pp. 439–455.

Kroon, A. M., E. Agsteribbe, and H. DeVries, 1972 Protein synthesis in mitochondria and chloroplasts. In *The Mechanism of Protein Synthesis and its Regulation,* edited by L. Bosch, American Elsevier, New York, pp. 539–582.

Kroon, A. M., A. J. Arendzen, and H. DeVries, 1974 On the sensitivity of mammalian mitochondrial protein synthesis to inhibition by the macrolide antibiotics. In *The Biogenesis of Mitochondria,* edited by A. M. Kroon and C. Saccone, Academic Press, New York, pp. 395–402.

Küntzel, H., 1969*a* Mitochondrial and cytoplasmic ribosomes from *Neurospora crassa:* Characterization of their subunits. *J. Mol. Biol.* **40:**315–320.

Küntzel, H., 1969*b* Proteins of mitochondrial and cytoplasmic ribosomes from *Neurospora crassa. Nature (London)* **222:**142–146.

Küntzel, H. and H. Noll, 1967 Mitochondrial and cytoplasmic polysomes from *Neurospora crassa. Nature (London)* **215:**1340–1345.

Kurland, C., 1960 Molecular characterization of ribonucleic acid from *Escherichia coli* ribosomes. I. Isolation and molecular weights. *J. Mol. Biol.* **2:**83–91.

Lamb, A. J., G. D. Clark-Walker, and A. W. Linnane, 1968 The biogenesis of mitochondria. 4. The differentiation of mitochondrial and cytoplasmic protein synthesizing systems *in vitro* by antibiotics. *Biochim. Biophys. Acta* **161:**415–427.

Laub-Kupersztejn, R. and J. Thirion, 1974 Existence of two distinct protein synthesis systems in the trypanosomatid *Crithidia luciliae. Biochim. Biophys. Acta* **340:**314–322.

Leaver, C. J. and M. A. Harmey, 1973 Ribosomal RNAs of higher plant mitochondria. In *Ribosomes and RNA Metabolism,* edited by J. Zelinka and J. Balan, Publishing House of the Slovak Academy of Sciences, pp. 407–417.

Lederman, M. and G. Attardi, 1973 Expression of the mitochondrial genome in HeLa cells. XVI. Electrophoretic properties of the products of *in vivo* and *in vitro* mitochondrial protein synthesis. *J. Mol. Biol.* **78:**275–283.

Leister, D. E. and I. B. Dawid, 1974 Physical properties and protein constituents of cytoplasmic and mitochondrial ribosomes of *Xenopus laevis. J. Biol. Chem.* **249:**5108–5118.

Leister, D. E. and I. B. Dawid, 1975 Mitochondrial ribosomal proteins in *Xenopus laevis/X. mulleri* interspecific hybrids. *J. Mol. Biol.* **96:**119–123.

Lengyel, P., 1974 The process of translation: a bird's eye view. In *Ribosomes,* edited by M. Nomura, A. Tissières and P. Lengyel, Cold Spring Harbor Laboratory, Cold Spring Harbor, N.Y., pp. 13–52.

Lietman, P. S., 1971 Mitochondrial protein synthesis: Inhibition by emetine hydrochloride. *Mol. Pharmacol.* **7:**122–128.

Linnane, A. W. and J. M. Haslam, 1970 The biogenesis of yeast mitochondria. In *Current Topics in Cellular Regulation,* edited by B. C. Horecker and E. R. Stadtman, Academic Press, New York, pp. 101–172.

Lizardi, P. M. and D. J. L. Luck, 1971 Absence of a 5 S RNA component in the mitochondrial ribosomes of *N. crassa. Nature (London) New Biol.* **229:**140–142.

Lizardi, P. M. and D. J. L. Luck, 1972 The intracellular site of synthesis of mitochondrial ribosomal proteins in *N. crassa. J. Cell Biol.* **54:**56–74.

Loening, U. E., 1968 Molecular weights of ribosomal RNA in relation to evolution. *J. Mol. Biol.* **38:**355–365.

Loening, U. E., K. W. Jones, and M. L. Birnstiel, 1969 Properties of the ribosomal RNA precursor in *Xenopus laevis*; comparison to the precursor in mammals and in plants. *J. Mol. Biol.* **45:**353–366.

Loomeijer, F. J. and A. M. Kroon, 1972 Synthesis of 6-^{14}C-ethidium bromide. *Anal. Biochem.* **49:**455–458.

Lyttleton, J. W., 1962 Isolation of ribosomes from spinach chloroplasts. *Exp. Cell Res.* **26:**312–317.

Mager, J., 1960 Chloramphenicol and chlortetracycline inhibition of amino acid incorporation into proteins in a cell-free system from *Tetrahymena pyriformis. Biochim. Biophys. Acta* **38:**150–152.

Mahler, H. R., 1973 Biogenetic autonomy of mitochondria. *CRC Crit. Rev. Biochem.* **1:**381–460.

Mahler, H. R. and R. N. Bastos, 1974 A novel reaction of mitochondrial DNA with ethidium bromide. *FEBS Lett.* **39:**27–34.

Mazumder, R., 1973 Effect of thiostrepton on recycling of *E. coli* initiation factor 2. *Proc. Natl. Acad. Sci. USA* **70:**1939–1942.

McConkey, E. H., 1974 Composition of mammalian ribosomal subunits: A re-evaluation. *Proc. Natl. Acad. Sci. USA* **71:**1379–1383.

McLean, J. C., G. L. Cohn, I. K. Brandt, and M. V. Simpson, 1958 Incorporation of labeled amino acids into the protein of muscle and liver mitochondria. *J. Biol. Chem.* **233:**657–663.

Millis, A. J. T. and Y. Suyama, 1972 Effects of chloramphenicol and cycloheximide on the biosynthesis of mitochondrial ribosomes in *Tetrahymena*. *J. Biol. Chem.* **247:**4063–4073.

Morimoto, H. and H. O. Halvorson, 1971 Characterization of mitochondrial ribosomes from yeast. *Proc. Natl. Acad. Sci. USA* **68:**324–328.

Morimoto, H., A. H. Scragg, J. Nekhorocheff, V. Villa, and H. O. Halvorson, 1971 Comparison of the protein synthesizing systems from mitochondria and cytoplasm of yeast. In *Autonomy and Biogenesis of Mitochondria and Chloroplasts,* edited by N. K. Boardman, A. W. Linnane, and R. M. Smillie, American Elsevier, New York, pp. 282–292.

Neupert, W., W. Sebald, A. J. Schwab, A. Pfaller, and T. Bucher, 1969 Puromycin sensitivity of ribosomal label after incorporation of ^{14}C-labeled amino acids into isolated mitochondria from *Neurospora crassa*. *Eur. J. Biochem.* **10:**585–588.

Nijhof, W. and A. M. Kroon, 1974 The interference of chloramphenicol and thiamphenicol with the biogenesis of mitochondria in animal tissues: A possible clue to the toxic action. *Postgrad. Med. J.* **50:**53–59.

O'Brien, T. W., 1971 The general occurrence of 55 S ribosomes in mammalian liver mitochondria. *J. Biol. Chem.* **246:**3409–3417.

O'Brien, T. W., 1972 Occurrence of 55 S miniribosomes in mitochondria of the shark. *J. Cell Biol.* **55:**191a.

O'Brien, T. W. and G. F. Kalf, 1967*a* Ribosomes from rat liver mitochondria. I. Isolation procedure and contamination studies. *J. Biol. Chem.* **242:**2172–2179.

O'Brien, T. W. and G. F. Kalf, 1967*b* Ribosomes from rat liver mitochondria. II. Partial characterization. *J. Biol. Chem.* **242:**2180–2185.

O'Brien, T. W., N. D. Denslow, and G. R. Martin, 1974 The structure, composition and function of 55 S mitochondrial ribosomes. In *The Biogenesis of Mitochondria,* edited by A. Kroon and C. Saccone, Academic Press, New York, pp. 347–356.

Oerter, D., R. Bass, H. J. Kirstaedter, and H. J. Merker, 1974 Effect of chloramphenicol and thiamphenicol on mitochondrial components. *Postgrad. Med. J.* **50:**65–68.

Ojala, D. and G. Attardi, 1972 Expression of the mitochondrial genome in HeLa cells. X. Properties of mitochondrial polysomes. *J. Mol. Biol.* **65:**273–289.

Olsnes, S., K. Refsnes, and A. Pihl, 1974 Mechanism of action of the toxic lectins abrin and ricin. *Nature (London)* **249:**627–631.

Parisi, B. and R. Cella, 1971 Origin of the ribosome specific factors responsible for peptide chain elongation in yeast. *FEBS Lett.* **14:**209–213.

Penman, S., M. Rosbash, and M. Penman, 1970 Messenger and heterogeneous nuclear RNA in HeLa cells: Differential inhibition by cordycepin. *Proc. Natl. Acad. Sci. USA* **67:**1878–1885.

Perlman, S. and S. Penman, 1970*a* Mitochondrial protein synthesis: Resistance to emetine and response to RNA synthesis inhibitors. *Biochem. Biophys. Res. Commun.* **40:**941–948.

Perlman, S. and S. Penman, 1970*b* Protein-synthesizing structures associated with mitochondria. *Nature (London)* **227:**133–137.

Pestka, S., 1971 Inhibitors of ribosome functions. *Annu. Rev. Microbiol.* **25:**487–562.

Pring, D. R., 1974 Maize mitochondria: Purification and characterization of ribosomes and ribosomal ribonucleic acid. *Plant Physiol.* **53:**677–683.

Pring, D. R. and D. W. Thornbury, 1975 Molecular weights of maize mitochondrial and cytoplasmic ribosomal RNAs under denaturing conditions. *Biochim. Biophys. Acta* **383:**140–146.

Quetier, F. and F. Vedel, 1974 Identification of mitochondrial rRNA from plant cells. *FEBS Lett.* **42:**305–308.

Rabbitts, T. H. and T. S. Work, 1971 The mitochondrial ribosome and ribosomal RNA of the chick. *FEBS Lett.* **14:**214–218.

Rawson, J. R. and E. Stutz, 1969 Isolation and characterization of *Euglena gracilis* cytoplasmic and chloroplast ribosomes and their ribosomal RNA components. *Biochim. Biophys. Acta* **190:**368–380.

Reijnders, L., P. Sloof, J. Sival, and P. Borst, 1973 Gel electrophoresis of RNA under denaturing conditions. *Biochim. Biophys. Acta* **324:**320–333.

Rendi, R., 1959 The effect of chloramphenicol on the incorporation of labeled amino acids into proteins by isolated subcellular fractions from rat liver. *Exp. Cell Res.* **18:**187–189.

Richter, D., 1971 Production of mitochondrial peptide-chain elongation factors in yeast deficient in mitochondrial deoxyribonucleic acid. *Biochemistry* **10:**4422–4425.

Richter, D. and F. Lipmann, 1970 Separation of mitochondrial and cytoplasmic peptide chain elongation factors from yeast. *Biochemistry* **9:**5065–5070.

Richter, D., L. Lin, and J. Bodley, 1971 Studies on translocation. IX. The pattern of action of antibiotic translocation inhibitors in eukaryotic and prokaryotic systems. *Arch. Biochem. Biophys.* **147:**186–191.

Richter, D., V. A. Erdmann, and M. Sprinzl, 1973 Specific recognition of GTψC loop (loop IV) of tRNA by 50 S ribosomal subunits from *E. coli. Nature* (*London*) *New Biol.* **246:**132–135.

Rifkin, M., D. Wood, and D. Luck, 1967 Ribosomal RNA and ribosomes from mitochondria of *Neurospora crassa. Proc. Natl. Acad. Sci. USA* **58:**1025–1032.

Robberson, D., Y. Aloni, G. Attardi, and N. Davidson, 1971 Expression of the mitochondrial genome in HeLa cells. VI. Size determination of mitochondrial ribosomal RNA by electron microscopy. *J. Mol. Biol.* **60:**473–484.

Sacchi, A., F. Cerbone, P. Cammarano, and U. Ferrini, 1973 Physiochemical characterization of ribosome-like (55 S) particles from rat liver mitochondria. *Biochim. Biophys. Acta* **308:**390–403.

Sager, R. and M. G. Hamilton, 1967 Cytoplasmic and chloroplast ribosomes of *Chlamydomonas*: Ultracentrifugal characterization. *Science* **157:**709–711.

Sala, F. and H. Küntzel, 1970 Peptide chain initiation in homologous and heterologous systems from mitochondria and bacteria. *Eur. J. Biochem.* **15:**280–286.

Schatz, G. and T. L. Mason, 1974 The biosynthesis of mitochondrial proteins. *Annu. Rev. Biochem.* **43:**51–87.

Schmitt, H., 1969 Characterization of mitochondrial ribosomes from *Saccharomyces cerevisiae. FEBS Lett.* **4:**234–238.

Schmitt, H., 1970 Characterization of a 72 S mitochondrial ribosome from *Saccharomyces cerevisiae. Eur. J. Biochem.* **17:**278–283.

Schmitt, H., 1971 Core particles and proteins from mitochondrial ribosomes of yeast. *FEBS Lett.* **15:**186–190.

Schmitt, H., 1972 Analysis and site of synthesis of ribosomal proteins from yeast mitochondria. *FEBS Lett.* **26:**215–220.

Schmitt, H., H. Grossfeld, J. S. Beckmann, and U. Z. Littauer, 1974 Biogenesis of mitochondria from *Artemia salina* cysts and the transcription *in vivo* of the DNA. In *The Biogenesis of Mitochondria,* edited by A. M. Kroon and C. Saccone, Academic Press, New York, pp. 135–146.

Scragg, A. H., 1971 Chain elongation factors of yeast mitochondria. *FEBS Lett.* **17:**111–114.

Scragg, A. H., H. Morimoto, V. Villa, J. Nekhorocheff, and H. Halvorson, 1971 Cell-free protein synthesizing system from yeast mitochondria. *Science* **171:**908.

Sperti, S., L. Montanaro, A. Mattioli, and F. Stierpe, 1973 Inhibition by ricin of protein synthesis *in vitro*: 60 S ribosomal subunit as the target of the toxin. *Biochem. J.* **136:**813–815.

Stegeman, W. J., C. S. Cooper, and C. J. Avers, 1970 Physical characterization of ribosomes from purified mitochondria of yeast. *Biochem. Biophys. Res. Commun.* **39:**69–76.

Stevens, B., J. Curgy, G. Ledoigt, and J. Andre, 1974 Analysis of mitoribosomes from *Tetrahymena* by polyacrylamide gel electrophoresis and electron microscopy. In *The Biogenesis of Mitochondria,* Academic Press, New York, pp. 327–335.

Stuart, K. D., G. S. Linden, and J. S. Hanas, 1973 Cycloheximide resistant protein synthesis in *Trypanosoma brucei. J. Cell Biol.* **59:**339a.

Stutz, E. and H. Noll, 1967 Characterization of cytoplasmic and chloroplast polysomes in plants: Evidence for three classes of ribosomal RNA in nature. *Proc. Natl. Acad. Sci. USA* **57:**774–781.

Swanson, R. F., 1971 Incorporation of high molecular weight polynucleotides by isolated mitochondria. *Nature (London)* **231:**31–34.

Swanson, R. F., 1973 Specificity of mitochondrial and cytoplasmic ribosomes and elongation factors from *Xenopus laevis. Biochemistry* **12:**2142–2146.

Swanson, R. F. and I. B. Dawid, 1970 The mitochondrial ribosome of *Xenopus laevis. Proc. Natl. Acad. Sci. USA* **66:**117–124.

Tait, A., 1972 Altered mitochondrial ribosomes in an erythromycin resistant mutant of *Paramecium. FEBS Lett.* **24:**117–120.

Taylor, M. M. and R. Storck, 1964 Uniqueness of bacterial ribosomes. *Proc. Natl. Acad. Sci. USA* **52:**958–965.

Tissières, A., J. Watson, D. Schlessinger, and B. Hollingworth, 1959 Ribonucleoprotein particles from *Escherichia coli. J. Mol. Biol.* **1:**221–234.

Tonnesen, T. and J. D. Friesen, 1973 The effects of daunomycin and ethidium bromide on *Escherichia coli. Mol. Gen. Genet.* **124:**177–186.

Towers, N. R., H. Dixon, G. M. Kellerman, and A. W. Linnane, 1972 Biogenesis of mitochondria. XXII. The sensitivity of rat liver mitochondria to antibiotics; a phylogenetic difference between a mammalian system and yeast. *Arch. Biochem. Biophys.* **151:**361–369.

Ts'o, P. O. P., J. Bonner, and J. Vinograd, 1958 Structure and properties of microsomal nucleoprotein particles from pea seedlings. *Biochim. Biophys. Acta* **30:**570–582.

Vasconcelos, A. C. L. and L. Bogorad, 1971 Proteins of cytoplasmic, chloroplast and mitochondrial ribosomes of some plants. *Biochim. Biophys. Acta* **228:**492–502.

Vazquez, D., 1974 Inhibitors of protein synthesis. *FEBS Lett.* **40:**S63–S84.

Verma, I. M., M. Edelman, M. Herzhog, and U. Littauer, 1970 Size determination of mitochondrial ribosomal RNA from *Aspergillus nidulans* by electron microscopy. *J. Mol. Biol.* **52:**137–140.

Vesco, C. and S. Penman, 1969 The cytoplasmic RNA of HeLa cells: New discrete species associated with mitochondria. *Proc. Natl. Acad. Sci USA* **62**:218–225.

Vignais, P. V., J. Huet, and J. Andre, 1969 Isolation and characterization of ribosomes from yeast mitochondria. *FEBS Lett.* **3**:177–181.

Vignais, P. V., B. J. Stevens, J. Huet, and J. Andre, 1972 Mitoribosomes from *Candida utilis*. Morphological, physical and chemical characterization of the monomer form and of its subunits. *J. Cell Biol.* **54**:468–492.

Wengler, G., G. Wengler, and K. Scherrer, 1972 Ribonucleoprotein particles in HeLa cells: The contamination of the postmitochondrial-cytoplasmic fraction from HeLa cells with ribonucleoprotein particles of mitochondrial origin. *Eur J. Biochem.* **24**:477–484.

Wilson, R. H., J. B. Hanson, and H. H. Mollenhauer, 1968 Ribosome particles in corn mitochondria. *Plant Physiol.* **43**:1874–1877.

Wittmann, H. G., G. Stöffler, E. Kaltschmidt, V. Rudloff, H. Janda, M. Dzionara, D. Bonner, K. Nierhaus, M. Cech, I. Hindennach, and B. Wittmann, 1970 Protein chemical and serological studies on ribosomes of bacteria, yeast and plants. *FEBS Symp.* **21**:33–46.

Wool, I. G. and G. Stöffler, 1974 Structure and function of eukaryotic ribosomes In *Ribosomes,* edited by M. Nomura, A. Tissières, and P. Lengyel, Cold Spring Harbor Laboratory, Cold Spring Harbor, N.Y., pp. 417–460.

Yu, R., R. Poulson, and P. R. Stewart, 1972*a* Comparative studies on mitochondrial development in yeasts I. Mitochondrial ribosomes from *Candida parapsilosis*. *Mol. Gen. Genet.* **114**:325–38.

Yu, R., R. Poulson, and P. R. Stewart, 1972*b* Comparative studies on mitochondrial development in yeasts. II. Mitochondrial ribosomes from *Saccharomyces cerevisiae*. *Mol. Gen. Genet.* **114**:339–349.

Yunis, A. A., G. K. Arimura, and D. R. Manyan, 1974 Comparative metabolic effects of chloramphenicol and thiamphenicol in mammalian cells. *Postgrad. Med. J.* **50**:60–65.

Yurina, N. P. and M. S. Odintsova, 1974 Buoyant density of chloroplast ribosomes in CsCl. *Plant Sci. Lett.* **3**:229–234.

Zylber, E., C. Vesco, and S. Penman, 1969 Selective inhibition of the synthesis of mitochondria-associated RNA by ethidium bromide. *J. Mol. Biol.* **44**:195–204.

PART T
MUTANT ENZYMES

17

Genetic Variants of Enzymes Detectable by Zone Electrophoresis

CHARLES R. SHAW AND RUPI PRASAD

The method of staining for a specific enzyme activity directly in the electrophoretic medium following electrophoresis of tissue extracts, the so-called zymogram technique developed by Hunter and Markert (1957), provided a valuable tool for a wide variety of biological researches. Its first application was in the study of isozymes, or multiple molecular forms of enzymes occurring within the same organism (Markert and Moller, 1959). The method soon led to the detection of a number of genetically variant enzymes which have been used as genetic markers in a great variety of studies. These include studies of genetic polymorphism, genetic difference between species, genetic mapping both by classical crossing techniques and by the newer technique of cultured cell hybridization, and in ontogenetic studies to determine time of activation of parental genes. Doubtless other applications will be devised and the importance of the technique will continue to expand.

The zymogram technique has certain unique advantages over many of the other methods for detecting of genetic markers. It analyzes directly the

CHARLES R. SHAW AND RUPI PRASAD—The University of Texas, M. D. Anderson Hospital and Tumor Institute, Texas Medical Center, Houston, Texas.

gene product, the protein, rather than traits such as morphology and color which are usually a number of steps removed from the gene. It employs crude tissue extracts, obviating the technical problems of purification of the protein. It requires a relatively small amount of material, so that a number of enzymes can be studied in such small quantities as the extract from a single *Drosophila* or from one bottle of cultured cells. Additionally, large numbers of organisms can be studied in a relatively short time by application of many extracts to a single slab of electrophoretic medium (starch gel and polyacrylamide gel are most commonly used). The slab technique, with samples applied across a straight row, permits precise comparison of migration rates so that slight differences are readily detectable.

By using a specific enzyme stain in the gel rather than a general protein stain, there is a high probability that the zones of activity being compared between organisms are probably homologous. This is especially true if the two organisms each contain only a single band of activity. Where multiple bands exist, analysis is more complicated and comparison is often difficult. However, a growing mass of information on the various isozyme systems assures increasing confidence in comparison and analysis of many of these enzymes. An example is the now well-understood tetrameric lactate dehydrogenases, which are composed of two different subunits associating randomly and resulting in five isozymes. It is common, in comparing two organisms, to find one of the two loci varying while the other is constant, and analysis is easy if one understands the molecular structure.

Specificities of the enzyme stains and of the enzymes themselves, of course, vary widely. Some are highly specific, such as the lactate dehydrogenases. Others are quite nonspecific, such as the staining system for the acetyl esterases. This employs a synthetic substrate, usually a naphthyl acetate, and a wide variety of hydrolytic enzymes will appear with such a stain. In Table 1, the esterases are all listed together as a single category except for cholinesterase, which has a special significance. No attempt has been made in this presentation to indicate the degree of specificity of the staining systems, but much of this is dealt with in the individual literature.

The zymogram techniques have been much improved over the years, producing better resolution of the bands and a great increase in the number of enzyme stains. Methods are now available for demonstrating some 50 or more enzyme activities.

Genetically variant enzymes were at first considered a relatively rare phenomenon, but as surveys of wild populations expanded through the 1960s it was demonstrated that such variation is common and widespread throughout the plant and animal kingdoms. Frequency of variants range from common polymorphisms to very rare alleles. It is now generally accepted that if one looks at sufficient numbers of individuals within a species a variant can be found of almost any enzyme.

The list of electrophoretic variants provided in the following table is intended to cover most of the known variants to date. In view of the large number, however, and the rapid development of the field, we cannot claim the list to be exhaustive. Nonetheless, it offers the reader a widely diverse source of organisms and enzymes which may be utilized for his specific research needs. The additional point is perhaps worth making that one can easily find his own markers in virtually any organisms he wishes to employ, because, based on a number of surveys most of which are published in the accompanying reference list, any wild population will be found to be polymorphic, on the average, at some 30% of loci.

Information on specific methods employed in the detection of variants listed has not been provided here. These are, for the most part, standard methods with individual modifications, most of which are described in Shaw and Prasad (1970). Experience has shown that some modification of recommended buffer systems is usually needed in different laboratories to obtain optimum resolution. Variable factors affecting migration include such things as the batch of starch used, water purity, dimensions of the gel, voltage, and gel temperature.

Most of the variants listed here are intraspecific. However, in certain cases, mention is made of differences between closely related species, particularly if they are hybridizable, as this obviously provides a useful system for genetic study. This point is so noted in the remarks column of the table. The references are not intended to be exhaustive, but to be the more pertinent ones. An exhaustive list would be of prohibitive length—e.g., in the case of the human glucose-6-phosphate dehydrogenase polymorphism, on which several thousand papers have been published.

Most of the interspecific variants probably differ by a single amino acid, but in practically no case has this been demonstrated. An exception is in certain of the rare variants of human erythrocyte carbonic anhydrase, extensively studied in Tashian's laboratory, where single amino acid substitutions were found. The main impediment to such studies, of course, is the relatively low concentration of most enzymes in most tissues, making preparation of sufficient amounts of pure material quite difficult. This is in contrast to the situation with hemoglobins as well as with certain serum proteins.

Not all amino acid substitutions produce alteration in electrophoretic mobility, since many of these result in no significant change in that charge of the molecule. Various estimates, i.e., Shaw (1965), indicate that only about one-third of amino acid substitutions are electrophoretically de-

(Text continued on p. 603.)

TABLE 1. Genetic Variants of Enzymes

Enzyme	*Organism and tissue*	*Remarks*	*References*
Acid phosphatase	Man, erythrocytes	Several alleles, differ in activity, population studies	Scott, 1966 Hopkinson *et al.*, 1963, 1964 Spencer *et al.*, 1964*a* Luffman and Harris, 1967
	Man, placenta and leukocytes	Five isozymes, all affected in variant	Beckman *et al.*, 1970
	Infrahuman primates, erythrocytes (many spp.)	Polymorphisms	Barnicot and Cohen, 1970
	Chicken (*Gallus*), liver	Polymorphism	Okada and Hachinohe, 1968
	Housefly (*Musca*)	Polymorphism	Ogita and Kassi, 1965
	Tetrahymena	Several alleles	Allen *et al.*, 1963
	Fungi (*Thamnidium*)	Inter- and intraspecific variation	Stout and Shaw, 1973
	Drosophila (several spp.)		Stone *et al.*, 1968 Gillespie and Kojima, 1969 Barnicot and Cohen, 1970
Adenosine deaminase	Human erythrocytes	Polymorphism, two alleles	Spencer *et al.*, 1967 Hopkinson and Harris, 1969
Adenylate kinase	Infrahuman primate erythrocytes (many spp.)	Inter- and intraspecific variations	Barnicot and Cohen, 1970
	Man, erythrocytes and other tissues	Polymorphism	Fildes and Harris, 1966 Bowman *et al.*, 1967*b* Bockelmann *et al.*, 1968 Brock, 1970

Alcohol dehydrogenase	*Drosophila*	Many alleles in several spp.	Johnson and Denniston, 1964
			Grell *et al.*, 1965
			Stone *et al.*, 1968
			Gillespie and Kojima, 1969
			Ayala *et al.*, 1970
			Wright and Shaw, 1970
	Myxomycetes	Interspecific variation	Shaw, 1970
	Maize	Two isozymes, both vary genetically	Scandalios, 1967
			Schwartz, 1969
			Efron, 1970
	Fungi (*Thamnidium*)	Polymorphism	Stout and Shaw, 1973
	Mouse (*Mus*)	Polymorphism	Selander *et al.*, 1969
	Man	Three loci, two are variant	Smith *et al.*, 1971
Aldehyde oxidase	*Drosophila*	Inter- and intraspecific variations	Courtright, 1967
	Mus musculus liver		Watson *et al.*, 1972
Alkaline phospharase	Mysomcetes		Shaw, 1970
	Man, serum and placenta	Polymorphism	Boyer, 1961
			Robson and Harris, 1967
			Beratis and Hirschhorn, 1972
	Teleost fish	Interspecific variation	Shaw, 1970
	Aspergillus		Dorn, 1965
	Bats	Polymorphism	Shaw, 1970
	Drosophila	Polymorphism	Beckman and Johnson, 1964*b*
			Hubby and Lewontin, 1966
	Chicken, serum	Dominant inheritance	Law and Munro, 1965
			Wilcox, 1966
			Law, 1967
	Maize	Activity altered	Tsai and Nelson, 1969

TABLE 1. Continued

Enzyme	*Organism and tissue*	*Remarks*	*References*
Amylase	*Drosophila*	Much polymorphism	Doane, 1965, 1967
	Mus musculus, saliva and pancreas	Different molecules in the two tissues both show genetic variation	Sick and Nielsen, 1964
	Barley	Two forms, α and β; only α varies	Frydenberg and Nielsen, 1966
	Man	Polymorphism	Kamaryt and Laxova, 1965
	Saliva	Polymorphism	Boettcher and de la Lande, 1969
	Saliva and pancreas	Polymorphism	Ogita, 1966
	Cattle (*Bos*), serum	Three alleles, polymorphic	Ashton, 1965
	Vole (*Clethrionomys*), salivary gland	Five alleles	Nielson, 1969
	Housefly (*Musca*)	Polymorphism	Ogita, 1962
Aryl sulfatase	Sea urchin	Interspecific variation	Fedecka-Bruner *et al.,* 1971
Aspartate aminotransferase	Herring (*Clupea*)	Polymorphism	Odense *et al.,* 1966
Carbonic anhydrase	Man, erythrocytes	Rare variants, several alleles	Shaw *et al.,* 1962
			Tashian *et al.,* 1963
			Shows, 1967
			Tashian, 1969
	Infrahuman primates, erythrocytes	Several polymorphisms in different spp.	Barnicot and Jolly, 1964
			Tashian, 1965
			Tashian *et al.,* 1968
			Tashian *et al.,* 1971

	Peromyscus, erythrocytes	Two alleles	Wilmont and Underhill, 1972
	Cattle (*Bos*), erythrocytes	Two alleles, polymorphic	Sartore *et al.*, 1969
	Buffalo (*Bison*), erythrocytes	Three alleles, polymorphic	Sartore *et al.*, 1969
	Horse (*Equus*), erythrocytes	Five alleles, some have quantitative differences	Sandberg, 1968
Catalase	Myxomycetes	Intrageneric differences	Shaw, 1970
	Enterobacteriaceae		Baptist *et al.*, 1969
	Drosophila	A and B forms, both vary	Kanapi and Wheeler, unpublished
	Maize		Beckman and Scandalios, 1964
	Man, erythrocytes		Baur, 1963
			Nance *et al.*, 1968
Ceruloplasmin	Man, serum	Rare variants, heterozygotes only	Shreffler *et al.*, 1967
			McCombs *et al.*, 1970
			Shokier and Shreffler, 1970
Cholinesterase	Man, serum	Complex allelism, with activity variation	Kalow and Staron, 1957
			Liddell *et al.*, 1962
			Harris *et al.*, 1963
			Horsfall *et al.*, 1963
			Simpson and Kalow, 1964
			Yoshida and Motulsky, 1969
			Altland and Goedde, 1970
	Acetylcholinesterase, human erythrocyte	Polymorphism	Coates and Simpson, 1972
	Horse, serum	Polymorphism	Oki *et al.*, 1964
Creatine kinase	*Peromyscus*, heart and skeletal muscle	Different enzymes in the two tissues, both polymorphic	Shaw, unpublished

TABLE 1. *Continued*

Enzyme	*Organism and tissue*	*Remarks*	*References*
Diaphorase, NADH	Man, erythrocytes	Polymorphism, at least five alleles	Hopkinson *et al.*, 1970
			Tariverdian and Wendt, 1970
Enolase	Fish (Salmonidae) muscle	Interspecific, differences	Tsuyki and Wold, 1964
	Yeast (*Saccharomyces*)	Induced mutations (with 5-fluorouracil)	Dave *et al.*, 1966
			Pfleiderer *et al.*, 1968
Esterase	Bacteria	Inter- and intraspecific	Norris, 1962
	Tetrahymena		Allen, 1960
			Allen and Weremiuk, 1971
	Paramecium	Intersyngenic variation	Tait, 1970
			Allen *et al.*, 1971
	Drosophila, many spp.	Many alleles, many loci	Wright and Macintyre, 1963
			Hybby and Lewontin, 1966
			McReynolds, 1967
			Stone *et al.*, 1968
			Gillespie and Kojima, 1969
			Ayala *et al.*, 1970
			Miziantz and Case, 1971
	Myxomycetes	Intrageneric	Shaw, 1970
	Peromyscus, erythrocytes and many tissues		Randerson, 1964
		Several loci, polymorphism	Selander *et al.*, 1971
	Mouse (*Mus*), several tissues		Popp and Popp, 1962
			Petras, 1963
			Ruddle and Roderick, 1965
			Petras and Sinclair, 1969
			Ruddle *et al.*, 1969

	Man, erythrocytes		Tashian and Shaw, 1962
			Tashian, 1969
	Cucurbita, seed	Interspecific	Wall and Whitaker, 1971
	Housefly (*Musca*)		Ogita, 1962
			van Asperen and van Mazijk, 1965
	American lobster, heart	Five loci, two show variants	Barlow and Ridgeway, 1971
	Butterfly (*Colias*)	Thirteen alleles, polymorphism	Burns and Johnson, 1967
	Butterfly (*Hemiargus*)	Fourteen alleles	Burns and Johnson, 1971
	Harvester ant	Two loci, both polymorphic	Johnson *et al.*, 1969
	Rabbit, erythrocytes	Three loci, all polymorphic	Grunder *et al.*, 1965
	Chicken, serum	Three loci, all polymorphic	Grunder, 1968
	Microtus		Semeonoff, 1972
	Rat (*Rattus*), serum	Variant inactive	Womack, 1972
	Ringneck dove, erythrocytes	Two alleles	Boehm and Irwin, 1971
	Marlin (*Tetrapturus*)	Polymorphism	Edmunds, 1972
	Tuna (*Thunnus*), several spp.	Polymorphism, both inter- and intraspecific	Sprague, 1967, 1970
			McCabe and Dean, 1970
	Fish (*Catostomus*)	Polymorphism, shows selective advantage	Koehn, 1969
	Fish (*Fundulus*)	Much polymorphism	Holmes and Whitt, 1970
Fumarase	Myxomycetes	Intrageneric	Shaw, 1970
Galactose-1-phosphate uridyltransferase	Man, erythrocytes		Mathai and Beutler, 1966

TABLE 1. Continued

Enzyme	*Organism and tissue*	*Remarks*	*References*
Glucosephosphate isomerase (also called phosphoglucose isomerase)	Man		Detter *et al.*, 1968
	Mouse (*Mus musculus*)	Two alleles	Carter and Parr, 1967
			DeLorenzo and Ruddle, 1969
	Fungi (*Thamnidium*)	Interspecific variation	Stout and Shaw, 1973
	Platyfish (*Xiphophorus*)	Two loci, both polymorphic	Siciliano *et al.*, 1973
	Rabbit		Welch *et al.*, 1970
	Chinchilla	Polymorphic	Carter *et al.*, 1972
	Cuis (*Cavia musteloides*)	Polymorphic	Carter *et al.*, 1972
	Guinea pig (*Cavia porcellus*)	Interspecific variation	Carter *et al.*, 1972
	Peromyscus	Polymorphic	Selander *et al.*, 1971
	Frog	Interspecific variation	Johnson and Chapman, 1971*b*
Glycose-6-phosphate dehydrogenase (G6PD)	Man, erythrocytes	Many alleles (~85), some with altered activity; sexlinked	Kirkman *et al.*, 1960
			Marks *et al.*, 1961
			Boyer *et al.*, 1962
			Shows *et al.*, 1964
			Dern *et al.*, 1969
			Yoshida *et al.*, 1971 (review)
			Chan *et al.*, 1972
	Dove (*Streptopelia*)	Autosomal	Cooper *et al.*, 1969*b*
	Enterobacteria		Bowman *et al.*, 1967*a*
			Williams and Bowden, 1968
			Baptist *et al.*, 1969

	American bison, erythrocytes	Sex linked	Naik and Anderson, 1970
	Drosophila	Polymorphism, X-linked strain differences	Young *et al.*, 1964 Wright and Shaw, 1970
	Myxomycetes	Interspecific	Shaw, 1970
	Mole rat (*Spalax*)	Polymorphic	Nevo and Shaw, 1972
	Fungi (*Thamnidium*)	Interspecific	Stout and Shaw, 1973
	Hare, erythrocytes	X-linked	Ohno and Poole, 1965
	Chicken, erythrocytes	Autosomal	Bhatnagar, 1969
	Pheasant, erythrocytes	Autosomal	Bhatnagar, 1969
	Quail (*Coturnix*)	Autosomal	Manwell and Baker, 1968
Glutamate dehydrogenase (GDH)	Enterobacteriaceae		Baptist *et al.*, 1969
	Man, leukocytes	Polymorphism, probably several alleles	Long, unpublished
	Fungi (*Thamnidium*)	Polymorphism	Stout and Shaw, 1973
Glutamate oxaloacetate transaminase (GOT)	Mouse, mitochondria	Strain differences	DeLorenzo and Ruddle, 1970
	Peromyscus	Polymorphism	Selander *et al.*, 1971
	Toad (*Xenopus*)	Soluble form variant	Johnson and Chapman, 1971*a*
Glutathione reductase	Man, erythrocytes	Many alleles, some with reduced activity associated with gout	Long, 1967, and 1970

TABLE 1. Continued

Enzyme	*Organism and tissue*	*Remarks*	*References*
Glyceraldehyde-3-phosphate dehydrogenase (G3PDH)	*Streptococcus*	Several alleles, strain differences	Williams, 1964 Williams and Bowden, 1968
	Platyfish (*Xiphophorus*)	Two loci, both polymorphic	Wright *et al.*, 1972 Siciliano *et al.*, 1973
α-Glycerophosphate dehydrogenase	*Drosophila* spp.	Polymorphism	Wright and Shaw, 1969 Ayala *et al.*, 1970
	Peromycus	Polymorphism	Selander *et al.*, 1971
	Ocean perch (*Sebastodes*)	Two alleles	Johnson *et al.*, 1970*b*
	Tuna (*Katsuwonus*)	Two alleles	McCabe *et al.*, 1970
	Horseshoe crab (*Limulus*)	Polymorphism	Selander *et al.*, 1970
Hexokinase	Myxomycetes	Interspecific variation	Shaw, 1970
	Drosophila	Two alleles, linkage data several loci, polymorphism	Grell, 1967 Knutsen *et al.*, 1969
Hexose-6-phosphate dehydrogenase (H6PD)	*Peromyscus*, many tissues	Polymorphism, several alleles, autosomal	Shaw and Barto, 1965 Shaw, 1966 Shaw and Koen, 1968
	Trout (*Salvelinus*), liver	Three alleles	Stegeman and Goldberg, 1971
	Mouse (*Mus*)	Two alleles, autosomal	Ruddle *et al.*, 1968 Selander *et al.*, 1969

Hypoxanthine-guanine phosphoribosyl transferase (HPGRT)	Man, erythrocytes		Der Kaloustian *et al.*, 1969
	Man–mouse hybrid cultured cells		Shin *et al.*, 1971
			Bakay and Nyhan, 1972
Isocitrate dehydrogenase (IDH)	Mouse, cytoplasm		Henderson, 1965
			McLaren and Tait, 1969
			Selander *et al.*, 1969
	Drosophila		Wright and Shaw, 1970
			Fox, 1971
	Myxomycetes	Interspecific variation	Shaw, 1970
	Escherichia coli		Baptist *et al.*, 1969
	Mole rat (*Spalax*)	Polymorphism	Nevo and Shaw, 1972
	Platyfish (*Xiphophorus*)	Polymorphism for both loci	Siciliano *et al.*, 1973
	Herring, liver and heart	Mictochondrial and supernatant	Wolf *et al.*, 1970
	Trout, liver and heart	Mitochondrial only	Wolf *et al.*, 1970
	Carp and goldfish	Polymorphism	Quiroz-Gutierrez and Ohno, 1970
	Horseshoe crab (*Limulus*)	Polymorphism	Selander *et al.*, 1970
	Snail (*Cepaea*)	Polymorphism	Manwell and Baker, 1968
Lactate dehydrogenase (LDH)	Man, erythrocytes and many other tissues	Care variants only, involving either of the two main loci	Boyer *et al.*, 1963
			Nance *et al.*, 1963
			Krause and Neely, 1964
			Vesell, 1965
			Das *et al.*, 1970
	Deermouse (*Peromyscus*)	B subunit polymorphism	Shaw and Barto, 1963
		A subunit	Cattanach and Perez, 1969
		Polymorphism, both subunits	Selander *et al.*, 1971

TABLE 1. Continued

Enzyme	*Organism and tissue*	*Remarks*	*References*
Lactate dehydrogenase (LDH) (contd)	Baboon	A subunit polymorphism	Syner and Goodman, 1966
	Pigeon	B subunit, rare variant	Zinkham *et al.*, 1965
	Pigeon, testis	C subunit polymorphism	Zinkham *et al.*, 1964
	Horse	B subunit, rare variant	Rauch, 1968
	Quail (*Coturnix*)	Polymorphism	Manwell and Baker, 1969
	Mink	B subunits vary	Saison, 1971
	Frog (*Rana*)	B subunits vary	Wright and Moyer, 1966 Salthe, 1969 Wright and Subtelny, 1971
	Myxomycetes	Interspecific variation	Shaw, 1970
	Mole rat (*Spalax*)	Polymorphism for both subunits	Nevo and Shaw, 1972
	Fungi (*Thamnidium*)	Interspecific variation	Stout and Shaw, 1973
	Horseshoe crab (*Libulus*)	Polymorphism	Selander *et al.*, 1970
	Cricket frog (*Acris*)	Polymorphism	Salthe and Nevo, 1969
	Trout (Salmonidae), several tissues	At least five loci, three are polymorphic	Morrison and Wright, 1966 Goldberg, 1966 Williscroft and Tsuyuki, 1970
	Bass (*Micropterus*)	Interspecific variation	Whitt *et al.*, 1971
	Blenny (*Anoplarchus*), muscle	*A* locus variant	M. Johnson, 1971
	Cod (*Gadua*), several tissues	*B* locus, four alleles	Odense *et al.*, 1969
	Salmon (*Onchorhynchus*) serum	*B* locus variant	Hodgins *et al.*, 1969
	Dace (*Rhinichthys*)	Two spp., variants at three alleles	Clayton and Gee, 1969
	Herring (*Clupea*)	*B* locus polymorphism	Odense *et al.*, 1966
	Whiting	*B* locus, three alleles	Markert and Faulhaber, 1965
	Rattail (*Coryphaenoids*)	Polymorphism	Whitt and Prosser, 1971

	Hake (*Merluccius*), several tissues	Polymorphism	Utter and Hodgins, 1969
	Killifish (*Fundulus*), eye	B subunits vary	Whitt, 1969
	Mackerel (*Scomber*), eye	B subunits vary	Whitt, 1969
	Teleosts (several spp.)	Much polymorphism	Whitt, 1970*a*
	Cattle (*Bos*), muscle	A subunits vary	Rauch, 1971
	Pig (*Sus*), sperm	C subunit polymorphism	Hyldgaard-Jensen and Moustgaard, 1967
Leucine aminopeptidase	*Drosophila* spp.	Polymorphism, at least two loci	Beckman and Johnson, 1964*a*
			Johnson and Sakai, 1964
			Hubby and Throckmorton, 1968
			Ayala *et al.*, 1970
			Kanapi and Wheeler, unpublished
	Myxomycetes	Interspecific variation	Shaw, 1970
	Cucurbita spp.	Interspecific variation	Wall and Whitaker, 1971
	Mole rat (*Spalax*)	Polymorphism	Nevo and Shaw, 1972
	Maize	Polymorphism	Beckman *et al.*, 1964
	Chicken, serum	Dominant, associated with alkaline phosphatase	Law, 1967
Lipase	Rabbit, adipose tissue	Polymorphism	Cortner and Schnatz, 1970
Malate dehydrogenase (MDH)	Mouse, cell cytoplasm	Strain differences	Henderson, 1966
	Mouse, mitochondria		Shows *et al.*, 1970
			Chapman *et al.*, 1971
	Peromyscus	Polymorphism	Selander *et al.*, 1971
	Trout (Salmonidae)		Bailey and Wilson, 1970
	Birds and reptiles	Polymorphism	Karig and Wilson, 1971

TABLE 1. *Continued*

Enzyme	*Organism and tissue*	*Remarks*	*References*
Malate dehydrogenase (MDH) (contd)	*Drosophila*	Polymorphism	Hubby and Lewontin, 1966
			Ayala *et al.*, 1970
	Enterobacteriaceae		Baptist *et al.*, 1969
	Ascaris	Polymorphism	Zee *et al.*, 1970
	Fungi (*Thamnidium*)	Interspecific variation	Stout and Shaw, 1973
	Platyfish (*Xiphophorus*)	Polymorphism	Siciliano *et al.*, 1973
	Marlin, blood	Polymorphism	Edmunds, 1972
	Walleye (*Strizostedion*), muscle	Three alleles	Clayton *et al.*, 1971
	Butterfly (*Colias*)	Polymorphism	G. Johnson, 1971
	Mussel, liver	Polymorphism	Koehn and Mitton, 1972
	Fundulus, muscle	Two loci, both polymorphic	Whitt, 1970*b*
	Sunfish (*Lipomis*)	Two loci, both polymorphic	Wheat *et al.*, 1972
	Mouse (*Mus*)	Mitochondrial form varies	Selander *et al.*, 1969
			Shows *et al.*, 1970
	Man	Variants of both mitochondrial and soluble forms	Davidson and Cortner, 1967*a*
			Davidson and Cortner, 1967*b*
			Blake and Kirk, 1970
	Horseshoe crab (*Limulus*)	Mitochondrial and soluble, both polymorphic	Selander *et al.*, 1970
	Frog (*Rana*)	Interspecific variation	Johnson and Chapman, 1971*b*
	Maize	Mitochondrial form	Longo and Scandalios, 1969
Methemoglobin reductase	Man, erythrocytes	Six alleles, most have reduced activity	Hsieh and Jaffi, 1971

Octanol dehydrogenase (ODH)	*Drosophila* spp.	Polymorphism, many alleles	Courtright *et al.*, 1966
			Stone *et al.*, 1968
			Pipkin, 1968
			Gillespie and Kojima, 1969
			Ogonji, 1971
Peptidase	Man, erythrocytes	Polymorphism, several alleles	Lewis and Harris, 1967
			Lewis *et al.*, 1968
			Lewis and Harris, 1969
			Lewis, 1971
			Chapman *et al.*, 1971
	Mouse (*Mus*), erythrocytes	Two alleles	Lewis and Truslove, 1969
	Maize	Two alleles	Melville and Scandalios, 1972
Phosphoglucomutase (PGM)	Man, erythrocytes and other tissues	Polymorphism, several alleles with variation of all three loci	Spencer *et al.*, 1964*b*
			Hopkinson and Harris, 1965
			Hopkinson and Harris, 1968
			Ishimoto and Yada, 1969
			McAlpine *et al.*, 1970
			Lewis, 1971
	Mouse (*Mus*)	Polymorphism, two loci	Selander *et al.*, 1969
			Shows *et al.*, 1969
			Chapman *et al.*, 1971
	Mole rat (*Spalax*)	Polymorphism	Nevo and Shaw, 1972
	Peromyscus	Polymorphism	Selander *et al.*, 1971
	Mosquito (*Aedes*)	Polymorphism two loci	Coluzzi *et al.*, 1971
	Ocean perch (*Sebastodes*), muscle	Two alleles at one locus only	Johnson *et al.*, 1971
	Eelpoint (*Zoarces*), brain	Three loci, all show variation	Hjorth, 1971
	Herring (*Clupea*), liver	Two loci, one shows polymorphism	Lush, 1969
	Trout (*Salmo*), muscle	One locus polymorphic	Roberts *et al.*, 1969
			Klose *et al.*, 1969

TABLE 1. Continued

Enzyme	Organism and tissue	Remarks	References
Phosphoglucomutase (PGM) (contd)	Salmon (*Oncorhynchus*), several tissues	One locus polymorphic	Utter and Hodgins, 1970
	Neurospora	Variants at two loci	Mishra and Tatum, 1970
	Enterobacteriaceae		Baptist *et al.*, 1969
	Primates, infrahuman, erythrocytes	Much polymorphism	Barnicot and Cohen, 1970
	Drosophila	Polymorphism	Trippa *et al.*, 1970
	Horseshoe crab (*Limulus*)	Polymorphism	Selander *et al.*, 1970
6-Phosphogluconate dehydrogenase (6PGD)	Man, erythrocytes and other tissues	Two alleles, polymorphism, with some quantitative variation	Fildes and Parr, 1963
			Dern *et al.*, 1966
			Parr, 1966
			Carter *et al.*, 1968
	Infrahuman primates, erythrocytes	Much polymorphism	Barnicot and Cohen, 1970
	Cyprinid fish (*Rutilus*)		Klose and Wolf, 1970
	Platyfish (*Xiphophorus*)		Siciliano *et al.*, 1973
	Enterobacteriaceae		Bowman *et al.*, 1967*a*
			Baptist *et al.*, 1969
	Fungi (*Thamnidium*)	Interspecific variation	Stout and Shaw, 1973
	Dove (*Streptopelia*)	Polymorphic in two spp.	Cooper *et al.*, 1969*a*
	Mole rat (*Spalax*)	Polymorphic	Nevo and Shaw, 1972
	Marsupial mouse (*Sminthopsis*)	Polymorphic	Cooper and Hope, 1971
	Drosophila	Strain differences	Wright and Shaw, 1970
	Peromyscus	Polymorphic	Shaw, 1965
			Selander *et al.*, 1971
	Marlin, blood	Three alleles	Edmunds, 1972
	Rat (*Rattus*)	Two alleles, polymorphic	Carter and Parr, 1969
	Pig (*Sus*), erythrocytes	Two alleles	Saison and Giblett, 1969

	Tuna (*Katsuwonus*)	Four alleles	McCabe *et al.*, 1970
	Mouse (*Mus*)	Polymorphic	Selander *et al.*, 1969
	Frog (*Rana*)	Interspecific variation	Johnson and Chapman, 1971*b*
	Snail (*Cepaea*)	Polymorphic	Manwell and Baker, 1968
	Quail (*Coturnix*)	Polymorphic	Manwell and Baker, 1969
Phosphoglycerate kinase (PGK)	Man, erythrocytes	Polymorphism, X-linked	Chen *et al.*, 1970
Phosphoglucose isomerase (PGI)		Eight different rare variants	Ritter and Wendt, 1971
Ribulose diphosphate carboxylase	Tomato	Several variants, some with altered activity	Andersen *et al.*, 1970
Sorbitol dehydrogenase	Goldfish, liver, kidney, and gonads	Polymorphic	Lin *et al.*, 1969
	Pig (*Sus*)	Two alleles	Op't Hof, 1969
Tetrazolium oxidase	Man, erythrocytes	Rare variant	Brewer, 1967
	Enterobacteriaeae		Baptist *et al.*, 1969
	Drosophila		Ayala *et al.*, 1970
	Fungi (*Thamnidium*)	Interspecific variation	Stout and Shaw, 1973
	Myoxmycetes	Interspecific variation	Shaw, 1970
	Trout (*Salmo*)	Two alleles	Utter, 1971
	Salmon (*Oncorhynchus*)	Two alleles	Utter, 1971
	Tuna (*Thunnus*)	Two alleles	Edmunds and Sammons, 1971
	Rockfish (*Sabostodes*)	Variation in 15 spp.	Johnson *et al.*, 1970*a*
	Dog	Two alleles	Baur and Schoor, 1969

TABLE 1. Continued

Enzyme	*Organism and tissue*	*Remarks*	*References*
Trypsin inhibitor	Soybean	Two alleles with different activities	Clark and Hymowitz, 1972
Tyrosinase	Mouse (*Mus*)	Multiple variants	Holstein *et al.*, 1971
Tyrosine aminotransferase	Mouse (*Mus*)	Some variation in activity between strains	Blake, 1970
Xanthine dehydrogenase (XDH)	*Drosophila*	Three loci affect activity; multiple alleles with different activities	Glassman *et al.*, 1968
	Myxomycetes	Interspecific variation	Shaw, 1970

tectable. This means that many of the estimates of frequencies based on electrophoretic studies miss two-thirds of the mutations.

A final point should be mentioned. This is the matter of neutral mutations, a subject presently of much interest and much controversy. Population geneticists (Crow, 1969; King and Jukes, 1969) have theorized that neutral or nearly neutral mutations are possible and that they can theoretically become incorporated into an evolving population. Since the variants detected by the zymogram retain enzyme activity, some of them probably qualify as neutral or nearly neutral, i.e., provide little selective advantage or disadvantage. Indeed, most of them may be of this type. But this is not known, and in most cases there has been no effort to detect any change in physiological activity. This question will doubtless continue to be the subject of considerable research (Shaw, 1970).

Acknowledgments

We are grateful to our colleagues Michael Siciliano, Daniel Stout, and David Wright for their help in compiling this chapter. Our special thanks to Gregory Whitt for providing the contents of his very extensive reprint files.

Work reported from this laboratory was supported in part by NIH Research Grant GM 15597.

Literature Cited

Allen, S. L., 1960 Inherited variations in the esterases of *Tetrahymena*. *Genetics* **45:**1051.

Allen, S. L. and S. L. Weremiuk, 1971 Intersyngenic variations in the esterases and acid phosphatases of *Tetrahymena pyriformis*. *Biochem. Genet.* **5:**119.

Allen, S. L., M. S. Misch, and B. M. Morrison, 1963 Genetic control of an acid phosphatase in *Tetrahymena*: Formation of hybrid enzyme. *Genetics* **48:**1635.

Allen, S. L., B. C. Byrne, and D. L. Cronkite, 1971 Intersyngenic variations in the esterases of bacterized *Paramecium aurelia*. *Biochem. Genet.* **5:**135.

Altland, K. and H. W. Goedde, 1970 Variability of pseudocholinesterase. *Humangenetik* **9:**241.

Andersen, W. R., G. F. Wildner, and R. S. Criddle, 1970 Ribulose diphosphate carboxylase. III. Altered forms of ribulose diphosphate carboxylase from mutant tomato plants. *Arch. Biochem. Biophys.* **137:**84.

Ashton, G. C., 1965 Serum amaylase (thread protein) polymorphism in cattle. *Genetics* **51:**431.

Ayala, F. J., C. A. Mourao, S. Perez-Salas, R. Richmond, and T. Dobzhansky, 1970 Enzyme variability in the *Drosophila willistoni* group I: Genetic differentiation among sibling species. *Proc. Natl. Acad. Sci. USA* **67:**225.

Bailey, G. S. and A. C. Wilson, 1970 Multiple forms of supernatant malate dehydrogenase in salmonid fishes: Biochemical, immunological, and genetic studies. *J. Biol. Chem.* **245:**5927.

Bakay, B. and W. L. Nyhan, 1972 Electrophoretic properties of hypoxanthine-guanine phosphoribosyl transferase in erythrocytes of subjects with Lesch-Nyhan syndrome. *Biochem. Genet.* **6:**139.

Baptist, J. N., C. R. Shaw, and M. Mandel, 1969 Zone electrophoresis of enzymes in bacterial taxonomy. *J. Bacteriol.* **99:**180.

Barlow, J. and G. J. Ridgeway, 1971 Polymorphisms of esterase isozymes in the American lobster (*Homarus americanus*). *J. Fish. Res. Board Can.* **28:**15.

Barnicot, N. A. and P. Cohen, 1970 Red cell enzymes of primates (Anthropoidea). *Biochem. Genet.* **4:**41.

Barnicot, N. A. and C. Jolly, 1964 A carbonic anhydrase variant in the baboon. *Nature (London)* **202:**198.

Baur, E. W., 1963 Catalase abnormality in a Caucasian family in the United States. *Science* **140:**816.

Baur, E. W. and R. T. Schoor, 1969 Genetic polymorphism of tetrazolium oxidase in dogs. *Science* **166:**1524.

Beckman, G., L. Beckman, and A. Tarnvik, 1970 A rare subunit variant shared by five acid phosphatase isozymes from human leukocytes and placentae. *Hum. Hered.* **20:**81.

Beckman, L. and F. M. Johnson, 1964*a* Genetic control of aminopeptidases in *Drosophila melanogaster. Hereditas* **51:**221.

Beckman, L. and F. M. Johnson, 1964*b* Variation in larval alkaline phosphatase controlled by *Aph* alleles in *Drosophila melanogaster. Genetics* **49:**829.

Beckman, L. and J. G. Scandalios, 1964 Catalase hybrid enzymes in maize. *Science* **146:**1175.

Beckman, L., J. G. Scandalios, and J. L. Brewbaker, 1964 Genetics of leucine aminopeptidase isozymes in maize. *Genetics* **50:**899.

Beratis, N. G. and K. Hirschhorn, 1972 Properties of placental alkaline phosphatase. III. Thermostability and urea inhibition of isolated components of the three common phenotypes. *Biochem. Genet.* **6:**1.

Bhatnagar, M. K., 1969 Autosomal determination of erythrocyte glucose-6-phosphate dehydrogenase in domestic chickens and ring-necked pheasants. *Biochem. Genet.* **3:**85.

Blake, R. L., 1970 Control of liver tyrosine aminotransferase expression: Enzyme regulatory studies on inbred strains and mutant mice. *Biochem. Genet.* **4:**215.

Blake, N. M. and R. L. Kirk, 1970 Genetic variants of soluble malate dehydrogenase in New Guinea populations. *Humangenetik* **11:**72.

Bockelmann, W., U. Wolf, and H. Ritter, 1968 Polymorphism of the phosphotransferases adenylate kinase and pyrucate kinase. *Humangenetik* **6:**78.

Boehm, L. G. and M. R. Irwin, 1971 Cellular esterases of the ringneck dove. *J. Hered.* **62:**44.

Boettcher, B. and F. de la Lande, 1969 Electrophoresis of human saliva and identification of inherited variants of amylase isozymes. *Aus. J. Exp. Biol. Med. Sci.* **47:**97.

Bowman, J. E., R. R. Brubaker, H. Frischer, and P. E. Carson, 1967*a* Characterization of enterobacteria by starch-gel electrophoresis of glucose-6-phosphate dehydrogenase and phosphogluconate dehydrogenase. *J. Bacteriol.* **94:**544.

Bowman, J. E., H. Frischer, F. Ajmar, P. E. Carson, and M. K. Gower, 1967*b* Population, family, and biochemical investigation of human adenylate kinase polymorphism. *Nature* (*London*) **214:** 1156.

Boyer, S. H., 1961 Alkaline phosphatase in human sera and placentae. *Science* **134:** 1002.

Boyer, S. H., I. H. Porter, and R. G. Weilbacher, 1962 Electrophoretic heterogeneity of glucose-6-phosphate dehydrogenase and its relationship to enzyme deficiency in man. *Proc. Natl. Acad. Sci. USA* **48:** 1869.

Boyer, S. H., D. C. Fainer, and E. J. Watson-Williams, 1963 Lactate dehydrogenase variant from human blood: Evidence for molecular subunits. *Science* **141:** 642.

Brewer, G. J., 1967 Achromatic regions of tetrazolium stained starch gels: Inherited electrophoretic variation. *Am. J. Hum. Genet.* **19:** 674.

Brock, D. J. H., 1970 Adenylate kinase isoenzyme patterns in tissues of people of different phenotype. *Biochem. Genet.* **4:** 617.

Burns, J. M. and F. M. Johnson, 1967 Esterase polymorphism in natural populations of a sulfur butterfly, *Colias eurytheme*. *Science* **156:** 93.

Burns, J. M. and F. M. Johnson, 1971 Esterase polymorphism in the butterfly *Hemiargus isola:* Stability in a variable environment. *Proc. Natl. Acad. Sci. USA* **68:** 34.

Carter, N. D. and C. W. Parr, 1967 Isoenzymes of phosphoglucose isomerase in mice. *Nature* (*London*) **216:** 511.

Carter, N. D. and C. W. Parr, 1969 Phosphogluconate dehydrogenase polymorphism in British wild rats. *Nature* (*London*) **224:** 1214.

Carter, N. D., R. A. Fildes, L. I. Fitch, and C. W. Parr, 1968 Genetically determined electrophoretic variation of human phosphogluconate dehydrogenase. *Acta Genet. Stat. Med.* **18:** 109.

Carter, N. D., M. R. Hill, and B. J. Weir, 1972 Genetic variation of phosphoglucose isomerase in some hystricomorph rodents. *Biochem. Genet.* **6:** 147.

Cattanach, B. M. and J. N. Perez, 1969 A genetically determined variant of the A-subunit of lactate dehydrogenase in the deer mouse. *Biochem. Genet.* **3:** 499.

Chan, T. K., D. Todd, and M. C. S. Lai, 1972 Glucose 6-phosphate dehydrogenase: Identity of erythrocytes and leukocyte enzyme with report of a new variant in Chinese. *Biochem. Genet.* **6:** 119.

Chapman, V. M., F. H. Ruddle, and T. H. Roderick, 1971 Linkage of isozyme loci in the mouse: *phosphoglucomutase-2* (*Pgm-2*), *mitochondrial NADP malate dehydrogenase* (*Mod-2*), and *dipeptidase-1* (*Dip-1*). *Biochem. Genet.* **5:** 101.

Chen, S. H., A. Yoshida, and E. R. Giblett, 1970 Polymorphism of phosphoglycerate kinase (PGK), an X-linked locus in man. *Am. J. Hum. Genet.* **22:** 14a.

Clark, R. W. and T. Hymowitz, 1972 Activity variation between and within two soybean trypsin inhibitor electrophoretic forms. *Biochem. Genet.* **6:** 169.

Clarke, B., 1970 Darwinian evolution of proteins. *Science* **168:** 1009.

Clayton, J. W. and J. H. Gee, 1969 Lactate dehydrogenase isozymes in longnose and blacknose dace (*Rhinichthys cataractae* and *R. atratulus*) and their hybrid. *J. Fish Res. Board Can.* **26:** 3049.

Clayton, J. W., D. N. Tretiak, and A. H. Kooyman, 1971 Genetics of multiple malate dehydrogenase isozymes in skeletal muscle of walleye (*Strizostedion vitreum vitreum*). *J. Fish. Res. Board Can.* **28:** 1005.

Coates, P. M. and N. Simpson, 1972 Genetic variation in human erythrocyte acetylcholinesterase. *Science* **175:** 1466.

Coluzzi, M., L. Bullini, and A. P. Bianchi Bullini, 1971 Phosphoglucomutase (PGM) allozymes in two forms of the *mariae* complex of the genus *Aedes*. *Biochem. Genet.* **5:** 253.

Cooper, D. W. and R. M. Hope, 1971 '6-Phosphogluconate dehydrogenase polymorphism in the marsupial mouse, *Sminthopsis crassicaudata. Biochem. Genet.* **5:**65.

Cooper, D. W., M. R. Irwin, and W. H. Stone, 1969*a* Inherited variation in the dehydrogenases of doves (*Streptopelia*). I. Studies on 6-phosphogluconate dehydrogenase. *Genetics* **62:**597.

Cooper, D. W., M. R. Irwin, and W. H. Stone, 1969*b* Inherited variation in the dehydrogenase of doves (*Streptopelia*). II. Autosomal inheritance of glucose-6-phosphatate dehydrogenase. *Genetics* **62:**607.

Cortner, J. A. and J. D. Schnatz, 1970 Alkaline lipolytic activity of rabbit adipose tissue: Genotypes and their inheritance. *Biochem. Genet.* **4:**529.

Courtright, J. B., 1967 Polygenic control of aldehyde oxidàse in *Drosophila. Genetics* **57:**25.

Courtright, J. B., R. B. Imberski, and H. Ursprung, 1966 The genetic control of alcoho' dehydrogenase and octanol dehydrogenase isozymes in *Drosophila. Genetics* **54:**1251–1260.

Crow, J. F., 1969 Molecular genetics and population genetics. *Proc. XII Int. Congr. Genet.* **3:**105.

Das, S. R., B. N. Mukherjee, and S. K. Das, 1970 LDH variants in India. *Humangenetik* **9:**107.

Dave, P. J., R. W. Kaplan, and G. Pfleiderer, 1966 The effect of mutations on the isozymes of the enolase of yeast and other glycolytic enzymes. *Biochem. Z.* **345:**440–453.

Davidson, R. G. and J. A. Cortner, 1967*a* Genetic variant of human erythrocyte malate dehydrogenase. *Nature* (*London*) **215:**761.

Davidson, R. G. and J. A. Cortner, 1967*b* Mitochondrial malate dehydrogenase: A new genetic polymorphism in man. *Science* **157:**1569.

DeLorenzo, R. J. and F. H. Ruddle, 1969 Genetic control of two electrophoretic variants of glucosephosphate isomerase in the mouse (*Mus musculus*). *Biochem. Genet.* **3:**151.

DeLorenzo, R. J. and F. H. Ruddle, 1970 Glutamate oxaloacetate transaminase (GOT) genetics in *Mus musculus*: Linkage, polymorphism, and phenotypes of the *Got-2* and *Got-1* loci. *Biochem. Genet.* **4:**259.

Der Kaloustian, V. M., R. Byrne, W. J. Young, and B. Childs, 1969 An electrophoretic method for detecting hypoxanthine-guanine phosphoribosyl transferase variants. *Biochem. Genet.* **3:**299.

Dern, R. J., G. J. Brewer, R. E. Tashian, and T. B. Shows, 1966 Hereditary variation of erythrocytic 6-phosphogluconate dehydrogenase. *J. Lab. Clin. Med.* **67:**255.

Dern, R. J., P. R. McCudy, and A. Yoshida, 1969 A new structural variant of glucose-6-phosphate dehydrogenase with a high production rate (G6PD hectoen). *J. Lab. Clin. Med.* **73:**283.

Detter, J. C., P. O. Ways, E. R. Giblett, M. A. Baughan, D. A. Hopkinson, S. Povey, and H. Harris, 1968 Inherited variations in human phosphohexose isomerase. *Ann. Hum. Genet.* **31:**329.

Doane, W. W., 1965 Disc electrophoresis of a α-amylase isozymes in *Drosophila melanogaster. Am. Zool.* **5:**346.

Doane, W. W., 1967 Quantitation of amylases in *Drosophila* separated by acrylamide gel electrophoresis. *J. Exp. Zool.* **164:**363.

Dorn, G., 1965 Phosphatase mutants in *Aspergillus nidulans. Science* **150:**1183.

Edmunds, P. H., 1972 Genic polymorphism of blood proteins from white marlin. Research Report 77, U.S. Fish and Wildlife Service.

Edmunds, P. H. and J. I. Sammons, 1971 Genic polymorphism of tetrazolium oxidase in bluefin tuna, *Thunnus thynnus,* from the western North Atlantic. *J. Fish. Res. Board Can.* **28:**1053.

Efron, Y., 1970 Alcohol dehydrogenase in maize: Genetic control of enzyme activity. *Science* **170:**751.

Efron, Y., 1971 Differences between maize inbreds in the activity level of the AP_1-controlled acid phosphatase. *Biochem. Genet.* **5:**33.

Fedecka-Bruner, B., M. Anderson, and D. Epel, 1971 Control of enzyme synthesis in early sea urchin development: Aryl sulfatase activity in normal and hybrid embryos. *Dev. Biol.* **25:**655.

Fildes, R. A. and H. Harris, 1966 Genetically determined variation of adenylate kinase in man. *Nature (London)* **209:**261.

Fildes, R. A. and C. W. Parr, 1963 Human red-cell phosphogluconate dehydrogenase. *Nature (London)* **200:**890.

Fox, D. J., 1971 The soluble citric acid cycle enzymes of *Drosophila melanogaster.* I. Genetics and ontogeny of NADP-linked isocitrate dehydrogenase. *Biochem. Genet.* **5:**69.

Frydenberg, O. and G. Nielsen, 1966 Amylase isozymes in germinating barley seeds. *Hereditas* **54:**123.

Gillespie, J. H. and K. Kojima, 1969 The degree of polymorphism in enzymes involved in energy production to that in nonspecific enzymes in two *Drosophila ananassae* populations. *Proc. Natl. Acad. Sci. USA* **61:**582.

Glassman, E., T. Shinoda, E. J. Duke, and J. F. Collins, 1968 Molecular forms of xanthine dehydrogenase and related enzymes. *Ann. N.Y. Acad. Sci.* **151:**263.

Goldberg, E., 1966 Lactate dehydrogenase of trout: Hybridization *in vivo* and *in vitro. Science* **151:**1091.

Grell, E. H., 1967 Electrophoretic variant of α-glycerophosphate dehydrogenase in *Drosophila melanogaster. Science* **158:**1319.

Grell, E. H., K. B. Jacobson, and J. B. Murphy, 1965 Alcohol dehydrogenase in *Drosophila melanogaster*: Isozymes and genetic variants. *Science* **149:**80.

Grunder, A. A., 1968 Inheritance of electrophoretic variance of serum esterases in domestic fowl. *Can. J. Genet. Cytol.* **10:**961.

Grunder, A. A., G. Sartore, and C. Stormont, 1965 Genetic variation in red cell esterases of rabbits. *Genetics* **52:**1345.

Harris, H., E. B. Robson, A. M. Glen-Bott, and J. A. Thornton, 1963 Evidence for non-allelism between genes affecting human serum cholinesterase. *Nature (London)* **200:**1185.

Henderson, N. S., 1965 Isozymes of isocitrate dehydrogenase: Subunit structure and intracellular location. *J. Exp. Zool.* **158:**263.

Henderson, N. S., 1966 Isozymes and genetic control of NADP-malate dehydrogenase in mice. *Arch. Biochem. Biophys.* **117(1):**28–33.

Hjorth, J. P., 1971 Genetics of *Zoarces* populations. I. Three loci determining the phosphoglucomutase isoenzymes in brain tissues. *Hereditas* **69:**233.

Hodgins, H. O., W. E. Ames, and F. M. Utter, 1969 Variants of lactate dehydrogenase isozymes in sera of sockeye salmon (*Oncorhynchus nerka*). *J. Fish. Res. Board Can.* **26:**15.

Holmes, R. S. and G. S. Whitt, 1970 Developmental genetics of the esterase isozymes of *Fundulus heteroclitus. Biochem. Genet.* **4:**471.

Holstein, T. J., W. C. Quevedo, Jr., and J. B. Burnett, 1971 Multiple forms of tyrosinase in rodents and lagomorphs with special reference to their genetic control in mice. *J. Exp. Zool.* **177:**173.

Hopkinson, D. A., and H. Harris, 1965 Evidence for a second "structural" locus determining human phosphoglucomutase. *Nature* (*London*) **208:**410.

Hopkinson, D. A. and H. Harris, 1968 A third phosphoglucomutase locus in man. *Ann. Hum. Genet.* **31:**359.

Hopkinson, D. A. and H. Harris, 1969 The investigation of reactive sulphydryls in enzymes and their variants by starch gel electrophoresis: Studies on red cell adenosine deaminase. *Ann. Hum. Genet.* **33:**81.

Hopkinson, D. A., N. Spencer, and H. Harris, 1963 Red cell acid phosphatase variants; a new human polymorphism. *Nature* (*London*) **199:**969.

Hopkinson, D. A., N. Spencer, and H. Harris, 1964 Genetical studies on human red cell acid phosphatase. *Am. J. Hum. Genet.* **16:**141.

Hopkinson, D. A., G. Corney, P. J. L. Cook, E. B. Robson, and H. Harris, 1970 Genetically determined electrophoretic variants of human red cell NADH diaphorase. *Ann. Hum. Genet.* **34:**1.

Horsfall, W. R., H. Ledmann, and D. Davies, 1963 Incidence of pseudocholinesterase variants in Australian aborigines. *Nature* (*London*) **199:**1115.

Hsieh, H. S. and R. Jaffi, 1971 Electrophoretic and functional variations NADH-methemoglobin reductase in hereditary methemoglobinemia. *J. Clin. Invest.* **50:**196.

Hubby, J. L. and R. C. Lewontin, 1966 A molecular approach to the study of genic heterozygosity in natural populations. I. The number of alleles at different loci in *Drosophia pseudoobscura. Genetics* **54:**577.

Hubby, J. L. and L. H. Throckmorton, 1968 Protein differences in *Drosophila.* IV. A study of sibling species. *Am. Nat.* **102:**193.

Hunter, R. L. and C. L. Markert, 1957 Histochemical demonstration of enzymes separated by zone electrophoresis in starch gels. *Science* **125:**1294.

Hyldgaard-Jensen, J. and J. Moustgaard, 1967 Three lactic dehydrogenase isoenzymes systems in pig spermatozoa and the polymorphism of subunits controlled by a third locus *C. Nature* (*London*) **216:**506.

Ishimoto, G. and S. Yada, 1969 Frequency of red cell phosphoglucomutase phenotypes in the Japanese population. *Hum. Hered.* **19:**198.

Johnson, A. G., F. M. Utter, and H. O. Hodgins, 1970*a* Interspecific variation of tetrazolium oxidase in *Sebastodes* (rockfish). *Comp. Biochem. Physiol.* **37:**1970*a*.

Johnson, A. G., F. M. Utter, and H. O. Hodgins, 1970*b* Electrophoretic variants of *l* alpha-glycerophosphate dehydrogenase in Pacific Ocean perch (*Sebastodes alutus*). *J. Fish. Res. Board Can.* **27:**943.

Johnson, A. G., F. M. Utter, and H. O. Hodgins, 1971 Phosphoglucomutase polymorphism in Pacific Ocean perch, *Sebastodes alutus. Comp. Biochem. Physiol.* **39B:**285.

Johnson, F. M. and C. Denniston, 1964 Genetic variation of alcohol dehydrogenase in *Drosophila melanogaster. Nature* (*London*) **204:**906.

Johnson, F. M. and R. K. Sakai, 1964 A leucine aminopeptidase polymorphism in *Drosophila buskii. Nature* (*London*) **203:**373.

Johnson, F. M., H. E. Schaffer, J. E. Gillaspy, and E. S. Rockwood, 1969 Isozyme genotype–environment relationships in natural populations of the harvester ant, *Poganomyrmex barbatus,* from Texas. *Biochem. Genet.* **3:**429.

Johnson, G. B., 1971 Analysis of enzyme variation in natural populations of the butterfly *Colias eurytheme. Proc. Natl. Acad. Sci. USA* **68:**997.

Johnson, K. E. and V. M. Chapman, 1971*a* Expression of the paternal genes for glutamate-oxaloacetate transaminase (GOT) during embryogenesis in *Xenopus laevis*. *J. Exp. Zool.* **178**:319.

Johnson, K. E. and V. M. Chapman, 1971*b* Expression of paternal genes during embryogenesis in the viable interspecific hybrid amphibian embryo *Rana pipiens* × *Rana palustris*: Electrophoretic analysis of five enzyme systems. *J. Exp. Zool.* **178**:313.

Johnson, M. S., 1971 Adaptive lactate dehydrogenase variation in the crested blenny, *Anoplarchus*. *Heredity* **27**:205.

Kalow, W. and N. Staron, 1957 On the distribution and inheritance of atypical forms of human serum cholinesterase as indicated by dibucaine numbers. *Can. J. Biochem. Physiol.* **35**:1305.

Kamaryt, J. and R. Laxova, 1965 Amylase heterogeneity: Some genetic and clinical aspects. *Humangenetik* **1**:579.

Kanapi, C. G. and M. R. Wheeler, 1970 Comparative isozyme patterns in three species of the *Drosophila nasuta* complex. *Tex. Rep. Biol. Med.* **28**:261–78.

Karig, L. M. and A. C. Wilson, 1971 Genetic variation in supernatant malate dehydrogenase of birds and reptiles. *Biochem. Genet.* **5**:211.

King, J. L. and T. H. Jukes, 1969 Non-Darwinian evolution. *Science* **164**:788.

Kirkman, H. N., H. D. Riley, Jr., and B. B. Crowell, 1960 Different enzymatic expressions of mutants of human glucose-6-phosphate dehydrogenase. *Proc. Natl. Acad. Sci. USA* **46**:938.

Klose, J. and U. Wolf, 1970 Transitional hemizygosity of the maternally derived allele at the 6PGD locus during early development of the cyprinid fish, *Rutilus rutilus*. *Biochem. Genet.* **4**:87.

Klose, J., H. Hitzeroth, H. Ritter, E. Schmidt, and U. Wolf, 1969 Persistance of maternal isozyme patterns of the lactate dehydrogenase and phosphoglucomutase system during early development of hybrid trout. *Biochem. Genet.* **3**:91.

Knutsen, C., C. F. Sing, and G. J. Brewer, 1969 Hexokinase isozyme variability in *Drosophila robusta*. *Biochem. Genet.* **3**:475.

Koehn, R. K., 1969 Esterase heterogeneity: Dynamics of a polymorphism. *Science* **163**:943.

Koehn, R. K. and J. B. Mitton, 1972 Population genetics of marine pelecypods. I. Ecological heterogeneity and evolutionary strategy at an enzyme locus. *Am. Nat.* **106**:47.

Kraus, A. P. and C. L. Neely, Jr., 1964 Human erythrocyte lactate dehydrogenase: Four genetically determined variants. *Science* **145**:595.

Law, G. R. J., 1967 Alkaline phosphatase and leucine aminopeptidase association in plasma of the chicken. *Science* **156**:1106.

Law, G. R. J. and S. S. Munro, 1965 Inheritance of two alkaline phosphatase variants in fowl plasma. *Science* **149**:1518.

Lewis, W. H. P., 1971 Polymorphism of human enzyme proteins. *Nature* (*London*) **230**:215.

Lewis, W. H. P. and H. Harris, 1967 Human red cell peptidases. *Nature* (*London*) **215**:351.

Lewis, W. H. P. and H. Harris, 1969 Peptidase D (prolidase) variants in man. *Ann. Hum. Genet.* **32**:317.

Lewis, W. H. P. and G. M. Truslove, 1969 Electrophoretic heterogeneity of mouse erythrocyte peptidases. *Biochem. Genet.* **3**:493.

Lewis, W. H. P., G. Corney, and H. Harris, 1968 Pep A 5-1 and pep A 6-1: Two new variants of peptidase A with features of special interest. *Ann. Hum. Genet.* **32:**35.

Liddell, J., H. Lehmann, D. Davies, and A. Sharih, 1962 Physical separation of pseudocholinesterase variants in human serum. *Lancet* **1:**463.

Lin, C., G. Schipmann, W. A. Kittrel, and S. Ohno, 1969 The predominance of heterozygotes found in wild goldfish of Lake Erie at the gene locus for sorbitol dehydrogenase. *Biochem. Genet.* **3:**603.

Long, W. K., 1967 Glutathione reductase in red blood cells: Variant associated with gout. *Science* **155:**721.

Long, W. K., 1970 Association between glutathione reductase variants and plasma uric acid concentration in a Negro population. *Am. J. Hum. Genet.* **22:**14*a*.

Longo, G. P. and J. G. Scandalios, 1969 Nuclear gene control of mitochondrial malic dehydrogenase in maize. *Proc. Natl. Acad. Sci. USA* **62:**104.

Luffmann, J. E. and H. Harris, 1967 A comparison of some properties of human red cell acid phosphatase in different phenotypes. *Ann. Hum. Genet.* **30:**387.

Lush, I. E., 1969 Polymorphism of a phosphoglucomutase isoenzyme in the herring (*Clupea harengus*). *Comp. Biochem. Physiol.* **30:**391.

Manwell, C. and C. M. A. Baker, 1968 Genetic variation of isocitrate, malate and 6-phosphogluconate dehydrogenase in snails of the genus *Cepaea*—Introgressive hybridization, polymorphism and pollution? *Comp. Biochem. Physiol.* **26:**195.

Manwell, C. and C. M. A. Baker, 1969 Hybrid proteins, heterosis and the origin of species. I. Unusual variation of polychaete *Hyalinoecia* "nothing dehydrogenases" and of quail *Coturnix* erythrocyte enzymes. *Comp. Biochem. Physiol.* **28:**1007.

Markert, C. L. and I. Faulhaber, 1965 Lactate dehydrogenase isozyme patterns in fish. *J. Exp. Zool.* **159:**319.

Markert, C. L. and F. Moller, 1959 Multiple forms of enzymes: Tissue, ontogenetic, and species specific patterns. *Proc. Natl. Acad. Sci. USA* **45:**753.

Marks, P. A., R. T. Gross, and J. Banks, 1961 Evidence for heterogeneity among subjects with glucose-6-phosphate dehydrogenase deficiency. *J. Clin. Invest.* **40:**1060.

Mathai, C. K. and E. Beutler, 1966 Electrophoretic variation of galactose-1-phosphate uridyltransferase. *Science* **154:**1179.

McAlpine, P. J., D. A. Hopkinson, and H. Harris, 1970 Thermostability studies on the isozymes of human phosphoglucomutase. *Ann. Hum. Genet.* **34:**61.

McCabe, M. M. and D. M. Dean, 1970 Esterase polymorphisms in the skipjack tuna. *Comp. Biochem. Physiol.* **34:**671.

McCabe, M. M., D. M. Dean, and C. S. Olson, 1970 Multiple forms of 6-phosphogluconate dehydrogenase and alpha-glycerophosphate dehydrogenase in the skipjack tuna, *Katsuwonus pelamis*. *Comp. Biochem. Physiol.* **34:**755.

McCombs, M. L., B. H. Bowman, and J. B. Alperin, 1970 A new ceruloplasmin variant: Cp Galveston. *Clin. Genet.* **1:**30.

McLaren, A. and A. Tait, 1969 Cytoplasmic isocitrate dehydrogenase variation within the C3H inbred strain. *Genet. Res.* **14:**93.

McReynolds, M. S., 1967 Homologous esterases in three species of the *virilis* groups of *Drosophila*. *Genetics* **56:**527.

Melville, J. C. and J. G. Scandalios, 1972 Maize endopeptidase: Genetic control, chemical characterization, and relationship to an endogenous trypsin inhibitor. *Biochem. Genet.* **7:**15.

Mishra, N. C. and E. L. Tatum, 1970 Phosphoglucomutase mutant of *Neurospora sitophila* and their relation to morphology. *Proc. Natl. Acad. Sci. USA* **66:**638.

Miziantz, T. J. and S. T. Case, 1971 Demonstration of genetic control of esterase-A in *Drosophila melanogaster. J. Hered.* **62:**345.

Morrison, W. J. and J. E. Wright, 1966 Genetic analysis of three lactate dehydrogenase isozyme systems in trout: Evidence for linkage of genes coding subunits A and B. *J. Exp. Zool.* **163:**259.

Naik, S. N. and D. E. Anderson, 1970 Study of glucose 6-phosphate dehydrogenase and 6-phosphogluconate dehydrogenase in the American buffalo (*Bison bison*). *Biochem. Genet.* **4:**651.

Nance, W. E., A. Claflin, and D. Smithies, 1963 Lactic dehydrogenase: Genetic control in man. *Science* **142:**1075.

Nance, W. E., J. Empson, T. Bennett, and L. Larson, 1968 Haptoglobin and catalase loci in man: Possible genetic linkage. *Science* **160:**1230.

Nevo, E. and C. R. Shaw, 1972 Genetic variation in a subterranean mammal, *Spalax ehrenbergi. Biochem. Genet.* **7:**235.

Nielsen, J. T., 1969 Genetic studies of the amylase isoenzymes of the bank vole, *Clethrionomys glareola. Hereditas* **61:**400.

Norris, J. R., 1962 Electrophoretic analysis of bacterial esterase systems—An aid to taxonomy. *J. Gen. Microbiol.* **28:**vii.

Odense, P. H., T. M. Allen, and T. C. Leung, 1966 Multiple forms of lactate dehydrogenase and aspartate aminotransferase in herring (*Clupea harengus harengus l.*). *Can. J. Biochem. Physiol.* **44:**1319.

Odense, P. H., T. C. Leung, T. M. Allen, and E. Parker, 1969 Multiple forms of lactate dehydrogenase in the cod, *Gadus morhua l. Biochem. Genet.* **3:**317.

Ogita, Z., 1962 Genetic variation of amylase and esterase in the housefly (*Musca domestica*). *Jpn. J. Genet.* **37:**518.

Ogita, Z., 1966 Genetico-biochemical studies of salivary and pancreatic amylase isozymes in the human. *Med. J. Osaka Univ.* **16:**271.

Ogita, Z. and T. Kassi, 1965 Genetic control of multiple molecular forms of the acid phosphomonoesterases in the housefly, *Musca domestica. Jpn. J. Genet.* **40:**185.

Ogonji, G. O., 1971 Bearing of genetic data on the interpretation of the subunit structure of octanol dehydrogenase of *Drosophila. J. Exp. Zool.* **178:**513.

Ohno, S. and J. Poole, 1965 Sex-linkage of erythrocyte glucose-6-phosphate dehydrogenase in two species of wild hares. *Science* **150:**1737.

Okada, I. and Y. Hachinohe, 1968 Genetic variation in liver acid phosphatases of chickens. *Jpn. J. Genet.* **43:**243.

Oki, Y., W. T. Oliver, and H. S. Funnell, 1964 Multiple forms of cholinesterase in horse plasma. *Nature (London)* **203:**605.

Op't Hof, J., 1969 Isoenzymes and population genetics of sorbitol dehydrogenase (E.C. 1.1.1.14) in swine (*Sus scrofa*). *Humangenetik* **7:**258.

Parr, C. W., 1966 Erythrocyte phosphogluconate dehydrogenase polymorphism. *Nature (London)* **210:**487.

Petras, M. L., 1963 Genetic control of a serum esterase component in *Mus musculus. Proc. Natl. Acad. Sci. USA* **50:**112.

Petras, M. L. and P. Sinclair, 1969 Another esterase variant in the kidney of the house mouse, *Mus musculus. Canad. J. Genet. Cytol.* **11:**97.

Pfleiderer, G., A. N. Kreiling, R. W. Kaplan, and P. Fortnagel, 1968 Biochemical, immunological, and genetic investigations of the multiple forms of yeast enolase. *Ann. N.Y. Acad. Sci.* **151:**78.

Pipkin, S. B., 1968 Genetics of octanol dehydrogenase in *Drosophila metzii*. *Genetics* **60:**81.

Popp, R. A. and D. M. Popp, 1962 Inheritance of serum esterases having different electrophoretic patterns (among inbred strains of mice). *J. Hered.* **53:**111.

Quiroz-Gutierrez, A. and S. Ohno, 1970 The evidence of gene duplication for S-form NADP-linked isocitrate dehydrogenase in carp and goldfish. *Biochem. Genet.* **4:**93.

Randerson, S., 1964 The inheritance of an erythrocytic esterase component in the deer mouse. *Genetics* **50:**277.

Rauch, N., 1968 A mutant form of lactate dehydrogenase in the horse. *Ann. N.Y. Acad. Sci.* **151:**672.

Rauch, N., 1971 Mutation at the *A* locus of lactate dehydrogenase in Holstein-Friesian cows. *Am. J. Vet. Res.* **32:**1439.

Richmond, R. C., 1970 Non-Darwinian evolution: A critique. *Nature* (*London*) **225:**1025.

Ritter, H. and G. G. Wendt, 1971 Population genetics of phosphoglucose isomerase (E.C. 5.3.1.9.). *Humangenetik* **13:**356.

Roberts, R. L., J. F. Wohnus, and S. Ohno, 1969 Phosphoglucomutase polymorphism in the rainbow trout, *Salmo gairdneri*. *Experientia* **25:**1109.

Robson, E. B. and H. Harris, 1967 Further studies on the genetics of placental alkaline phosphatase. *Ann. Hum. Genet.* **30:**219.

Ruddle, F. H. and T. H. Roderick, 1965 The gentic control of three kidney esterases in C57BL/6J and RF/J mice. *Genetics* **51:**445.

Ruddle, F. H., T. B. Shows, and T. H. Roderick, 1968 Autosomal control of an electrophoretic variant of glucose-6-phosphate dehydrogenase in the mouse (*Mus musculus*). *Genetics* **58:**599.

Ruddle, R. H., T. B. Shows, and T. H. Roderick, 1969 Esterase genetics in *Mus musculus:* Expression, linkage, and polymorphism of locus *Es-2*. *Genetics* **62:**393.

Saison, R., 1971 A genetically controlled lactate dehydrogenase variant at the *B* locus in mink. *Biochem. Genet.* **5:**27.

Saison, R. and E. R. Giblett, 1969 6-phosphogluconic dehydrogenase polymorphism in the pig. *Vox Sang.* **16:**514.

Salthe, S. N., 1969 Geographic variation of the lactate dehydrogenases of *Rana pipiens* and *Rana palustris*. *Biochem. Genet.* **2:**271.

Salthe, S. N. and E. Nevo, 1969 Geographic variation of lactate dehydrogenase in the cricket frog, *Acris crepitans*. *Biochem. Genet.* **3:**335.

Sandberg, K., 1968 Genetic polymorphism in carbonic anhydrase from horse erythrocytes. *Hereditas* **60:**411.

Sartore, G., C. Stormont, B. G. Morris, and A. A. Grunder, 1969 Multiple electrophoretic forms of carbonic anhydrase in red cells of domestic cattle (*Bos taurus*) and American buffalo (*Bison bison*). *Genetics* **61:**823.

Scandalios, J. G., 1967 Genetic control of alcohol dehydrogenase isozymes in maize. *Biochem. Genet.* **1:**1–8.

Schwartz, D., 1969 Alcohol dehydrogenase in maize: Genetic basis for multiple isozymes. *Science* **164:**585.

Scott, E. M., 1966 Kinetic comparison of genetically different acid phosphatases of human erythrocytes. *J. Biol. Chem.* **241(13):**3049–3052.

Selander, R. K., W. G. Hunt, and S. Y. Yang, 1969 Protein polymorphism and genetic heterozygosity in two European subspecies of the house mouse. *Evolution* **23:**379.

Selander, R. K., S. Y. Yang, R. C. Lewontin, and W. E. Johnson, 1970 Genetic variation in the horseshoe crab (*Limulus polyphemus*), a phylogenetic "relic." *Evolution* **24:**402.

Selander, R. K., M. H. Smith, S. Y. Yang, W. E. Johnson, and J. B. Gentry, 1971 Biochemical polymorphism and systematics in the genus *Peromyscus.* I. Variation in the old field mouse (*Peromyscus polionotus*). Univ. Texas Publ. *Stud. Genet.* **6:**7103.

Semeonoff, R., 1972 Esterase polymorphisms in *Microtus ochrogaster:* Interaction and linkage. *Biochem. Genet.* **6:**125.

Shaw, C. R., 1965 Electrophoretic variation in enzymes. *Science* **149:**936.

Shaw, C. R., 1966 Glucose-6-phosphate dehydrogenase: Homologous molecules in deer mouse and man. *Science* **153:**1013.

Shaw, C. R., 1970 How many genes evolve? *Biochem. Genet.* **4:**275.

Shaw, C. R. and E. Barto, 1963 Genetic evidence for the subunit structure of lactate dehydrogenase isozymes. *Proc. Natl. Acad. Sci. USA* **50:**211.

Shaw, C. R. and E. Barto, 1965 Autosomally determined polymorphism of glucose-6-phosphate dehydrogenase in *Peromyscus. Science* **148:**1099.

Shaw, C. R. and A. L. Koen, 1968 Glucose 6-phosphate dehydrogenase and hexose 6-phosphate dehydrogenase of mammalian tissues. *Ann. N.Y. Acad. Sci.* **151:**149.

Shaw, C. R. and R. Prasad, 1970 Starch gel electrophoresis of enzymes—A compilation of recipes. *Biochem. Genet.* **4:**297.

Shaw, C. R., F. N. Syner, and R. E. Tashian, 1962 New genetically-determined molecular form of erythrocyte esterase in man. *Science* **138:**31.

Shin, S., P. M. Khan, and P. R. Cook, 1971 Characterization of hypoxanthine-guanine phosphoribosyl transferase in man–mouse somatic cell hybrids by an improved electrophoretic method. *Biochem. Genet.* **5:**91.

Shokier, M. H. K. and D. C. Shreffler, 1970 Two new ceruloplasmin variants in Negroes—Data on three populations. *Biochem. Genet.* **4:**517.

Shows, T. B., 1967 The amino acid substitution and some chemical properties of a variant human erythrocyte carbonic anhydrase: Carbonic anhydrase $Id_{Michigan}$. *Biochem. Genet.* **1(2):**171–195.

Shows, T. B., R. E. Tashian, and G. J. Brewer, 1964 Erythrocyte glucose-6-phosphate dehydrogenase in Caucasians: New inherited variant. *Science* **145:**1056.

Shows, T. B., F. Ruddle, and T. Roderick, 1969 Phosphoglucomutase electrophoretic variants in the mouse. *Biochem. Genet.* **3:**25.

Shows, T. B., V. M. Chapman, and F. H. Ruddle, 1970 Mitochondrial malate dehydrogenase and malic enzyme: Mendelian inherited electrophoretic variants in the mouse. *Biochem. Genet.* **4:**707.

Shreffler, D. C., G. J. Brewer, J. C. Gall, and M. S. Honeyman, 1967 Electrophoretic variation in human serum ceruloplasmin: A new genetic polymorphism. *Biochem. Genet.* **1:**101.

Siciliano, M., D. Wright, S. George, and C. R. Shaw, 1973 Inter- and intraspecific genetic distances among teleosts. *Proc. 17th Int. Congr. Zool.* pp. 1–24, Theme 5, Etudes Moleculaires des Différences entre les Espèces, Paris, Masson.

Sick, K. and J. T. Nielsen, 1964 Genetics of amylase isozymes in the mouse. *Hereditas* **51:**291–296.

Simpson, N. E. and W. Kalow, 1964 The "silent" gene for serum cholinesterases. *Am. J. Hum. Genet.* **16:**180.

Smith, M., D. A. Hopkinson, and H. Harris, 1971 Developmental changes and polymorphism in human alcohol dehydrogenase. *Ann. Hum. Genet.* **34:**251.

Spencer, N., D. A. Hopkinson, and H. Harris, 1964*a* Quantitative differences and gene dosage in the human red cell acid phosphatase polymorphism. *Nature* (*London*) **201:**299.

Spencer, N., D. A. Hopkinson, and H. Harris, 1964*b* Phosphoglucomutase polymorphism in man. *Nature* (*London*) **204:**742.

Spencer, N., D. A. Hopkinson, and H. Harris, 1967 Adenosine deaminase polymorphism in man. *Ann. Hum. Genet.* **32:**9.

Sprague, L. M., 1967 Multiple molecular forms of serum esterase in three tuna species from the Pacific Ocean. *Hereditas* **57:**198.

Sprague, L. M., 1970 The electrophoretic patterns of skipjack tuna tissue esterases. *Hereditas* **65:**187.

Stegeman, J. J. and E. Goldberg, 1971 Distribution and characterization of hexose 6-phosphate dehydrogenase in trout. *Biochem. Genet.* **5:**579.

Stone, W. S., M. R. Wheeler, F. M. Johnson, and K. Kojima, 1968 Genetic variation in natural island populations of members of the *Drosophila nasuta* and *Drosophila ananassae* subgroups. *Proc. Natl. Acad. Sci. USA* **59:**102.

Stout, D. and C. R. Shaw, 1973 Comparative enzyme patterns in two species of *Thamnidium. Mycologia* **LXV:**803.

Syner, F. N. and M. Goodman, 1966 Polymorphism of lactate dehydrogenase in gelada baboons. *Science* **151:**206.

Tait, A., 1970 Enzyme variation between syngens in *Paramecium aurelia. Biochem. Genet.* **4:**461.

Tariverdian, G. H. R. and G. G. Wendt, 1970 Genetisch kontrollierte Varianten der NADH-diaphorase. *Humangenetik* **11:**75.

Tashian, R. E., 1965 Genetic variation and evolution of the carboxylic esterases and carbonic anhydrases of primate erythrocytes. *Am. J. Hum. Genet.* **17:**257.

Tashian, R. E., 1969 The esterases and carbonic anhydrases of human erythrocytes. In *Biochemical Method in Red Cell Genetics,* edited by J. Yunis, Academic Press, New York.

Tashian, R. E. and M. W. Shaw, 1962 Inheritance of an erythrocyte acetylesterase variant in man. *Am. J. Hum. Genet.* **14:**295.

Tashian, R. E., C. C. Plato, and T. B. Shows, Jr., 1963 Inherited variant of erythrocyte carbonic anhydrase in micronesians from Guam and Saipan. *Science* **140:**53.

Tashian, R. E., D. C. Shreffler, and T. B. Shows, 1968 Genetic and phylogenetic variation in the different molecular forms of mammalian erythrocyte carbonic anhydrases. *Ann. N.Y. Acad. Sci.* **151:**64.

Tashian, R. E., M. Goodman, V. E. Headings, J. DeSimone, and R. H. Ward, 1971 Genetic variation and evolution in the red cell carbonic anhydrase isozymes of macaque monkeys. *Biochem. Genet.* **5:**183.

Trippa, G., C. Santolamazza, and R. Scozzari, 1970 Phosphoglucomutase (PGM) locus in *Drosophila melanogaster:* Linkage and population data. *Biochem. Genet.* **4:**665.

Tsai, C.-Y. and O. E. Nelson, 1969 Mutations at the shrunken 4 locus in maize that produce three altered phosphorylases. *Genetics* **61:**813.

Tsuyuki, H. and F. Wold, 1964 Enolase: Multiple molecular forms in fish muscle. *Science* **146:**535.

Utter, F. M., 1971 Tetrazolium oxidase phenotype of rainbow trout (*Salmo gairdneri*) and Pacific salmon (*Oncorhynchus* spp.). *Comp. Biochem. Physiol.* **39B:**891.

Utter, F. M. and H. O. Hodgins, 1969 Lactate dehydrogenase isozymes of Pacific hake (*Merluccius productus*). *J. Exp. Zool.* **172**:59.

Utter, F. M. and H. O. Hodgins, 1970 Phosphoglucomutase polymorphism in sockeye salmon. *Comp. Biochem. Physiol.* **36**:195.

van Asperen, K. and M. E. van Mazijk, 1965 Agar gel-electrophoretic esterase patterns in houseflies. *Nature (London)* **205**:1291.

Vesell, E. S., 1965 Polymorphism of human lactate dehydrogenase isozymes. *Science* **148**:1103.

Wall, J. R. and T. W. Whitaker, 1971 Genetic control of leucine aminopeptidase and esterase isozymes in the interspecific cross *Cucurbita ecuadorenis* × *C. maxima*. *Biochem. Genet.* **5**:223.

Watson, J. G., T. J. Higgins, P. B. Collins, and S. Chaykin, 1972 The mouse liver aldehyde oxidase locus (*Aox*). *Biochem. Genet.* **6**:195.

Welch, S. G., L. I. Fitch, and C. W. Parr, 1970 A variant of phosphoglucose isomerase in the rabbit. *Biochem. J.* **117**:525.

Wheat, T. E., G. S. Whitt, and W. F. Childers, 1972 Linkage relationships between the homologous malate dehydrogenase loci in teleosts. *Genetics* **70**:337.

Whitt, G. S., 1969 Homology of lactate dehydrogenase genes: *E* gene function in the teleost nervous system. *Science* **166**:1156.

Whitt, G. S., 1970*a* Developmental genetics of the lactate dehydrogenase isozymes of fish. *J. Exp. Zool.* **175**:1.

Whitt, G. S., 1970*b* Genetic variation of supernatant and mitochondrial malate dehydrogenase isozymes in the teleost *Fundulus heteroclitus*. *Experientia* **26**:734.

Whitt, G. S. and C. L. Prosser, 1971 Lactate dehydrogenase isozymes, cytochrome oxidase activity, and muscle ions of the rattail (*Coryphaenoides* sp.). *Am. Zool.* **11**:503.

Whitt, G. S., W. F. Childers, and T. E. Wheat, 1971 The inheritance of tissue—specific lactate dehydrogenase isozymes in interspecific bass (*Micropterus*) hybrids. *Biochem. Genet.* **5**:257.

Wilcox, F. H., 1966 A recessively inherited electrophoretic variant of alkaline phosphatase in chicken serum. *Genetics* **53**:799.

Williams, R. A. D., 1964 Location of glyceraldehyde-3-phosphate dehydrogenase in starch gels. *Nature (London)* **203**:1070.

Williams, R. A. D. and E. Bowden, 1968 The starch-gel electrophoresis of glucose-6-phosphate dehydrogenase and glyceraldehyde-3-phosphate dehydrogenase of *Streptococcus faecalis, S. faecium,* and *S. durans*. *J. Gen. Microbiol.* **50**:329.

Williscroft, S. N. and H. Tsuyuki, 1970 Lactate dehydrogenase systems of rainbow trout—Evidence for polymorphism in liver and additional subunits in gills. *J. Fish Res. Board Can.* **27**:1563.

Wilmont, P. L. and D. K. Underhill, 1972 Carbonic anhydrase polymorphism in a New Jersey population of the white-footed mouse *Peromyscus leucopus*. *Genetics* **71**:315.

Wolf, U., W. Engel, and J. Faust, 1970 Zum Mechanismus der Diploidisierung in der Wirbeltierevolution: Koexistenz von tetrasomen und disomen Genloci der Isocitrat-Dehydrogenasen bei der Regenbogenforelle (*Salmo iridens*). *Humangenetik* **9**:150.

Womack, J. E., 1972 Genetic control of the major electrophoretic component of rat plasma esterase. *J. Hered.* **63**:41.

Wright, D. A. and F. H. Moyer, 1966 Parental influence on lactate dehydrogenase in the early development of hybrid frogs in the genus *Rana*. *J. Exp. Zool.* **163**:215.

Wright, D. A. and C. R. Shaw, 1969 Genetics and ontogeny of α-glycerophosphate dehydrogenase isozymes in *Drosophila melanogaster*. *Biochem. Genet.* **3**:343.

Wright, D. A. and C. R. Shaw, 1970 Time of expression of genes controlling specific enzymes in *Drosophila* embryos. *Biochem. Genet.* **4:**385.

Wright, D. A. and S. Subtelny, 1971 Nuclear and cytoplasmic contributions to dehydrogenase phenotypes in hybrid frog embryos. *Dev. Biol.* **24:**119.

Wright, D. A., M. J. Siciliano, and J. N. Baptist, 1972 Genetic evidence for the tetramer structure of glyceraldehyde 3-phosphate dehydrogenase. *Experientia* **28:**889.

Wright, T. R. F. and R. J. Macintyre, 1963 A homologous gene–enzyme system, esterase 6, in *Drosophila melanogaster* and *D. simulans. Genetics* **48:**1717.

Yoshida, A. and A. Motulsky, 1969 A pseudocholinestera variant (E cynthiana) associated with elevated plasma enzyme activity. *Am. J. Hum. Genet.* **21:**486.

Yoshida, A., E. Beutler, and A. G. Motulsky, 1971 Human glucose-6-phosphate dehydrogenase variants. *Bull. World Health Org.* **45:**243.

Young, W. J., J. E. Porter, and B. Childs, 1964 Glucose-6-phosphate dehydrogenase in *Drosophila:* X-linked electrophoretic variants. *Science* **143:**140.

Zee, D. S., H. Isensee, and W. H. Zinkham, 1970 Polymorphism of malate dehydrogenase in *Ascaris suum. Biochem. Genet.* **4:**253.

Zinkham, W. H., A. Blanco, and L. Kupchyk, 1964 Lactate dehydrogenase in pigeon testes: Genetic control by three loci. *Science* **144:**1353.

Zinkham, W. H., L. Kupchyk, A. Blanco, and H. Isensee, 1965 Polymorphism of lactate dehydrogenase isozymes in pigeons. *Nature* (*London*) **208:**284.

Errata for Previous Volumes

Volume 1

p. 233. The portion of the *S. typhimurium* histidine operon shown as Fig. 3G duplicates Fig. 3F. It should be replaced by the figure reproduced on the page following this one (p. 618).

p. 354. The text on lines 19 to 22

germspores are derived . . . of cross 2.

should read

germspores are derived from primordia which generally contain only one nucleus. These must divide mitotically to produce the multinucleated germspores. Such uninucleated primordia have been observed by P. Reau and myself in a germsporangium of cross 2.

Volume 2

p. 495. (Locus *mt,* line 4) mt^E, Mt^E should be mt^+, mt^+.

523. (Locus *fna,* line 9) *fAA** should be *fA*.

528. (Table 5) Genes *nd 146, nd 163, nd 203,* and *nd 1183* should be deleted.

540. (Table 8, row 6) scl^{c1} should be scl^{c6}.

572. (Fig. 1) In the horizontal row "Species 5," the "±" should be under Species 7, Mating type O.

Volume 3

p. 681. (Fig. 1) The label "retinula cell receptor depolarization" should read "retinula cell re- or depolarization."

Figure 3G

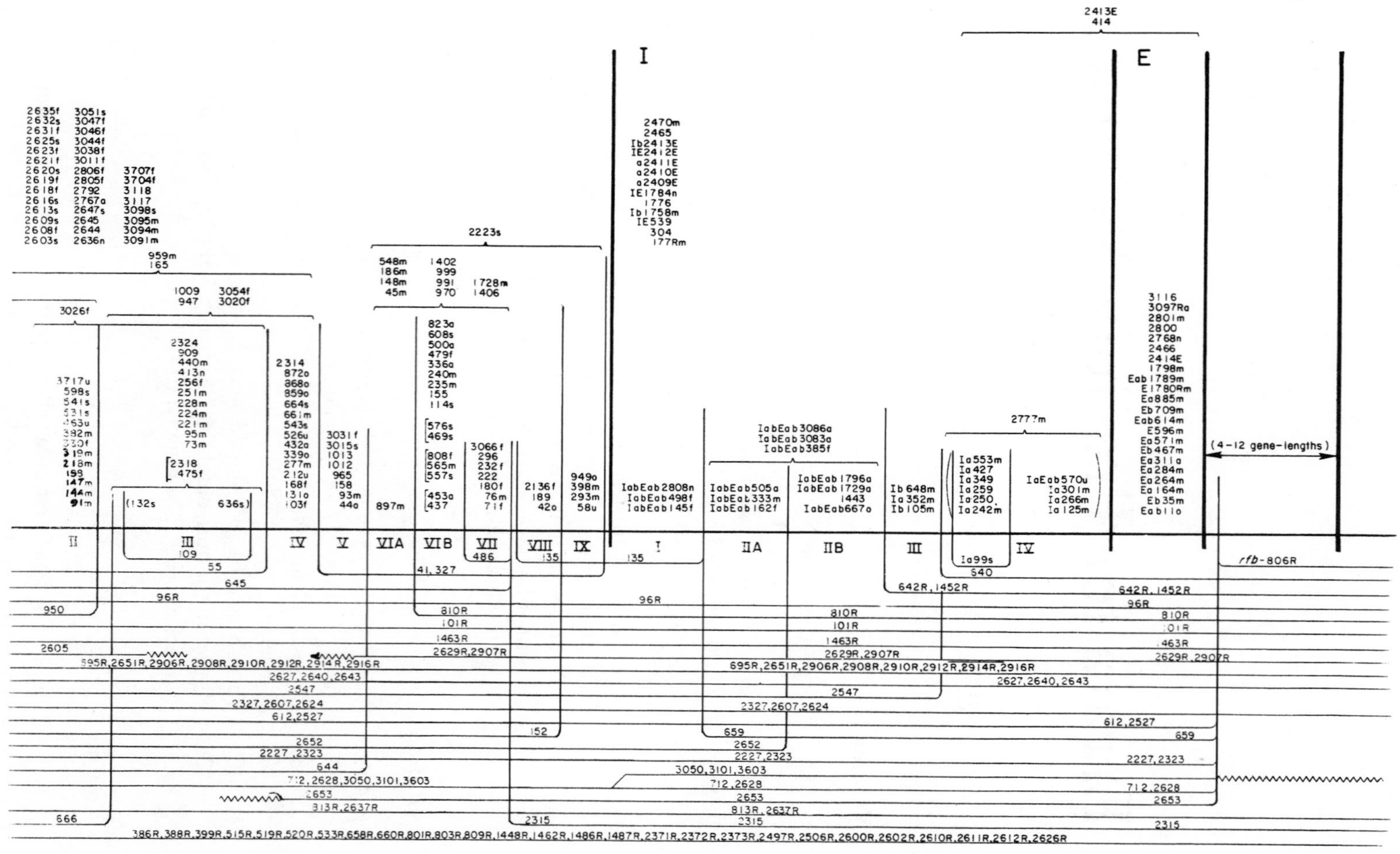

2635f 3051s
2632s 3047f
2631f 3046f
2625s 3044f
2623f 3038f
2621f 3011f
2620s 2806f 3707f
2619f 2805f 3704f
2618f 2792 3118
2616s 2767a 3117
2613s 2647s 3098s
2609s 2645 3095m
2608f 2644 3094m
2603s 2636n 3091m
959m
165
3026f
1009 3054f
947 3020f
2223s
548m 1402
186m 999
148m 991 1728m
45m 970 1406
I
2470m
2465
Ib2413E
IE2412E
a2411E
a2410E
a2409E
IE1784n
1776
Ib1758m
IE539
304
177Rm
2413E
414
E
3116
3097Ra
2801m
2800
2768n
2466
2414E
1798m
Eab1789m
E1780Rm
Ea885m
Eb709m
Eab614m
E596m
Ea571m
Eb467m
Ea311a
Ea284m
Ea264m
Ea164m
Eb35m
Eab11a
(4–12 gene-lengths)
2777m
Ia553m
Ia427
Ia349
Ia259
Ia250
Ia242m
IaEab570u
Ia301m
Ia266m
Ia125m
Ib648m
Ia352m
Ib105m
IabEab3086a
IabEab3083a
IabEab385f
IabEab505a
IabEab333m
IabEab162f
IabEab1796a
IabEab1729a
1443
IabEab667a
IabEab2808n
IabEab498f
IabEab145f
II III IV V VIA VIB VII VIII IX I IIA IIB III IV
rfb-806R
950
2605
666
96R
645
55
109
41,327
486
135
642R,1452R
810R
101R
1463R
2629R,2907R
695R,2651R,2906R,2908R,2910R,2912R,2914R,2916R
2627,2640,2643
2547
2327,2607,2624
612,2527
152
659
2652
2227,2323
644
712,2628,3050,3101,3603
3050,3101,3603
712,2628
2653
813R,2637R
2315
386R,388R,399R,515R,519R,520R,533R,658R,660R,801R,803R,809R,1448R,1462R,1486R,1487R,2371R,2372R,2373R,2497R,2506R,2600R,2602R,2610R,2611R,2612R,2626R

Author Index

Subject Index

Contents of Other Volumes

Volume 1: Bacteria, Bacteriophages, and Fungi

Volume 2: Plants, Plant Viruses, and Protists

Volume 3: Invertebrates of Genetic Interest

Mosquitoes and Flies of Genetic Interest

The Genus *Drosophila*

Drosophila melanogaster

Volume 4: Vertebrates of Genetic Interest